Geotechnical Practice for Waste Disposal

Geotechnical Practice for Waste Disposal

Edited by

David E. Daniel

Professor of Civil Engineering
The University of Texas at Austin
USA

CHAPMAN & HALL

London · Glasgow · New York · Tokyo · Melbourne · Madras

Published by Chapman & Hall, 2–6 Boundary Row, London SE1 8HN

Chapman & Hall, 2–6 Boundary Row, London SE1 8HN, UK

Blackie Academic & Professional, Wester Cleddens Road, Bishopbriggs, Glasgow G64 2NZ, UK

Chapman & Hall Inc., 29 West 35th Street, New York NY10001, USA

Chapman & Hall Japan, Thomson Publishing Japan, Hirakawacho Nemoto Building, 6F, 1-7-11 Hirakawa-cho, Chiyoda-ku, Tokyo 102, Japan

Chapman & Hall Australia, Thomas Nelson Australia, 102 Dodds Street, South Melbourne, Victoria 3205, Australia

Chapman & Hall India, R. Seshadri, 32 Second Main Road, CIT East, Madras 600 035, India

First edition 1993

Typeset in 10/12 pt Palatino by Best-set Typesetter Ltd, Hong Kong
Printed in Great Britain by St Edmundsbury Press, Bury St Edmunds

ISBN 0 412 35170 6

A catalogue record for this book is available from the British Library

Library of Congress Cataloging-in-Publication data available

Geotechnical practice for waste disposal / edited by David E. Daniel.
– 1st ed.
p. cm.
Includes index and bibliographical references
1. Waste disposal in the ground. 2. Environmental geotechnology.
I. Daniel, David E. (David Edwin), 1949–
TD795.7.G455 1993
628.4'456 – dc20 92–40842
CIP

Contents

Contributors

Morton Barlaz, Department of Civil Engineering, North Carolina State University, Raleigh, North Carolina.

Richard C. Benson, Technos Inc., Miami, Florida.

Gaylen R. Brubaker, Remediation Technologies Inc., Chapel Hill, North Carolina.

Keros Cartwright, Illinois State Geological Survey, Champaign, Illinois.

David E. Daniel, Department of Civil Engineering, The University of Texas at Austin, Texas.

Jeffrey C. Evans, Department of Civil Engineering, Bucknell University, Lewisburg, Pennsylvania.

Lorne G. Everett, Geraghty and Miller, Inc., and University of California at Santa Barbara, Santa Barbara, California.

Robert K. Ham, Department of Civil and Environmental Engineering, The University of Wisconsin, Madison, Wisconsin.

Mark P. Hemingway, Southwestern Laboratories, Austin, Texas.

Bruce R. Hensel, Illinois State Geological Survey, Champaign, Illinois.

Larry A. Holm, CH2M Hill, Oak Ridge, Tennessee.

David L. Jordan, Hydrosystems Inc., Sterling, Virginia.

Bob Kent, Geomatrix Consultants, Inc., Santa Anna Heights, California.

Robert M. Koerner, Professor of Civil Engineering, Director, Geosynthetic Research Institute, Drexel University, Philadelphia, Pennsylvania.

Raymond C. Loehr, Department of Environmental and Water Resources Engineering, The University of Texas at Austin, Texas.

Perry Mann, International Technology Corp., Austin, Texas.

Issa S. Oweis, Converse Consultants East, Parsippany, New Jersey.

R. Lee Peyton, Department of Civil Engineering, University of Missouri, Columbia, Missouri.

Roman Z. Pyrih, GEOCHEM Div., Terra Vac, Lakewood, Colorado.

Charles O. Riggs, Sverdrup Environmental, Inc., St Louis, Missouri.

Jim V. Rouse, GEOCHEM Div., Terra Vac, Lakewood, Colorado.

Charles D. Shackelford, Department of Civil Engineering, Colorado State University, Fort Collins, Colorado.

Paul R. Schroeder, US Army Engineer, Waterways Experiment Station, Vicksburg, Mississippi.

Lyle R. Silka, Hydrosystems Inc., Sterling, Virginia.

Paul B. Trost, Waste-Tech Services Inc., Golden, Colorado.

Dirk Van Zyl, Golder Associates, Inc., Lakewood, Colorado.

Preface

Earth scientists and geotechnical engineers are increasingly challenged to solve environmental problems related to waste disposal facilities and cleanup of contaminated sites. The effort has given rise to a new discipline of specialists in the field of environmental geotechnology. To be effective, environmental geotechnologists must not only be armed with the traditional knowledge of fields such as geology and civil engineering, but also be knowledgeable of principles of hydrogeology, chemistry, and biological processes. In addition, the environmental geotechnologist must be completely up to date on the often complex cadre of local and national regulations, must comprehend the often complex legal issues and sometimes mind-boggling financial implications of a project, and must be able to communicate effectively with a host of other technical specialists, regulatory officials, attorneys, local land owners, journalists, and others. The field of environmental geotechnology will no doubt continue to offer unique challenges.

The purpose of this book is to summarize the current state of practice in the field of environmental geotechnology. Part One covers broadly applicable principles such as hydrogeology, geochemistry, and contaminant transport in soil and rock. Part Two describes in detail the underlying principles for design and construction of new waste disposal facilities. Part Three covers techniques for site remediation. Finally, Part Four addresses the methodologies for monitoring.

The topics of 'waste disposal' and 'site remediation' are extraordinarily broad. This book is written for geologists, hydrogeologists, geotechnical engineers, environmental engineers, soil scientists, and others with similar backgrounds. The book does not attempt to go beyond the immediate capabilities of the environmental geotechnologist. For example, the book does not cover incineration of waste because the environmental geotechnologist would not logically design an incinerator; a combustion specialist would.

In preparing this book, the editor has assembled a group of exceptionally knowledgeable contributors to whom the editor is most grateful. The editor thanks the numerous scientists and engineers who have shaped his thinking over the years and expresses particular appreciation to his students for innumerable thoughtful questions and many hours of sacrifice and hard work. Finally, many thanks to Cindy Symington for her help in preparing this book.

David E. Daniel

PART ONE

General Principles

CHAPTER 1

Introduction

David E. Daniel

1.1 EARLY DAYS OF ENVIRONMENTAL GEOTECHNOLOGY

In the 1940s to 1960s, the industrialized countries of the world underwent an enormous expansion in capacity for manufacturing goods, processing petroleum, and making new chemicals. Geotechnical engineers and earth scientists played an important role in that expansion by identifying mineral and petroleum resources, investigating subsurface stratigraphy and soil conditions, designing foundations for buildings and machinery, and developing earthwork specifications.

The early roots of the field of geotechnical engineering can be traced to the 1800s. Legendary contributions were made by Coulomb, Rankine, Terzaghi, and many others over the past two centuries. Although the International Society for Soil Mechanics and Foundations Engineering held its first international conference in 1936, the modern practice of soil mechanics and foundation engineering blossomed throughout the world after World War II and underwent particularly intensive growth and maturity in the 1950s and 1960s. Most of the modern journals in geotechnical engineering, such as *Géotechnique*, the *Canadian Geotechnical Journal*, and the American Society of Engineer's (ASCE's) *Journal of the Soil Mechanics and Foundation Division* (now *Journal of Geotechnical Engineering*) published their first issues in the 1950s.

Despite the advances made in the field that we now call geotechnical engineering in the 1950s and 1960s, there was practically no emphasis on environmental matters in geotechnical engineering practice during this period. The geotechnical engineer's role in environmental decisions, however, was relatively minor until the late 1970s. The

Geotechnical Practice for Waste Disposal.
Edited by David E. Daniel.
Published in 1993 by Chapman & Hall, London. ISBN 0 412 35170 6

author recalls working as a junior engineer at the San Francisco firm of Woodward-Clyde Consultants from the period 1974–1977. In the beginning of this experience, geotechnical engineers generally shunned projects that involved waste materials or contamination. The author recalls that the firm had one engineer who was particularly experienced in evaluation of 'garbage dumps'. This individual's special capability in this area was not viewed as particularly valuable by his fellow engineers: there were far more glamorous projects than evaluation of garbage dumps and preventing environmental damage from garbage dumps. By the end of the 1970s, though, as will be seen in the next subsection, the situation began to change markedly and irrevocably.

Perhaps a typical example of the pre-1970s situation is the following project from the author's project files. In the late 1950s, a petroleum refinery in Texas underwent a major expansion, which required new structures as well as a 600 ha 'evaporation pond' for waste water. Waste water containing benzene (a carcinogen) was to be pumped into the pond and allowed to evaporate. At the time, evaporation was considered a viable 'disposal' method; today in the US, evaporation is not considered a suitable technique for treatment or disposal of hazardous waste.

A well known and very capable geotechnical engineering company was hired to develop geotechnical recommendations for the plant expansion. Approximately 20 exploratory borings were drilled to a depth of approximately 15 m. Various laboratory tests (mostly strength and compressibility tests to support foundation design recommendations) were performed, and several hydraulic conductivity tests were conducted on subsoils from the area where the evaporation pond was proposed. The hydraulic conductivity tests indicated hydraulic conductivities less than 1×10^{-8} cm/s.

The evaporation pond was discussed in the final report that the geotechnical engineer prepared, but only one paragraph was devoted to the pond and to potential groundwater contamination issues. The subsoils were described as 'impermeable' in the report. Implicitly, groundwater contamination was not considered to be a relevant issue. Construction began a short time after the geotechnical report was issued, and the unlined evaporation pond was put into service in about 1960.

In 1980, a regulatory agency directed the company to install a groundwater monitoring system. An enormous pool of contaminated liquid was found beneath the pond (Fig. 1.1). Fortunately, the uppermost aquifer, which was located nearly 100 m beneath the ground surface, had not yet been impacted, thanks to a low-hydraulic-conductivity stratum located about 75 m below the surface (Fig. 1.1). Nevertheless, the pond was taken out of service and remediation was

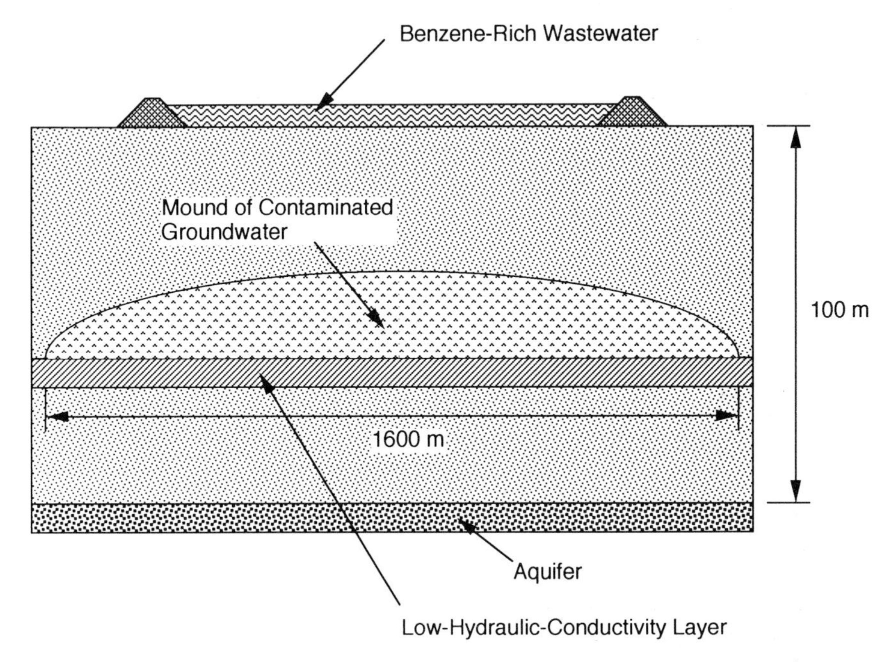

Fig. 1.1 Mound of contaminated liquid that formed over a period of 20^+ years from unanticipated seepage of waste water into soil beneath evaporation pond.

initiated by installing approximately 25 recovery wells. Remediation will probably take decades.

Where did the original geotechnical specialists err? Subsequent studies showed that the actual hydraulic conductivity of the subsoils located 0–15 m below the surface was not the value of $< 1 \times 10^{-8}$ cm/s reported by the geotechnical engineer but instead was approximately 1×10^{-4} cm/s. The soils beneath the evaporation pond were far from impermeable. The problem was that the original testing program was restricted to laboratory hydraulic conductivity tests on small samples of soil. The soils in the field contained numerous cracks (probably caused by cyclic wetting and drying) that were not included in the small laboratory test specimens. Today's specialist in environmental geotechnology knows that small-scale laboratory hydraulic conductivity tests do not always give representative in situ values, but the engineer of the 1950s had little clue that laboratory tests could produce misleading results.

The problem just described is not untypical. In the 1950s and 1960s, there was little understanding of the long-term consequences of waste disposal to the land. Clayey soils were thought to be 'impermeable'. The mere existence of such materials was thought to be enough to ensure protection of the environment. There was practically no understanding by geotechnical engineers of the relationship between disposal of chemicals in the ground and long-term ground water impacts. Many of the realities that we now know to exist were simply not recognized as problems before the 1970s. Even today, the uneducated engineer or scientist can easily repeat the common mistakes of the past if he or she does not possess the appropriate knowledge and understanding of recent experience and discoveries.

1.2 ORIGINS OF ENVIRONMENTAL GEOTECHNICS

The field of practice now called 'environmental geotechnics' evolved over a period of about two decades starting in the 1970s. In the US, the birth of environmental geotechnology can be traced largely to the nuclear power industry. In the 1970s, an enormous amount of geotechnical engineering work was focused on design of nuclear power plants. One part of the process of constructing a nuclear power plant in the US was development of an 'environmental impact statement'. No nuclear power facility could be granted a construction permit until an environmental impact statement was completed. Because geotechnical engineers were the lead technical specialists in most siting investigations, it was natural that geotechnical engineers manage and coordinate preparation of environmental impact statements. Many geotechnical engineering firms developed the capability for preparing such statements and hired environmental specialists, e.g., biologists and botanists, to conduct the technical work. For many geotechnical engineers, this was their first exposure to concern for the environment in their technical work and their first opportunity to work with diverse environmental specialists. These early experiences were critical in the development of an understanding among geotechnical engineers that geotechnical engineering principles, along with other scientific and engineering specialties, could be brought to bear in solving important technical problems involving the environment.

Concerns about the safety of nuclear power plants have always been voiced, but in the mid to late 1970s, one of the most frequently expressed concerns was over the ultimate disposal of high-level radioactive waste. Where would such wastes go? The wastes remain dangerous for hundreds of thousands of years; how can they be safely contained? To address these questions, numerous investigations were performed, primarily in the US and Europe, to study the proper disposal of high-level radioactive wastes. Geotechnical engineers played an important role in these early studies through investigation and characterization of suitable host rocks for waste repositories, analysis of long-term performance of the earth materials under realistic temperatures and pressures, evaluation of probable ground water impacts, and assessment of potential risks. For many large geotechnical engineering companies, these types of investigations represented the first significant attempt to perform a comprehensive technical analysis of a waste containment facility.

In the US, the next significant event in the development of environmental geotechnology was the widely publicized contamination at a site in New York called Love Canal. At this site, dangerous chemical wastes had been buried in an old canal and covered with clayey soils. The chemicals slowly seeped out of the canal over a period of many years until, in the mid 1970s, suspected health impacts were identified. The area was eventually evacuated, and unprecedented publicity was given to the dangers posed to the residents of the area from the wastes and to the enormous health impacts from hazardous waste in general. For the first time, national attention was focused on the adverse impacts of improper disposal and management of chemical wastes. Also, in the late 1970s and early 1980s, significant discussion began to occur concerning appropriate design standards for waste containment facilities. Geotechnical engineers at this time began to be brought into the process of designing waste containment facilities.

The 1980s brought an explosive expansion of environmental geotechnology, particularly in the US, where two significant processes were at work. First, stringent regulations drove waste containment systems to unprecedented levels of sophistication and complexity. Double liner systems, leak detection layers, geomembrane liners, and a host of new concepts were not only discussed but were written into regulations. Regulatory agencies, rather than engineers, drove the evolution of increasingly sophisticated containment designs. Environmental geotechnologists were called upon to develop and implement the required designs and containment systems. The regulatory agencies and environmental geotechnologists literally pushed containment technology to unprecedented levels of sophistication.

The second process stemmed from promulgation by the US Congress in 1980 of the Comprehensive Environmental Response, Compensation, and Liability Act (CERCLA), which became commonly referred to as 'Superfund'. Superfund charted a course in the US for cleanup of contaminated sites, such as the Love Canal site, and provided monies for cleanup if the responsible parties could not be identified or were incapable of paying for the cleanup. The requirement that contaminated sites be cleaned up was the single most important boost to environmental geotechnology in its relatively short history. Tens of thousands of engineers are actively engaged in investigation, design, and actual cleanup of contaminated sites in the US, and similar priorities are being established in many other countries, as well.

Because of the complexity of environmental contamination, environmental geotechnologists are as diverse as the problems they tackle.

Table 1.1 Rates of generation of chemical wastes in selected countries (from van Veen and Mensink, 1985)

country	*population* $(\times 10^6)$	*waste production* $(\times 10^6$ *tonnes)*	*per capita waste production (kg/person)*
Switzerland	6.4	0.1	16
Denmark	5.1	0.1	20
Canada	23.9	1	42
Sweden	8.3	0.5	60
Netherlands	14.1	1	71
Italy	57	5	88
United Kingdom	56	5	89
France	53.7	5	93
West Germany	61.6	6	97
Belgium	9.9	1	101
United States	227.7	40	176

Environmental geotechnologists include civil engineers (especially those schooled in soil and rock mechanics, geotechnical engineering, sanitary engineering, and environmental engineering), soil scientists, geologists, hydrogeologists, geohydrologists, engineering geologists, geological engineers, mining engineers, and agricultural engineers. This book is written from the point of view of the civil engineer who must not only examine and define the scope of environmental problems but, most importantly, solve those problems in practical, reliable, cost-effective ways. The challenge is immense.

Those who assume that the field of environmental geotechnology is restricted to one or two countries could not be more incorrect. All of the industrialized countries of the world generate significant quantities of waste, as the data in Table 1.1 indicate. Those countries with relatively recently-developed industrialized bases frequently have less mature and sophisticated environmental controls. Many of the past mistakes made in the industrialized countries of the world can be avoided if engineers in industrially-emerging nations will take the time to learn from those mistakes and apply the technologies (especially the most cost effective ones) that have been found to work in the highly developed countries. The field of environmental geotechnology is truly a field of world-wide importance.

1.3 REGULATORY REQUIREMENTS

No book on waste disposal would be complete without some discussion of regulatory requirements. However, the focus of this book is technology, not regulation. Thus, the following brief discussion will serve to summarize the regulatory framework in one country (the US), where the regulatory climate is the most complex. This discussion will be helpful in following some of the material in later chapters. As stated earlier, waste containment technology has been driven by regulation, not engineering. One simply cannot function as an environmental geotechnologist without some appreciation for how regulations drive technology.

In the US, two key pieces of legislation, and amendments to the two legislative benchmarks, drive the environmentalist's technical approach to problems. Disposal of newly-generated solid waste is regulated under the Resource Conservation and Recovery Act (RCRA) and the Hazardous and Solid Waste Amendments (HSWA) to RCRA. Hazardous waste is covered in Subtitle C of RCRA, and non-hazardous waste is regulated in Subtitle D.

Cleanup of old waste disposal sites is governed by legislation mentioned earlier, viz., Comprehensive Environmental Response, Compensation, and Liability Act (CERCLA), which is also known as Superfund. The original CERCLA legislation has expired and has been superseded by the Superfund Amendments and Reauthorization Act (SARA).

The spirit and details of RCRA and SARA are quite different. In RCRA, stringent, explicit requirements are set forth for nearly all aspects of hazardous and nonhazardous waste management, including treatment, storage, and disposal of waste. With cleanup (Superfund), the problem is different because each site is unique and must be evaluated individually. Thus, SARA is far more flexible than RCRA. However, RCRA requirements may have to be considered in implementation of cleanup under Superfund under the principle of ARARs, which stands for Applicable or Relevant and Appropriate Regulations. If RCRA applies to the waste (e.g., if the waste was disposed of after RCRA went into effect), RCRA requirements are applicable and must generally be followed. Even if RCRA does not apply to the waste (e.g., the waste was disposed of before RCRA took effect), RCRA requirements may still be relevant and appropriate, and if so, typically must at least be considered.

Hazardous wastes are widely perceived to be incredibly dangerous waste materials that require the highest degree of care through regulation. However, a fundamental problem exists in defining what constitutes a hazardous waste. If one assumes that a hazardous waste is any waste that is potentially dangerous to humans, practically all wastes would be considered hazardous because practically everything is potentially harmful to humans. Ordinary table salt (NaCl), for example, can be lethal if ingested in excessive doses. A health-based definition of hazardous waste would be logical but awkward to apply due to the many complexities involved. Ultimately, any definition that is selected for hazardous waste must be easy to understand and must rely upon reproducible, widely-accepted test methods.

Under RCRA, a waste must first be a solid waste to be considered a hazardous waste. Hazardous wastes are, by statutory specification, a subset of solid wastes. A solid waste, by definition, is any material, other than those which are specifically excluded, that is:

1. disposed of or abandoned in lieu of disposal;
2. burned, incinerated, or recycled; or
3. considered inherently waste-like.

The definition of solid waste was evidently intentionally made very broad and inclusive.

A solid waste, unless specifically excluded, is considered a hazardous waste if any one of the following four criteria are met.

1. The waste is specifically listed as a hazardous waste (RCRA provides an extensive list of waste constituents that are by definition hazardous – these wastes are called **listed wastes**).
2. The waste is a mixture containing hazardous waste (i.e., any waste mixed with a hazardous waste is also a hazardous waste).
3. The waste is derived from the treatment, storage, or disposal of hazardous waste, i.e., once a hazardous waste, always a hazardous waste (but see exemptions discussed later).
4. The waste exhibits any one of four characteristics of hazardous waste:
   (a) ignitability (the waste presents a fire hazard during routine management and either has a flash point $<60\,^{\circ}C$ or is not a liquid and is capable of causing fire, e.g., by spontaneous chemical changes, that creates a hazard);
   (b) corrosivity ($pH \leqslant 2$ or $pH \geqslant 12.5$, or corrodes steel at a rate greater than 6 mm/yr at $58\,^{\circ}C$);
   (c) reactivity (unstable waste that undergoes violent reaction); or
   (d) toxicity (based on the Toxicity Characteristic Leaching Procedure, or TCLP, test).

The TCLP test is a revision of the US Environmental Protection Agency's (EPA's) earlier test procedure known as the Extraction Procedure (EP, or EP toxicity test). In the original EP toxicity test, the pulverized solid waste was mixed with a leachant at a liquid:solids ratio of 20:1. The liquid and solid were mixed in a sealed container, and then the solid and liquid phases were separated by centrifugation and filtration. The liquid was analyzed for 14 compounds (8 metals, 4 insecticides, and 2 herbicides). Concentrations were compared to published values; if the published values were exceeded, the waste was defined as hazardous. If not, the waste was not considered a

Table 1.2 Toxicity characteristic constituents and regulatory levels

TCLP constituents/regulatory levels		*old EP constituents/regulatory levels*	
Benzene	0.5 mg/L	Arsenic	5.0 mg/L
Carbon tetrachloride	0.5 mg/L	Barium	100.0 mg/L
Chlordane	0.03 mg/L	Cadmium	1.0 mg/L
Chlorobenzene	100.0 mg/L	Chromium	5.0 mg/L
Chloroform	6.0 mg/L	Lead	5.0 mg/L
m-Cresol	200.0 mg/L*	Mercury	0.2 mg/L
o-Cresol	200.0 mg/L	Selenium	1.0 mg/L
p-Cresol	200.0 mg/L	Silver	5.0 mg/L
1,4-Dichlorobenzene	7.5 mg/L	Endrin	0.02 mg/L
1,2-Dichloroethane	0.5 mg/L	Lindane	0.4 mg/L
1,1-Dichloroethylene	0.7 mg/L	Methoxychlor	10.0 mg/L
2,4-Dinitrotoluene	0.13 mg/L	Toxaphene	0.5 mg/L
Heptachlor (and its hydroxide)	0.008 mg/L	2,4 D	10.0 mg/L
Hexachloro-1,3-butadiene	0.5 mg/L	2,4,5T-Silvex	1.0 mg/L
Hexachlorobenzene	0.13 mg/L		
Hexachloroethane	3.0 mg/L†		
Methyl ethyl ketone	200.0 mg/L		
Nitrobenzene	2.0 mg/L		
Pentachlorophenol	100.0 mg/L		
Pyridine	5.0 mg/L†		
Tetrachloroethylene	0.7 mg/L		
Trichloroethylene	0.5 mg/L		
2,4,5-Trichlorophenol	400.0 mg/L		
2,4,6-Trichlorophenol	2.0 mg/L		
Vinyl Chloride	0.2 mg/L		

* If o-, m-, and p-Cresol concentrations cannot be differentiated, the total cresol concentration is used. The regulatory level for total cresol is 200.0 mg/L.
† Quantitation limit is greater than the calculated regulatory level. The quantitation limit, therefore, becomes the regulatory level.

characteristic hazardous waste and would have to meet one of the other criteria to be considered a hazardous waste.

In 1984, EPA was directed to reconsider the EP toxicity test. In 1986, EPA proposed to revise and expand the EP text. After much deliberation, collaboration, and revision, EPA issued the final procedure for the revised test (i.e., the TCLP) in 1989. The TCLP procedure varies slightly depending on the nature of possible contamination and the form of the waste. Basically, though, the procedure still involves mixing the solid waste with a leachant at a liquid: solid ratio of 20:1. The compounds analyzed in the TCLP test, and trigger values for definition as a hazardous waste, are listed in Table 1.2.

A number of waste types are specifically excluded from categorization as a hazardous waste. The two most notable exclusions are household refuse and small-quantity hazardous waste.

In the US, it is illegal to dispose of untreated hazardous waste. All hazardous waste must first be treated with best demonstrated available technology (BDAT) prior to disposal. For this reason, the amount of hazardous waste disposed of in landfills has declined while the amount destroyed, e.g., via incineration, has increased markedly.

In the US, the volume of household waste (often termed **municipal solid waste**, or MSW) that must be landfilled has remained relatively constant or increased slightly. Increased recycling efforts have served more to slow the rate of growth of MSW that must be landfilled rather than to reduce the amount going to landfills. Landfills are, and will continue to be for the foreseeable future, the primary means for disposal of MSW. The need for landfills for all categories of waste will remain because society can never fully rid itself of all solid residual material.

1.4 WASTE CHARACTERISTICS

Hazardous waste cover a broad spectrum of materials, particularly when one considers newly generated, treated wastes and old, untreated wastes buried in the ground some years in the past. Figure 1.2 shows a distribution of industrial waste constituents in the mid 1980s in the US. Many wastes are mixtures of materials. Any attempt to characterize such materials would be pointless; virtually all waste forms can be found. The wastes range from strongly acidic to neutral to strongly

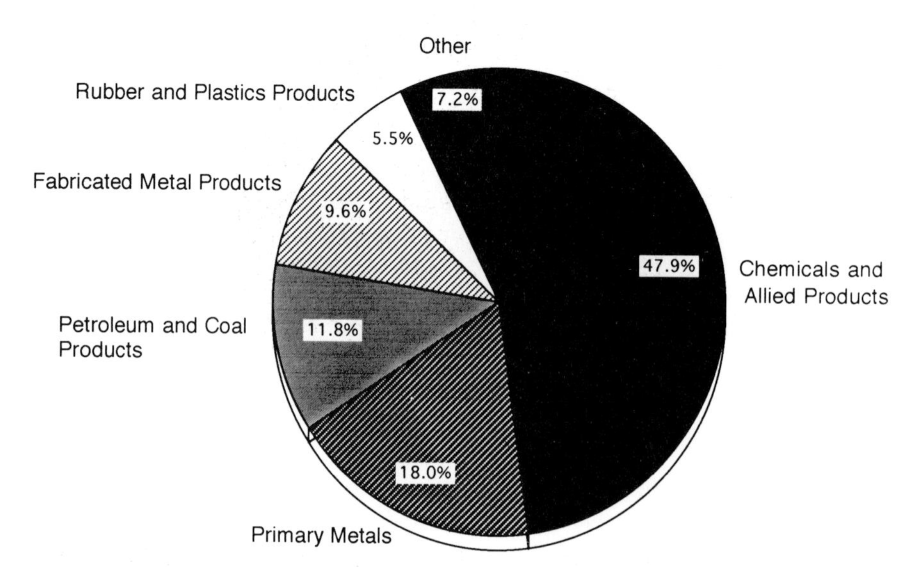

Fig. 1.2 Distribution of industrial waste in the US in the mid 1980s.

alkaline. Some wastes are rich in metals, some are rich in organics, and some are mixtures that contain both metals and organics.

One special type of waste deserves mention at this point to introduce terminology. Organic liquids may be divided into two groups: water soluble liquids (e.g., ethyl alcohol) and non-water soluble liquids (e.g., trichloroethylene). Actually, virtually all organic liquids are soluble to some extent in water, but with many organic liquids the solubilities are measured in parts per million or parts per billion of organic constituent that can be dissolved in water. Non-water-soluble organic liquids are called **non-aqueous-phase liquids** (NAPLs). There are two types of non-aqueous-phase organic liquids; dense non-aqueous phase liquids (DNAPLs) are heavier than water and light non-aqueous phase liquids (LNAPLs) are lighter than water. One would expect to find a LNAPL (e.g., gasoline) floating on the water table and a DNAPL (such as the chlorinated solvent perchloroethylene) perhaps below the water table at the interface between an aquifer and a lower-hydraulic-conductivity formation. Cleanup of LNAPLs and especially DNAPLs poses unique challenges, as discussed later in this book.

Municipal solid waste (MSW) is somewhat more consistent than industrial waste. The distribution of MSW produced in the US is shown in Fig. 1.3 (by weight) and Fig. 1.4 (by volume). As recycling becomes more prevalent, the distribution can be expected to change. An important feature of MSW is that the waste decomposes and

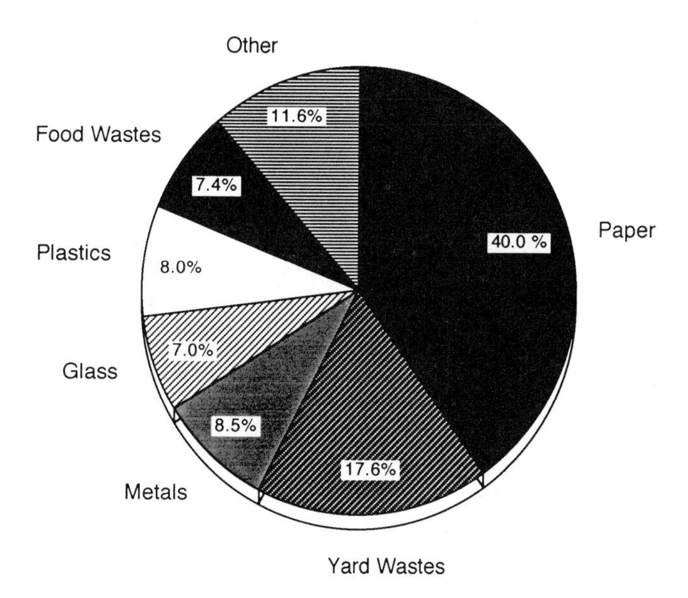

Fig. 1.3 Distribution of municipal solid waste generated in the US by weight (USEPA, 1990).

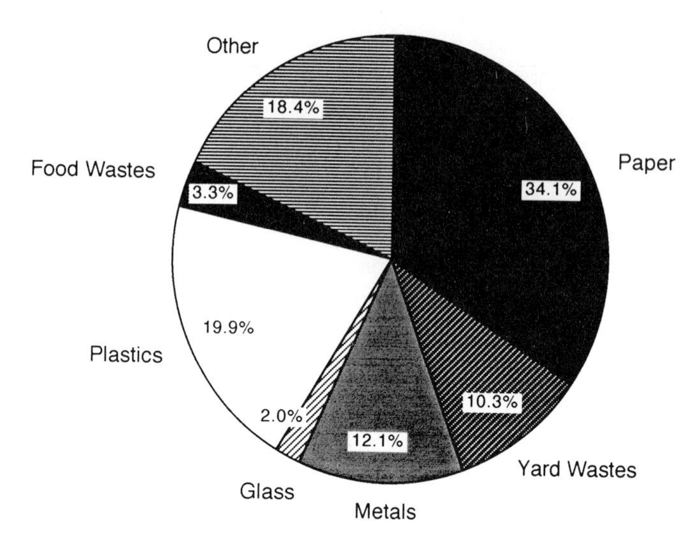

Fig. 1.4 Distribution of municipal solid waste in US landfills by volume (USEPA, 1990).

produces gas, including methane, that can be very dangerous if improperly controlled.

The liquid that is derived from waste is called leachate. Leachate can be produced directly from buried liquid wastes or consolidation of fluid-bearing wastes, by decomposition or chemical reactions, or by the leaching action of water moving through the waste. The control of leachate and gas is the single most important design requirement for new waste disposal facilities.

REFERENCES

USEPA (1990) *Characterization of municipal solid waste in the United States: 1990 update*, Office of solid waste and emergency response, Washington, DC, EPA/530-SW-90-042.

van Veen, F. and J.A. Mensink (1985) Brief survey of legislation and arrangements for the disposal of chemical waste in a number of industrialized countries, *Haz. Waste & Haz. Mat.*, **2**(3), 333–53.

CHAPTER 2

Geochemistry

Jim V. Rouse and Roman Z. Pyrih

2.1 INTRODUCTION

Not so many years ago when the subject of geochemistry was mentioned, topics such as crustal abundance of the various elements, radiochemical age dating, or computer modeling of stable chemical states immediately came to mind. Most engineers in the geotechnical professions equated geochemistry with analytical services. However, there is a rapidly emerging side to the discipline of geochemistry which relates directly to geotechnical practices in waste disposal.

This relationship between the two disciplines is based on the principle that an interdisciplinary approach to problem solving has the best chance of success. This is especially true when dealing with waste disposal. It is essential for the geotechnical professional to be able to communicate with the geochemist, to ask the right questions, and to understand the significance of geochemical data. At the same time, it is the obligation of the geochemist to provide data in usable form, to voice alerts and anticipate problems, and, most importantly, to integrate the principles of geochemistry into the engineering and design of waste disposal facilities, soil cleanup strategies, and ground-water restoration programs. When integrated with sound engineering design, geochemistry can provide an additional level of environmental safeguards and control.

The purpose of this chapter is to introduce the geotechnical professional to two very useful concepts of geochemistry: the concept of attenuation and the idea of phased ground-water monitoring. When integrated into the design of a waste-disposal facility, attenuation introduces one additional environmental safeguard against the migration of potential ground-water contaminants. Likewise, taking a phased

Geotechnical Practice for Waste Disposal.
Edited by David E. Daniel.
Published in 1993 by Chapman & Hall, London. ISBN 0 412 35170 6

approach to the monitoring of chemical constituents during and after the operational life of the facility is not only cost-effective, but is more environmentally protective than the approach of collecting and analyzing for 'everything'.

The discussion of the basic geochemical principles involved in these concepts is not intensive; it is provided so that the geotechnical engineer can relate the concepts to everyday problems. Geochemical literature cited throughout the chapter will provide additional insight into various aspects of environmental geochemistry, for those who wish to pursue the subject further. The examples of geochemical processes which are referenced in the chapter focus on geochemical and biogeochemical interactions between ground water and the natural geological material through which it flows. As such, the examples illustrate the effect of geochemical and biogeochemical interactions on chemical constituents present in seepage solutions and ground water.

2.2 GEOCHEMICAL ATTENUATION

Natural geological material has the ability to interact geochemically with chemical constituents of ground water. The results of this interaction can lead to the partial or total immobilization of potential ground-water contaminants. This process of immobilizing and retarding the chemical constituents from moving with ground-water flow is called attenuation.

The ability of natural geological material such as soil to immobilize potential ground-water contaminants and attenuate their movement is well known. Under the right conditions, soils and sediments can act as geochemical 'traps' and prevent the movement of various chemical constituents. Early in the development of geotechnical practices in waste disposal, Griffin *et al.* (1976, 1977) described and quantified the ability of clay-bearing material to remove heavy metals and other trace elements from seepage solutions resulting from landfill leachate. The authors of this early effort made the innovative suggestion that overall pollution from landfills could be reduced if liners of natural earth material were designed for higher hydraulic conductivities. Higher hydraulic conductivities would allow for contaminant attenuation and avoid the development of 'bathtub' conditions. Unfortunately, recent regulatory reliance on specific design criteria runs counter to this often reasonable and technically-sound approach.

As described by Rouse and Pyrih (1985), a number of naturally occurring geochemical processes are effective in removing ground-water contaminants. These mechanisms can include cation and anion exchange with clays, adsorption of cations and anions on hydrous oxides of iron and manganese, sorption on organic matter, direct precipitation of ions from solution, and co-precipitation by adsorption. In addition, volatilization and biodegradation can be effective mechanisms for attenuating the movement of organic contaminants (Dragun, 1988).

Most geological material in contact with ground water contains some percentage of clay minerals. Such geological material usually consists of complex mixtures of clay minerals, iron and manganese hydrous oxides, and organic matter. Under the proper conditions, the clay minerals, the hydrous oxides and the organic matter impart to natural earth material the ability to scavenge and to concentrate cations and anions from seepage solutions or migrating ground water (Dragun, 1988; Hornick, 1976).

A mechanism with the potential to affect significant attenuation is cation and anion exchange between clay minerals and ions in solution. Ion-exchange or ion-replacement reactions can occur to some extent in all clay minerals (Grim, 1968). Analogous reactions resulting in ion-exchange from solution can occur not only with clay minerals but also at the surface of iron and manganese hydrous-oxides (Jenne, 1968) and organic matter (Schmidt-Collerur, 1978) that can be found associated with natural geological materials.

Other mechanisms can be activated as seepage solution or ground water contacts geological materials. Precipitation and coprecipitation can remove chemical constituents from solution as insoluble precipitates. Often, precipitation reactions are initiated by changes in chemical parameters such as pH and Eh. As a general rule, neutralization of pH optimizes conditions for geochemical removal of many constituents from solutions. Geochemical attenuation mechanisms are most active in a pH range between 5 and 8.

The extent to which geological materials will function as a geochemical trap and attenuate the movement of potential ground-water contaminants will depend upon:

1. the chemical composition of the seepage solution or ground water;
2. the geochemical and mineralogical properties of the geological material; and
3. the pH and Eh conditions that are established during contact of the water with the geological material.

2.2.1 Examples of geochemical attenuation

The environmental fate and geochemical behavior of chromium and selenium have been under intensive investigation (Rouse and Pyrih 1989; Rouse, 1988; Dragun, 1988; Hornick, 1976). Griffin *et al.* (1976) reported that clay-bearing material could remove trace elements including chromium and selenium from seepage solutions resulting from landfill leachate. Above values of pH 6, trivalent chromium was immobilized as a result of precipitation. Below pH 4, the trivalent species was adsorbed by both kaolinite and montmorillonite clays. Between pH 4 and 6, the combination of adsorption and precipitation mechanisms rendered trivalent chromium immobile. The removal of selenium (as the selenite ion) from leachate solution by kaolinite and montmorillonite clays reached a maximum in the pH range of 2 to 3, but quickly decreased as the pH increased. The adsorption of selenium became negligible as the pH approached 10. Fritz and Hall (1988) established that selenium adsorption occurs not only on clays but on hydrous oxides and organic matter. Maximum adsorption of selenite occurred between pH 4 and 6; above pH 8.5 a significant decrease in selenite adsorption was observed.

Experiences with the subsurface migration of arsenic have shown that this potential ground-water contaminant can be and is attenuated by geochemical mechanisms (Pyrih and Rouse, 1989). Under slightly acidic pH conditions, arsenic will readily anion-exchange with clay minerals such as illite and smectite. In addition, arsenic is sorbed by hydrous oxides of iron and manganese and precipitated as an insoluble arsenate by metals such as iron, copper, or zinc. Iron/arsenic and iron/chromium reactions are the basis for newly-evolving process technology designed to remove traces of arsenic and chromium from natural water systems (Peck, 1990).

The environmental fate of free cyanide and of metal-cyanide complexes has also been under intensive investigation. Recent laboratory studies and field investigations of cyanide attenuation by natural geological material indicate that free cyanide and metal-cyanide complexes move only short distances through soil before being immobilized. This attenuation of cyanide is due to geochemical and biogeochemical processes that naturally occur in the subsurface. Several geochemical mechanisms can account for the removal of cyanide from solutions. These mechanisms include sorption on mineral surfaces or organic detritus; precipitation of insoluble metal-cyanide complexes typically found with iron; chemical conversion to thiocyanate; and chemical oxidation to cyanate. Furthermore, cyanide can be altered by biological mechanisms and metabolized by plants. Under aerobic conditions occurring in near-surface soils and ground water, cyanide will be decomposed to ammonia, nitrogen or even nitrate, and to carbon dioxide. Under anaerobic conditions, cyanide will be decomposed to ammonium ion, nitrogen, thiocyanate, and carbon dioxide. (University of California at Berkeley, 1988; Smith, 1988; Rouse and Pyrih, 1988; Chatwin and Trepanowski, 1987; Simovic *et al.*, 1984; Schmidt *et al.*, 1981; Knowles, 1976).

Volatilization of free cyanide is an important physicochemical mechanism that can limit the mobility of free cyanide and weak metal-cyanide complexes in the near-surface environment. Free cyanide volatilization becomes quite significant when the pH in soil or ground water is reduced below 9.4. This mechanism is especially important since most soils are capable of neutralizing alkaline seepage and lowering the solution pH. Photodecomposition of metal-cyanide complexes accelerates the volatilization process by breaking down metal-cyanide complexes in the presence of sunlight. The extent of sunlight penetration will determine the degree of this photodecomposition. (Smith, 1988; Huiatt *et al.*, 1983; Schmidt *et al.*, 1981).

Moderately strong and strong metal-cyanide complexes such as ferro-and ferri-cyanide complexes are exceptionally stable and have a geochemistry that is totally different from the free forms of cyanide, such as molecular hydrogen cyanide and simple ionic cyanide. Such iron-cyanide complexes are better able to interact geochemically and biogeochemically with soil and aquifer materials. These interactions lead to immobilization of the cyanide complexes and eventual decomposition by microbial activity (Smith and Struhsacker, 1988; Scott, 1984; Huiatt *et al.*, 1983).

2.2.2 Quantifying attenuation capacities

The effectiveness and capacity of natural geological material to attenuate the movement of chemical constituents that may be present in seepage solutions or ground water can be quantified in the laboratory. To quantify attenuation capacities in a geological material, a phased geochemical program of testing and analyses should be designed that includes the following steps:

1. analysis of the earth material for mineralogical and geochemical properties;
2. conduct of sequential batch-contact measurements to quantity attenuation capacities;
3. conduct of long-term column experiments to confirm attenuation capacity; and
4. field demonstration.

Geochemical properties

As an initial effort to quantify the geochemical attenuation capacity of geological material, representative samples should be analyzed by X-ray diffraction to determine bulk mineralogy and the mineralogy of the clay-size fractions. In addition, samples should be analyzed for a list of geochemical parameters which experience has shown can affect the degree of attenuation taking place in natural earth material. This list of geochemical parameters includes at least the following:

- clay content
- cation-exchange capacity and exchangeable cations
- iron and manganese hydrous-oxide content
- organic carbon content
- soil pH
- acid and base neutralizing potential

A good source of information on procedures for measuring the parameters listed above is Klute (1986).

Laboratory evaluation

After data are developed on the geochemical character of site soils and the degree of variability of the geological material, the next task is to evaluate the material's attenuation capacity. The geochemical reactions which are likely to occur when seepage solutions percolate through soils or ground water flows through geological material can be simulated in the laboratory. Two procedures have been shown to be useful in demonstrating these reactions and evaluating the effectiveness of the attenuation mechanisms, namely:

- sequential batch-contact tests
- column percolation tests

Each procedure has certain advantages and disadvantages. Sequential batch-contact tests are useful in evaluating attenuation properties on a small sample of geological material and obtaining estimates of attenuation capacity. The testing can be completed relatively quickly and is ideally suited for determining attenuation characteristics on a number of samples at one time. However, the results from column testing generally describe attenuation processes more truly. Unfortunately, column tests may require several months to generate useful data for materials with low hydraulic conductivity.

2.2.3 Laboratory methodology and procedures

Sequential batch-contact testing

Sequential batch-contact testing is often selected as the laboratory technique to obtain preliminary indications of attenuation capacities. The sequential batch procedure simulates seepage solution percolating through columns of soil material wherein successive solution comes in contact continuously with fresh material. This testing procedure can simulate years of potential field seepage or ground-water flow in a few days of laboratory testing. The methodology and procedures for sequential batch-contact testwork are described in detail by Houle and Long (1980).

Sequential batch-contact testing (SBT) of a soil's attenuation property proceeds by placing weighed amounts of a bulk sample in polyethylene containers and adding measured volumes of representative seepage solution or ground water containing the potential contaminant. The proportions of solution to bulk sample are controlled in order to achieve the desired liquid to solid ratios. The resulting slurries are gently agitated over a period of 24 hours. At the conclusion of each 24-hour period, the slurry is decanted and filtered. The moist solids and the appropriate volume of leachate solution from the previous test are advanced in sequence according to the matrix illustrated in Fig. 2.1. Liquids are advanced left to right; solids are advanced from top to bottom. Care is taken to transfer all solids including any slimes adhering to the filter paper. The remainder of the leachate solution from each test is saved for chemical analysis. Inorganic leachate samples for laboratory analyses are usually filtered through a 0.45 micron filter, split into appropriate sample portions, and preserved as required. In principle, the same procedures can be used for organics with the following precautions:

1. materials must be non-absorbent (e.g., glass);
2. there must be no opportunity for volatilization (i.e., no air in the mixing chamber); and
3. samples must be handled and transferred in a way that ensures no losses of volatile organics.

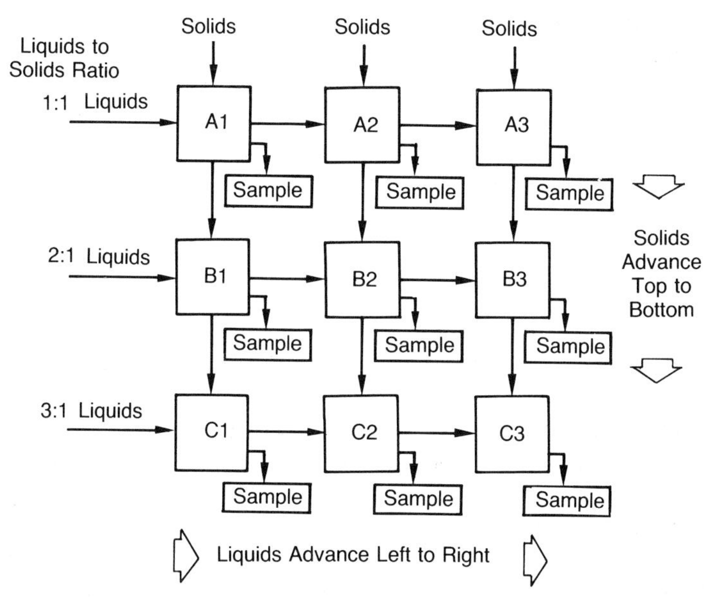

Fig. 2.1 Schematic of a typical sequential batch-contact test.

The liquid to solid ratios that are selected for the sequential batch tests control the volume of solution that a given weight of soil or geological material would contact. These ratios can be related to pore volume and, based on these ratios, an estimate of equivalent pore volume throughput can be calculated. At the liquid to solid ratios specified in the schematic of Fig. 2.1, leachate sample A3 would be equivalent to 0 to approximately 2 pore volumes of seepage throughput. At the other extreme of the schematic, leachate sample C1 would be equivalent to approximately 20 through 40 pore volumes of seepage throughput.

The attenuation capacity of the soil or geological material for a chemical constituent present in the solution or ground-water sample can be directly determined from the SBT data. The attenuation capacity is calculated by summing the amount of the constituent removed from solution during each contact with solids material, and by dividing this amount by the weight of the soil or geological sample in contact with the solution. This calculation is repeated for each contact of batch A, B, and C, and the results are summed over the entire test sequence. In contacts where a constituent was obviously desorbed from the earth material sample, the attenuation capacity is adjusted downward by subtracting the amount which was dissolved. The results of the calculations are reported as milligrams (mg) of constituent attenuated per kilogram (kg) of earth material.

A simpler form of the sequential batch-contact procedure can also be used to determine parameters useful for modeling contaminant transport in the ground. In a single-batch extraction test, leachate is mixed with soil for 24 hours. The concentration of contaminant constituents in the leachate is measured before and after 24 hours of gentle agitation. If the constituent does not tend to partition from the aqueous phase to the solid phase, the concentration in solution (aqueous phase) will be the same before and after agitation. Such a constituent is said to be 'conservative', 'non-reactive', or 'non-sorbed'.

With most waste constituents, the solution concentration will be lower after 24 hours of gentle mixing because some of the constituent will be sorbed by the soil. Such constituents are said to be 'sorbed', 'reactive', or 'attenuated' species. The mass of constituent sorbed, M, is easily calculated from the difference in solution concentration before and after agitation, and from the volume of leachate. A graph is plotted of the mass of chemical constituent sorbed per mass of soil (S) versus the concentration of the chemical constituent (C) in the solution. An example of such a plot is shown in Fig. 2.2. This type of graph is called an 'adsorption isotherm'. The relationship between S and C is sometimes linear, particularly for very low concentrations of metals and organics in aqueous solutions. If the isotherm is linear, the slope, R_d, is

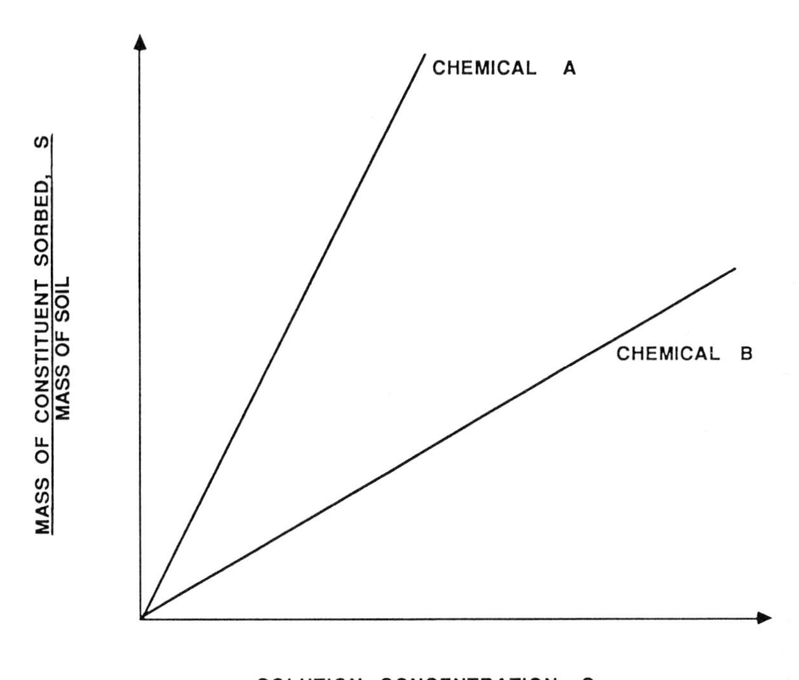

Fig. 2.2 Linear adsorption isotherms.

$$R_d = \frac{S}{C} \tag{2.1}$$

where R_d is denoted 'distribution ratio' or 'partition coefficient'. In Fig. 2.2, the R_d of chemical A is greater than R_d of chemical B, indicating that chemical A is more strongly sorbed by soil than chemical B.

If the soil and solution are in equilibrium after 24 hours of agitation, then R_d is equivalent to K_d, the 'distribution coefficient':

$$K_d = R_d = \frac{S}{C} \tag{2.2}$$

The distribution coefficient, K_d, is often an input parameter to contaminant transport models.

With many chemical/soil combinations, and particularly for batch tests performed on leachates containing high concentrations of solutes, the adsorption isotherms are non-linear. A finite sorption capacity results in the typical nonlinear adsorption isotherm shown in Fig. 2.3. A common model used to describe such isotherms is the Freundlich equation

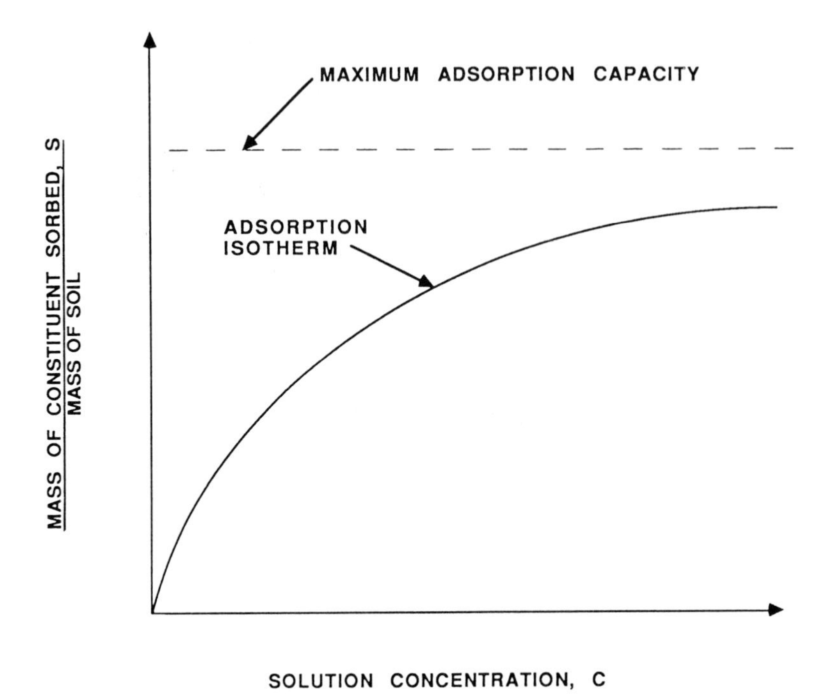

Fig. 2.3 Non-linear adsorption isotherm.

$$S = K_d C^n \tag{2.3}$$

where S is the mass of chemical constituent sorbed per mass of soil, C is the concentration of the chemical constituent in the solution and K_d and n are coefficients that depend upon conditions of the system.

If one uses a contaminant transport model that assumes a linear isotherm, but the isotherm is in fact non-linear, the appropriate concentration range must be selected to estimate the value of K_d to imput to the model.

The reader is reminded of the many factors, such as pH, that can affect attenuation. These factors must be carefully controlled in batch testwork, and must accurately reflect field conditions.

Column percolation testing

Column tests are at the other end of available laboratory methodology. Column tests generally provide the most realistic information on the attenuation properties of natural geological materials. If properly designed, the column test procedure can more accurately describe attenuation processes by simulating conditions that will occur in the field. Unlike the sequential batch-contact test, the column test is

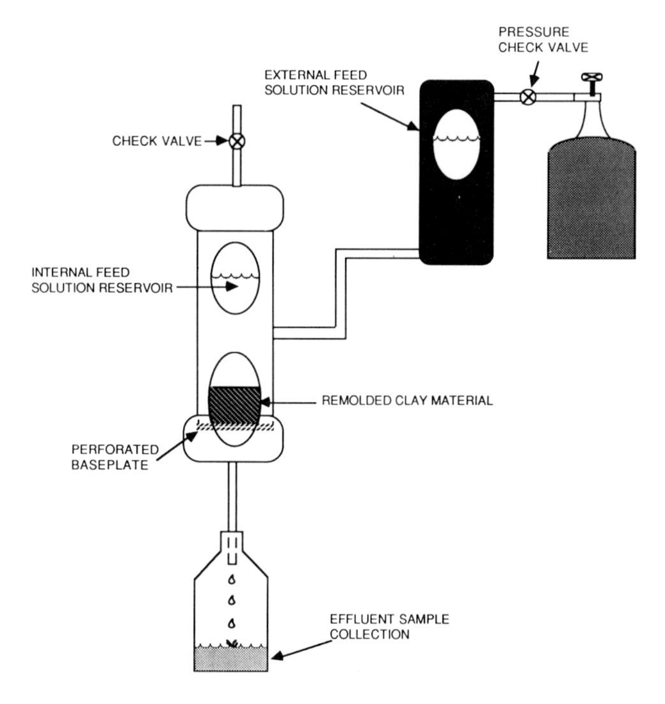

Fig. 2.4 Operation of a typical column test.

usually a long-term procedure often taking several months to complete, depending upon the permeability of the clay-bearing material that is being evaluated.

The objective of the column testing is to force representative solutions that contain chemical constituents of concern through columns packed with samples of earth material. During operation of the columns, effluent samples are periodically sampled for chemical analysis. By determining the chemical composition of the effluent that has percolated through the column material and comparing the results to the starting 'feed' solution, the attenuation properties of the soil or geological material can be quantified. Such tests frequently are performed in flexible-wall permeameters or in relatively simple rigid-wall columns fabricated from large diameter sections of plastic tubing or pipe. The operation of a typical down-flow column test performed on clay-bearing material is illustrated in the generalized sketch of Fig. 2.4.

It is not uncommon for column tests to show decreased hydraulic conductivities over time. This decrease in hydraulic conductivity is usually due to the formation of secondary minerals in interstitial voids of the clay-bearing material or to transformations in clay-mineral phases.

As with sequential batch-contact testing, attenuation capacity can be calculated directly from analyses of column effluent samples. Attenuation capacities are calculated by summing the amount of the constituent that was removed from solution during each pore volume of column throughput. If effluent analyses indicate an increase in constituent concentration (suggesting desorption), the attenuation capacity is adjusted downward by subtracting the amount that was desorbed. The results of the calculations are reported as milligrams (mg) of constituent attenuated per kilograms (kg) of natural geological material packed in the column and contacted by the column 'feed' solution.

Biogeochemical testing

Biogeochemical degradation has been shown to be extremely effective in attenuating the movement of many organic ground-water contaminants. Several laboratory testing programs have been developed to evaluate the effectiveness of these biogeochemical processes (Brubaker and Crockett, 1986). The purpose of such laboratory scale testwork is four-fold, namely:

1. to determine the availability of indigenous organisms and their capacity to survive in the subsurface environment;
2. to ascertain if these organisms have the ability to biodegrade soil and ground-water contaminants;
3. to evaluate critical factors limiting the growth of such organisms; and
4. to determine if enhancing environmental conditions for biodegradation might mobilize other contaminants or reaction products.

Biological degradation is discussed further in later chapters in this book.

2.2.4 Field methodology

Soil water samplers, commonly referred to as pressure/vacuum lysimeters, are field sampling devices that have been used successfully for monitoring water quality in the unsaturated zone. In the past, such lysimeters have been used in the mining industry to monitor the unsaturated zone for solution losses from heap leach facilities and for leachate seeping from tailing ponds (Pyrih, 1991). In more recent applications, lysimeters have been installed at various depths in ore heaps in order to evaluate the efficiency of the leaching process and to determine the completeness of heap neutralization. Soil water samplers provide a means for field verification of geochemical attenuation.

The pressure/vacuum lysimeter consists of a cylindrical chamber made of PVC and a low permeability porous ceramic cup. Pore water

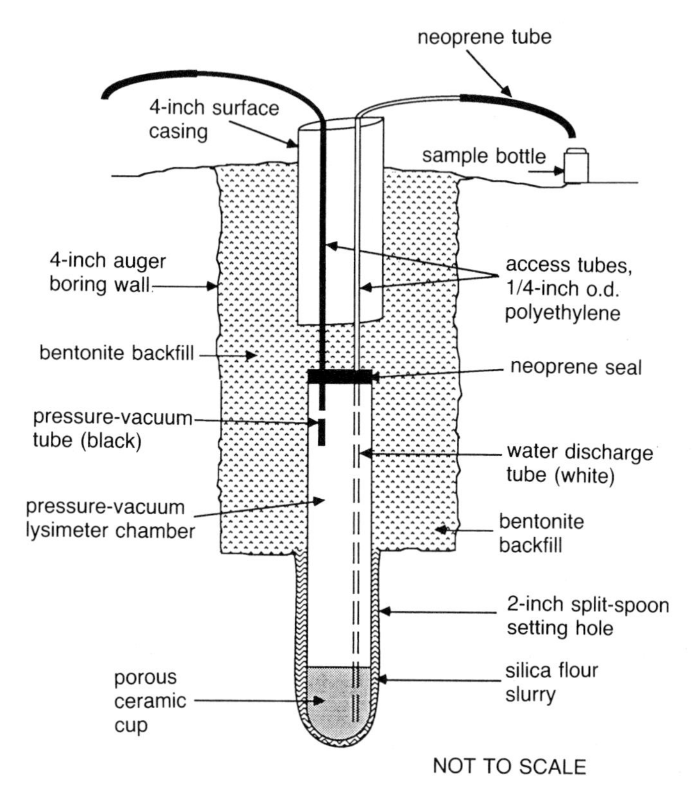

Fig. 2.5 Schematic of a typical pressure/vacuum lysimeter.

in the form of soil moisture (for example, seepage) is drawn by vacuum into the chamber through the porous ceramic and then retrieved through access tubes to the surface. The construction of a typical pressure/vacuum lysimeter is illustrated in Fig. 2.5.

The ability of a soil in the unsaturated zone to attenuate the movement of potential ground-water contaminants can be confirmed and quantified in field percolation tests using pressure/vacuum lysimeters. The lysimeters are used to sample the soil moisture as the seepage containing the contaminants of concern migrates downward. Field installation and operation of a 'nest' of soil moisture lysimeters is illustrated in Fig. 2.6. Further information on lysimeters is given in Chapter 24, Vadose Zone Monitoring.

2.3 PHASED GROUND-WATER MONITORING

The fact that geochemical attenuation is predictable and quantifiable can be used to advantage in the design of a phased ground-water

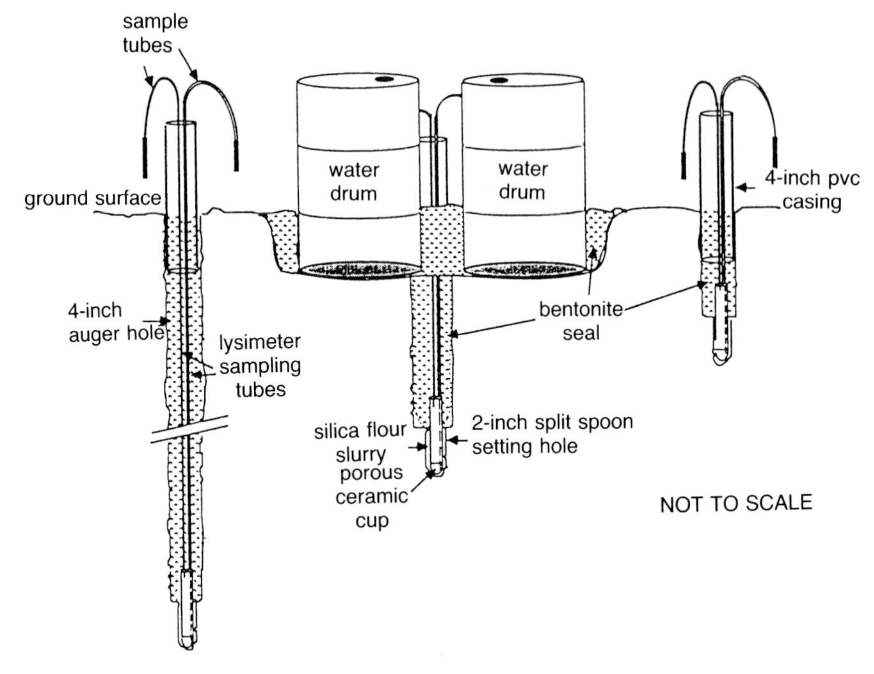

Fig. 2.6 Field installation of a nest of pressure/vacuum lysimeters.

monitoring program. Such a phased monitoring program is more cost effective and environmentally protective than the more common approach of collecting and analyzing samples for esoteric, nonconservative parameters such as heavy metals or complex organics. The phased monitoring system makes use of the fact that most environmentally sensitive contaminants are retarded during migration through geologic material, relative to the more-common major-ion components.

Experience has shown that plumes of migrating ground-water contamination can be divided into three zones, as illustrated in Fig. 2.7 (Cherry, Sheperd and Morin, 1982; Rouse and Pyrih, 1990). The 'core' zone, located closest to the source, is a zone in which the geochemical attenuation capacity of the subsurface material has been depleted. The ground water in the core zone is similar to that of the source. The 'active' zone is the site of intense geochemical interaction between subsurface materials and migrating ground-water contaminants. The active zone is the zone in which the previously-described reactions primarily occur, and marks the site of active precipitate formation. The downgradient 'neutralized' zone contains only conservative components which generally do not enter into geochemical attenuation reactions. These can include most of the major-ion components of the migrating seepage plume, but may include oxidized species of chromium or selenium.

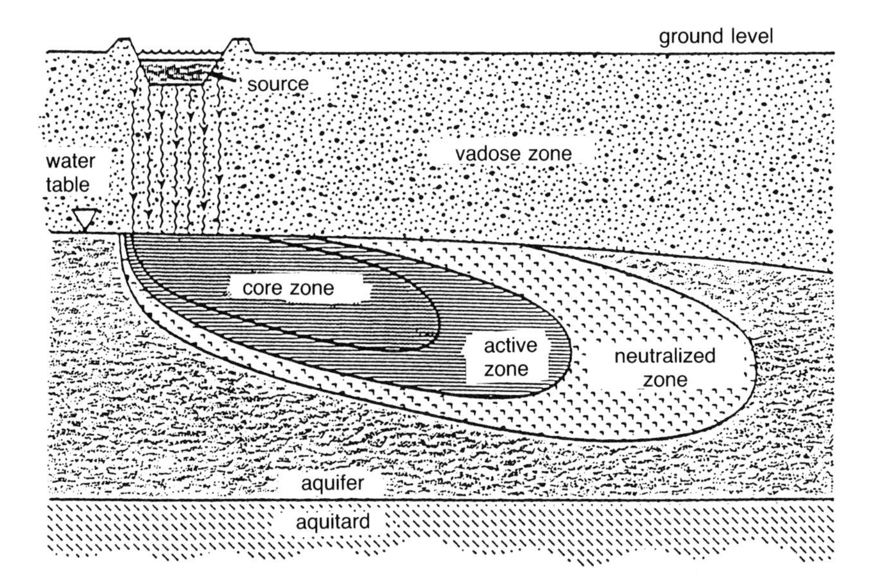

Fig. 2.7 Conceptual geochemical model of zones in a contaminant plume.

This conceptual model can be used to advantage in the design, operation, and interpretation of a phased monitoring system. A well located downgradient of a migrating seepage plume will yield ground water essentially identical in chemical composition to ambient ground water. The onset of the arrival of contamination will be characterized by a shift in major-ion chemistry and the detection of other conservative contaminants. Initial monitoring parameters should include only major ions (Ca, Mg, Na, K, HCO_3, CO_3, SO_4, Cl) and those parameters which geochemical testing has shown to be present in the seepage solution that are not attenuated by geochemical mechanisms. Graphical techniques such as the trilinear plot (Hill, 1940) or the Stiff diagram (Stiff, 1951) are useful in detecting such shifts in major-ion composition. There is no advantage to sampling for non-conservative ions during the initial monitoring phase, since such ions will not migrate as rapidly as the more-conservative major ions.

After major-ion sampling has demonstrated the arrival of the neutralized zone at a well, monitoring should be intensified to include those parameters which geochemical testing has shown to be least attenuated. The order of arrival of non-conservative contaminants will be inversely proportional to the degree of parameter attenuation, that is, those parameters showing relatively little attenuation will arrive before those demonstrating significant attenuation.

The phased monitoring approach has environmental advantages in addition to the obvious economic advantage of reduced analytical costs. By focusing attention on a specific suite of parameters and using graphical techniques which consider the overall percentage composition in a water sample, it is possible to detect subtle changes which might be overlooked when viewed in terms of an extensive suite of parameters. The detection of such a shift can, in turn, lead to the initiation of more intensive sampling. The authors have seen several investigations lead to a false sense of security by focusing on non-conservative parameters as the indication of the presence or absence of contamination. Major-ion analysis documented a more extensive plume in which geochemical attenuation had been effective in retarding the spread of non-conservative contaminants. Examples of such errors in interpretations include the attenuation and biotransformation of cyanide to nitrate and the reduction of nitrate to ammonia followed by geochemical attenuation of the ammonia.

2.4 SUMMARY

The movement of potential ground-water contaminants in the subsurface is controlled by geochemical processes which take place when chemical constituents of ground water interact with soils, sediments and bedrock. Under the right conditions, the mobility of the constituents is retarded by these interactions with the result that the contaminants become fixed in geochemical 'traps' below the surface. This behavior can be described by a relatively simple geochemical/geohydrological conceptual model.

To be effective, ground-water monitoring programs and cleanup strategies need to consider these geochemical interactions. Cost effective and environmentally sound on-site and in-place treatment technologies must be based on the geochemistry of the contaminants of concern.

Geochemical attenuation is a predictable, dependable, and quantifiable mechanism which can be used to advantage in the design of a waste-handling facility and in the operations of a monitoring system. Geotechnical engineers can more effectively design waste disposal systems when they understand and utilize geochemical principles as an integral part of the overall system design.

REFERENCES

Brubaker, G.R. and Crockett, E.L. (1986) *In situ* aquifer remediation using enhanced bioreclamation, in *Proceedings of the Second Annual Hazardous Materials Management Conference West*, Long Beach, CA.

Chatwin, T.D. and Trepanowski, J.J. (1987) Utilization of soils to mitigate cyanide releases, in *Proc., 3rd Western Reg. Conf. on Precious Metals, Coal, and Environ.*, Rapid City, SD, September 23–26, 151–70.

Cherry, J.A., Shepherd, T.A. and Morin, K.A. (1982) *Chemical composition and geochemical behavior of contaminated ground water of uranium tailings impoundments*, Preprint 82-114, SME-AIME, February 1982.

Dragun, J. (1988) *The Soil Chemistry of Hazardous Materials*, Hazardous Materials Control Research Institute, Silver Spring, MD.

Freeze, R.A. and Cherry, J.A. (1979) *Groundwater*, Prentice Hall, Englewood Cliffs, NJ.

Fritz, S.J. and Hall, S.D. (1988) Efficacy of various sorbic media in attenuation of selenium, in *J. Environ. Qual.*, **17**(3), 480–4.

Griffin, R.A., Frost, R.R., Au, A.K., Robinson, G.D. and Shimp, N.F. (1977) *Attenuation of Pollutants in Municipal Landfill Leachate by Clay Minerals: Part 2 – Heavy Metal Adsorption*, Environmental Geology Notes, Illinois State Geological Survey (79) April.

Griffin, R.A., Keros C., Shrimp, N.F., Steel, J.D., Ruch, R.R., White, W.A., Hughes, G.M. and Gilkeson, R.H. (1976) *Attenuation of Pollutants in Municipal Landfill Leachate by Clay Minerals: Part 1 – Column Leaching and Field Verification*, Environmental Geology Notes, Illinois State Geological Survey (78), November.

Grim, R.E. (1968) *Clay Mineralogy*, McGraw-Hill Book Co., New York.

Hill, R.A. (1940) Geochemical patterns in Coachella Valley, California, in *Trans. Amer. Geophys. Union*, **21**.

Hornick, S.B. (1976) The interaction of soils with waste constituents, in *Soil Chemistry, Volume A: Basic Elements*, (eds G.H. Bolt and M.G.M. Bruggenwert) Elsevier Scientific Publishing Company, New York, 4–19.

Houle, J.J. and Long, D.E. (1980) Interpreting results from serial batch extraction tests of wastes and soils, *Proc., 6th Ann. Res. Symp.*, EPA-600/9-80-010, 60–81.

Huiatt, J.L., Kerrigan, J.E., Ferron, A.O. and Potter, G.L. (1983) Cyanide from mineral processing, *Proc., Nat. Sc. Foundation Workshop*, Salt Lake City, UT, February 2–3, 1982.

Jenne, E.A. (1968) Controls on Mn, Fe, Co, Ni, Cu, and Zn concentrations in soils and water: the significant role of hydrous Mn and Fe oxides, *Adv. in Chem.* (73), 337–87.

Klute, A. (1986) *Methods of Soil Analysis*, Amer. Soc. of Agronomy, Madison, WI, Parts 1 and 2.

Knowles, C.J. (1976) Microorganisms and cyanide, *Bact. Rev.*, Sept., 652–781.

Peck, G.A. (1990) Surface treatment processes for heavy metals removal, in *Proc., Environ. Haz. Conf. and Exposition*, Seattle, WA, May 15–17.

Pyrih, R.Z. (1991) Use of lysimeters to monitor the effects of tailings dewatering on pore-water chemistry, *Proc., Randol Gold Forum CAIRNS '91*, April.

Pyrih, R.Z. and Rouse, J.V. (1989) Attenuation processes: a viable regulatory alternative, in *Proc., Environ. Haz. Conf. and Exposition*, Seattle, WA.

Rouse, J.V. (1988) Copper, chromium and arsenic in the environment: natural concentrations and geochemical attenuation, in *Proc., Amer. Wood Preservers' Assoc.*, **84**, 110–3.

Rouse, J.V. and Pyrih, R.Z. (1985) Natural geochemical attenuation of contaminants contained in acidic seepage, in *Proc., Int. Conf. on New Frontiers for Haz. Waste Management*, EPA/600/9-85/025, 192–9.

Rouse, J.V. and Pyrih, R.Z. (1988) Natural geochemical attenuation of trace elements in migrating precious-metal process solutions, in *Proc., Perth Int. Gold Conf.*, Oct. 77–81.

Rouse, J.V. and Pyrih, R.Z. (1989) In-place cleanup of chromium contamination of soil and ground water, *Proc., Environ. Haz. Conf.*, Houston, TX.

Rouse, J.V. and Pyrih, R.Z. (1990) In-place cleanup of heavy metal contamination of soil and ground water at wood preservation sites, in *Proc., Amer. Wood Preserver's Assoc.*, **86**.

Schmidt, J.W., Simovic, L. and Shannon, E. (1981) *Natural degradation of cyanide in gold milling effluents*, Industry Seminar, Waste Water Technology Centre, Technology Development Board, Water Pollution Control Directorate, Environment Canada.

Schmidt-Collerur, J.J. (1978) *Investigations of the relationship between organic matter and uranium deposits*, Bendix Field Engineering Corporation, final report, subcontract no. 76-030-E, Denver Research Institute.

Scott, J.S. (1984) An overview of cyanide treatment methods for gold mill effluents, in *Proc., Conf. on Cyanide and the Environ.*, Tucson, AZ, Dec. 11–14, 307–30.

Simovic, L., Snodgrass, W.J., Murphy, K.L. and Schmidt, J.W. (1984) Development of a model to describe the natural degradation of cyanide in gold mill effluents, in *Proc., Conf. on Cyanide and the Environ.*, Tucson, AZ, Dec. 11–14, pp. 413–32.

Smith A. (1988) Cyanide degradation and detoxification in a heap leach, in *Introduction to Evaluation, Design, and Operation of Precious Metal Heap Leaching Projects*, Society of Mining Engineers, Littleton, CO, pp. 293–305.

Smith, A. and Struhsacker, D.W. (1988) Cyanide geochemistry and detoxification regulations, in *Introduction to Evaluation, Design, and Operation of Precious Metal Heap Leaching Projects*, Society of Mining Engineers, Littleton, CO, pp. 275–92.

Stiff, H.A., Jr. (1951) The interpretation of chemical water analysis by means of patterns, *J. Petroleum Tech.*, **3**(10), 15–71.

University of California at Berkeley (1988) *Mining Waste Study, Final Report*, prepared by Mining Waste Study Team of the University of California at Berkeley, commissioned by the California State Legislature and funded by State Water Resources Control Board, the Department of Health Services, and the California Department of Conservation.

CHAPTER 3

Contaminant transport

Charles D. Shackelford

3.1 INTRODUCTION

The purpose of this chapter is to present the basic concepts for describing contaminant transport in and through porous materials. The presentation is based on the premise that the reader has little or no formal background in contaminant transport theory. As a result, the material covered is limited to one-dimensional transport of miscible contaminants (i.e., solutes) in saturated porous media. Several references are recommended for additional study.

3.2 TRANSPORT PROCESSES

3.2.1 Advection

Advection is the process by which solutes are transported along with the flowing fluid or solvent, typically water, in response to a gradient in total hydraulic head. Due to advection, non-reactive solutes (i.e., solutes which are not subject to chemical or biological reactions) are transported at an average rate equal to the seepage velocity of the fluid, or

$$v_s = \frac{v}{n} \qquad (3.1)$$

where v_s is the seepage or average linear velocity of the water (solvent), n is the total porosity of the porous material, and v is the flux of water (i.e., quantity of flow per unit area per unit time). Equation (3.1) presupposes that all of the voids in a porous material are equally

Geotechnical Practice for Waste Disposal.
Edited by David E. Daniel.
Published in 1993 by Chapman & Hall, London. ISBN 0 412 35170 6

effective in conducting flow. The flux is given by Darcy's law which, for one-dimensional flow, can be written as

$$v = \frac{Q}{A} = -K\frac{\partial h}{\partial x} = Ki \tag{3.2}$$

where Q is the volumetric flow rate of the water, A is the total cross-sectional area (solids plus voids) perpendicular to the direction of flow, K is the hydraulic conductivity, h is the total hydraulic head, x is the direction of flow, and i is the dimensionless hydraulic gradient. The seepage velocity, v_s, reflects the fact that the fluid actually can flow only through the void space of the porous material whereas the flux, v, represents the volumetric flow of fluid through the total cross-sectional area of the material.

The time required for a nonreactive solute to migrate through a saturated soil of thickness L, known as the solute transit time, due to advection can be estimated using the seepage velocity as follows.

$$t = \frac{L}{v_s} = \frac{nL}{Ki} \tag{3.3}$$

where t is the transit time. When Eq. (3.3) is used to estimate the transit time of the solute, it is implied that all solute constituents are carried along with the fluid at the velocity v_s. Under these conditions, the advective mass flux of a particular chemical species can be calculated as

$$J_A = vc = Kic = nv_sc \tag{3.4}$$

where J_A is the advective mass flux (mass flowing through a unit cross-sectional area in a unit of time) and c is the concentration of the solute in the liquid phase of the porous material based on the volume of solution in the porous material (i.e., mass of solute per unit volume of mixture).

With some porous materials, e.g., fractured soil or rock, most of the advective flow occurs through only part of the total void space of the material. For these materials, an 'effective porosity', n_e, is defined as the volume of fluid conducting pores divided by the total volume (pores plus solids) of the material. For materials with $n_e < n$, n_e should be substituted for n in Eqs. (3.1), (3.3), and (3.4). For more discussion of effective porosity, the reader is advised to consult Freeze and Cherry (1979), Horton *et al.* (1985), and Peyton *et al.* (1986).

3.2.2 Diffusion

Diffusion may be thought of as a transport process in which a chemical or chemical species migrates in response to a gradient in its concentration, although the actual driving force for diffusive transport is the gradient in chemical potential of the solute (Robinson and Stokes, 1959). A hydraulic gradient is not required for transport of contaminants by diffusion.

The fundamental equation for diffusion is Fick's first law which, for one-dimensional transport, can be written as

$$J_D = -D_0 \frac{\partial c}{\partial x} \tag{3.5}$$

where J_D is the diffusive mass flux, x is the direction of transport, and D_0 is the 'free-solution' diffusion coefficient. Equation (3.5) represents the one-dimensional form of Fick's first law describing diffusion in aqueous or free solution (i.e., no porous material).

For diffusion in saturated porous material, a modified form of Fick's first law is used

$$J_D = -\tau D_0 n \frac{\partial c}{\partial x} \tag{3.6}$$

or

$$J_D = -D^* n \frac{\partial c}{\partial x} \tag{3.7}$$

where τ is a dimensionless tortuosity factor and D^* is the 'effective' diffusion coefficient. The porosity term is required in Eq. (3.7) since the diffusive flux, J_D, is defined with respect to the total cross-sectional area of the porous medium. The tortuosity factor accounts for the increased distance of transport and the more tortuous pathways experienced by solutes diffusing through porous media. Tortuosity is expressed as

$$\tau = \left(\frac{L}{L_e}\right)^2 \tag{3.8}$$

where L is the macroscopic, straight-line distance between two points defining the flow path, and L_e is the actual, microscopic or effective distance of transport between the same two points. Since $L_e > L$, $\tau <$ 1.0 and $D^* < D_0$. Therefore, mass transport due to diffusion in porous materials is slower than mass transport due to diffusion in free or aqueous solution. Typical values of τ are reported in the range 0.01 to 0.67 (Perkins and Johnston, 1963; Freeze and Cherry, 1979; Daniel and Shackelford, 1988; Shackelford, 1989; Shackelford and Daniel, 1991). Factors affecting the determination of τ are described by Shackelford and Daniel (1991). In some instances, the porosity term in Eq. (3.7) is included in the definition of the effective diffusion coefficient (Shackelford, 1988a; Shackelford and Daniel, 1991; Shackelford, 1991).

3.2.3 Coupled flow processes

In general, water can flow through porous media not only in response to a gradient in total hydraulic head but also in response to gradients in chemical composition (chemico-osmosis), electricity (electro-osmosis), and temperature (thermal osmosis). Also, in the absence of advective flow, solutes can migrate through porous materials not only in response to a concentration gradient but also in response to an electrical gradient (electro-phoresis) and a thermal gradient (thermal diffusion). Solutes also may be filtered (ultrafiltration) by some porous materials. These additional flow processes typically are referred to as coupled flow processes in order to distinguish them from the direct flow processes represented by advection and diffusion. The description of coupled flow processes is based on the theory of the thermodynamics of irreversible processes. A description of this theory is beyond the scope of this chapter, but an excellent review of the relation of the theory to coupled and direct flow processes can be found in Bear (1972). Mitchell (1976) provides examples of the application of coupled flow processes to typical geotechnical engineering problems.

In the formulation of the contaminant transport equations, coupled flow processes often are neglected or are considered to be insignificant for at least two reasons. First, coupled flow processes are significant only at the relatively low flow rates which typically are associated with flow through fine-grained soils (e.g., clays and silts). Since most of the applications of contaminant transport theory traditionally have been associated with problems pertaining to the transport of contaminants through highly permeable, granular materials (e.g., aquifers), the effects of coupled flow processes typically have been assumed to be negligible. Nevertheless, coupled flow processes may be significant in materials with high activity and/or low void ratio (Olsen, 1969; Greenberg *et al.*, 1973). Second, the determination of the phenomenological coefficients associated with the equations describing coupled flow processes is difficult, at best. As a result, such determinations usually can be performed only under highly controlled conditions in the laboratory and extrapolation of the results to field situations is uncertain.

3.2.4 Mechanical dispersion

In traditional contaminant transport theory, a mechanical dispersive flux, J_M, is added to the total mass flux of the solute to account for the spreading of the solute due to variations in the seepage velocity, v_s, which occur during transport in and through porous materials. On a microscopic level, these variations are thought to be related to the three different effects illustrated in Fig. 3.1 (Fried, 1975; Bear, 1979; Freeze

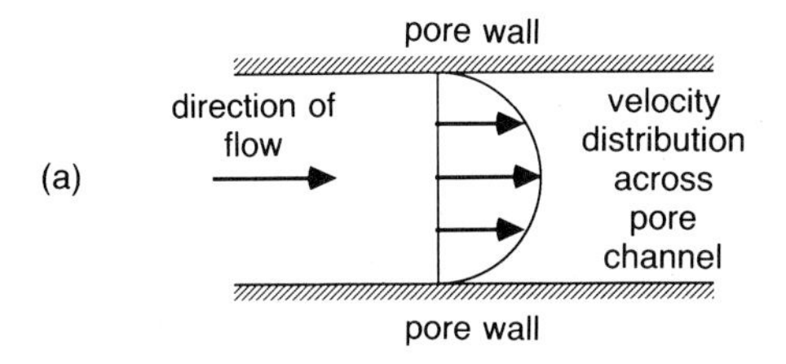

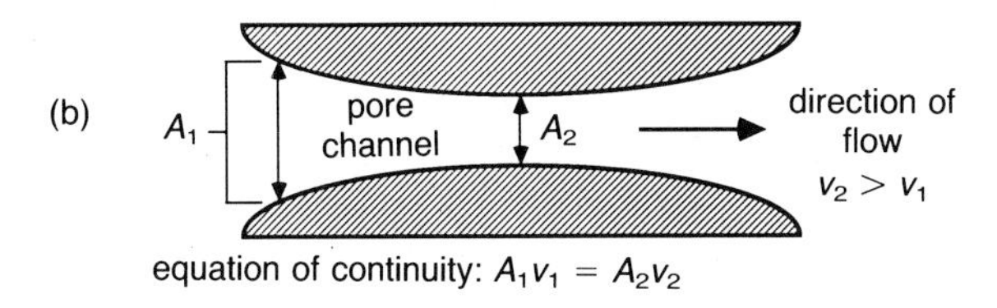

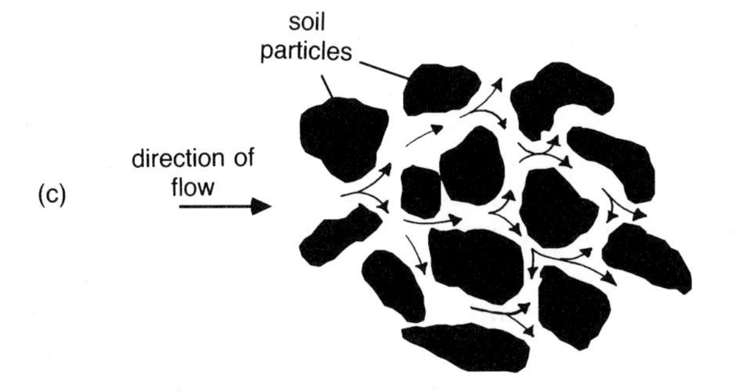

Fig. 3.1 Microscopic dispersion in soil: (a) effect of velocity distribution across single pore; (b) effect of variation in pore sizes: and (c) effect of tortuous nature of flow paths [after Freeze and Cherry (1979); and Fried (1975)].

and Cherry, 1979). First, the velocity of flow across any given pore channel within the material will be greater in the middle of the pore channel than it is near the walls of the pore channel (Fig. 3.1a). This is the same effect which is known to occur in fluid flow through pipes and in rivers, streams, and channels. Second, the equation of continuity predicts that the flow velocity across a smaller pore opening will be greater than that across a larger pore opening as illustrated in Fig. 3.1b. Finally, velocity variations will result due to the tortuous nature of the flow paths existing in nearly all porous materials (Fig. 3.1c).

On a larger scale, mechanical dispersion is thought to be caused by the different flow rates resulting from heterogeneities that typically are encountered whenever transport occurs over relatively large areas or regions. This type of dispersion is illustrated in Fig. 3.2 where flow through the bulk sand medium is shown to be interrupted by the existence of the low-hydraulic-conductivity clay lenses dispersed

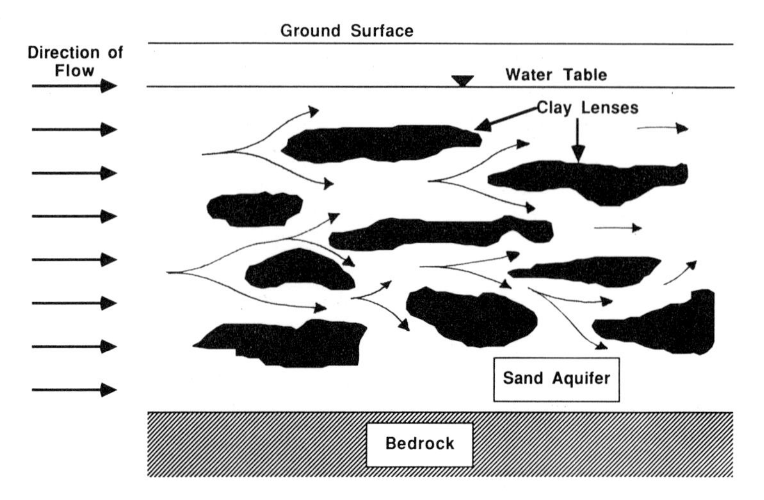

Fig. 3.2 Mechanical dispersion on large or regional scale.

throughout the sand layer. Due to the existence of the clay lenses, the transport proceeds around rather than through the clay resulting in a situation which is analogous to that shown in Fig. 3.1c for microscopic dispersion. Therefore, the dispersion illustrated in Fig. 3.2 may be thought to be due to a 'macroscopic tortuosity effect'.

The mechanical dispersive flux typically is assumed to be a Fickian process which, for one dimension, can be expressed as

$$J_M = -D_m n \frac{\partial c}{\partial x} \tag{3.9}$$

where J_M is the mechanical dispersive flux and D_m is the mechanical dispersion coefficient. Since mechanical dispersion results from variations in the magnitude of the seepage velocity, the mechanical dispersion coefficient often is assumed to be a function of the seepage velocity, or

$$D_m = \alpha_L v_s^{\beta} \tag{3.10}$$

where α_L is the longitudinal dispersivity of the porous medium in the direction of transport, and β is an empirically determined constant between 1 and 2 (Freeze and Cherry, 1979). In most applications, the exponent, β, is assumed to be unity, i.e., D_m is assumed to be a linear function of v_s in Eq. (3.10). However, β may be greater than unity in many situations (Anderson, 1979, 1984; Bear and Verruijt, 1987). Also, the dispersivity, α_L, is probably scale dependent with larger values for α_L being associated with greater transport distances (Pickens and Grisak, 1981). For example, values of α_L reported from the results of field studies may be as much as four to six orders-of-magnitude greater than the corresponding laboratory measured values which commonly are found to range between 0.1 and 10 mm (Freeze and Cherry, 1979). Regardless of the appropriate values for α_L and β, the relationship for D_m in Eq. (3.10) indicates that the effect of mechanical dispersion increases as the seepage velocity increases and vice versa. As a result, for typical problems involving low-hydraulic-conductivity barrier materials, mechanical dispersion does not appear to be significant due to the relatively low flow rates (Rowe, 1987).

3.2.5 Combined transport

In the absence of coupled flow processes, the total mass flux, J, of the solute or contaminant species is the sum of the advective, diffusive, and dispersive fluxes, or

$$J = J_A + J_D + J_M \tag{3.11}$$

Based on Eqs. (3.4), (3.7), and (3.9), the total mass flux for one-dimensional transport in saturated porous material is

$$J = nv_s c - D^* n \frac{\partial c}{\partial x} - D_m n \frac{\partial c}{\partial x} \tag{3.12}$$

or

$$J = nv_s c - D_h n \frac{\partial c}{\partial x} \tag{3.13}$$

where D_h is the hydrodynamic dispersion coefficient given by

$$D_h = D^* + D_m = D^* + \alpha_L v_s \tag{3.14}$$

The exponent β in Eq. (3.10) has been assumed to be unity in the formulation of Eq. (3.14). The hydrodynamic dispersion coefficient accounts for dispersion of the solute due to both diffusion and mechanical dispersion.

3.3 TRANSIENT TRANSPORT

3.3.1 Conservation of mass

The transient transport of a chemical species through saturated material is based on the conservation of mass for a representative elementary volume (REV) of soil, or (Freeze and Cherry, 1979)

net rate of mass increase within the REV	=	mass flux into the REV	−	mass flux out of the REV	±	increase (or decrease) in mass due to chemical reactions occurring within the REV

A REV is the minimum volume that will allow for the application of the continuum approach to flow or transport through porous media (Bear, 1972, 1979; Bear and Verruijt, 1987). In mathematical terminology, the conservation of mass is reflected by the continuity equation, or

$$\frac{\partial m}{\partial t} = -\nabla \cdot \mathbf{J} \pm R \pm \lambda m \tag{3.15}$$

where m is the total (adsorbed plus liquid phase) mass of solute per unit volume of soil, λ is a general rate constant used to describe such reactions as radioactive and/or biological decay and R is a general term representing all other chemical and biological reactions. The positive signs (+) is Eq. (3.15) are used for concentration source terms (e.g., mineral dissolution) whereas the (−) signs are used for terms representing concentration sinks (e.g., precipitation).

3.3.2 Advection-dispersion equation

A number of simplifying assumptions often are required in practice to reduce Eq. (3.15) to a more usable form. For example, the porous medium often is assumed to be homogeneous, isotropic, and non-deformable, and transport is assumed to be governed by steady flow of an incompressible fluid. Also, only trace concentrations of solutes typically are considered so that changes in fluid density due to changes in solute concentrations can be neglected. Coupled flow processes usually are neglected and only equilibrium exchange reactions (e.g., reversible sorption reactions) are included routinely in the modelling of the transport of reactive solutes through porous media. When these assumptions are acceptable, and transport is assumed to occur only in one direction (say, the x-direction), Eq. (3.13) can be used for the flux term, J, and Eq. (3.15) can be reduced to (Freeze and Cherry, 1979)

$$\frac{\partial c}{\partial t} = \frac{D_h \partial^2 c}{R_d \partial x^2} - \frac{v_s \partial c}{R_d \partial x} \tag{3.16}$$

where R_d is the dimensionless retardation factor. In the study of contaminant transport, Eq. (3.16) commonly is referred to as the one-dimensional advection-dispersion equation. When the seepage velocity is sufficiently low such that mechanical dispersion is negligible, the advection-dispersion equation (3.16) effectively reduces to an advection-diffusion equation since the hydrodynamic dispersion coefficient from Eq. (3.14) is dominated by the effective diffusion coefficient (i.e., $D_h \approx D^*$).

The retardation factor represents the relative rate of fluid flow to the transport rate of a reactive solute (Freeze and Cherry, 1979), or

$$R_d = \frac{v_s}{v_R} \tag{3.17}$$

where v_R is the transport rate, or velocity, for the center of mass of the reactive solute. For nonreactive (nonadsorbing) solutes, such as chloride (Cl^-), the retardation factor is unity and the solute is transported at the rate of the seepage velocity in accordance with Eq. (3.17). As a result, Eq. (3.16) may be written as

$$\frac{\partial c}{\partial t} = D_h \frac{\partial^2 c}{\partial x^2} - v_s \frac{\partial c}{\partial x} \tag{3.18}$$

However, for reactive (adsorbing solutes), the retardation factor is greater than unity, i.e., $R_d > 1$. Therefore, Eq. (3.17) indicates that adsorbing solutes are transported at a reduced rate, v_R, relative to nonadsorbing solutes. In such cases, it often is convenient to rewrite the advection-dispersion equation (3.16) for adsorbing solutes by utilizing Eq. (3.17) as

$$\frac{\partial c}{\partial t} = D_R \frac{\partial^2 c}{\partial x^2} - v_R \frac{\partial c}{\partial x} \tag{3.19}$$

where $D_R = D_h/R_d$.

The value of R_d typically is determined in the laboratory from the results of either column tests (discussed later) or batch equilibrium tests (discussed in Chapter 2) using the relationship

$$R_d = 1 + \frac{\rho_b}{n} K_p \tag{3.20}$$

where ρ_b is the bulk (dry) density of the soil and K_p is the 'partition coefficient'. The partition coefficient relates the change in the adsorbed concentration relative to a change in the equilibrium concentration of the chemical species as

$$K_p = \frac{dS}{dc} \tag{3.21}$$

where S is the adsorbed concentration, or mass of solute adsorbed per mass of soil. Equation (3.21) represents the slope of a plot of S versus c, also known as an adsorption isotherm. The term 'partition coefficient' applies when the adsorption isotherm is nonlinear and, therefore, when K_p is dependent on the equilibrium concentration, i.e., $K_p = f(c)$. Therefore, the partition coefficient of a nonlinear adsorption isotherm is equal to the tangential slope of the adsorption isotherm evaluated at a specific value for the equilibrium concentration, c. When the adsorption isotherm is linear, the partition coefficient is constant and is called the 'distribution coefficient, K_d'.

In general, there are two types of nonlinear adsorption isotherms – concave and convex – as well as the linear adsorption isotherms which are of interest in contaminant transport (Melnyk, 1985; Shackelford and Daniel, 1991). In chemical engineering, concave and convex adsorption

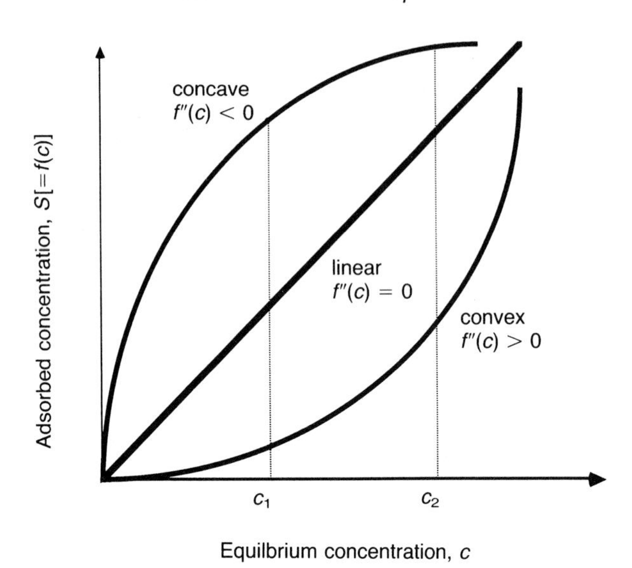

Fig. 3.3 General types of adsorption isotherms [after Melnyk (1985); and Shackelford and Daniel (1990)].

isotherms often are referred to as favorable and unfavorable isotherms, respectively (Weber and Smith, 1987). Concave isotherms are the more common type of nonlinear isotherms for transport in soil. The difference in the three types of isotherms is illustrated schematically in Fig. (3.3). In a concave isotherm, the slope of the isotherm at a lower equilibrium concentration is greater than it is at a higher equilibrium concentration, whereas the opposite is true for a convex isotherm (compare points for c_1 and c_2 in Fig. 3.3). Since the rate of transport for a reactive solute (v_R) is inversely proportional to the retardation factor and, therefore, the partition coefficient (compare Eqs. (3.17), (3.20), and (3.21)), a reactive solute will be transported faster at a higher concentration for a concave isotherm, whereas the opposite is true for a convex isotherm (see Figs. 3.4a and 3.4b). As a result, the concentration profile described by a convex isotherm (Fig. 3.4b) tends to spread out more during transport and forms what is known as a 'concentration wave'. However, since it is physically impossible for a higher concentration of a given solute to reach a specified distance before a lower concentration of the same solute reaches the same distance, the concentration profile shown in Fig. 3.4a for concave adsorption behavior tends to form what is known as a 'concentration step'. As a result, the limiting partition coefficient for reactive solutes described by concave isotherms is based on the use of secant lines

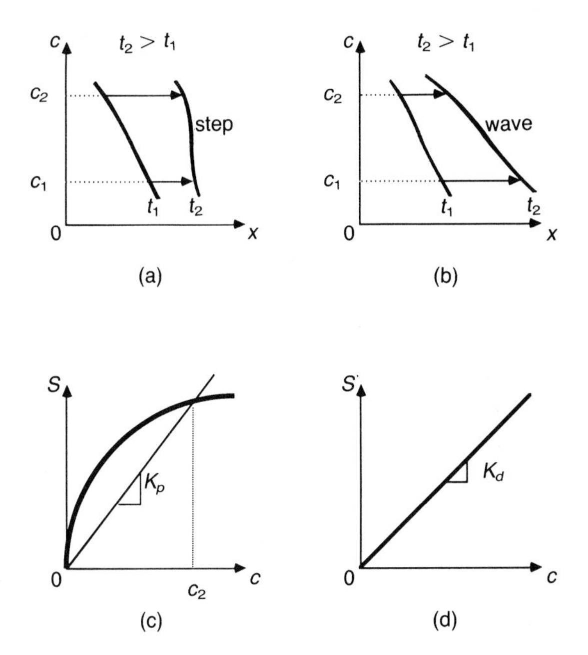

Fig. 3.4 Effects of different types of adsorption isotherms: (a) concentration-distance curve for concave adsorption behavior; (b) concentration-distance curve for convex adsorption behavior; (c) secant formulation for concave isotherm; and (d) distribution coefficient for linear adsorption isotherm.

instead of tangent lines, as shown in Fig. 3.4c (Melnyk, 1985). The use of secant lines forces all equilibrium concentrations of a solute up to the value at the intersection of the secant line and the adsorption isotherm (i.e., c_2 in Fig. 3.4c) to have the same value for the partition coefficient and, therefore, the same transport rate. In essence, the slope of the secant line is analogous to the slope associated with a constant, distribution coefficient for a linear isotherm. The above concepts are described in more detail by Melnyk (1985).

In most applications, nonlinear adsorption behavior is described by either a Freundlich or a Langmuir isotherm equation, as

$$S = K_f c^N \tag{3.22}$$

or

$$S = \frac{K_L M c}{1 + K_L c} \tag{3.23}$$

where S is the adsorbed concentration, K_f and N are the Freundlich isotherm equation parameters (K_d and n, respectively, in Eq. (2.3)), and K_L and M are the Langmuir isotherm equation parameters.

For the cases of a nonlinear, convex adsorption isotherm described by a Freundlich or a Langmuir isotherm, the retardation factor becomes, respectively

$$R_d = 1 + \frac{\rho_b}{n} K_f N c^{N-1} \tag{3.24}$$

or

$$R_d = 1 + \frac{\rho_b}{n} \frac{K_L M}{(1 + K_L c)^2} \tag{3.25}$$

When the adsorption behavior is described by nonlinear, concave isotherms, the limiting partition coefficient can be evaluated using a secant formulation, or

$$K_p = \left.\frac{\Delta S}{\Delta c}\right|_{c_0} \tag{3.26}$$

where c_0 is the equilibrium concentration which defines the particular slope of the secant line used for the evaluation of K_p (e.g., c_2 in Fig. 3.4c). For secant lines passing through the origin of the adsorption isotherm, the limiting retardation factors based on the Freundlich and Langmuir isotherm equations are, respectively

$$R_d = 1 + \frac{\rho_b}{n} K_f {c_0}^{N-1} \tag{3.27}$$

and

$$R_d = 1 + \frac{\rho_b}{n} \frac{K_L M}{1 + K_L c_0} \tag{3.28}$$

Retardation factors based on Eqs. (3.27) or (3.28) can be used to estimate the extent of migration of the center of mass of a contaminant plume in the case of nonlinear, concave adsorption behavior. However, it is unlikely that accurate estimates of the distribution of contaminants in the soil can be made using retardation factors based on Eqs. (3.27) or (3.28) due to the approximation of the nonlinear adsorption behavior by linear secant formulations. Davidson *et al.* (1976) illustrate the application of Eq. (3.27) to describe the migration of high concentrations of pesticides from hazardous waste landfill sites.

For the case of a linear adsorption isotherm (Fig. 3.4d), the retardation factor is written simply as

$$R_d = 1 + \frac{\rho_b}{n} K_d \tag{3.29}$$

As a result, all concentrations have the same transport rate, i.e., the slope of the isotherm is constant at all equilibrium concentrations.

3.3.3 Transport of reactive (adsorbing) organic compounds

The adsorption of hydrophobic organic contaminants at relatively low concentrations also can be described adequately by Eq. (3.29) provided the distribution coefficient is defined as

$$K_d = f_{oc} \cdot K_{oc} \tag{3.30}$$

where f_{oc} is the fraction (by weight) of organic carbon in the soil and K_{oc} is the organic carbon partition coefficient (Karichoff *et al.*, 1979). The organic carbon partition coefficient has been correlated empirically with a number of parameters, particularly the octanol-water partition coefficient, K_{ow}. These empirical correlations are covered in more detail by Griffin and Roy (1985).

3.3.4 Transport of radioactive chemical species

In the disposal of nuclear wastes, the migration of radioactive chemical species, or radioisotopes, is a primary environmental concern. In addition, radioisotopes often are used as tracers in the study of solute transport through porous materials. In such circumstances, the decay term, λm, in Eq. (3.15) should be evaluated. If the half-life of the radioactive isotope is short relative to the time frame of transport, the decay term should be included in the general formulation of the advection-dispersion equation (3.16) to give

$$\frac{\partial c}{\partial t} = \frac{D_h \partial^2 c}{R_d \partial x^2} - \frac{v_s \partial c}{R_d \partial x} - \lambda c \tag{3.31}$$

where λ is the decay constant of a linear or first-order decay reaction given by

$$\frac{\mathrm{d}m}{\mathrm{d}t} = -\lambda m \tag{3.32}$$

In the formulation of Eqs. (3.31) and (3.32), the decay constant, λ, of the solute in both the liquid and the solid phases of the soil has been assumed to be the same. Based on Eq. (3.32), the decay constant can be evaluated as follows:

$$\int_{m}^{0.5m} \frac{\mathrm{d}m}{m} = -\lambda \int_{0}^{t_{\frac{1}{2}}} \mathrm{d}t \tag{3.33}$$

or

$$\lambda = \left(\frac{\ln (2)}{t_{\frac{1}{2}}}\right) = \frac{0.693}{t_{\frac{1}{2}}} \tag{3.34}$$

where $t_{\frac{1}{2}}$ is the half-life of the radioisotope, i.e., the time required for half of the radioisotope to decay. Equation (3.34) indicates that as $t_{\frac{1}{2}}$ increases, the decay constant λ decreases. Therefore, for first-order decay, the decay term in Eq. (3.15) can be neglected without incurring significant error in the analysis when the radioactive solutes have long half-lives.

3.3.5 Fick's second law

At low values for the seepage velocity, diffusion is more significant relative to advection. In the limit, i.e., as $v_s \rightarrow 0$, Eqs. (3.16) and (3.18) reduce to the following two expressions for Fick's 2nd law for diffusion of adsorbing and nonadsorbing solutes, respectively:

$$\frac{\partial c}{\partial t} = \frac{D^* \partial^2 c}{R_d \partial x^2} = D_A^* \frac{\partial^2 c}{\partial x^2} = D_s \frac{\partial^2 c}{\partial x^2} \tag{3.35}$$

and

$$\frac{\partial c}{\partial t} = D^* \frac{\partial^2 c}{\partial x^2} \tag{3.36}$$

where D_A^* in Eq. (3.35) is referred to as an 'apparent diffusion coefficient' (Li and Gregory, 1974) and D_s is known as the 'effective diffusion coefficient of the reactive solute' (Gillham *et al.*, 1984; Quigley *et al.*, 1987). The result of using D_A^* or D_s in Eq. (3.35) is that only one unknown must be solved instead of the two unknowns (since $D_A^* = D_s = D^*/R_d$). Shackelford and Daniel (1991) and Shackelford (1991) discuss the use of D_A^* or D_s in application and summarize several other expressions for Fick's 2nd law of diffusion which exist due to differences in the definitions of D^* and differences in the expression of the solute concentration.

3.4 EFFECTS OF MATERIAL PROPERTIES

The effects of several material properties (R_d, D_m, D^*, K, and n) on the transport of solutes in saturated porous media can be illustrated with the aid of solute 'breakthrough curves'. Breakthrough curves represent the temporal variation in the concentration of a solute at the effluent end of a column of porous material. Breakthrough curves can be measured using laboratory columns by:

1. establishing steady-state fluid flow conditions;
2. continuously introducing at the influent end of the column a liquid containing a solute at concentration c_0; and
3. monitoring the solute concentration at the effluent end.

A schematic representation of a solute breakthrough curve for a continuous source concentration for a column of length L is presented in Fig. 3.5a. The effluent concentration, c, is expressed for convenience as a relative concentration, c/c_0.

With respect to Fig. 3.5a, the breakthrough curve can be divided into three distinct zones. Zone 1 represents the time required (t_0) for the solute to reach the effluent end of the column. Zone 3 represents the steady-state transport condition with respect to the solute which occurs after time t_f where the effluent concentration is the same as the influent concentration, c_0. It is important to note that it takes some time before steady-state solute transport occurs whereas steady-state fluid (solvent) flow is a pre-condition for the column test. Zone 2 is a transient, transitional zone wherein the effluent concentration is gradually rising from zero to c_0. Due to the spreading effect of the solute front in zone 2, there are an infinite number of possible transit times depending on the particular choice of the relative concentration, c/c_0. However, the 'usual' practice has been to define the transit time with respect to a relative concentration of 0.5, which is time t_L in Fig. 3.5a (Bowders *et al.*, 1985). The concentration-distance curves for this case are illustrated

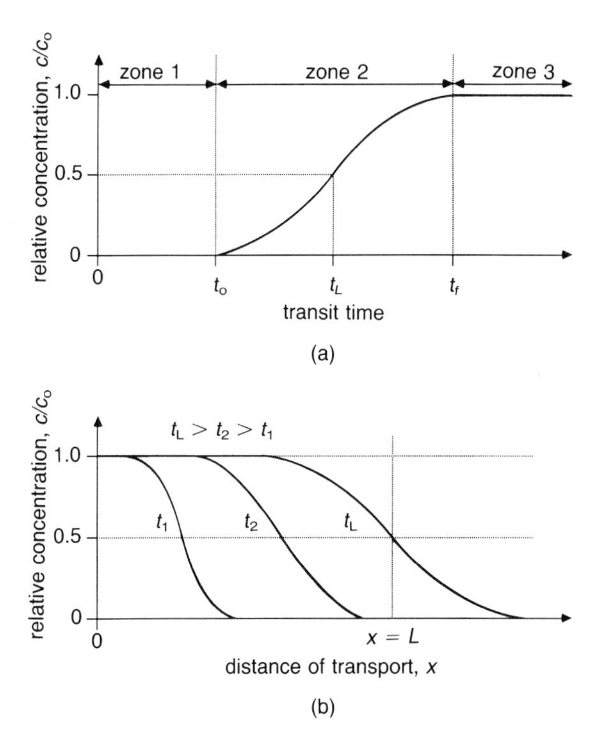

Fig. 3.5 Solute breakthrough curve for constant source concentration: (a) concentration-time profile; and (b) concentration-distance profiles for different times.

in Fig. 3.5b for three different elapsed times after introduction of the solute at the influent end of the column.

3.4.1 Effect of retardation

The effect of retardation on the transport of a solute is illustrated in Fig. 3.6a. The abscissas for the breakthrough curves in Fig. 3.6 have been defined in terms of pore volumes of flow instead of time. The number of pore volumes of flow (T) is equal to the cumulative volume of flow through the sample divided by the volume of void space in the material, or

$$T = \frac{vAt}{LAn} = \frac{vt}{Ln} = \frac{v_s}{(L/t)} = \frac{v_s}{v_R} \tag{3.37}$$

where A is the cross-sectional area of the material and L is the length of the column. Under steady-state fluid flow conditions, Eq. (3.37) indicates that the number of pore volumes of flow is directly proportional to time. For purely advective transport of nonreactive solutes, the velocity of the solute, v_R, is the same as the seepage velocity, v_s, and (provided $n_e = n$) complete breakthrough of the solute will occur at one pore volume of flow (curve 1 in Fig. 3.6a). This ideal condition often is referred to as 'piston' or 'plug' flow since there is no spreading of the solute front.

For solutes subject to reversible sorption reactions, i.e., reactive solutes, $v_R < v_s$ and the solute will be retarded. This retardation effect is manifested as an offset in the solute breakthrough curve of the reactive solute relative to the nonreactive solute (A in Fig. 3.6a). With respect to the advective transport of an adsorbing solute, v_R ($=L/t$) in Eq. (3.35) represents the 'velocity' of the retarded solute. The transit time based solely on the advective transport of adsorbing solutes can be estimated from the retarded solute velocity as

$$t = \frac{L}{v_R} = \frac{LR_d}{v_s} = \frac{nLR_d}{Ki} \tag{3.38}$$

Since $R_d > 1.0$ for retarded solutes, the transit time based on Eq. (3.38) for adsorbing solutes will be greater than the transit time for nonadsorbing solutes (Eq. 3.3).

Again, it is important to distinguish clearly between steady-state

Fig. 3.6 (opposite) Effect of material properties on solute breakthrough curves for constant source concentration: (a) effect of retardation-piston flow; (b) effect of effective porosity-piston flow; (c) effect of mechanical dispersion; (d) effect of diffusion. [Note: curves 1 and 3 – nonreactive solutes; curves 2 and 4 – reactive solutes].

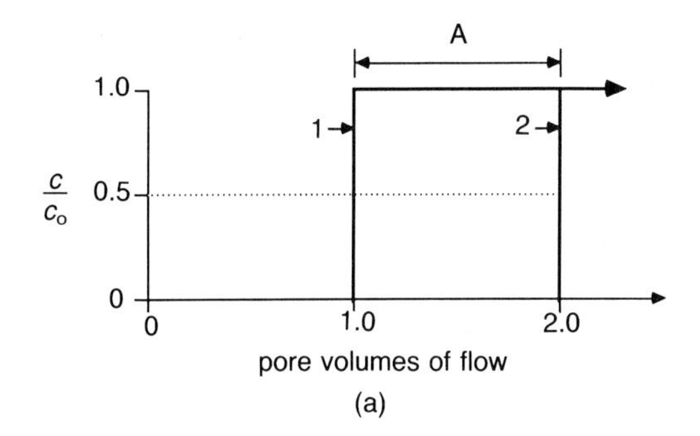
A
1.0
1
2
$\frac{c}{c_o}$
0.5
0
0
1.0
2.0
pore volumes of flow
(a)

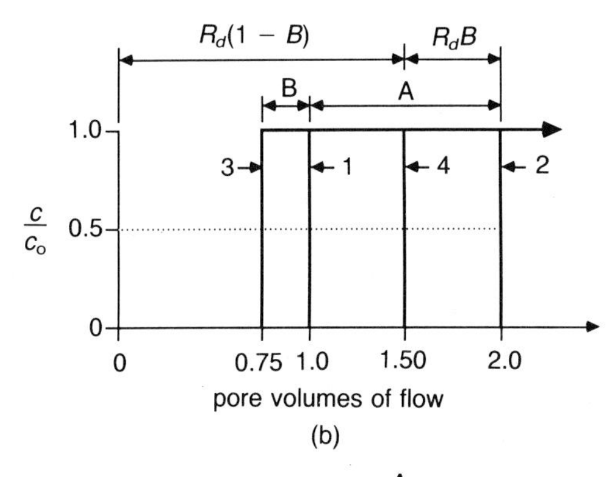
$R_d(1 - B)$
R_dB
B
A
1.0
3
1
4
2
$\frac{c}{c_o}$
0.5
0
0
0.75
1.0
1.50
2.0
pore volumes of flow
(b)

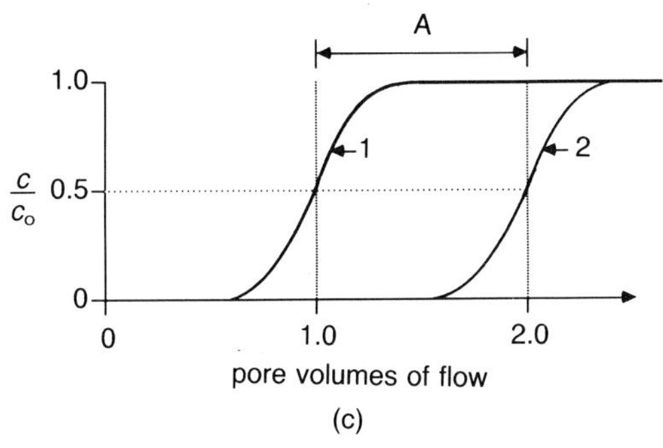
A
1.0
1
2
$\frac{c}{c_o}$
0.5
0
0
1.0
2.0
pore volumes of flow
(c)

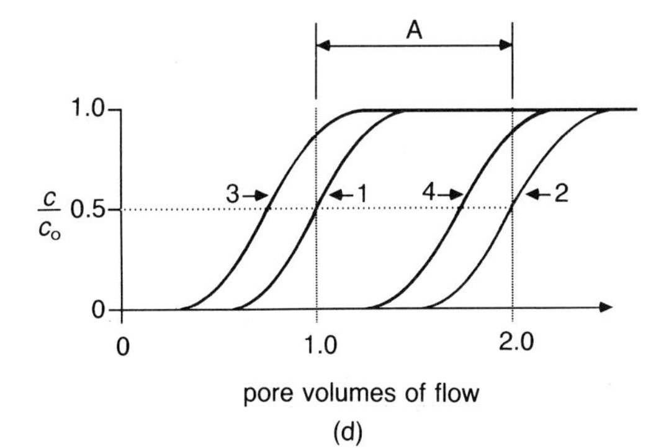
A
1.0
3
1
4
2
$\frac{c}{c_o}$
0.5
0
0
1.0
2.0
pore volumes of flow
(d)

fluid flow and steady-state solute transport. The retardation factor, R_d, as defined by Eq. (3.17) is a measure of how much slower a retarded solute moves relative to the steady-state movement of the fluid, or solvent. Once complete breakthrough of the solute has occurred, the solute is no longer retarded (as the term is used here) since 'steady-state' solute transport, by definition, means that the influent and effluent mass fluxes of the solute are equal. As a result, steady-state transport is achieved only after the liquid-phase concentration of the solute is in equilibrium with the solid-phase concentration of the solute throughout the entire length of the column.

A comparison of Eqs. (3.17) and (3.37) reveals that the retardation factor, R_d, is equal to the number of pore volumes of flow, T, it takes for the solute to break through the effluent end of the column. This relationship can be utilized to determine the value for the retardation factor for an adsorbing solute. For example, the breakthrough curve in Fig. 3.6a for the retarded solute is given by curve 2, so the retardation factor is 2.0, i.e., $R_d = T = 2.0$.

3.4.2 Effective porosity

In fine-grained soils, some of the fluid in the pore space may be immobile due to dead-end pores and/or attraction of fluid molecules to the surface of the soil particles (Bear, 1972, 1979). As a result, the portion of the pore space available for solute transport may be significantly less than the total pore space. In such cases, an effective porosity, n_e, should be used instead of the total porosity, n, in Eq. (3.1) to determine the seepage velocity. Effective porosities on the order of 2–100% of the total porosity have been measured in compacted soils (Horton *et al.*, 1985; Liao and Daniel, 1989).

The effect of effective porosity on purely advective (piston) transport of solutes is illustrated by the offset, B, in Fig. 3.6b. Since $n_e < n$, the seepage velocity is greater than that predicted using the total porosity, and actual breakthrough of the solutes occurs earlier than expected. For the case shown in Fig. 3.6b, B represents 0.25 pore volumes of flow and, therefore, only 75% of the total pore space is effective in conducting the fluid flow (i.e., $n_e = 0.75n$). As a result, breakthrough of nonreactive (nonadsorbing) solutes will occur after only 0.75 pore volumes of flow (curve 3) rather than after one pore volume of flow (curve 1). On the other hand, complete breakthrough (i.e., assuming piston transport) of reactive solutes occurs after $R_d(1 - B)$ pore volumes of flow, or R_dB pore volumes sooner than expected. For example, R_d for the reactive solute in Fig. 3.6 is 2 and, therefore, breakthrough of the reactive solute is expected at 2 pore volumes of flow (curve 2). However, due to the effective porosity effect, breakthrough of the reactive solute actually occurs after only 1.5 pore volumes of flow (curve 4).

3.4.3 Mechanical dispersion

In general, complete breakthrough of a solute will not occur instantaneously. Instead, there is a gradual rise in the solute concentration from $c = 0$ to $c = c_0$ due to the spreading effect caused, in part, by mechanical dispersion. This spreading effect gives the solute breakthrough curve its characteristic S-shape shown in Fig. 3.5a. The effects of mechanical dispersion on the breakthrough curves illustrated in Fig. 3.6a are shown in Fig. 3.6c. At relatively high seepage velocities, mechanical dispersion dominates the mixing process and the solute breakthrough curves, including the effects of mechanical dispersion, intersect those for piston flow at a relative concentration, c/c_0, of 0.5, as shown in Fig. 3.6c.

3.4.4 Effect of diffusion

The breakthrough curves illustrated in Fig. 3.6c represent spreading of the solute front primarily due to mechanical dispersion, i.e., spreading due to diffusion is negligible. This is the case commonly referred to in groundwater hydrology textbooks because the primary concern is with contaminant migration in aquifers, i.e., coarse-grained, water-bearing strata subject to relatively high seepage velocities. Curves 3 and 4 in Fig. 3.6d represent breakthrough curves for nonreactive and reactive solutes, respectively, when the seepage velocity is sufficiently low such that the effect of diffusion is not masked by the effect of advection, including mechanical dispersion. For comparison, the breakthrough curves in Fig. 3.6c also are provided in Fig. 3.6d. The spreading effect is still noticeable in Fig. 3.6d, but curves 1 and 3 and curves 2 and 4 do not intersect each other at $c/c_0 = 0.5$. Instead, curves 3 and 4 are displaced to the left of curves 1 and 2, respectively. This displacement due to diffusion can result in transit times (at $c/c_0 = 0.5$) which, in some cases, can be much less than those predicted by considering only advection. This effect previously has been recognized analytically by De Weist (1965) and experimentally by Biggar and Nielsen (1960).

The decrease in the transit time, illustrated in Fig. 3.6d, due to diffusion is a function of the magnitude of the seepage velocity. This dependence is illustrated in Fig. 3.7 where the transit time for purely advective, purely diffusive, and advective-dispersive transport is illustrated as a function of the logarithm of the seepage velocity. Although no data are provided in Fig. 3.7, the general trends are apparent from previous analyses (Shackelford, 1988b, 1989). The horizontal distance between the purely advective and the advective-dispersive curve in Fig. 3.7 represents the offset distance in Fig. 3.6d for a given seepage velocity. The independence of pure diffusion on hydraulic conductivity is indicated by the vertical line. From the trends presented in Fig. 3.7, it is apparent that the effect of diffusion on

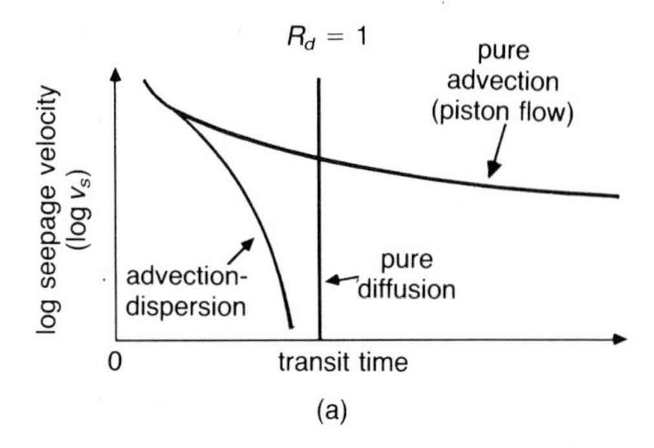

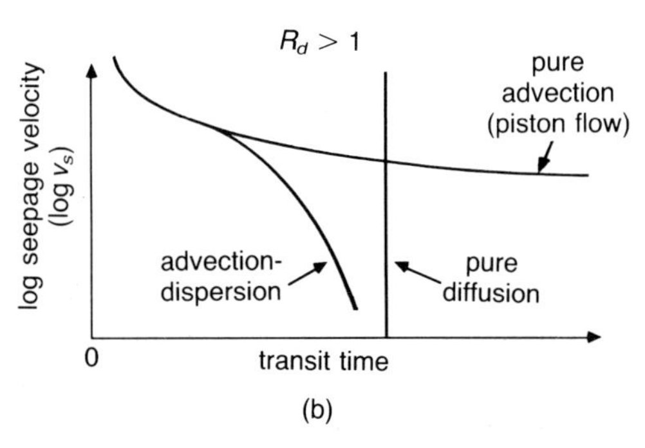

Fig. 3.7 Schematic relationship between transit time and seepage velocity for a transport distance, L, and a constant source concentration: (a) nonreactive solutes; and (b) reactive solutes [after Shackelford (1988b, 1989, 1991)].

transit time increases as the seepage velocity decreases. The seepage velocity at which diffusion begins to dominate solute transport is indicated by the inflection point in the advective-dispersive curve. Available evidence indicates that diffusion can be a significant transport process when the seepage velocity is in the range 0.064–0.09 m/yr (Rowe *et al.*, 1988), and a dominant transport process when v_s is $\leqslant$ 0.005 m/yr (Gillham *et al.*, 1984; Shackelford, 1988b, 1989).

The effect of diffusion on solute breakthrough curves is similar to the effect of effective porosity (compare Figs. 3.6b and 3.6d). Whereas the effect of diffusion is noticeable only at relatively low seepage velocities, the effect of effective porosity is expected to be independent of the seepage velocity. However, there is some indication that the value of n_e can be a function of the hydraulic gradient (and, therefore, v_s) imposed in the study, with lower values of n_e being observed at lower hydraulic gradients (Liao and Daniel, 1989). Therefore, attempts to discern the effective porosity of a soil, e.g., by performing column tests at elevated seepage velocities, may not prove successful.

3.5 APPLICATIONS FOR ONE-DIMENSIONAL PROBLEMS

3.5.1 Waste disposal scenarios

Analytical solutions may be used, for example, to estimate both the flux and the concentration of contaminants at the effluent end of a column of porous material or at the bottom of a containment barrier. The appropriate analytical solution to be used depends on the susceptibility of the contaminant to geochemical reactions (i.e., non-reactive versus reactive solutes) and the overall flow and boundary conditions. In general, there are three possible scenarios or cases to consider for containment systems in which mechanical dispersion is negligible (Shackelford, 1989):

1. diffusion without advection (i.e., pure diffusion);
2. diffusion with positive advection; and,
3. diffusion with negative advection.

Each of these three cases is illustrated in Fig. 3.8.

For the case shown in Fig. 3.8a, the hydraulic gradient across a clay barrier is zero, so there is no advective flow. However, since the concentration of the contaminants in the leachate is greater than that in the natural soil, a concentration gradient is established across the barrier and contaminants will diffuse through the barrier. Whereas the diffusive flux of contaminants is described by Eq. (3.7), the transient diffusive transport of contaminants can be modelled using analytical solutions to Eq. (3.35), i.e., analytical solutions to Eq. (3.16) with v_s equal to zero.

For the case depicted in Fig. 3.8b, a hydraulic gradient has been established across the barrier such that advective transport of contaminants occurs in the same direction as the diffusive transport. The advective transport is termed 'positive' since it results in an increase in the contaminant concentration in the natural soil (Shackelford, 1989). This case commonly is described in texts on contaminant transport. (Bear, 1972, 1979; Freeze and Cherry, 1979; Bear and Verruijt, 1987). For this case, the total mass flux of the contaminant is given by Eq. (3.12) where D_h equals D^* (i.e., $D_m = 0$). The time-dependent concentration of contaminants occurring at the barrier bottom, assuming a homogeneous soil, can be determined with analytical solutions to the advective-dispersive equation, e.g., Eq. (3.16).

In some instances, the direction of advective transport may be opposite to that of diffusive transport, as shown in Fig. 3.8c. This situation may occur in practice, for example, when the barrier (either natural or man-made) is located over a confined aquifer under artesian pressure, or when vertical barriers, such as slurry walls, are used to isolate a contaminated area (see Fig. 3.9). In this case, the advective

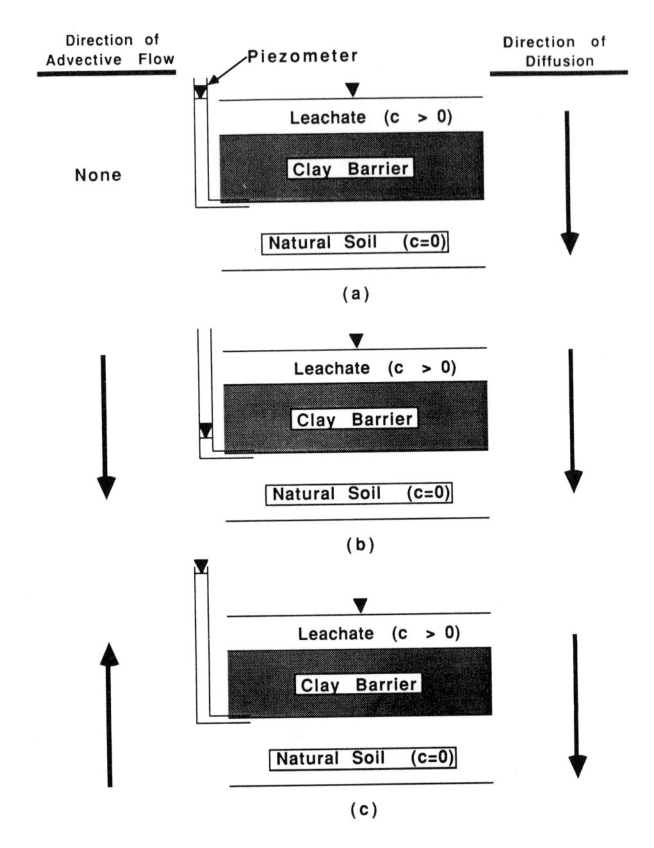

Fig. 3.8 Waste disposal scenarios: (a) pure diffusion; (b) diffusion with positive advection; and (c) diffusion with negative advection [after Shackelford (1989)].

flow is termed 'negative' since it works to prevent the escape of pollutants from the contaminated area (Shackelford, 1989). Gray and Weber (1984) analyzed this case using Eq. (3.16) simply by substituting a '$-v_s$' for 'v_s'. However, the boundary and flow conditions for this case can be complex. For example, if the concentration, c_2, in Fig. 3.9 is initially zero, the initial flow of water into the containment area may dilute the concentration c_1 with time and, therefore, reduce the concentration gradient for outward (positive) diffusive transport. The contaminant migration front, if any, will depend on the offsetting effects of the negative advective and positive diffusive fluxes. The situation for reactive solutes is more complex. Regardless, significant diffusive transport of contaminants may still result if a relatively thin barrier is built to contain relatively high concentrations of pollutants over extended periods (Gray and Weber, 1984).

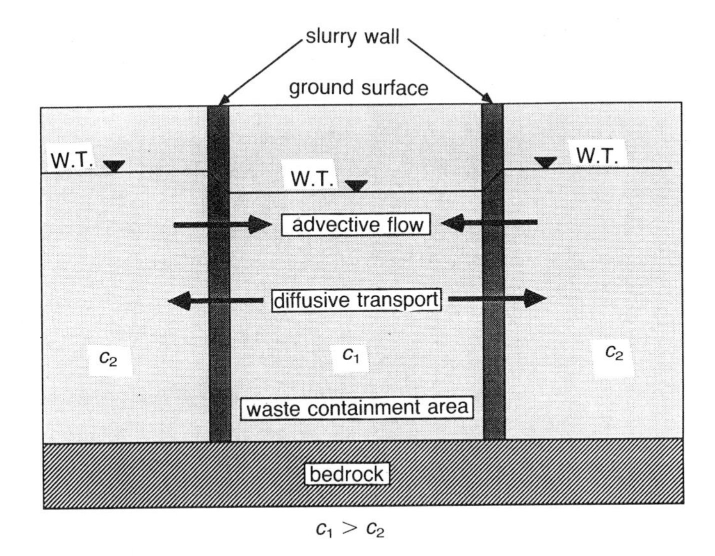

Fig. 3.9 Slurry wall scenario illustrating case for diffusion against advection [after Gray and Weber (1984); and Shackelford (1989)].

3.5.2 Steady-state flux determination

The steady-state (long-term) flux of a contaminant through a barrier of thickness L due to purely diffusive transport (i.e., Fig. 3.8a) is given by the following form of Eq. (3.7).

$$J_D = -D^* n \frac{\Delta c}{\Delta x} = -D^* n \frac{\Delta c}{L}\bigg|_{\Delta x = L} \tag{3.39}$$

Here the differentials in Eq. (3.7) have been replaced by difference operators to indicate steady-state conditions. The implicit assumption in the use of Eq. (3.39) is that the contaminant exiting the bottom of the barrier is instantaneously and continuously flushed so that the concentration gradient across the barrier is maintained at the steady-state value, i.e., so that $\Delta c/\Delta x \neq f(t)$.

For example, consider the clay barrier scenario shown in Fig. 3.10a. The clay barrier, with a thickness L (=1 m) and porosity n (=0.4), is used to contain a leachate with a constant concentration, c_0 (=10 mg/L), of a chemical species (e.g., Cl^-). The barrier is underlain by a leachate collection system for collection and removal of leachate which migrates through the barrier. For the case of pure diffusion, the flow rate of the leachate through the barrier is considered to be negligible so that diffusion is the dominant mechanism of contaminant transport through the barrier (see Fig. 3.7). Under these conditions, the contaminant concentration profiles in the barrier as a function of time will resemble those shown in Fig. 3.10b (Quigley *et al.*, 1987; Daniel and Shackelford,

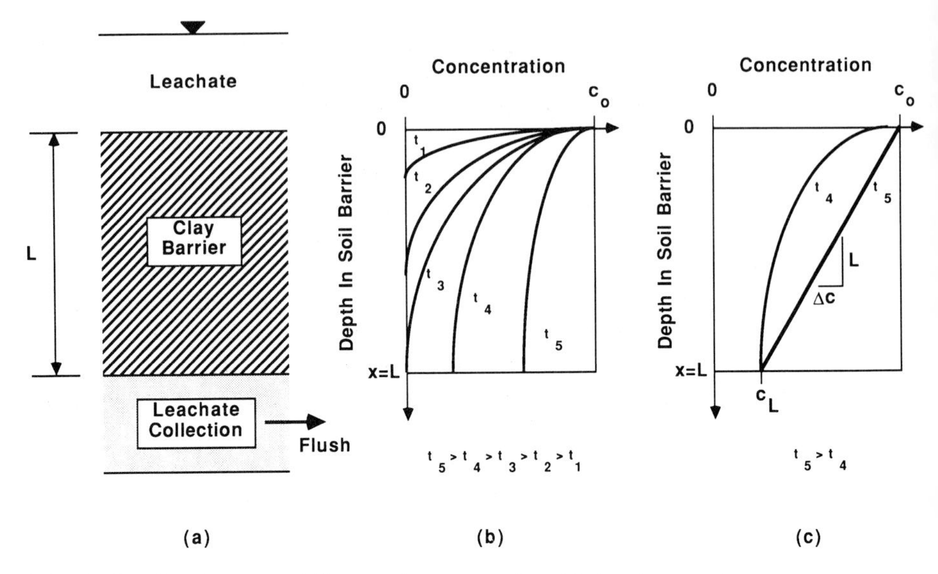

Fig. 3.10 Steady-state diffusive flux scenario: (a) barrier configuration; (b) concentration profiles as a function of time; and (c) establishment of steady-state concentration gradient across barrier.

1988). For example, in Fig. 3.10b, the contaminant species reaches the bottom of the barrier after an elapsed time t_3, and the concentration of the contaminant species in the leachate collection system continues to rise as a function of time. If none of the contaminant is removed from the leachate collection system, the concentration at the bottom of the barrier will eventually reach the level, c_0, at which time migration via diffusion will stop since the concentration gradient across the barrier will be zero (i.e., $\Delta c = c_0 - c_0 = 0$). However, if the leachate collection system is continuously flushed so as to maintain a constant concentration, c_L (=2 mg/L), in the leachate collection system, as shown in Fig. 3.10c, then a steady-state concentration gradient eventually will be established across the barrier (e.g., after time t_5 in Fig. 3.10c), and the steady-state diffusive flux can be determined using Eq. (3.39).

$$J_D = -D^* n \frac{c_L - c_0}{L}$$

$$= -\left(1 \times 10^{-9} \frac{\text{m}^2}{\text{s}}\right)(0.4) \frac{\left(2\frac{\text{mg}}{L} - 10\frac{\text{mg}}{L}\right)}{1\,\text{m}} \left(1000 \frac{L}{\text{m}^3}\right)\left(31.536 \times 10^6 \frac{\text{s}}{\text{yr}}\right)$$

$$= 101 \left(\frac{\text{mg}}{\text{m}^2\text{yr}}\right) \tag{3.40}$$

Here a value of $1.0 \times 10^{-9}\,\text{m}^2/\text{s}$ has been used for the effective diffusion coefficient, D^*, for chloride, a nonreactive chemical species

(Shackelford, 1988a). An additional example of the use of Eq. (3.39) to estimate the flux of organic contaminants at the bottom of a natural clayey barrier is given by Johnson *et al.* (1989).

The situation for the case of diffusion with 'positive' advection (Fig. 3.8b) is not as straightforward as that for the case of pure diffusive transport. When the concentration of contaminant species in the leachate is assumed to be constant (e.g., c_0), breakthrough of the contaminant species through the bottom of the barrier will resemble the solute breakthrough curve shown in Fig. 3.5a, and the concentration profiles within the barrier as a function of time will resemble those shown in Fig. 3.5b. In accordance with Fig. 3.5a, steady-state transport conditions are established only after an elapsed time t_f. At this time, the concentration of the specified chemical species at the barrier bottom reaches the value c_0 corresponding to the concentration in the leachate (i.e., $c/c_0 = 1$), and the concentration gradient across the barrier is zero (i.e., $\Delta c = c_0 - c_0 = 0$). Therefore, there is no diffusive flux after time t_f, and the steady-state flux of the contaminant is due solely to the advective component of flux given by Eq. (3.4). Even if an attempt is made to maintain a reduced concentration of the contaminant species in the leachate collection system by flushing, steady-state diffusive flux conditions still will not be established because time-dependent concentration profiles (i.e., $c = f(x, t)$) within the barrier still exist due to the advective component of transport. Due to these time-dependent concentration profiles, the differential concentration gradient in Eq. (3.7) cannot be replaced by the difference concentration gradient in Eq. (3.39) (i.e., $\partial c/\partial x \neq \Delta c/\Delta x$) and, therefore, steady-state diffusive flux conditions cannot exist (i.e., since $\partial c/\partial x = f(t)$).

3.5.3 Analytical solutions

Mathematical solutions to the transport equations for the complex geometries, heterogeneous transport parameters, and realistic boundary conditions characteristic of many practical problems typically require the application of numerical techniques (e.g., finite differences or finite elements). However, the application of numerical methods to the solution of practical contaminant transport problems is beyond the scope of this chapter. An excellent review of the application of numerical techniques to the solution of groundwater problems can be found in Huyakorn and Pinder (1983).

Although restricted in use, analytical solutions can be beneficial when utilized for preliminary design, rudimentary evaluation of remedial action, and/or verification of more sophisticated numerical solutions. Analytical solutions to the various forms of the one-dimensional advection-dispersion equation (i.e., (3.16), (3.18), and (3.31)) can be found in Lapidus and Amundson (1952), Bastian and

Lapidus (1956), Ogata and Banks (1961), Brenner (1962), Lindstrom *et al.* (1967), Gershon and Nir (1969), Ogata (1970), Bear (1972, 1979), Selim and Mansell (1976), Van Genuchten (1981), Van Genuchten and Alves (1982), and Javendel *et al.* (1984). Semi-analytical solutions are provided by Javendel *et al.* (1984) and Rowe and Booker (1985). Three analytical solutions to Eq. (3.31) are presented in Table 3.1 (Van Genuchten, 1981; Van Genuchten and Alves, 1982). The second analytical solution in case A is required for the situation in which the decay constant, λ, is zero since division by zero in the first solution is mathematically impossible.

Table 3.1 Analytical solutions to the one-dimensional advection-dispersion equation with simultaneous reversible sorption and linear, first-order decay reactions (after Van Genuchten, 1981; and Van Genuchten and Alves, 1982)

Case A: Initial and Boundary Conditions

Initial condition: $c = 0 \qquad (x \geqslant 0;\ t = 0)$

1st boundary condition: $\left(-D_h \frac{\partial c}{\partial x} + v_s c\right)\Big|_{x=0^+} = v_s c_0\Big|_{x=0^-} \qquad (x = 0;\ t > 0)$

2nd boundary condition: $\frac{\partial c}{\partial x} = 0 \qquad (x = \infty;\ t > 0)$

Solution 1: $(R_d \geqslant 1;\ \lambda > 0)$*

$$\frac{c(x,t)}{c_0} = \left(\frac{v_s}{v_s + U}\right)\exp\left[\frac{(v_s - U)x}{2D_h}\right]\operatorname{erfc}\left[\frac{R_d x - Ut}{2\sqrt{D_h R_d t}}\right] + \left(\frac{v_s}{v_s - U}\right)\exp\left[\frac{(v_s + U)x}{2D_h}\right]\operatorname{erfc}\left[\frac{R_d x + Ut}{2\sqrt{D_h R_d t}}\right] + \left(\frac{v_s^2}{2D_h R_d \lambda}\right)\exp\left(\frac{v_s x}{D_h} - \lambda t\right)\operatorname{erfc}\left(\frac{R_d x + Ut}{2\sqrt{D_h R_d t}}\right)$$

Solution 2: $(R_d \geqslant 1;\ \lambda = 0)$

$$\frac{c(x,t)}{c_0} = \frac{1}{2}\operatorname{erfc}\left[\frac{R_d x - v_s t}{2\sqrt{D_h R_d t}}\right] + \left(\frac{v_s^2 t}{\pi D_h R_d}\right)^{\frac{1}{2}}\exp\left(-\frac{(R_d x - v_s t)^2}{4 D_h R_d t}\right) - \frac{1}{2}\left(1 + \frac{v_s x}{D_h} + \frac{v_s^2 t}{D_h R_d}\right)\exp\left(\frac{v_s x}{D_h}\right)\operatorname{erfc}\left(\frac{R_d x + v_s t}{2\sqrt{D_h R_d t}}\right)$$

Case B: Initial and Boundary Conditions

Initial condition $c = 0 \qquad (x \geqslant 0;\ t = 0)$

1st boundary condition: $c = c_0 \qquad (x = 0;\ t > 0)$

2nd boundary condition: $\frac{\partial c}{\partial x} = 0 \qquad (x = \infty;\ t > 0)$

Solution: $(R_d \geqslant 1;\ \lambda \geqslant 0)$*

$$\frac{c(x,t)}{c_0} = \frac{1}{2}\exp\left[\frac{(v_s - U)x}{2D_h}\right]\operatorname{erfc}\left[\frac{R_d x - Ut}{2\sqrt{D_h R_d t}}\right] + \frac{1}{2}\exp\left[\frac{(v_s + U)x}{2D_h}\right]\operatorname{erfc}\left[\frac{R_d x + Ut}{2\sqrt{D_h R_d t}}\right]$$

* $U = (v_s^2 + 4D_h R_d \lambda)^{\frac{1}{2}}$ [Note: when $\lambda = 0$, $U = v_s$]

Both cases in Table 3.1 correspond to a constant source concentration for semi-infinite porous media. However, the first boundary condition for each of the two cases is different. In case A, the first boundary condition represents the conservation of mass required at the liquid-soil interface. The first boundary condition for case B represents an approximation of the first boundary condition for case A. In reality, the first boundary condition for case B is impossible in that it represents a discontinuity in the concentration at the liquid-soil interface (Kreft and Zuber, 1978; Parker and Van Genuchten, 1984). However, Gershon and Nir (1969) compared breakthrough curves based on one-dimensional analytical solutions derived from the initial and boundary conditions represented by cases A and B and found that the approximate boundary condition in case B resulted in differences of up to only 5% in the region of $c/c_0 = 0.5$. Therefore, the analytical solution represented by case B in Table 3.1 should be sufficient for most practical purposes (i.e., in lieu of the analytical solution presented for case A).

The 'erfc' appearing in the analytical solutions in Table 3.1 stands for the 'complementary error function' which, for any argument z, is given by

$$\text{erfc}\,(z) = 1 - \text{erf}\,(z) \tag{3.41}$$

and erf (z) is the 'error function' of the argument, z. Values of erf(z) and erfc(z) are tabulated in several texts (Carslaw and Jaeger, 1959; Abramowitz and Stegun, 1972; Crank, 1975; Freeze and Cherry, 1979). As an alternative, values of erf(z) may be determined with the following series solution (Carslaw and Jaeger, 1959; Abramowitz and Stegun, 1972), or

$$\text{erf}\,(z) = \frac{2}{\sqrt{\pi}} \sum_{m=0}^{\infty} \frac{(-1)^m z^{2m+1}}{m!\,(2m+1)} \tag{3.42}$$

A small computer program (~100 lines) can be written to perform the series summation. Experience indicates that fifty terms ($m = 50$) are required in Eq. (3.42) to achieve an accuracy to six decimal places for values of erf(z) (Shackelford, 1990). For situations in which the quantity $v_s x \gg D_h$ and/or t is small, the series summation represented by Eq. (3.42) may not be appropriate for convergence of the analytical solutions in Table 3.1 (Brenner, 1962; Cadena, 1989). In such cases, other approximations can be used for evaluation of erf(z) (Abramowitz and Stegun, 1972).

The analytical solutions in Table 3.1 may be used to derive many of the analytical solutions given in other references. For example, in the absence of decay (i.e., $\lambda = 0$), the analytical solution given by case B reduces to the well-known Ogata and Banks (1961) solution (also see Lapidus and Amundson, 1952; Ogata, 1970; Bear, 1972, 1979; Freeze and Cherry, 1979; Van Genuchten, 1981; and Van Genuchten and Alves, 1982), or

$$\frac{c(x, t)}{c_0} = \frac{1}{2}\left[\operatorname{erfc}\left(\frac{R_d x - v_s t}{2\sqrt{R_d D_h t}}\right) + \exp\left(\frac{v_s x}{D_h}\right)\operatorname{erfc}\left(\frac{R_d x + v_s t}{2\sqrt{R_d D_h t}}\right)\right] \quad (3.43)$$

Dimensionless parameters

The arguments for analytical solutions often can be expressed for convenience in terms of dimensionless parameters. For example, Eq. (3.43) can be expressed as

$$\frac{c(x, t)}{c_0} = \frac{1}{2}\left\{\operatorname{erfc}\left(\frac{1 - T_R}{2\sqrt{T_R/P_L}}\right) + \exp(P_L)\operatorname{erfc}\left(\frac{1 + T_R}{2\sqrt{T_R/P_L}}\right)\right\} \quad (3.44)$$

where

$$T_R = \frac{v_s t}{R_d x} = \frac{v_R t}{x} = \left.\frac{v_R t}{L}\right|_{x=L} \quad (3.45)$$

and

$$P_L = \frac{v_s x}{D_h} = \left.\frac{v_s L}{D_h}\right|_{x=L} \quad (3.46)$$

Equation (3.45) represents the number of pore volumes, T, divided by the retardation factor, R_d, whereas P_L is the Peclet number defined with respect to the barrier thickness, L (Bear, 1972). Equation (3.44) can be presented graphically as a plot of c/c_0 versus T_R for various values of P_L (Ogata and Banks, 1961; Ogata, 1970; Goldman *et al.*, 1986; Shackelford, 1990).

3.5.4 Transit time determinations

Several performance criteria can be used to provide a definition for transit time for the design of waste containment barriers (Goldman *et al.*, 1986). Two specific transit-time analyses are considered here. First, the time required for the concentration of a specific leachate component to reach a specific value at the barrier bottom is illustrated. Second, the time required to reach a specified flux of a particular contaminant at the barrier bottom is considered.

Specified leachate concentration

Equation (3.43) can be used directly to determine the concentration, c, of a specified leachate constituent at the bottom of a clay barrier of thickness, L, for any given time t. For given values of c, c_0, v_s, D_h, and L $(=x)$, all variables in Eq. (3.43) are defined except for the time, t. An iterative procedure must be used with Eq. (3.43) to solve for t. Shackelford (1990) illustrates the iterative procedure using Eq. (3.44) to estimate the barrier thickness required to provide the necessary transit time for the concentration of a specified leachate constituent. The iterative procedure involves the following steps:

1. assume a liner thickness, L;
2. calculate P_L from Eq. (3.46);
3. determine T_R from a graph of T_R versus c/c_0 for the desired value of c/c_0 and the calculated value of P_L; and
4. use Eq. (3.45) to determine the transit time, t.

This procedure is repeated until t is greater than or equal to the design life of the containment barrier.

Specified leachate flux

A similar approach to the one presented above for a specified leachate concentration can be developed for a specified leachate flux by using the analytical solution for the concentration, $c(x, t)$, in the determination of the advective, diffusive, and dispersive fluxes using Eqs. (3.4), (3.7), and (3.9). For example, if Eq. (3.44) is used to represent the solute concentration, then the total flux, J, of the solute at a distance L and any time t can be represented by a dimensionless flux number, F_N, as follows (Shackelford, 1990):

$$F_N = \frac{JL}{nc_0D_h} = \frac{1}{2}\{P_LQ_1 + Q_2\} \tag{3.47}$$

where

$$Q_1 = \text{erfc}\left(\frac{1 - T_R}{2\sqrt{T_R/P_L}}\right); \qquad Q_2 = \frac{2\exp\left[-\left(\frac{1 - T_R}{2\sqrt{T_R/P_L}}\right)^2\right]}{\sqrt{\frac{\pi T_R}{P_L}}} \tag{3.48}$$

A graphical solution to Eq. (3.47) is presented by Shackelford (1990) for the case where $D_h \approx D^*$. The graphical solution of Eq. (3.47) may be used to estimate the barrier thickness required to provide the necessary transit time for a specified contaminant flux by substituting F_N for c/c_0 in the procedure outlined for the specified leachate concentration.

3.6 CASE HISTORIES

Several case histories illustrating the migration of contaminants beneath landfills have appeared in the literature within the past 15 years or so. For example, Goodall and Quigley (1977) and Quigley *et al.*

(1984) evaluated measured concentration profiles below the Confederation Road sanitary landfill site near Sarnia, Ontario, using Eq. (3.43). They concluded that diffusion is an important transport mechanism of solutes in landfill leachate under low-flow situations.

Johnson *et al.* (1989) evaluated the migration of chloride (Cl^-) and organic compounds beneath a hazardous waste disposal site located 15 km southwest of Sarnia, Ontario. The results indicated that the influence of advection was negligible and that diffusion was the dominant mechanism of transport in the natural clay. Johnson *et al.* (1989) contend that the study has important implications with regard to the design of relatively thin (1 m thick) clay liners, or even double liners, with low hydraulic conductivities. They concluded that diffusion of contaminants through clay liners can result in relatively short breakthrough times, and the resulting fluxes may be significant especially for the common priority pollutants.

An excellent and detailed case study describing the effect of longitudinal dispersion on the migration of contaminants from beneath an abandoned landfill in Ontario, Canada, is presented in a series of seven papers in a special issue of the *Journal of Hydrology* (Cherry, 1983). Landfill-derived contamination was delineated by monitoring elevated distributions of chloride, sulfate, and electrical conductance in an underlying sand aquifer. The extent of the contaminant plume resulting from migration of the landfill leachate was detected to be of the order of 600 m in width, 700 m in length, and 20 m in depth beneath the landfill. Irregular concentration profiles beneath and near the landfill which gradually became smoother downgradient from the landfill, where maximum concentrations were much lower, were attributed to the strong influence of longitudinal dispersion in the sand aquifer.

REFERENCES

Abramowitz, M. and Stegun, I.A. (1972) *Handbook of Mathematical Functions.* Dover Publishers, Inc., New York, N.Y.

Anderson, M.P. (1979) Using models to simulate the movement of contaminants through groundwater flow systems, *Crit. Rev. of the Environ. Controls*, CRC, **9**(2), 97–156.

Anderson, M.P. (1984) Movement of contaminants in groundwater: groundwater transport – advection and dispersion, in *Groundwater Contamination*, Studies in Geophysics, National Academy Press, Washington, D.C., pp. 37–45.

Bastian, W.C. and Lapidus, L. (1956) Longitudinal diffusion in ion exchange and chromatographic columns. Finite column, *J. Phys. Chem.*, **60**, 816–7.

Bear, J. (1972) *Dynamics of Fluids in Porous Media.* American Elsevier Publishing, Inc., New York, N.Y.

Bear, J. (1979) *Hydraulics of Groundwater*. McGraw-Hill Book Co., New York, N.Y.

Bear, J. and Verruijt, A. (1987) *Theory and Applications of Transport In Porous Media*, D. Reidel Publ. Co., Dordrecht, Holland.

Biggar, J.W. and Nielson, D.R. (1960) Diffusion effects in miscible displacement occurring in saturated and unsaturated porous materials, *J. Geophys. Res.* **65**(9), 2887–95.

Bowders, J.J., Daniel, D.E., Broderick, G.P. and Liljestrand, H.M. (1985) Methods for testing the compatibility of clay liners with landfill leachate, *Haz. and Solid Waste Testing*, in *Proc. 4th Symp.*, STP 886, Amer. Soc. for Testing and Materials, Philadelphia, PA.

Brenner, H. (1962) The diffusion of longitudinal mixing in beds of finite length: numerical values, *Chem. Eng. Sc.*, **17**:229–43.

Carslaw, H.S. and Jaeger, J.C. (1959) *Conduction of Heat in Solids*, 2nd edn. Oxford University Press, Oxford, England.

Cadena, F. (1989) Numerical approach to solution of pollutant transport models using personal computers, *Computers in Education*, Division of ASEE, **9**(2), 34–6.

Cherry, J.A. (guest ed) (1983) Migration of contaminants in groundwater at a landfill: a case study, *J. of Hydrology*, **63**, 1–192.

Crank, J. (1975) *The Mathematics of Diffusion*, 2nd edn. Oxford University Press, London, England.

Daniel, D.E. and Shackelford, C.D. (1988) Disposal barriers that release contaminants only by molecular diffusion, *Nuclear and Chemical Waste Management*, **8**, 299–305.

Davidson, J.M., Ou, L.-T. and Rao, P.S.C. (1976) Behavior of high pesticide concentrations in soil water systems, Residual management by land disposal, *Proc., Haz. Waste Res. Symp.*, USEPA, Cincinatti, Ohio, EPA-600/9-76-015, 206–12.

De Weist, R.J.M. (1965) *Geohydrology*, John Wiley and Sons, Inc., New York, N.Y.

Freeze, R.A. and Cherry, J.A. (1979) *Groundwater*, Prentice-Hall, Inc., Englewood Cliffs, N.J.

Fried, J.J. (1975) *Groundwater Pollution*, Elsevier Scientific Publishers, Amsterdam.

Gershon, N.D. and Nir, A. (1969) Effects of boundary conditions of models on tracer distribution in flow through porous mediums, *Water Resources Res.*, **5**(4), 830–9.

Gillham, R.W., Robin, M.J.L., Dytynyshyn, D.J. and Johnston, H.M. (1984) Diffusion of nonreactive and reactive solutes through fine-grained barrier materials, *Canadian Geotech. J.*, **21**, 541–50.

Goldman, L.J., Truesdale, R.S., Kingsbury, G.L., Northeim, C.M. and Damle, A.S. (1986) *Design, construction, and evaluation of clay liners for waste management facilities*, EPA Draft Technical Resource Document, Hazardous Waste Engineering Laboratory, Washington, DC, EPA/530-SW-86-007 [NTIS PB86-184496].

Goodall, D.C. and Quigley, R.M. (1977) Pollutant migration from two sanitary landfill sites near Sarnia, Ontario, *Canadian Geotech. J.*, **14**, 223–36.

Gray, D.H. and Weber, W.J., Jr. (1984) Diffusional transport of hazardous waste leachate across clay barriers, *Proc., 7th Ann. Madison Waste Conf.*, Sept 11–12, Univ. of Wisconsin, Madison, 373–89.

Greenberg, J.A., Mitchell, J.K. and Witherspoon, P.A. (1973) Coupled salt and water flows in a groundwater basin, *J. Geophys. Res.*, **78**(27), 6341–53.

Griffin, R.A. and Roy, W.R. (1985) *Interaction of organic solvents with saturated soil-water systems*, Environmental Institute for Waste Management Studies, University of Alabama, Tuscaloosa, Alabama.

Horton, R., Thompson, M.L. and McBride, J.F. (1985) Estimating transit times of noninteracting pollutants through compacted soil materials, *Proc., 11th Ann. Solid Waste Res. Symp.*, Apr 29–May 1, Cincinnati, Ohio, EPA/600/9-85/013, 275–82.

Huyakorn, P.S. and Pinder, G.F. (1983) *Computational Methods in Subsurface Flow*, Academic Press, Inc., Orlando, Florida.

Javendel, I., Doughty, C. and Tsang, C.-F. (1984) *Groundwater Transport: Handbook of Mathematical Models.* Water Resources Monograph Series 10, American Geophysical Union, Washington, DC.

Johnson, R.L., Cherry, J.A. and Pankow, J.F. (1989) Diffusive contaminant transport in natural clay: a field example and implications for clay-lined waste disposal sites, *Environ. Sc. and Tech.*, **23**, 340–9.

Karichoff, S.W., Brown, D.S. and Scott, T.A. (1979) Sorption of hydrophobic pollutants on natural sediments, *Water Research*, **13**, 241–8.

Kreft, A. and Zuber, A. (1978) On the physical meaning of the dispersion equation and its solutions for different initial and boundary conditions, *Chem. Eng. Sc.*, **33**, 1471–80.

Lapidus, L. and Amundson, N.R. (1952) Mathematics of adsorption in beds. VI. The effect of longitudinal diffusion in ion exchange and chromatographic columns, *J. Phys. Chem.*, **56**, 984–8.

Li, Y-H. and Gregory, S. (1974) Diffusion of ions in sea water and in deep-sea sediments, *Geochimica et Cosmochimica Acta*, **38**, 703–14.

Liao, W.P. and Daniel, D.E. (1989) Time of travel of contaminants through soil liners, *Proc., 12th Ann. Madison Waste Conf.*, Sept 20–21, Univ. of Wisconsin, Madison, 367–76.

Lindstrom, F.T., Haque, R., Freed V.H. and Boersma, L. (1967) Theory on the movement of some herbicides in soils, linear diffusion and convection of chemicals in soils, *Environ. Sc. and Tech.*, **1**(7), 561–5.

Melnyk, T.W. (1985) *Effects of sorption behavior on contaminant migration*, Atomic Energy of Canada Limited (AECL-8390), Whiteshell Nuclear Research Establishment, Pinawa, Manitoba R0E 1L0.

Mitchell, J.K. (1976) *Fundamentals of Soil Behavior*, John Wiley and Sons, Inc., New York, N.Y.

Ogata, A. (1970) Theory of dispersion in granular medium, *US Geol. Surv. Prof. Paper 411-I.*

Ogata, A. and Banks, R.B. (1961) A solution of the differential equation of longitudinal dispersion in porous media, *US Geol. Surv. Prof. Paper 411-A.*

Olsen, H.W. (1969) Simultaneous fluxes of liquid and charge in saturated kaolinite, *Proc., Soil Sc. Soc. of Amer.*, **33**(3), 338–44.

Parker, J.C. and Van Genuchten, M. Th. (1984) Flux-averaged and volume-averaged concentrations in continuum approaches to solute transport, *Water Resources Res.*, **20**(7), 866–72.

Perkins, T.K. and Johnston, O.C. (1963) A review of diffusion and dispersion in porous media. *J. of Soc. of Petrol. Engin.*, **3**, 70–83.

Peyton, G.R., Gibb, J.P., LeFaivre, M.H., Ritchey, J.D., Burch, S.L. and Barcelona, M.J. (1986) Effective porosity of geologic materials, Land disposal, remedial action, incineration, and treatment of hazardous waste, *Proc., 12th Ann. Res. Symp.*, Cincinnati, Ohio, EPA/600/9-86/022, 21–8.

Pickens, J.F. and Grisak, G.E. (1981) Scale-dependent dispersion in a stratified granular aquifer, *Water Resources Res.*, **17**(4), 1151–211.

Quigley, R.M., Crooks, V.E. and Fernandez, F. (1984) Engineered clay liners (an overview report). Preprints of the Seminar on the Design and Construction of Municipal and Industrial Waste Disposal Facilities, June 6–7, Toronto, Ontario, Canada, 115–34.

Quigley, R.M., Yanful, E.K. and Fernandez, F. (1987) Ion transfer by diffusion through clayey barriers, *Geotech. Practice for Waste Disposal*, ASCE special publication No. 13, (ed. R.D. Woods), 137–58.

Robinson, R.A. and Stokes, R.H. (1959) *Electrolyte Solutions*, 2nd edn. Butterworths Scientific Publications, London, England.

Rowe, R.K. (1987) Pollutant transport through barriers, *Geotech. Practice for Waste Disposal*, ASCE special publication No. 13, (ed. R.D. Woods), 159–81.

Rowe, R.K. and Booker, J.R. (1985) 1-D pollutant migration in soils of finite depth, *J. Geotech. Eng.*, ASCE, **111**(4), 479–99.

Rowe, R.K., Caers, C.J. and Barone, F. (1988) Laboratory determination of diffusion and distribution coefficients of contaminants using undisturbed clayey soil, *Canadian Geotech. J.*, **25**, 108–18.

Selim, H.M. and Mansell, R.S. (1976) Analytical solution of the equation for transport of reactive solutes through soils, *Water Resources Res.*, **12**(3), 528–32.

Shackelford, C.D. (1988a) Diffusion of inorganic chemical wastes in compacted clay, Ph.D. Dissertation, University of Texas, Austin, TX.

Shackelford, C.D. (1988b) Diffusion as a transport process in fine-grained barrier materials, *Geotech. News*, **6**(2), 24–7.

Shackelford, C.D. (1989) Diffusion of contaminants through waste containment barriers, *Transportation Research Record 1219*, Transportation Research Board, National Research Council, Washington, DC, 169–82.

Shackelford, C.D. (1990) Transit-time design of earthen barriers, *Eng. Geol.*, Elsevier Publ., Amsterdam, **29**, 79–94.

Shackelford, C.D. (1991) Laboratory diffusion testing for waste disposal-A review, *J. Contaminant Hydrology*, Elsevier Publ., Amsterdam, **7**(3), 177–217.

Shackelford, C.D. and Daniel, D.E. (1991) Diffusion in saturated soil: I. Background, ASCE *J. Geotech. Eng.*, **117**(3), 467–84.

Van Genuchten, M. Th. (1981) Analytical solutions for chemical transport with simultaneous adsorption, zero-order production, and first-order decay, *J. Hydrology*, **49**:213–33.

Van Genuchten, M. Th. and Alves, W.J. (1982) Analytical solutions of the one-dimensional convective-dispersive solute transport equation, US Department of Agriculture, Technical Bulletin No. 1661.

Weber, W.J., Jr. and Smith, E.H. (1987) Simulation and design models for adsorption processes, *Environ. Sc. and Tech.*, **21**(11), 1040–50.

CHAPTER 4

Hydrogeology

Keros Cartwright and Bruce R. Hensel

4.1 INTRODUCTION

The basic theory and concepts of hydrogeology will be discussed in this chapter. Sections 4.2–4 will deal primarily with the definitions and equations of ground-water flow in the saturated and unsaturated zone. Section 4.5 will examine the geology of hydrogeology, and is especially important because the movement of fluid through the subsurface and the equations that describe this movement are dictated by the geology of the subsurface materials. Geologic characteristics such as permeability, porosity, anisotropy, and homogeneity, are functions not only of the type of rock or soil material, but also of the depositional environment, diagenesis, and tectonic processes such as folding and faulting that may have occurred after deposition. This information is needed to build a geologic framework in which hydrogeologic principles may be applied before ground-water flow and contaminant transport can be evaluated.

4.2 BASIC PRINCIPLES

4.2.1 Definition of geologic materials

Two basic types of geologic materials will be discussed in this chapter. **Soil** is used in the engineering sense to describe nonlithified and/or unconsolidated sand, silt, and/or clay deposits that overlay the bedrock surface. **Rock** is used to describe all igneous, metamorphic, and lithified sedimentary rocks. When these materials have like permeability and porosity characteristics, they make-up a hydrostratigraphic unit that will either allow or be a barrier to ground-water flow.

Geotechnical Practice for Waste Disposal.
Edited by David E. Daniel.
Published in 1993 by Chapman & Hall, London. ISBN 0 412 35170 6

4.2.2 Definition of ground water

Ground water is defined as water that occurs in the voids of saturated rock and soil material (Freeze and Cherry, 1979; Todd, 1980). Water found in unsaturated materials is defined as **vadose water** (Todd, 1980). There is no distinction between the two types of water other than occurrence, and water can move from the unsaturated to saturated zone and vice versa. Therefore, for the purpose of this chapter, the term ground water will be used to identify all water that occurs in voids beneath the ground surface.

4.2.3 Porosity

The voids in which ground water are found can have many shapes and forms (Fig. 4.1). The **primary porosity** of a material is the ratio of the volume of intergranular pore space relative to the total volume of the material. This value may be quite high, 45–50% in clays, or quite low, 0–5% in crystalline rocks. This type of porosity is often referred to as **intergranular** porosity. The **secondary porosity** of a material refers to the porosity formed subsequent to deposition. Types of secondary porosity are fractures, weathered rock or soil zones, solution openings,

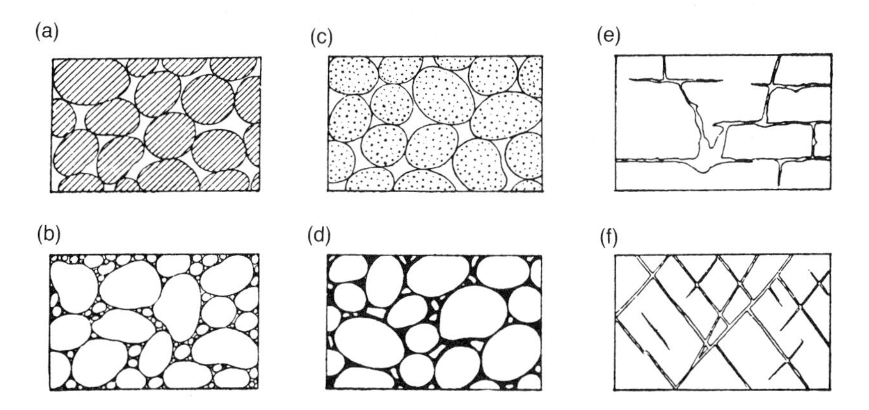

Fig. 4.1 Types of porosity commonly found in geologic materials. A–D are examples of primary porosity, the primary porosity of case D has been diminished by digenetic processes, cases E and F are examples of secondary porosity, from Meinzer (1923). Diagram showing several types of rock interstices and the relation of rock texture to porosity: A, Well-sorted sedimentary deposit having high porosity; B, poorly sorted sedimentary deposit having low porosity; C, well-sorted sedimentary deposit consisting of pebbles that are themselves porous, so that the deposit as a whole has a very high porosity; D, well-sorted sedimentary deposit whose porosity has been diminished by the deposition of mineral matter in the interstices; E, rock rendered porous by solution; F, rock rendered porous by fracturing.

and macropores caused by animal burrowing, root action, and dehydration of soils. This type of porosity is sometimes referred to as **fracture** porosity. The total porosity of a soil or rock is the sum of the primary and secondary porosity. Table 4.1 lists porosity characteristics for flow through common geologic materials.

An important concept in the evaluation of contaminant transport is that of **effective porosity**. Whereas total porosity is the ratio of void volume to sample volume in a rock or soil material, effective porosity is the percentage of void volume that is interconnected, allowing a fluid to pass through the material. The effective porosity of a soil or rock material must, by definition, be equal to or less than the total porosity of the material. In general, the effective porosity of coarse-grained materials such as sand approaches the total porosity. The effective porosity of fine-grained materials such as clay may be 10–20% of total porosity (Horton, Thompson, and McBride, 1988).

A term related to porosity that is important in the analysis of unsaturated flow is **moisture content**. The moisture content, defined here as a volumetric term, is the ratio of water filled pores to total sample volume. Another related term, **degree of saturation**, or

Table 4.1 Porosity characteristics of ground-water flow through various geologic material

type of porosity	*geologic characteristic*		
	sedimentary		*igneous and metamorphic*
	lithified	*unlithified*	
intergranular		gravel sand silt loess peat outwash	
intergranular and fractures	weather limestone/ dolomite oolitic limestone chalk sandstone shale coal	clay-marine/ lacustrine till	tuff weathered basaltic rocks weathered granitic rocks
fracture	limestone dolomite marble		basaltic rocks granitic rocks quartzite gneiss

moisture content, is the ratio of water filled pores to total porosity. For a saturated material, the moisture content will be equal to the total porosity and the degree of saturation will be equal to 1.0.

4.2.4 Hydraulic head

Freeze and Cherry (1979) noted that ground-water flow, in its simplest form, is a mechanical process. When water moves through a porous material, mechanical energy is transformed to thermal energy as the water molecules encounter the frictional resistance of the matrix. Therefore, the direction of flow must be away from areas where the mechanical energy per unit mass of a fluid is high, and towards areas where mechanical energy is low. This energy concept is the basis for the definition of hydraulic head, in that the fluid potential, or head, is the mechanical energy per unit mass of fluid.

Hydraulic head has three components: the elevation head, the pressure head, and the velocity head (Freeze and Cherry, 1979). In most situations, ground-water velocity is low and the velocity head is negligible compared to pressure and elevation head. The **elevation head** is a measure of the potential energy of the water. It describes the potential for downward movement of water in response to gravity. **Pressure head** describes the fluid pressure and is measured as the height a column of water at the measurement point will rise or fall to in a manometer (Fig. 4.2).

4.2.5 Saturated and unsaturated materials

When a rock or soil material is saturated, the pressure head will be zero or a positive value. All of the void spaces in a saturated material are usually filled with water. Ground water in saturated materials will flow into an opening, such as a well, under conditions of normal atmospheric pressure.

In an unsaturated or tension-saturated material the pressure head will have a negative value (Fig. 4.2). **Tension**, or suction, is the absolute value of pressure head in unsaturated material. Void spaces in an unsaturated material will often be filled with both water and gasses. Ground water in such materials will not flow into an opening unless a pressure lower than the pressure head is applied to the opening. Restating the last sentence in soil physics terms, soil water will not flow into an opening unless a suction greater than the tension is applied to the opening.

In many soils, particularly in fine grained soils, a capillary zone may exist where most or all of the pore spaces are saturated with water even though the pressure head is negative. In both the unsaturated

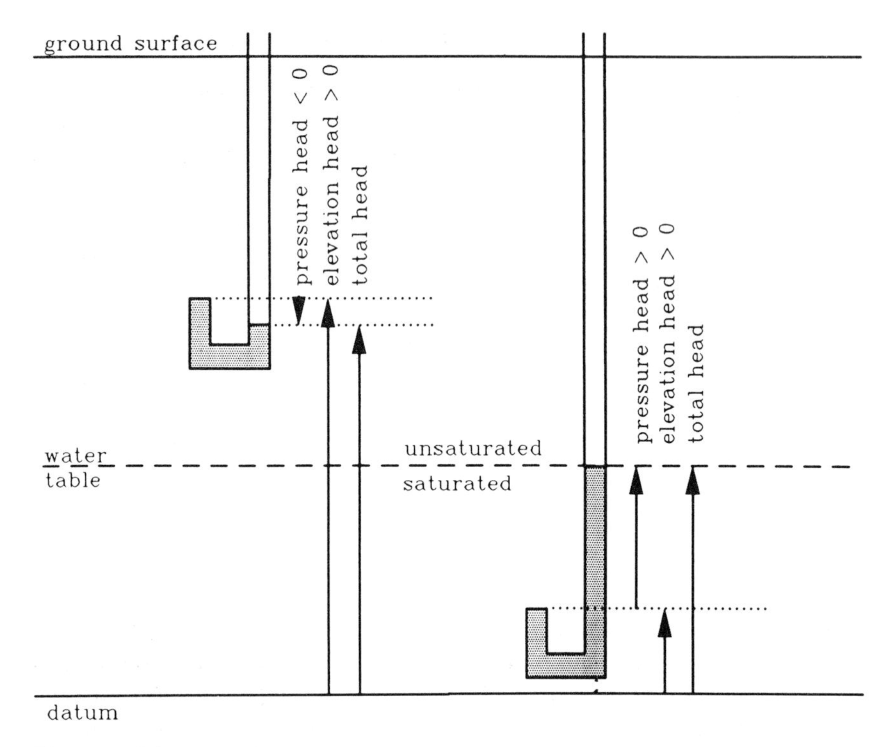

Fig. 4.2 Schematic drawing illustrating head in the saturated and unsaturated zone. Relative head value is indicated by length of line, positive values are indicated by upward pointing arrow, negative values by downward pointing arrows.

and the capillary zones, ground water does not flow to an opening at atmospheric pressure because the capillary forces hold the water molecules to the soil particles.

4.2.6 Conservation of mass

Ground water is subject to the principle of conservation of mass. That is, water is not created or consumed in the subsurface (except as a byproduct of certain chemical reactions). Therefore for any volume of soil or rock material

$$V_{in} = V_{out} + S \tag{4.1}$$

Where V_{in} is the volume of water that enters the system, V_{out} is the volume of water that exits the system and S is the change in the volume of water stored in the system. This is a simplified version of the **continuity equation**. The concept of continuity is the basis of the ground-water flow equation that describes the value of head at any point within a three-dimensional flow field.

4.2.7 Hydraulic conductivity and intrinsic permeability

The hydraulic conductivity of a rock or soil material describes the ease with which a particular fluid, usually water, may flow through that material. Hydraulic conductivity is neither a rock or fluid property, it is dependent on the density and viscosity of the fluid as well as the intrinsic permeability of the rock or soil material. The intrinsic permeability of a material is dependent upon its grain size, packing, shape, and pore distribution; thus it is a property related to the soil or rock material. Freeze and Cherry (1979, p. 26–9) present a comprehensive discussion on this topic. The terms hydraulic conductivity and permeability are often interchanged, with intrinsic permeability used to describe the material property that affects fluid movement. In this chapter, hydraulic conductivity will be used to describe the movement of water in the subsurface and permeability will hereafter be used in place of intrinsic permeability.

4.2.8 Infiltration and recharge

Almost all ground water originates as surface water (Todd, 1980). Infiltration describes the process of water entering the soil at the ground surface. The source of this water may be rainfall, melting snow, seepage from surface water bodies, or irrigation water. Recharge describes the process of water entering the saturated zone. Figure 4.3 illustrates the difference between recharge and infiltration. **Discharge** occurs when ground water exits the flow system as baseflow, evapotranspiration, springs, or through wells. Most discharge is baseflow to surface water bodies (Todd, 1980) such as oceans, lakes, and rivers.

Before water can be made available for recharge, it must infiltrate into the soil. Freeze and Cherry (1979) define infiltration as 'entry into the soil of water made available at the ground surface, together with associated flow away from the ground surface within the unsaturated zone'. Not all water made available at the ground surface infiltrates and not all water which infiltrates into the soil will recharge ground water. Water which infiltrates may take several courses: it may be drawn back to the surface by evaporation or through plant roots as transpiration; it may redistribute through the unsaturated soil column replenishing soil-moisture deficits without recharging ground water; for certain geologic and topographic conditions it may flow to a nearby surface water body as subsurface storm flow; or it may seep to the saturated zone as recharge water. Infiltration will occur for almost any event where water is applied to an unsaturated ground surface; but only water that is not returned to the atmosphere as evapotranspiration, not discharged to surface water bodies through the unsaturated zone, and is not used to replenish soil-moisture deficits is available for recharge.

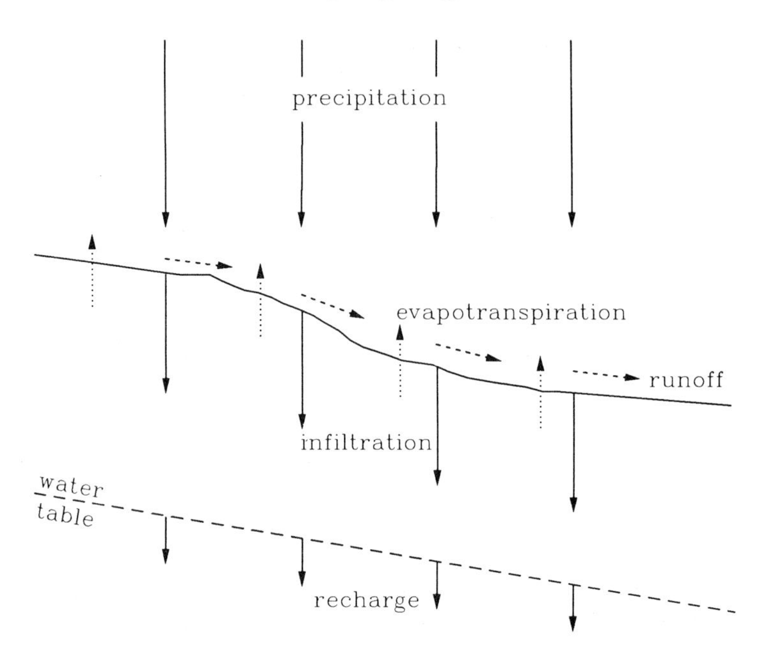

Fig. 4.3 Infiltration and recharge. Infiltration is water that percolates into the ground. Recharge is water that reaches and flows downward from the water table. Only that portion of infiltrating water that is not used to replenish soil-moisture deficits and is not discharged back to the atmosphere through evapotranspiration is available for recharge.

Ground-water recharge is defined as 'entry into the saturated zone of water made available at the water table, together with the associated flow (of ground water) away from the water table within the saturated zone' (Freeze and Cherry, 1979). If the water table is at the ground surface and flow is downward, then water made available at the ground surface can recharge the ground-water flow system. However, if flow is upwards from the water table, discharge, rather than recharge will occur. Recharge has also been defined as the amount of water that reaches an aquifer. Strictly speaking, this definition is incorrect. When this definition of recharge is used, it should be termed as recharge to an aquifer.

Numerous field studies have shown that, in mid-latitude regions, ground water recharge is most likely in the spring, at which time snow melts, precipitation is high and evapotranspiration is low; so that the water table is high and antecedent soil moisture is high (Walton, 1965; Rehm, Groenewold and Peterson, 1982; Steenhuis *et al.*, 1985). Steenhuis *et al.* (1985) measured recharge on Long Island and estimated that most (75–90%) precipitation from mid-October to mid-May becomes recharge, while very little precipitation occurring during late spring, summer and early fall infiltrates to the water table to replenish

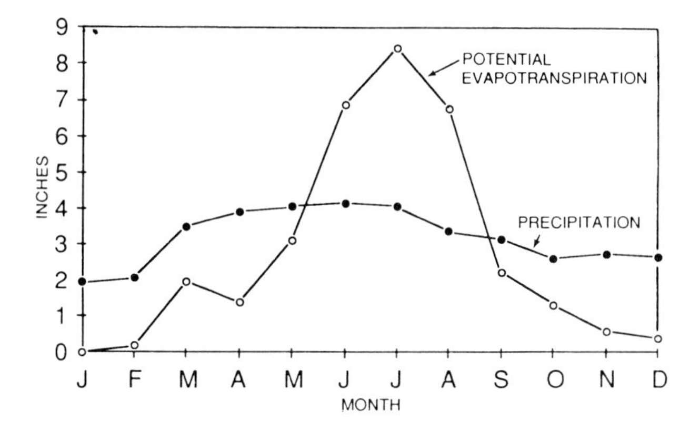

Fig. 4.4 Precipitation vs. evapotranspiration in the midwestern US from Bowman and Collins (1987).

ground water. In general, during the summer months, evapotranspiration and soil moisture requirements use all infiltrating water (Fig. 4.4) except during periods of excessive rainfall (Walton, 1965).

4.2.9 Hydrostratigraphic units

Geologic materials may be classified into three hydrogeologic types: aquifers, aquitards, and aquifuges. **Aquifers** are hydrogeologic units that are capable of sustaining a water supply. **Aquitards** will not sustain a water supply. The distinction between aquifers and aquitards is subject to scale and interpretation. A hydrogeologic unit that supplies adequate water to a domestic well, and is considered an aquifer by the well owner, may be considered a portion of an aquitard to the owner of a municipal water supply system. An **aquifuge** is a unit with no interconnected porosity (effective porosity is essentially 0.0). Aquitards and aquifuges are also commonly referred to as **confining units**.

Seaber (1988) formally defines a hydrostratigraphic unit as a volume of geologic material distinguished by its porosity and permeability. An example of a formal hydrostratigraphic classification is one proposed in Illinois by Cartwright (1983). In this classification there are three aquisystems based on the nature of the geologic material: the Crystalline Rock Aquisystem consisting of the PreCambrian basement rocks; the Indurated Rock Aquisystem consisting of all Paleozoic sedimentary rocks; and the Non-Indurated Rock Aquisystem consisting of unconsolidated Mesozoic and Cenozoic materials. The aquisystems are then further subdivided into aquigroups, with the top of an aquigroup being a major aquitard. Aquigroups then are subdivided into aquifers and minor aquitards.

4.3 FLOW IN UNSATURATED MATERIALS

4.3.1 Definitions

The **infiltration rate** in a soil is the volume of water flowing into a soil profile per unit of ground surface area and unit time. The infiltration rate is therefore the flux of water entering the ground. Horton (1940) defined **infiltration capacity** as the maximum infiltration rate that occurs when the rate of water made available at the surface exceeds the ability of the soil to adsorb water. Hillel (1971) defined the term **infiltrability** for the infiltration rate when water at atmospheric pressure is made freely available at the ground surface. When water is ponded at the surface, the infiltration rate can be expected to exceed the infiltrability because the ponded water will have a head greater than atmospheric pressure. When water is supplied to the surface slowly, so that ponding does not occur, the infiltration rate will be less than or equal to the infiltrability. In the first case, the infiltration capacity of the soil will dictate the infiltration rate. In the second case, water will infiltrate as fast as it is applied so the supply rate dictates the infiltration rate (Hillel, 1980).

Field capacity is the moisture content in a soil after the initial rapid stage of downward gravity drainage (redistribution) through the soil column. This redistribution of water is a continual process (Hillel, 1980) that is relatively rapid in wet soils and slow in dry soils. As the soil moisture content of a wet soil approaches field capacity, the rate of downward redistribution exponentially decreases (Richards *et al.*, 1956). Foth and Turk (1972) suggest that soils commonly reach field capacity when about half the pore space is occupied by water. A soil with a moisture content below field capacity has a **soil moisture deficit**. When water infiltrates into a soil with a moisture deficit it is drawn into the pores, and little is available for downward redistribution until the moisture content of the soil is replenished to a state greater than the soil's field capacity.

4.3.2 Equations of unsaturated flow

Lehr (1988) presented a discussion on the microscopic processes that affect the movement of water in the unsaturated zone. Movement in the unsaturated zone is in response to suction forces and gravity. The suction forces consist of **adhesion**, where water molecules are attracted to rock or soil particles and **cohesion**, where water molecules are attracted to other water molecules. Collectively, these forces are also referred to as **capillary forces**. The capillary forces are a result of hydrogen bonding between the water molecules and other water/soil/rock molecules. Capillary forces tend to be stronger in fine-grained materials than in coarse-grained materials because the fine-grained materials have a greater surface area allowing more molecules for adhesion.

Capillary forces dominate in dry soils; however, as the soil wets, gravity becomes the more dominant force. When pressure head is negative (unsaturated conditions), water movement will only occur in the liquid phase through continuous water films around particles. As the moisture content increases, small pores are filled with water before larger pores, the number and volume of potential flow paths increases, and the overall hydraulic conductivity of the unsaturated material increases.

The mathematical representation of unsaturated flow on a macroscopic scale is derived from Darcy's law and the continuity equation. This equation, known as the **Richards Equation**, is presented below using the notation of Freeze and Cherry (1979).

$$\frac{\partial}{\partial x}\left(K_{(\psi)}\frac{\partial h}{\partial x}\right) + \frac{\partial}{\partial y}\left(K_{(\psi)}\frac{\partial h}{\partial y}\right) + \frac{\partial}{\partial z}\left(K_{(\psi)}\frac{\partial h}{\partial z}\right) = \frac{\partial \theta}{\partial t} \tag{4.2}$$

$K_{(\psi)}$ is the unsaturated hydraulic conductivity that varies with respect to pressure head (ψ), h is total hydraulic head, and θ is the moisture content of the soil at a given pressure head. As can be seen in Eq. (4.2), flow in the unsaturated zone is a function of the moisture content and the pressure head.

4.3.3 Moisture content, tension, and hydraulic conductivity

The hydraulic conductivity in an unsaturated material is lower than the hydraulic conductivity of that material when it is saturated. Hydraulic conductivity in an unsaturated material is dependent upon the moisture content and tension in the material. For any given soil and condition of wetting or drying, there is a unique relationship between moisture content and tension. A **soil moisture characteristic curve** is a graphical representation of this relationship. Figure 4.5 shows a hypothetical soil characteristic curve. When tension is zero or greater, the moisture content is equal to porosity (n) and the material's hydraulic conductivity is equal to its saturated (maximum) hydraulic conductivity. As the soil begins to dry, tension increases; but the moisture content and unsaturated hydraulic conductivity decrease little because capillary forces continue to hold water in the pores. At some tension, the pores begin to drain and moisture content and hydraulic conductivity decrease rapidly.

If the dry soil is re-wetted, moisture content and hydraulic conductivity increase; however, for any given point on the wetting curve both moisture content and hydraulic conductivity will be lower than if the soil were being dried. This effect, called **hysteresis**, is

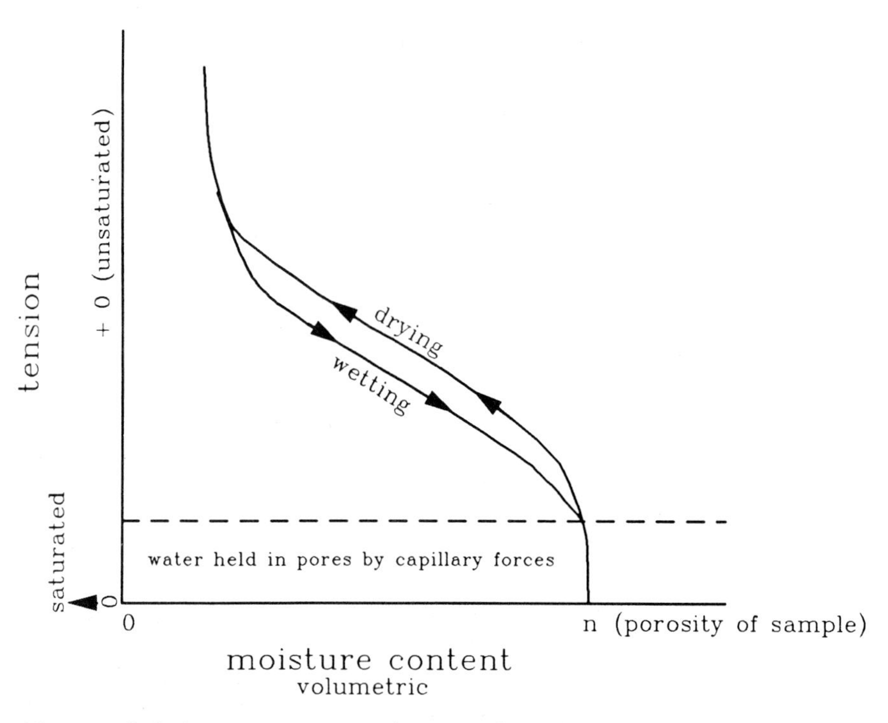

Fig. 4.5 Soil characteristic curve showing effect of hysteresis.

attributed to several processes. One of the most important processes is termed by Hillel (1982) as the 'ink bottle' effect. This term describes the phenomena where the tension at which a saturated pore will drain is greater than the tension at which it will fill with water. A pore in a drying soil will drain when the tension reaches some specific value; however when the soil is rewetting that same pore will fill when tension reaches a value lower than the value at which the pore drained. Thus, in a given soil at a specific value of tension, there will be more pores filled with water during drying than during wetting, so the moisture content of a drying soil is greater than a wetting soil at that specific value of tension. Other processes effecting hysteresis are the contact angle of the meniscus, entrapped air, and soil shrinking/ swelling (Hillel, 1982).

One example of the potential importance of the hysteresis effect could be the case of characteristic curves for a compacted soil liner. If this curve is obtained by laboratory analysis for drying, the lab-estimated unsaturated hydraulic conductivity may be higher at any given tension than the field values because the field liner would likely be wetting.

4.4 FLOW IN SATURATED MATERIALS

4.4.1 Gradient

Ground water in the saturated zone moves from areas of high energy to areas of low energy. The change in energy over a given distance is defined as the gradient. There are several types of gradients which can drive ground-water flow in the saturated zone. A temperature gradient would cause water to flow from areas of high temperature to areas of low temperature, an electrical gradient can cause movement of water from areas of high voltage to areas of low voltage, a density gradient could cause ground-water flow from zones of high salinity to zones of low salinity, and a chemical gradient may cause water to flow from areas of low salinity to areas of high salinity. However, the **hydraulic gradient**, where water flows from high head to low head, is generally the dominant driving force of ground-water flow. The hydraulic gradient is the maximum change in head over distance from a given point.

The hydraulic gradient is a vector, therefore it has magnitude and direction. The magnitude is the change in head over distance, and the direction is that of the maximum change in head over distance. Hydraulic gradient cannot be determined with two points, or with any number of points that lie on a straight line. Hydraulic gradient must be determined from a surface, either the water table or the potentiometric surface, so that the steepest value may be obtained. On a potentiometric surface, hydraulic gradient is perpendicular to lines of equal potential. In isotropic materials, flow lines will be parallel to the hydraulic gradient. However, in non-isotropic materials, flow lines will not necessarily be parallel to hydraulic gradient – see sections 4.4.3 (Homogeneity and isotropy) and 4.5.3 (Aquifers with flow in secondary pore openings).

4.4.2 Equations of saturated flow

The mathematical equation that describes ground-water flow is Darcy's Law. This empirical equation, often given in this simplified form for one-dimensional flow in isotropic media,

$$Q = -KiA \tag{4.3}$$

defines the volumetric flux of fluid (Q) through a given cross-sectional area (A) for a material with hydraulic conductivity (K) under a hydraulic gradient (i). A more general form of Darcy's law is

$$v = -Ki \tag{4.4}$$

where v is defined by Freeze and Cherry (1979) as the **specific discharge**. They make this distinction because v is a flux, not a velocity. Darcy's law describes flow through a continuum at a macroscopic scale, and v describes the time it would take for a particle of water to traverse a distance assuming a porosity value of 1.0. In reality, ground water must flow more rapidly than the specific discharge would suggest because a significant portion of the continuum is occupied by a solid matrix. Just as a river's water velocity must increase when the channel is constricted, ground-water velocity must increase when the cross-sectional area available for flow is decreased. An approximation of macroscopic ground-water velocity is given by the equation

$$\bar{v} = \frac{-Ki}{n_e} \tag{4.5}$$

where n_e is effective porosity and $\bar{v}$ is the **average linear velocity** or seepage velocity. Because n_e is always less than 1.0, $\bar{v}$ is always greater than the specific discharge (v). The average linear velocity ($\bar{v}$) approximates the velocity of ground water at a macroscopic scale. However, because water molecules must take a tortuous path around the solid matrix particles, $\bar{v}$ is not a true measure of the actual movement of ground-water molecules on a microscopic scale. However, it is sufficient for evaluation of ground-water flow and contaminant transport problems that are also evaluated on a macroscopic scale.

The equation of ground-water flow for steady state conditions is derived by combining a three-dimensional form of the Darcy equation with the continuity equation

$$\frac{\partial}{\partial x}\left(K_x\frac{\partial h}{\partial x}\right) + \frac{\partial}{\partial y}\left(K_y\frac{\partial h}{\partial y}\right) + \frac{\partial}{\partial z}\left(K_z\frac{\partial h}{\partial z}\right) = 0 \tag{4.6}$$

where K_x, K_y, and K_z are the hydraulic conductivity values in the x, y, and z directions. This equation describes the head distribution throughout a three-dimensional flow field for steady state conditions in which the cartesian coordinates (x, y, z) are parallel to the principal hydraulic conductivity axes. A similar equation for a transient flow system would be

$$\frac{\partial}{\partial x}\left(K_x\frac{\partial h}{\partial x}\right) + \frac{\partial}{\partial y}\left(K_y\frac{\partial h}{\partial y}\right) + \frac{\partial}{\partial z}\left(K_z\frac{\partial h}{\partial z}\right) = S_s\frac{\partial h}{\partial t} \tag{4.7}$$

where S_s is the specific storage, h is hydraulic head, and t is time. This equation describes the head distribution throughout a three-dimensional flow field at any given point in time and space. The left sides of Eqs. (4.6) and (4.7) are identical; however the right sides differ.

Equation (4.6) defines **steady state flow** where head is constant over time with no change in storage, while Eq. (4.7) defines **transient flow** where head changes with respect to time.

4.4.3 Homogeneity and isotropy

Equations (4.6) and (4.7) define head distribution in an anisotropic, heterogeneous material. If the material was isotropic then K would be constant in any direction so that $K_x = K_y = K_z$. If the material was also homogeneous then K would be constant at any place in the material. For an isotropic, homogeneous material, with steady state flow, Eq. (4.6) reduces to Laplace's equation.

$$\frac{\partial^2 h}{\partial x^2} + \frac{\partial^2 h}{\partial y^2} + \frac{\partial^2 h}{\partial z^2} = 0 \tag{4.8}$$

Homogeneous and isotropic conditions are often assumed for many analytical solutions to ground-water problems. However, for most geologic materials, particularly for materials deposited in fluvial environments, heterogeneous and anisotropic conditions are more

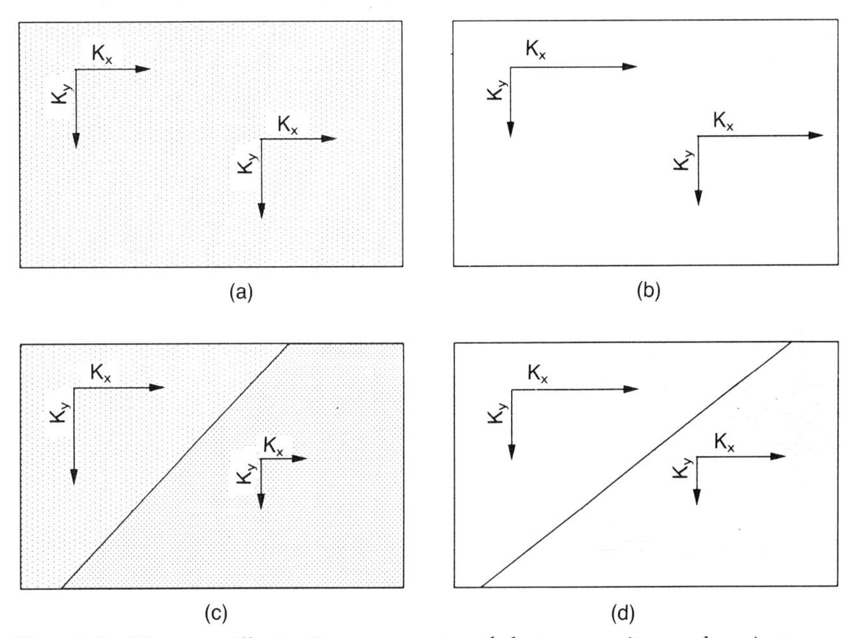

Fig. 4.6 Diagram illustrating concepts of heterogeneity and anisotropy. (a) Homogeneous, isotropic in a uniform sand, (b) homogeneous, anisotropic in a stratified sand, (c) heterogeneous, isotropic, uniform sands, facies change, (d) heterogeneous, anisotropic, stratified sands, facies change.

common. Figure 4.6 illustrates the concepts of homogeneity and isotropy. Both homogeneity and isotropy are subject to scale. For example, a sandstone layer may be isotropic when viewed by itself; but as one layer in a sequence of sandstone layers with slightly different hydraulic conductivities it would be a portion of a vertically anisotropic flow system. Similarly, the hydraulic conductivity of a till may vary within a couple of meters; but on a large scale the range of values may have a normal distribution so that the unit may be considered to be homogeneous.

An example of an anisotropic, heterogeneous flow system might be an alluvial valley fill aquifer. The uneven distribution of sand, silt, and clay deposits commonly found in such deposits would mean changes in hydraulic conductivity from point to point within the flow system that could vary by orders of magnitude. Horizontal stratification of sands and silts in such deposits results in horizontal hydraulic conductivities that are commonly 2–10 times larger than vertical conductivities (Revelle, 1941), with factors of 100 or more being possible.

An example of a homogeneous, isotropic system could be a sandstone deposited in an aeolian environment. The sorting of grains by wind action would tend to remove irregularities that could cause anisotropy and heterogeneity.

4.5 GROUND-WATER FLOW REGIMES

4.5.1 Flow systems

The route ground water takes to a discharge point is known as a flow path. A flow system is a set of flow paths with common recharge and discharge areas (Cartwright and Hunt, 1981). Toth (1963) developed the concept of flow systems to distinguish flow in local, intermediate, and regional systems. Water in a **local flow system** will flow to a nearby discharge area, such as a pond or stream. Water in a **regional flow system** will travel a greater distance than the local flow system, and often will discharge to oceans, large lakes, or rivers (Fig. 4.7).

Flow systems are dependent on both the hydrogeologic characteristics of the soil/rock material and landscape position. Areas of steep relief tend to have dominant local flow systems, and areas of flat relief tend to have dominant intermediate and regional flow systems (Freeze and Witherspoon, 1967). Geological heterogeneities affect the interrelationship between regional and local flow systems, the pattern of recharge and discharge areas, and the quantity of ground water that flows through the systems (Freeze and Cherry, 1979).

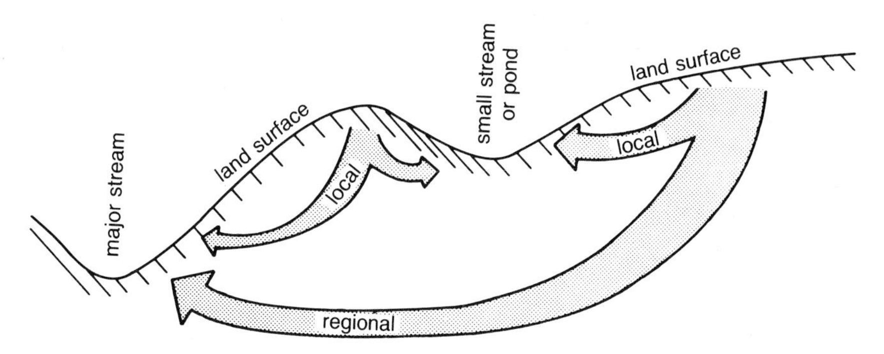

Fig. 4.7 Local and regional flow systems and relationship to recharge and discharge areas (from Cartwright and Sherman, 1969).

4.5.2 Recharge and discharge areas

In a typical watershed, the recharge area will be larger than the discharge area (Freeze and Witherspoon, 1967), with discharge occurring over only 5–30% of the watershed (Freeze and Cherry, 1979). In general, ground water is recharged in highlands and discharges in lowland areas; however geology, topography, vegetation, and land usage can complicate the determination of recharge and discharge areas.

Rates of recharge are affected by the infiltrability of the soil, the hydraulic conductivity of the unsaturated materials, depth to the water table, ambient soil-moisture content, landscape slope and position, land use, and availability of recharge water. In general, if recharge water is available, recharge rates will be more rapid in coarse-grained materials than in fine-grained materials because coarse-grained materials have higher hydraulic conductivity and lower field capacity than fine-grained materials.

4.5.3 Types of aquifers

There are many ways to classify the materials through which ground water flows. One basic division that has already been discussed is that of an aquifer versus an aquitard. Aquifers can further be divided by hydraulic characteristics (confined vs. unconfined), porosity characteristics (primary vs. secondary), and geologic characteristics. Figure 4.8 and Table 4.2 lists ranges of hydraulic conductivity for many of these materials.

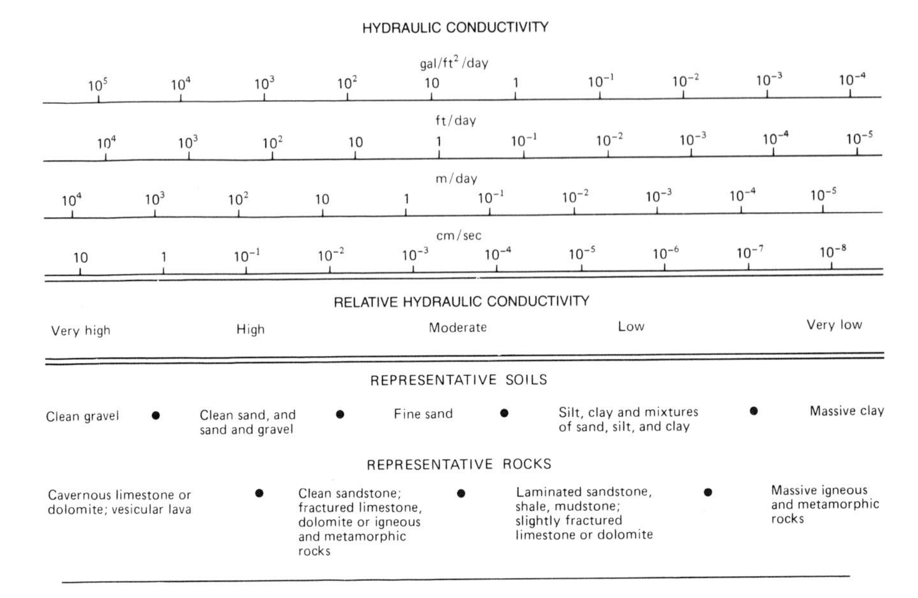

Fig. 4.8 Range of hydraulic conductivity values in selected soil and rock materials (modified from US Department of Interior, 1981).

Confined vs unconfined aquifers

A confined aquifer is overlain by an aquitard and contains water under pressure so that the water level in a penetrating well rises above the top of the aquifer. The elevation to which water in a confined aquifer will rise in a well is called the **potentiometric surface** (Fig. 4.9). Freeze and Cherry (1979) note that a map of the potentiometric surface is really a map of head using only two or three dimensions. If there is a vertical component of flow in the system then any interpretations based on a potentiometric surface may be in error.

Unconfined aquifers generally occur at relatively shallow depths. The top of the saturated zone in an unconfined aquifer is within the rock or soil strata of the aquifer. This surface is the **water table** (Fig. 4.9). Unlike a confined aquifer, an unconfined aquifer is not necessarily bounded at the top by an aquitard. Using a hydrostratigraphic definition, the upper boundary of an unconfined aquifer would be top of the hydrostratigraphic unit that makes up the aquifer. Using an older definition, the upper boundary of an unconfined aquifer is the water table. Under the latter definition, the upper boundary of an unconfined aquifer is not fixed in space or time because the water table rises and falls in response to seasonal and man-induced stresses. In general, the water table will not coincide with the potentiometric surface in areas where both types of aquifer are present (Freeze and Cherry, 1979).

Table 4.2 Typical hydraulic conductivity, porosity, and effective porosity values

geologic material	*hydraulic conductivity (m/s)*	*porosity total*	*porosity effective*	*sources*
Igneous/Meta.		0–5		5, 11
fractured	10^{-10}–10^{-13}	0–10		5
weathered	10^{-4}–10^{-8}	40–50		11
Basalt				
dense	10^{-9}–10^{-13}	1–8		2, 4
permeable	10^{-2}–10^{-7}	5–50	5–10	2, 5, 6
lava	10^{-1}–10^{-17}			12
Tuff	10^{-3}–10^{-8}	10–40		2, 9, 11, 12
Granite	10^{-9}–10^{-13}	0–2		4
weathered	10^{-4}–10^{-5}	0–45	0–8	2, 6, 9
Quartzite		1		4
Gneiss	10^{-9}–10^{-13}	0–2		4
Sedimentary rocks				
Limestone/dolomite	10^{-6}–10^{-9}	0–20		1, 5
fractured	10^{-4}–10^{-9}	0–20	0–18	1, 6
Limestone				
oolitic	10^{-6}–10^{-7}	1–25		1
karst	10^{+2}–10^{-6}	5–50		1, 5
marble	10^{-5}–10^{-8}	0–2		1
chalk	10^{-5}–10^{-9}	8–45		1, 4, 6
weathered	10^{-3}–10^{-5}			4
Sandstone/siltstone	10^{-4}–10^{-10}	10–50	10–48	3, 4, 6, 9, 11
Shale	10^{-9}–10^{-13}	0–10		3, 5, 9
Coal	10^{-6}–10^{-11}			11
Unconsolidated sediments				
Gravel	10^{-1}–10^{-4}	25–60	7–40	2, 5, 6, 9, 11
Sand				
clean	10^{-2}–10^{-6}	25–55	24–40	2, 5, 6, 9, 11
silty	10^{-3}–10^{-7}			
Silt	10^{-6}–10^{-10}	35–60		2, 5, 9, 11
Loess	10^{-5}–10^{-11}			2, 9, 10, 11
Lacustrine silt/clay	10^{-9}–10^{-13}	40–70		5, 10
fractured	10^{-8}–10^{-11}			10
Marine clay	10^{-9}–10^{-12}	40–70		5
Clay/silt compacted	10^{-9}–10^{-11}	30–50	<0.1	7, 8
Peat	10^{-6}–10^{-9}	60–92		9, 11
Glacial deposits				
Outwash	10^{-3}–10^{-7}		35	6, 10
Till/basal	10^{-7}–10^{-11}			5
weathered	10^{-6}–10^{-9}			10
fractured	10^{-5}–10^{-9}			10
Till/supraglacial	10^{-5}–10^{-9}			10
weathered	10^{-5}–10^{-9}			10
fractured	10^{-5}–10^{-9}			10

Sources: 1. Brahana *et al.* (1988); 2. Davis (1969) – K values assume water at 20 °C; 3. Davis (1988); 4. de Marsily (1986); 5. Freeze and Cherry (1979); 6. Gelhar *et al.* (1985); 7. Horton, Thompson and McBride (1988); 8. Krapac *et al.* (1991); 9. Morris and Johnson (1967); 10. Stephenson, Fleming and Mickelson (1988); 11. Walton (1985); 12. Wood and Fernandez (1988). Note: 1, 2, 3, 6 and 12 are compilations of data from other sources.

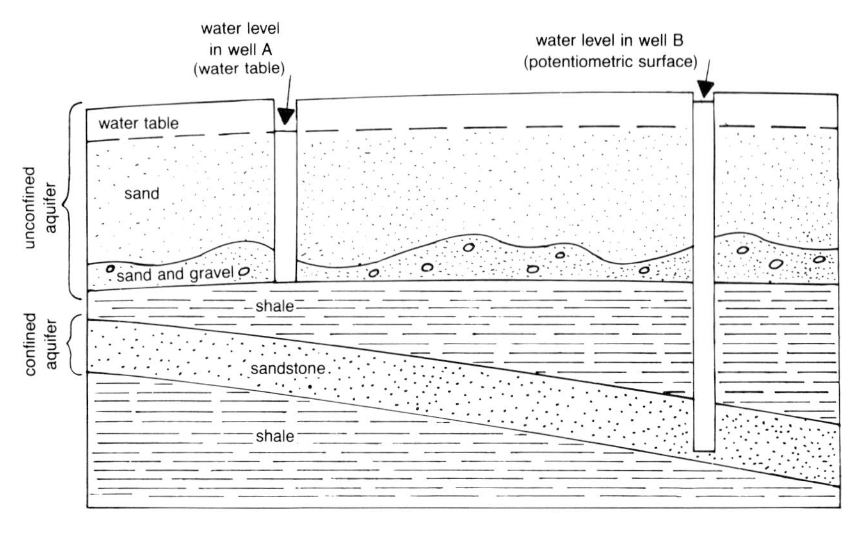

Fig. 4.9 Confined and unconfined aquifers, from Cartwright and Hunt (1981).

Aquifers with intergranular flow

Aquifers in which water flow is through intergranular (primary) pore openings may consist of lithified sedimentary deposits such as sandstone and conglomerate, or unlithified sand and gravel deposits. The porosity, permeability, anisotropy and heterogeneity of these deposits are functions of the depositional environment and the aquifer material.

Darcy's law is generally assumed to be valid in aquifers with intergranular flow. Ground water moves through the pore openings in response to head differentials. The pore openings in many porous media aquifers are sufficiently large and interconnected so that effective porosity approaches total porosity.

Flow is tortuous on a microscopic scale as the water weaves its way between the grains of the aquifer matrix. However, flow is fairly uniform on a macroscopic scale, and the direction of flow will be fairly uniform from recharge areas to discharge areas unless there is a disturbance in the flow system. Disturbances may be caused by addition or extraction of water at sources or sinks such as wells and small lakes and streams; or by changes in the geology/permeability of the aquifer such as a facies change, fault, or intrusive body.

Contaminant transport in porous media aquifers will be affected primarily by the hydraulic conductivity and effective porosity of the aquifer, and by dispersion (Fig. 4.10). As a plume moves downgradient, it may move more rapidly in zones of higher hydraulic conductivity and may take on a fingering configuration if the zones of

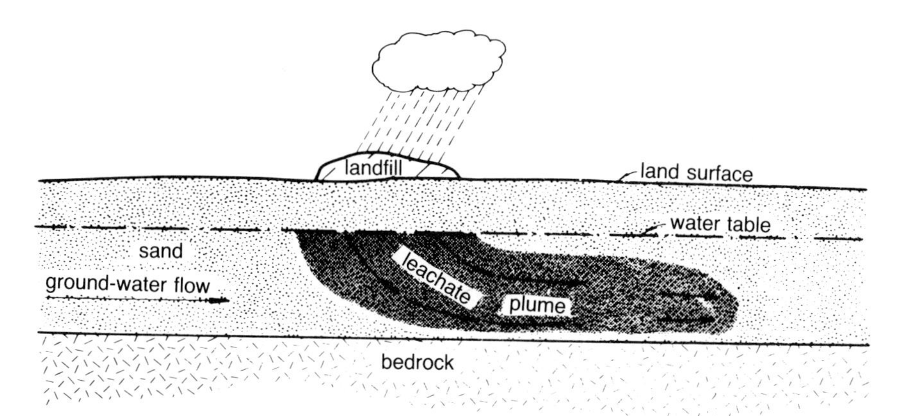

Fig. 4.10 Contaminant plume in an aquifer with intergranular flow, from USEPA (1980).

higher hydraulic conductivity are extensive. On a microscopic scale, molecules of contaminant and water are taking a tortuous flow path around the sediment grains, causing mixing with non-contaminated water so the area of the plume increases as the concentration decreases.

Examples of aquifers with intergranular flow are given below. Each represents a different type of depositional environment.

Example 1

Large alluvial aquifers can be found in the valleys of many major river systems. These aquifers consist of sands and gravels deposited in either a fluvial or glaciofluvial environment. Examples of this type of aquifer would be the sand and gravel deposits of the Ohio, Missouri, and Mississippi River Valleys. These aquifers often can be subdivided into a lower substratum where hydraulic conductivity is usually high and the flow system is often heterogeneous and isotropic (Foreman and Sharp, 1981) and an upper top-stratum where hydraulic conductivity is variable and tends to be greatest in the direction of the paleocurrent of a given channel deposit (Sharp, 1988). The direction of ground-water flow in alluvial valley aquifers is generally towards the river although some flow in the downstream direction is also common. Recharge may come from underlying and adjacent bedrock formations, precipitation, and seepage from tributary streams, and discharge is commonly to the river.

Example 2

Sands and gravels within glacial deposits occur in different forms related to different glacial environments. Common types of glacial deposits that may contain sand and gravel are till, ice-contact, and outwash.

Till is deposited directly from glacial ice and is usually poorly sorted (Davis and DeWiest, 1966). The textural and hydrogeologic properties of tills vary greatly, and are related to the source rock of the till and the distance that the material has been transported by ice. In the mid-continental US, till is largely derived from shale and carbonate rocks, furthermore this material has been transported great distances. These tills are typically fine-grained and have low hydraulic conductivity, and are considered to be confining beds. In areas where tills are derived from nearby precambrian bedrock or younger sandstones, such as in the Canadian Shield area or New England, the tills are often coarse-grained with high hydraulic conductivity and may be valuable aquifers (Stephenson, Fleming and Mickelson, 1988).

Ice-contact deposits are sediments that were glaciofluvially deposited in contact with ice (Stephenson, Fleming and Mickelson, 1988). They may range in texture from silt and clay to sand and gravel and may be well or poorly sorted. Ice contact sands and gravels may be discontinuous and of limited extent (100s to 10s of meters or less), in which case they may only yield sufficient water for domestic purposes. However, when they are laterally extensive, these formations may be productive aquifers.

Outwash describes glaciofluvial sediments deposited by streams flowing away from the ice (Stephenson, Fleming and Mickelson, 1988). These types of deposits are generally coarse-textured and well sorted. Some alluvial aquifers described above may be partially or wholly made up of outwash deposits. Outwash from older glaciations may be buried by younger glacial deposits, as is the case with some highly productive buried bedrock valley aquifers such as the Mahomet Aquifer of central Illinois (Stephenson, 1967).

Outwash deposits tend to be more extensive and more homogeneous than ice contact deposits. Tills are usually finer-grained and with lower hydraulic conductivity than glaciofluvial deposits. Some sandy tills and outwash deposits may be fairly isotropic; ice contact deposits will tend to be vertically and horizontally anisotropic.

Sand and gravel deposits within glacial materials are often bounded by clayey till deposits with hydraulic conductivities that may be four or more orders of magnitude lower than the aquifer (Stephenson, Fleming and Mickelson, 1988), although fractures in the till may decrease that ratio considerably. Recharge to these aquifers may be in the form of precipitation to surficial aquifers or seepage through overlying till units to buried aquifers. Discharge may be to a lake or stream if the aquifer is hydraulically connected, or to the surrounding till unit if the aquifer is of limited extent. Buried bedrock valley aquifers and other deposits in contact with the underlying bedrock may be hydraulically connected if both units are aquifers.

Example 3

Clastic sedimentary rock aquifers commonly are comprised of sandstone or conglomerate. These aquifers are important on a world-wide basis because 25% of the sedimentary rocks in the world are estimated to be sandstone (Freeze and Cherry, 1979). Sandstone can be deposited in aeolian, fluvial, shoreline, or shelf and slope environments. The aeolian deposits are relatively homogeneous, compared to fluvial deposits, and as isotropic as any deposits found in nature (Freeze and Cherry, 1979). The fluvial, shoreline, and shelf deposits are commonly horizontally stratified and vertically isotropic; thus horizontal hydraulic conductivity may exceed vertical hydraulic conductivity by as much as three- to 10-fold (Davis, 1988). However, vertical fracturing can result in increased vertical hydraulic conductivity compared to horizontal. Sandstone formations commonly have permeability values that are lower than the sands from which they formed because cementing materials, primarily silicate and carbonate minerals, precipitate between the grains and because compaction reduces the porosity of the formation during lithification (Freeze and Cherry, 1979; Davis, 1988). However, if the porosity of clastic rocks is low (<0.15) then permeability may be controlled more by secondary fractures than by porosity, which may result in hydraulic conductivity as high as or higher than the source sands (Davis, 1988).

Flow systems in large sandstone aquifers will typically fit the model described by Toth (1963). Recharge to regional aquifers will primarily occur at outcrop areas, although some recharge will also occur as leakage through overlying geologic units where the aquifer is buried. Regional discharge will usually be to large lakes and rivers, although local discharge to smaller surface water bodies is also likely.

Aquifers with flow in secondary pore openings

Some geologic formations that lack significant primary porosity can still be valuable aquifers because ground water is able to flow through fractures and solution openings in the rock. In fact, many such formations constitute major regional aquifers. Flow through secondary pore openings is most common in sedimentary carbonate and in igneous/metamorphic rocks.

For highly fractured rocks, it may be valid to assume that the continuum approach of Darcy's law can be used to describe ground-water flow on a large scale. For rock materials with widely spaced fractures or very large cave-like openings, the continuum approach generally is not valid.

An example of an aquifer where the continuum approach is valid on a large scale is the weathered portion of the Silurian dolomite aquifer in northeastern Illinois and eastern Wisconsin. An example of an aquifer where the continuum approach is not valid is the karst carbonate aquifer of central Kentucky.

Most fracture flow aquifers fall between these extremes; so the common approach is to assume a large scale continuum. In other words, local ground-water flow in a fractured aquifer on the scale of a landfill may not be Darcian; but flow for the entire aquifer, which may underlie several counties, may be treated as Darcian.

Geologic controls are the rule for aquifers with secondary porosity. A tectonically folded aquifer may have greater permeability at the crest of the anticline and in the synclinal trough than on the flanks (Davis and DeWiest, 1966) because added stress at those points may have caused a greater degree of fracturing.

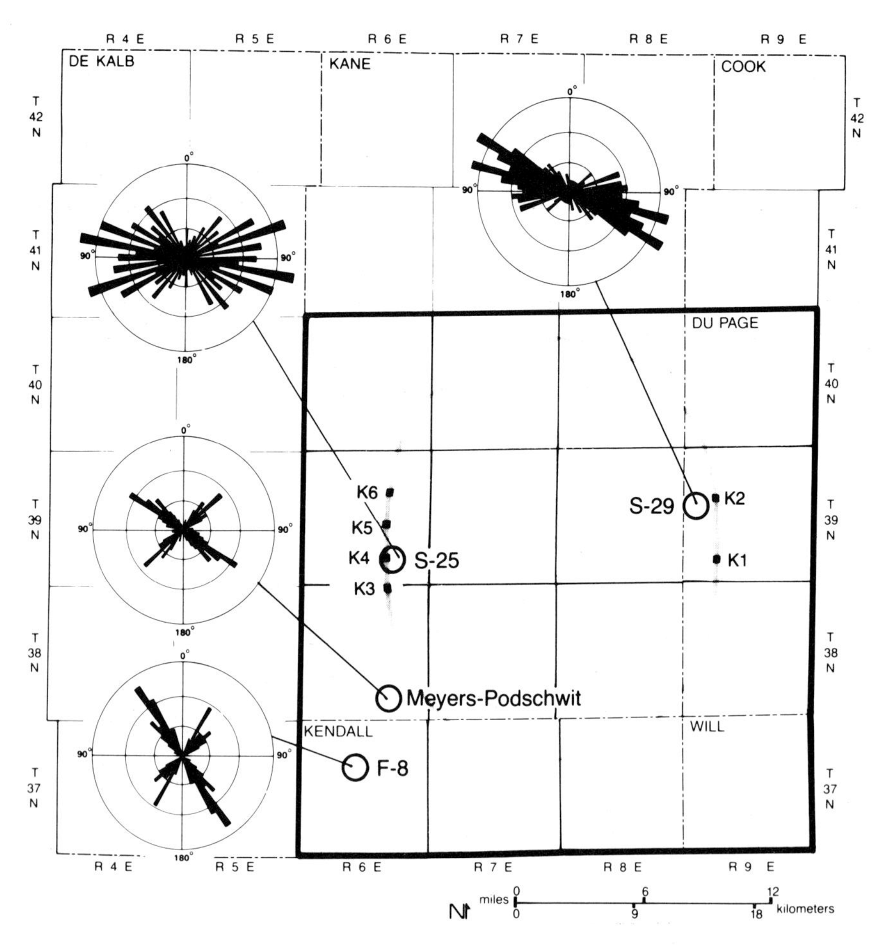

Fig. 4.11 Rose diagram for fractured carbonate rocks, from Curry *et al.* (1988).

Fractured aquifers may be highly anisotropic, particularly in the horizontal orientation. That is because geologic formations commonly have a vertical fracture pattern where there is a primary and secondary fracture orientation (Fig. 4.11). The effect of this fracture orientation on ground-water flow is that hydraulic conductivity is greatest in the direction of primary fracturing, lower in the direction of secondary fracturing, and least in orientations where there are few fractures. Thus, if the hydraulic gradient is tangential to the primary directions of fracturing, flow will not be orthogonal to lines of equal potential and parallel to the hydraulic gradient; but rather in a direction that is highly controlled by anisotropic hydraulic conductivities dictated by fracturing. It may be almost 90 degrees to the gradient in extremely anisotropic conditions.

Contaminant transport in fractured aquifers can be extremely complex because of the extreme anisotropy and heterogeneity of the aquifer (Fig. 4.12), and because transport rates can be rapid. The effective fracture porosity of fractured aquifers may be significantly less than 1%. As can be seen from Eq. (4.5), formations with very low effective porosities can have very rapid ground-water velocities. Thus, contaminant transport in a fractured aquifer may be very rapid due to rapid fluid movement through the fractures, particularly through large fractures.

An extreme example of contaminant transport through large openings is in karst aquifers. In this hydrogeologic environment, the best method to monitor a waste disposal/storage site is to locate and monitor all of the discharge points of the karst system underlying the potential source of contaminants (Quinlan and Ewers, 1985).

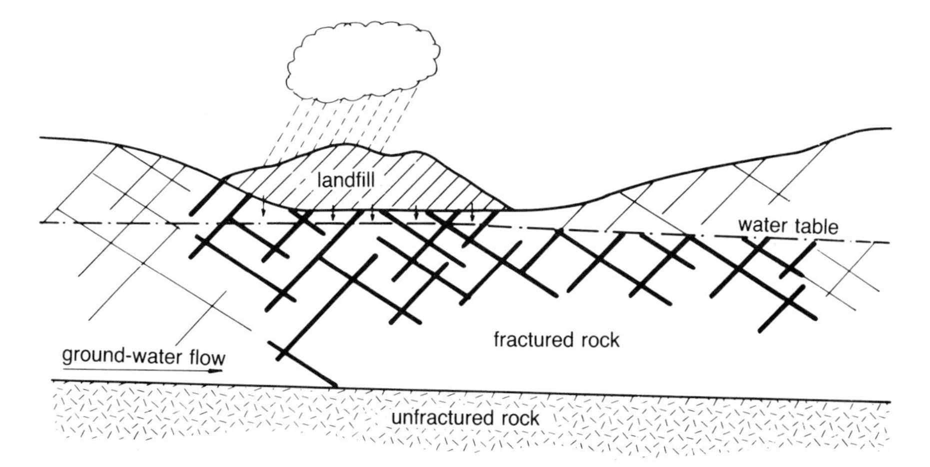

Fig. 4.12 Contaminant plume in an aquifer with flow through fractures, from USEPA (1980).

4.5.4 Flow in aquitards

Aquitards are hydrostratigraphic units that transmit ground water less readily than aquifers. The definition of an aquitard is use dependent. However, for characterizing ground-water flow systems, an aquitard may be considered as any continuous hydrogeologic unit that has a hydraulic conductivity 2 or more orders of magnitude less than overlaying and/or underlying aquifers. Units comprised of shale or clay are commonly aquitards, although massive, non-fractured carbonates and crystalline rocks can also be aquitards. However, it is difficult to classify a particular rock type as an aquitard without extensive field testing because unexpected geologic conditions may allow fairly rapid transmission of water.

An example of a rock that may not necessarily be an aquitard would be shale. Davis (1988) notes that fractures may cause shales at shallow depths to be permeable enough to transmit water to a well or away from waste disposal trenches; however shales at deep depths may be as close to being absolutely impermeable as any geologic material within a few kilometers of the Earth's surface.

The permeability and hydraulic conductivity of aquitard materials tends to be low because fracture and pore openings tend to be too small or too poorly connected to allow significant fluid movement. In essence, the effective porosity of these materials approaches zero. The distinction between effective porosity and total porosity is important because the total porosity of some aquitard materials such as clay and some volcanic rocks may be an order of magnitude or more greater than the effective porosity. As noted earlier, the permeability of all of these units can be increased considerably by open fractures.

When aquitards are underlain by aquifers, it is common to have a situation where flow is vertical through the aquitard; so that flow is downward in recharge areas and upward in discharge areas. This condition may be enhanced if vertical fractures are present.

Clays and shales are commonly touted as confining materials to contaminant transport. Stephenson, Fleming and Mickelson (1988) compiled ground-water velocities for clay-rich glacial tills and found intergranular ground water velocities of 0.005–11 mm/yr. Other studies have shown that such velocities may be 300 mm/yr (Hughes *et al.*, 1971). At these low ground-water velocities diffusion becomes an important contaminant transport mechanism (Freeze and Cherry, 1979). However, the data ranges compiled by Stephenson, Fleming and Mickelson also showed that sand lenses or fractures in the till can result in overall velocities as great as 40 m/yr.

4.6 PRACTICAL IMPLICATIONS

When characterizing ground-water flow at a waste disposal site, care must be used when defining the confining layer. A hydrostratigraphic unit cannot be classified as a confining layer based only on its rock or soil type. Nor can this distinction be made based on limited, small-scale testing. For example, the hydraulic conductivity of a glacial till may have been determined *in situ* as 1×10^{-11} m/s, and isotope analysis on water extracted from cores from the boring for that well may show that the unit contains ancient water. But, fractures within meters of the well may be rapidly transmitting modern water to an underlying aquifer. Characterizing a confining layer for waste disposal requires detailed knowledge of the hydrogeology of the entire unit, which means one must understand the geologic processes that have affected the unit during and after deposition.

The movement of ground water is dictated by the physical characteristics of the hydrostratigraphic units within the flow system. Therefore, it is necessary to establish a good geologic framework before characterizing any ground-water flow problem. If this framework is established, so that the character, depositional environment, and post-depositional history of the geologic materials is understood, it is then possible to apply hydrogeologic principles to characterize ground-water movement.

REFERENCES

Bowman, J.A. and Collins, M.A. (1987) Impacts of irrigation and drought on groundwater resources, *Illinois State Water Survey Report of Investigations* **109**.

Brahana, J.V., Thrailkill, J., Freeman, T. and Ward, W.C. (1988) Carbonate rocks in *Hydrogeology: The Geology of North America* (eds W. Back, J.S. Rosenshein and P.R. Seaber) V. O-2, Geological Society of America, pp. 333–52.

Cartwright, K. (1983) Classification of hydrostratigraphic units in Illinois, Illinois State Geological Survey draft memorandum, open file.

Cartwright, K. and Hunt, C.S. (1981) Hydrogeologic aspects of coal mining in Illinois: an overview, *Illinois State Geological Survey EGN 90*.

Cartwright, K. and Sherman, F.B. (1969) Evaluating sanitary landfill sites in Illinois, *Illinois State Geological Survey EGN 27*.

Curry, B.B., Graese, A.M., Hasek, M.J., Vaiden, R.C., Bauer, R.A., Schumacher, D.A., Norton, K.A. and Dixon, W.G. Jr. (1988) Geological-geotechnical studies for siting the superconducting super collider in Illinois: results of the 1986 drilling program, *Illinois State Geological Survey EGN 122*.

Davis, S.N. (1969) Porosity and permeability of natural materials, in *Flow through porous media* (ed. R.J.M. De Wiest) Academic Press, pp. 53–89.

Davis, S.N. (1988) Sandstones and shales, in *Hydrogeology: The Geology of North*

America, (eds W. Back, J.S. Rosenshein and P.R. Seaber) V. O-2, Geological Society of America, pp. 323–32.

Davis, S.N. and DeWiest, R.J.M. (1966) *Hydrogeology*, John Wiley and Sons, Inc. de Marsily, G. (1986) *Quantitative hydrogeology*, Academic Press.

Foreman, T.L. and Sharp, J.M. Jr. (1981) Hydraulic properties of a major alluvial aquifer; an isotropic, inhomogeneous system, *J. Hydrology*, **53**, 247–58.

Foth, H.D. and Turk, L.M. (1972) *Fundamentals of Soil Science*, 5th edn, John Wiley and Sons, Inc.

Freeze, R.A. and Cherry, J.A. (1979) *Groundwater*, Prentice-Hall, Inc.

Freeze, R.A. and Witherspoon, P.A. (1967) Theoretical analysis of regional groundwater flow, 2: effect of water-table configuration and subsurface permeability variation. *Water Resources Res.*, **3**, 623–34.

Gelhar, L.W., Mantoglou, A., Welty, C. and Rehfeldt K.R. (1985) *A review of field-scale physical solute transport processes in saturated and unsaturated porous media*, Electric Power Research Institute EA-4190.

Hillel, D. (1971) *Soil and water, physical principles and processes*, Academic Press.

Hillel, D. (1980) *Applications of soil physics*, Academic Press.

Hillel, D. (1982) *Introduction to soil physics*, Academic Press.

Horton, R., Thompson, M.L. and McBride, J.F. (1988) *Determination of effective porosity of soil materials*, USEPA No. EPA/600/2-88/045, NTIS PB88-242391.

Horton, R.E. (1940) An approach toward a physical interpretation of infiltration capacity, *Proc., Soil Sc. Soc. of Amer.*, **5**, 399–417.

Hughes, G.M., Landon, R.A. and Farvolden, R.N. (1971) *Hydrogeology of solid waste disposal sites in northeastern Illinois*, USEPA Solid Waste Management Series SW-12d.

Krapac, I.G., Cartwright, K., Hensel, B.R., Herzog, B.L., Larson, T.H., Panno, S.V., Risatti, J.B., Su, W.J. and Rehfeldt, K.R. (1991) Construction, Monitoring, and Performance of Two Soil Liners, *Illinois State Geological Survey Environmental Geology 141*.

Lehr, J.H. (1988) The misunderstood world of unsaturated flow, *G. Water Monit. Rev.*, **8**(2), pp. 4–6.

Meinzer, O.E. (1923) *The occurrence of groundwater in the United States*, USGS Water Supply Paper 489.

Morris, D.A. and Johnson, A.I. (1967) *Summary of hydrologic and physical properties of rock and soil materials*, as analyzed by the Hydrologic Laboratory of the U.S. Geological Survey 1948–1960, USGS Water Supply Paper 1839-D.

Quinlan, J.F. and Ewers, R.O. (1985) Ground water flow in limestone terrains: strategy rationale and procedure for reliable, efficient monitoring of ground water in karst areas, in *Proc., 5th Natl. Symp. and Exp. on Aquifer Rest. and G. Water Monit.*, National Water Well Association, pp. 197–234.

Rehm, B.W., Groenewold, G.H. and Peterson, W.M. (1982) *Mechanisms, distribution, and frequency of ground water recharge in an upland area of western North Dakota*, North Dakota Geological Survey Report of Investigations 75.

Revelle, R. (1941) Criteria for recognition of sea waters in ground-waters, in *Trans. Amer. Geophys. Union*, **22**, 593–7.

Richards, L.A., Gardner, W.R. and Ogata, G. (1956) Physical Processes determining water loss from soil, in *Proc., Soil Sc. Soc. of Amer.*, **20**, pp. 310–4.

Seaber, P.R. (1988) Hydrostratigraphic units, in *Hydrogeology: The Geology of North America*, (eds W. Back, J.S. Rosenshein and P.R. Seaber) V. O-2, Geological Society of America, pp. 9–14.

Sharp, J.M. Jr. (1988) Alluvial aquifers along major rivers, in *Hydrogeology: The Geology of North America* (eds W. Back, J.S. Rosenshein and P.R. Seaber) V. O-2, Geological Society of America, pp. 273–82.
Steenhuis, T.S., Jackson, C.D., Kung, S.K.J. and Brutsaert, W. (1985) Measurement of recharge on eastern Long Island, New York, USA, *J. Hydrology*, **79**, 145–69.
Stephenson, D.A. (1967) *Hydrogeology of glacial deposits of the Mahomet Bedrock Valley in east-central Illinois*, Illinois State Geological Survey Circular 409.
Stephenson, D.A., Fleming, A.H. and Mickelson, D.M. (1988) Glacial deposits, in *Hydrogeology: The Geology of North America* (eds W. Back, J.S. Rosenshein and P.R. Seaber) V. O-2, Geol. Soc. of Amer., pp. 301–14.
Todd, D.K. (1980) *Groundwater Hydrology*, 2nd edn, John Wiley and Sons, Inc.
Toth, J.A. (1963) A theoretical analysis of ground-water flow in small drainage basins, *J. Geophys. Res.*, **68**, pp. 4795–812.
US Department of Interior (1981) *Groundwater manual*, John Wiley and Sons.
USEPA (1980) *Procedures manual for groundwater monitoring at solid waste disposal facilities, Solid Waste Management Series report SW-611*, Cincinnati, OH.
Walton, W.C. (1965) *Ground-water recharge and runoff in Illinois*, Illinois State Water Survey Report of Investigations 48.
Walton, W.C. (1985) *Practical aspects of ground water modeling*, 2nd edn, National Water Well Association, Worthington, OH.
Wood, W.W. and Fernandez, L.A. (1988) Volcanic rocks, in *Hydrogeology: The Geology of North America* (eds W. Back, J.S. Rosenshein and P.R. Seaber) V. O-2, Geological Society of America, pp. 353–65.

PART TWO

New Disposal Facilities

CHAPTER 5

Landfills and impoundments

David E. Daniel

5.1 INTRODUCTION

Landfills are the final repositories for unwanted or unusable wastes. Until the middle of this century, nearly all wastes were discarded in open, unengineered dumps. Waste was often burned to conserve space. Topographical anomalies that lended themselves naturally to dumping were typically selected for dump sites. The most common waste dumps were natural depressions (creeks, low-lying areas, and flood plains) that were otherwise of little use and mined-out areas, e.g., sand or gravel quarries. The practice of open dumping changed little, until a few decades ago.

The **sanitary landfill** began to become commonplace shortly after World War II. A sanitary landfill consists of a refuse disposal area in which the waste is disposed of in cells that range in thickness up to about 5 m. Within each cell, waste is covered with a 150- to 300-mm-thick layer of soil (called **daily cover**) at the end of each working day. Today, several foams are available that in some cases have been substituted for daily soil cover. The configuration of cells is sketched in Fig. 5.1 for a waste containment unit that contains a liner. In fact, however, early sanitary landfills had no man-made liner; liners did not become common in sanitary landfills until the 1970s, and even today, not all sanitary landfills have liners.

The sanitary landfill represented a dramatic improvement over the open dump. Controlled placement of waste in sanitary landfills (particularly daily covering) greatly reduced the number of rodents and insects, dramatically reduced public health risks, and generally contributed to major aesthetic improvements in waste disposal.

Engineered liners for waste disposal facilities did not become routine until the 1970s. The author recalls first seeing a diagram of a double

Geotechnical Practice for Waste Disposal.
Edited by David E. Daniel.
Published in 1993 by Chapman & Hall, London. ISBN 0 412 35170 6

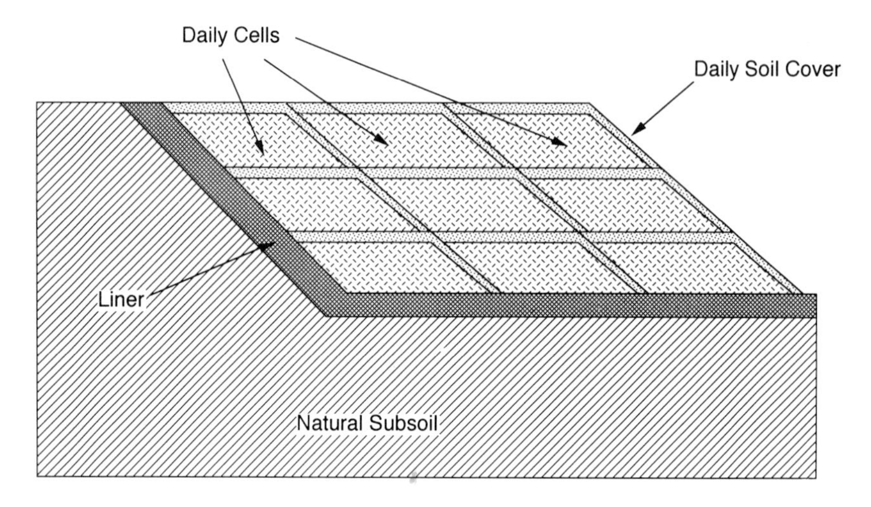

Fig. 5.1 Daily cells in a landfill.

liner for a landfill in about 1978. Sophisticated waste containment units with multiple liners and fluid collection systems did not become common in the US until they were mandated by the US Environmental Protection Agency (EPA) for hazardous waste landfills in the early 1980s.

As mentioned in Chapter 1, regulations have driven landfill practices in the US and most other countries. Requirements for landfills vary with the type of waste and other factors such as the hydrogeology of a site, climate, and type of waste to be buried. In this chapter, general principles of landfill design will be discussed. In the remaining chapters that comprise Part II of the book, specific details on the design of various landfill components will be discussed.

5.2 OBJECTIVES OF WASTE DISPOSAL FACILITIES

The objective of a waste disposal facility are frequently not spelled out in clear terms that can be agreed upon by all parties involved in the development of a facility. Further, people frequently have different understandings about the objective of a waste disposal facility. For an owner, the objective may be to dispose of waste that poses a public health risk at a cost that can be borne by the local citizens. For the engineer, the objective is frequently to design a facility that optimizes the tradeoff between adequate waste containment and cost. For an individual living near a landfill, the objective is for the facility to be operated with zero impact upon the individual or the individual's property.

A critical question that is often asked is: Will the landfill leak? The answer to this question depends upon the definition that one selects for the term 'leak'. If one assumes that a leak is an unanticipated discharge, then the landfill need not leak. If one assumes that a leak is any release, then it is clear that all landfills either leak or have the potential to leak. No liner material known to mankind is forever impermeable to all chemicals. Some chemicals will eventually migrate by advection or diffusion through all liner materials unless an inward gradient producing a chemical flux toward the containment unit is maintained. The objective of the landfill design should not be to stop the release of all chemicals from the facility for an infinite period of time. Such an objective is unrealistic. The real issue with respect to containment of waste in the disposal facility revolves around how much waste material in **leachate** (the liquid that seeps from the landfill) will be released from the containment facility over a given period of time and what the environmental impact of the release will be. For well designed facilities, the quantities of chemicals released from the waste disposal facility are minute and environmental impacts (short and long term) are negligible.

The objective of a waste disposal facility is to contain the waste in a manner that is protective of human health and the environment. Because no endeavor of mankind can be undertaken without some risk, there is always a risk that a landfill will fail to perform up to expectations. The best that can humanly be expected of designers is to ensure that the risk posed by a disposal facility is extremely low. Monitoring systems are installed around landfills to determine if the facility is performing in an unexpected manner.

Regulations dictate the minimum technology that is required to keep risks associated with waste containment facilities small. Usually, the owner's and engineer's objectives are to comply with regulatory requirements. It is assumed by most engineers that so long as the minimum regulatory requirements are met, the environmental risk will be low. However, the author encourages the engineer to think about and evaluate environmental risks independently of regulatory requirements and to satisfy himself or herself that the risks are suitably low with the design that is ultimately selected.

5.3 SITING OF LANDFILLS

An entire book could be written on site selection for waste containment units. Some of the factors that must be considered include proximity to waste generators, availability of transportation systems for moving waste to the site, climate, geology, hydrogeology, surface hydrology, proximity of airports, demographics, land use, impacts on the local community, and others. It is widely understood that landfills are perceived to an undesirable addition to a community. An attitude of NIMBY (not in my back yard) is often used to describe the attitude of residents of local communities. Citizens of an area will typically accept that landfills are needed (although many suffer from the misconception that waste minimization and recycling can eliminate the need for landfills), but not in their community. The problem, of course, is that the landfill must be located in someone's community. The process of getting local citizens to accept the siting of a waste disposal facility in their community is frequently a more significant problem than dealing with technical issues. Most experienced landfill siting specialists have found that effective communication at a very early stage is usually the most effective way to deal with NIMBY attitudes. Many citizens completely misunderstand what a modern waste disposal facility is like; visions of smoldering wastes in an open dump are common. Educating people who have never had the opportunity, or taken the time, to learn how a modern waste disposal disposal facility is designed and operated is usually a critical prerequisite in gaining community support for a new waste disposal facility.

Another common approach for dealing with the NIMBY problem is to build new waste containment units on top of (vertical expansion) or next to (lateral expansion) an existing landfill. There may still be strong local opposition to expansion of a landfill (e.g., because of concerns over truck traffic), but there are fewer NIMBY obstacles to overcome in expanding a facility compared to siting a new facility. Another approach is to build a new landfill in an area that already has one or more landfills.

The specific geotechnical problems related to site selection are site characterization and analysis. The primary issues are definition of subsurface stratigraphy, identification of groundwater conditions (including identification of aquifers and aquitards, measurement of piezometric levels in aquifers, and characterization of background water quality and geochemical conditions), determination of the properties of major subsurface units (with particular emphasis on hydraulic conductivity, geochemical attenuation characteristics, and strength), evaluation of the availability of construction materials (e.g., clay for liners and granular material for drains), and analysis of the probable performance of the site, including assessment of probable groundwater impacts and analysis of the stability of the facility.

An ideal site is one that is located reasonably close to the source of the waste, has convenient transportation access, is not situated in a low-lying area or floodplain, is underlain by suitably strong materials, and has favorable hydrogeological characteristics. Some sites are so complex that their characterization is at best very difficult and, in some cases, impossible. If the site is so complex that it cannot be adequately characterized, the site is not a favorable one to host a waste disposal facility.

5.4 CONTAINMENT TECHNOLOGY

One of the first questions that the landfill designer must address is whether the waste containment units will be above or below ground. As indicated in Fig. 5.2, the waste can be disposed of either above the existing surface, below the surface, or both above and below. Above-ground disposal is particularly attractive for sites with shallow water tables. Above-ground landfills have the advantage that leachate can be drained by gravity, the facility is conspicuous and therefore is not easily forgotten and ignored, and construction of liner and drainage-

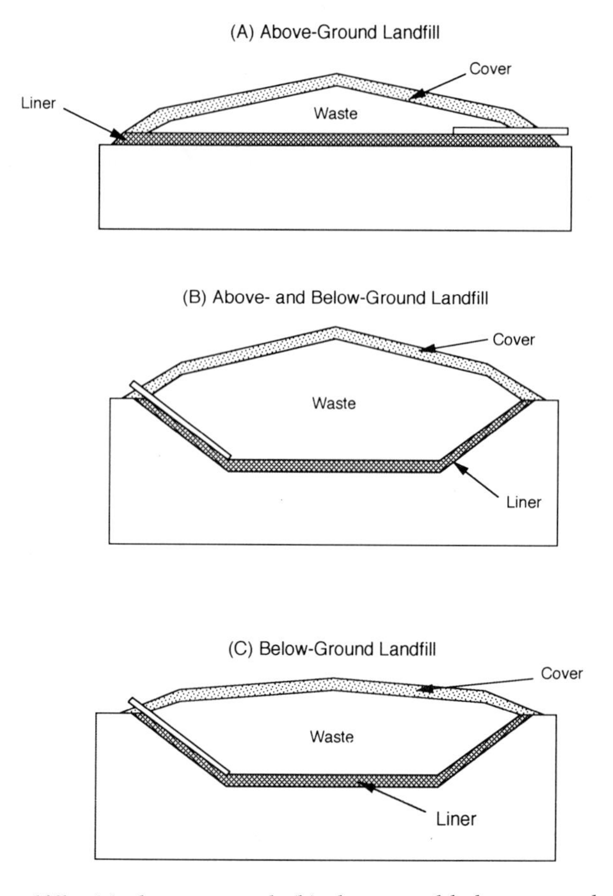

Fig. 5.2 Landfills: (a) above-ground; (b) above- and below-ground; (c) below-ground.

system components occurs on more or less level ground, which simplifies construction. Vertical expansions of existing facilities essentially constitute above-ground landfills. Disposal units that are partly below ground have the advantage of allowing more waste to be disposed of in a given area, offer more efficient use of construction materials (excavated material can be used for daily cover or other purposes), and potentially could allow for productive use of the land (e.g., for a park) if the final surface is relatively flat.

Several systems can be employed to control release of waste constituents. The primary objectives of control systems are:

1. to minimize infiltration of water into the waste through the cover or sidewalls of the facility;
2. to collect and remove gas and leachate; and
3. to minimize release of leachate to the subsurface.

In the past, natural soils were often used as liners (Fig. 5.3a), but this practice is becoming less common because it is very difficult to prove

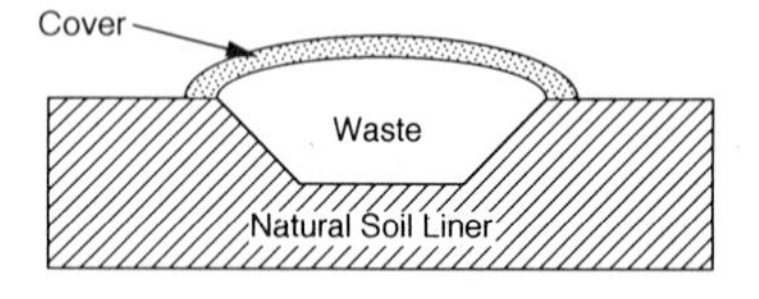

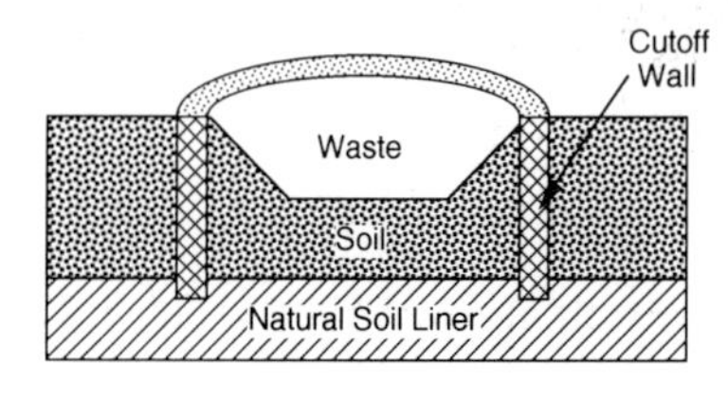

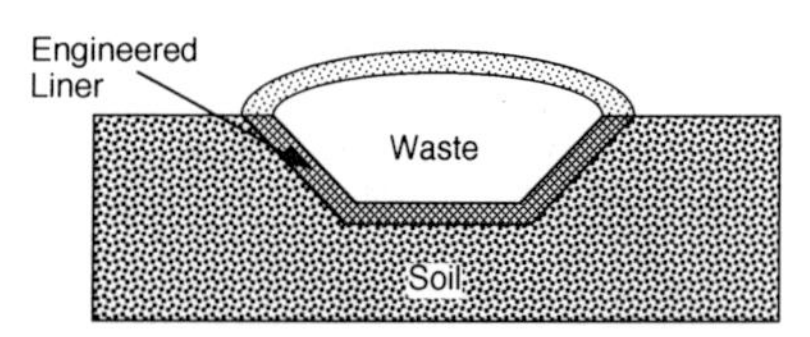

Fig. 5.3 Lining of waste disposal units with (a) natural soils, (b) vertical cutoff walls, and (c) engineered liners.

that the natural liner is free of high-hydraulic-conductivity zones or secondary features such as cracks and joints. Vertical cutoff walls are occasionally used in conjunction with a natural soil liner (Fig. 5.3b), but this practice is unusual for new disposal facilities. (However, surrounding a contaminated area with a cutoff wall is a common practice for site remediation, as discussed in Chapter 17). A more typical situation is shown in Fig. 5.3c in which an engineered liner is employed to control movement of contaminants out of the disposal unit.

In the US, Subtitles C and D of the Resource Conservation and Recovery Act give the USEPA the authority to establish minimum containment requirements for hazardous and nonhazardous solid waste, respectively. Figure 5.4 illustrates the minimum containment requirements that are usually imposed. For hazardous waste landfills (Fig. 5.4a), a double liner system is required with a leachate collection layer located above the primary liner and a leak detection layer located between the two liners. The Subtitle D requirements (Fig. 5.4b) were promulgated a few weeks before final revisions were made to this chapter and do not go into effect until 1993. Nevertheless, the minimum liner consists of a single composite liner with a leachate collection system. Recommended European designs for waste containment

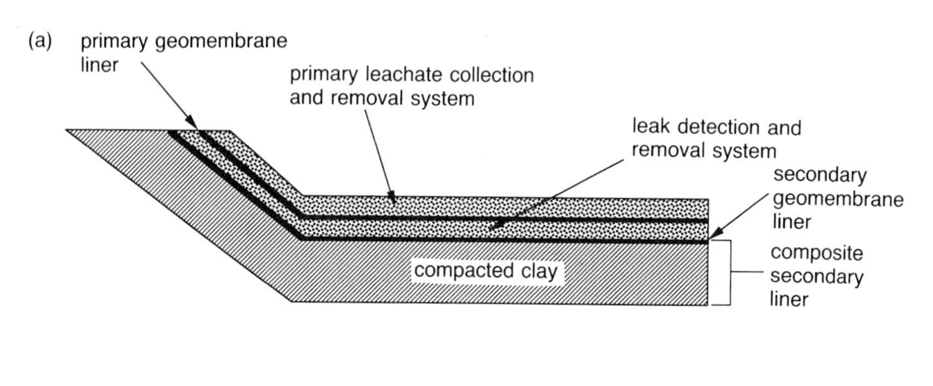

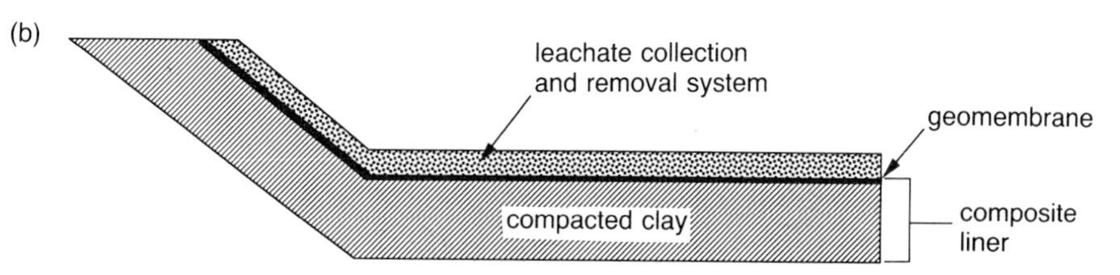

Fig. 5.4 Minimum liner requirements of the US Environmental Protection Agency: (a) for hazardous waste landfills (b) for non-hazardous waste landfills.

facilities call for a single composite liner with a leachate collection system (Fig. 5.4b).

For a hazardous waste facility, the EPA requires that the compacted soil liner be at least 0.9 m thick and have a hydraulic conductivity $\leqslant 1 \times 10^{-7}$ cm/s. Drainage layers are typically required to have a hydraulic conductivity $\geqslant$1 cm/s, and the leak detection system must be capable of detecting a leak within 24 hours. Geosynthetic drainage materials (typically geonets) are commonly used for the leak detection layer (especially on side slopes). Geomembranes must be at least 0.76 mm thick. For non-hazardous waste facilities, the requirements are similar, but the minimum thickness of the compacted soil liner is 0.6 m.

The final cover system for a waste disposal facility may consist of several layers (Fig. 5.5). A surface layer controls erosion and, for sites where vegetation grows on the surface, supports vegetative growth. The surface layer usually consists of topsoil that is seeded with appropriate vegetation. Occasionally, at arid sites, gravel or cobbles are placed at the surface. A protective layer isolates the surface environment from the underlying waste and components of the cover system. The protective layer also stores water that infiltrates through the surface layer; later, the stored water can be returned to the atmosphere by evaporation or evapotranspiration from plants. At sites with substantial precipitation, a drainage layer is often placed in the cover. The drainage layer serves three purposes:

1. to drain water away quickly so that the water is not available for percolation through the cover;

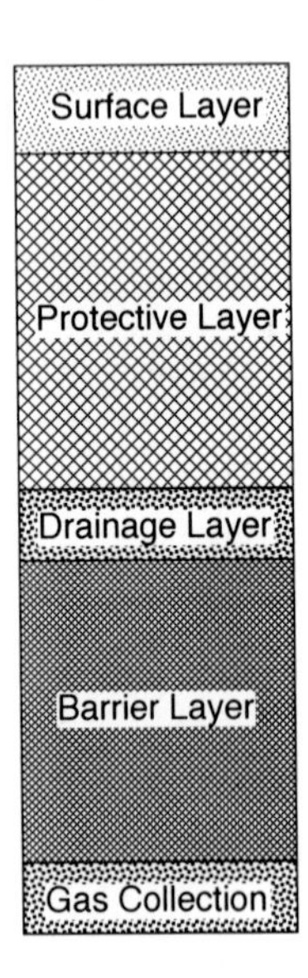

Fig. 5.5 Components of a cover system.

2. to minimize the head of water on the liner, which minimizes percolation; and
3. to reduce water pressures that could trigger instability of the cover.

Nearly all engineered cover systems (except perhaps those at arid sites) have a barrier layer. The barrier layer is generally either a single geomembrane, a layer of low-hydraulic-conductivity compacted soil, or a composite geomembrane-soil liner. A new material, termed a **geosynthetic clay liner**, is experiencing increased usage in cover systems. Many waste disposal regulations require that the barrier layer in a cover system be no more permeable than the liner system in the disposal facility. The reason for this requirement is to ensure that the containment facility does not fill with liquid should the leachate collection cease to be operated or malfunction. A gas collection layer is recommended for wastes that produce significant quantities of potentially harmful or dangerous gas. More details on cover systems and their components are given in Chapters 10 and 18.

In the US, an extra component is usually added to the minimum liner requirements for hazardous waste landfills. At most facilities, the primary liner consists of a geomembrane/clay **composite liner**. For example, one possible configuration of liner and cover system components is shown in Fig. 5.6. The purpose of adding the clay component to the primary liner is to minimize the flux of liquid into the leak detection layer. The compacted clay component of the primary liner system is shown in Fig. 5.6 to cover the floor and sidewalls of the disposal unit. In reality, the clay is usually not placed on the sideslopes (just on the base of the unit) because compaction of clay on a side slope directly above geosynthetic materials can be very difficult and risks damaging the geosynthetic components. A thin, manufactured

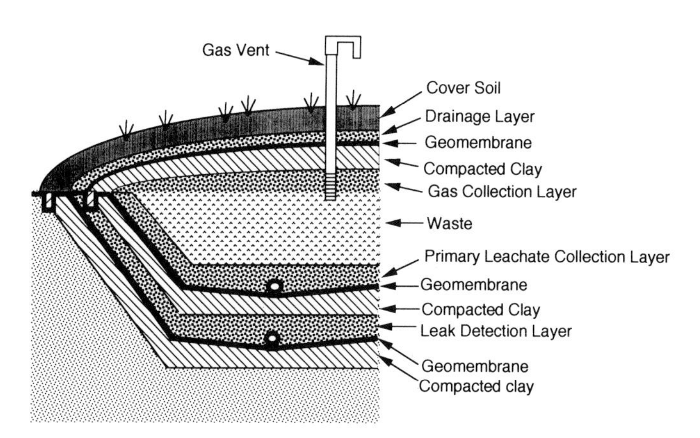

Fig. 5.6 Double composite liner system and multiple-component cover system in a waste containment unit.

clay blanket (called geosynthetic clay liner) is recommended by many as the clay component of the primary liner. A design recommended by Daniel and Koerner (1991) is sketched in Fig. 5.7. The geosynthetic clay liner (discussed more in Chapter 7) has the advantages that the liner can be placed with lightweight equipment, which minimizes risk of damage to underlying components, and the dry clay blanket does not tend to yield water due to consolidation (which a compacted clay liner does when it is loaded). Consolidation water entering the leak detection layer is usually misinterpreted as possible liner leakage.

Because composite liners are critical components of many waste disposal units, it is appropriate to discuss in more detail the rationale behind the composite liner. Defect-free geomembranes are practically impermeable: water passes through an intact geomembrane via diffusion at an extremely slow rate. The problem with geomembranes is that defects, such as pinholes or tiny flaws in seams, can occur. The size and number of holes is controlled by the construction methodology and degree of construction quality assurance. Darilek, Laine, and Parra (1989) used an electrical resistivity method to survey 28 completed liners for leaks. They located 542 leaks (0–79 leaks/site) with an average areal density of 26 leaks/ha. Leaks were found at all sites, except for two small geomembrane-lined tanks. Of the leaks detected, 18% were in parent material and 82% were in seams and at details such as sumps and pipe protrusions, which also include seams.

Giroud and Bonaparte (1989a) independently evaluated leaks in geomembrane liners. They concluded that 1 defect per 10 m of seam can be expected for geomembranes installed without independent quality assurance and that an average of 1 defect per 300 m of field

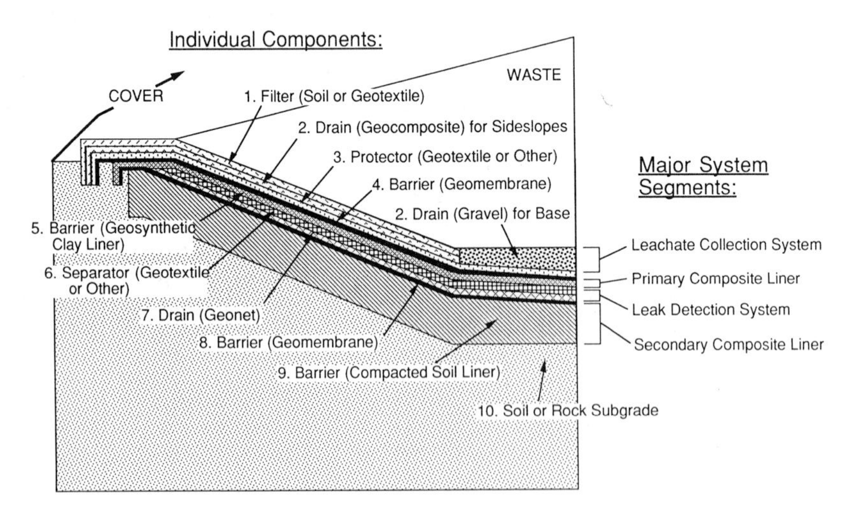

Fig. 5.7 Lining system recommended by Daniel and Koerner (1991).

seam can be expected with reasonably good installation practice and independent quality assurance. For typical panel widths, seam defects are likely to result in 3–5 leaks/ha with good quality assurance.

Bonaparte and Gross (1990) examined leakage rates measured in the leak detection layer of double liner systems at actual landfills and impoundments. In 19 geomembrane-lined landfills, leakage occurred through the geomembrane liner in all 19 cells. Of the 6 surface impoundments, leakage occurred in only 2. However, with the impoundments, liners were leak tested by filling the cells with water prior to operation – any leaks found during testing were repaired.

A composite liner overcomes the major problem (occasional defects) with a single geomembrane liner. The way that a composite works is sketched in Fig. 5.8 and is contrasted with individual geomembranes and soil liners. If there is a hole in a geomembrane liner, liquid will easily move through the hole, assuming the subgrade soil does not impede seepage. With a soil liner alone, seepage takes place over the entire area of the liner. With a composite liner, liquid moves easily through any hole in the geomembrane but will then encounter low-permeability soil. The low-permeability soil impedes further migration of the liquid. Thus, leakage through a hole in a geomembrane is minimized by placing a low-permeability soil beneath the membrane. Similarly, leakage through a soil liner is reduced by placing it in contact with a geomembrane which, despite occasional holes or defects in seams, greatly reduces the area of flow through the soil liner and thereby significantly decreases the rate of flow through the soil liner.

To achieve good composite action, the geomembrane must be placed against the soil with good hydraulic contact (often called **intimate contact**). One would normally not separate the geomembrane and soil liner with a highly permeable material, such as a bed of sand or a geotextile, because intimate hydraulic contact would be jeopardized (Fig. 5.9). If stones of a size and shape that could puncture the geomembrane exist in the soil liner, the stones must be removed or a

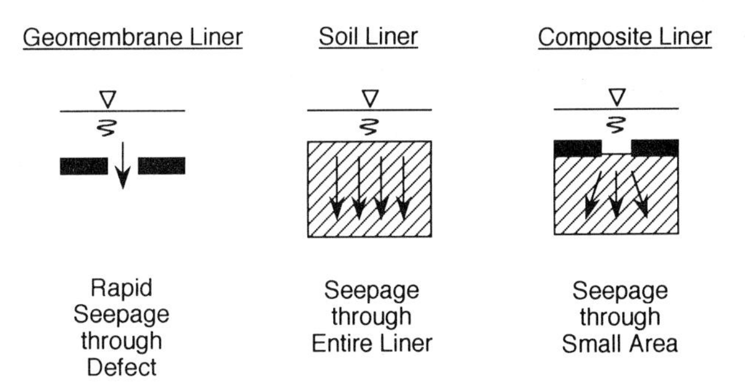

Fig. 5.8 Seepage patterns through geomembrane, soil, and composite liners.

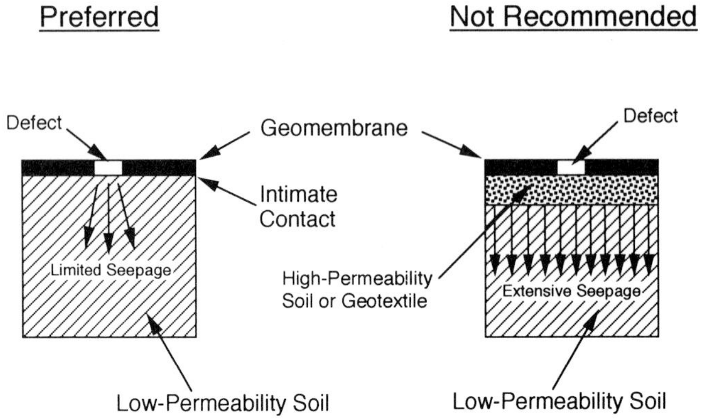

Fig. 5.9 Proper design of composite liner for intimate hydraulic contact between geomembrane and compacted soil.

separate layer of select soil liner material must be placed in contact with the geomembrane. To achieve intimate contact, the surface of a compacted soil liner on which the geomembrane is placed should be smooth-rolled with a steel-drum roller. Also, the geomembrane should be placed and backfilled in a way that minimizes wrinkles.

Seepage rates through geomembrane liners, soil liners, and composite liners may be calculated using equations published by Giroud and Bonaparte (1989a) and Giroud *et al.* (1989). The following example calculations are presented to compare calculated flow rates through different lining systems. Assumptions made for the calculations are that the head of liquid, h, is 300 mm; the hydraulic conductivity of the soil liner, K, is 10^{-8} cm/s (best case), 10^{-7} cm/s (average case), or 10^{-6} cm/s (worst case); the geomembrane contains holes with an area of 0.1 cm^2; and the number of holes per hectare is 2 (best case), 20 (average case), or 60 (worst case). Computed flow rates are summarized in Table 5.1.

Calculated flow rates through the composite liner are typically at least 100 times less than through the geomembrane or soil liner alone. Further, even if the soil liner is relatively permeable (10^{-6} cm/s) and there are 20 holes/ha in the geomembrane, the calculated flow rate through the composite liner is far less than the flow rate through relatively good-quality soil liners or geomembrane liners on their own.

A compelling attribute of composite liners is that neither the clay soil nor the geomembrane component needs to be constructed perfectly to realize excellent performance of the composite liner. This fact is critical to designers who cannot always be assured that the liner will be built to the desired high standards.

The performance of composite liners to date has generally been very good. Bonaparte and Gross (1990) report leakage rates measured in

Table 5.1 Example calculations of flow rates

type of liner	*flow rate (L/ha/day)*		
	best case	*average case*	*worst case*
Geomembrane alone	2 500	25 000	75 000
holes/ha	2	20	60
Compacted soil alone	115	1 150	11 500
K (cm/s)	10^{-8}	10^{-7}	10^{-6}
Composite	0.8	47	770
holes/ha	2	20	60
K (cm/s)	10^{-8}	10^{-7}	10^{-6}
contact	poor	poor	poor

leak detection layers for double-liner systems in landfills and impoundments. Analysis of the data is complicated by the fact that most, if not all, of the liquid collected initially in a leak detection system beneath a composite primary liner is the result of consolidation of the clay-liner component of the primary composite liner. For example, if a 0.6 m thick layer of saturated soil compresses 3% in thickness over a period of two years, the average flow rate due to consolidation would be 270 L/ha/day, which is likely to be far greater than the long-term leak rate.

Bonaparte and Gross (1990) report that geosynthetic clay liners (GCLs) were used as the lower component of the primary composite liner in seven liner systems. For these systems, there was no consolidation water produced and interpretation of the leak rate through the composite liner was unambiguous. No flow was detected in the leak detection system of any of the seven composite liners. For this reason, many designers in the US are beginning to employ a geomembrane/geosynthetic clay liner as a composite liner for a double liner systems.

One difficulty with composite liners is interfacial shear between the geomembrane and soil liner. The interfacial shear resistance at this surface is often low and can be lowered even more if positive water pressure develops at the interface. Engineers should measure the geomembrane-soil friction angle with adequately large shear boxes and employ appropriate geotechnical methods for slope analysis (see Chapter 11). Anchorage may be needed for the geomembrane, or operational constraints during filling of a waste cell may be necessary to minimize development of shear stresses along the interface.

Another potential problem is that the temperature of an exposed geomembrane reach up to 65 °C during summer days. This high temperature can dry out and crack the underlying clay liner. When the geomembrane cools, moisture can condense beneath the membrane and produce a thin layer of water at the clay/geomembrane interface.

Condensate can migrate down-slope and pool at the toe of a slope. The recommended solution is to cover the geomembrane as soon as possible. ↑

5.5 DISPOSAL UNIT

Nearly all landfills (except perhaps those that contain huge quantities of mining wastes) consist of several units or cells rather than one massive landfill. The reasons for dividing the disposal area into waste containment units or cells are many. For the owner, the process of obtaining a waste disposal permit can be lengthy and expensive. Development of a site for waste disposal is generally not worth the owner's effort unless permission can be granted to dispose of waste at a site for many years. In the US, waste containment facilities with a double composite lining system are constructed at a cost on the order of \$150,000 to \$300,000 per acre (\$370,000 to \$740,000 per ha). To limit the initial financial investment, it is wise to develop a portion of the site initially rather than the whole site. Also, construction logistics favor tractable disposal units rather than one massive landfill. In addition, some of the components of a landfill cannot be exposed to extreme weather conditions or ultraviolet light for extended periods; the simplest way to deal with these problems is to cover the materials as soon as possible, which is a practice that favors many small cells rather than one massive landfill.

Landfills that are at least partly below ground usually contain an earthen access ramp to allow trucks and other traffic to drive down into the disposal cell. The ramp may be placed on top of the liner system, or the liner system (all or part) may be placed on top of the ramp. An access road leads to the earthen ramp. One of the major forces that can act on a lining system is the force from a heavy truck that brakes while moving down the earthen ramp. It is beneficial to place a curve in the access road, just before the earthen ramp that leads down into a disposal cell, to ensure that trucks slow down before entering the disposal cell. The curve should ideally be designed to allow the truck driver to get a good view of the ramp prior to driving onto the ramp.

5.6 OPERATIONS

A number of operational details have to be carefully planned from a geotechnical standpoint. The first layer of waste placed on top of a leachate collection layer is usually **select waste**. Select waste is not so much waste that is selected to contain controlled material as it is waste that is screened to make sure that it does not contain large objects that could puncture a underlying liner or contain fine materials (like industrial sludges) that could plug the leachate collection system. In a few landfills, the first layer of waste is a genuine select waste: shredded tires. The tires provide very large hydraulic conductivity and protect the leachate collection system and underlying liner from puncture.

The waste must be placed in a way that does not create shearing stresses that could trigger instability. The slippage at the Kettleman Hills facility (Seed *et al.*, 1990) occurred in part as a result of the way in which wastes were placed. The facility was located on a gentle slope in a valley (such units are often called valley landfills), and the facility was filled in a way that created a large mound of waste and a significant slope. Sometimes the stability of a facility may hinge on the way in which the disposal cell is filled. If so, the filling procedures must be spelled out clearly by the designers.

Daily cover serves a critically important function of keeping insects and rodents out of the waste. One problem that daily cover can create is hydraulic isolation of one cell from another. If the daily cover consists of relatively impermeable soil, water cannot migrate uniformly through the waste. Instead, water will be channeled in the landfill. Some cells may be saturated with water and others may be virtually dry. Wide variation in moisture conditions leads to problems with differential settlement and leachate collection. If leachate will be reintroduced to the disposal unit (**leachate recirculation**), it is particularly important that daily cover has a high hydraulic conductivity. A common manifestation of low-hydraulic-conductivity daily cover is the appearance of leachate seeps on landfill covers: leachate flows laterally along the surface of daily cover rather than infiltrating downward, until the leachate 'daylights' on the sloping cover of a landfill.

In cold climates, it is sometimes necessary to spread waste over the surface of the leachate collection layer to provide adequate depth of cover over an underlying compacted clay liner to prevent the clay liner from freezing. Designers of the leachate collection layer must bear in mind the need to drain not only leachate but precipitation that falls in the cell prior to placement of waste. For large cells, it is sometimes cost effective to divide the leachate collection layer into separate areas so that rainfall that falls in an area where waste has not yet been placed can be collected and released without the need for treatment.

An important operations consideration is the cleanout of leachate collection pipes. If pipes will be cleaned, the design must accommodate access points for the cleanout equipment.

REFERENCES

Bonaparte, R. and Gross, B.A. (1990) Field behavior of double liner systems, in *Waste Containment Systems: Construction, Regulation, and Performance*, (ed. R. Bonaparte), American Society of Civil Engineers, New York, pp. 52–83.

Daniel, D.E. and Koerner, R.M. (1991) Landfill liners, *Civil Eng.*, **61**(12), 46–9.
Darilek, G.T., Laine, D.L. and Parra J.O. (1989) The electrical leak location method for geomembrane liners: development and applications, *Geosynthetics '89*, Industrial Fabrics Association International, Minneapolis, Minn., **2**, 456–66.
Giroud, J.P. and Bonaparte, R. (1989a) Leakage through liners constructed with geomembranes – Part I: geomembrane liners, *Geotextiles and Geomembranes*, **8**, 27–67.
Giroud, J.P. and Bonaparte, R. (1989b) Leakage through liners constructed with geomembranes – Part II: composite liners, *Geotextiles and Geomembranes*, **8**, 71–111.
Giroud, J.P., Khatami, A. and Badu-Tweneboah, K. (1989) Evaluation of the rate of leakage through composite liners, *Geotextiles and Geomembranes*, **8**, 337–40.
Seed, R.B., Mitchell, J.K. and Seed H.B. (1990) Kettleman hills waste landfill slope failure II: stability analyses, *J. Geotech. Eng.*, **116**(4), 669–90.

CHAPTER 6

Leachate and gas generation

Morton A. Barlaz and Robert K. Ham

6.1 INTRODUCTION

The generation of leachate and gas from landfills is a well documented phenomenon which impacts landfill design and operation. Leachate and gas production are both influenced by the contents of the landfill, the local climate, and the manner in which the facility is operated. Gas composition and volume will be discussed in this chapter as will leachate composition. The quantity of leachate produced from a landfill is a function of site surface hydrology, which in turn is a function of climate, the presence of groundwater and the numerous factors affecting infiltration of surface water into the landfill. Leachate quantity is discussed in a later chapter.

The central theme for this chapter is leachate and gas generation from landfills containing municipal solid waste and other non-hazardous waste. This theme is consistent with present US regulations governing the generation, treatment and disposal of wastes classified as hazardous. Under present regulations, the burial of untreated hazardous wastes in landfills is almost completely prohibited, particularly the burial of materials in a physical state from which they could exert a significant influence on gas and leachate generation. Nonetheless, it must be recognized that there are numerous landfills in which hazardous wastes were buried together with municipal solid wastes in the years before promulgation of the aforementioned regulations. Hazardous wastes also enter municipal solid waste landfills from unregulated small quantity sources. In addition, co-disposal of municipal solid wastes with specific hazardous wastes is permitted in other countries and is considered desirable by some. While the discussion to follow will concentrate on gas and leachate as produced

Geotechnical Practice for Waste Disposal.
Edited by David E. Daniel.
Published in 1993 by Chapman & Hall, London. ISBN 0 412 35170 6

Table 6.1 Composition of municipal refuse by component

*component**	*percent by wet weight*[†]
Paper	40.0
Yard	17.6
Metal	8.5
Plastics	8.0
Food	7.4
Glass	7.0
Other	11.6[‡]

* As reported by EPA (1990) for municipal solid waste in 1988.
[†] The moisture content of refuse with this composition, based on typical moisture contents for each component of the refuse, is 19.8% (wet weight) (Tchobanoglous *et al.*, 1977).
[‡] Includes rubber, leather, textiles, wood, and miscellaneous wastes.

from municipal solid waste landfills, comments on the impacts of hazardous waste will be inserted as appropriate.

6.1.1 Refuse composition

Traditionally, municipal refuse has been classified according to sortable categories such as glass, paper, metals, etc. (Table 6.1). While such data are needed for recycling studies and overall solid waste management planning, data on the chemical composition of refuse are more directly applicable to a discussion of refuse decomposition and its impact on gas and leachate characteristics. Recently, data on the chemical composition of refuse have been published and these data have been used to calculate a methane potential for each chemical constituent (Barlaz *et al.*, 1989a). Chemical constituent data are presented in Table 6.2. The data in Table 6.2 show that the principal biodegradable constituents of refuse are cellulose and hemicellulose. These two constituents account for 91% of the methane potential of refuse (Barlaz *et al.*, 1989a). The remainder of the methane potential of refuse consists of protein (8.3%), and soluble sugars (0.5%). Lignin, the other major organic component of refuse, does not decompose to any significant extent under anaerobic conditions (Young and Frazer, 1989). Knowing that refuse contains a significant biodegradable fraction, one can begin to understand the manner in which refuse decomposes and how this affects leachate and gas production.

Table 6.2 Composition and methane potential of municipal refuse by chemical constituent

chemical constituent	*% dry weight**	*methane potential*†
Cellulose	51.2	73.4
Hemicellulose	11.9	17.1
Protein‡	4.2	8.3
Lignin	15.2	0
Starch	0.5	0.7
Pectin§	<3.0	–
Soluble sugars	0.35	0.5
Total volatile solids‖	78.6	¶

* Adopted from Barlaz (1988).
† Data expressed as a percentage of the total methane potential based on the cellulose, hemicellulose, protein, sugar and starch data. The methane contribution of pectin was not calculated because of the uncertainty associated with the pectin concentration in refuse. Methane potential was calculated from the stoichiometry given by Parkin and Owen (1986) on the basis of 100% conversion of a constituent to carbon dioxide and methane. It should be recognized that 100% of the degradable constituents will not be decomposed as some fraction is surrounded by lignin and is not accessible for anaerobic degradation.
‡ Determined by multiplication of the total Kjeldahl nitrogen by 6.25. The actual protein content of refuse is probably lower than the value given here because some of the Kjeldahl nitrogen is actually nitrogen contained in humic materials and structural proteins which are not easily degradable (Burns and Martin, 1986).
§ Actual value is probably less than 3% but could not be quantified.
‖ An independent measure of volatile solids based on weight loss on ignition at 550°C. Not a total of previous entries in the table.
¶ The volatile solids concentration is presented to illustrate that the chemical constituent analyses account for 110% of the volatile solids. No methane potential is calculated for volatile solids because this measurement includes both degradable and non-degradable carbon.

6.2 LANDFILL MICROBIOLOGY

The decomposition of refuse to methane in sanitary landfills is a microbially mediated process which requires the coordinated activity of several trophic groups of bacteria. As discussed above, the principal substrates which decompose to methane in landfills are cellulose and hemicellulose. Similar substrates fuel the production of methane in other ecosystems including the rumen, marshes, rice paddies and sludge digesters (Wolfe, 1979). In the first part of this section the general pathway for anaerobic decomposition, as it has been documented to occur in other anaerobic ecosystems, is reviewed. Implications of this pathway for sanitary landfills are also presented. Following this general pathway review, refuse decomposition is characterized in four phases defined by different gas and leachate characteristics.

6.2.1 The microbiology of anaerobic decomposition

Three trophic groups of anaerobic bacteria are required for the production of methane from biological polymers (cellulose, hemicellulose, and protein) as illustrated in Fig. 6.1 (Wolfe, 1979; Zehnder *et al.*, 1982). The first group of microorganisms, referred to as the Hydrolytic and Fermentative microorganisms in Fig. 6.1, is responsible for the hydrolysis of biological polymers. The initial products of polymer hydrolysis are soluble sugars, amino acids, long chain carboxylic acids and glycerol. Hydrolytic and Fermentative microorganisms then ferment these initial products to short-chain carboxylic acids, alcohols, carbon dioxide and hydrogen. Acetate, a direct precursor of methane is also formed. The second group of bacteria active in the conversion of biological polymers to methane is the obligate proton-reducing acetogens. They oxidize the fermentation products of the first group of microorganisms to acetate, carbon dioxide and hydrogen. The conversion of fermentation intermediates like butyrate, propionate and ethanol is only thermodynamically favorable at very low hydrogen concentrations. Thus, these substrates are only utilized when the obligate proton-reducing acetogenic bacteria can function in syntrophic association with a hydrogen scavenger such as a methane producing or sulfate reducing organism. The importance of acetogenic bacteria which convert hydrogen plus carbon dioxide to acetate has not been established in the refuse ecosystem. This microbiological activity probably competes weakly with the hydrogenophilic methanogens for hydrogen.

The third group of bacteria necessary for the production of methane are the methanogens. The methanogens can utilize only a limited number of substrates including formate, methanol, methylamines, hydrogen plus carbon dioxide, and acetate (Wolin and Miller, 1985). In sludge digesters, it is estimated that 70% of the methane produced originates from acetate (Zeikus, 1980; Mah *et al.*, 1978). This value has not been investigated in the landfill ecosystem. The production of methane from acetate yields only 31 kJ/mole CH_4 produced. This is barely enough energy for the generation of adenosine triphosphate

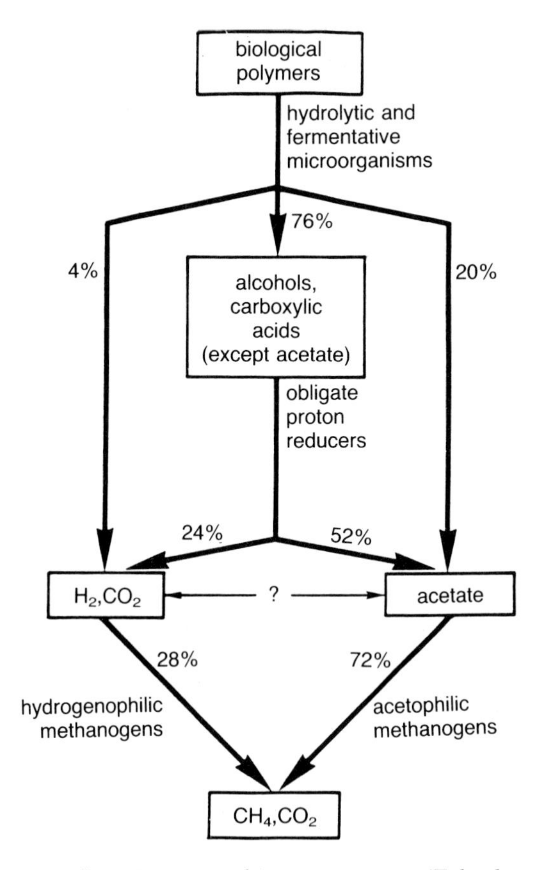

Fig. 6.1 Substrate flow in anaerobic ecosystems (Zehnder *et al.*, 1982). Complete multistep methanogenesis as it occurs in lacustrine and sulfate depleted marine sediments, bogs, marshes, trees and digesting sludge (reprinted by permission from author).

(ATP) which requires 30.6 kJ/mole, so the growth of methanogens on acetate is relatively slow. The conversion of hydrogen and carbon dioxide to methane yields 135.6 kJ/mole CH_4 produced. Thus, the latter reaction is energetically more favorable. The methanogens are most active in the pH range 6.8 to 7.4 (Zehnder, 1978).

The importance of the methanogens in anaerobic digestion has been summarized by Zeikus (1980). As a group, the methanogens:

1. control the pH of their ecosystem by the consumption of acetate;
2. regulate the flow of electrons by the consumption of hydrogen, creating thermodynamically favorable conditions for the catabolism of alcohols and acids; and
3. excrete organic growth factors including vitamins and amino acids which are used by other heterotrophic bacteria in the ecosystem.

Should the activity of the fermentative organisms exceed that of the acetogens and methanogens, there will be an imbalance in the ecosystem. Carboxylic acids and hydrogen will accumulate and the pH of the system will fall, thus inhibiting methanogenesis.

6.2.2 Microbiology of refuse decomposition

When refuse is placed in a landfill, biological decomposition resulting in methane formation as described in the previous section does not occur immediately. A period ranging from months to years is necessary for the proper growth conditions and the required microbiological system to become established. Refuse decomposition phases ranging in number from three to six or more have been identified by different investigators depending on the data base and purposes of each study.

A four-phase characterization of refuse decomposition describing chemical and microbiological characteristics of decomposition is summarized here (Barlaz *et al.*, 1989b). Refuse decomposition is described in an aerobic phase, an anaerobic acid phase, an accelerated methane production phase, and a decelerated methane production phase as follows and as summarized in Fig. 6.2. This description is based on data from laboratory scale lysimeters. The relationship between these data and field scale landfills will be discussed.

Phase 1 – aerobic phase

Oxygen is unavoidably present in the void space when refuse is landfilled. This oxygen, plus oxygen dissolved in the refuse-associated moisture triggers aerobic decomposition. In the aerobic phase, both oxygen and nitrate are consumed, with soluble sugars serving as the carbon source for microbial activity. All of the trophic groups required for refuse methanogenesis are present in fresh refuse (cellulolytics, acetogens, and methanogens), although there is little change in their populations. Leachate strength will be relatively low during the aerobic phase and gas composition will be nearly 100% CO_2.

The aerobic phase in newly placed refuse in a full-scale landfill is likely to last for only a few days. During the aerobic phase, refuse is typically below field capacity so any leachate produced from freshly buried refuse will likely have flowed through a channel in the refuse. In addition, since refuse is likely to be buried on top of older material and leachate is generally collected at the bottom of a landfill, leachate characteristic of the aerobic phase may not be observed. Instead, liquid percolating through fresh and then older refuse will reflect the characteristics of the older refuse, which may be in any of the other three phases of decomposition described below. Leachate generated from freshly buried refuse has been analyzed from test landfills and

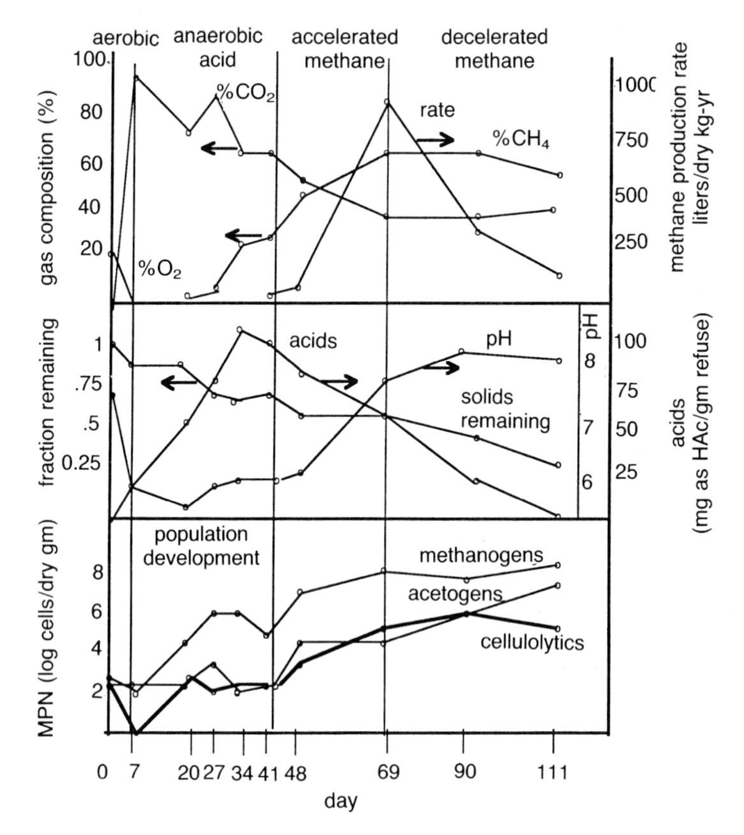

Fig. 6.2 Summary of observed trends in refuse decomposition with leachate recycle. Gas volume data were corrected to dry gas at standard temperature and pressure. The acids are expressed as acetic acid equivalents. Solids remaining is the ratio of the cellulose plus hemicellulose removed from a container divided by the weight of cellulose plus hemicellulose added to the container initially. Methanogen MPN data are the log of the average of the acetate- and H_2/CO_2- utilizing populations. (Reprinted from Barlaz *et al.*, 1989b)

laboratory cells. This leachate has sometimes been observed to be very strong, with chemical oxygen demand (COD) values of tens of thousands to one hundred thousand mg/l. Such numbers reflect primarily squeezings from compaction of the refuse.

Phase 2 – anaerobic acid phase

Eventually, oxygen entrained with the refuse is depleted by aerobic microbial activity. As a given refuse mass becomes covered by other refuse, replenishment of oxygen becomes insignificant and the system becomes anaerobic. In the anaerobic phase, carboxylic acids accumulate, the pH decreases, there is some cellulose and hemicellulose decomposition, and some methane may be detected in landfill gas. The first part of the acid phase is characterized by the rapid accumulation of carboxylic acids and a decrease in the pH of the refuse ecosystem. This pH decrease can be attributed to the accumulation of acidic end-products of sugar fermentation. In addition to carboxylic acids, it is likely that a wide variety of organic intermediates are produced. High CO_2 concentrations, first reported by Farquhar and Rovers (1973), result from fermentative activity, as does the accumulation of hydrogen gas. Both CO_2 and H_2 are products of sugar fermentation. No methane is detected early in the acid phase of refuse decomposition.

Carboxylic acid concentrations increase and the pH decreases throughout the acid phase. The data presented in Fig. 6.2 are based on a system in which the leachate was collected from the bottom of laboratory-scale landfills, externally neutralized and added back to the top of the test system, i.e. the leachate was recycled. Leachate recycle and neutralization is not typically practiced in field-scale landfills. The acid phase explains the long lag time between refuse burial and the onset of methane production in sanitary landfills in that low pH inhibits refuse methanogenesis. In a field-scale landfill, acids leaching from refuse in the acid phase of decomposition would be rapidly consumed by refuse in decomposition phases three or four described below.

With progression through the acid phase, the carbon dioxide concentration decreases as the methane concentration increases and newly produced gas purges carbon dioxide from the landfill. Cellulose and hemicellulose hydrolysis is not consistent in the acid phase and there is relatively little solids hydrolysis.

Measurable methane production denotes the beginning of the third phase of decomposition, the accelerated methane production phase. Leachate was externally neutralized in the research reported in Fig. 6.2. The resulting increased pH stimulated microbial activity and allowed decomposition to proceed to the third phase. In a full-scale landfill without leachate neutralization, it may take months to years for the microbiological system to progress to the third phase of decomposition. Acid concentrations may be reduced and the pH increased by slow microbial acid consumption, or by rinsing the acids to lower levels of the landfill or to the leachate collection system by water infiltration. It should be stressed that the low pH which results from an acid accumulation, and not high acid concentrations alone, limits progression of refuse decomposition to the accelerated methane production phase.

Phase 3 – accelerated methane production phase

In the accelerated methane production phase there is a rapid increase in the rate of methane production to some maximum value. Methane concentrations of 50–70% are typical of this phase with the balance of the gas being carbon dioxide. There is a decreased accumulation of carboxylic acids in the third phase of refuse decomposition. Carboxylic acids are consumed faster than they are produced as the methane production rate increases. Subsequently, the pH of the refuse ecosystem increases. There is little solids hydrolysis during this phase of decomposition (Fig. 6.2). Increases in the populations of cellulolytic, acetogenic and methanogenic bacteria are observed.

The data presented in Fig. 6.2 show an increasing methane production rate and decreasing carboxylic acid concentration in the accelerated methane production phase. However, it must be remembered that these data were collected from laboratory-scale lysimeters. Under field conditions, the measured methane production rate may represent an average of refuse in numerous states of decomposition. The sharp increase and decrease illustrated in Fig. 6.2 will be considerably dampened. Similarly, the decrease in carboxylic acid concentrations is a relative effect. Thus, one measurement from a landfill will probably not be sufficient to identify the state of refuse decomposition. Nevertheless, decreasing carboxylic acid concentrations typically correlate with increased rates of methane production. Estimation of the methane potential of a landfill by the use of solids analysis, in concert with measurement of the methane production rate, will be discussed in a later section.

Phase 4 – decelerated methane production phase

The final phase of refuse decomposition may be described as the decelerated methane production phase. It is characterized by a decrease in carboxylic acid concentrations to concentrations below 100 mg/l. With the depletion of carboxylic acids there is a further increase in the pH of the ecosystem. Though the acids are depleted, there is still some COD exerted by the leachate. While relatively low, it is likely to consist of less degradable compounds, including humic materials.

The rate of methane production decreases even though the methane and carbon dioxide concentrations remain constant at about 60% and 40%, respectively. The rate of cellulose and hemicellulose decomposition in the decelerated methane production phase is higher than that exhibited in any other phase of refuse decomposition. Solids hydrolysis controls the rate of methane production in that there is no longer an accumulation of carboxylic acids to serve as soluble substrate.

In earlier phases of refuse decomposition, the hydrolysis of cellulose and hemicellulose leads to accumulations of carboxylic acids. In the fourth phase of refuse decomposition, where the rate of polymer hydrolysis exceeds that exhibited earlier, no accumulations are observed. The difference may be explained by the makeup of the refuse ecosystem; specifically by increases in methanogenic and acetogenic activity. It is interesting to observe that the rate of methane production and the concentration of carboxylic acids decreased in parallel. Though there has been speculation that high concentrations of carboxylic acids inhibit methane production, acetate and butyrate concentrations of 9753 mg/l and 6956 mg/l, respectively, did not inhibit methane production in the containers described by Fig. 6.2 (Barlaz *et al.*, 1989c). In the second and third phases of refuse decomposition it is the utilization of carboxylic acids which limits the onset and rate of methane production. After consumption of the initial accumulation of carboxylic acids, it is polymer hydrolysis which limits the rate of methane production.

The identification and description of these four phases of refuse decomposition were developed based on data from a laboratory experiment. The characteristics of refuse decomposition described here represent what would happen under ideal circumstances for a given volume of newly placed refuse. The data presented in Fig. 6.2 have time as the abscissa. The times shown on the abscissa should not be applied to other conditions. These times were influenced by the frequency of leachate recycle and neutralization and the incubation temperature. Neither leachate recycle nor shredded refuse are typically used in landfills. However, no external additions of bacteria were made to the refuse used by Barlaz *et al.* (1989b). It is believed that the major differences in refuse decomposition under different conditions will be the length of time required for the different phases of refuse decomposition to evolve, the methane production rate, and possibly the methane yield.

6.2.3 Implications of microbial activity on gas and leachate characteristics

Several observations concerning refuse decomposition and the landfill ecosystem can be made based on an understanding of the groups of organisms involved. The groups of organisms involved in methane production obtain energy by fermentation only. Under aerobic conditions, the organic polymers will be oxidized to carbon dioxide. When present, nitrates or nitrites will serve as electron acceptors and be reduced to nitrogen as the biodegradable components of the refuse are converted to carbon dioxide. Like oxygen, there can be no methane production in the presence of nitrate or nitrite. If sulfate is present in the ecosystem, it is reduced by the sulfate reducing bacteria to hydrogen sulfide in the presence of hydrogen. Sulfate is not reduced by bacteria in the presence of oxygen. Hydrogen is also used by the methanogenic bacteria to reduce carbon dioxide to methane. Competition between sulfate reducing and methanogenic bacteria for hydrogen has been reported for other ecosystems but appears less important in the refuse ecosystem (Barlaz *et al.*, 1989b).

The methanogens are most active in the pH range 6.8 to 7.4 (Zehnder, 1978). Should the activity of the fermentative organisms exceed that of the acetogens and methanogens, there will be an imbalance in the ecosystem. Carboxylic acids and hydrogen will accumulate and the pH of the system will fall, thus inhibiting methanogenesis. The potential for a rapid accumulation of acids is particularly acute in the refuse ecosystem. Food and garden wastes make up about 26% of the solid waste stream (Table 6.1). These materials contain soluble sugars. In addition, there will be some spontaneous lysis of plant cells under anaerobic conditions. This lysis results in the release of additional soluble sugars (Greenhill, 1964a,b,c). These sugars will be converted to carbon dioxide in the presence of oxygen or nitrate, or carboxylic acids in the absence of oxygen and nitrate.

Based on the measured concentrations of soluble sugar (3.46 mg/dry gm) and nitrate (0.015 mg/dry gm) in refuse, and an assumed porosity of 25%, Barlaz *et al.* (1989a) calculated that the amounts of oxygen and nitrate present in freshly buried refuse were sufficient to support the oxidation to carbon dioxide of only 10% of fresh refuse soluble sugars. This means that 90% of the sugars are fermented to carboxylic acids which accumulate, reducing the pH of the refuse ecosystem and inhibiting methanogenesis.

When the fermentative bacteria are active, large quantities of soluble organics are produced which may percolate from the landfill in leachate. As the methanogens become active, these soluble organics are metabolized. Thus, the chemical oxygen demand (COD) or total organic carbon (TOC) of landfill leachate is typically highest before the onset of steady state methanogenesis (Ham and Bookter, 1982; Barlaz *et al.*, 1989b). Essentially all organic chemicals exert a COD. Prior to the decelerated methane production phase carboxylic acids account for 60% to 90% of the COD of leachate. Thus, high organic acid concentrations and high COD may be considered equivalent. The high organic acid concentrations plus dissolved carbon dioxide causes acidic pH levels in leachate. This in turn causes more dissolution of inorganic constituents. A further cause of high inorganic concentrations is complex formation between organic and inorganic constituents, holding higher concentrations of some metals in solution than would otherwise be expected.

6.3 THE RELATIONSHIP BETWEEN LABORATORY AND FIELD-SCALE DATA ON LEACHATE AND GAS CHARACTERISTICS

6.3.1 Gas production and characteristics

There is general agreement that landfill gas will contain between 50% and 70% methane with carbon dioxide and trace constituents making up the balance. Similar results have been reported from both field and laboratory-scale tests. Trace constituents which have been identified in landfill gas include light petroleum compounds (benzene, toluene, xylene), chlorinated compounds, hydrogen, and hydrogen sulfide. Vinyl chloride, a known carcinogen, has also been detected in landfill gas. While the volume of trace constituents is seemingly insignificant, they may present a problem for downstream gas processing equipment. In some landfills, concentrations of hydrogen as high as 20% have been observed, indicative of phase two decomposition somewhere in the landfill. This is generally a transient condition, and once methane production is established little hydrogen should be observed.

Comparison of methane production rate data between field-scale landfills and laboratory experiments is difficult because there is essentially no data in the open literature on methane production rates in field-scale facilities. Data from field-scale landfills is complicated by questions regarding the volume or mass of refuse responsible for the production of a measured volume of gas. There is a wider body of methane production data collected under laboratory conditions. However, the laboratory data are not perfectly comparable in that experimental conditions (moisture, particle size, temperature, etc.) are not uniform between studies. In addition, most laboratory experiments were conducted to explore techniques for enhancing methane production. The enhanced methane production rates would not be expected at field-scale landfills until certain of these techniques are employed in the field.

Methane yields of 42–120 liters CH_4/dry kg refuse have been reported in laboratory tests conducted with leachate recycle and neutralization (Buivid *et al.*, 1981; Barlaz *et al.*, 1987; Kinman *et al.*, 1987; Barlaz 1988). These studies show significant variation in methane production rate and methane yield. Certain of the differences can be explained by differences in experimental design. For example, the data reported by Barlaz *et al.* (1987) and Barlaz (1988) differ in reactor volume (100 vs. 2 liters), temperature (25 °C vs. 41 °C) and the rate of leachate recycle. Buivid *et al.* (1981) used refuse with an abnormally high paper content.

Methane yields were measured in field-scale tests cells as part of the Controlled Landfill Project in Mountain View California (Pacey, 1989).

Yields of 38.6–92.2 liters CH_4/dry kg refuse were measured after 1597 days of monitoring. However, mass balance data suggest that significant volumes of methane were not measured in certain test cells. The value of 92.2 liters CH_4/dry kg refuse is for a cell with buffer addition only and appears to be accurate based upon good correlation between the measured volume and mass balance data. The measured yield in the control cell was 76.6 liters CH_4/dry kg refuse and it is likely that not all methane produced was actually measured. Biochemical methane potential tests indicated a methane potential of 92.2 liters CH_4 @ STP/dry kg refuse remained in the control cell. Thus, the total yield for the control cell may be as high as 168.8 liters CH_4/dry kg refuse (92.2 + 76.6) if decomposition proceeds to completion.

A number often used as an estimate of methane production rate in field scale landfills is 0.1 ft^3 CH_4/wet lb-yr. Assuming refuse buried at 20% moisture, this converts to 7.8 liters CH_4/dry kg-yr, a number comparable to some of the lower values reported in the literature.

6.3.2 Leachate composition characteristics

Concentrations of leachate constituents can vary by orders of magnitude between different field-scale landfills. This has been illustrated by Halvadakis *et al.* (1983). For example, the COD at twelve sites varied from 0 to 89 520 mg/l. The composition of the soluble phase of the refuse ecosystem is influenced by the state of refuse decomposition as explained in a previous section. However, leachate produced from a field-scale landfill is not likely to be representative of refuse in one state of decomposition, but rather is more representative of an average leachate derived from several different refuse cells which may be in different states of decomposition. The wide range of concentrations makes the design of on-site leachate treatment facilities difficult. It is thus important to provide storage for flow and leachate strength equalization.

Perhaps the most comprehensive compilation of leachate quality data applicable to full-scale landfills was provided by Ehrig (1988). He obtained leachate composition data from 15 West German landfills ranging in age from 0 to 12 years. These data are summarized in Tables 6.3 and 6.4. Wide ranges in concentration data are apparent. Constituents for which the solubility is relatively unaffected by pH, such as chlorides, show no clear shift in going from acidic leachate, characteristic of the anaerobic acid phase, to the methane producing phases. Constituents which are more soluble under acidic conditions, such as iron, show a clear decrease in concentration in progressing from the acid phase to the methane producing phases. Ehrig (1988) also measured the COD in numerous landfill leachates. He reported wide variation in COD concentrations (700–30 000 mg/l) for landfills less

Table 6.3 Leachate analysis (parameters with differences between acetic and methanogenic phase) after Ehrig (1988) (Reprinted by permission of Springer-Verlag)

	average	*range*
acetic phase		
pH (−)	6.1	4.5 – 7.5
BOD_5 (mg/l)	13 000	4 000 – 40 000
COD (mg/l)	22 000	6 000 – 60 000
BOD_5/COD (−)	0.58	–
SO_4 (mg/l)	500	70 – 1 750
Ca (mg/l)	1 200	10 – 2 500
Mg (mg/l)	470	50 – 1 150
Fe (mg/l)	780	20 – 2 100
Mn (mg/l)	25	0.3 – 65
Zn (mg/l)	5	0.1 – 120
methanogenic phase		
pH (−)	8	7.5 – 9
BOD_5 (mg/l)	180	20 – 550
COD (mg/l)	3 000	500 – 4 500
BOD_5/COD (−)	0.06	–
SO_4 (mg/l)	80	10 – 420
Ca (mg/l)	60	20 – 600
Mg (mg/l)	180	40 – 350
Fe (mg/l)	15	3 – 280
Mn (mg/l)	0.7	0.03– 45
Zn (mg/l)	0.6	0.03– 4

than eight years old. As the biological and mass transport systems in the landfills 'matured', the COD stabilized at about 8000 mg/l. This occurred at about eight years.

6.3.3 Impact of hazardous wastes on leachate and gas characteristics

There are a wide variety of inorganic and organic wastes which are classified as hazardous. Thus it is only possible to make general statements on potential effects of hazardous wastes on leachate and gas characteristics. During the aerobic phase, there is an increase in temperature due to the waste heat of microbial metabolism. This, combined with vigorous CO_2 production during the aerobic and initial part of the anaerobic acid phase will enhance gas stripping of volatile chemicals.

Gas stripping does not necessarily mean release of an organic compound to the environment. As volatile organic compounds pass through soil separating the refuse from the atmosphere, there is an additional

Table 6.4 Leachate analysis (no difference between phases could be observed) after Ehrig (1988) (Reprinted by permission of Springer-Verlag)

	average	*range*
Cl (mg/l)	2100	100 – 5 000
Na (mg/l)	1350	50 – 4 000
K (mg/l)	1100	10 – 2 500
alkalinity (mg $CaCO_3$/l)	6700	300 –11 500
NH_4 (mg N/l)	750	30 – 3 000
orgN (mg n/l)	600	10 – 4 250
total N (mgN/l)	1250	50 – 5 000
NO_3 (mg N/l)	3	0.1– 50
NO_2 (mg N/l)	0.5	0 – 25
total P (mg P/l)	6	0.1– 30
AOX (μg Cl/l)*	2000	320 – 3 500
As (μg/l)	160	5 – 1 600
Cd (μg/l)	6	0.5– 140
Co (μg/l)	55	4 – 950
Ni (μg/l)	200	20 – 2 050
Pb (μg/l)	90	8 – 1 020
Cr (μg/l)	300	30 – 1 600
Cu (μg/l)	80	4 – 1 400
Hg (μg/l)	10	0.2– 50

* adsorbable organic halogen

opportunity for degradation by soil microorganisms. Whether this is a significant sink for volatile organics buried in landfills has not been studied. In addition to gas stripping, there are other potential fates of organics in landfills. Refuse in an active state of methane production represents a healthy ecosystem which may be capable of decomposing substrates other than carbohydrates and proteins. Thus some hazardous organics may be anaerobically degraded in sanitary landfills although there is very little data in this area. The existing data do show that phenol is readily degraded by refuse microorganisms. Watson-Craik and Senior (1989) reported on the conversion of phenol to methane in laboratory-scale lysimeters. Leachate recycle was necessary for complete degradation of the 188 mg/l phenol solution. Halogenated solvents such as perchloroethylene and trichloroethylene have been shown to undergo a biologically mediated reductive dehalogenation reaction under methanogenic conditions (Fathepure *et al.*, 1987). One end-product of reductive dehalogenation is vinyl chloride. The presence of vinyl chloride in landfill gas suggests that chlorinated solvents were previously buried in the landfill. Organics may also preferentially sorb to the non-degraded organic fraction of the decomposed refuse, thus reducing or eliminating transport.

The pH of refuse decreases during the anaerobic acid phase and this is likely to increase metals solubility and therefore metals mobility. However, this does not necessarily mean that metal concentrations will increase in leachate. Once refuse is in the accelerated methane production phase, indicative of active anaerobic metabolism, sulfate is reduced to hydrogen sulfide. Most metal sulfides are extremely insoluble and their formation would reduce metal mobility. As refuse decomposition proceeds to the decelerated methane production phase, humic materials are produced. These materials can be expected to behave as natural chelating agents, enhancing metal mobility. Carboxylic acids also act as chelating agents (Francis and Dodge, 1986). Pohland and Gould (1986), in an evaluation of the co-disposal of metal sludges with municipal solid waste, reported an increase in metal concentrations in leachate as the refuse became well decomposed.

6.4 FACTORS LIMITING THE ONSET OF METHANE PRODUCTION

Research on the decomposition of solid waste in sanitary landfills was first reported by Merz and Stone (1962). Since then, numerous researchers have tried to enhance refuse methanogenesis by manipulation of the landfill ecosystem (Barlaz *et al.*, 1990). Many factors influence the onset and rate of methane production including moisture content, pH, nutrient concentrations, and temperature among others (Farquhar and Rovers, 1973). It is important to note that all of the microorganisms necessary for the conversion of refuse to methane have been detected in fresh refuse (Barlaz *et al.*, 1989b). As a result of a review of data on both laboratory- and field-scale tests, it appears that the two variables most important to refuse methanogenesis are moisture content and pH.

The moisture content of fresh refuse ranges from 15% to 45% and is typically about 20% on a wet weight basis. While there is no definitive answer as to either the minimum or optimal moisture content required for refuse decomposition, 20% is clearly low. Wujcik and Jewell (1980) studied the effect of moisture content on the batch fermentation of wheat straw and dairy manure – compounds which are analogous in chemical composition to refuse. Methane yields decreased at moisture contents below 70% and the yield at 30% moisture was 22% of the yield at 70% moisture. In many laboratory-scale studies on the effects of moisture on refuse decomposition, methane yields have been too low for quantification of yield as a function of moisture content.

The broadest data sets where moisture content can be evaluated are those of Emberton (1986) and Jenkins and Pettus (1985). Emberton evaluated methane production rate data for landfills across the US and categorized the landfills based on annual precipitation. Jenkins tested the effect of moisture content in refuse sampled from landfills. In both studies, the methane production rate exhibited an upward trend with increasing moisture contents; confounding factors such as density, refuse age and refuse composition notwithstanding.

A second key factor influencing the rate and onset of methane production is pH. As discussed in the section on anaerobic microbiology, microorganisms responsible for the conversion of refuse to methane are quite sensitive to pH. Their pH optimum is between 6.8 and 7.4 and methane production rates decrease sharply at pH values below about 6.5. Theoretically, refuse pH is an excellent indicator of potential methanogenic activity. Unfortunately, as discussed previously, leachate composition data at a field-scale landfill may well not be indicative of the decomposition state of the refuse. Leachate with a pH below 5 has been observed in landfills actively producing methane. Such leachate has most likely been in contact with some refuse in the anaerobic acid phase as well as refuse actively producing methane. Hence, leachate pH is not always a useful indicator of methanogenic activity.

Leachate recycle and neutralization has been shown to enhance the onset and rate of methane production (Pohland, 1975; Buivid *et al.*, 1981; Barlaz *et al.*, 1987). Given that moisture and pH are reported to be the two most significant factors limiting methane production, the stimulatory effect of leachate recycle and neutralization is logical. Recycling neutralized leachate back through a landfill increases refuse moisture content, substrate availability, and provides a degree of mixing in what may otherwise be an immobilized batch reactor. Neutralization of the leachate provides a means of externally raising the pH of the refuse ecosystem. There is limited field experience with leachate recycle systems and more is needed to fully document its value in a field-scale situation.

Moisture content is the factor which most often limits methane production. This is to be expected in dry climates where there is little opportunity for infiltration, and in wet climates where biological activity may be limited because landfills are typically designed to minimize water infiltration. Other conclusions have been reached in laboratory-scale experiments where moisture content was not limiting (Barlaz *et al.*, 1990). Prior to the onset of methane production, the rate at which the methanogenic bacteria convert acetate to methane limits the onset of methane production. In laboratory-scale work, seeding fresh refuse with anaerobically decomposed refuse reduced the time to the onset of methane production significantly (Barlaz *et al.*, 1987). With the onset of methane production and depletion of the soluble substrates, the rate at which polymers (cellulose, hemicellulose and protein) are hydrolyzed limits the rate of methane production.

6.4.1 Regulatory factors influencing methane production

Another factor limiting the onset and rate of methane production is regulatory policy and philosophy. Historically, municipal solid waste was dumped in holes and occasionally covered with whatever material was available. This often led to groundwater contamination and the release of contaminated stormwater to surface water. Subsequently, states began to regulate the burial of municipal solid waste, bringing about the modern sanitary landfill, complete with liners and leachate collection as described in Chapter 1.

Sanitary landfill design standards were developed to minimize the amount of moisture which came in contact with refuse, thus minimizing leachate production. However, at the time that the philosophy of a dry landfill was adopted, methane recovery for energy was in its infancy. The design of landfills to enhance methane production, by allowing the moisture content of fresh refuse to increase by surface water infiltration, was not given serious consideration. More recently, some solid waste regulations recognize that leachate recycle and neutralization for enhancement of methane production may be advantageous. In addition to energy recovery, enhanced methane production and recovery as an energy source offers other advantages.

1. With the onset of methane production, there is a reduction in leachate COD, thus reducing leachate treatment costs and the potential for groundwater contamination.
2. Refuse settles as it decomposes and the resulting settlement necessitates significant maintenance of the landfill cover to repair cracks. More complete refuse decomposition prior to placement of the final cover would minimize settlement and long term maintenance costs.
3. Enhanced methane production would make energy recovery projects more economical, thus more would be implemented. This would reduce emissions of methane to the atmosphere. Methane is a gas which has a greenhouse effect at least twenty times more damaging than carbon dioxide on a volume basis.
4. Enhanced refuse decomposition reduces the impact of future leachate and gas emissions, most likely reducing long-term monitoring and care requirements.
5. Enhanced decomposition means that most of the gaseous and leachate products of decomposition are released during the period when the gas and leachate control systems are most likely to function as designed, and when responsible parties are present to monitor and repair the site as necessary.

Landfills designed and operated in a mode of leachate recycle and neutralization may be recognized as more desirable in the future. It will then be necessary to design landfills not only with liners and leachate collection systems similar to those used today, but also with systems for distributing leachate over refuse without creating odor or side slope seepage problems.

6.5 MASS BALANCE ANALYSIS OF REFUSE DECOMPOSITION

A knowledge of the chemical composition of refuse remaining in a landfill at any time makes it possible to estimate the volume of methane which may yet be produced, remaining reserves of degradable matter, and the potential for loss of refuse materials and settlement as the refuse decomposes. The methane potential of refuse, combined with an estimate of the current rate of methane production as measured in a pump test, is useful for evaluation of the economics of landfill gas recovery projects.

The mass of methane which would be produced if all of a given constituent were converted to carbon dioxide, methane and ammonia may be calculated from Eq. (6.1) (Parkin and Owen, 1986).

$$C_nH_aO_bN_c + \left(n - \frac{a}{4} - \frac{b}{2} + \frac{3c}{4}\right)H_2O \rightarrow \left(\frac{n}{2} - \frac{a}{8} + \frac{b}{4} + \frac{3c}{8}\right)CO_2 + \left(\frac{n}{2} + \frac{a}{8} - \frac{b}{4} - \frac{3c}{8}\right)CH_4 + cNH_3 \quad (6.1)$$

Cellulose ($C_6H_{10}O_5$) and hemicellulose ($C_5H_8O_4$) comprise 91% of the methane potential of fresh refuse (Table 6.2). These two constituents, along with carboxylic acids will comprise at least 90% of the methane potential of refuse in virtually any state of decomposition. Based on Eq. (6.1), 415 and 424 liters of methane at S.T.P. would be expected for every kilogram of cellulose and hemicellulose degraded, respectively.

The methane potential of carboxylic acids can be calculated from the reactions governing the conversion of valerate to propionate and acetate, butyrate and propionate to acetate and hydrogen; hydrogen and carbon dioxide to methane; and acetate to methane and carbon dioxide (McInerney and Bryant, 1981). Calculated methane potentials are 720.8, 643.7, 537.0 and 373.0 liters CH_4/kg of valerate, butyrate, propionate and acetate, respectively.

Barlaz *et al.* (1989a) performed mass balances on shredded refuse incubated in laboratory-scale lysimeters with leachate recycle. Carbon recoveries of 64–111% were obtained. After correcting for explainable error, this range improved to 87–111%. Mass balances were useful for documenting the decomposition of specific chemical constituents and demonstrating the importance of cellulose and hemicellulose to methane production.

Mass balances may be used to estimate the methane potential remaining in a landfill by sampling the refuse, performing the appropriate chemical analyses and calculating the methane potential. Ideally, the chemical composition and methane potential of the refuse at burial would also be known, in which case comparison of the initial methane potential of the refuse with that at the time of sampling will provide information on the fraction of the refuse which has been degraded. It must be recognized that representative sampling of a full-scale sanitary landfill is not realistic; however, it is possible to obtain multiple samples at apparently representative locations within the landfill and get an idea of the range and extent of loss of decomposable materials. Samples should be as large as can reasonably be handled and be reduced by proven techniques.

6.5.1 The final state of refuse composition

As described here, the calculated methane potential assumes complete loss of degradable organics. Some fraction of these constituents is surrounded by lignin and not readily available for anaerobic decomposition. Thus, the actual methane potential of a refuse sample will be less than that calculated. Bookter and Ham (1982) measured cellulose concentrations as low as 8.2% for nine year old shredded refuse which was well decomposed as evidenced by its humus like appearance. As part of this same study various landfills around the country were sampled and cellulose concentrations as low as 6.6% were reported for non-shredded refuse. Refuse sampled from a landfill in Wisconsin had cellulose concentrations of 10.0%, 6.1% and 3.1% for refuse buried in 1948, 1954 and 1957, respectively. Kinman *et al.* (1989) found between 1.6% and 62.0% paper in 20 year-old refuse excavated from the Mallard North landfill in Chicago, IL. The average paper concentration was 32.0% and no cellulose concentration data were reported.

It is not possible at present to make a general statement as to the maximum amount of cellulose and hemicellulose decomposition which can be expected in a landfill. Only a few researchers have measured this parameter and there is little data in the open literature. Complete cellulose and hemicellulose disappearance is not expected though the data of Bookter and Ham (1982) suggest that values below 10% are possible. However, excavated refuse may be diluted with cover soil, thus diluting measured cellulose concentrations. Closely controlled laboratory experiments and additional data on the composition of refuse several years after burial is needed to estimate the fraction of the cellulose and hemicellulose in effect not available for anaerobic decomposition.

6.6 CASE STUDY

The 36 ha Omega Hills Landfill, owned and operated by Waste Management of Wisconsin, Inc. (WMWI), is located 25 kms from downtown Milwaukee. Since commencing operation in 1972, approximately 9 million cubic meters of municipal, industrial, and commercial wastes have been deposited, mostly from the Milwaukee metropolitan area. Average refuse depth is 37 m feet with a maximum depth of 64 m.

A monitoring program in 1984 at the landfill documented gas migration and vegetative stress in the final clay cover. In addition, odor from escaping gas was an aesthetic concern. In response to these site conditions, a gas extraction system was designed and construction began in the spring of 1985. The gas extraction system was designed to provide a zero pressure envelope throughout most of the landfill by withdrawing gas at a rate equal to its production.

Sixty-five gas extraction points are located throughout the Omega Hills Landfill, consisting of 45 wells, 11 trenches, and 9 leachate collection system connections. An average well spacing of 92 m was used to provide zone of influence overlap. The gas collectors are connected by a 5.8-km-long gas collection header system designed to handle up to 200 000 m^3 of gas per day. The header system consists of high density polyethylene pipe (HDPE) ranging from 100–600 mm in diameter. A flexible coupling connects each well assembly to the header and allows for differential settlement between the two components.

The header system is designed to handle both gas and liquid flow by use of minimum slopes to facilitate drainage of condensate to six low points. Barometric drip legs at the low points allow gravity drainage of condensate from the header without air intrusion. Condensate is discharged into the existing leachate control piping system which flows to the on-site leachate pretreatment facility and from there to the sanitary sewer. To minimize head loss, the header lines generally were sized for maximum gas velocities of 9–11 m/sec.

Landfill gas is withdrawn from the wells and is transported through the header to the 150 m^2 gas compressor building. In the compressor building, the gas first flows through water wash scrubbers for particulate removal and then to compressors. The wash water is treated along with the leachate. From the compressor building, the gas is transported to the 290 m^2 turbine building where it fuels the gas fired turbine engines which drive the electrical generators. A gas flaring station is used only as a backup should the electrical generation system be down for any length of time.

The compressors were manufactured by Hall Systems, Inc., of Tulsa, Oklahoma. The first stage of each compressor consists of a centrifugal blower capable of producing up to a 1.5 m water column head. The centrifugal blower provides the head and flow rates required to withdraw landfill gas from the collection system. The gas then enters two additional stages of compression, with each stage followed by cooling of the gas. After compression, gas at a delivery pressure of 28 kPa is fed to underground piping and transported to the turbine generator sets. To prevent liquid condensation between the compressors and the turbines, a portion of the gas bypasses the final cooling stage and heats the rest of the gas. Each compressor is driven by a 450 kW electric motor. The compressor package, which includes the centrifugal blower, a piston type, two-stage compressor, motor, gas scrubbers, and gas chiller unit, is skid mounted.

The two Centaur turbine generator sets were obtained from Solar Turbine, Inc. Each set is skid mounted and weighs approximately 18 tonnes. Each turbine generator set can produce approximately 3.3 Mw of electricity. At this loading the turbine requires approximately 45 standard cubic meters per minute of landfill gas at 50% methane. Electricity is generated at 4160 volts and is stepped up to 26.4 kV and purchased by the Wisconsin Electric Power Company.

Special attention was paid to the acoustical treatment of the facilities because much of the equipment produces significant noise levels which cannot be permitted to adversely impact adjacent communities.

The system went on-line December 17, 1985, and has since been operating continuously with no obvious trends in gas generation rates or composition. Gas flow meets the quantity and quality requirements of the two turbines, which are basically operated at capacity. The plant has exceeded the estimated on-line time with only minor maintenance and other downtime periods having been experienced in the first four and a half years of operation. The owner estimates a payback period of seven to eight years.

In the fall of 1988, a third Centaur turbine-generator set was placed on-line, increasing the plant capacity by 50%. There have been periods when insufficient gas was available to operate all three turbines simultaneously at capacity, largely because of liquid accumulations in gas wells which limits their productivity.

REFERENCES

Barlaz, M.A. (1988) Microbial and chemical dynamics during refuse decomposition in a simulated sanitary landfill, Ph.D. thesis, Dept. of Civil and Environmental Engineering, Univ. of Wisconsin, Madison, WI.

Barlaz, M.A., Ham, R.K. and Milke, M.W. (1987) *Waste Management and Res.*, **5**, 27.

Barlaz, M.A., Ham, R.K. and Schaefer, D.M. (1989a) *J. Environ. Eng.*, ASCE, **115**(6), 1088–102.

Barlaz, M.A., Schaefer, D.M. and Ham, R.K. (1989b) *Appl. Env. Microbiol.*,

55(1), 55–65.

Barlaz, M.A., Schaefer, D.M. and Ham, R.K. (1989c) *Appl. Biochem. Biotechnol.*, **20/21**, 181–205.

Barlaz, M.A., Ham, R.K. and Schaefer, D.M. (1990) *CRC Critical Reviews in Environ. Control*, **19**(6), 557–84.

Bookter, T.J. and Ham, R.K. (1982) *J. Environ. Eng.*, ASCE, **108**(EE6), 1089.

Buivid, M.G., Wise, D.L., Blanchet, M.J. *et al.* (1981) *Resource Recovery and Conservation*, **6**, 3.

Burns, R.G. and Martin, J.P. (1986) Biodegradation of organic residues in soil, in *Microflora and Faunal Interactions in Natural and Agro-Ecosystems*, (eds M.J. Mitchell and J.P. Nakas) Martinus Nighoff/Dr. W. Junk Publ., Bordrecht, The Netherlands.

Ehrig, H.J. (1988) Water and element balances of landfills, in *Lecture Notes in Earth Sciences*. (ed. P. Baccini) Springer-Verlag, Berlin.

Emberton, J.R. (1986) The biological and chemical characterization of landfills, *Proc., Energy from Landfill Gas*, Solihull, West Midlands, UK, Oct. 30–1.

EPA (1990) *Characterization of Municipal Solid Waste in the United States; 1990 Update*, EPA/530-SW-90-042, USEPA, Washington, DC, PB 90-215112.

Fathepure, B.Z., Nengu, J.P. and Boyd, S.A. (1987) *Appl. Env. Microbiol.*, **53**(11), 2671–4.

Farquhar, G.J. and Rovers, F.A. (1973) *Water, Air, and Soil Pollution*, **2**, 483.

Francis, A.J. and Dodge, C.J. (1986) *Arch. Environ. Contam. Toxicol.*, **15**, 611–16.

Greenhill, W.L. (1964a) *J. British Grasslands Soc.*, **19**, 30–7.

Greenhill, W.L. (1964b) *J. British Grasslands Soc.*, **19**, 231–6.

Greenhill, W.L. (1964c) *J. British Grasslands Soc.*, **19**, 336–9.

Halvadakis, C.P., Robertson, A.P. and Leckie, J.O. (1983) Landfill Methanogenesis: Literature Review and Critique, *Technical Report No. 271*, Department of Civil Engineering, Stanford University.

Ham, R.K. and Bookter, T.J. (1982) *J. Environ. Eng.*, ASCE, **108**(EE6), 1147.

Jenkins, R.L. and Pettus, J.A. (1985) The use of *in vitro* anaerobic landfill samples for estimating gas generation rates, in *Biotechnological Advances in Processing Municipal Wastes for Fuels and Chemicals* (ed. A.A. Antonopoulos) Argonne Natl. Lab. Report ANL/CNSV–TM–167, p. 419.

Kinman, R.N., Nutini, P.L., Walsh, J.L. *et al.* (1987) *Waste Management and Res.*, **5**, 13.

Kinman, R.N., Rickabaugh, J. and Berg, G. (1989) Analysis of 20 year-old refuse from the Mallard North landfill in Chicago, Illinois, *Proc. Purdue Ind. Waste Conf., May 9–11*, West Lafayette, IN.

Mah, R.A., Smith, M.R. and Baresi, L. (1978) *Appl. Environ. Microbiol.*, **35**(6), 1174.

McInerney, M.J. and Bryant, M.P. (1981) Basic principles of bioconversions in anaerobic digestion and methanogenesis, in *Biomass Conversion Processes for Energy and Fuels*, (eds S.S. Sofar and O. Zaborsky) Plenum Publishing Corp., New York.

Merz, R.C. and Stone, R. (1962) *Public Works*, **93**(9), Sept., 103.

Pacey, J. (1989) Enhancement of degradation: large-scale experiments, in *Sanitary Landfilling: Process Technology and Environmental Impact*, (eds T. Christensen, R. Cossu and R. Stegmann) Academic Press, London, pp. 103–19.

Parkin, G.F. and Owen, W.F. (1986) *J. Environ Eng.*, ASCE, **112**(5), 867.

Pohland, F.G. (1975) Sanitary Landfill Stabilization with Leachate Recycle and Residual Treatment, Georgia Institute of Technology, EPA Grant No. R-801397.

Pohland, F.G. and Gould, J.P. (1986) *Wat. Sci. Tech.*, **18**(12), 177.
Tchobanoglous, G., Theisen, H. and Eliassen, R. (1977) *Solid Wastes*, McGraw-Hill, Inc., N.Y., 334.
Watson-Craik, I.A. and Senior, E. (1989) *Water Res.*, **13**(10), 1293–303.
Wolfe, R.S. (1979) Methanogenesis, in *Microbial Biochemistry, Int. Rev. of Biochemistry*, **21**, (ed. J.R. Quayle) University Park Press, Baltimore, MD.
Wolin, M.J. and Miller, T.L. (1985) in *Biology of Industrial Microorganisms*, (eds A.L. Demsin and N.A. Solomon) Benjamin/Cumings Publ. Co. Inc. Menlo Park, CA., pp. 189–221.
Wujcik, W.J. and Jewell, W.J. (1980) *Biotech. Bioeng. Symp. No. 10*, 43–65.
Young, L.Y. and Frazer, A.C. (1987) *Geomicrobiology J.*, **5**, 261.
Zehnder, A.J.B. (1978) Ecology of methane formation, in *Water Pollution Microbiology*, **2**, (ed. Ralph Mitchell) John Wiley & Sons, N.Y. 1978, p. 349.
Zehnder, A.J.B., Ingvorsen, K. and Marti, T. (1982) Microbiology of methane bacteria, in *Anaerobic Digestion*, (ed. D.E. Hughes) Elsevier Biomedical Press B.V., Amsterdam, p. 45.
Zeikus, J.G. (1980) Microbial populations in digesters, in *Anaerobic Digestion*, (ed. D.A. Stafford) Appl. Sci. Publishers, London, p. 61.

CHAPTER 7

Clay liners

David E. Daniel

7.1 INTRODUCTION

Low-hydraulic-conductivity soil liners are given various names, including **soil liner** and **clay liner**. In this chapter, the term clay liner is used even though other minerals in the liner material (e.g., sand) may be present in larger quantities than clay. The term *clay* is emphasized because clay is largely responsible for the low hydraulic conductivity of earthen liners. Attention in this chapter is focused on three types of clay liners:

1. naturally-occurring clay liners;
2. compacted clay liners; and
3. geosynthetic clay liners.

A clay liner serves as a hydraulic barrier to flow of fluids. Clay liners are used to minimize infiltration of water into buried waste (cover systems) or to control release of leachate from the waste (liner systems). To meet these objectives, clay liners must have low hydraulic conductivity over long periods of time. Further, one must be able to verify that the hydraulic conductivity will be suitably low, which is often the most difficult problem to be resolved. In addition, clay liners are expected to attenuate the movement of leachate, to prolong release of chemicals in leachate, and to serve other site-specific functions.

7.2 NATURAL CLAY LINERS

Natural clay liners are naturally-occurring formations of low-hydraulic-conductivity, clay-rich soil. Waste can be buried above or within a

Geotechnical Practice for Waste Disposal.
Edited by David E. Daniel.
Published in 1993 by Chapman & Hall, London. ISBN 0 412 35170 6

natural liner. Natural liners normally contain significant amounts of clay minerals and have hydraulic conductivities less than or equal to 1×10^{-6} to 1×10^{-7} cm/s. Natural liners more typically serve as a back-up to engineered liners, but occasionally (for old landfills or, where regulations allow, for new landfills), a natural liner may represent the only liner at a waste disposal facility.

The continuity and hydraulic conductivity of natural liner materials are critical issues. To function effectively, the natural liner must be continuous and be free from major hydraulic imperfections such as fractures, joints, and holes.

An evaluation of a liner's continuity begins with a geologic evaluation and includes a careful study of local and regional hydrogeology (see Chapter 4). Exploratory borings are an essential investigatory tool; surface and borehold geophysics often provide a wealth of valuable information. Analysis of radioisotopic concentrations of certain constituents in ground water can lead to the determination of the age of ground water; knowledge of the age of ground water can help to establish that a soil liner hydrogeologically isolates one aquifer from another. Another useful way to investigate the ability of a natural liner to isolate one stratum from another is to pump from a well in an underlying aquifer and to observe changes in water levels in wells installed in an overlying aquifer, or vice versa.

Hydraulic conductivity of a natural liner should be studied with a combination of laboratory and *in situ* hydraulic conductivity tests. Laboratory test are usually performed with flexible-wall cells used to permeate samples obtained by pushing a thin-walled tube into soil that underlies a borehole (Daniel *et al.*, 1984). The results of such tests should be viewed with suspicion because if the liner contains hydraulic defects (e.g., cracks, fissures, slickensides, or root holes), the defects will probably be missed and the measured hydraulic conductivity will be too low (Olson and Daniel, 1981; Daniel, Trautwein, and McMurtry, 1985; Keller, van der Kamp, and Cherry, 1986; Bradbury and Muldoon, 1990).

In situ hydraulic conductivity tests are an essential part of a credible effort to characterize the hydraulic conductivity of natural soil liners. The normal type of test involves a single cased borehole and either a constant or a falling head. Several equations may be used to compute hydraulic conductivity; see Olson and Daniel (1981) and Chapuis (1989) for details. The number of tests that are needed to characterize the hydraulic conductivity of a natural deposit of soil or rock varies from site to site and depends upon hydrogeologic complexity, the required accuracy to which the hydraulic conductivity needs to be determined, available funding, and other factors. Typically, of the order of 10–20 *in situ* hydraulic conductivity tests are appropriate. Great care must be taken to seal the boreholes so that the hydraulic integrity of the liner is not compromised. Sealing techniques are discussed in Part 4 of this book.

It is extremely difficult and expensive to prove that a naturally-occurring stratum of soil or rock uniformly possesses low hydraulic conductivity. For this reason, use of a natural soil liner as the sole means for protecting ground water from contamination is not normally recommended. An exception might be the case of an extraordinarily uniform, massive, and well-characterized stratum of material, but such strata are rare. Another exception might be site remediation cases in which ground water flow and contamination patterns are monitored and contingencies are made in case strata are more permeable than expected.

The interested reader is encouraged to consult Keller, van der Kamp, and Cherry (1986) and Bradbury and Muldoon (1990) for examples of well-conceived evaluations of natural soil liners.

7.3 COMPACTED CLAY LINERS

7.3.1 Introduction

Compacted clay liners are constructed primarily from natural soil materials, although the liner may contain processed materials such as bentonite or even synthetic materials such as polymers. Clay liners are constructed in layers called **lifts**. On side slopes, the lifts can be horizontal or parallel to the slope, although parallel lifts are not recommended for side slopes steeper than 2.5 to 3 on 1 (horizontal to vertical). As suggested by Fig. 7.1, lifts parallel to the side slopes are preferred because the effect of a zone of poor material, or imperfect bonding of lifts, is less with parallel lifts. If horizontal lifts are used, it may help to place a slight inward inclination on the lift interfaces (Fig. 7.1) to minimize the tendency for leachate to flow along lift interfaces.

7.3.2 Compaction requirements

The objective of compaction is to remold chunks (clods) of soil into a homogeneous mass that is free of large, continuous interclod voids. If this objective is accomplished with suitable soil materials, low hydraulic conductivity ($\leq 1 \times 10^{-7}$ cm/s) will result.

Experience has shown that the water content of the soil, method of compaction, and compactive effort have a major influence on the hydraulic conductivity of compacted soil liners. Laboratory studies have demonstrated that low hydraulic conductivity is easiest to achieve when the soil is compacted wet of optimum water content with a high level of kneading-type compactive energy (Mitchell, Hooper, and

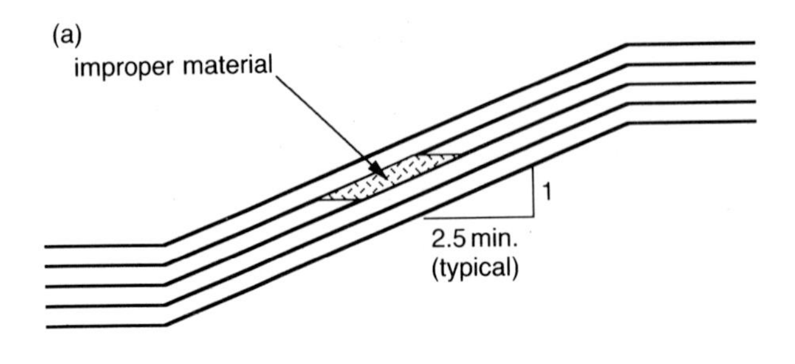

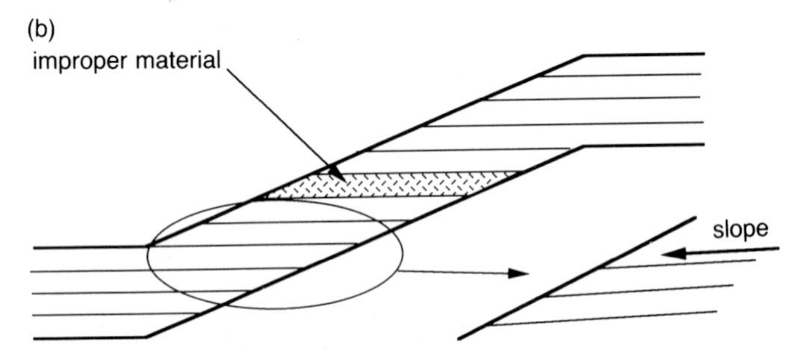

Fig. 7.1 Side slopes constructed with (a) parallel and (b) horizontal lifts.

Campanella, 1965). Figure 7.2 illustrates the influence of molding water content and compactive energy upon hydraulic conductivity. The soil must be sufficiently wet so that, upon compaction, clods of clayey soil will mold together, eliminating large inter-clod pores. Kneading the soil during compaction with a high level of compactive energy helps to remold clods and to eliminate large pore spaces.

Studies have also demonstrated lifts of soil must be bonded together to eliminate, to the extent possible, highly permeable zones at lift interfaces. The idea is illustrated in Fig. 7.3; if permeable inter-lift zones are eliminated, hydraulic connection between 'defects' in each lift is destroyed and a low overall hydraulic conductivity is achieved.

7.3.4 Materials

The minimum requirements recommended to achieve a hydraulic conductivity $\leqslant 1 \times 10^{-7}$ cm/s for most soil liners materials are as follows:

percentage fines:	$\geqslant$20–30%
plasticity index:	$\geqslant$7–10%
percentage gravel:	$\leqslant$30%
maximum particle size:	25–50 mm

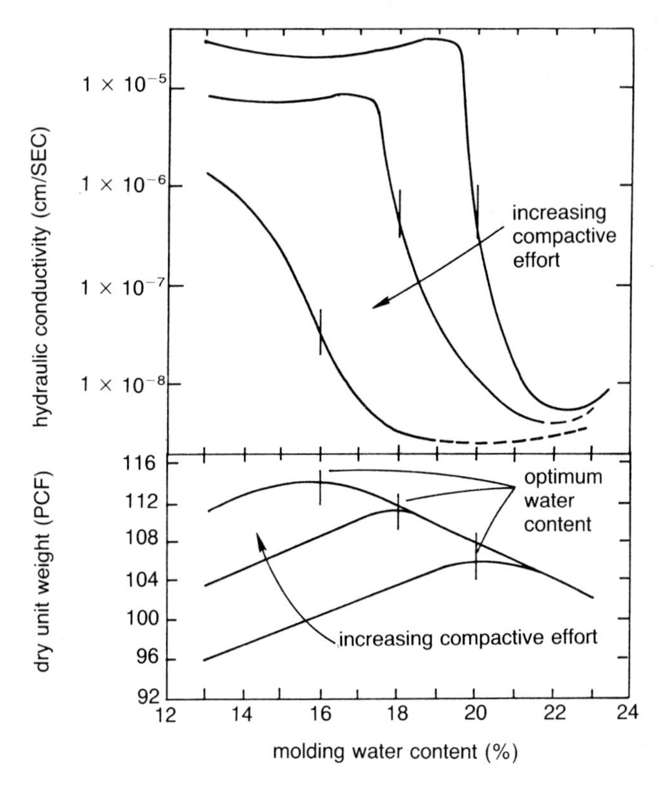

Fig. 7.2 Effect of molding water content and compactive energy on hydraulic conductivity, from Mitchell *et al.* (1965).

Percentage fines is defined as the percent by dry weight passing the US No. 200 sieve, which has openings of 75 μm. Plasticity index may be determined by ASTM D4318. Percentage gravel is defined as the percent by dry weight retained on a No. 4 sieve (4.76 mm openings). Local experience may dictate more stringent requirements, and, for some soils, more restrictive criteria may be appropriate. However, if the criteria tabulated above are not met, it is unlikely that a natural soil liner material will be suitable without additives such as bentonite.

Recent work conducted to evaluate gravel content bears mentioning. Shelley (1991) mixed kaolinite and mine spoil with varying percentages of gravel (maximum particle size: 20 mm), moistened the soil to a few percent wet of optimum, compacted the soil/gravel mixtures using standard Proctor compaction procedures (ASTM D698), and then permeated the compacted specimens. Results are summarized in Fig. 7.4. Shelley found that the soil could contain up to 50–60% gravel without a detrimental impact upon hydraulic conductivity. Shelley reported that at gravel percentages ≤50–60%, clay particles plugged the voids between the gravel particles. Shakoor and Cook (1990) report

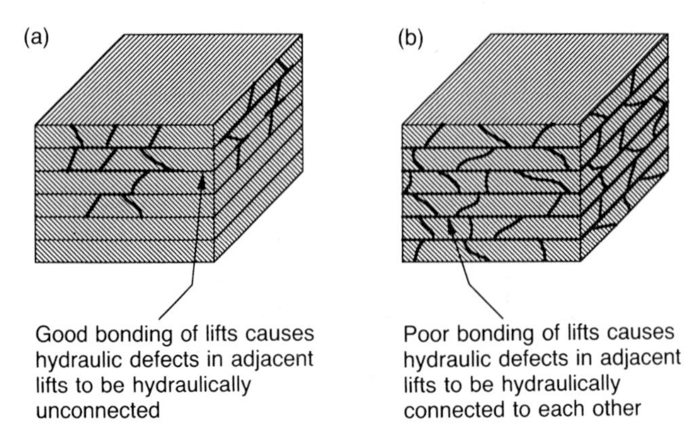

Fig. 7.3 Effect of (a) good and (b) poor bonding of lifts on the performance of a compacted clay liner.

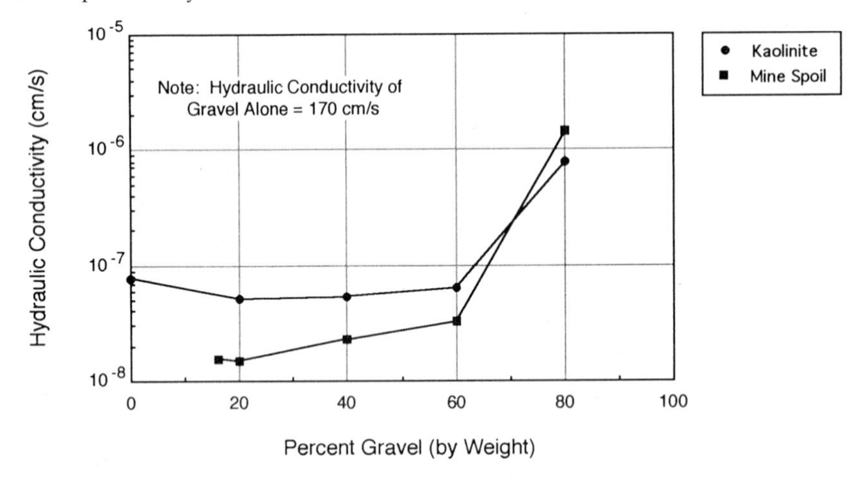

Fig. 7.4 Effect of percent of gravel in clay on hydraulic conductivity, after Shelley (1991).

similar results. However, in both investigations, the soil and gravel were carefully and uniformly mixed in the laboratory; in the field, the mixing will be less perfect. The main issue is not necessarily how much gravel is present (assuming the gravel content is ≤50–60%) but, rather, with the likelihood that pockets of gravel (segregation of gravel) can occur during construction. The potential for gravel segregation to occur depends on the soil material and construction procedures. The author recommends that the gravel content not exceed about 30%, but notes that this value should be increased or decreased as appropriate for a given material and construction process.

If suitable materials are unavailable locally, local soils can be blended with commercial clays, e.g., bentonite, to achieve low hydraulic con-

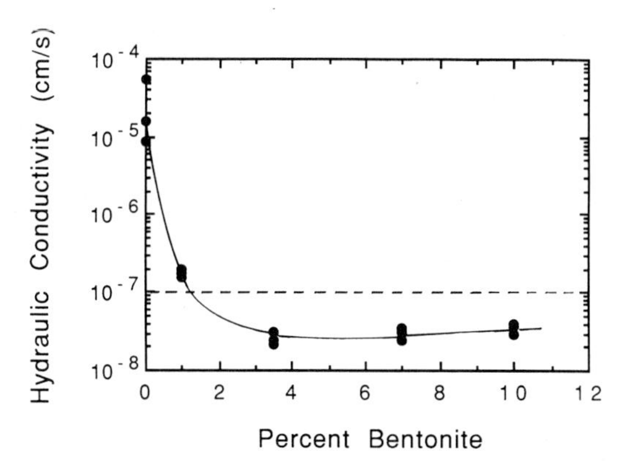

Fig. 7.5 Effect of percent of bentonite on hydraulic conductivity.

ductivity. A relatively small amount of bentonite can lower hydraulic conductivity as much as several orders of magnitude (Fig. 7.5).

One should be cautious about using highly plastic soils (soils with plasticity indices >30–40%) because these materials form hard clods when the soil is dry and are very sticky when the soil is wet. Highly plastic soils, for these reasons, are difficult to work with in the field.

7.3.5 Construction

Processing – Summary.

Some liner materials need to be processed to break down clods of soil (Benson and Daniel, 1990; Shackelford and Javed, 1991), to sieve out stones and rocks, to moisten the soil, or to incorporate additives. Clods of soil can be broken down with tilling equipment. Stones can be sieved out of the soil with large vibratory sieves or mechanized 'rock pickers' passed over a loose lift of soil. Road reclaimers can process soil in a loose lift and crush stones or large clods.

If the soil must be wetted or dried more than 2–3 percentage points in water content, the soil should be processed by spreading it in a loose lift about 300 mm thick. Water can be added and mixed into the soil with a tiller, or the soil can be disced or tilled to allow it to dry uniformly. It is essential that time be allowed for the soil to wet or dry uniformly. At least 1–3 days is usually needed for adequate hydration or dehydration. Frozen soil should never be used to construct a soil liner.

Additives such as bentonite can be introduced in two ways. One technique is to mix soil and additive in a pugmill. Water can also be added in a pugmill either concurrently with bentonite or in a separate processing step. Alternatively, the soil can be spread in a loose lift

200–300 mm thick, the additive spread over the surface, and rototillers used to mix the materials. Several passes of the tiller over a given spot are usually needed. Water can be added in the tiller during mixing or later, after mixing is complete. The pugmill is more reliable in providing thorough, controlled mixing, but, done carefully, the other method can provide adequate mixing. For more information on bentonite and soil/bentonite testing, the reader is referred to Alther (1983), Alther (1987), Chapuis (1990).

Surface preparation

It is crucial that each lift of a soil liner be effectively bonded to the overlying and underlying lifts (Fig. 7.3). The surface of a previously-compacted lift must be rough rather than smooth. If the surface has been smoothed, e.g., with a smooth steel-drummed roller, the surface should be excavated to a depth of 20–30 mm with a disc or other suitable device.

Soil placement

Soil is placed in a loose lift that is no thicker than about 230 mm (9 in.). If grade stakes are used to gauge thickness, the stakes must be removed and the hole left by the stakes sealed. Other techniques, e.g., use of lasers, are preferable for control of elevations. After the soil is placed, a small amount of water may be added to offset evaporative losses, and the soil may be tilled one last time prior to compaction.

Compaction

Heavy, footed compactors with large feet that fully penetrate a loose lift of soil (Fig. 7.6) are ideal. Recommended specifications include:

minimum weight: 18 000 kg (40 000 lbs);
minimum foot length: 180–200 mm; and
minimum number of passes: 5.

More passes may sometimes be needed. A 'pass' is defined as one pass of the compactor, not just an axle, over a given area, and the recommended minimum of five passes is for a vehicle with front and rear drums. In the US, the Caterpillar 815B and 825C are examples of equipment in widespread use that have led to satisfactory results in most cases.

Statically operated compactors are preferred over vibratory compactors for soil liners. The weight of the compactor must be compatible with the soil; relatively dry soils with firm clods require a very heavy

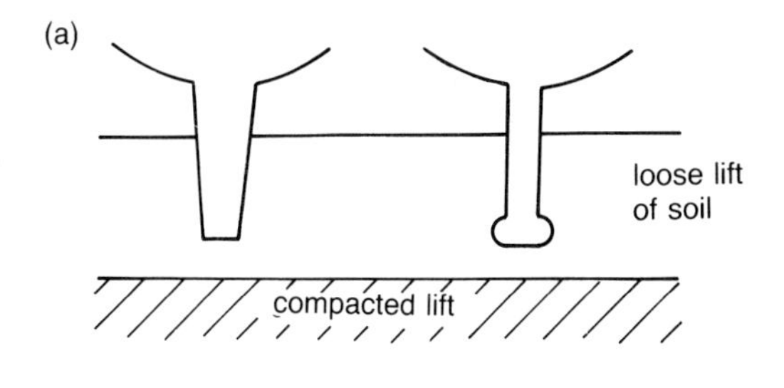

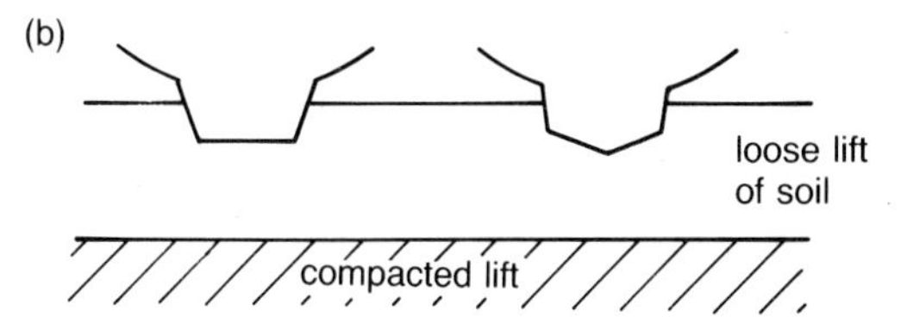

Fig. 7.6 (a) Fully and (b) partly penetrating feet on a footed roller.

compactor whereas relatively wet soils with soft clods require a roller that is not so heavy that it becomes bogged down in the soil.

Protection

After compaction of a lift, the soil must be protected from desiccation and freezing. Desiccation can cause cracking of the clay (Boynton and Daniel, 1985; Kleppe and Olson, 1985). Freeze-thaw changes the structure and fabric of compacted clay in a way that increases hydraulic conductivity (Fig. 7.7) (Chamberlain and Gow, 1979; Zimmie and La Plante, 1990; Kim and Daniel, 1992). Desiccation can be minimized in several ways: the lift can be temporarily covered with a sheet of plastic (but one must be careful that the plastic does not heat excessively and itself dry the clay), the surface can be smooth-rolled to form a relatively impermeable layer at the surface, or the soil can be periodically moistened. The compacted lift can be protected from damage by frost by avoiding construction in freezing weather or by temporarily covering the lift with an insulating layer of material. The protective measures discussed in this section apply to each lift as well as to the completed liner or cover barrier.

Quality control tests

A critical component in construction quality assurance are quality control (QC) tests. For soil liners, the tests fall into two categories:

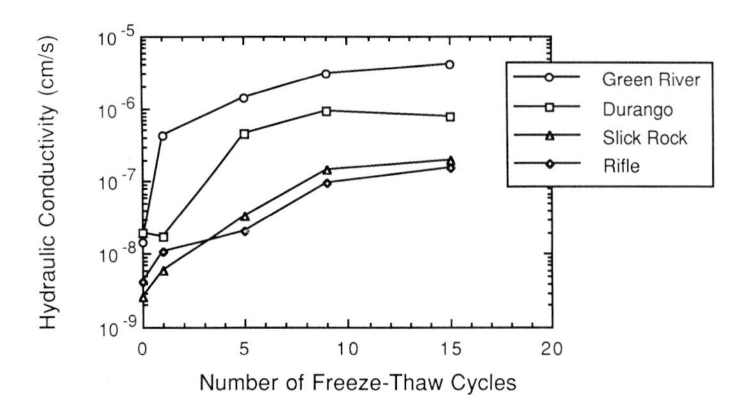

Fig. 7.7 Effect of freeze-thaw on hydraulic conductivity of compacted clay, after Chamberlain *et al.* (1990).

1. tests to verify that the materials of construction are adequate; and
2. tests and observation to verify that the compaction process is adequate.

Recommended tests and minimum testing frequencies are summarized by Daniel (1990).

7.3.6 Water content and dry unit weight

A critical step in design of a compacted soil liner is determination of the range of acceptable water content and dry unit weight of the soil. If the soil is too dry at the time of compaction, suitably low hydraulic conductivity may be unachievable. If the soil is too wet, a variety of problems may ensue, e.g., problems with construction equipment operating on soft, weak soils and potential slope instability caused by low strength of the soil.

Once an acceptable water content range has been selected, the soil must be compacted with adequate compactive energy to compress large voids and to remold clods of soil into a homogeneous, relatively impermeable mass. The dry unit weight of the soil can be a useful indicator of the effectiveness of compaction.

One problem confronting the designer is that both the water content of the soil and the compactive energy delivered to soil during construction of a soil liner vary. Further, precise duplication of field compaction in the laboratory is impossible. Accordingly, the recommended approach (Daniel and Benson, 1990) for establishing water content and dry unit weight requirements during the design stage is to utilize a range in water content that more than spans the range anticipated in the field, and to compact the soil with three compactive energies

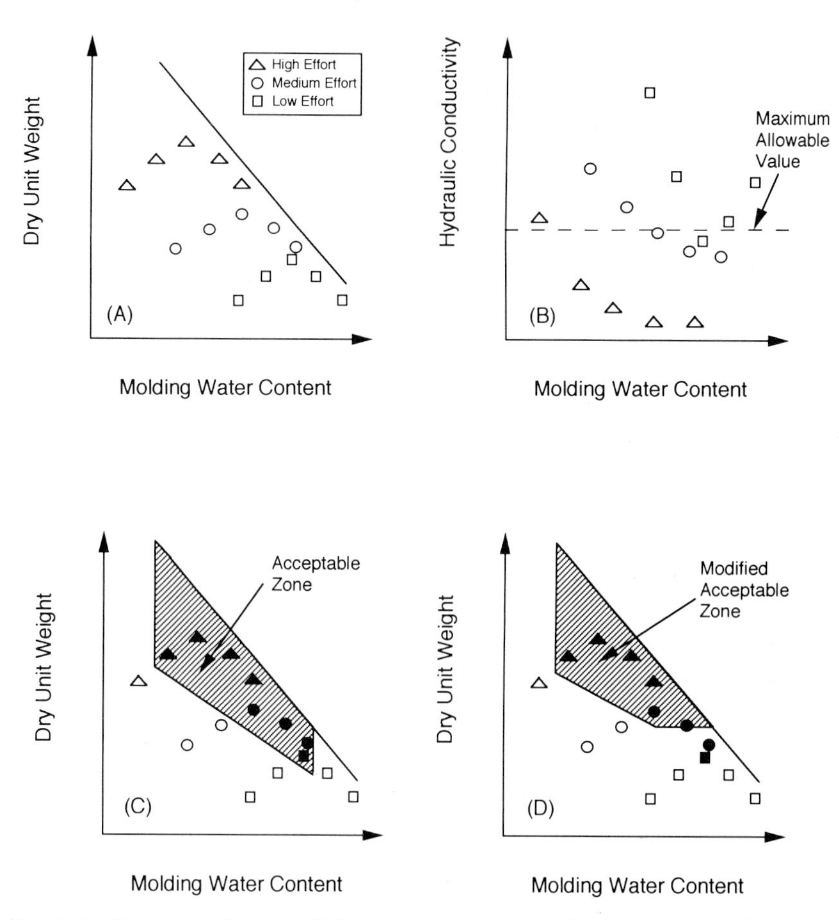

Fig. 7.8 Procedure for establishing acceptable zone for water content and dry unit weight, after Daniel and Benson (1990).

that represent estimates of the lowest compactive energy anticipated, average energy, and highest compactive energy. The author recommends standard and modified Proctor (ASTM D698 and D1557) for the average and high compactive energies, respectively, and 'reduced Proctor' for the lowest energy. 'Reduced Proctor' is the same as standard Proctor but with only 15 drops of the compactive ram per lift rather than the usual 25 drops.

The recommended procedure is illustrated in Fig. 7.8. One compacts 5–6 samples of soil with three different compactive efforts and plots three compaction curves (Fig. 7.8a). Next, the compacted soils are permeated and hydraulic conductivity is measured (Fig. 7.8b). The compaction points are replotted (Fig. 7.8c) with solid symbols used for test specimens that had adequately low hydraulic conductivity and open symbols used for test specimens that were too permeable. An

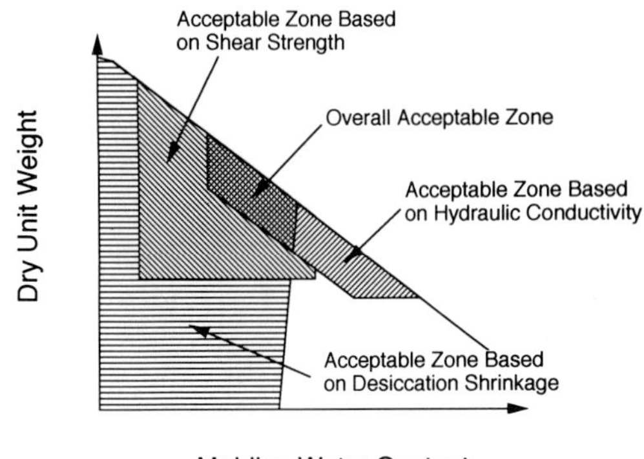

Fig. 7.9 Overall acceptable zone based on hydraulic conductivity, shear strength, and shrinkage upon desiccation.

'acceptable zone' is drawn (usually with some judgement applied) that encompasses the solid points. Finally, the acceptable zone is modified to account for any other relevant factors, e.g., shear strength considerations or local construction practices (Fig. 7.8d). Figure 7.9 illustrates how an acceptable zone can be defined from hydraulic conductivity, shear strength, and desiccation shrinkage criteria.

7.3.7 Test pads

The construction of a test pad prior to building a full-sized liner has many advantages. By constructing a test pad, one can experiment with compaction water content, construction equipment, number of passes of the equipment, lift thickness, etc. Most importantly, though, one can conduct extensive testing, including quality control testing and *in situ* hydraulic conductivity testing, on the test pad.

It is usually recommended that the test pad have a width of at least three construction vehicles (>10 m), and an equal or greater length. The pad should ideally be the same thickness as the full-sized liner, but the trial pad may be thinner than the full-sized liner. (The full-thickness liner should perform at least as well as, and probably better than, a thinner test section because defects in any one lift become less important as the number of lifts increases.) The *in situ* hydraulic conductivity may be determined in many ways. The large sealed double-ring infiltrometer is usually the best large-scale test (Daniel, 1989; Sai and Anderson, 1990), although the Boutwell test (Daniel, 1989, and references therein) is enjoying increased popularity due to its ease of operation and relatively short testing times.

One problem with *in situ* tests on test pads is that the test pad is subjected to essentially zero overburden stress. Hydraulic conductivity

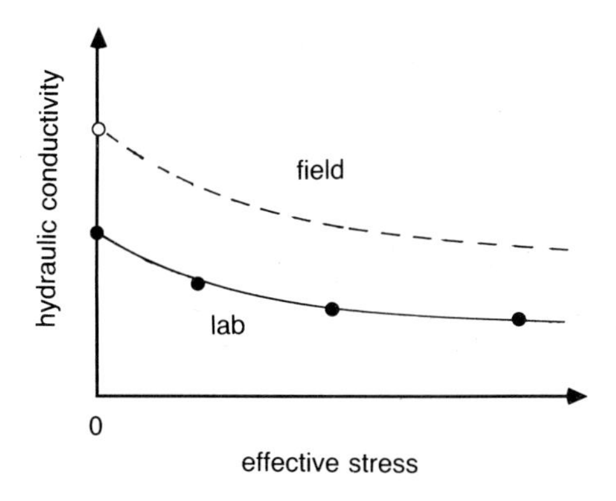

Fig. 7.10 Recommended procedure for adjusting the hydraulic conductivity measured in the field on a test pad for the influence of compressive stress.

decreases with increasing compressive stress. The author recommends that the hydraulic conductivity measured on a test pad with *in situ* methods be corrected for the effects of overburden stress based on results of laboratory hydraulic conductivity tests performed over a range in compressive stress (Fig. 7.10).

7.3.8 Chemical attack by waste

Waste liquids may attack and effectively destroy earthen liners. It is convenient to consider acids and bases, neutral inorganic liquids, neutral organic liquids, and leachates separately. Testing protocols have been described by Bowders *et al.* (1986).

Acids and bases

Strong acids and bases can dissolve solid material in the soil, form channels, and increase hydraulic conductivity. Some acids, e.g., hydrofluoric and phosphoric acid, are particularly aggressive and dissolve soil readily. Concentration of acid, duration of reaction, liquid-solid ratio, type of clay, and temperature are also important variables (Grim, 1953). Leachates with pH <3 or >11 are usually of the most concern.

When concentrated acid is passed through clayey soil, hydraulic conductivity often declines initially but later increases. Soils have a high capacity to buffer acid; many pore volumes of flow are usually needed before the full effect of the acid is observed. Examples include Nasiatka *et al.* (1981), Peterson and Gee (1986), and Bowders and Daniel (1987). Soils that are composed primarily of sand, with a small amount of bentonite, are particularly susceptible to attack by acids because the small mass of bentonite is readily dissolved (Nasiatka *et al.* 1981).

Neutral, inorganic liquids

The effects of neutral, inorganic liquids may be evaluated with the Gouy–Chapman theory (Mitchell, 1976), which states that the thickness (T) of the diffuse double layer varies with the dielectric constant of the pore fluid (D), the electrolyte concentration (n_0), and the cation valence (v) as follows:

$$T \propto [D/(n_0 v^2)]^{\frac{1}{2}} \tag{7.1}$$

For solutions containing mainly water, the dielectric constant of the liquid is relatively constant, and thus the main parameters are n_0 and v. As the diffuse double layer of adsorbed water and cations expands, hydraulic conductivity decreases because flow channels become constricted. Attempts to validate quantitatively the effect of D, n_0, and v on conductivity have generally failed. Qualitatively, however, the Gouy–Chapman theory explains the observed patterns. Aqueous solutions with few electrolytes, e.g., distilled water, tend to expand the double layer and to produce low hydraulic conductivity. A strong (high n_0) solution containing polyvalent cations tends to produce the largest conductivity. Further details and supporting data are reported by Fireman (1944) and McNeal and Coleman (1966).

Neutral, organic liquids

Most organic chemicals have lower dielectric constants than water. Low D tends to cause low T, as shown in Eq. (7.1) and thus high hydraulic conductivity. In addition, low-dielectric-constant liquids cause clay particles to flocculate and cause the soil to shrink and to crack (Anderson, 1982; Fernandez and Quigley, 1985, 1988). This phenomenon is called 'syneresis' and the cracks 'syneresis cracks'. Numerous studies have shown that organic chemicals can cause large increases in hydraulic conductivity (Anderson, 1982; Acar *et al.*, 1985; Fernandez and Quigley, 1985; Foreman and Daniel, 1986; and others).

High compressive stress causes the soil to compact when an organic solvent passes through the soil rather than to crack (Broderick and Daniel, 1990; Fernandez and Quigley, 1991). Thus, soil liners perform much better at high compressive stress than low stress when they are permeated with organic liquids.

Dilute organic liquids do not tend to alter hydraulic conductivity significantly (Bowders and Daniel, 1987, and references therein). If a small amount of low-dielectric-constant liquid, e.g., trichlorethylene (D

= 3), is mixed with water, the dielectric constant of the mixture is only slightly less than that of water (80). Tests have indicated that the dielectric constant must be less than 30–50 for hydraulic conductivity to increase. Experience indicates that soil liners are not attacked by an organic liquid if

1. the solution consists of at least 50% water, and
2. there is no separation of phases, i.e., all of the organic liquid is dissolved in the water and none exists as a separate phase.

A recent, comprehensive review is provided by Budhu *et al.* (1991).

7.3.9 Reliability of compacted clay liners

Many examples can be cited of soil liners that failed to function effectively as hydraulic barriers. Soil liners are not the problem; inadequate investigations (natural soil liners) or inadequate construction or quality control (compacted soil liners) are the main causes of problems. Gordon *et al.* (1989), Cartwright and Krapac (1990), Johnson *et al.* (1990), and Reades *et al.* (1990) provide examples of well-built, full-scale compacted soil liners that have *in situ* hydraulic conductivities $\leqslant 1 \times 10^{-7}$ m/s. Good-quality soil liners that will perform effectively can be constructed.

7.4 GEOSYNTHETIC CLAY LINERS

A relatively new type of manufactured clay liner is receiving widespread attention as a potential hydraulic barrier in liner and cover systems at waste disposal facilities. The liner consists of a thin layer of clay sandwiched between two geotextiles or glued to a geomembrane. Four companies currently manufacture these types of materials, and new products of a similar design are expected to appear on the market in the next year or two.

Various terms have been used to describe these materials, e.g., clay mat, bentonite matting, bentonitic clay liner, prefabricated clay blanket, prefabricated clay liner, etc. Recently, the term geosynthetic clay liner (GCL) has received widespread use for these materials and is used in the remainder of this chapter.

7.4.1 Types of geosynthetic clay liners

Four GCLs are currently manufactured: Bentofix®, Bentomat®, Claymax®, and Gundseal. The configurations, summarized in Fig. 7.11, fall into 2 categories:

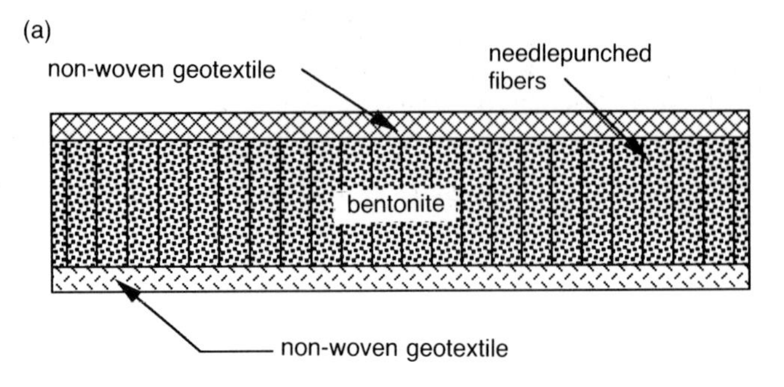

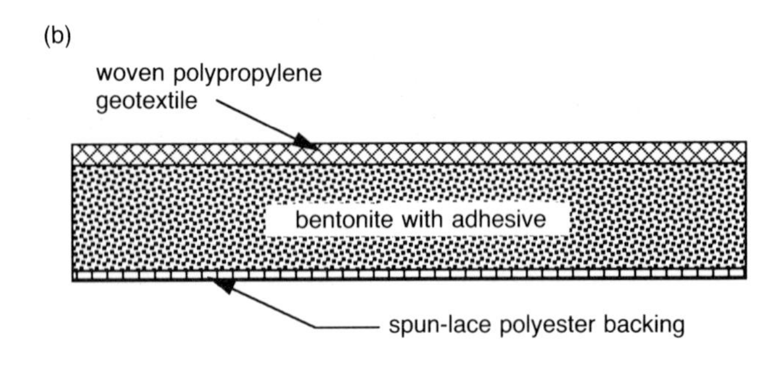

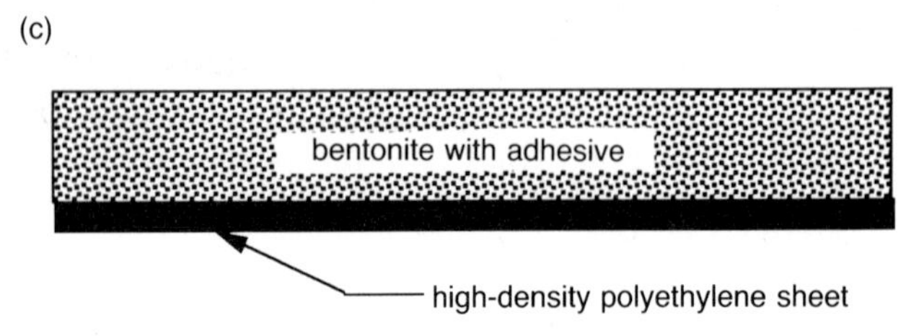

Fig. 7.11 Geosynthetic clay liners: (a) Bentofix Ⓡ and Bentomat Ⓡ; (b) Claymax Ⓡ; (c) Gundseal.

1. bentonite sandwiched between two geotextiles (Bentofix®, Bentomat®, and Claymax®); and
2. bentonite mixed with an adhesive and glued to a geomembrane (Gundseal).

All the GCLs contain approximately 5 kg/m^2 of bentonite. The materials are manufactured in panels with widths of approximately 4–5 m and lengths of 25–60 m.

The panels are placed on rolls at the factory, stored, shipped to the construction site, and unrolled in their final location. All four GCL's are said to be self-sealing at overlaps between panels (Fig. 7.12): when

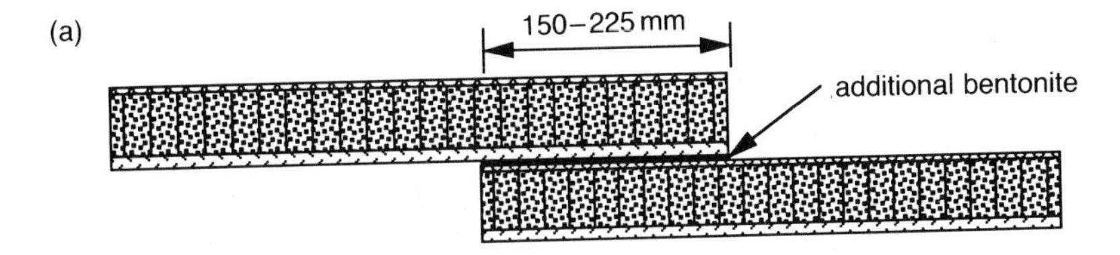

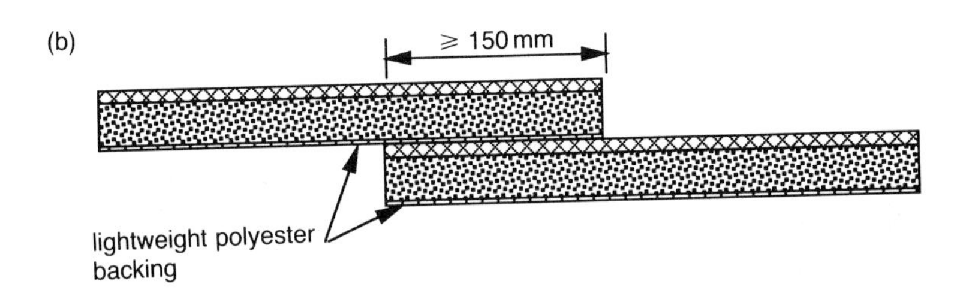

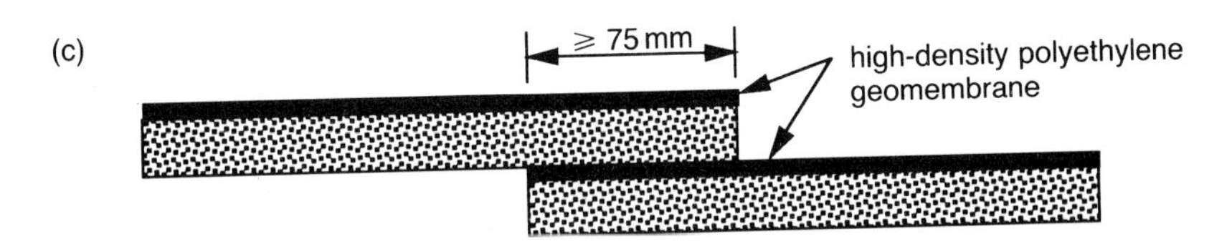

Fig. 7.12 Overlapped geosythetic clay liners: (a) Bentofix ® and Bentomat ® seam; (b) Claymax ® seam; (c) Gundseal seam.

water hydrates the clay in the GCL, the clay swells and automatically seals the overlap. Because no mechanical seaming of joints between panels is needed, GCLs can be installed very rapidly.

After the GCL is placed, it must be covered immediately. If the GCL becomes wet, e.g., from a rainstorm, the clay swells in an uneven manner and the material generally may not properly self-seam. Thus, the GCL cannot be installed when precipitation is threatening, and the GCL cannot be left exposed.

Bentofix® is manufactured in Germany and Canada by Naue-Fasertechnik. Sodium-activated bentonite (German product) is placed between two thick, nonwoven, high density polyethylene (HDPE) geotextiles and then the geotextiles are needlepunched together. The lower and upper geotextiles weigh 800 g/m^2 and 400 g/m^2, respectively. The purpose of needlepunching the geotextiles together is twofold:

1. to hold the GCL together during handling and deployment; and
2. to provide increased in-plane shear strength after deployment.

In the field, the Bentofix® sheets are unrolled and overlapped. Additional granular bentonite is placed between overlapped panels to make the material self-sealing at overlaps. Further information is available from Scheu *et al.* (1990).

Bentomat® is manufactured by American Colloid Company in Villa Rica, Georgia. Several geotextiles and grades of bentonite can be used in making Bentomat®. The usual configuration consists of regular or contaminant-resistant sodium bentonite sandwiched between two geotextiles (each 170 g/m^2) that are needlepunched together. Conceptually, the product is similar to Bentofix®, but Bentomat® is made with thinner geotextiles and a natural, sodium bentonite.

At overlapped sections of Bentomat®, 0.4 kg/m of additional bentonite is placed between the GCL panels to assist in self-sealing upon hydration.

Claymax® is manufactured by the James Clem Corporation in Fairmont, Georgia. With this GCL, bentonite is mixed with a water-soluble adhesive and sandwiched between two geotextiles. The upper geotextile usually consists of a 119 g/m^2 woven polypropylene needle-punched with a 17 g/m^2 nylon fabric to form a 136 g/m^2 primary geotextile for the GCL. The other geotextile is a 25 g/m^2 open-weave, spun-lace polyester backing. The adhesive mixed with the bentonite serves to hold the GCL together during manufacture and installation. After Claymax® has been installed, the adhesive no longer serves a purpose; hence, the solubility of the adhesive is insignificant in terms of long-term performance.

Claymax® panels are overlapped in the field, but no extra bentonite is required; the bentonite is said to ooze through the openings in the geotextiles (especially the open-weave polyester backing) to self-seal the overlapped panels when the bentonite is hydrated.

Further information about Claymax® may be obtained from Schubert (1987), Shan (1990), and Shan and Daniel (1991).

Gundseal is manufactured by Gundle Lining Systems in Spearfish, South Dakota. The material consists of sodium bentonite that is mixed with an adhesive and attached to a high density polyethylene (HDPE) geomembrane through a calendaring process. The geomembrane is usually a 0.5 mm smooth HDPE sheet, but textured geomembranes and very low density polyethylene (VLDPE) geomembranes can be used.

Gundseal is unrolled in the field with the HDPE facing either up or down. Overlapped areas are said to be self-sealing at the bentonite/polyethylene contact; no mechanical seaming is necessary, although the polyethylene sheets could be welded together, if desired.

If the bentonite is facing downward, Gundseal serves as a composite geomembrane/clay liner with the geomembrane on the top. If one wished to place a conventional geomembrane on top of Gundseal to form a composite liner, the bentonite in the Gundseal would face upward as shown in Fig. 7.1; the composite liner would consist of a separate geomembrane underlain by the bentonite component of Gundseal underlain by the polyethylene geomembrane component of Gundseal.

7.4.2 Engineering properties

Hydraulic conductivity

Various organizations have measured the hydraulic conductivity of GCLs. The tests have been performed with flexible-wall permeameters over a range in effective stress. The tests on Gundseal were performed by punching holes in the geomembrane (otherwise the geomembrane would restrict cross-plane flow). Figure 7.13 summarizes the available hydraulic conductivity data. Information about the tests and sources of data may be found in Daniel and Estornell (1990), Scheu *et al.* (1990), and Estornell (1991).

The hydraulic conductivity to water varies between approximately 1×10^{-10} and 1×10^{-8} cm/s, depending on compressive stress. There are some differences in hydraulic conductivity between the various GCLs, but those differences are difficult to sort out given the very

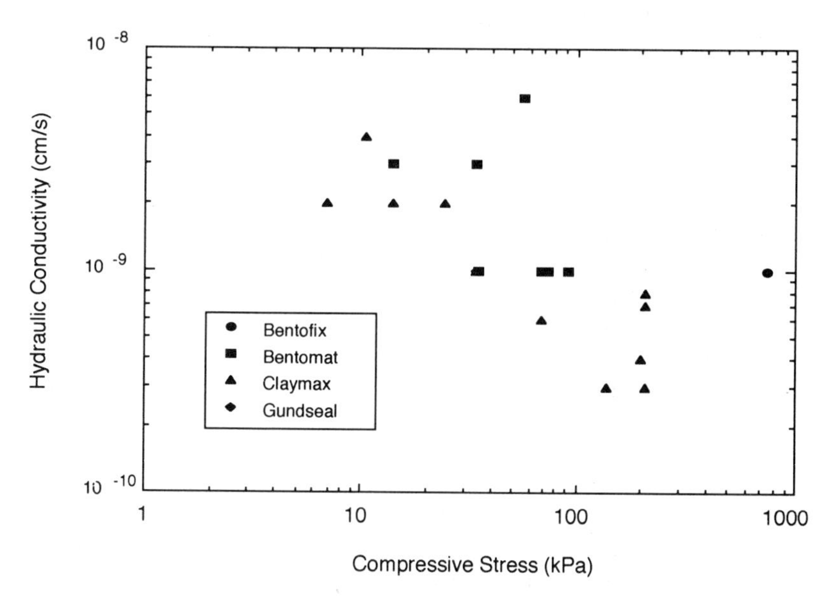

Fig. 7.13 Data on hydraulic conductivity of geosynthetic clay liners.

limited data base for some of the materials and differences in testing procedures employed to measure hydraulic conductivity. The engineer who compares the hydraulic conductivity of one GCL to that of another GCL should be careful to compare hydraulic conductivities determined at the same compressive stress and similar test conditions.

Estornell (1991) built several tanks to evaluate the hydraulic integrity of overlapped seams. The tanks measured 1.2 m in width by 2.4 m in length. Samples of three GCLs (Bentomat®, Claymax®, and Gundseal) were placed on a drainage layer on the bottom of the tank and then covered with 300–600 mm of gravel. The GCLs were flooded with water to a depth of 300–600 mm, and the flow rate in the drainage layer located beneath the GCL was monitored. Tests were performed with control samples containing no overlapped seam and with overlapped samples. Estornell found that the fluxes through the control samples and the overlapped samples were consistently low and were about the same. The overlapped sections did in fact self-seal under in these large-scale, controlled tests.

Shan and Daniel (1991) describe tests in which one geosynthetic clay liner was permeated with a variety of chemicals. Shan and Daniel found that the liner maintained low hydraulic conductivity to a broad range of chemicals when the bentonite was fully hydrated with fresh water prior to introduction of the chemical. However, when the dry GCL was permeated directly with an organic chemical, the bentonite did not hydrate, did not swell, and did not attain a low hydraulic conductivity. The designer should be careful to utilize the results of hydraulic conductivity tests performed under the most critical conditions of hydration that can be expected in the field.

Shan and Daniel (1991) also describe tests aimed at documenting the ability of one GCL to self heal when damaged by puncture, desiccation, or freeze/thaw. Shan and Daniel found that if the dry material is punctured, the bentonite will swell when hydrated and fill small punctures. If the GCL is hydrated and dried, severe desiccation cracks can form. However, Shan and Daniel found that when the GCL was rewetted, the bentonite swelled, and the hydraulic conductivity returned to the original, low value for the undamaged material. Several cycles of wet/dry did not cause a permanent change in the hydraulic conductivity of the hydrated material under controlled laboratory test conditions. Similarly, freeze/thaw cycles did not cause an increase in hydraulic conductivity of one GCL investigated. The very high swelling capacity of bentonite was presumably responsible for the self healing capability that was observed.

One topic of interest concerning geosynthetic clay liners is the behavior of composite geomembrane/geosynthetic clay liner. The question is the following: if a geomembrane is placed on a geosynthetic clay liner with a geotextile separating the geomembrane from the bentonite (Fig. 7.14a), will the highly transmissive geotextile transmit water laterally (in which case the composite action is less than ideal), or will bentonite seal the geotextile (Fig. 7.14b) such that lateral flow is minimal? Estornell (1991) used the large tanks mentioned earlier to study this issue for two GCLs that had geotextiles on both the upper and lower surface and one GCL that had no geotextile on one surface. Estornell placed a geomembrane that contained several punctures directly on the GCLs and found that lateral wicking of water through the geotextile in two of the GCLs was significant. For the GCL that did not contain a geotextile on one surface of the material, no lateral flow at the geomembrane/bentonite interface was noted (in fact, the bentonite was only wetted to a distance of 50–75 mm from the puncture in the geomembrane after about three months of wetting). These tests illustrate that composite action with a geomembrane may be affected

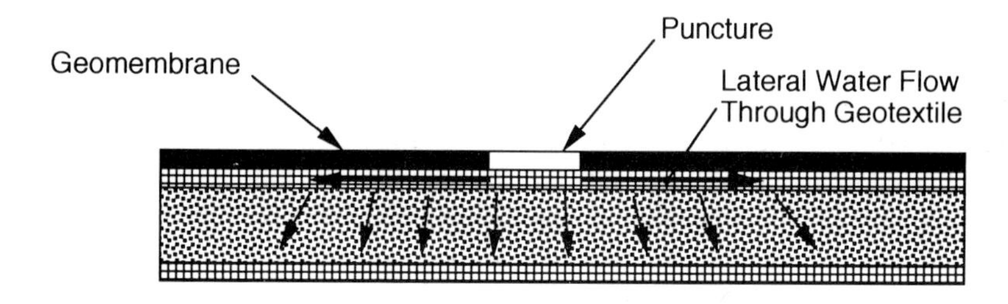

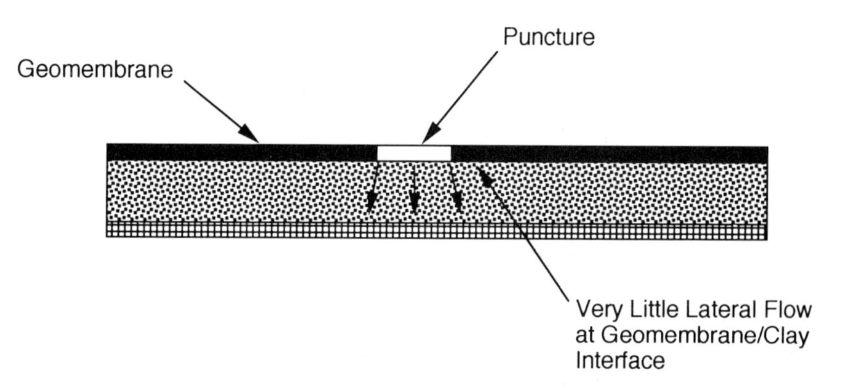

Fig. 7.14 (a) Poor and (b) good composite action between geomembrane and geosynthetic clay liner.

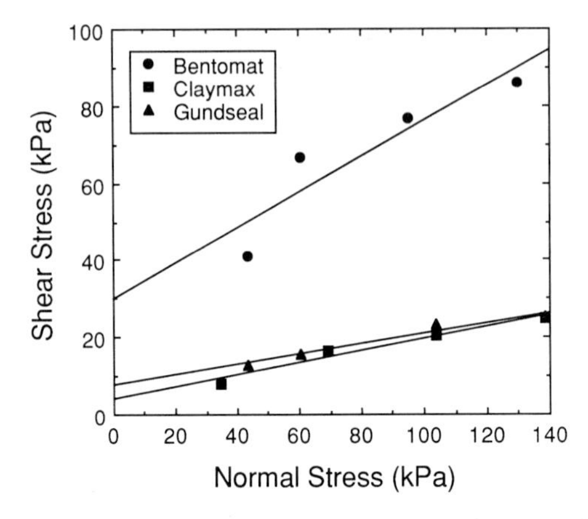

Fig. 7.15 Drained direct shear data on fully hydrated geosynthetic clay liners.

by a geotextile that separates the geomembrane from the clay. Further study of this phenomenon is in progress, and at least one manufacturer is attempting to add agents to the geotextile in contact with a geomembrane that will reduce the in-plane transmissivity of the geotextile. Also, the reader should realize the effectiveness of the composite action between a geomembrane and a GCL depends on overburden stress; the higher the overburden stress, the less the lateral flow along the interface.

Shear strength

Fully-drained direct shear tests have been performed on Bentomat®, Claymax®, and Gundseal. The tests were performed on 60-mm-diameter samples that were subjected to a normal stress, hydrated, and then sheared very slowly. Separate consolidation tests were performed to obtain the parameters needed to determine the time to failure required to ensure fully drained conditions. Times to failure were typically 4–8 days, although the tests were continued to residual conditions over a shearing period of about four weeks.

Results of the tests are plotted in Fig. 7.15. The straight lines shown in the figure were determined from linear regression; regression coefficients are given in Table 7.1.

7.4.3 Current applications

To date, there have been two primary applications of GCLs for waste containment applications:

Table 7.1 Regression coefficients

geosynthetic clay liner	*effective cohesion (kPa)*	*effective angle of internal friction (degrees)*
Bentomat®	30	26
Claymax®	4	9
Gundseal	8	8

1. as the clay component of a primary composite liner for double liner systems; and
2. as the clay component of a composite liner in final cover systems for landfills or site remediation projects.

Geosynthetic clay liners have also been used by themselves as a barrier layer in liners and covers, although to a lesser extent. Secondary containment structures have also been lined with GCLs (Bruton, 1991).

7.4.4 Advantages and disadvantages of GCLs

Geosynthetic clay liners are usually viewed as an alternative to a compacted clay liner (Grube, 1991; Grube and Daniel, 1991). Hence, the following discussion is addressed at the relative advantages of GCLs compared to CCLs.

Table 7.2 contrasts the differences between geosynthetic clay liners and compacted clay liners. The main advantages of GCLs are that GCLs can be installed much more quickly, lightweight construction equipment can be used (which is especially important when the clay liner is placed on top of other components of a liner system that might be punctured or damaged by heavy equipment), GCLs are installed dry and are therefore not as vulnerable to damage from desiccation during construction, and dry GCLs do not produce water upon loading (wet, compacted clay liners consolidate when loaded and the consolidation water can be misinterpreted as leakage if the liner is placed on top of a leak detection layer). The main disadvantages of GCLs are a general lack of experience, the vulnerability of a thin GCL to puncture, questionable composite behavior of some GCLs with an adjacent geomembrane, less leachate attenuation capacity than a thick liner, and questions about stability of hydrated bentonite.

GCLs and CCLs both enjoy advantages and disadvantages. The engineer should weigh the pluses and minuses for each project and make a decision about which type of clay liner is most appropriate based on such an evaluation.

Table 7.2 Comparison of geosynthetic clay liners with conventional compacted clay liners

compacted clay liner (CCL)	*geosynthetic clay liner (GCL)*
thick (0.6–1.5 m)	thin (≤10 mm)
field constructed	manufactured
hard to build correctly	easy to build (unroll and place)
impossible to puncture	possible to damage and puncture
constructed with heavy equipment	light construction equipment can be used
often requires test pad at each site	repeated field testing not needed
site-specific data on soils needed	manufactured product; data available
large leachate-attenuation capacity	small leachate-attenuation capacity
relatively long containment time	shorter containment time
large thickness takes up space	little space is taken
cost is highly variable	more predictable cost
soil has low tensile strength	higher tensile strength
can desiccate and crack	can't crack until wetted
difficult to repair	not difficult to repair
vulnerable to freeze/thaw damage	less vulnerable to freeze/thaw damage
performance is highly dependent upon quality of construction	hydraulic properties are less sensitive to construction variabilities
slow construction	much faster construction

REFERENCES

Acar, Y.B., Hamidon, A., Field, S.D. and Scott, L. (1985) The effect of organic fluids on hydraulic conductivity of compacted kaolinite, *Hydraulic Barriers in Soil and Rock, ASTM STP 874*, (eds A.I. Johnson, R.K. Frobel, N.J. Cavalli and C.B. Pettersson), Am. Soc. for Testing and Materials, Philadelphia, pp. 171–87.

Alther, G.R. (1983) The methylene blue test for bentonite liner quality control, *Geotech. Testing J.*, **6**(3), 128–32.

Alther, G.R. (1987) The qualifications of bentonite as a soil sealant, *Eng. Geol.*, **23**, 177–91.

Anderson, D.C. (1982) Does landfill leachate make clay liners more permeable?, *Civ. Eng.*, **52**(9), 66–9.

Benson, C.B. and Daniel, D.E. (1990) The influence of clods on the hydraulic conductivity of a compacted clay, *J. Geotech. Eng.*, **116**(8), 1231–48.

Bowders, J.J., Daniel, D.E., Broderick, G.P. and Liljestrand, H.M. (1986) Methods for testing the compatibility of clay liners with landfill leachate, *Haz. and Ind. Solid Waste Testing, 4th Symp., ASTM STP 886*, (eds J.K. Petros, W.J. Lacey and R.A. Conway), Am. Soc. for Testing and Materials, Philadelphia, pp. 233–50.

Bowders, J.J. and Daniel, D.E. (1987) Hydraulic conductivity of compacted clays to dilute organic chemicals, *J. Geotech. Eng.*, **113**(12), 1432–48.

Boynton, S.S. and Daniel, D.E. (1985) Hydraulic conductivity tests on compacted clay, *J. Geotech. Eng.*, **111**(4), 465–78.

Bradbury, K.R. and Muldoon, M.A. (1990) Hydraulic conductivity determinations in unlithified glacial and fluvial materials, in *Ground Water and Vadose Zone Monitoring, STP 1053,* (eds D.M. Nielsen and A.I. Johnson), Am. Soc. for Testing and Materials, Philadelphia, pp. 138–51.

Broderick, G.P. and Daniel, D.E. (1990) Stabilizing compacted clay against chemical attack, *J. Geotech. Eng.*, **116**(10), 1549–67.

Bruton, D. (1991) Bentonite mats meet secondary containment standards, *Geotech. Fabrics Rep.*, **9**(4), 26–7.

Budhu, M., Giese, R.F., Campbell, G. and Baumgrass, L. (1991) The permeability of soils with organic fluids, *Canadian Geotech. J.*, **28**(1), 140–7.

Cartwright, K. and Krapac, I.G. (1990) Construction and performance of a long-term earthen liner experiment, in *Waste Containment Systems: Construction, Regulation, and Performance,* (ed R. Bonaparte), ASCE, New York, pp. 135–55.

Chamberlain, E.J. and Gow, A.J. (1979) Effects of freezing and thawing on the permeability and structure of soil, *Eng. Geol.*, **13**, 73–92.

Chamberlain, E.J., Iskander, I. and Hunsicker, S.E. (1990) Effect of freeze-thaw cycles on the permeability and macrostructure of soils, *Proc., Int. Symp. on Frozen Soil Impacts on Agricultural, Range, and Forest Lands, CRREL Special Report 90–1,* U.S. Army Corps of Engineers Cold Regions Research & Engineering Laboratory, Hanover, Ne Hampshire, 145–55.

Chapuis, R.P. (1989) Shape factors for permeability tests in boreholes and piezometers, *Ground Water*, **27**(5), 647–54.

Chapuis, R.P. (1990) Soil–bentonite liners: predicting permeability from laboratory tests, *Canadian Geotech. J.*, **27**(1), 47–57.

Daniel, D.E. (1984) Predicting hydraulic conductivity of clay liners, *J. Geotech. Eng.*, **110**(2), 285–300.

Daniel, D.E. (1989) *In situ* hydraulic conductivity tests for compacted clay, *J. Geotech. Eng.*, **115**(9), 1205–26.

Daniel, D.E. (1990) Summary review of construction quality control for compacted soil liners, in *Waste Containment Systems: Construction, Regulation, and Performance,* (ed R. Bonaparte), Am. Soc. of Civil Engineers, New York, pp. 175–89.

Daniel, D.E. and Benson, C.H. (1990) Water content-density criteria for compacted soil liners, *J. Geotech. Eng.*, **116**(12), 1811–30.

Daniel, D.E. and Estornell, P.M. (1990) *Compilation of information on alternative barriers for liner and cover systems,* USEPA, EPA 600/2-91/002, Cincinnati, Ohio.

Daniel, D.E., Trautwein, S.J., Boynton, S.S. and Foreman, D.E. (1984) Permeability testing with flexible-wall permeameters, *Geotech. Testing J.*, **7**(3), 113–22.

Daniel, D.E., Trautwein, S.J. and McMurtry, D. (1985) A case history of leakage from a surface impoundment, in *Seepage and Leakage from Dams and Impoundments,* (eds R.L. Volpe and W.E. Kelly), Am. Soc. of Civil Engineers, New York, pp. 220–35.

Estornell, P.M. (1991) Bench-scale hydraulic conductivity tests of bentonitic blanket materials for liner and cover systems, M.S. Thesis, Univ. of Texas, Austin, TX.

Fernandez, F. and R.M. Quigley (1985) Hydraulic conductivity of natural clays permeated with simple liquid hydrocarbons, *Canadian Geotech. J.*, **22**(2), 205–14.

Fernandez, F. and Quigley, R.M. (1988) Viscosity and dielectric constant controls on the hydraulic conductivity of clayey soils permeated with water-soluble organics, *Canadian Geotech. J.*, **25**(4), 582–9.

Fernandez, F. and Quigley, R.M. (1991) Controlling the destructive effects of clay – organic liquid interactions by application of effective stress, *Canadian Geotech. J.*, **28**(3), 388–98.

Fireman, M. (1944) Permeability measurements on disturbed soil sample, *Soil Sc.*, **58**, 337–55.

Foreman, D.E. and Daniel, D.E. (1986) Permeation of compacted clay with organic chemicals, *J. Geotech. Eng.*, **112**(7), 669–81.

Gordon, M.E., Huebner, P.M. and Miazga, T.J. (1989) Hydraulic conductivity of three landfill clay liners, *J. Geotech. Eng.*, **15**(8), 1148–60.

Grim, R.E. (1953) *Clay Mineralogy*, McGraw-Hill, New York.

Grube, W.E. Jr. (1991) Soil barrier alternatives, in *Remedial Action, Treatment, and Disposal of Hazardous Waste*, USEPA, Cincinnati, Ohio, EPA/600/9-91/002, pp. 436–44.

Grube, W.E., Jr. and Daniel D.E. (1991) Alternative barrier technology for landfill liner and cover systems, presented at the 84th Ann. Meeting & Exhib. of the Air & Waste Management Assoc., Vancouver, British Columbia, Paper 91-5.9.

Johnson, G.M., Crumbley, W.S. and Boutwell, G.P. (1990) Field verification of clay liner hydraulic conductivity, in *Waste Containment Systems: Construction, Regulation, and Performance*, (ed R. Bonaparte), ASCE, New York, pp. 226–45.

Keller, C.K., van der Kamp, G. and Cherry, J.A. (1986) Fracture permeability and groundwater flow in clayey till near Saskatoon, Saskatchewan, *Canadian Geotech. J.*, **23**, 229–40.

Kim, W.H. and Daniel, D.E. (1992) Effects of freezing on the hydraulic conductivity of a compacted clay, *J. Geotech. Eng.*, **118**(7).

Kleppe, J.H. and Olson, R.E. (1985) Desiccation cracking of soil barriers, in *Hydraulic Barriers in Soil and Rock, ASTM STP 874*, Am. Soc. for Testing and Materials, Philadelphia, pp. 263–75.

McNeal, B.L. and Coleman, N.T. (1966) The effect of solution of soil hydraulic conductivity, *Proc., Soil Sc. Soc. of Amer.*, **30**(3), 308–12.

Mitchell, J.K. (1976) *Fundamentals of Soil Behavior*, John Wiley & Sons, New York.

Mitchell, J.K., Hooper, D.R. and Campanella, R.G. (1965) Permeability of compacted clay, *J. Soil Mech. and Foundations Div.*, ASCE, **91**(SM4), 41–65.

Nasiatka, D.M., Shepherd, T.A. and Nelson, J.D. (1981) Clay liner permeability in low pH environments in *Proc., Symp. on Uranium Mill Tailings Management*, Colorado State Univ., Fort Collins, Colorado, pp. 627–45.

Olson, R.E. and Daniel, D.E. (1981) Measurement of the hydraulic conductivity of fine-grained soils, in *Permeability and Ground Water Contaminant Transport, ASTM STP 746*, (eds T.F. Zimmie and C.O. Riggs), Am. Soc. for Testing and Materials, Philadelphia, pp. 18–64.

Peterson, S.R. and Gee, G.W. (1986) Interactions between acidic solutions and clay liners: permeability and neutralization, in *Hydraulic Barriers in Soil and Rock, ASTM STP 874*, Am. Soc. for Testing and Materials, Philadelphia, pp. 229–45.

Reades, D.W., Lahti, L.R., Quigley, R.M. and Bacopoulos, A. (1990) Detailed case history of clay liner performance, in *Waste Containment Systems: Construction, Regulation, and Performance*, (ed R. Bonaparte), ASCE, New York, pp. 156–74.

Sai, J.O. and Anderson, D.C. (1990) Field hydraulic conductivity test for compacted soil liners, *Geotech. Testing J.*, **13**(3), 215–25.

Scheu, C., Johannben, K. and Saathoff, F. (1990) Non-woven bentonite fabrics

– a new fibre reinforced mineral liner system, in *Geotextiles, Geomembranes and Related Products*, (ed D. Hoet), Balkema Publishing, Rotterdam, pp. 467–72.

Schubert, W.R. (1987) Bentonite matting in composite lining systems, in *Geotechnical Practice for Waste Disposal '87*, (ed R.D. Woods), ASCE, New York, pp. 784–96.

Shackelford, C.D. and Javed, F. (1991) Large-scale laboratory permeability testing of a compacted clay soil, *Geotech. Testing J.*, **14**(2), 171–91.

Shakoor, A. and Cook, B.D. (1990) The effect of stone content, size, and shape on the engineering properties of a compacted silty clay, *Bul. Assoc. of Eng. Geol.*, **XXVII**(2), 245–53.

Shan, H.Y. (1990) Laboratory tests on bentonitic blanket, M.S. Thesis, Univ. of Texas, Austin, TX.

Shan, H.Y. and Daniel, D.E. (1991) Results of laboratory tests on a geotextile/ bentonite liner material, in *Geosynthetics 91*, Industrial Fabrics Assoc. Int., St. Paul, Minnesota, **2**, pp. 517–35.

Shelley, T.L. (1991) Effect of gravel on hydraulic conductivity of compacted soil liners, M.S. Thesis, Univ. of Texas, Austin, Texas.

Zimmie, T.F. and La Plante, C. (1990) The effects of freeze/thaw cycles on the permeability of a fine-grained soil, in *Proc., 22nd Mid-Atlantic Ind. Waste Conf.*, Drexel Univ., Philadelphia.

CHAPTER 8

Geomembrane liners

Robert M. Koerner

As defined by the American Society for Testing and Materials (ASTM), geomembranes are 'very low permeability synthetic membrane liners or barriers used with any geotechnical engineering related material so as to control fluid migration in a man-made project, structure or system'. They are used exclusively as liquid or vapor barriers and have wide application in broad areas of environmental and transportation engineering practice, as well as in geotechnical engineering. Within the environmental applications area is the containment of solid waste materials which is the focus of this book. Geomembranes, which are the focus of this particular chapter, form an essential part of the overall *liner system*.

The outline of the chapter will be the following sections.

8.1 Historical perspective of polymeric liners
8.2 Current use in waste disposal practice
8.3 Geomembrane types and manufacture
8.4 Test methods and properties
8.5 Design models and examples
8.6 Construction quality control and assurance
8.7 Durability and other concerns
8.8 Future trends and growth

8.1 HISTORICAL PERSPECTIVE OF POLYMERIC LINERS

The genesis of modern liners, which in the context of geosynthetics are called geomembranes, is intrinsically tied to the development and growth of the polymer industry. According to Staff (1984) the earliest type of thin prefabricated polymer sheets placed on a prepared soil

Geotechnical Practice for Waste Disposal.
Edited by David E. Daniel.
Published in 1993 by Chapman & Hall, London. ISBN 0 412 35170 6

base as currently practiced, was the use of polyvinyl chloride as swimming pool liners in the early 1930s. Bell and Yoder (1957) describe transportation related trials that were practiced by the Bavarian Highway Department using low density polyethylene seepage barriers in the late 1930s. For large fabricated sheets, however, it seems as though the synthetic thermoset materials made the first significant impact. Potable water reservoirs sealed with butyl rubber liners are reported by Staff (1969) in the late 1940s. A testing and design oriented approach toward the lining of water canals was undertaken by the US Bureau of Reclamation in the 1950s (Hickey, 1957). The research was performed by Luritzen (1957) who worked closely with manufacturers and resin suppliers. A bevy of worldwide activity using a wide range of polymeric material ensued. As seen in Fig. 8.1, polyvinyl chloride canal liner installations were made in Canada, Russia, Taiwan, and in Europe throughout the 1960s and 1970s. Chlorosulfonated polyethylene was developed in the USA and Europe and created a major impact. Polyethylene liners were developed in West Germany and spread to all of Europe, Africa, Australia and North America. South Africa was very much involved in polyethylene liner development which it exported and then further developed in North America.

Today, polymeric geomembranes are indeed worldwide in both their availability and applicability. They are truly a subset of geosynthetics

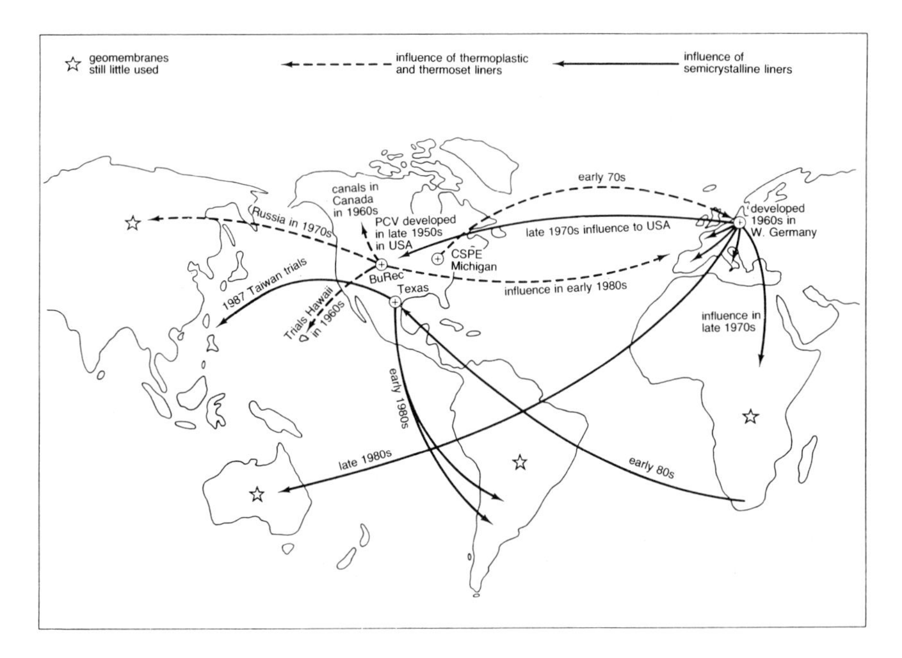

Fig. 8.1 Initial development and movement of geomembranes on a worldwide basis, after Koerner (1990a).

and a major category in their own right. In the applications of interest in this book, i.e., solid waste landfill liners, they are a primary focus of attention for the liner system containing the waste. Note that the liner system referred to is often a composite liner consisting of a geomembrane placed in intimate contact with a clay soil of the type described in Chapter 7 or with a geosynthetic clay liner of the type described in Chapter 18. In such cases that other geosynthetics may be used for drainage, filtration, separation or reinforcement (such as geotextiles, geonets, geogrids and geocomposites) they are auxiliary to the proper and long-term functioning of the liner system.

8.2 CURRENT USE IN WASTE DISPOSAL PRACTICE

Perhaps the most significant research effort leading to the mandated use of geomembranes in waste disposal facilities is that of Anderson and Brown (Anderson *et al.*, 1981). They presented hydraulic conductivity data of different clay soils using various organic solvents which resulted in extremely high values over permeation of the same clays with water. While this data was somewhat flawed by virtue of their experimental procedures (e.g., use of 100% solvents and fixed wall permeameters), it resulted in immediate action by the USEPA which issued the following regulation in July, of 1982.

> Prevention (via geomembranes), rather than minimization (via clay soil liners), of leachate migration similarly produces better environmental results in the case of surface impoundments (landfills) used to dispose of hazardous wastes. A liner that prevents rather than minimizes leachate migration provides added assurance that environmental contamination will not occur.

This regulation set in motion a number of different liner systems which consisted of single geomembranes, redundant geomembranes, composite geomembrane/clay liners, redundant composite liners, and other variations which appeared on almost an annual basis (Table 8.1). As a result of this progression of events, the type of liner system listed as a double geocomposite appears is shown in Fig. 8.2. It is illustrated with a variety of other geosynthetic materials such as geotextiles, geonets, geocomposites and geogrids, since many solid waste facilities have some or all of these polymeric materials contained within them. The drainage and filtration geosynthetic materials will be described in Chapter 9. Also illustrated in Fig. 8.2 is the cover system above the waste with its own composite liner plus a variety of other possible geosynthetic materials. Included in this cover system is the possible use of geosynthetic clay liners. These aspects are covered in detail in Chapters 7 and 18. This chapter, however, will concentrate completely on geomembrane liners. Other references, e.g., Koerner (1990a), are available on the entire gambit of geosynthetic materials.

8.3 GEOMEMBRANE TYPES AND MANUFACTURING

There are three categories of polymers that can be used to make geomembranes; thermoset elastomers, thermoplastics and bituminous types. At this point in time, thermoset elastomers are rarely used in waste containment due primarily to the difficulty in making field seams. An adhesive tape is used which generally does not show good durability in accelerated aging tests. Also, bituminous geomembranes

Table 8.1 Genesis of liner systems used in the USA, after Koerner (1990b)

approx. date	*type of liner systems*	*PLCR system*	*P-GM*	*LDCR system*	*S-GM*
pre-1982	single clay	soil/pipe	clay	none	none
1982	single GM	soil/pipe	GM	none	none
1983	double GM	soil/pipe	GM	soil/pipe	GM
1984	single GM, single GC	soil/pipe	GM	soil/pipe	GM/clay
1985	single GM, single GC	soil/pipe	GM	GN	GM/clay
1987	double GC	soil/pip	GM/clay	GT/GN	GM/clay
1990	double GC	GT/GC or GT/GN/GN	GM/GCL	GT/GN	GM/clay

Abbreviations: GC = geocomposite liner or drain; GCL = geosynthetic clay liner; GM = geomembrane barrier; GN = geonet drain; GT = geotextile filter; LDCR = leak detection, collection and removal system; PLCR = primary leachate collection and removal system; P-CM = primary geomembrane barrier; S-GM = secondary geomembrane barrier.

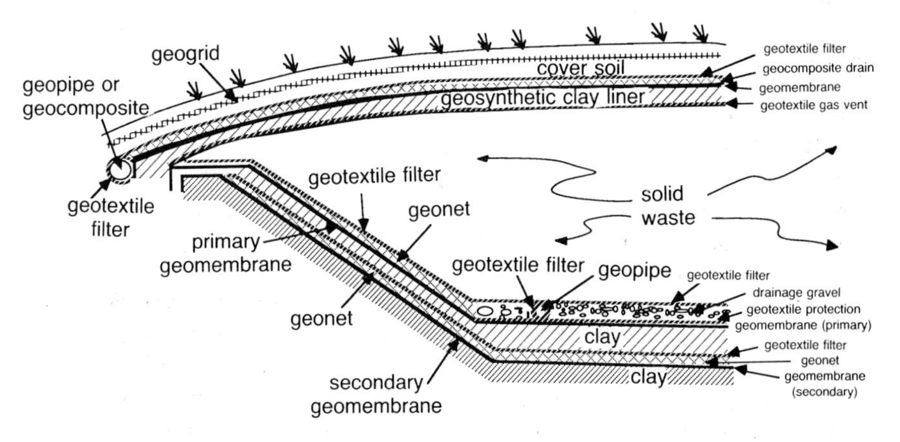

Fig. 8.2 Cross-section of solid waste landfill liner system and closure system illustrating the major use of geosynthetics.

Table 8.2 Typical range of formulations for thermoplastic geomembranes

*type of geomembrane**	*resin (%)*	*plasticizer (%)*	*carbon black and filler (%)*	*additive*† *(%)*
HDPE	96–97	0	2–2.5	0.5–1
VLDPE	96–97	0	2–2.5	0.5–1
CPE or CPE–R	60–75	10–15	20–30	3–5
CSPE–R	45–50	2–5	45–50	2–4
EIA–R	50–65	10–20	20–30	3–5
PVC	45–50	35–40	10–15	3–5

* the 'R' designation refers to fabric reinforcement
† includes antioxidant, processing aids and lubricants

are rarely used in North America, but still see some use in Europe, particularly in France. Thus essentially all of the geomembranes being used today in North America are in the thermoplastic category. By thermoplastic it is meant that the polymer can be melted and with the removal of heat will revert to its original structure (or nearly so). Currently, the most commonly used thermoplastic geomembranes are the following.

- high density polyethylene (HDPE) – either smooth or textured
- very low density polyethylene (VLDPE) – either smooth or textured
- chlorinated polyethylene (CPE) – either nonreinforced or reinforced with a fabric scrim
- chlorosulfonated polyethylene (CSPE) – generally reinforced with a fabric scrim
- ethylene interpolymer alloy (EIA) – always reinforced with a fabric
- polyvinyl chloride (PVC)

Note, however, that within the above generic categories various additives, fillers and/or plasticizers are used with the base resin which gives the geomembrane its name (Table 8.2).

Thus it is seen that the type of geomembrane selected is actually a compound, which the geomembrane formulator makes into the final sheet product. The manufacturing can be done in one of three ways; extrusion, calendering or spread coating (Fig. 8.3).

Both HDPE and VLDPE geomembranes are made by extrusion processes. A continuous screw extruder accepts the formulated compound from a hopper, melts it under vacuum and delivers it through rollers in either a wide flat sheet final product form, or in the form of a circular disc. The circular disc is carried upward in the form of a large cylinder, brought over a nip bar, cut, unfolded and rolled into the final sheet form. This latter technique is called the 'blown film' process. Widths of up to 10.7 m can be made by the various types of extrusion processes.

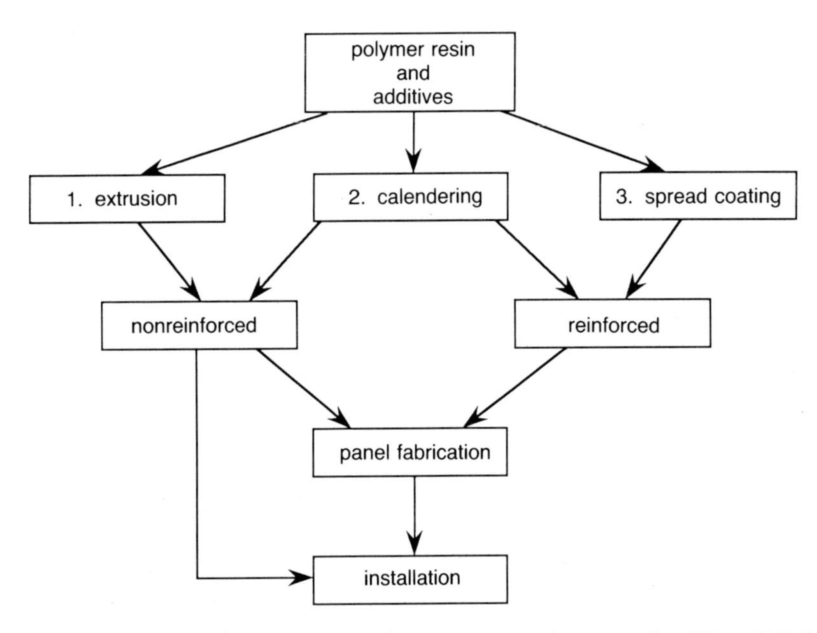

Fig. 8.3 Various processes to manufacture geomembranes, after Haxo (1986).

They are sent directly to the construction site in rolls for final placement and field seaming.

CPE, CPE–R, CSPE–R and PVC are made by the calendering process. The formulated compound is blended in a high shear-flow mixer into a viscous form and fed into the calender which consists of counter rotating sets of rollers forming the individual plys of the geomembrane. Multiple laminates can be built up using an 'inverted-L' calender and often a fabric scrim is embedded between the individual plys. In such cases, the geomembrane is said to be scrim reinforced and carries an 'R' designation. Calendered geomembranes are usually made into 1.8 m wide rolls which are then factory seamed into wider panels, consisting of 5 or 6 roll widths, in a fabrication facility. They are then accordion folded in both their width and their length directions, placed on wooden pallets, and sent to the construction site for final placement and field seaming.

The spread coating process of manufacturing geomembranes is the least widely practiced. It consists of the formulated compound being screeded over a fabric reinforcing substrate. It is often placed on both sides of the fabric, as with EIA–R geomembranes, or can be on one-side only as with certain geomembrane/geotextile composite liners. The finished product may be factory seamed into wider panels or sent directly to the site in rolls for final placement and field seaming.

A closing comment in this section should be made as to quality control of geomembrane manufacturing in a factory environment. As might be expected, manufacturing quality control can be, and usually is, very good. This is to be expected, since factory temperature is generally controlled, harsh weather conditions are eliminated, and the work force is stable and (hopefully) dedicated to a single task.

8.4 TEST METHODS AND PROPERTIES

The individual properties of the polymeric geomembranes just described can be grouped into categories as shown below. Regarding test methods for geomembranes *per se*, an ASTM subcommittee under the Soil and Rock Committee, designated as D18.20, Impermeable Barriers, was formed in 1982. In 1985, however, a special subcommittee of D-35 began considering geomembranes. Today, D-35 has been renamed as 'Committee D35 on Geosynthetics'. Many standards are completed within this committee and still others are in various stages of development. Note that there are many additional standards with geomembrane applicability in the plastics, rubber, roofing, and water canal divisions of ASTM, the National Sanitation Foundation (NSF) and the Geosynthetic Research Institute (GRI). Whenever possible the ASTM D35 test method will be referenced. Other ASTM standards will have the next priority, followed by the other groups that were mentioned. Lastly, references from the open literature will be cited if no standardized test procedures exist. This testing section is subdivided into the following categories, as per Koerner (1990b):

- physical properties
- mechanical properties
- chemical properties
- biological properties
- thermal properties

8.4.1 Physical properties

Included in this grouping are the thickness of the geomembrane, via ASTM D751, the density (or specific gravity of the polymer), via ASTM D792 or D1505, and the mass per unit area, via ASTM D1910. Also the water vapor transmission, via ASTM E96, may be considered in this category as well as solvent vapor transmission (Haxo, 1988).

8.4.2 Mechanical properties

Tensile behavior of geomembranes is obviously important and there are a large number of index tests including ASTM D412, D638, and D882. The test specimens are narrow width or dumbbell shaped, in these particular tests. For a wide width performance test, ASTM D4885 is recommended. In this case the test specimen is 200 mm wide in an attempt to simulate plane strain conditions. Three dimensional, out-of-plane, tension tests are just emerging and GRI GM-4 can be used as an interim guide until a more formalized test method appears. Seam tests in both shear and peel modes are important and NSF No. 54 (1991) should be consulted for details which depend upon the particular type of geomembrane being evaluated.

Other index strength tests for geomembranes include tear tests via ASTM D2263, D1004, D751, D1424, D2261 or D1938, which all cover the general topic. Impact is handled via ASTM D1709, D3029 or D3998. Puncture resistance is measured via ASTM D2582 or D3787, but also Federal Test Method No. 101C is often referenced in the literature.

The important test evaluation of interface friction between geomembranes and other geosynthetics or soil has no prescribed test procedure. Most testing laboratories use a variation of the direct shear soil test which is covered under ASTM D-3080. There are many unanswered questions, however, in using this test for geomembranes such as the following.

- length and width of the shear box
- method of geomembrane fixity to shear box
- saturation conditions of test specimens
- shear rate of conducting the test
- interpretation of an adhesion value when the results indicate a vertical axis intercept

Geomembrane anchorage behavior has been briefly evaluated and can be addressed by GRI Method GM-2.

Stress cracking of semi-crystalline geomembranes (such as HDPE) is conventionally evaluated by the bent strip test, ASTM D1693, but one should also consider a notched constant load test such as GRI Method GM-5. The current recommendation in this test is to require at least 100 hours transition time between the ductile and brittle behavior of the geomembrane test specimens (Halse *et al.*, 1989).

8.4.3 Chemical resistance

The chemical resistance of geomembranes to water is evaluated indirectly using swelling resistance via ASTM D570, ozone resistance via ASTM D1149 and ultraviolet light resistance via ASTM D3334. In this latter regard there are a number of accelerated laboratory testing methods covered under ASTM G23, G26 and G53.

Chemical resistance of geomembranes to leachate is generally performed using EPA Method 9090 (USEPA, 1985). Note, however, that ASTM has a task group on this activity and should have a test method finalized in the very near future. In the evaluation procedure one takes replicate samples of the candidate geomembrane and incubates them in the site specific leachate at a constant temperature. Two sets of tests are necessary, one at 23 °C, the other at 50 °C. The samples are sequentially retrieved at 30, 60, 90 and 120 days. A suite of index strength tests are then performed to note any changes that may have occurred from the original, as-received, material. This requires a subjective assessment of the data as to the degree of chemical resistance of the candidate geomembrane. The assessment procedure is not covered by either the EPA or the ASTM (proposed) test method. Note should also be mentioned that the 50 °C incubation is generally the more aggressive situation and shows the greater variation from the as-received geomembrane material.

8.4.4 Biological resistance

Geomembrane resistance to fungi can be evaluated via ASTM G21 and resistance to bacteria via ASTM G22. Quite clearly there is little concern for biological degradation of the resin (with its very high molecular weight), or of the inert fillers that may be present. Questions may be raised, however, as far as plasticizers (recall Table 8.2), but no authoritative information is available to the author's knowledge.

Burrowing animals are another matter. If an animal with sharp and hard teeth wants to penetrate a geomembrane there is a high likelihood that this may occur. The possible exception being thick HDPE geomembranes. Why an animal would want to do this is another matter entirely. Bio-barriers, consisting of large rocks have been designed to thwart the animals in this regard.

8.4.5 Thermal properties

The behavior of geomembranes to extreme swings in temperature can be evaluated for high temperatures via ASTM D794 and for cold temperatures via simulated testing conditions.

The coefficient of linear expansion/contraction is evaluated by means of ASTM D2102 and D2259 for contraction and D1042 for expansion and dimensional change. It should be mentioned that polyethylene geomembranes have the highest values of thermal expansion/contraction of the currently manufactured products.

8.5 DESIGN MODELS AND EXAMPLES

The essence of design for any engineering material is to compare an allowable (or test) property with a required (or design) property for a resulting factor-of-safety. This is conventionally viewed as follows:

$$\text{factor-of-safety } (FS) = \frac{\text{allowable (test) property}}{\text{required (design) property}} \tag{8.1}$$

Since we have just covered the major available test methods from which the properties are obtained, focus now shifts to various design methods and models. For geomembranes, a number of these models are available. Among the more important ones are the following:

(a) thickness design due to bending
(b) strength design due to subsidence
(c) strength design due to unbalanced shear stresses
(d) anchorage design due to embedment

for which the free-body diagrams for each situation are shown on Figs. 8.4a–d respectively. The requisite design equations are as follows:

for bending thickness

$$t_{\text{reqd}} = \frac{p}{\cos\beta}\,\frac{x}{\sigma_{\text{allow}}}\,(\tan\delta_U + \tan\delta_L) \tag{8.2}$$

$$FS = \frac{t_{\text{act}}}{t_{\text{reqd}}} \tag{8.3}$$

for subsidence strength

$$\sigma_{\text{reqd}} = \frac{2DL^2p}{3t(D^2 + L^2)} \tag{8.4}$$

$$FS = \frac{\sigma_{\text{allow}}}{\sigma_{\text{reqd}}} \tag{8.5}$$

for tensile strength

$$T_{\text{reqd}} = [(c_{aU} - c_{aL}) + p\cos\omega\,(\tan\delta_U - \tan\delta_L)]\,LW \tag{8.6}$$

$$FS = \frac{T_{\text{allow}}}{T_{\text{reqd}}} \tag{8.7}$$

for anchorage design

$$\sigma_{\text{allow}}(t) = p\tan\delta\,(L_{RO}) + 2(K_O\sigma_{v\,\text{ave}})(\tan\delta)(d_{AT}) \tag{8.8}$$

which is solved implicitly for L_{RO} and d_{AT} and a suitable multiplier is then used for the desired factor-of-safety value. The above formulations are derived (along with illustrative problems) in Koerner (1990b) and Koerner and Hwu (1991). The following constants are defined for these equations.

t = geomembrane thickness
p = mobilizing pressure, e.g., the full weight of waste

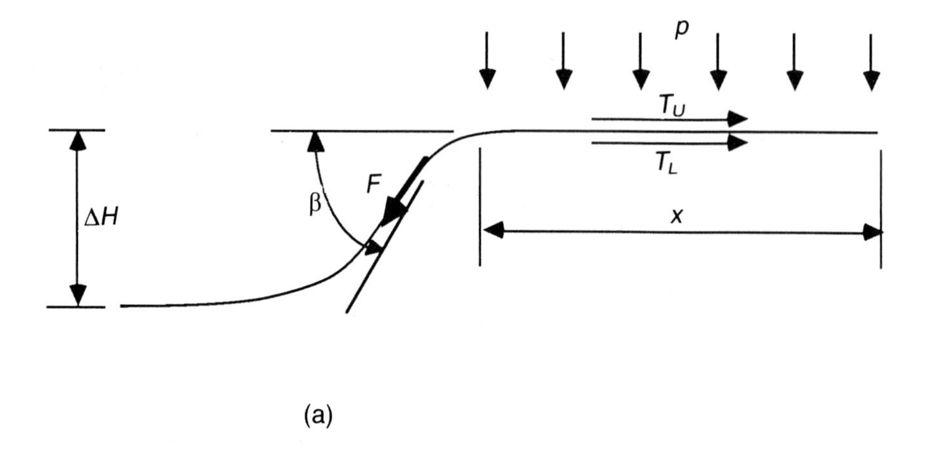

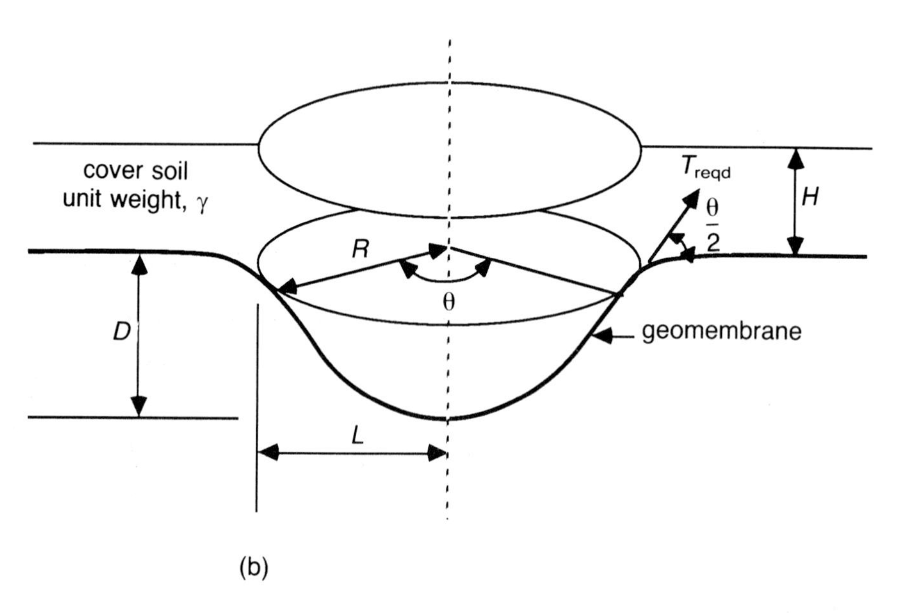

Fig. 8.4 Free-body diagrams for various geomembranes design scenarios: (a) thickness design due to bending; (b) strength design due to subsidence; (c) strength design due to unbalance shear; and (d) anchorage design due to embeddment.

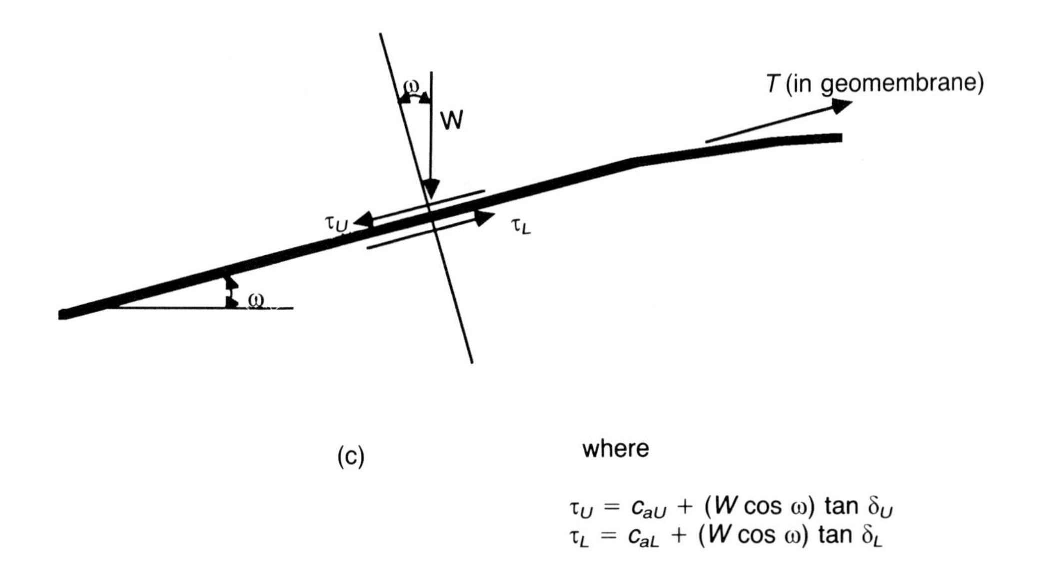

(c)

where

$$\tau_U = c_{aU} + (W \cos \omega) \tan \delta_U$$
$$\tau_L = c_{aL} + (W \cos \omega) \tan \delta_L$$

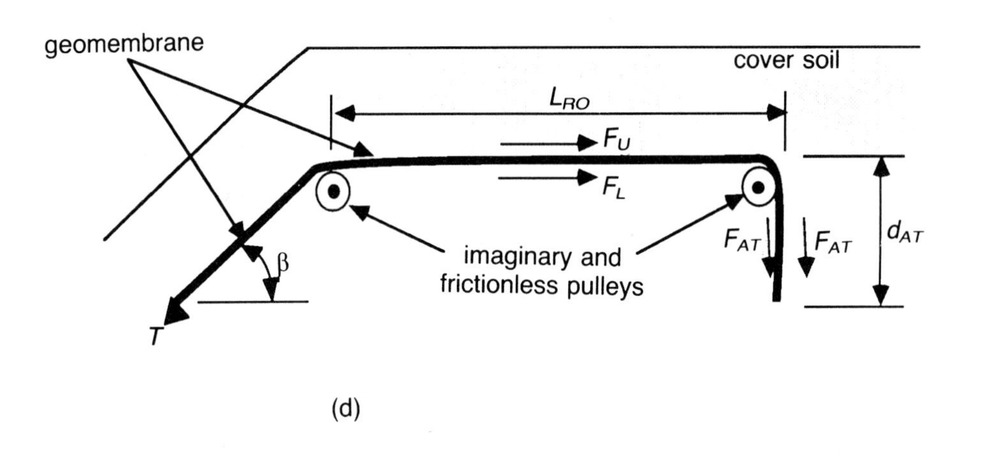

(d)

σ_{allow}	= allowable geomembrane stress; which is yield for HDPE, fabric break for CPE–R, CSPE–R and EIA–R and the stress at 50% strain for CPE, VLDPE and PVC
σ_{reqd}	= required geomembrane stress
T_{allow}	= allowable geomembrane strength; which is yield for HDPE, fabric break for CPE–R, CSPE–R and EIA–R and the strength at 50% strain for CPE, VLDPE and PVC (Note: $T_{\text{allow}} = \sigma_{\text{allow}}t$)
T_{reqd}	= required geomembrane strength
x	= anchorage mobilization distance
β, ω	= various slope angles
δ_U and δ_L	= friction angles (upper and lower) with respect to the geomembrane
c_{aU} and c_{aL}	= adhesion values (upper and lower) with respect to the geomembrane
D, L, W,	= various dimensional values
L_{RO}	= length of runout
d_{AT}	= depth of anchor trench

In performing the above calculations for factor-of-safety it must be emphasized that the test value for allowable properties of thickness, stress, strength, etc., must simulate the *in situ* behavior of the material. In this regard, partial factors-of-safety might be considered for laboratory generated results which do not simulate the anticipated field performance, e.g., for degradation and aging of the geomembrane.

8.6 CONSTRUCTION QUALITY CONTROL AND ASSURANCE (CQC AND CQA)

Construction quality control of geomembranes during manufacturing is an important consideration and most manufacturers and many large owner/operators have excellent guides in this regard. They should be consulted for the many details that go beyond this brief coverage of the subject.

Storage at the manufacturing facility, shipment to the site and storage at the construction site are important considerations. Roll or panel deployment is the first field activity for CQC operations and while much can be said, one aspect is very important, i.e., protection from wind uplift damage. Clearly, large areas of exposed geomembrane are very dangerous since the number of sandbags required to hold geomembranes down under the wind uplift forces becomes huge as wind velocities increase (Wayne and Koerner, 1988). The geomembrane should be covered immediately after field seaming and the acceptance of these seams by the parties involved. This acceptance procedure often goes beyond CQC and involves a separate inspection team, called the construction quality assurance (CQA) auditor. This group is an added level of inspection to see that the construction activities are indeed being undertaken according to the project's plans and specifications.

Regarding the field seaming of geomembranes, there are three major categories; they are the *extrusion welding, thermal fusion or melt bonding* and *solvent/adhesive* seaming methods. Each will be explained along with their specific variations so as to give an overview of field seaming technology, for additional detail see Frobel (1984) and Koerner (1990d).

Extrusion welding is used exclusively on geomembranes made from polyethylene. It is a direct parallel of metallurgical welding in that a ribbon of molten polymer is extruded over the edge of, or in between, the two surfaces to be joined. The surfaces upon which the extrudate is to be placed must be ground to remove oxides and waxes prior to the extrudate being placed. The hot extrudate causes the surfaces of the sheet material to become molten and the entire mass then cools and fuses together. The technique is called extrusion *fillet* welding when the extrudate is placed over the leading edge of the seam, and is called extrusion *flat* welding when the extrudate is placed between the two sheets to be joined. It should be noted that extrusion fillet welding is essentially the only method for seaming polyethylenes that are used for patching and in poorly accessible areas like sump bottoms and around pipes.

There are three seaming methods based on **thermal fusion**, or **melt bonding**, that can be used on both semicrystalline and flexible thermoplastic geomembranes. In each of them, portions of the opposing surfaces are melted without the addition of new polymer. This being the case, temperature, pressure and seaming rate all play important roles in that too much melting weakens the geomembrane and too little melting results in poor seam strength. The **hot air method** makes use of a device consisting of a resistance heater, a blower, and temperature controls to blow hot air between two sheets to melt the opposing surfaces. Usually, temperatures of the 'gun' greater than 250 °C (480 °F) are required. Immediately following the melting of the surfaces, pressure is applied by counter-rotating knurled rollers on the top and bottom of the seamed area. The **hot wedge or hot shoe method** is the widest practiced method for joining polyethylene geomembranes. It consists of an electrically heated resistance element in the shape of a wedge that travels between the two sheets to be sealed. As it melts the surfaces, a shear flow occurs across the upper and lower surfaces of the wedge and then roller pressure is applied. Hot wedge units are automated as far as temperature, amount of pressure applied and travel rate is concerned. A standard hot wedge creates a single uniform width seam, while a dual hot wedge (or 'split' wedge) forms two parallel seams with a uniform unbonded space between them. This space can be used to evaluate seam quality and continuity by pressurizing it with air and monitoring any drop in pressure that may signify a leak in the seam. A newest variation of thermal fusion seaming of geomembranes is the **ultrasonic seaming method**. Here a vibrating knurled horn is passed between the two geomembranes to be bonded. The vibrational energy (about 40 000 Hz) causes melting of the surfaces which when forced together by pressure from rollers results in the completed seam.

Regarding the solvent and/or adhesive seam types; **solvent seams** make use of a liquid solvent squirted from a squeeze bottle between the two geomembrane sheets to be joined. After a few seconds, pressure is applied to make complete contact. As with any of the solvent-seaming processes to be described, a portion of the two adjacent materials to be bonded is truly dissolved into a liquid phase. Too much solvent will weaken the adjoining sheets, and too little solvent will result in a weak seam. Therefore great care is required in providing the proper amount of solvent with respect to the particular type and thickness of geomembrane. Care must also be exercised in allowing the proper amount of time to elapse before contacting the two surfaces, and in applying the proper pressure and duration of rolling. **Bodied solvent seams** are similar to solvent seams except that 5–15% of the parent lining compound is dissolved in the solvent which is then used to make the seam. The purpose of this procedure is to compensate for the lost material while the seam is in a liquid state and to fill in surface dimples or undulations on reinforced liners. The viscous liquid is brushed onto the two opposing surfaces to be bonded. Pressure is necessary, and the use of heat guns or radiant heaters accelerates the curing process. **Solvent adhesive seams** make use of an adherent, within a bodied solvent solution, which is left after dissipation of the solvent. The adherent thus becomes an additional element in the system. Brushing is used to apply the three-part liquid (solvent, parent liner compound, and adhesive) onto the two opposing surfaces. Sufficient pressure must be used and heat guns or radiant heaters are often used to accelerate the curing process. **Contact adhesive seams** have the widest applicability to all geomembrane types but are mainly focused on the thermoset elastomeric geomembranes such as EPDM, butyl, nitrile, etc. The solution is applied to both mating surfaces by brush and then rolled. After reaching the proper degree of tackiness, the two sheets are placed on top of one another, followed by roller pressure. The adhesive forms the bond and is an additional element in the system. Due to the lack of current use of thermoset geomembranes, the technique is not used at this time.

In order to give an overview as to which seaming methods are used

Table 8.3 Most commonly used field seaming methods for various geomembranes

type of seaming method	*types of geomembranes*						
	HDPE	*VLDPE*	*CPE*	*CPE–R*	*CSPE–R*	*EIA–R*	*PVC*
extrusion fillet	P	P	n/a	n/a	n/a	n/a	n/a
extrusion flat	P	n/a	n/a	n/a	n/a	n/a	n/a
hot air	S	P	S	S	S	P	S
hot wedge	P	P	S	S	S	P	S
solvent	n/a	n/a	S	S	S	n/a	P
bodied	n/a	n/a	P	P	P	n/a	P
solvent adhesive	n/a	n/a	S	S	S	n/a	S

Note: P = primary method used; S = secondary method used; n/a = method is 'not applicable'.

for the various thermoplastic geomembranes that are currently being used (Table 8.3).

Upon fabrication of the geomembrane field seams, selective destructive test samples should be taken and from these, shear and peel tests performed. Typical sampling frequency is one per 150–300 m. The sample is often 1 m in length, with $\frac{1}{3}$ being evaluated on the job site by the CQC personnel, $\frac{1}{3}$ being sent to the CQA firm for laboratory testing and the last $\frac{1}{3}$ being archived for future reference, if necessary. The shear and peel tests should be assessed on the basis that the sheet material adjacent to the seam area fails without the seam delaminating. This is called a 'film tear bond' (Haxo, 1988). Additionally, the strength of the seam in shear should be 70–90% of the strength of the percent sheet. For the seam tested in peel, this value is from 50–70% for polyethylene and above 1.7 kN/m for other geomembranes (NSF 54, 1991).

Lastly, all geomembrane field seams in between the locations where destructive samples are removed, should be evaluated by a suitable nondestructive test method. Table 8.4 gives an overview of these test methods, but see references in Koerner (1990c) for additional details. A conference proceedings on the seaming of geosynthetics with twenty papers on geomembrane seaming and inspection is also available, Koerner (1990d).

8.7 DURABILITY AND OTHER CONCERNS

The degradation of polymer materials is known to occur via a number of different phenomena and mechanisms. Table 8.5 gives an overview of the situation where it can be seen that for most cases proper

Table 8.4 Overview of nondestructive seam tests, after Koerner (1990c)

nondestructive test method	*primary user*			*general comments*					
	installation contractor (CQC)	*project design engineer (CQA)*	*third-party auditor (CQA)*	*cost of equipment*	*speed of tests*	*cost of tests*	*type of result*	*recording method*	*operator dependency*
air lance	yes	–	–	$200	fast	nil	yes-no	manual	very high
pick test	yes	–	–	nil	fast	nil	yes-no	manual	very high
electric wire	yes	yes	–	$500	fast	nil	yes-no	manual	high
dual seam (positive pressure)	yes	yes	–	$200	fast	mod.	yes-no	manual	low
vacuum chamber (negative pressure)	yes	yes	–	$1,000	slow	very high	yes-no	manual	very high
ultrasonic pulse echo	–	yes	yes	$5,000	moderate	high	yes-no	automatic	moderate
ultrasonic impedance	–	yes	yes	$7,000	moderate	high	qualitative	automatic	unknown
ultrasonic shadow	–	yes	yes	$5,000	moderate	high	qualitative	automatic	moderate
electric field	yes	yes	yes	$20,000	slow	high	yes-no	manual and automatic	low
acoustic sensing	yes	yes	yes	$1,000	fast	nil	yes-no	manual	moderate

Table 8.5 Degradation phenomena in geomembranes (from a geosynthetics engineering perspective), after Koerner, Halse and Lord (1991)

degradation classification	*degradation mechanism*	*initial change in material*		*subsequent change in material*		*preventative measure*
		*laboratory**	*field†*	*laboratory‡*	*field§*	
ultraviolet	• chain scission • bond breaking	• mol. wt. • stress crack resist.	• color • crazing	• elongation • modulus • strength	• color • cracking	• screening agent • antidegradient • cover the geomembrane
radiation	• chain scission	• mol. wt. • stress crack resist.	• color • crazing	• elongation • modulus • strength	• color • cracking	• cover for β and α-rays • shield for neutrons • reduce dosage for γ rays
chemical	• reaction with structure • reaction with additives	• carbonyl index • IR • mol. wt.	• texture • color • crazing • TGA	• elongation • modulus • strength	• texture • cracking • reactions	• proper resin • proper additives
swelling	• liquid absorption	• TGA	• thickness • color • texture	• thickness • modulus • strength	• thickness • softness	• proper resin • proper manufacturing
extraction	• additive expulsion	• TGA • IR	• texture • color • thickness	• elongation • modulus • thickness	• texture • color	• proper compounds • proper manufacturing

Table 8.5 Degradation phenomena in geomembranes (from a geosynthetics engineering perspective), after Koerner, Halse and Lord (1991)

degradation classification	*degradation mechanism*	*initial change in material*		*subsequent change in material*		*preventative measure*
		*laboratory**	*field†*	*laboratory‡*	*field§*	
delamination	• adhesion loss	• thickness • edge effects	• thickness • edge effects	• thickness • ply adhesion	• layer separation • thickness	• proper manufacturing • protect geomembrane edges
oxidation	• reaction with structure	• IR • carbonyl index	• color • crazing • strength	• elongation • modulus	• color • cracking	• anti-oxidant • cover with soil • cover with liquid
biological	• reactions with additives	• mol. wt. • carbonyl index • IR	• texture • surface film	• elongation • modulus • strength	• texture • surface layer • cracking	• avoid sensitive additives • biocide

* Initial laboratory changes are generally sensed by chemical fingerprinting methods: mol. wt. = molecular weight; IR = infrared spectroscopy; TGA = thermal geometric analysis; OIT = oxidation induction time.
† Initial field changes are generally sensed on a qualitative basis.
‡ Subsequent laboratory changes can be sensed by numerous physical and mechanical tests. Listed in the table are those considered to be the most sensitive parameters.
§ Subsequent field changes are a continuation of the initial changes until physical and mechanical properties begin to visually change.

materials, construction, and timely covering will eliminate many of the potential problems. This, however, is not the case for oxidation degradation which will eventually bring the geomembrane from its normally ductile behavior into a brittle mode. Thus the geomembrane will eventually become sensitive to ground movements, subsidence, excessive stresses, etc. It should be noted that elevated temperature over a normal background level greatly increases the reaction and its negative effects.

Thus it is necessary to provide insight into different predictive methods for assessing lifetime. There are at least three such methods:

1. the **rate process method**, using ductile-to-brittle experimental curves, to extrapolate the response into the site-specific temperatures and minimum acceptable stress levels;
2. **Arrhenius modeling** (again using elevated temperature) obtaining a reaction rate which is assumed to simulate time at the site-specific temperature; and
3. **Hoechst multiparameter approach** which superimposes ductile-to-brittle behavior, stress relaxation behavior and field obtained strain behavior to form the basis of predictive technique.

These methods are discussed in Koerner *et al.* (1991) and also in a recent seminar on the durability and aging of geosynthetics (Koerner, 1989). It is perhaps interesting to comment on the author's personal observation that geomembranes are not easy to place properly, but will generally serve their intended function for time frames far longer than most people suspect.

8.8 FUTURE TRENDS AND GROWTH

Projections into the future status of geomembranes are difficult at best. Furthermore, they are usually biased on the part of the person making the prediction by reason of personal knowledge of certain specific situations. The position from the author's point of view is that of a very strong environmental awareness in the area of groundwater pollution. Solid waste containment (of the type described in this book) is clearly powering the geomembrane market in North America. There is every indication that this 'mind-set' will continue and be spread worldwide. Thus environmental applications will continue to be strong with a long-term extended growth pattern.

With an ever-increasing worldwide population growth, water supplies will become more scarce and require storage reservoir and surface impoundment liners. Similarly, canal liners for the transportation of this water to areas of their primary use is an important application. Recent concerns of potable water reservoir pollution from the atmosphere has ushered in the concept of floating geomembrane covers.

With industrialization on a worldwide basis comes the necessity of canal liners to transport chemicals and waste waters. The application here is among the most demanding of all geomembrane applications.

With continued applications of geomembranes in the above mentioned environmentally related categories will come an awareness on the part of many transportation and geotechnical engineers for the use of geomembranes in many applications where a liquid or vapor barrier is necessary. Certainly rehabilitation of our infrastructure such as remediation of concrete, masonry and soil/rock dams is an area which can use geomembranes of a wide variety of types (Cazzuffi, 1987). With familiarity, comes additional use and even forays into areas where geomembranes have not previously been used.

The flow chart of Fig. 8.5 of potential use and opportunities (along

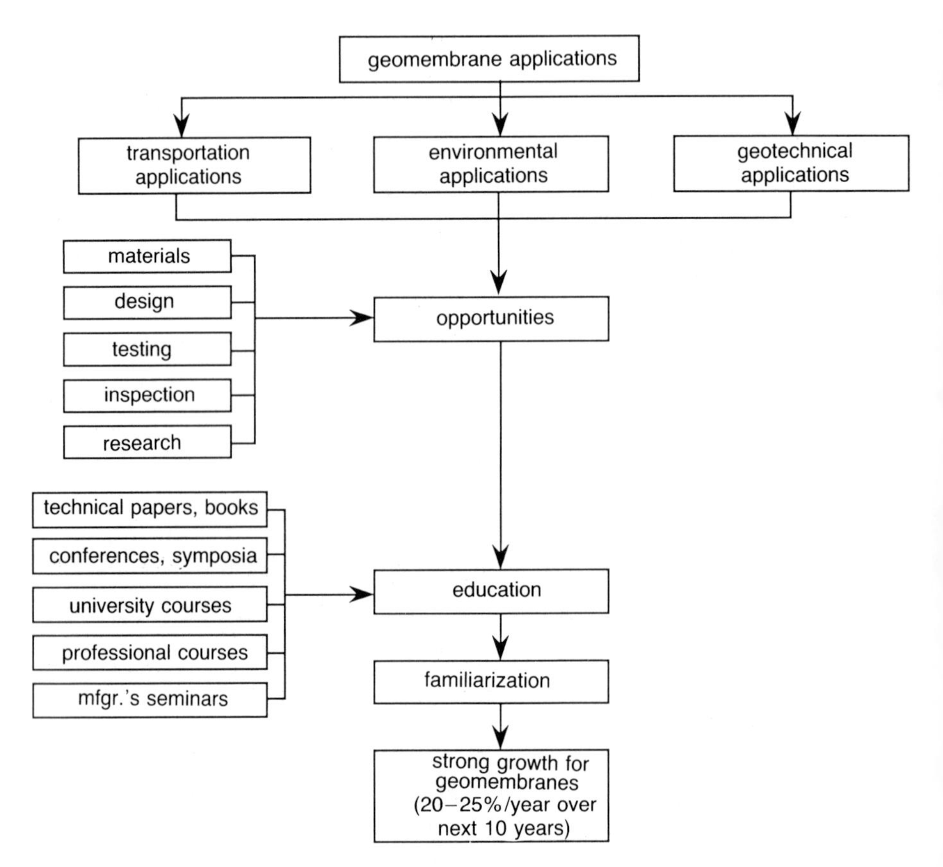

Fig. 8.5 Future growth and trends in the use of geomembranes.

with education) should bring about a strong growth for geomembranes in the near term, e.g., 20–25% over the next 10 years, and, quite possibly, will be continued even beyond that time frame.

REFERENCES

Anderson, D.C., Brown, K.W. and Green, J. (1981) Organic leachate effects on the permeability of clay liners, *Proc. Natl. Conf. on Mgmt. of Uncontrolled Hazardous Waste Substances*, HMCRI, Silver Spring, MD, pp. 223–9.

Bell, J.R. and Yoder, E.J. (1957) Plastic moisture barrier for highway subgrade protection, *Proc. Highway Research Board*, Washington, DC, **36**, 713–35.

Cazzuffi, D. (1987) The use of geomembranes in Italian dams, *Intl. J. Water Power and Dam Construction*, March, 44–52.

Frobel, R.K. (1984) Methods of constructing and evaluating geomembrane seams, *Proc. Intl. Conf. on Geomembranes*, Denver, CO, IFAI, 359–64.

Halse, Y.H., *et al.* (1989) Methods to evaluate stress crack resistance of HDPE FML sheets and seams, *Proc. 1990 EPA Research Seminar*.

Haxo, H.E., Jr. (1986) Quality assurance of geomembranes used as linings for hazardous waste containment, *J. Geotextiles and Geomembranes*, Elsevier Appl. Sci. Publ. Ltd., England, 225–47.

Haxo, H.E. (1988) *Lining of waste containment and other impoundment facilities*, USEPA, 600/2-88/052.

Hickey, M.E. (1957) Evaluation of plastic films as canal lining materials, *US BuRec Lab. Report B-25*.

Koerner, R.M. (ed.). (1989) *Durability and Aging of Geosynthetics*, Elsevier Appl. Sci. Publ., London.

Koerner, R.M. (1990a) Geomembrane overview, significance and background, in *Geomembranes: Identification and Performance Testing*, (eds A.L. Rollin and J.-M. Rigo) RILEM TC-103.

Koerner, R.M. (1990b) *Designing with Geosynthetics, 2nd edn*, Prentice Hall, Englewood Cliffs, NJ.

Koerner, R.M. (1990c) Preservation of the environment via geosynthetic containment systems, *Proc. 4th Intl. Conf. on Geotextiles and Geomembranes*, The Netherlands, June 1990 (reprinted in *Geotech. Fabrics Rep.* Sept/Oct. 1990 and Nov/Dec. 1990 issues).

Koerner, R.M. (ed.). (1990d) *The Seaming of Geosynthetics*, Elsevier Appl. Sci. Publ., London, 1990.

Koerner, R.M., Halse, Y.H. and Lord, A.E., Jr. (1991) Long-term durability and aging of geomembranes, in *Waste Containment Systems: Construction, Regulation and Performance* (ed. R. Bonaparte), American Society of Civil Engineers, New York, pp. 106–34.

Koerner, R.M. and Hwu, B-L. (1991) Stability and Tension Considerations Regarding Cove Soils on Geomembrane Lined Slopes, *J. Geotextiles and Geomembranes*, **10**(4), 335–55.

Luritzen, C.W. (1957) Seepage control with plastic films, in *Irrigation Engineering and Maintenance*, pp. 18–33.

National Sanitation Foundation Standard No. 54, (1991), Ann Arbor, Michigan.

Staff, C.E. (1969) Seepage prevention with impermeable membranes, *Civ. Eng.*, **37**(2).

Staff, C.E. (1984) The foundation and growth of the geomembrane industry in the United States, *Proc. Intl. Conf. on Geomembranes*, Denver, CO, IFAI, 5–8.

USEPA (1985) *Compatibility Test for Wastes and Membrane Liners*, Method 9090, Washington, DC.

Wayne, M.H. and Koerner, R.M. (1988) Effect of wind uplift on liner systems, *Geotech. Fabrics Rep., IFAI*, **6**(4), 26–9.

CHAPTER 9

Collection and removal systems

Robert M. Koerner

The collection of liquids in waste containment systems and their proper removal represents an important element in the successful functioning of these important facilities. Focus in this chapter is on primary and secondary (leak detection) leachate collection systems beneath the waste and on surface water removal in the cover system above the waste. The gas collection system in the closure system will also be addressed. The individual sections are as follows.

9.1 Overview and background
9.2 Hydraulic concepts, definitions and test methods
9.3 Primary leachate collection and removal system
9.4 Secondary leachate collection and removal system (i.e., leak detection)
9.5 Surface water collection and removal system
9.6 Methane gas collection and removal system

9.1 OVERVIEW AND BACKGROUND

The general area of 'liquid management' is the focus of this chapter. This refers to the timely and efficient removal of liquids out of, and away from, the waste containment facility where they can be properly treated and then disposed. The liquid itself comes primarily from rainfall and snowmelt, but some is also contained in the solid waste as it is brought to the landfill. Taken collectively, these liquids combine to form leachate which has a wide variation of characteristics.

After solid waste is placed in the landfill, the leachate flows gravita-

Geotechnical Practice for Waste Disposal.
Edited by David E. Daniel.
Published in 1993 by Chapman & Hall, London. ISBN 0 412 35170 6

tionally downward until it reaches the primary leachate collection system (PLCS) as shown in Fig. 9.1. This layer consists of a drainage material (usually gravel) protected by a filter layer (sand or geotextile). The leachate must gravitationally flow within this drainage material to a sump area where it is removed by a submersible pump within a manhole or large diameter pipe. Taken together, the drain, filter, sump and removal manhole/pipe is called the primary leachate collection and removal system, or PLCRS.

When dealing with solid waste liner systems having double liners, i.e., primary and secondary, a secondary leachate collection system (SLCS) must be included between the two liners (Fig. 9.1). This layer is also called a leak detection layer or witness drain. It consists of either gravel or a geonet when the primary liner is a geomembrane by itself. When the primary liner is a composite liner (geomembrane with clay beneath), the leak detection drainage material must be protected by a filter/separator consisting of sand (for gravel) or a geotextile (for a geonet). The leachate passing through the primary liner, i.e., the 'leaking liquid', flows gravitationally downward to its own sump area where it is collected and removed. Removal is generally by small diameter pipes since, hopefully, the quantities involved are very low, or even nonexistent. Taken together this drain, filter, sump and removal piping is called the secondary leachate collection and removal system, or SLCRS.

Above the waste in the closure system, concern must be focused on a surface water collection system (SWCS) above the cap geomembrane as shown in Fig. 9.1. The drainage material is either gravel or a geocomposite sheet drain. Whatever the choice, the material must be protected by an appropriate filter since cover soil is placed above it. If the drainage media is gravel the filter will be sand or a geotextile, while

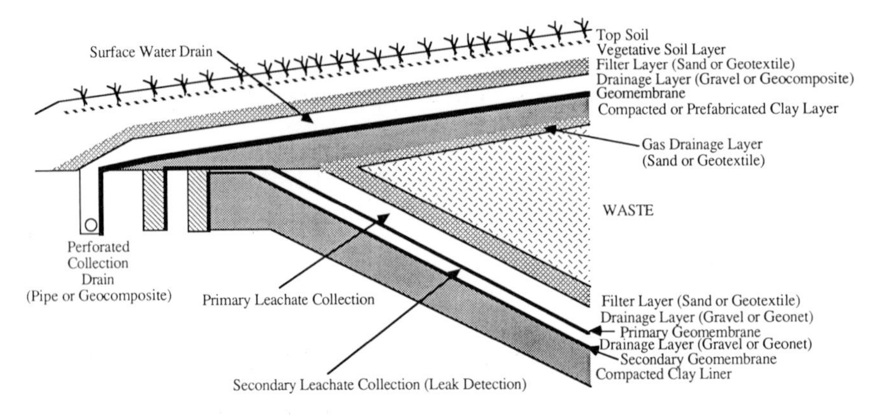

Fig. 9.1 Cross-section of solid waste facility illustrating various drainage layers involved in a liquid management program.

if it is a geocomposite the filter will be a geotextile. The surface water which reaches this drainage layer flows gravitationally to the perimeter of the facility where it is removed by a perforated pipe or a geocomposite edge drain. Taken together the drain, filter and removal system is called the surface water collection and removal system, or SWCRS.

Lastly, one must consider landfill gases which can rise within the waste to the lower side of the cover system and will require a gas drainage layer and venting system. This drainage layer is usually sand, but can also be a thick geotextile. Gases in this drainage layer flow up-gradient beneath the clay/geomembrane barrier and must be removed by a penetration through the cap barrier system in the form of a vent, flare, or gas collection and utilization piping system.

In the above description of the four aspects of liquid/gas collection and removal systems, natural soils were always counterpointed with their geosynthetic alternative. (The different geosynthetic alternatives are shown in Fig. 9.2, i.e., geonets, geocomposite sheet drains and geocomposite edge drains.) This is done in recognition that most regulations allow for alternates to natural soils, if they can be shown to be technically equivalent. Thus design of the geosynthetic alternative is necessary, whereas the natural soil is often mandated by minimum technology guidance (MTG). Some of the practical and economic trade-offs in considering natural soil versus the geosynthetic alternative are given in Table 9.1.

Regarding a cost comparison between natural soils and their much thinner geosynthetic alternatives, it is important to note that with vertical 'air' space in landfills at such a premium it certainly behooves an owner/operator/designer to consider a geosynthetic replacement. For example, at a $10/kN waste tipping fee (the 1992 price in northeast USA), the savings of a 7 mm geonet drain over 30 cm of gravel in air space alone is approximately $2 000 000 for a 20 ha site.

9.2 HYDRAULIC CONCEPTS, DEFINITIONS AND TEST METHODS

The design of the various liquid management drainage and filtration components just described is covered by hydraulic concepts included in the areas of hydraulics and hydraulic systems. Many textbooks used in geotechnical engineering cover the particulars for natural soils in flow through porous media, but relatively few include information on geosynthetics. In this latter regard see Fluet (1987) and Koerner (1990a,b).

It is important to note that with some of the high flow geosynthetic materials, like geonets, sheet drain geocomposites and edge drain

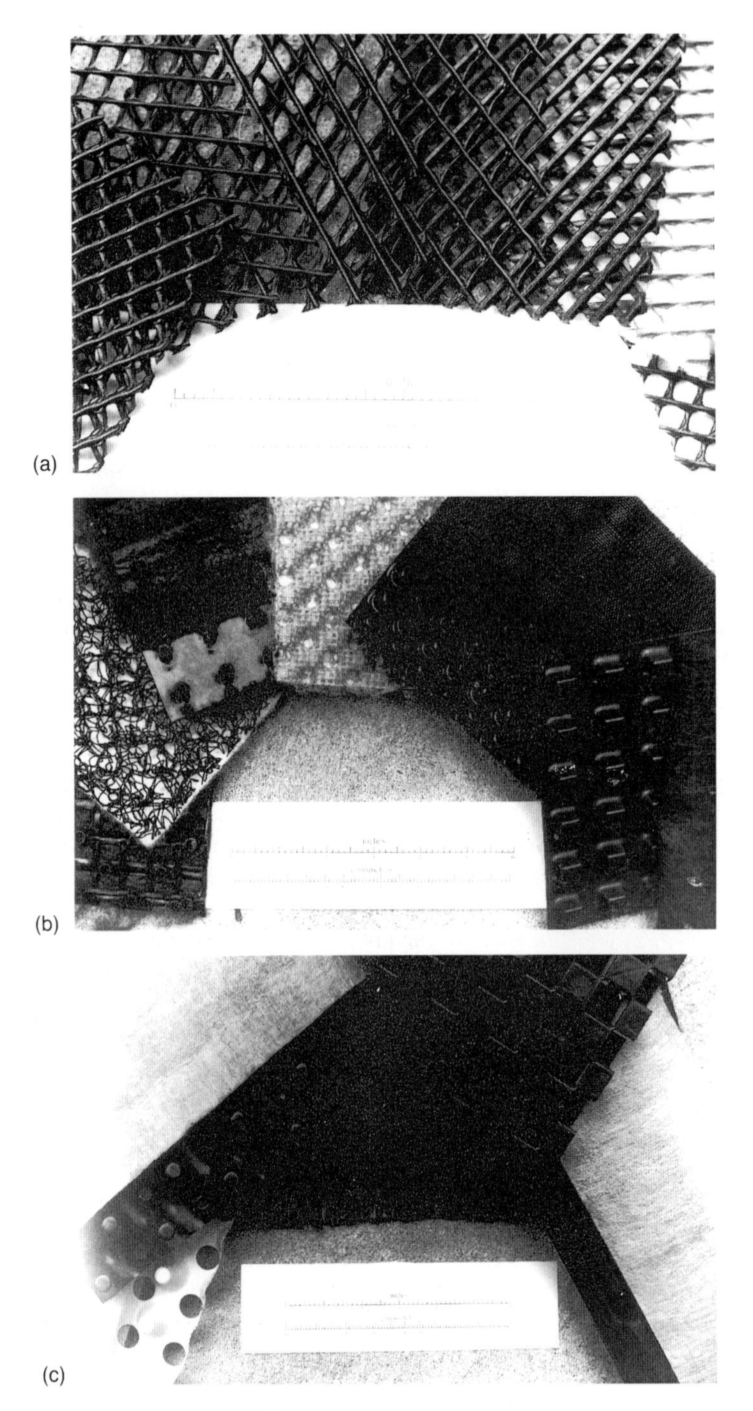

Fig. 9.2 Types of geosynthetic drainage materials: (a) geonets; (b) geocomposite sheet drains; and (c) geocomposite edge drains.

Table 9.1 Selected comments on natural soils *vis-à-vis* geosynthetics used in landfill drainage and filtration applications

type	*advantages*	*disadvantages*
Drainage media		
pipe (perforated)	common usage traditional design rapid transmission not likely clogged	takes vertical space creep deformation needs filter
soil (gravel)	common usage traditional design durable	takes vertical space slow transmission particulate clogging biological clogging needs filter soil moves under load
geonet	saves vertical space rapid transmission not likely clogged does not move	intrusion creep deformation needs geotextile filter new technology
geocomposite (sheet drain or edge drain)	saves vertical space rapid transmission not likely clogged does not move	intrusion creep deformation needs geotextile filter very new technology
Filtration media		
soil (sand)	common usage traditional design durable	takes vertical space particulate clogging biological clogging moves under load
geotextile	saves vertical space easy placement does not move	particulate clogging biological clogging installation survivability new technology

geocomposites (Fig. 9.2), laminar flow conditions do not exist. This being the case one cannot work with Darcy's formula in the conventional sense of considering a permeability coefficient, or hydraulic conductivity, value. In these cases where turbulent flow exists, flow rate at a given hydraulic gradient will have to be explicitly stated.

For the various **drainage** designs, flow is assumed to be within the plane of the material as shown in Fig. 9.3(a). If *laminar* conditions exist as in a soil or geotextile;

$$q = kiA = ki(W \times t) = (kt)iW = (\theta)iW$$
$$Q' = q/W = \theta i \qquad (9.1)$$

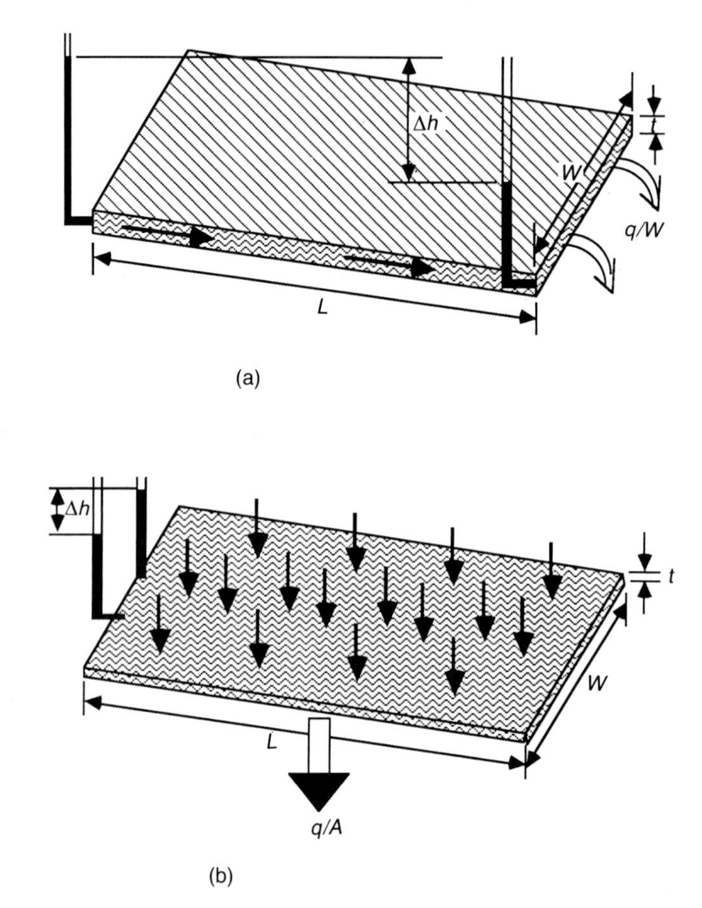

Fig. 9.3 Various hydraulic concepts illustrating (a) in-plane drainage conditions and (b) cross-plane filtration conditions.

where q = flow rate, Q' = flow rate per unit width, k = hydraulic conductivity (or 'permeability coefficient'), i = hydraulic gradient, W, L, t are various geometric properties, and $\theta = (kt)$ = transmissivity.

Note that the flow rate per unit width, Q', is equal to the transmissivity, θ, if and only if, the hydraulic gradient is unity. If the flow is *turbulent* as in a geonet or drainage geocomposite, however, it is recommended to work directly with the flow rate per unit width, Q', at a given value of hydraulic gradient (and at a given value of applied normal stress).

For the **filtration** portion of the various designs, flow is assumed to be across the plane of the material as shown in Fig. 9.3(b). Since laminar conditions usually exist in filter media, e.g., in sands or geotextiles, Darcy's formula can be used directly.

$$q = kiA = k\frac{\Delta h}{t}A = \frac{k}{t}\Delta hA$$

$$\frac{k}{t} = \frac{q}{\Delta hA}$$

$$Q'' = \psi(\Delta h) \tag{9.2}$$

where Q'' = flow rate per unit area, and $\psi = k/t$ = permittivity.

If there is concern regarding laminar flow conditions and the use of permittivity, ψ, or if bulk flow is being considered, one can alternatively work with the term, Q'', which is flow rate per unit area, or 'flux'.

Using these hydraulic definitions one can design against a factor-of-safety concept in the traditional manner:

for drainage problems

$$FS = Q'_{\text{allow}}/Q'_{\text{reqd}} \tag{9.3}$$

or

$$FS = \theta_{\text{allow}}/\theta_{\text{reqd}}$$

where FS = factor-of-safety, Q' = flow rate per unit width, and θ = transmissivity.

for filtration problems

$$FS = Q''_{\text{allow}}/Q''_{\text{reqd}} \tag{9.4}$$

or

$$FS = \psi_{\text{allow}}/\psi_{\text{reqd}}$$

where FS = factor-of-safety, Q'' = flow rate per unit area (or flux), and ψ = permittivity.

In the above relationships the values for allowable properties are obtained from an appropriate test method. Clearly, ASTM is the leading standardization group in this regard. Committee D-18 has standards for soil properties and Committee D-35 has standards for geosynthetic properties. The most important test methods for use in the type of problems to be presented in this chapter are the following:

- D2434 Permeability of soil
- D2412 Strength of plastic pipe
- D4716 Transmissivity of geosynthetics
- D4491 Permittivity of geotextiles
- D4751 Apparent opening size of geotextiles

These tests methods (and others) will be referenced further in their appropriate design sections.

9.3 PRIMARY LEACHATE COLLECTION AND REMOVAL SYSTEM

The primary leachate collection and removal system (PLCRS) is located above the primary geomembrane as shown in Fig. 9.1. It consists of a drainage layer which is usually gravel, perforated pipe collector system, sump collection area, removal system (manhole or large diameter pipes) and a protective filter layer. Each topic will be described in this section along with selected comments on long-term filter performance.

9.3.1 Drainage considerations

Immediately after a landfill cell is constructed, the PLCRS serves as a construction dewatering system. After waste begins to be placed, attenuation of much of the precipitation occurs and the flow rate significantly decreases. The 1985 EPA minimum technology guidance (MTG) for hazardous waste landfills calls for the following properties of the PLCRS.

- 30 cm thick layer
- hydraulic conductivity of 0.01 cm/sec or higher, (note that new regulations *may* increase this value to 1 cm/sec, or higher)
- slopes of greater than, or equal to 2%
- must include a perforated pipe removal system
- must include filter soil above the drainage soil
- must cover both bottom and sidewalls of the landfill facility

In all primary leachate collection systems known to the author, natural soil is used as per the above stated MTG or state mandated requirement, whichever takes precedence. Problems are encountered, however, in holding these sands and gravels on side slopes, particularly when using low friction geomembranes as the primary liner. Many problems of surface sliding have occurred. Options in this regard are the following:

- use a sufficiently low side slope so as to assure drainage soil stability;
- use high friction or roughened geomembrane sheets to increase the frictional stability;
- reinforce the PLCS with a geogrid or geotextile in the form of veneer stability;
- construct the PLCS as the waste is being placed, i.e., one lift at a time; and
- use a geocomposite sheet drain or geonet on the side slopes for the PLCS.

Of these options, the roughened geomembranes and/or veneer stability techniques are being used most frequently, see Koerner and Hwu (1990) for design details. The use of a geosynthetic drainage system (geonet or sheet drain) will be discussed later in the SLCRS (leak detection) section.

9.3.2 Collection considerations

If one considers using the above minimum values for the drainage stone design, the mound equation, after Moore (1983), can be used to obtain the perforated collector pipe spacing as per Fig. 9.4.

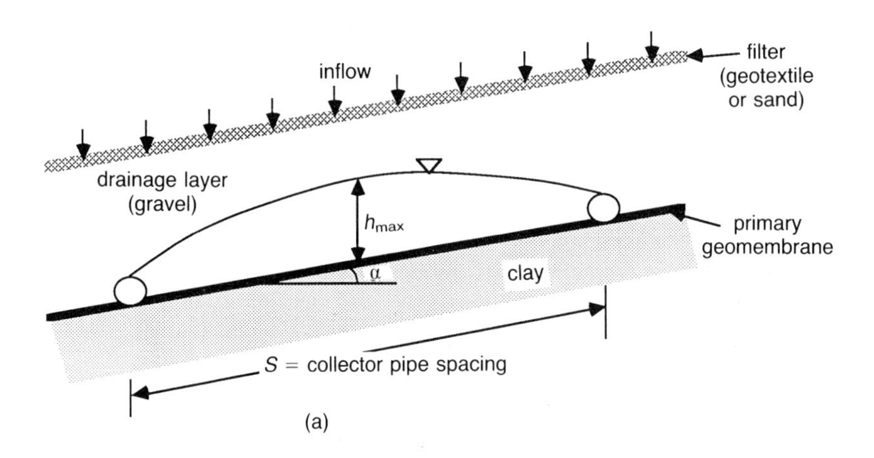

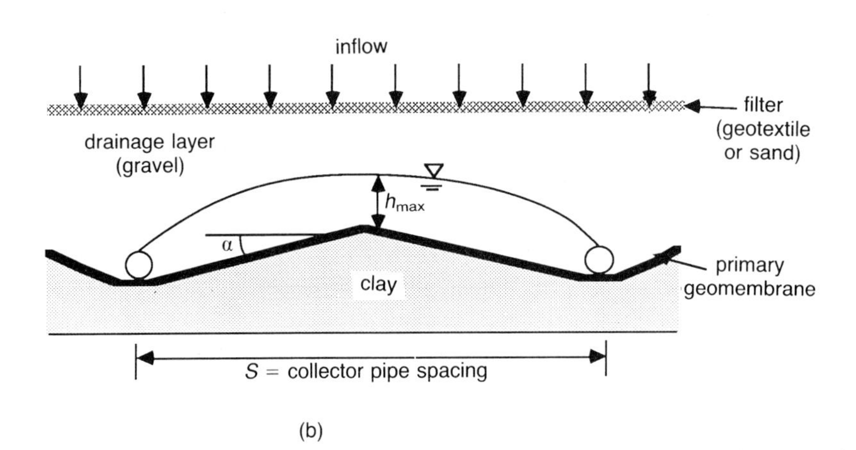

Fig. 9.4 Illustrations of leachate 'mounds' occurring between collection pipes: (a) uniform slope between pipes; (b) crested slope between pipes.

$$h_c = \frac{S\sqrt{c}}{2}\left[\frac{\tan^2 \alpha}{c} + 1 - \frac{\tan \alpha}{c}\sqrt{\tan^2 \alpha + c}\right] \quad (9.5)$$

where h_c = maximum height of leachate between adjacent collector pipes, S = spacing between adjacent pipes, α = slope of ground surface between adjacent pipes, $c = q/k$ (dimensionless), q = infiltration intensity, and k = hydraulic conductivity (permeability) of collection stone.

In this formulation, the value of q is most significant. Immediately after construction of the primary liner system, the PLCRS must act as a construction dewatering system. Here the infiltration intensity, or flow rate, will be at a maximum and site specific weather data must be used. Once waste is placed, precipitation will be greatly attenuated and a water balance method can be used; see Chapter 10. Flow rates are proportionately reduced as additional waste is placed in the facility.

Next, the size (diameter) of the piping must be determined. This is usually done using Manning's equation (Gupta, 1989), which, in SI units, is as follows:

$$V = \frac{1}{n} R_H{}^{0.66} S^{0.50}$$
$$Q = VA \quad (9.6)$$

where V = velocity of flow, A = cross-section area, R_H = hydraulic radius (= flow area/wetted perimeter) = $D/4$, for pipes flowing full, D = inside diameter of the pipe, S = slope of the pipe, and n = Manning's roughness coefficient (dimensionless).

Table 9.2 gives typical values of roughness coefficient n after Fox and McDonald (1985). With the current tendency toward use of plastic pipe in landfills, it should be mentioned that smooth interior PVC or HDPE pipe has a Manning coefficient varying from 0.009 to 0.010, while a

Table 9.2 Values of Manning roughness coefficient, n, for representative pipe interior surfaces, after Fox and McDonald (1985)

type of pipe surface	*representative n value*
Lucite, glass, or plastic	0.010
Wood or finished concrete	0.013
Unfinished concrete, well-laid brickwork, concrete or cast iron pipe	0.015
Riveted or spiral steel pipe	0.017
Smooth, uniform earth channel	0.022
Corrugated flumes, typical canals, river free from large stones and heavy weeds	0.025
Canals and rivers with many stones and weeds	0.035

profiled pipe interior has a Manning coefficient ranging from 0.018 to 0.025. The Manning formula is greatly aided by use of a nomograph which is available in many hydraulic's textbooks (Fig. 9.5).

The load carrying capability of the pipe network must obviously be assured. Here the modified Iowa formula is generally used (Spangler, 1971; Moser, 1990).

$$\Delta X = \frac{D_L K W_c}{(EI/r^3)(0.061E')} \tag{9.7}$$

where ΔX = horizontal deflection or change in diameter of the pipe, D_L = deflection lag factor (varies from 1.0 to 1.5), K = bedding constant (varies from 0.83 to 0.110), W_c = prism load per unit length of pipe, E = modulus of elasticity of the pipe material, I = moment of inertia of the pipe wall per unit length, r = mean radius of the pipe, and E' = modulus of soil reaction.

In this equation the value of W_c is calculated as the weight of waste above the pipe and E' is usually taken from the US Bureau of Reclamation studies (Howard, 1977). The horizontal deflection is assumed to be equal to the vertical shortening (as per ASTM D-2412) and should be

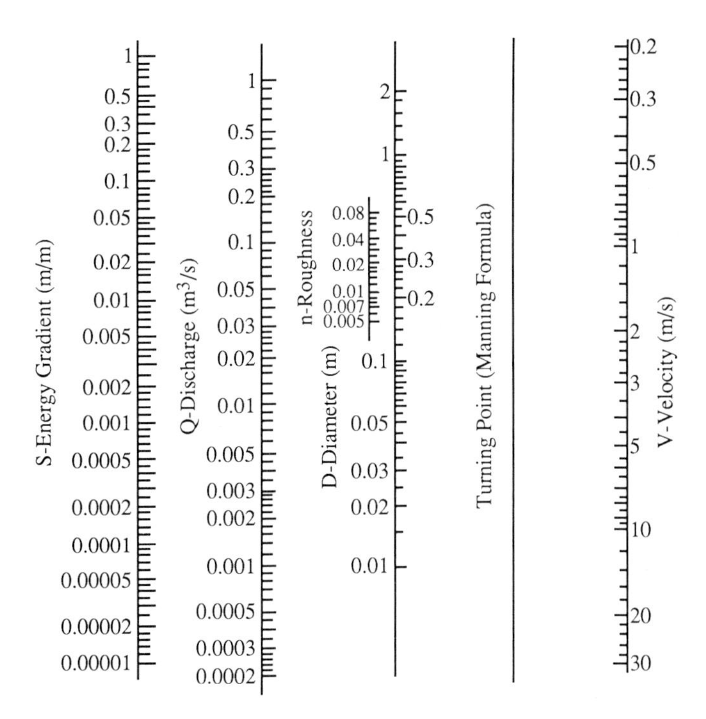

Fig. 9.5 Solution chart for the Manning formula $V = \frac{1}{n} R_H^{0.66} S^{0.50}$, after Hwang (1981).

limited to approximately 5% of the pipe diameter. Note that plastic pipe is governed by a limiting deflection rather than an ultimate crush strength as is rigid pipe.

9.3.3 Removal considerations

There are many possible configurations for the collection and removal network on the bottom of the cell (Fig. 9.6). All of them, however,

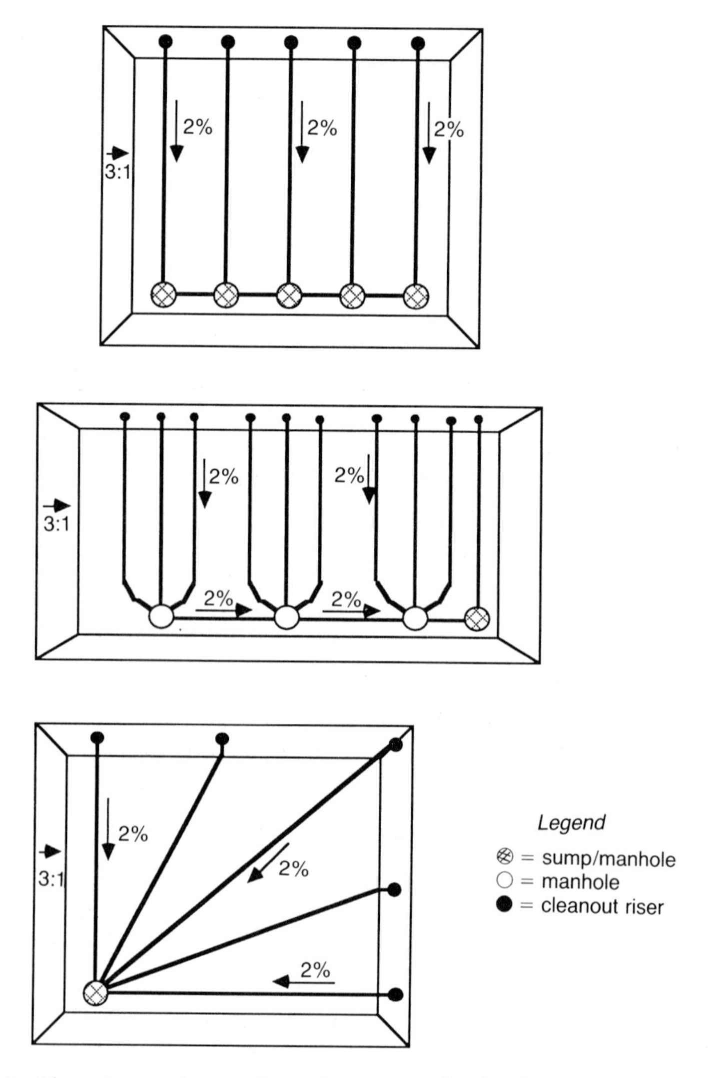

Fig. 9.6 Plan views of several configurations for leachate removal systems, after Jordan (1984).

gravitationally drain to a sump area where a leachate removal facility is located. Here the variations are enormous with some options being as follows:

- manhole extending vertically through the waste (and final cover) for leachate removal;
- pipe risers extending up the sidewall slope located above the primary geomembrane; and
- sloped pipe extending through the liner system to a location beyond the landfill cell area.

Of these options, the current tendency is toward side slopes risers to avoid the downdrag problems associated with vertical manholes and the liner penetrations associated with continuous piping systems (Richardson and Koerner, 1988). Except in very arid areas with dry waste materials, most owner/operators of landfills use submersible pumps in the sump area to pump the collected leachate to trucks, storage tanks, reservoirs or directly to treatment facilities.

9.3.4 Filtration considerations

The granular filter above the leachate collection drainage stone is required by MTG to be at least 15 cm (6.0 in) of soil. This soil must be appropriately designed considering the overlying waste and underlying drainage soil. In this regard, conventional soil filter design is used such as the following.

$$\frac{d_{15}\text{ (of filter)}}{d_{85}\text{ (of soil)}} < 4 \text{ to } 5 < \frac{d_{15}\text{ (of filter)}}{d_{15}\text{ (of soil)}} \tag{9.8}$$

As stated by Cedergren (1989), the left half of the inequality is rationalized as follows.

> The 15% size (d_{15}) of the filter material must be not more than four or five times the 85% size (d_{85}) of the protected soil. The ratio of d_{15} of a filter to d_{85} of a soil is called the **piping ratio**.

The right half of the inequality is rationalized as follows.

> The 15% size (d_{15}) of a filter material should be at least four or five times the 15% size (d_{15}) of the protected soil. The intent is to guarantee sufficient permeability to prevent the buildup of large seepage forces and hydrostatic pressures.

The above design conditions are usually performed on a graph of the various particle size distributions of the soils involved and are explained in all geotechnical engineering textbooks.

When considering placement of a soil filter around slotted or perforated pipe, the US Army, *et al.* (1971) criteria is generally used. These criteria are as follows.

for slotted pipe

$$\frac{d_{85}\text{ (of filter)}}{\text{slot width}} > 1.2 \qquad (9.9a)$$

for perforated pipe

$$\frac{d_{85}\text{ (of filter)}}{\text{hole diameter}} > 1.0 \qquad (9.9b)$$

The option of using a geotextile filter above the drainage media or around a slotted or perforated pipe is very compelling since the savings in air space represents additional revenue to the owner of the facility and in many situations is cost effective in its own right. The design of a geotextile filter must be assured in the same manner as with a soil filter, i.e., adequate soil retention and sufficient permeability. The approach uses factor-of-safety concepts as follows; see Koerner (1990a) for additional details.

for soil retention

$$O_{95} < (2\text{ to }3)\ d_{85} \qquad (9.10)$$

for adequate permeability

$$FS = \frac{\psi_{allow}}{\psi_{reqd}} \qquad (9.11)$$

where O_{95} = 95% opening size of the geotextile (as per ASTM D4751), d_{85} = 85% size of the protected soil (or waste), ψ_{allow} = permittivity of the geotextile (as per ASTM D4491), and ψ_{allow} = required permittivity as per the inflowing liquid (estimated precipitation or via a water balance procedure; see Chapter 10).

9.3.5 Filter clogging

The long-term flow performance of the filter above the PLCRS (either natural soil or geotextile) must be assured for the length of the landfill filling operation and for the post-closure care period which might include leachate recycling procedures. This is typically for 30 to 40 years. Thus a question must be raised as to potential clogging of the filter by the precipitates in the leachate, by biological micro-organisms, or by a synergistic combination of both. Considering the characteristics of leachate from domestic waste landfills (not necessarily from hazardous waste or ash monofils), the tendency for such clogging is certainly possible.

In a recent study by Koerner and Koerner (1990), six domestic landfill leachates have been evaluated for their tendency to clog both geotextile and natural soil filters. Using ninety-six, 100 mm diameter test columns, the effects of aerobic and anaerobic conditions, four different geotextiles, with soil (a #40 mesh Ottawa sand) and without soil, were evaluated. Flow tests of a falling head variety were conducted and within six months the following trends were observed.

no clogging	(0%–25% flow reduction)	6 of 96 columns = 7%
minor clogging	(25%–50% flow reduction)	4 of 96 columns = 4%
moderate clogging	(50%–75% flow reduction)	37 of 96 columns = 38%
major clogging	(75%–95% flow reduction)	35 of 96 columns = 36%
severe clogging	(95%–100% flow reduction)	14 of 96 columns = 15%

Clearly the situation of filter clogging must be addressed and should be done in the design stage. A number of possibilities exist. One option that a designer has is to provide access to the removal piping system as per cleanout risers of the type shown in the sketches of Fig. 9.6. With such cleanout capability one might consider a leachate backflush, a water backflush, a nitrogen gas backflush or a vacuum extraction. (Trials are currently ongoing for each of these procedures where the water backflush appears to be the most promising.) In this case, a very open geotextile could be used. Another option might be not to use a filter at all. While this will generally require approval of the permitting agency, it might be a viable option when using high permeability drainage stone, e.g., hydraulic conductivity greater than 1.0 cm/sec. With adequate gradient, the fine particulates and the accompanying micro-organisms should be carried along with the leachate to the sump and be pumped out of the withdrawal manhole. The last option might be to not worry about filter clogging at all. If the drainage material beneath the filter is not saturated (which is probably the case), the hydraulic head of leachate buildup is on the filter and is not on the geomembrane. Thus the MTG restriction of less than 30 cm head on the liner is not violated. In such a case the leachate mound will then build up in the waste until it becomes sufficiently large to cause breakthrough of the clogged filter. At breakthrough the leachate will flow rapidly into the underlying PLCRS until further clogging occurs in the filter and the process repeats itself.

Whatever the option taken in regard to potential filter clogging it should be done in an open and candid manner.

9.4 SECONDARY LEACHATE COLLECTION AND REMOVAL SYSTEM (I.E., LEAK DETECTION)

The secondary leachate collection and removal system (SLCRS) is located between the primary and secondary liner systems and serves as a leak detection layer, or witness drain. It must gravitationally flow to a sump area located at the low point of the landfill cell where it is periodically sampled so as to assess the adequacy of the primary liner against leakage of leachate.

9.4.1 Drainage considerations

The 1985 EPA minimum technical guidance (MTG) for hazardous waste landfills calls for the following:

- 30 cm thick layer
- hydraulic conductivity of 0.01 cm/sec, or higher
- must cover both bottom and side walls
- must have a response time from liner leak to detection of less than 24 hours

This layer, when constructed of natural soils (as it certainly can be), presents some very real challenges. Side slope stability and pipe installation, along with placement of the primary liner system above it, are all difficult to construct. Furthermore, the response time criterion might be difficult to meet; see Richardson and Koerner (1988) for such calculations.

When considering that the amount of leachate the SLCRS must handle is only a small fraction of PLCRS, one should seriously consider use of a geonet (Fig. 9.2a). The savings in air space by itself is considerable (recall the calculation in section 9.1) plus the above difficulties are avoided, e.g., no pipe system need be used with a geonet. Thus this section will consider the SLCS to be a geonet, recognizing that natural soil can be used if so desired.

The design of the geonet can be based on transmissivity, but it is straightforward (and technically more correct) to proceed with flow rate per unit width at the project specific hydraulic gradient and applied normal stress, i.e.,

$$FS = \frac{Q'_{allow}}{Q'_{reqd}}$$

where Q'_{allow} = allowable flow rate per unit width of the candidate geonet as per ASTM D4716, and Q'_{reqd} = required flow rate per unit width coming from the leaking primary liner.

Regarding the ASTM test for Q'_{allow}, a cross-section and typical data set is given in Fig. 9.7. Note, however, that this type of data is usually

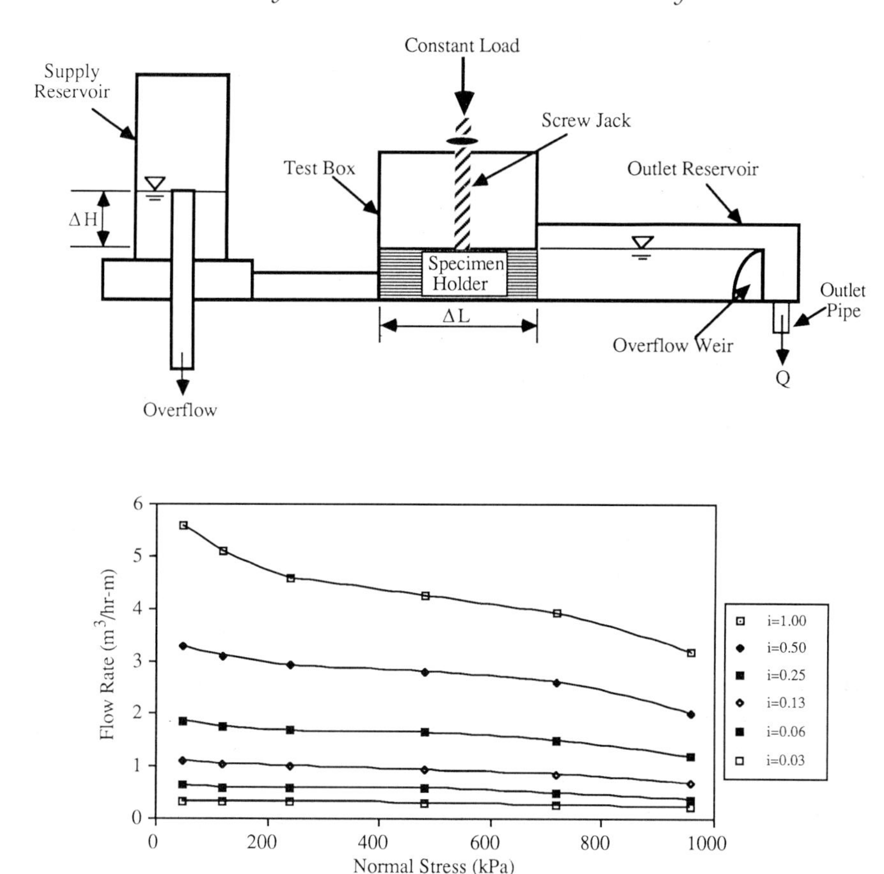

Fig. 9.7 Schematic diagram of lateral flow rest device following ASTM D4716 and typical geonet flow rate data between rigid platens.

presented for the geonet placed between two rigid geomembranes. When the primary liner is a composite liner this is certainly not the field simulated situation. In this case, a geotextile (functioning as a filter/separator) must be placed over the geonet and under the clay component of the overlying composite liner. The stress on the clay will cause an intrusion of the geotextile into the apertures of the geonet reducing a portion of its planar flow capability. Hwu *et al.* (1990) have given insight into the amount of flow reduction for a wide range of conditions, soils and geotextiles. If the reduction in flow rate becomes excessive (for example, from very large landfill heights), it may be necessary to consider a double geonet (Koerner and Hwu, 1989).

Regarding the quantity Q'_{reqd}, one must estimate the flow rate of leachate coming through the primary liner. At the minimum one must consider the vapor diffusion rates for the 'ideal' liner system. This amount varies according to type and thickness of primary geomembrane, but is of the order of 10 l/ha-day. Beyond this one must estimate the number and size of holes that may occur in the primary geomembrane, and whether or not it is a single geomembrane or a composite geomembrane/clay primary liner. Guidance in this regard can be found in Giroud and Bonaparte (1989a,b).

9.4.2 Collection considerations

When using a geonet as a SLCRS there is no network of perforated removal pipes that are necessary. The response time from liner leak to sump detection is surely less than 24 hours since flow in the geonet's apertures is essentially uninhibited.

9.4.3 Removal considerations

Flow within the SLCS is gravitational and usually follows the same grading scheme as the overlying PLCS, recall the patterns of Fig. 9.6. Since both systems are draining to the same down gradient location in the landfill cell, the removal sumps are usually immediately adjacent to one another. What differs, however, is the amount of leachate in the SLCRS versus the PLCRS. Instead of requiring submersible pumps, one should have need for no more than well samplers and, at that, using small diameter riser pipes. Typically 100–150 mm smooth, solid-wall plastic pipes coming up the side slope between the primary and secondary liners are adequate. Sampling frequency is intermittent and done in accordance with the permit for the facility. It is important to recognize that if liquid is found in the SLCRS it might not be from a leaking primary liner. Gross *et al.* (1990) gives insight as to other possible liquids, for example:

- compression water if a moist soil drainage layer is used;
- consolidation water from the saturated clay component if a composite primary liner is used;
- infiltration water from the secondary (bottom) liner in areas of high water table; or
- penetration water from the termination area around the periphery of the site, i.e., between the anchorages of the primary and secondary liner systems.

Thus one should always chemically analyze the liquid in the SLCRS via mass spectroscopy/gas chromatography (MS/GC) methods before assuming that the primary liner is leaking.

In closing, it should be mentioned that response action plans (RAPs) will soon be mandated for all landfill facilities and will require feedback of leakage rates found in the SLCRS. At that time, design values should become more realistic than they are currently.

9.5 SURFACE WATER COLLECTION AND REMOVAL SYSTEM

Above the completed landfill facility will be either a temporary or a permanent liner system (see Chapter 18), and above this liner will generally be a surface water collection and removal system (SWCRS). This might not be the case in extremely arid regions of the southwest USA where the nominal amount of moisture that does occur should be preserved, rather than drained away from the site. In general, however, the water that does penetrate the cover soil should be efficiently and expeditiously removed off of the liner and away from the site; recall Fig. 9.1.

9.5.1 Drainage considerations

The 1989 EPA minimum technology guidance (MTG) for final covers on hazardous waste landfill calls for the SWCRS to include the following.

- 30 cm thick layer
- hydraulic conductivity of 0.01 cm/sec, or higher
  (or geosynthetic material with equivalent performance)
- minimum final slope of 3%

Clearly, natural soils can be, and have been, used for surface water collection systems. Savings of air space is not a consideration since minimum cover thickness over the geomembrane/clay liner can easily accommodate the 30 cm thickness requirement. Indeed most designs respond to the above stated MTG using granular soil drainage layers, taking into consideration the conditions of stability against the cover soil sliding off of the geomembrane. On steep side slopes of aboveground landfills this consideration may be troublesome requiring high-friction geomembranes or some type of geosynthetic reinforcement (Koerner and Hwu, 1990).

The alternate geosynthetic material referenced in the MTG makes reference to either geonets or geocomposite sheet drains as shown in Fig. 9.2. Since geonets were just described, the following comments will be focused on geocomposite sheet drains. There are many commercially available products in this regard; see Table 9.3. All are designed as were the geonets described earlier, i.e.,

$$FS = \frac{Q'_{\text{allow}}}{Q'_{\text{reqd}}}$$

The allowable flow rate is obtained from an ASTM D4716 test, recall Fig. 9.7, and compensated for the intrusion of the geotextile filter/separator. The required flow rate is obtained from a suitable water balance method considering site-specific precipitation, runoff, evapotranspiration and infiltration. The infiltration is, of course, the liquid

Table 9.3 Summary of geocomposite sheet drainage systems

manufacturer/agent	*product name*	*core structure*	*core polymer*	*geotextile**	*roll size (ft) width/length*	*thickness (mils)*
Akzo	Enkamat 7010	3-D web	nylon 6	–	3.2/492	400
	Enkamat 7020	3-D web	nylon 6	–	3.2/330	800
	Enkamat 9010	3-D web	nylon 6	polyester	3.0/99	400
	Enkamat 9120	3-D web	nylon 6	polyester	3.0/99	800
American Wick Drain Corp.	Amerdrain 480 mat	Nippled core	polyethylene	polypropylene	4/104	375
Burcan Industries	Hitek 6c	cuspated	polyethylene	polypropylene	3.0/450	255
	Hitek 20c	cuspated	polyethylene	polypropylene	3.6/125	785
	Hitek 40c	cuspated	polyethylene	polypropylene	3.5/80	1575
Exxon	Tiger Drain	cuspated	polyethylene	polypropylene	4/100	600
Exxon	Battle Drain	cuspated	composite	polypropylene	4/100	600
Greenstreak	Sheet Drain	raised core	polystyrene	nonwoven-PP	4′ × 10′ panels	375
	Deck Drain	" "	"	woven-PP	"	375
Geotech Systems	Geotech Drain Board	EPS panel	–	–	4.0/4.0	1000
Huesker Synthetic	HaTe-Drainmatte		polypropylene	PES	13/328	260
JDR Enterprises	J-Drain 100	extruded rib	polyethylene	polypropylene		250
	J-Drain 200	extruded rib	polyethylene	polypropylene		250
Mirafi	Miradrain 6000	raised core	polystyrene	polypropylene	4/8, 25, 50	380
	Miradrain 6200	raised core	polystyrene	polypropylene + backup film	4/8, 25, 50	380
	Miradrain 9000	raised core	polystyrene	polypropylene	4/8, 25, 50	380
	Miradrain 8000	raised core	polyvinyl chloride	polypropylene	4/8, 25, 50	380
	Miradrain 5000	raised core perforated	polystyrene	polypropylene	4/8, 25, 50	380
	Miradrain 4000	cuspated	polystyrene	polypropylene	4/8, 25, 50	750

Table 9.3 *Continued*

manufacturer/agent	*product name*	*core structure*	*core polymer*	*geotextile**	*roll size (ft) width/length*	*thickness (mils)*
Monsanto	Hydraway Drain	raised cyl. tubes	polyethylene	polypropylene	12, 18/400	1000
Nilex	Nudrain A		polyethylene	polypropylene	1.6/49, 98	1575
	Nudrain C		polyethylene	polypropylene	3.6/98	787
NW Fabrics	Permadrain	cuspated	polyethylene	polypropylene	any size	–
Pro Drain Systems	PDS 20	cuspated	polyethylene	polypropylene	3.7/10–500	750
	PDS 40	cuspated	polyethylene	polypropylene	3.3/10–250	1500
Rhone-Poulenc	Megadrain	3-D web	nylon 6	polyester	–	375

* All geotextile filters are nonwoven; either needle-punched or heat-bonded.

that must be carried in the drainage network. The HELP computer code is well suited for this determination; see Chapter 10.

9.5.2 Collection considerations

As shown in Fig. 9.1, the surface water drainage system gathers and transports the infiltrating water to a collection system around the periphery of the cell or landfill. This collector system is usually a perforated pipe and is designed similar to a highway underdrain system (Cedergren, 1989). However, as with highway underdrains, geocomposite edge drains are seeing use due in large part to their low cost and ease of placement (Fig. 9.2c). Table 9.4 gives an overview of the available products. In the design of the collector system it must be remembered that the required flow rate is cumulative over the length of the landfill cover and over the length of pipe or edge drain to the outlet.

9.5.3 Filter considerations

The 1989 cover MTG calls for a 150 mm thick filter soil over the drainage soil. This soil is designed as with any other soil filter as was described in Section 9.3. If a geocomposite sheet drain is used, the filter will be a geotextile which is also designed as was described in Section 9.3.

It should be mentioned that the design of the surface water drain and its filter is much more straightforward than with the PLCRS drain since water is the permeant and not leachate. Furthermore, the normal stresses in the landfill cover are much less than beneath the waste so that crushing of the core and intrusion of the geotextile are easier to meet with a wide variety of geosynthetic products.

9.6 METHANE GAS COLLECTION AND REMOVAL SYSTEM

When dealing with domestic waste landfills, the anaerobic decomposition of the waste will generate numerous gases. The two principal types are carbon dioxide (CO_2) and methane (CH_4). Carbon dioxide is heavier than air and will migrate to the bottom of the waste facility and will be removed with the leachate. Methane, however, is lighter than air and will rise to the top of the waste (or to the sides) where it will meet the cover liner system and be trapped beneath it (Fig. 9.1). Thus, it is necessary to have a gas collector system beneath the liner and have it uniformly graded to the high point of the cover. Here it is either vented, flared, or captured and used for energy production. Additionally, certain chemicals within the landfill may also emit gases

Table 9.4 Properties of commercially available highway edge drains as listed in manufacturers literature

company	*trademark*	*core characteristics*					*geotextile type*
		material	*shape*	*thickness*	*strength**	*flow rate**	
Monsanto	Hydraway 2000	PE	straight column	1.00	13 700	27 gpm/ft. @ 210 psf i = 0.2 27 gpm/ft. @ 2090 psf i = 0.2	PET nonwoven needle-punched
Contech	Stripdrain 75	PE	tapered column	0.75	5 800	26 gpm/ft. @ 500 psf i = 1.0 22 gpm/ft. @ 1500 psf i = 1.0	PET nonwoven needle-punched
	Stripdrain 100	PE	tapered column	1.00	8 000	15 gpm/ft. @ 1440 psf i = 0.1 3.2 gpm/ft. @ 1500 psf i = 0.01 4.7 gpm/ft. @ 1500 psf i = 0.02	PET nonwoven needle-punched
Pro Drain Systems	PDS-20	PE	tapered column	n/a	n/a	n/a	n/a
	PDS-30	PE	tapered column	1.20	n/a	n/a	PP nonwoven needle-punched
ICE	Akwadrain B	PE	double cuspated	1.00	8 000	12 gpm/ft. @ 1000 psf i = 1.0 12 gpm/ft. @ 2000 psf i = 1.0	PP nonwoven needle-punched
	Akwadrain 125	PE	double cuspated	1.25	n/a	n/a	PP nonwoven needle-punched
Adv. Drainage Systems (ADS)	Advan Edge	PE	corrugated with columns	1.60	4 320	7 gpm/ft. @ 10 psi i = 0.1	PP nonwoven melt bonded
Nilex	Nudrain™ 'A'	PE	double cuspated	1.60	2 700	5.9 l/sec/250 mm width	PP nonwoven melt-bonded
Greenstreak	Pave Drain™	PE	tapered column (tapered)	1.10	8 500	8 gpm/ft. @ i = 0.05 38 gpm/ft. @ i = 0.5	PP nonwoven needle-punched
Drysol	Drainage Cell	PP	cellular	1.20	n/a	n/a	PP nonwoven needle-punched

* The values of flow rate are assumed to be at a hydraulic gradient of 1·0, in which case it is numerically equal to transmissivity. The values, however, are taken directly from manufacturers' literature where considerable variation in test method, manner of presentation of results, and concepts involved, tend to vary.
n/a = not available

or vapors that require venting. The collection layer itself is not covered in detail by MTG and is clearly site specific and waste specific. Design proceeds as with any drainage system in that a factor-of-safety is formulated as follows:

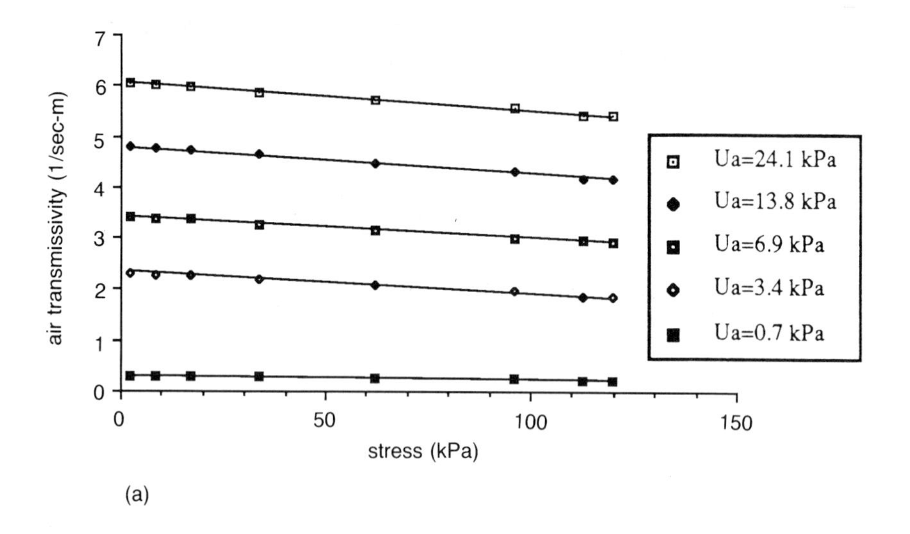

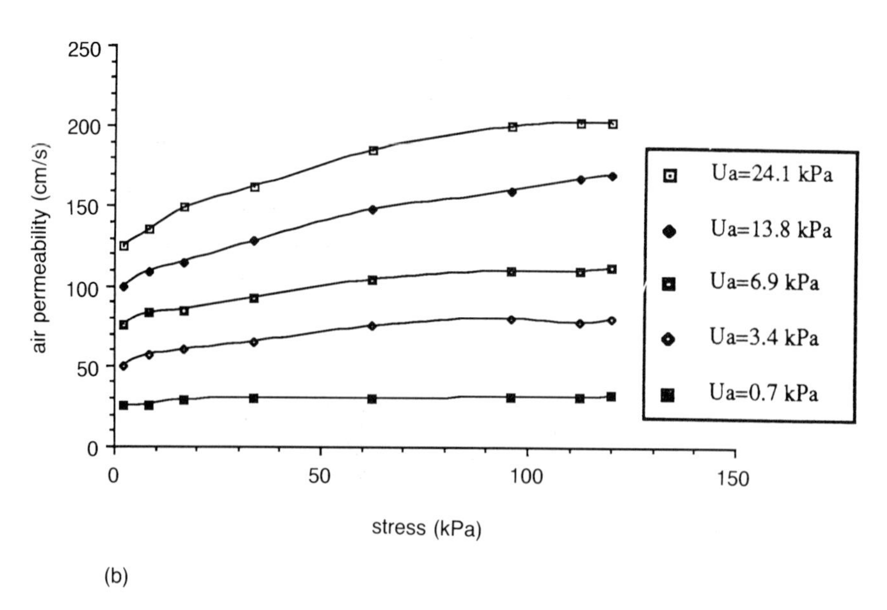

Fig. 9.8 (a) Air transmissivity and (b) in-plane coefficient of air permeability, versus applied normal stress for one layer of 600 g/m^2 needle-punched nonwoven geotextile, after Koerner and Bove (1984).

$$FS = \frac{Q'_{\text{allow}}}{Q'_{\text{reqd}}}$$

The allowable flow rate, Q'_{allow}, is not at issue when using natural soils since gas flow-to-water flow is in rough proportion to the differences in viscosity, i.e., air flows about 1000 times easier than does water. Most soil layers for gas venting are 15–30 cm in thickness. Flow rate is of

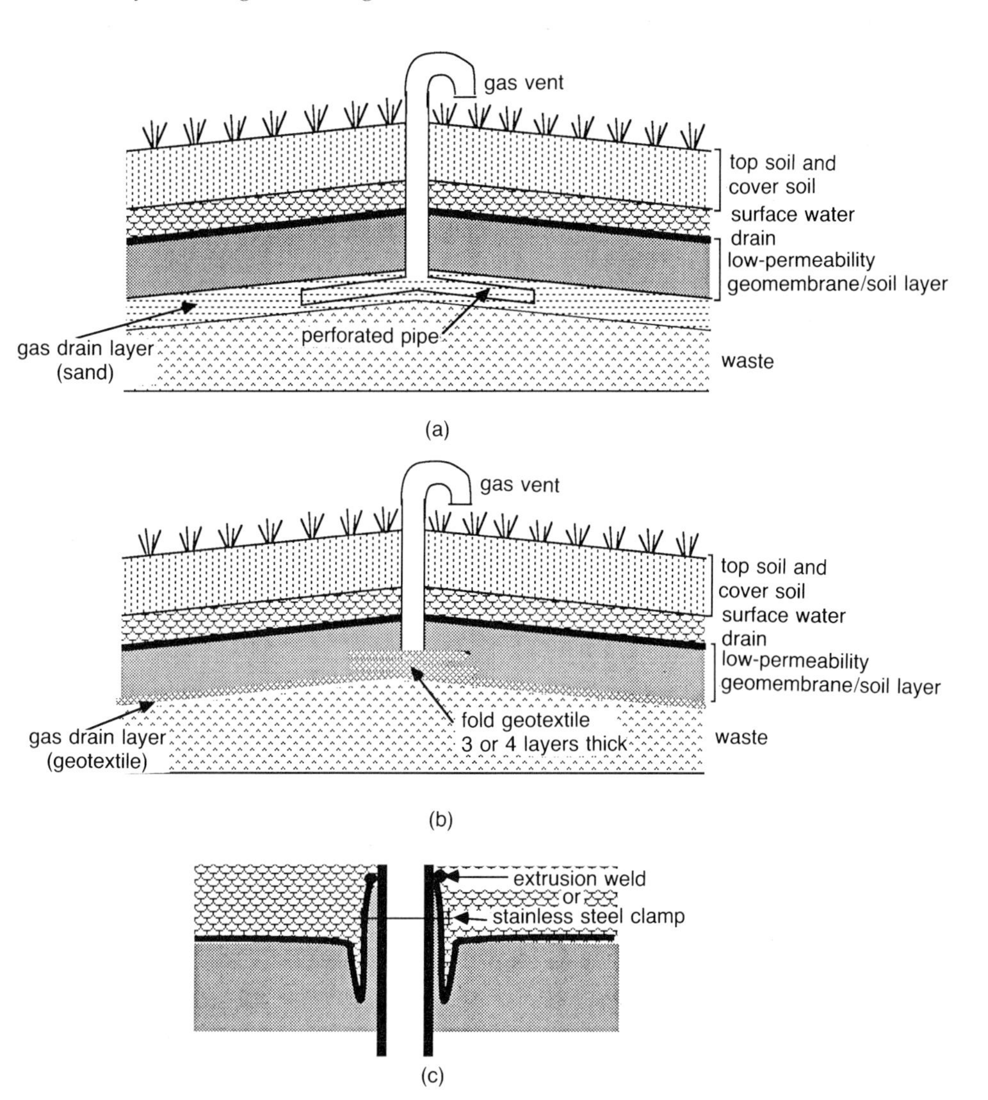

Fig. 9.9 Schemes for methane gas removal though final cover liner system: (a) natural soil vent system, after EPA (1989); (b) geotextile vent system; (c) suggested bellows at geomembrane penetration.

issue, however, if one is considering the use of a geotextile gas drain. Since this layer represents a savings of air space (it is within the landfill enclosure), it should be considered and properly investigated. Figure 9.8 gives air transmissivity and air permeability in the plane of a $600\,g/m^2$ needle punched nonwoven geotextile. Here the data is seen to be sensitive to the incoming gas flow rate and the applied normal stress.

Concerning the required flow rate, Q'_{reqd}, Emcon (1980) has determined gas recovery rates from actual field measurements at six landfills. Using data from recovery projects or gas control programs, annual gas recovery rates from 0.27 to 0.77 cubic meters per kilometer of waste have been estimated.

Of particular note is the extremely long time frames for decomposition of the waste (along with the accompanying methane gas generation) to occur. Baron *et al.* (1981) estimate the following.

- 50% decomposition (i.e., half-life) = 23 yrs
- 90.5% decomposition = 80 yrs
- 99.1% decomposition = 160 yrs

Thus the detail of methane gas collection and removal is important to the proper long-term functioning of the closure system.

Regarding removal of the collected gas at the high point of the closure (or along the flow path at regular intervals), the cover barrier system must be penetrated. Several designs are possible, but differential settlement of the underlying waste must be considered. Figure 9.9 gives some ideas as to different removal schemes with a granular venting layer or with a geotextile. The idea of a bellows arrangement at the geomembrane penetration is attractive. Prefabricated geomembrane boot attachments are recommended in this regard.

REFERENCES

Baron, J.L. *et al.* (1981) *Landfill methane utilization technology workbook*, US DOE, Contract No. 31-109-38-5686, Washington, DC.

Cedergren, H.R. (1989) *Seepage, Drainage and Flow Nets*, J. Wiley and Sons, NY, NY.

Emcon Associates (1980) *State-of-the-art of methane gas enhancement in landfills*, Project 343.1.1, San Jose, CA.

Fluet, J.E. (1987) *Geotextile testing and the design engineer*, ASTM STP 952, Philadelphia, PA.

Fox, R.W. and McDonald, A.T. (1985) *Introduction to Fluid Mechanics*, 3rd edn, J. Wiley and Sons, NY, NY.

Giroud, J.-P. and Bonaparte, R. (1989a) Leakage through liners constructed with geomembranes – Part I: geomembrane liners, *J. Geotextiles and Geomembranes*, Elsevier Appl. Sci. Publ., London, **8**(1), 27–68.

Giroud, J.-P. and Bonaparte, R. (1989b) Leakage through liners constructed with geomembranes – Part II: composite liners, *J. Geotextiles and Geomembranes*, Elsevier Appl. Sci. Publ., Long, **8**(2), 71–112.

Gross, B.A., Bonaparte, R. and Giroud, J.-P. (1990) Evaluation of flow from landfill leachate detection layers, *Proc. 4th Intl. Conf. on Geosynthetics*, A.A. Balkema Publ., Rotterdam, The Netherlands, 481–6.

Gupta, R.A. (1989) *Hydrology and Hydraulic Systems*, Prentice-Hall, Inc., Englewood Cliffs, NJ.

Howard, A.K. (1977) Soil reaction for buried flexible pipe, *J. Geotech. Eng. Div., ASCE*, 33–43.

Hwang, H.C. (1981) *Fundamentals of Hydraulic Engineering Systems*, Prentice-Hall, Inc., Englewood Cliffs, NJ.

Hwu, B-L., Koerner, R.M. and Sprague, C.J. (1990) Geotextile intrusion into geonets, *Proc. 4th Intl. Geosynthetic Conf.*, The Hague, Netherlands, 351–6.

Jordan, E.C. Co. (1984) *Performance standard for evaluating leak detection*, US EPA Contract No. 68-01-6871, Washington, DC.

Koerner, G.R. and Koerner, R.M. (1990) *Biological activity and potential remediation involving geotextile landfill leachate filters*, ASTM STP 1081 (ed. R.M. Koerner), Philadelphia, PA.

Koerner, R.M. (1990a) *Designing with Geosynthetics*, Prentice-Hall Publ., Co., Englewood Cliffs, NJ.

Koerner, R.M. (1990b) *Geosynthetic testing for waste containment applications*, ASTM STP 1081, Philadelphia, PA.

Koerner, R.M., Bove, J.A. and Martin, J.P. (1984) Water and air transmissivity of geotextiles, *J. Geotextiles and Geomembranes*, Elsevier Appl. Sci. Publ., **3**(1), 57–75.

Koerner, R.M. and Hwu, B-L. (1989) Behavior of double geonet drainage systems, *Geotech. Fabrics Rep.*, Sept./Oct., 39–44.

Koerner, R.M. and Hwu, B-L. (1990) Stability and tension considerations regarding cover soils on geomembrane lined slopes, *Intl. J. Geotextiles and Geomembranes*, Elsevier Appl. Sci. Publ., London.

Moore, C. (1983) *Landfill and surface impoundment performance evaluation*, US EPA, SW-869, Washington, DC.

Moser, A.P. (1990) *Buried Pipe Design*, McGraw-Hill, NY, NY.

Richardson, G.N. and Koerner, R.M. (1988) *Geosynthetic design guidance for hazardous waste landfill cells and surface impoundments*, US EPA Contract No. 68-03-3338, Cincinnati, OH.

Spangler, M.G. (1971) *Soil Engineering*, 2nd edn, International Textbook Co., Scranton, PA.

US Army, Navy and Air Force (1971) *Dewatering and groundwater conrol in deep excavations*, TM5-818-5, NAVFAC P-418, AFM 88-5, Ch. 6, Washington, DC.

US EPA (1985) *Minimum technology guidance on double liner systems for landfills and surface impoundments: design, construction and operations*, EPA/530-SW-84-014.

US EPA (1989) *Final covers on hazardous waste landfills and surface impoundments*, EPA/530-SW-89-047, OSW, Washington, DC.

CHAPTER 10

Water balance for landfills

R. Lee Peyton and Paul R. Schroeder

10.1 INTRODUCTION

A key step in the design of a solid waste disposal facility is the execution of a 'water balance' or 'water budget' analysis. A water balance is an accounting of the final disposition of precipitation falling on a site. Water balance analysis can be used to estimate the potential leachate production and liner/drain system performance and to compare the relative effectiveness of alternative cover and liner/drain designs. Knowledge of the possible range of leachate production is important for sizing the leachate collection system (e.g., pipes) and in making decisions about how to manage treatment of the leachate. Similarly, prediction of liner leakage and the depth of leachate buildup in a drain layer is important in the selection of liner and drain materials and in the design of collection pipe spacing and liner slope.

A water balance analysis accounts for the effects of many hydrologic processes on water movement at a site. Once precipitation strikes a site, it may either:

- be intercepted by vegetation and subsequently evaporated;
- be stored temporarily on the surface in the form of snow or ice if the temperature is freezing and later evaporated or melted;
- run off the surface of the cover system; or
- infiltrate into the cover.

Water infiltrating into the cover may:

- be transpired to the atmosphere via plants that draw the water from the ground through their roots;
- evaporate from the surface soil into the atmosphere;
- be stored in the cover soil; or

Geotechnical Practice for Waste Disposal.
Edited by David E. Daniel.
Published in 1993 by Chapman & Hall, London. ISBN 0 412 35170 6

- percolate downward through the cover soil and past the evaporative zone.

Water percolating through past the evaporative zone may:

- be collected in the cover if a lateral drainage layer is present;
- be stored in the underlying waste or soil; or
- contribute to the leachate production which is either collected in the leachate collection system or lost as leakage through the liner system.

This chapter presents a description of hydrologic processes affecting leachate generation, and methods for estimating the water balance. In addition, several case studies comparing the measured water balance at several facilities with predictions generated by the Hydrologic Evaluation of Landfill Performance (HELP) model are presented. Finally, the sensitivity of various landfill design parameters on the water balance components is shown using the HELP model to predict the effects.

10.2 WATER BALANCE CALCULATIONS

Potential pathways of water movement are shown in Fig. 10.1. The two principal methods of analyzing water routing are a hand method (Thornthwaite method) and a computer method (HELP model).

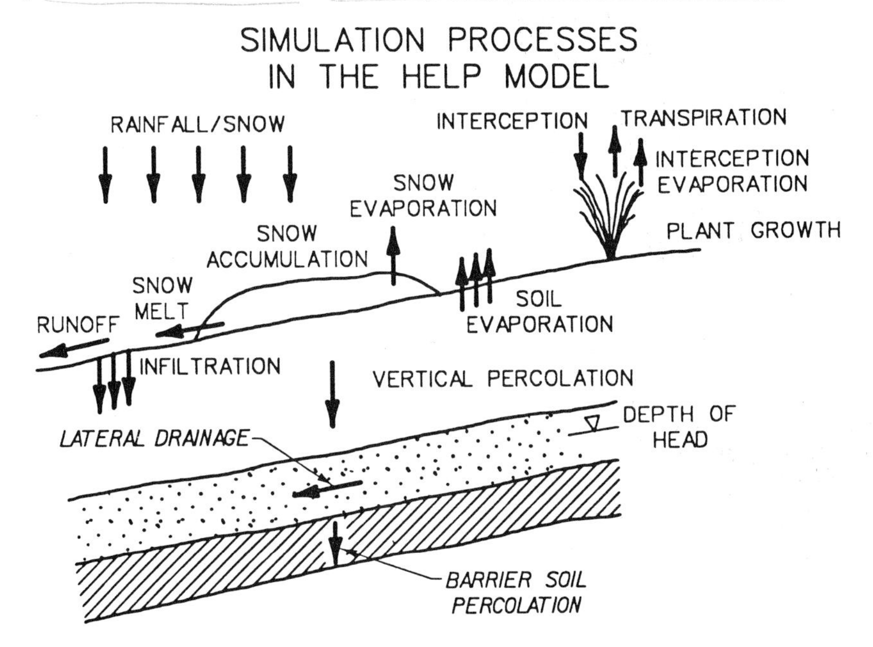

Fig. 10.1 Water pathways.

10.2.1 Thornthwaite method

The earliest, comprehensive method for water balance analysis was developed by Thornthwaite and Mather (1957). Fenn *et al.* (1975) developed for the US Environmental Protection Agency (EPA) a water balance method for predicting leachate generation for solid waste disposal sites based on the Thornthwaite method, proposing runoff coefficients for landfills and moisture storage values for municipal waste. The water balance method employs average monthly values of precipitation and other climatic parameters. To determine monthly infiltration (IN) into the cover, one subtracts monthly runoff (R) from monthly precipitation (P).

$$IN = P - R \tag{10.1}$$

Runoff can be calculated from precipitation, as follows

$$R = CP \tag{10.2}$$

where C is a runoff coefficient that can be estimated from the guidance provided by Fenn *et al.* (1975):

description of the grass-covered soil	slope of ground surface	runoff coefficient (C)
sandy soil	flat (<2%)	0.05–0.10
	mild (2 – 7%)	0.10–0.15
	steep (>7%)	0.15–0.20
clayey soil	flat (<2%)	0.13–0.17
	mild (2 – 7%)	0.18–0.22
	steep (>7%)	0.25–0.35

Potential evapotranspiration (PET), which depends on mean temperature, heat index, and hours of sunlight, can be calculated from tables provided by Thornthwaite and Mather (1957). The cumulative monthly infiltration minus potential evapotranspiration ($IN - PET$) is calculated. A negative number for $IN - PET$ indicates that the cover has a tendency to dry out; a positive value indicates that there is a tendency for the cover soil to become wetter. If $IN - PET$ is negative, water may evapotranspire from the cover soil, but if the soil is already dry, no further drying will occur. The amount of drying depends not only on $IN - PET$ but also on the water content of the soil.

If $IN - PET$ is positive, water may be stored in the cover soil (which would produce an increase in water content). However, if the water content is already very high, the soil can store no additional water and water will percolate downward through the cover soil. The field capacity of the soil is the maximum water content that a soil can attain without draining water by gravity. Field capacity can be determined by allowing a saturated soil to drain by gravity; the water content when equilibrium is reached is the field capacity. Thornthwaite and Mather (1957) and Fenn *et al.* (1975) recommended values of field capacity for various soils. One must allocate $IN - PET$ to either soil storage or, when field capacity would be exceeded, to downward percolation.

An excellent description of the Thornthwaite method as applied by Fenn *et al.* (1975) including its use and limitations is given by Kmet (1982). Hand calculations are easily performed with this technique, which is particularly suitable for analysis with spreadsheet software. The main disadvantage is that numerous simplifying assumptions must be made.

10.2.2 Computer analysis with program HELP

The state of the art in water balance analysis for landfills is a computer program known as HELP. The HELP model was developed by the US Army Engineer Waterways Experiment Station for the US EPA and is described in detail by Schroeder *et al.* (1984a,b). Version 3 of the model is being released in 1993 by the US EPA Risk Reduction Engineering Laboratory.

The HELP model was adapted from the Hydrologic Simulation Model for Estimating Percolation at Solid Waste Disposal Sites of the US EPA (Perrier and Gibson, 1980; Schroeder and Gibson, 1982), the Chemical Runoff and Erosion from Agricultural Management Systems (CREAMS) (Knisel, 1980), and Simulator for Water Resources in Rural Basins (SWRRB) (Williams *et al.*, 1985). The following subsections summarize the main features and assumptions of the model. The User's Manuals (Schroeder *et al.*, 1984a,b; 1988a,b) should be consulted for further detail.

The HELP model is a quasi-two-dimensional, gradually varying, deterministic, computer-based water budget model. It is termed quasi-two-dimensional because it contains a one-dimensional vertical drainage model and a one-dimensional lateral drainage model coupled at the base of lateral drainage layers or the tops of liners incorporated in the cover system or located below the waste. The program computes free downward vertical drainage to the top of a liner, at which point the liner restricts drainage and a zone of saturation can develop. The models for lateral drainage and leakage or percolation through the liner then use the height of saturated material above the liner to compute simultaneously the rates of lateral drainage to collection systems and vertical leakage through the liner, respectively. The model is termed gradually varying because the simulation progresses through time using analyses that are assumed steady for each time period. Version 2 of the model (the latest version at the time of this writing) uses a time period of 6 hours (Schroeder *et al.*, 1988b). The model is deterministic and quantitative. Finally, the HELP model is a computer-based water budget model that runs on a personal computer; the model uses a computer to apportion the precipitation and initial water content of soil layers into estimates of the following water budget components: surface runoff, evapotranspiration, changes in snow storage, changes in water content of the soil, lateral drainage collected in each drain system, and leakage or percolation through each liner in the system (Fig. 10.2). The HELP program can generate daily, monthly, annual, and long-term average water budgets.

The HELP model is a tool developed specifically to aid analysis in the evaluation and comparison of alternative landfill designs. A secondary goal in development of the model was to provide an accurate prediction of water budget components. Testing, verification studies, and refinements are ongoing to improve the accuracy of HELP. In general, the accuracy and precision of the model is limited by uncertainty and variability in the properties of material existing in landfills.

Infiltration

Daily infiltration into the landfill is determined indirectly from a surface water balance. Infiltration equals the sum of rainfall and snowmelt, minus the sum of runoff and surface evapotranspiration. Runoff and surface evaporation are in part a function of interception. Precipitation on days having a mean temperature below 0 °C is treated as snowfall and is added to the surface snow storage. Decreases in snow storage occur by snowmelt and surface evaporation.

Snowmelt is computed using a slightly modified version of the simple degree-day method. The analysis permits a small quantity of snowmelt to occur at mean daily temperatures between −5 °C and 0 °C to account for the variation in temperature during a day. It also accounts for the fact that landfills often have higher soil temperatures because of heat generated from biodegradation of waste. Snowmelt contributes to runoff, evaporation, and infiltration.

Interception is modeled in the method of Horton (1919). Interception approaches a maximum value exponentially as the rainfall increases to about 5 mm. The maximum interception is a function of the quantity of aboveground biomass of leaf area index and is limited to a maximum of 1.3 mm. The interception evaporates from the surface.

The HELP model uses the US Soil Conservation Service (SCS) curve number method for estimating surface runoff (US Department of Agriculture, 1972). The SCS curve number method is an empirical procedure developed for small watersheds (12–200 hectares) with mild slopes (about 3–7%). The method correlates daily runoff with daily rainfall for watersheds with a variety of soils, types of vegetation, land

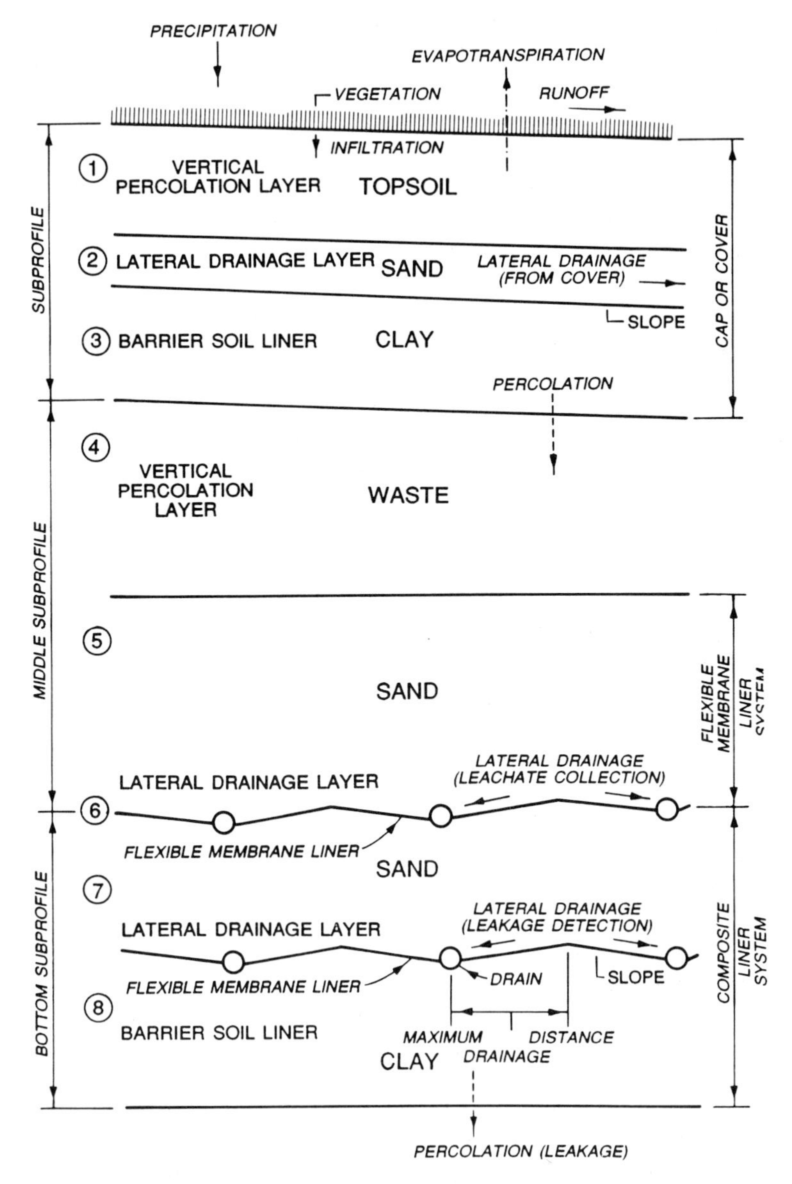

Fig. 10.2 Water balance that can be analyzed in program HELP.

management practices, and antecedent moisture conditions (level of prior rainfall). Version 3 of the model (scheduled for release in 1993) will include a procedure to adjust the curve number as a function of surface slope since surface slopes greater than 20% can produce significantly greater runoff.

Several limitations merit mentioning.

- The SCS curve number method is applied to landfills, which are much smaller than the watersheds for which the method was developed. Verification studies have shown good agreement between the predicted and observed cumulative annual volume of runoff.
- Cumulative volume of runoff is independent of rainfall duration and intensity since over a long simulation period, a variety of precipitation events will occur. The predicted value represents an average of the measured runoff for the typical variety of rainfall events.
- No surface run-on from surrounding areas is permitted by the model.
- Estimates of runoff greater than predicted by the SCS curve number method are produced when the surface soils are saturated or limit infiltration due to very low hydraulic conductivity.

Evapotranspiration

Evapotranspiration consists of evaporation of water from the surface, from the soil, and from plants. Each component is calculated separately.

Evaporation of water from the surface is limited to the smaller of the potential evapotranspiration and the sum of snow storage and interception. The HELP model uses a modified Penman method to compute evapotranspiration (Ritchie, 1972). The potential evapotranspiration is a function of vegetative cover, daily temperature, and daily solar radiation.

The HELP model uses Ritchie's (1972) method of evaporation from soil. The method involves a two-stage square-root-of-time routine. In stage one, the soil evaporation equals the evaporative demand placed on the soil. Demand is based on energy and is equal to the potential evapotranspiration discounted for surface evaporation and shading from ground cover. A vegetative growth model is used to compute the total quantity of vegetation, both active and dormant, which provides shading. In stage two, evaporation from the soil is limited by low soil moisture and low rates of water vapor transport to the surface. Stage two soil evaporation is a function of the square root of the length of time that the soil has been in dry condition.

The HELP model estimates plant transpiration by assuming that the potential plant transpiration is a linear function of the potential evapotranspiration and the active leaf area index.

Key limitations are listed below.

- The potential evapotranspiration is a function solely of energy available at the surface and, therefore, is not affected by energy produced in the landfill, by wind, or by humidity.
- A constant value is used for the albedo (i.e., the fraction of incident

solar radiation that is reflected). The value is typical for brown soils and grasses and is modified only when the surface is covered with snow.

- The program uses a constant evaporative zone depth. This depth is the maximum depth to which soil suction can draw water to the surface. This depth is a function of soil properties, design, vegetation, and climatic conditions.
- The vegetative growth model produces representative leaf area indices and biomass estimates that are sufficient to estimate interception, surface shading, and plant transpiration.

Subsurface water routing

Subsurface water routing includes vertical unsaturated drainage, percolation through saturated soil liners, leakage through geomembranes, and lateral drainage in drainage layers. In modeling these processes, the soil moisture of each layer (with the exception of barrier layers) is computed by sequential analysis. The soil moisture controls the rate of subsurface water movement, but the rate of movement also affects the moisture content. Consequently, an iterative analysis procedure is used.

The HELP model simulates unsaturated vertical drainage by assuming unit hydraulic gradient. The unsaturated hydraulic conductivity is calculated by the Campbell (1974) equation which is based on the Brooks and Corey (1964) model. The model does not allow drainage from one layer at a rate greater than the maximum infiltration rate of the underlying layer. This allows the simulation of a layer having a lower hydraulic conductivity below a layer of higher hydraulic conductivity without treating the lower layer as a liner.

Vertical drainage through soil liners is termed percolation in the HELP model. Soil liners are assumed to remain saturated; however, percolation occurs only when there is a zone of saturation directly above the liner. Percolation is calculated from Darcy's law.

Leakage through geomembranes is modeled as a reduction of the cross-sectional area of flow through the subsoil below the liner. The rate of flow through the leaking subsoil is computed as the percolation rate through a saturated barrier soil liner. This method provides good results for composite liners but is not very good for just a geomembrane. Version 3 will include an improved leakage model for geomembranes based on the works of Brown *et al.* (1987) and Giroud *et al.* (1989a,b,c).

The HELP model simulates lateral drainage using a steady-state analytical approximation of the Boussinesq equation (McEnroe and Schroeder, 1988). The analytical approximation was developed by converting the Boussinesq equation to a nondimensional form and solving it for the extremes in nondimensional average saturated soil depth. These two solutions were then fitted with an approximation that covers the rest of the range of nondimensional depths. The approximation involves less than 1% error.

Vegetative growth

The HELP model accounts for seasonal variation in active and dormant aboveground biomass and leaf area index through a general vegetative growth model. The maximum value of leaf area index depends on type of vegetation, soil fertility, climate, and management factors.

Input data

Required climatic input data include daily precipitation, daily mean temperature, daily solar evaporation, maximum leaf area index, growing season, and evaporative zone depth. The computer program includes default daily weather data for 102 US cities and can synthetically generate weather data for 183 US cities, if desired.

The input soil data required for each layer include porosity, field capacity, wilting point, and saturated hydraulic conductivity. Default data are included for 15 soil types as well as solid waste. Design data include SCS runoff curve number (defaults are provided), surface area, and miscellaneous control parameters such as number of layers in the profile and thicknesses of the layers.

Output data

The output from the HELP consists primarily of percolation or leakage through each liner and depth of saturation on the surface of liners (e.g., in drainage layers). Incremental and cumulative qualities of water budget for the various components are computed and printed.

10.3 CASE STUDIES

Available data on landfill leachate production are very limited, especially for periods of record that extend significantly beyond the initial water-balance equilibration period, which may last up to several years. Available data on the important facets of the water balance, such as runoff, evaporation, rainfall, soil moisture, leachate ponding depths, percolation rates and detailed soil characteristics, are even more limited. This may be affected to some degree by the reluctance of waste disposal operators to subject their facilities to public scrutiny. Also, this type of data collection can be costly. However, the verification of water balance models and the understanding of processes involved in leachate production and migration are highly dependent on obtaining this information.

This section presents published water balance data for 17 landfill cells at six sites. Leachate drainage rates were measured from all cells. Runoff was measured from 11 cells at two sites. Daily rainfall was measured at one site. Soil moisture, leachate ponding depths and barrier-soil percolation rates were not adequately or consistently measured in any of the cells. Most sites had at least limited data on the hydraulic conductivity of the clay liner, although the testing methods, and therefore the meaning of the test results, varied widely. Data describing cover soils or the extent of surface vegetation were generally lacking.

10.3.1 University of Wisconsin lysimeters

From 1970 to 1977, eight large lysimeters filled with either shredded or unprocessed refuse were monitored for surface runoff and leachate production at the University of Wisconsin at Madison (Ham 1980). Each cell was 18 m long by 9 m wide. The depth of refuse was either 1.2 m or 2.4 m. Refuse was underlain with a 100 mm layer of crushed granite over a polyethylene barrier. Bottom slopes, of approximately 3%, directed leachate to a collection box at the center of the cell where it was periodically pumped and the volume of leachate was measured. Four cells were covered with 150 mm of sandy silt soil; the remaining four cells were left uncovered. The top surfaces of all cells were sloped at 3% toward one of the 18 m long walls where surface runoff was collected and measured.

Climatological conditions at this location are as follows:

1. average temperature is 7 °C;
2. average annual precipitation is 790 mm;
3. minimum daily temperature falls below freezing on 163 days per year; and
4. average daily solar radiation is 330 langleys.

The vegetative cover was mixed volunteer vegetation, comparable to meadow grass, which became established on both covered and uncovered cells over a several year period. This vegetation grew more quickly and more densely on the uncovered cells.

The water balance measurements are shown in Table 10.1 expressed in terms of percent precipitation based on US National Oceanic and Atmospheric Administration (NOAA) precipitation records for a gage located approximately two miles from the landfill site. It is important to highlight the variability in the measured results between similar cells. The measured runoff varied from 71–129% of the mean for the

Table 10.1 Cumulative measurements of water balance components for landfill cells at University of Wisconsin at Madison

cell number	*period of measurement* (yr)*	*runoff (% precip.)*	*ET + ΔS[†] (% precip.)*	*leachate drainage (% precip.)*
Covered cells				
1	4	5.4	67.7	26.9
2	4	6.0	67.1	26.9
3	5	9.1	73.6	17.3
8	3	9.8	67.5	22.7
Mean		7.6	68.9	23.5
Standard deviation		2.2	3.1	4.6
Uncovered cells				
4	5	3.6	79.2	17.2
5	2	3.4	90.1	6.5
6	2	3.9	92.6	3.5
7	3	2.8	73.5	23.7
Mean		3.4	83.9	12.7
Standard deviation		0.5	9.1	9.4

* Excludes first two years of landfill operation.
[†] ET + ΔS = evapotranspiration + change in landfill moisture storage
= 100 − runoff % − leachate drainage %

covered cells and 82–115% for the uncovered cells. The measured drainage varied from 74–114% of the mean for the covered cells and 28–187% for the uncovered cells. The difference between the cell designs (besides the presence of a soil cover) was depth of the waste layer and presence of shredded refuse, neither of which should significantly affect the water balance after a two-year equilibration period. Overall, evapotranspiration accounted for 69–84% of the water balance and was 2.9–6.6 times as large as the leachate drainage volume. Runoff was the smallest component of the water balance ranging from 3–8% of precipitation.

10.3.2 Sonoma County test cells

A solid-waste stabilization project was conducted in Sonoma County, California from 1971 to 1974 to determine the effect of applying water, septic tank pumpings, or recycled leachate to waste in landfills (EMCON Associates 1975). Each of the five cells studied was 15 m long by 15 m wide and 2.4 m deep. The landfill contained a soil liner; no flexible membrane liner was present. The soil liner material was native sandy clay. Pervious lenses were encountered during excavation and were replaced by a 0.6 m thickness of sandy clay. The landfill cover material

Table 10.2 Cumulative measurements of water balance components for landfill cells at Sonoma County, California

cell	*period of measurement* (yr)*	*runoff (% precip.)*	*ET + ΔS + PERC*[†] *(% precip.)*	*leachate drainage (% precip.)*
A	2	59.3	39.5	1.26
B	2	64.1	32.7	3.21
C	2	–[‡]	–[‡]	314.49
D	2	–[‡]	–[‡]	996.53
E	2	59.7	35.9	4.35
Mean[§]		61.0	36.0	2.94
Standard deviation[§]		2.7	3.4	1.56

* Excludes first year of landfill operation.
[†] $ET + \Delta S + PERC$ = evaporation + change in landfill moisture storage + percolation through bottom soil liner
= 100 − runoff % − leachate drainage %
[‡] No measurement
[§] Computed only for cells A, B and E.

was a 0.6 m thick layer of sandy clay. There was no documentation of surface vegetation.

Cells A, B and E were constructed and operated as typical sanitary landfills except that the moisture content of the refuse was brought to field capacity with water prior to capping in cell B and with septic tank pumpings in cell E. Cells C and D were constructed with an inflow pipe network placed on top of a 300 mm thick sand/gravel distribution medium installed between the waste layer and the soil cover. Water was continuously added to the waste through this network in cell C, whereas landfill leachate was continuously recycled through the waste layer in cell D.

This test site was located 45 miles north of San Francisco. The mean temperature is 14 °C, with the daily minimum temperature falling below freezing on 39 days per year. Mean annual precipitation is 790 mm. Ninety-five per cent of this precipitation occurs from October to April. The mean daily solar radiation is approximately 410 langleys. Daily precipitation was measured at the landfill site.

The water balance measurements are shown in Table 10.2. In this case, the largest component of the water balance was runoff at 61% of precipitation. The evapotranspiration variable accounted for 36% of the precipitation, approximately 12 times the leachate drainage percentage. The leachate drainage from cell C was 314% of precipitation compared to inflow through the pipe network of 388%. For cell D, the leachate drainage was 997% of precipitation compared to inflow of 1018%.

10.3.3 Boone County test cell

A test cell was studied from 1971 to 1980 in Boone County, Kentucky, to evaluate volume and characteristics of the leachate, composition of gases, internal temperature, settlement, and clay liner efficiency (EMCON Associates, 1983; Wigh, 1984). The cell consisted of a 9-m wide by 45-m long trench with vertical walls. The bottom of the trench at both ends was sloped toward the center at 14%. The middle 15 m were sloped at 7% to the transverse centerline. A 0.8 mm synthetic liner, 9 m wide by 15 m long, was centered over the base of the cell. A percolation collection pipe embedded in gravel was placed on the synthetic liner at the bottom of the cell. A 450-mm thick compacted clay liner was placed over the synthetic liner and collection pipe. This liner was found to have an average in-place hydraulic conductivity of 4×10^{-7} cm/s at the conclusion of the cell study. A second pipe was embedded in a gravel-filled section of the clay liner directly above the lower pipe to collect lateral leachate drainage above the clay liner. A polyethylene sheet was placed beneath this pipe to prevent leachate from short-circuiting to the lower pipe. Residential refuse was placed and compacted above this liner system, and a 0.6 m layer of cover soil was deposited onto the completed waste layer.

At the end of the study, a field assessment was conducted. The cover soil was classified as CL according to the Unified Soil Classification System (USCS) and was found to have an average in-place hydraulic conductivity of 5×10^{-5} cm/s. Also, secondary openings were found in the soil cover through which relatively rapid infiltration could occur.

The test site is located approximately 30 km south of Cincinnati, Ohio. The mean annual temperature is 12 °C, with the daily minimum temperature falling below freezing on 111 days per year. The mean annual precipitation is 1.1 m. The mean daily solar radiation is approximately 360 langleys. Precipitation at the test site was recorded once or more per week throughout the study period. These values were used to adjust daily precipitation records from the nearest NOAA weather station, located approximately 24 kms away at Covington, Kentucky.

At this landfill, leachate drainage amounted to 28.8% of the precipitation over a 7-yr measurement period. Runoff was not measured. A negligible amount of percolation was measured from the percolation collection pipe.

10.3.4 Wisconsin County landfills

Three landfills located in Brown, Eau Claire and Marathon Counties, Wisconsin were monitored for periods ranging from 3 to 7 yr (Gordon *et al.*, 1984). The Brown County landfill began in 1976 with a planned size of 23 ha. The landfill is located near Green Bay, Wisconsin, where the mean annual temperature is 7 °C. The daily minimum temperature falls below freezing on 163 days per year. The mean annual precipitation is 690 mm, and the mean daily solar radiation is approximately 330 langleys. Daily precipitation data were measured at the NOAA weather station at Green Bay.

The base of the Brown County landfill consists of a 1.2-m thick compacted clay liner and a leachate collection system. Phases 1 through 3 (approximately 7 ha) were designed with a 1% base slope, a leachate flow distance of 91 m, and a 300-mm thick sand blanket on the base and sidewalls; the leachate flow distance was shortened to 30 m for Phase 4 (approximately 3 ha). In-place hydraulic conductivity tests resulted in an average hydraulic conductivity of the liner of 7×10^{-8} cm/s. The average depth of waste was 23 m. The waste initially consisted of municipal and commercial refuse. Later, industrial waste consisting of flyash and water treatment plant sludge was added. Daily cover is a 150 mm thickness of silty clay soil (USCS classification CL). During the period of leachate monitoring presented here, much of the landfill area was overlain only with daily cover. Final cover consists of 0.6 m of compacted clay and 150 mm of topsoil.

The Eau Claire County landfill opened in December 1978 in the vicinity of Eau Claire, Wisconsin. The data presented here covers the period of landfill expansion up to 6 ha.

The mean annual temperature is 6 °C with 172 days per year experiencing a minimum temperature below freezing. The mean annual precipitation is 740 mm, and the mean daily solar radiation is approximately 330 langleys.

The base of the landfill consists of a 1.2 m thick compacted clay liner overlaid with a 300-mm thick sand blanket and a leachate collection system. The liner slope is 1%, and the maximum leachate flow distance along the base is approximately 40 m. The waste is primarily municipal and commercial refuse with minor amounts of industrial wastes. Daily cover at this landfill is a 150 mm layer of sand. A 750 mm interim layer of papermill sludge covered a large percentage of the site during the period considered here. Final capping will include a 300 mm sand blanket over the sludge and then 150 mm of topsoil. The ultimate maximum fill thickness will be approximately 15 m.

The Marathon County landfill was opened in December 1980 in central Wisconsin near the city of Wausau. Approximately 6 ha were under development at the end of the data collection period presented in this section. The mean annual temperature is 6 °C with 170 days per year experiencing a minimum temperature below freezing. The mean annual precipitation is 790 mm, and the mean daily solar radiation is approximately 330 langleys.

The facility is designed with a 1.2-m-thick compacted clay liner and a leachate collection system. The liner is sloped at 1% toward 200-mm-diameter perforated PVC pipes. The pipes are embedded in a 300-mm-deep trench oriented at approximately 45 degrees to the slope of the liner. The clay liner thickness increases to 1.5 m in the vicinity of these trenches. The maximum leachate flow distance is about 75 m in Phases 1 and 2 (3.6 ha total). Phase 3 (2 ha) was designed for a maximum flow distance of 30 m. Following construction of the clay liner and installation of the collection pipe, a minimum 300-mm-thick silty sand drainage blanket was placed over the base and sidewalls. The waste is 75% municipal refuse and 25% papermill sludge. The thickness of the waste layer ranges from 15 to 25 m. The daily cover is 150 to 300 mm of sand. The final cover is 0.6 m of clay covered by 150 mm of silty sand and 150 mm of topsoil.

The leachate drainage results for these three landfills are shown in Table 10.3. The leachate collected ranged from 3.1 to 7.7% of precipitation. Runoff was not measured. It is possible that steady-state moisture conditions had not been reached during the period of these measurements because the landfills remained active during this period and because of their large depth. If this was the case, the drainage percentages shown in Table 10.3 are probably lower than the long-term steady state values.

10.4 SENSITIVITY OF WATER BALANCE TO LANDFILL DESIGN PARAMETERS

10.4.1 Introduction

This section examines the sensitivity of a landfill water balance to numerous landfill design variables using the HELP model. This information is useful in a variety of ways. It can aid the design engineer in selecting preliminary design alternatives for municipal or hazardous waste landfills. It can serve as a basis for regulatory agencies to establish and evaluate technical guidelines. It can also provide additional insight

Table 10.3 Cumulative measurements of water balance components for Wisconsin County landfills

landfill	*period of measurement** *(yr.)*	*leachate drainage* *(% precip.)*
Brown	5	3.1
Eau Claire	5	7.7
Marathon	3	6.9

*Excludes the first year of landfill operation, except for Marathon which includes the first year.

on the importance and interaction of specific design variables on the water balance. Finally, it can assist in evaluating the suitability of methodologies used in the computer model. The analyses include examination of cover systems and lateral drainage/liner systems.

10.4.2 Comparison of typical cover designs

Design parameters

Three locations were studied to determine the effect of various climatological regimes on cover performance – Santa Maria, California; Schenectady, New York; and Shreveport, Louisiana. These locations represent a wide range in levels of precipitation, temperature, and solar radiation.

Two cover designs were examined as shown in Fig. 10.3. One is typical of newer RCRA landfills where 600 mm of topsoil overlies a 300-mm-thick lateral drainage layer having a saturated hydraulic conductivity of 3×10^{-2} cm/sec, a slope of 0.03 m/m and a maximum

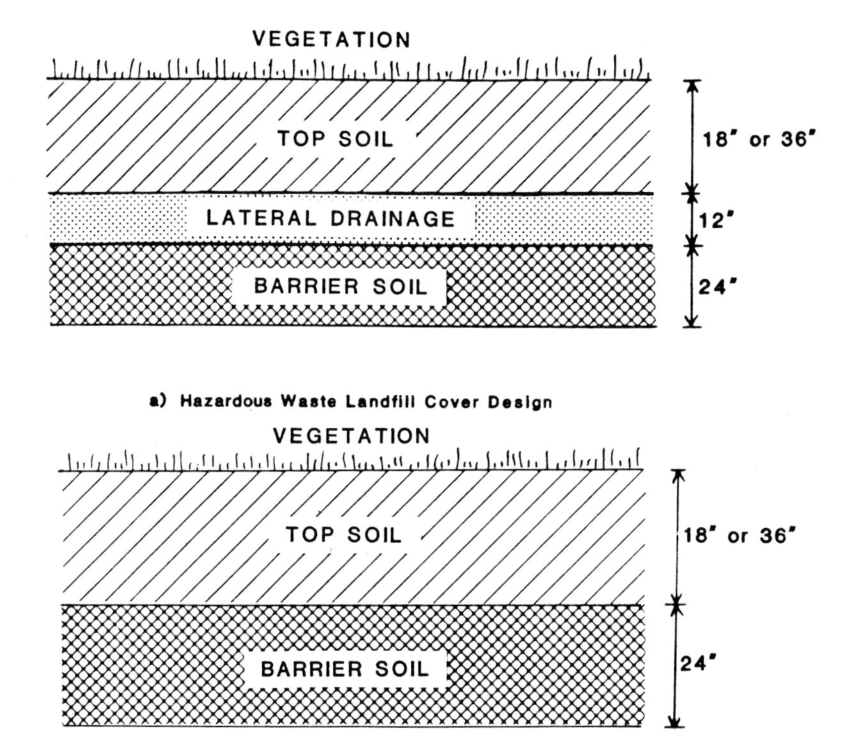

Fig. 10.3 Typical cover profiles: (a) hazardous waste landfill cover design; and (b) municipal landfill cover design.

drainage length of 60 m. The drainage layer is underlain by a 0.6-m-thick soil liner having a saturated hydraulic conductivity of 1×10^{-7} cm/sec. The other design is typical of older municipal sanitary landfills where a topsoil layer overlies a 0.6-m-thick soil liner having a saturated hydraulic conductivity of 1×10^{-6} cm/sec.

Two types of topsoil were considered in the cover designs: sandy loam and silty clayey loam. The topsoil type was used to select soil porosity, field capacity, wilting point and hydraulic conductivity, besides influencing the selection of the runoff curve number. In addition to two types of topsoil, two thicknesses of topsoil were examined: 450 and 900 mm.

The vegetative cover was designated as being either a good stand of grass or a poor stand of grass. This selection dictated the values for leaf area index, evaporative depth and runoff curve number, and influenced the value used for the saturated hydraulic conductivity of the topsoil.

Effects of vegetation

A good stand of grass represents three times the quantity of vegetation as that of a poor stand. Table 10.4 presents the water balance results for both cover systems at all three sites as a function of level of vegetation. The results are given in units of percent of the precipitation.

Vegetation reduces surface runoff and increases evapotranspiration.

Table 10.4 Effects of climate and vegetation

	two layer cover design location*			*three layer cover design† location*		
	CA	*LA* (% *precipitation*)	*NY*	*CA*	*LA* (% *precipitation*)	*NY*
Poor grass						
Runoff	5.6	4.6	5.5	3.0	4.4	2.2
Evapotranspiration	51.8	53.0	52.1	51.6	51.9	50.3
Lateral drainage	–	–	–	41.2	40.6	44.0
Percolation	42.6	42.4	42.4	4.2	3.1	2.5
Good grass						
Runoff	3.1	0.2	3.5	0.0	0.2	0.0
Evapotranspiration	55.0	57.2	55.3	52.6	53.0	51.0
Lateral drainage	–	–	–	43.2	43.7	45.5
Percolation	42.9	42.6	41.2	4.2	3.1	2.5

* 900 mm of sandy loam topsoil and 0.6 m of 10^{-6} cm/sec clay liner.
† 450 mm of sandy loam topsoil, 300 mm of 0.03 cm/sec sand with 60 m drain length at 3% slope, and 0.6 m of 10^{-7} cm/sec clay liner.

Evapotranspiration is greater because of the plant demand for moisture and a greater quantity of water is available for evapotranspiration due to greater infiltration and a greater evaporative zone. Runoff is less because vegetation increases the minimum infiltration rate, drying rate, interception, and surface roughness resulting in a decrease in the runoff curve number. The influence of surface vegetation on the volume of lateral drainage and percolation or leakage from the cover is varied. However, the quantity of vegetation tends to have very little effect on the percolation or leakage through the cover system. For the cover with lateral drainage, the increase in infiltration with good grass was greater than the increase in evapotranspiration, resulting in a larger volume of lateral drainage and a negligible change in percolation. For the cover without lateral drainage, the increase in infiltration yielded high heads or depths of saturation above the liner that permitted greater evapotranspiration by maintaining higher moisture contents in the evaporative zone. Consequently, the increase in evapotranspiration was greater than the increase in infiltration. This resulted in a trend toward a small decrease in percolation for a higher level of vegetation. The opposite trend may occur for vegetative layers having a lower saturated hydraulic conductivity and a higher plant available water capacity (field capacity minus wilting point). The results were similar at all three sites despite quite different climates.

In summary, vegetation decreases runoff and increases evapotranspiration but tends to have little effect on the rest of the water balance. The magnitude of the effects is design dependent and to a lesser degree climate dependent. The main function of vegetation is to control erosion.

Effects of topsoil thickness

Two topsoil thicknesses were examined: 450 and 900 mm. Table 10.5 presents the water balance results for the two layer cover system at all three sites as a function of topsoil thickness. The results are given in units of percent of the precipitation. The cover system with lateral drainage was not used in this analysis because lateral drainage would negate the effects by preventing or minimizing the intrusion of the saturated zone above the liner into the evaporative zone.

Significant differences existed between the 450- and 900-mm topsoil depth simulations in the absence of lateral drainage. The effects were similar at all three sites. Runoff and evapotranspiration were greater for the shallower depth to the liner, indicating that the head above the barrier soil layer maintained higher moisture contents in the evaporative zone. The percolation was consequently less than the cases with greater topsoil thickness. The 900-mm depth to the liner permits larger heads and longer sustaining heads since a greater thickness of material

Table 10.5 Effects of climate and topsoil thickness

	45.7 cm of topsoil location			*91.4 cm of topsoil location*		
	CA	*LA* (*% precipitation*)	*NY*	*CA*	*LA* (*% precipitation*)	*NY*
Runoff	11.2	7.5	13.4	5.6	4.6	5.5
Evapotranspiration	51.9	56.9	54.5	51.8	53.0	52.1
Percolation	36.9	35.6	32.1	42.6	42.4	42.4

Sandy loam topsoil with a poor stand of grass underlain by 0.6 m of 10^{-6} cm/sec clay liner.

is below the evaporative zone and is free from abstraction of water by evapotranspiration. The larger heads provide a greater pressure gradient to increase the leakage rate through the cover system.

In general, the effects of topsoil thickness would vary greatly as the thickness increases from several centimeters to several meters. Throughout the transition, the quantity of runoff should continue to decrease until the depth to the liner becomes sufficiently great so as to prevent the zone of saturation from ever climbing into the evaporative zone. Similarly, the percolation through the liner should continue to increase until there is no interaction between the saturation zone and the evaporative zone. The evapotranspiration is expected to increase initially as the available storage in the evaporative zone increases, that is until the depth to the liner equals the maximum depth that evapotranspiration can reach. At greater depths the evapotranspiration should continue to decrease until the depth to the liner is sufficient to prevent any further interactions between the evaporative and saturation zones.

While percolation increases with topsoil thickness given identical properties for all layers in the cover system, adequate thickness must be provided in a design to insure the integrity of the cover system. A small topsoil thickness would not provide adequate water storage to support vegetation, maintain soil stability and control erosion. Similarly, a shallow depth to the liner would promote desiccation or freezing of the liner which may greatly increase its permeability and therefore the percolation.

Effects of topsoil type

Two topsoil types were examined: sandy loam and silty clayey loam. Table 10.6 presents the water balance results for the three layer cover system at all three sites as a function of topsoil type. The results are given in units of percent of the precipitation. The cover system without

Table 10.6 Effects of climate and topsoil types

	*three layer cover design**					
	sandy loam location			*silty clayey loam location*		
	CA	*LA (% precip.)*	*NY*	*CA*	*LA (% precip.)*	*NY*
Runoff	3.0	4.4	2.2	21.6	22.3	19.2
Evapotranspiration	51.6	51.9	50.3	61.2	64.4	58.6
Lateral drainage	41.2	40.6	44.0	15.0	11.3	20.3
Percolation	4.2	3.1	2.5	2.2	2.0	1.9

*450 mm of topsoil with poor stand of grass, 300 mm of 0.03 cm/sec sand with 60 m drain length at 3% slope, and 0.6 m of 10^{-7} cm/sec clay liner.

lateral drainage was not used in this analysis because the intrusion of the saturated zone above the liner into the evaporative zone would decrease the magnitude of the effects.

The results show that the clayey topsoil significantly increased both runoff and evapotranspiration, which in turn greatly decreased lateral drainage and percolation. The results were similar at all three sites. Runoff increased from about 3% to 20% of the precipitation, due primarily to the larger runoff curve number selected for the clayey soil based on its lower minimum infiltration rate. Evapotranspiration increased approximately from 51% to 61% of precipitation, due to the lower hydraulic conductivity of the clayey soil and, more importantly, the larger plant available water capacity. The lower hydraulic conductivity of the clayey soil slowed the drainage rate, maintaining moisture contents above field capacity for longer periods of time and allowing greater evapotranspiration. The larger plant available water capacity of the clayey soil provided a larger moisture reservoir available for evapotranspiration after gravity drainage ceased. The lateral drainage was reduced from about 42% to 16% of the precipitation, and the percolation was reduced from about 3% to 2% of precipitation.

Use of lateral drainage layer

In general, the use of a lateral drainage layer would be expected to decrease the height of the saturation zone above the liner by draining some of the infiltrated water from the cover system. As such, percolation through clay liners would decrease slightly. In addition, runoff and evapotranspiration would also tend to decrease but the magnitude of the change would be design dependent. Topsoil thickness, topsoil type, vegetation and climate would have impacts.

Effects of climate

The effects of climate were examined in each of the previous sections. Climate affects the absolute magnitude, in millimeters, of the water budget components. However, Tables 10.4 through 10.6 show that climate has a much smaller effect on the relative magnitude of the water budget components in terms of percent of the precipitation. The relative proportions of the water budget components are primarily design dependent while the magnitudes are strongly dependent on the magnitude of the precipitation.

10.4.3 Liner/drain systems

This section examines the effects of liner/drain system design on the performance of the drain system under conditions typical of cover systems and leachate collection systems in open and closed landfills. Performance was determined by the apportionment of the drainage into the drain layer between lateral drainage and percolation through the liner. In addition, the effect of design on the resulting depth of saturation was also examined. For the cover system or open landfill the drainage into the drain layer was 1.3 m/yr, distributed temporally in accordance with the precipitation at Shreveport, LA. For the closed landfill the drainage into the drain layer was distributed uniformly through time at a rate of 200 mm/yr.

Clay liner/drain systems

Saturated hydraulic conductivities

The liner/drain system used in this analysis is shown in Fig. 10.4a. The value of KD (the saturated hydraulic conductivity of the drain layer) ranged from 0.001 to 1 cm/sec while the value of KP (the saturated hydraulic conductivity of the clay liner) ranged from 10^{-8} to 10^{-5} cm/sec. The slope of the liner surface toward the drainage collector was 3%, and the maximum drainage length to the collector was 23 m. The results of the drainage efficiency determinations for the various combinations of KD and KP are shown in Fig. 10.5 where the average annual volumes of lateral drainage and percolation expressed as a percentage of annual inflow are plotted.

Summarizing the results shown in Fig. 10.5, the saturated hydraulic conductivity of the liner is the primary control of leakage through a clay liner. At hydraulic conductivities below about 10^{-6} cm/sec the leakage is nearly proportional to the value of KP; that is, an order of magnitude decrease in the value of KP yields nearly an order of magnitude decrease in percolation. The value of KD has only a small effect on the leakage through liners having a KP of 10^{-7} cm/sec or less.

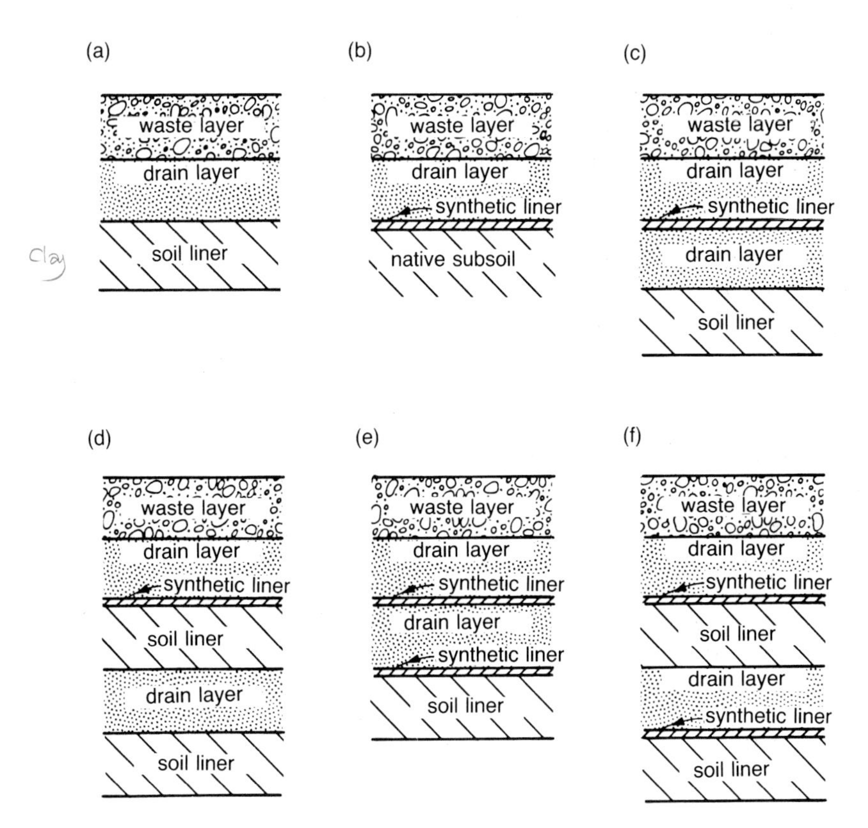

Fig. 10.4 Alternative designs considered.

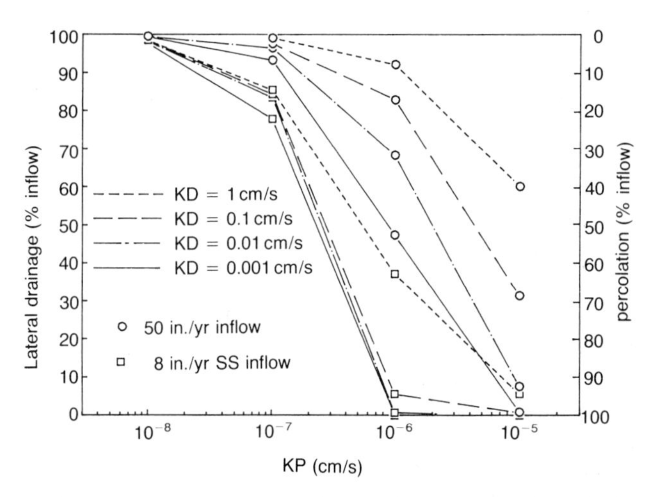

Fig. 10.5 Effect of saturated hydraulic conductivity on lateral drainage and percolation.

Changing the value of KD by three orders of magnitude when using these low permeability liners yields much less than an order of magnitude change in percolation.

Figures 10.6 and 10.7 show that percolation tends to dominate at ratios of KD/KP below 10^6. This is particularly true as the depth of saturation or inflow decreases. When heads remain constant, the ratio of lateral drainage to percolation is a linear function of KD/KP. Using the maximum head allowed by RCRA of 300 mm and the current minimum KD/KP ratio implied by RCRA of 10^5, a percolation of 2.3% of inflow results; however, an unusually large steady-state inflow of 2 m/yr is required to achieve this condition. Therefore when using the RCRA guidance design, the peak and steady-state average heads will be considerably smaller than 300 mm at virtually all locations.

Slope and drainage length

The volumes of lateral drainage and percolation vary little with changes in slope and drainage length under both steady and unsteady inflows. A ninefold increase in slope reduced the percolation by a maximum of

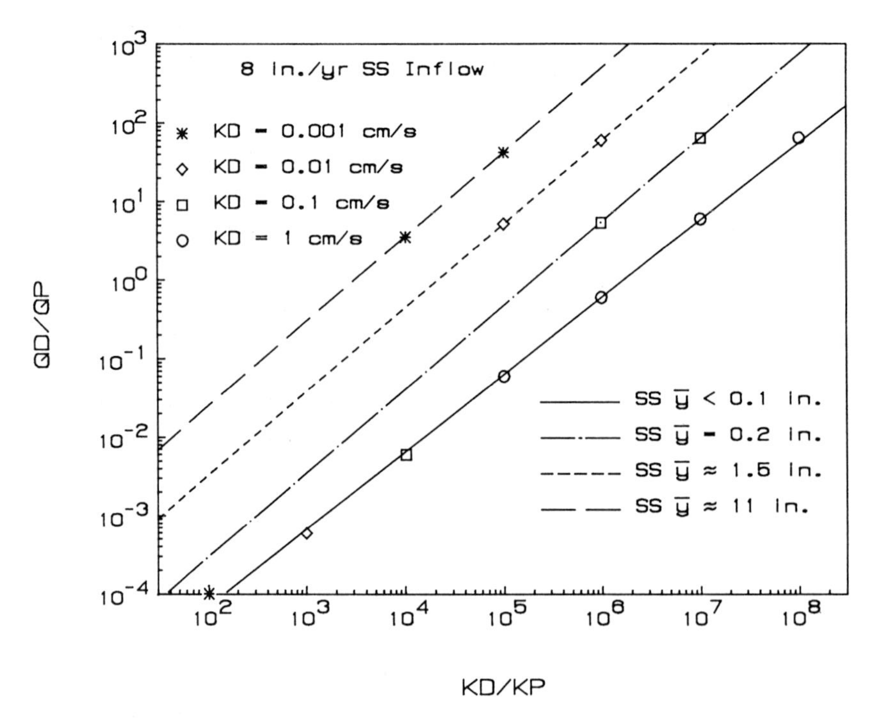

Fig. 10.6 Effect of ratio of drainage layer saturated hydraulic conductivity to soil liner saturated hydraulic conductivity on ratio of lateral drainage to percolation for a steady-state (SS) inflow of 8 in/yr (200 mm/yr).

25% for the unsteady inflow and 13% for the steady inflow. As the drainage length reduces and the slope increases, the lateral drainage rate increases. As a result, the head decreases and is maintained at smaller depths for shorter durations. Consequently, the percolation decreases since it is a function of the head on the liner. A ninefold decrease in drainage length reduced the percolation by a maximum of 50% for the unsteady inflow and 25% for the steady inflow. A ninefold increase in slope and decrease in length decreased the percolation by about 60% for the unsteady inflow and about 30% for the steady inflow.

The head in the drain layer varies greatly with changes in slope and drainage length. For a steady inflow the average head increases linearly with an increase in drainage length and an increase in the inverse of the slope. A similar relationship exists between the peak average head during the simulation and the ratio L/S (drainage length divided by drainage slope) for unsteady inflow. The average head is slightly influenced by the product of the slope and drainage length when the head is similar to this product.

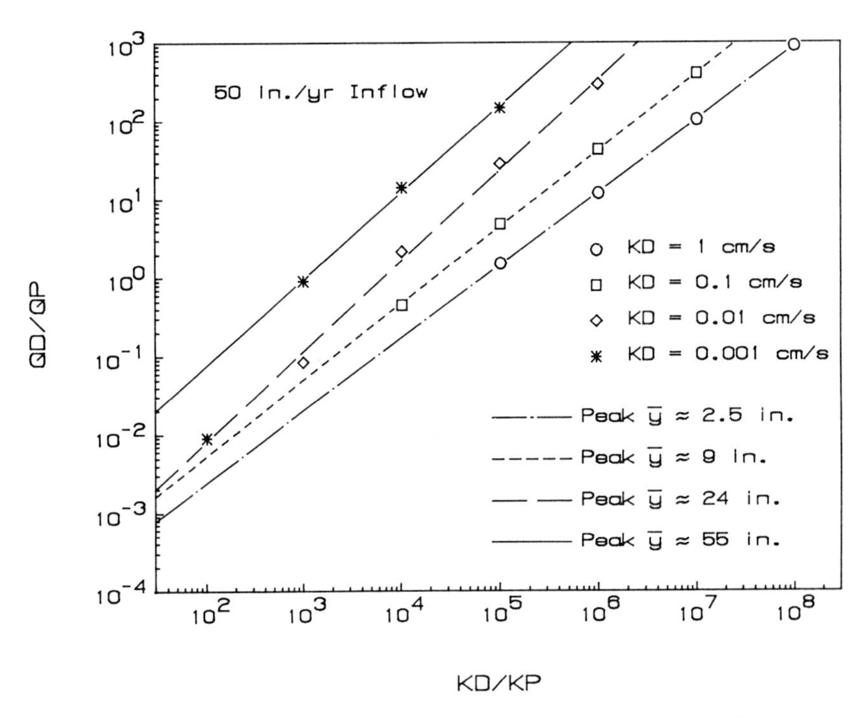

Fig. 10.7 Effect of ratio of drainage layer saturated hydraulic conductivity to soil liner saturated hydraulic conductivity on ratio of lateral drainage to percolation for an unsteady inflow of 50 in/yr (1.3 m/yr).

Geomembrane/drain suptems

A single synthetic liner under a drain layer as shown in Fig. 10.4 is examined in this section. It is assumed that the synthetic liner was laid directly on a 3 m-thick layer of native subsoil. The drainage layer had a saturated hydraulic conductivity of 10^{-2} cm/sec, a slope of 3% and a drainage length of 23 m. This case will be used to demonstrate the influence of the synthetic liner leakage fraction and the saturated hydraulic conductivity of the native subsoil on the liner system performance. The properties of the subsoil ranged from sand to clay in the analysis.

Liner leakage fraction

Brown *et al.* (1987) conducted laboratory experiments and developed predictive equations to quantify leakage rates through various size holes in synthetic liners over soil. They assumed that the measured leakage rates corresponded to a uniform vertical percolation rate equal to the saturated hydraulic conductivity through a circular cross-sectional area of the soil liner directly beneath the hole. Using the data relating leakage and cross-sectional area of flow, Brown *et al.* (1987) developed predictive equations for the radius or area of this flow cross section as a function of hole size, depth of leachate ponding, and saturated hydraulic conductivity of the soil. Figure 10.8 presents their results. The radius of saturated flow through the subsoil was significantly greater than the radius of the hole in the synthetic liner. The cross-sectional area of saturated flow was multiplied by the number of holes per unit area of synthetic liner to compute the synthetic liner leakage fraction. Liner leakage fraction is simply defined as the total horizontal area of saturated flow through the subsoil beneath all of the liner holes divided by the horizontal area of the liner.

Liner leakage fraction is a function of many parameters, some quantitatively defined and others qualitatively. Liner leakage fraction increases linearly with increases in the number of holes of the same size and shape. Shape also has a strong effect on the leakage; tears have larger leakage than punctures. Increasing the size of circular holes yields only a slight increase in the leakage while increasing the length of a tear or bad seam increases the leakage nearly linearly. Leakage also increases nearly linearly with increases in head or depth of saturation above the liner. The leakage fraction is also affected by the gap width between the liner and the subsoil. Gap width is a measure of the seal between the liner and the subsoil. The smaller the gap the better the seal. The seal is a function of the subsoil, installation, liner placement and subsoil preparation. Installation of the liner on coarse-grained subsoil, clods, debris or filter fabric provide a poor seal as will wrinkles in the liner. Coarse-grained subsoils decrease the leakage fraction while

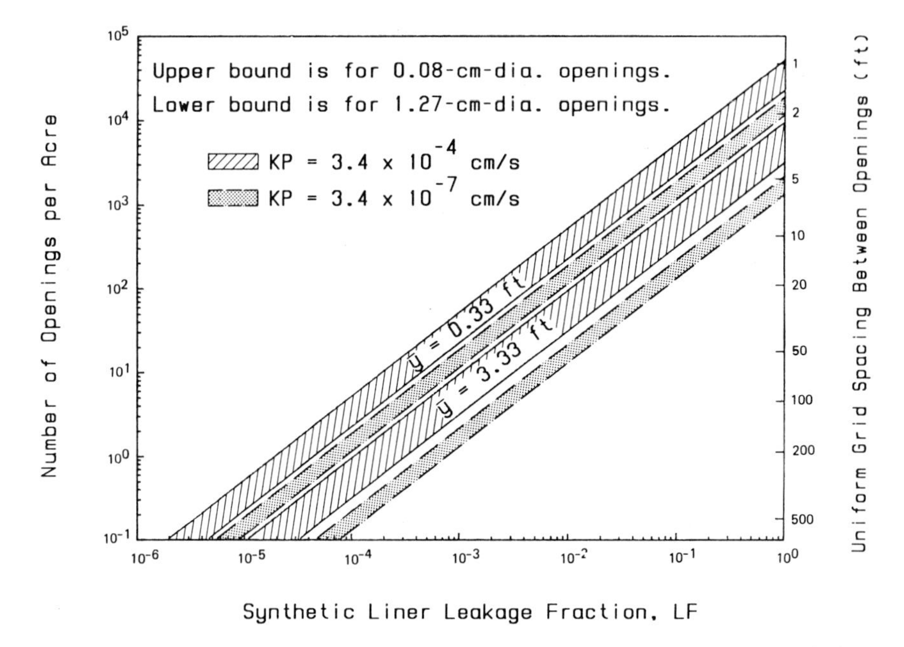

Fig. 10.8 Synthetic liner leakage fraction as a function of density of holes, size of holes, head on the liner, and saturated hydraulic conductivity of the liner.

greatly increasing the leakage. The greater permeability of coarse materials allows greater flow through a smaller area of saturated flow, reducing the spreading required to accommodate the leakage through the liner.

System performance

The percolation rate through a leaking synthetic liner is a linear function of the leakage fraction for a given subsoil when the average head on the liner is constant. The percolation rate expressed as a percentage of inflow rate is shown graphically in Fig. 10.9 as a function of the leakage fraction. This relationship is shown for a range of values of the average head and for a steady inflow rate of 200 mm/yr. Figure 10.9 emphasizes the significant influence of average head or inflow on controlling the distribution of the inflow between vertical percolation and lateral drainage. This figure shows that to maintain the vertical percolation rate at less than 1% of the inflow rate for heads greater than 2.5 mm, the leakage fraction for a clay subsoil (KP = 10^{-6} cm/sec) must be less than 5×10^{-4} and for a sandy subsoil (KP = 10^{-3} cm/sec) must be less than 5×10^{-7}. The overall effectiveness of a flexible membrane liner is equivalent to a soil liner having a saturated hydraulic conductivity equal to the product of the leakage fraction and the saturated hydraulic conductivity of the subsoil when the conductivity

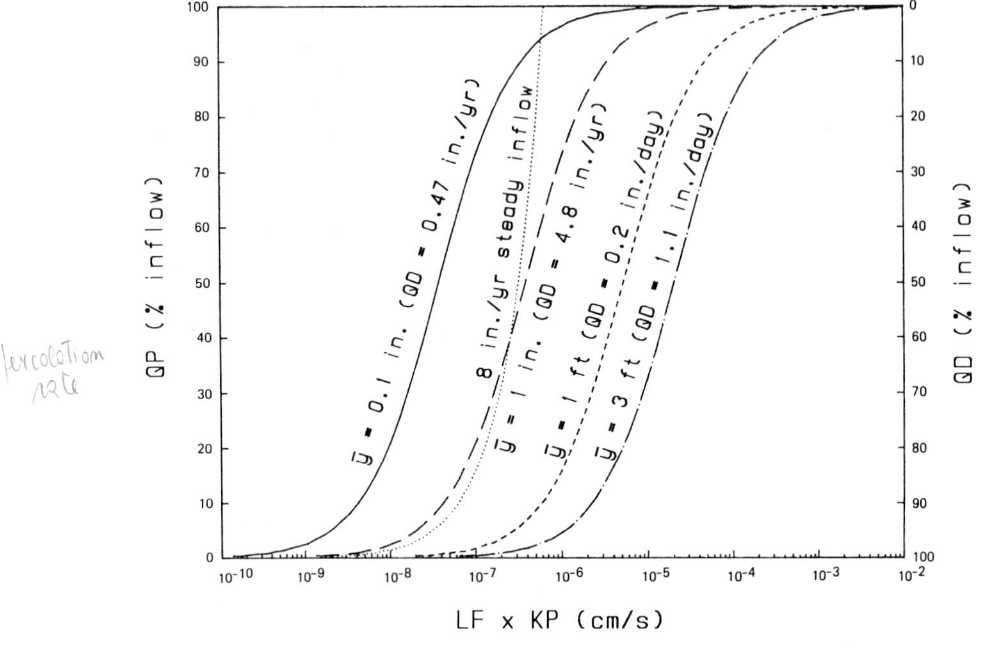

Fig. 10.9 Effect of liner leakage fraction on system performance.

of the subsoil is equal to or less than the conductivity of the material above the liner.

Double liner systems

A comparison of designs in Fig. 10.4c–f was made by Schroeder and Peyton (1990). Their results are summarized here. They found that the Design (f) in Fig. 10.4f is the most effective in detecting the earliest leaks with the least amount of vertical leakage through the primary liner and also through the bottom soil liner. Design (d) yields the same quantity of leakage through the primary liner; however, leakage in design (d) would probably never be detected or collected. Therefore, the bottom liner in design (d) is not functional. Designs (d) and (e) yield the same leakage through the bottom liner but design (e) detects leakage through the primary liner at the lowest leakage fraction. Design (c) also detects leaks at very small leakage fractions but allows significant vertical percolation through the bottom soil liner before detection. The leakage through the primary liner in designs (c) and (e) is large even at low leakage fractions. Therefore, synthetic membranes placed on highly permeable subsoils are ineffective except for very low inflows and for very low leakage fractions. Synthetic membranes are best used in conjunction with a low-permeability soil as a composite liner. Comparison of designs (b) and (c) demonstrates this point. Both designs are composed of one synthetic membrane and one soil liner, but the leakage from the composite liner (design b) is much lower than the leakage from the double liner system (design c).

It is interesting to compare the single liner performance of design (b) to the double liner performance of design (d), assuming the soil liner saturated hydraulic conductivity in design (b) is the same as design (d). The vertical percolation leaving the system in design (b) is essentially the same as that leaving the secondary liner in design (d). The secondary liner in design (d) is nonfunctional since the percolation rate of the second soil liner is generally equal to or greater than the leakage rate.

10.4.4 Summary of sensitivity analysis

The interrelationship between variables influencing the hydrologic performance of a landfill cover is complex. It is difficult to isolate one parameter and exactly predict its effect on the water balance without first placing restrictions (sometimes severe restrictions) on the values of the remaining parameters. With this qualification in mind, the following general summary statements are made.

The primary importance of the topsoil depth is in controlling the extent or existence of overlap between the evaporative depth and the head in the lateral drainage layer. The greater this overlap, the greater will be evapotranspiration and runoff. Surface vegetation has a significant effect on evapotranspiration from soils with long flow-through travel times and large plant available water capacities; otherwise, the effect of vegetation on evapotranspiration is small. The general influence of surface vegetation on lateral drainage and percolation is difficult to predict outside the context of an individual cover design. Clay soils increase runoff and evapotranspiration and decrease lateral drainage and percolation. Landfills in colder climates and in areas of lower solar radiation are likely to show less evapotranspiration and greater lateral drainage and percolation.

The ratio of lateral drainage to percolation is a positive function of the ratio of KD/KP and the average head above the liner. However, the average head is a function of lateral drainage, percolation and L/S. The quantity of lateral drainage, and therefore also the average head, is in turn a function of the infiltration. Therefore, the ratio of lateral drainage to percolation increases with increases in infiltration and the ratio of KD/KP for a given drain and liner design. The ratio of lateral drainage to percolation for a given ratio of KD/KP increases with increases in infiltration and the term S/L. The percolation and average head above the liner is a positive function of the term L/S.

Leakage through a flexible membrane liner increases with the number and size of holes, the depth of water buildup on the liner, the hydraulic conductivity of the subsoil and the gap between the liner and the subsoil. Flexible membrane liners reduce leakage through liner systems by reducing the area of saturated flow through the subsoil. The overall effectiveness of a flexible membrane liner system is equivalent to a soil liner having a saturated hydraulic conductivity equal to the product of the saturated hydraulic conductivity of the subsoil and the ratio of the reduced area of flow through the subsoil to the area of the liner. Composite liners provide the best reduction in leakage. Drain systems which yield low head buildup on the FML improve the performance of an FML system.

REFERENCES

Brooks, R.H. and Corey, A.T. (1964) *Hydraulic properties of porous media*, Hydrology Paper No. 3, Colorado State University.

Brown, K.W., Thomas, J.C., Lytton, R.L., Jayawikrama, P. and Bahrt, S.C. (1987) *Quantification of leak rates through holes in landfill liners*, EPA/600/S2-87-062, USEPA, Cincinnati, OH.

Campbell, G.S. (1974) A simple method for determining unsaturated hydraulic conductivity from moisture retention data, *Soil Sc.*, **117**(6), 311–4.

EMCON Associates (1975) *Sonoma County Solid Waste Stabilization Study*. EPA 530–SW–65d.1, US Environmental Protection Agency, Washington DC, 283pp.

EMCON Associates (1983) *Field Assessment of Site Closure, Boone County, Kentucky*. EPA 600/S2–83–058, US Environmental Protection Agency, Cincinatti, Ohio, 6pp.

Fenn, D.G., Hanley, K.J. and DeGeare, T.V. (1975) *Use of the water balance method for predicting leachate generation from solid waste disposal sites*, EPA/530/SW-168, USEPA, Cincinnati, OH.

Giroud, J.P. and Bonaparte, R. (1989a) Leakage through liners constructed with geomembranes – Part I. Geomembrane liners, *Geotextiles and Geomembranes*, **8**(1), 27–67.

Giroud, J.P. and Bonaparte, R. (1989b) Leakage through liners constructed with geomembranes – Part II. Composite liners, *Geotextiles and Geomembranes*, **8**(2), 71–111.

Giroud, J.P., Khatami, A. and Badu-Tweneboah, K. (1989) Evaluation of the rate of leakage through composite liners, *Geotextiles and Geomembranes*, **8**(4), 337–40.

Gordon, M.E., Huebner, P.M. and Kmet, P. (1984) *An evaluation of the performance of four clay-lined landfills in Wisconsin*. Seventh Annual Madison Waste Conference, University of Wisconsin, Madison Extension, Madison, Wisconsin, 62pp.

Ham, R.K. (1980) *Decomposition of residential and light commercial solid waste in test lysimeters*. SW–190c, US Environmental Protection Agency, Washington DC, 103pp.

Horton, R.E. (1919) Rainfall interception, *Monthly Weather Rev., US Weather Bureau*, **47**(9), 603–23.

Kmet, P. (1982) *EPA's water balance method – its use and limitations.* Wisconsin DNR, Bureau of Solid Waste Management, Madison, WI.

Knisel, W.G. (ed.) (1980) *CREAMS: a field scale model for chemical runoff and erosion from agricultural management systems,* Vols. I, II, and III, USDA-SEA, AR, Cons. Res. Report 24.

McEnroe, B.M. and Schroeder, P.R. (1988) Leachate collection in landfills: steady case, *J. Envir. Eng.*, ASCE, **114**(5), 1052–62.

Perrier, E.R. and Gibson, A.C. (1980) *Hydrologic simulation on solid waste disposal sites,* EPA-SW-868, USEPA, Cincinnati, OH.

Ritchie, J.T. (1972) A model for predicting evaporation from a row crop with incomplete cover, *Water Resources Res.*, **8**(5), 1204–13.

Schroeder, P.R. and Gibson, A.C. (1982) Supporting documentation for the hydrologic simulation model for estimating percolation at solid waste disposal sites (HSSWDS), draft report, USEPA, Cincinnati, OH.

Schroeder, P.R. and Peyton, R.L. (1990) Evaluation of landfill-liner designs. *J. Environ. Engin. Div.*, ASCE, **116**(3), 421–37.

Schroeder, P.R., Morgan, J.M., Walski, T.M. and Gibson, A.C. (1984a) *The hydrologic evaluation of landfill performance (HELP) model: Vol. I. User's Guide for Version 1.* Technical Resource Document, EPA/530-SW-84-009, USEPA, Cincinnati, Ohio.

Schroeder, P.R., Gibson, A.C. and Smolen, M.D. (1984b) *The hydrologic evaluation of landfill performance (HELP) model: Vol. II. Documentation for Version 1.* Tech. Res. Document, EPA/530-SW-84-010, USEPA, Cincinnati, Ohio.

Schroeder, P.R., Peyton, R.L., McEnroe, B.M. and Sjostrom, J.W. (1988a) The hydrologic evaluation of landfill performance (HELP) model: Volume III. User's Guide for Version 2, Internal Working Document, USAE Waterways Experiment Station, Vicksburg, MS.

Schroeder, P.R., McEnroe, B.M., Peyton, R.L. and Sjostrom, J.W. (1988b) The hydrologic evaluation of landfill performance (HELP) model: Volume IV. Documentation for Version 2, Internal Working Document, USAE Waterways Experiment Station, Vicksburg, MS.

Thornthwaite, C.W. and Mather, J.R. (1957) Instruction and tables for computing potential evapotranspiration and the water balance, *Publication in Climatology,* **10**(3), 185–311, Drexel Institute of Technology, Centerton, NJ.

USDA, Soil Conservation Service (1972) *National Engineering Handbook,* section 4, Hydrology. US Government Printing Office, Washington, DC, 631pp.

Wigh, R.J. (1984) *Landfill research at the Boone County field site.* EPA 600/2–84–050, US Environmental Protection Agency, Cincinatti, Ohio, 116pp.

Williams, J.R., Nicks, A.D. and Arnold, J.G. (1985) Simulator for Water Resources in Rural Basins, *J. Hydraulic Eng.*, ASCE, **111**(6), 970–86.

CHAPTER 11

Stability of landfills

Issa S. Oweis

11.1 INTRODUCTION

The stability of landfills is controlled in broad terms by the following factors:

1. the properties of the supporting soil;
2. the strength characteristics and weight of refuse;
3. inclination of the slope;
4. leachate levels and movements within the landfill;
5. type of cover; and
6. cover resistance to erosion.

In all cases, the presence of water acts as a destabilizing agent in reducing the strength and increasing the destabilizing force.

Assessment of the stability of solid waste landfills is somewhat less reliable than for soil embankments. The unit weight of refuse and its strength are difficult to determine and could vary over a wide range. Assessment of these variables is largely based on case histories and site-specific investigations (Oweis and Khera, 1990). Because of the extreme variability in refuse composition, the usual soil sampling and testing of soils on relatively small samples is not applicable for refuse such as typical municipal waste.

The purpose of this chapter is to present procedures for stability evaluation. Potential instability could occur in the foundation soil, the refuse, or the cover. In all cases the safety margin is expressed in terms of the factor of safety, F, and is defined as

$$F = \frac{\text{Available strength along the potential failure surface}}{\text{Mobilized strength along the potential failure surface}} \qquad (11.1)$$

Geotechnical Practice for Waste Disposal.
Edited by David E. Daniel.
Published in 1993 by Chapman & Hall, London. ISBN 0 412 35170 6

11.2 TYPES OF INSTABILITY

Figures 11.1 and 11.2 illustrate two stability situations. In Fig. 11.1 a landfill rests on a firm base such as rock or compact sand. The landfill has a liner and a leachate collection system and no leachate mound is expected to develop. The refuse mass in this case could slide along the liner (potential failure surface abc), or if the slopes are steep, the refuse could fail along a failure surface such as de. Modern landfills are designed with typical slopes of 3:1 horizontal to vertical and in a dry condition, refuse is known to stand almost vertically for cuts as deep as 6 m or more. From observations of the Global landfill failure, the steep chasm in refuse at the top of the failure zone was about 12 m high (Oweis *et al.*, 1985). If the leachate collection system fails, a leachate mound will develop, resulting in a lower factor of safety.

Figure 11.2 depicts an old landfill on soft ground. It is desired to increase the height of the landfill from level ab to cd. Because no leachate collection system is present, a leachate mound should be expected unless a water balance analysis indicates otherwise. In this

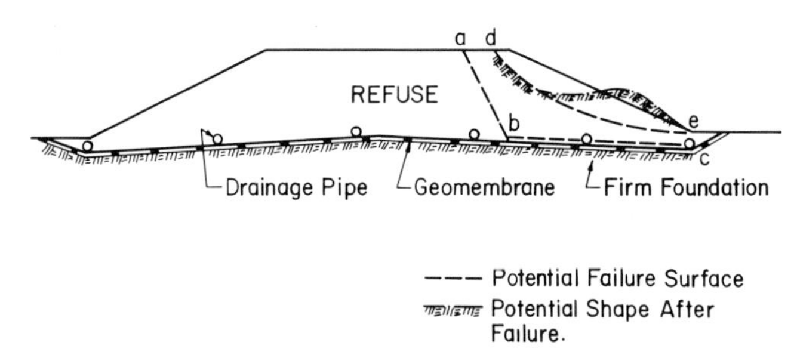

Fig. 11.1 Stability of a landfill on firm base.

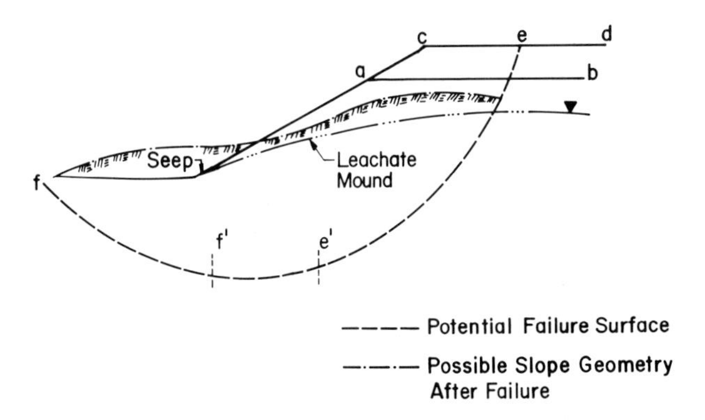

Fig. 11.2 Stability of a landfill expansion on soft base.

case, then, the potential failure surface could include the weak foundation of the landfill.

If shear stresses are a large percentage of the strength of soft foundation soils, lateral movements at the toe and beyond could impact cutoff walls, perimeter leachate collection systems, or surface water drains.

The cover of a landfill could also fail by sliding along a plane parallel to slope. An initially safe, dry cover may fail at a later time because of saturation. Saturation increases the weight of the cover and decreases the frictional resistance along the interface. The source of saturation is usually precipitation, cracking of the cover and poor drainage, but saturation could also result from leachate seeps from the side slopes of the landfill.

In addition failure can occur by slippage between components of a liner or cover system. Shear along the contact between two surfaces is often called 'interfacial shear'. The interface between a geomembrane and geonet, and between a geomembrane and a wet compacted clay, are especially critical. A failure at the Kettleman Hills facility is an example of instability associated with interfacial shear (Seed *et al.*, 1990).

11.3 TOTAL AND EFFECTIVE STRESSES

In soil mechanics, changes in strength and volume are considered to be due solely to the changes in effective stress. In a saturated soil mass, the normal stress on any plane is due to the sum of two components: a) pressure carried by water in the pore space and b) effective stress carried by the soil skeleton. The total stress is produced by the overburden and stresses produced by surface loading. The effective stress (σ') is the total stress (σ) less the porewater pressure u

$$\sigma' = \sigma - u \tag{11.2}$$

Because soils contain water in the void space, any volume change will require either a decrease or increase in porewater pressure. For a saturated soil to experience compression, water would have to be expelled from the voids, whereas for the soil to expand the effective stress would have to decrease and water drawn into the soil. Consider a deposit of soft saturated clay subjected to loading from a sanitary landfill. The clay tends to compress over time. Because of the low hydraulic conductivity of the clay, the rate at which the water is expelled is much slower than the rate at which the load is applied. This results in excess porewater pressure generated in the clay. For saturated clay, the excess porewater pressure could be estimated based on Skempton (1954):

$$\delta u = B(\delta\sigma_3 + A(\delta\sigma_1 - \delta\sigma_3)) \tag{11.3}$$

where $\delta\sigma_1$ and $\delta\sigma_3$ are the additional major and minor principal stresses, respectively, produced by the loading, δ_u is the change in excess porewater pressure, and B and A are porewater pressure coefficients. If the soil is saturated, B is essentially equal to 1 for compressible soils. The parameter A at failure depends on the degree of overconsolidation and ranges from about -0.5 for highly overconsolidated clays to 3 for very sensitive to quick clays. The range for normally consolidated clay is 0.7–1.3 (Leonards, 1962).

The *in situ* effective stresses are those responsible for the consolidation of the soil prior to load application. A soil profile with a horizontal surface is shown in Fig. 11.3. The *in situ* vertical effective stress (σ'_v) and horizontal (σ'_h) effective stress are also principal stresses, σ'_1 and σ'_3, respectively, because the horizontal and vertical planes are free from shear stresses. σ'_h could be larger than σ'_v for highly overconsolidated clays, and in such cases, the major principal stress would be σ'_h. The total vertical stress is

$$\sigma_v = z_1\gamma_m + z_2\gamma \tag{11.4}$$

where γ_m and γ are total unit weights for zones shown in Fig. 11.3. The effective vertical stress is

$$\sigma'_v = \sigma_v - u_0 \tag{11.5}$$

where u_0 is the initial (natural) porewater pressure equal to $\gamma_w z_2$, where γ_w is the unit weight of water.

For a soil element along a slope with a slope angle of i with the horizontal, the vertical and horizontal stresses are no longer principal stresses. In this case, the normal total, effective, and shear stresses are respectively:

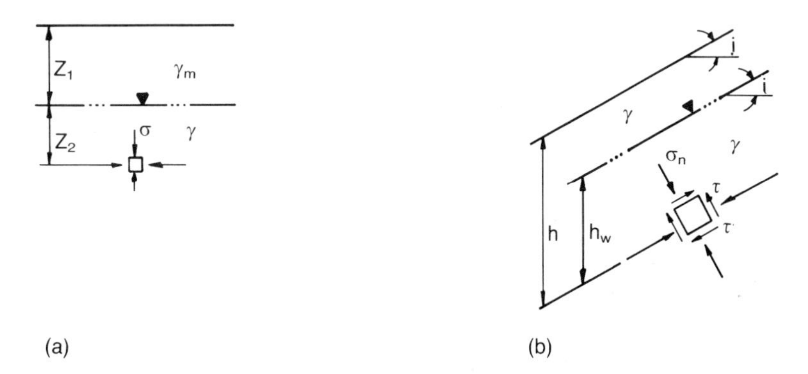

Fig. 11.3 Total stresses beneath (a) a level, i.e. horizontal ground surface and (b) within a slope.

$$\sigma_n = \gamma h \cos^2 i \tag{11.6a}$$

$$\sigma'_n = \sigma_n - \gamma_w h_w \cos^2 i \tag{11.6b}$$

$$\tau = \gamma h \sin i \cos i \tag{11.6c}$$

The effective initial horizontal stress for a sedimentary deposit is computed based on:

$$\sigma'_h = K'_0 \sigma'_v \tag{11.7}$$

where K'_0, the earth pressure coefficient at rest, depends on the properties of the clay and the stress history. For normally consolidated soils K'_0 can be estimated based on

$$K'_0 \approx 1 - \sin \phi' \tag{11.8}$$

where ϕ' is the effective angle of shearing resistance (effective friction angle).

The increases in total principal stresses $\delta\sigma_1$ and $\delta\sigma_3$ are computed using stress distribution theories. The changes in effective principal stresses are

$$\delta\sigma'_1 = \delta\sigma_1 - \delta u \tag{11.9a}$$

$$\delta\sigma'_3 = \delta\sigma_3 - \delta u \tag{11.9b}$$

In cohesionless soils under normal static loading, the excess porewater pressure is dissipated as the load is applied, δu is taken as zero, and the increase in total stress produces an immediate equivalent increase in effective stress. In clays, the development of excess porewater pressure δu creates a hydraulic gradient and this excess porewater pressure dissipates over time. The rate of dissipation depends on the hydraulic conductivity of the clay and other parameters (Lambe and Whitman, 1969). After full dissipation of the excess porewater pressure, the porewater pressure returns to natural porewater pressure, u_0. This process will produce a strength increase.

11.4 FAILURE CRITERION

The revised Coulomb equation (Terzaghi and Peck, 1967) is used to describe failure conditions in soils. The criterion can be expressed as follows:

$$\tau_f = c' + (\sigma'_n) \tan \phi' \tag{11.10}$$

where τ_f = shear resistance (strength) along the plane of failure, c' = effective cohesion, σ'_n = effective stress normal to the plane of failure, and ϕ' is the effective angle of shearing resistance. For initial principal stresses σ'_1 and σ'_3, failure would occur when the effective stress circle becomes tangent to the failure envelope as a result of the changes in the principal stresses generated by loading the soil. In soft, saturated clay, the effective cohesion component is usually small (Bishop, 1971).

In order to apply effective stress relationships to field conditions, excess pore pressures generated by loading must be known. If effective stress parameters are to be used to assess the stability of saturated clays, then a monitoring program is required to verify the calculated porewater pressures. However, to avoid the need for field pore pressure measurements, the stability of cohesive soils may be evaluated by using the undrained shear strength ($\phi = 0$ analysis). The undrained shear strength, s_u, as determined in the triaxial test, is usually taken as one-half the principal stress difference ($\sigma_1 - \sigma_3$) at failure:

$$s_u = \frac{1}{2}(\sigma_1 - \sigma_3)_f \tag{11.11}$$

In the unconfined compression test, σ_3 is zero and s_u is half the unconfined compressive strength, q_u.

An empirical relationship in terms of the plasticity index has been found to be applicable to a wide range of soft, sedimented clays and may be used for a rough evaluation of the undrained shear strength as determined from the unconfined compression or vane tests (Terzaghi and Peck, 1967). The relationship (Skempton, 1954) is

$$\frac{s_u}{\sigma'_v} = 0.11 + 0.0037\ \mathrm{PI} \tag{11.13}$$

The undrained shear strength is sensitive to the type of testing procedure used, the orientation of the failure planes, and the direction of shear stresses. The triaxial compression test usually provides undrained strengths higher than the triaxial extension test where the horizontal stress is the major principal stress. The direct simple shear test provides values somewhere in between (Ladd *et al.*, 1977; Mayne, 1988). Mayne (1988) made a statistical evaluation of published data on many clays and found that the ratio s_u/σ'_v varies from 0.25 to 0.55 under isotropically consolidated, undrained triaxial conditions (CIUC) with an average of 0.35. The range and the average under conditions of simple shear (DSS) are (0.15–0.35) and 0.23, respectively.

The choice of the type of the strength test to be used requires judgement and experience. For reasonably homogeneous clays, the simple direct shear test is applicable along the base of the failure plane (e′f′ in Fig. 11.2) where the shape of the failure plane is nearly horizontal. The triaxial extension undrained strength is applicable in the region of the failure plane beyond the toe of the (ff′) slope. The triaxial compression is applicable for the portion of the potential failure surface in the same quadrant as the slope (ee′).

The undrained shear strength of overconsolidated (OC) clays depends on the clay type and stress history. The ratio $(s_u/\sigma'_v)_{oc}$ is defined as

$$\left(\frac{s_u}{\sigma'_v}\right)_{oc} = \left(\frac{s_u}{\sigma'_v}\right)_{nc} (\mathrm{OCR})^m \tag{11.14}$$

where OCR is the overconsolidation ratio (Ladd *et al.*, 1977) determined in the consolidation test and the subscripts $_{oc}$ and $_{nc}$ indicate overconsolidated and normally consolidated, respectively. The exponent m typically varies from 0.7 to 0.85, depending on the type of test used (Mayne, 1988).

The effective stress parameters (σ' and ϕ') are used for cohesionless soils under normal static loading conditions. In most projects, the angle of shearing resistance ϕ, is estimated based on the soil classification and estimated *in situ* relative density based on field penetration tests (Gibbs and Holtz, 1957), such as the Standard Penetration Test (ASTM D1586). Under seismic loading conditions, the same problem is faced in assessing the developed excess porewater pressure, and consequently, the undrained shear strength for cohesionless soils is used for evaluating the potential for liquefaction. Liquefaction occurs when the porewater pressure reduces the effective stress in a sand deposit to practically zero during ground shaking, leading to loss of strength which transforms the liquified layer into a viscous fluid. The excess porewater pressures in saturated cohesionless soils develop during shock or earthquake loading for the same reason as fine grained soils during a much slower loading rate. The soil tends to compress, and if the drainage rate is not fast enough, excess porewater pressures develop. The s_u/σ'_v for cohesionless soils has been correlated with the Standard Penetration Test N value, the amount of fines in the soil, and earthquake duration, as characterized by earthquake magnitude (Seed and De Alba, 1986; Seed, 1987).

11.5 CHARACTERIZATION OF REFUSE

Unlike soils, refuse is not a geologic material and the composition varies depending on the type of waste generated. Through experience, however, the mechanical behavior of refuse is expressed in terms of an apparent frictional parameter ϕ_a and cohesion c_a. The apparent cohesion is a frictional parameter characterizing the lateral reinforcement present in the waste, which is analogous to the apparent cohesion in reinforced earth. The parameters ϕ_a and c_a are empirical in nature and may be defined in laboratory tests at an arbitrary strain level that is considered by the user to be excessive. Alternatively, these parameters are backcalculated from actual cases of failure or cases where large deformations in refuse have occurred (Oweis *et al.*, 1985), or conservatively backcalculated from slopes that have not failed, by assuming a factor of safety of 1.0. The apparent reinforcement effect in refuse allows the material to withstand shear strains substantially larger than those for soils at peak strength. This could be significant when considering refuse fills on weak clays where the factor of safety using peak strength is marginal. The soil along the potential failure surface may develop its post peak or residual strength without any obvious manifestations such as cracks on the top of the slope. For embankments on clays, a factor of safety of 1.5 is usually considered adequate. Such a factor may not be applicable for refuse because refuse properties could vary over a wider range.

In order to conduct a stability analysis, data on the composition, unit weight of refuse and leachate mounding are needed.

11.5.1 Refuse composition

The composition of municipal refuse was discussed in Chapter 1. The unit weight of refuse could be estimated from the composition of various components, their compactability, and the contribution of daily cover and moisture adsorption of refuse (Oweis and Khera, 1990). Table 11.1 shows some published values for the unit weight. It should be noted that with more recycling, the unit weights will be larger

Table 11.1 Unit weight of municipal refuse

municipal solid waste (MSW) form	*unit weight*	
	kN/m^3	*pcf*
MSW, moderate to good compaction	4.7–6.3	30–40
MSW, good to excellent compaction	8.6–9.4	55–60
MSW, baled waste	8.6–14.1	55–90
MSW, active landfill with a leachate mound	6.6	42
MSW, old landfill	9.7	62
MSW, from test pits	8.9–16.2	
MSW, after degredation and settlement	9.9–11	63–70
Incinerator residue as received	7.2–12.7	46–81
Incinerator residue (maximum dry density–standard compaction)	13.5	86
Incinerator residue–old and recent residue (in-place densities)	14.9–16.6	94.7–105.7
Wood chips (dry)	2.3	14.6
Wood chips (with 64% moisture)	3.8	23.9
Leaf compost as delivered	2.6	16.7
In windows (40% moisture)	5.2	33.3
Final product	3.5	22.2

Source: NSWMA (1985), Landva and Clark (1986), Schoenberger and Fungaroli (1971), Poran and Ali (1989), Oweis and Khera (1990).

because the light waste (e.g. paper, aluminum cans, etc.) will be taken out of the waste stream. A baling operation substantially reduces the refuse volume and increases the unit weight up to $11\,kN/m^3$, (Tchobanoglous *et al.*, 1977). For submerged refuse, a saturated unit weight of $10.3–11.1\,kN/m^3$ could be used. In the absence of site specific data, the unit weights in Table 11.1 could be assumed for average municipal waste in communities with minimal recycling.

Old landfills that have undergone substantial settlement are expected to have larger unit weight. The long-term unit weight could be estimated knowing the age of the fill and deformation characteristics. For example, if an old landfill eventually settles 30% of its original volume by the time it is completed (Oweis and Khera, 1986), and undergoes additional long term settlement of 10%, the unit weight of refuse would increase by 67%, which is consistent with reported values (see Table 11.1). In communities with an active recycling program, an approximate estimate could be made for the unit weight based on the composition of the waste stream (Oweis and Khera, 1990). If, for example, paper products, plastics, and glass are excluded from the waste stream and replaced by food waste, the unit weight may increase by about 30% or more, which is significant in the stability assessment.

Incineration is used in many locations to reduce the volume of combustible components of municipal refuse which, is often about 70% of the waste stream. Well-burned residue from a modern incinerator may weigh about $16\,kN/m^3$ (100 pcf) or more after compaction. For existing landfills, the unit weight could be established by field density measurements.

11.5.2 Shear strength of waste

The modified Coloumb failure criterion could be used to characterize the shear strength of waste. Considering an apparent friction angle ϕ_a and apparent cohesion c_a, the shear strength τ along a given plane is a function of the effective stress σ'_n normal to that plane. The laboratory determination of strength is usually given at a strain level of 15–20% because the samples do not actually collapse. Fang *et al.* (1977) used the double punch test for determining the tensile resistance of compacted refuse, and together with the unconfined compression test, values of c_a and ϕ_a of 63 kPa and 19°, respectively, were derived. Drained direct shear tests on specimens that measured 287 mm by 434 mm reported by Landva and Clark (1990) indicated the following results.

Shredded refuse: $c_a = 23\,kPa$ (480 psf) $\phi_a = 24°$
Old refuse: $c_a = 16\,kPa$ (334 psf) $\phi_a = 33°$

Artificial refuse: $c_a = 0$ $\phi_a = 27°$ to $41°$
Fresh artificial refuse: $c_a = 0$ $\phi_a = 36°$

These results are consistent with the c_a and ϕ_a combinations inferred from the failure of the Global landfill and field testing of refuse described by Oweis *et al.* (1985). Triaxial tests on anistopically consolidated samples of 2-year-old milled refuse yielded ϕ_a of 40° (Stoll, 1971), and 15° was recommended to forestall lateral spreading.

Essentially all tests on refuse are affected by the soil content, and results can vary over a wide range, especially when small samples are tested. Direct shear tests on 106-mm-diameter samples of mixed refuse yielded inferred friction angles from 39° to 81° (Siegel *et al.*, 1990). Refuse with more granular soil typically exhibits higher friction.

The age of the refuse is expected to affect the field strength. As the refuse ages it becomes denser and hence, stronger. Decomposition, however, may produce the opposite effect. It is wise, therefore, to use conservative strength parameters for stability calculations. Another factor is the expected variation in the character of refuse as the proportion of recyclables becomes less. With more garbage waste, the refuse would be expected to be heavier and the strength is expected to be less than current data indicate.

Higher strengths were reported in the literature for incinerator residue. A friction angle of 45° was reported by Schoenberger and Fungaroli (1971) on a residue with a density of $15.4\,kN/m^3$ using the direct shear test. The high angularity of the particles was cited as a probable reason for the high strength. Poran and Ali (1989) used unconsolidated-undrained triaxial tests to determine the strength of partially saturated residue with about 20% by weight passing No. 200 sieve (0.074 mm openings) and reported apparent friction angles of 43° to 45° at a maximum dry density of $13.5\,kN/m^3$ and an optimum moisture content of 23.5% (60–70% saturation). It is probable that such apparently high friction angles are partially due to the development of negative porewater pressures. If waste will eventually become saturated in the landfill, tests for stability assessment should be based on saturated samples. Unless supported by laboratory tests, the angle of shearing resistance for stability assessment should not exceed an equivalent granular soil.

11.6 SHEAR ALONG INTERFACES

In modern landfill designs for liners and covers, geomembranes, geotextiles, drainage nets and reinforcement may be used in conjunction with earthen materials. The stability is expressed in terms of the safety factor against sliding along the interface. The interface friction angle, δ, depends on the type of the materials on both sides of the interface, moisture conditions along the interface, resin type, surface texturing, stiffness of the geotextile or the geomembrane, settlement of the subgrade and other factors related to field placement and quality control.

The value is usually measured using a direct shear test that is not standard. The size of the shear box used varies from 101.6 mm square (Martin *et al.*, 1984) to 300 mm square (Williams and Houlihan, 1987). Takasumi *et al.* (1991) summarize all the testing techniques and shear boxes that have been used.

The material to be tested is placed on the lower (fixed) portion of the box and the other material is made to slide over it. The reported value for a geomembrane on sand ranges from 17° to 36°, whereas for geotextiles on sand the reported range is 22° to 40°. The δ value for geotextiles on fine-grained soils is in the same order as the angle of the shearing resistance of the soil. The soil particles tend to be lodged between the yarns forcing the failure plane above the interface (Williams and Houlihan, 1987). The reported friction angle of the interface between geotextiles and geomembranes varied over a very wide range of 6° to 24°, which is roughly the same range as geomembranes over drainage nets. The problem with possible low values led some designers to tack weld the drainage net to the geomembrane (Duplancic *et al.*, 1987). Seed *et al.* (1990) carried out interface tests to conduct an analysis for a landfill slope failure. Based on the tests, they used a residual friction angle of 8 · 5° between an HDPE liner and compacted clay, with the clay at its as-compacted water content. When the interface was moistened, a residual friction of 8 · 0° was used. For a wet HDPE/clay interface, an interface shear strength of 43 kPa was found to represent a more critical failure mechanism than a residual friction of 8 · 0°.

Textured geomembranes with a rough surface are now produced and used. The manufacturers claim an improvement in the interface friction of 60% to more than 100%. Such improvement is significant, especially in cover design, where side slopes can be steep.

If the interface frictional characteristics are not adequate to ensure stability, cover soils can be reinforced with geogrids or high-strength geotextiles.

11.7 STRENGTH OF COMPACTED COHESIVE SOIL

Compacted cohesive soil is frequently used as a barrier in a liner or a cover structure. The undrained shear strength is sensitive to the moisture content and the method of compaction (Fig. 11.4). Clays compacted dry of the optimum are initially stronger than if compacted wet of the optimum. The loss of strength upon soaking, however, can be substantial.

In terms of total stresses, the undrained strength of the clay liners and covers is difficult to assess because of the difficulty of predicting the field degree of saturation. In terms of effective stresses, the strength is expressed in terms of the effective strength parameters c' and ϕ'. The failure envelope is usually curved showing essentially no effective cohesion at low effective stresses. For a rough estimate, ϕ' may be estimated from correlations with the plasticity index (Terzaghi and Peck, 1967).

11.8 METHODS OF STABILITY ANALYSIS

11.8.1 General

Analysis of slope stability in soils is usually based on limit equilibrium concepts. At failure, a slip surface is assumed to occur with simultaneous mobilization of the shear strength along that surface. The mass above the slip surface is assumed to move as a rigid body. In waste disposal sites, the driving force causing the development of a slip surface is the weight of the waste and leachate. The resisting (restoring) forces are due to the shear strength of the waste and soil and the weights of soil, water and waste located near the toe of the slope. Seismic forces could affect the stability by adding to the disturbing forces and reduction of shear strength. The techniques for stability assessment differ in the shape of the slip surface assumed and conditions of equilibrium considered.

The assumptions inherent in the traditional methods of stability analysis are rarely satisfied. Deformations do occur along the slip surface and the mass above it prior to failure even in very sensitive clay (Clausen *et al.*, 1984). It is possible to monitor subsoil movements and pore pressures to control filling to avoid failure when the factor of safety is marginal. The assumption of simultaneous mobilization of shear strength along the entire failure surface is likely to be violated. Most soils exhibit a peak strength and lesser residual strength at larger deformations. The failure is generally progressive in nature (Terzaghi and Peck, 1967). The highly stressed zones reach their peak strength first while others would be at less than peak. At failure, some zones are at peak strength mobilization while others may have weakened to below peak.

The safety of the slope is expressed in terms of the factor of safety F which is the ratio of the available shear strength to the mobilized strength given in Eq. (11.1). The factor of safety for the frictional

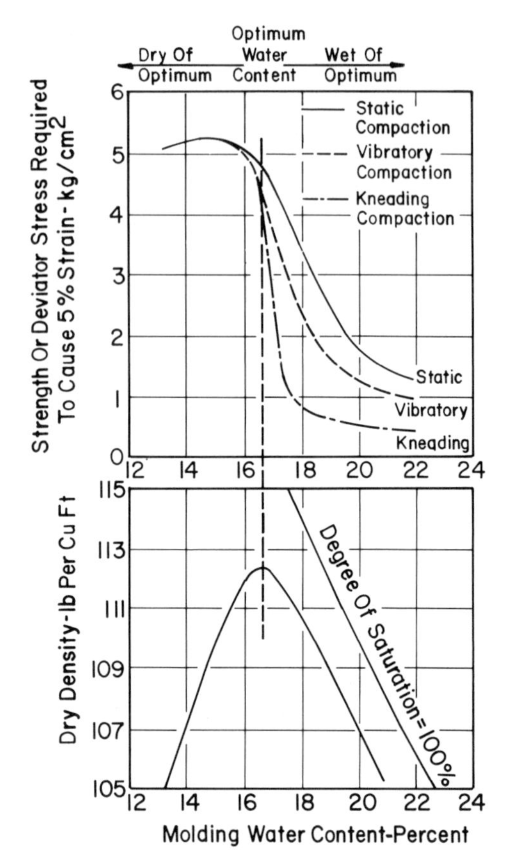

Fig. 11.4 Strength of compacted silty clay (Seed *et al.*, 1960).

component of the strength is assumed to be the same as the cohesive component. Thus

$$c'_m = c'/F \tag{11.15a}$$

$$\tan \phi'_m = \tan \phi'/F \tag{11.15b}$$

where c'_m and ϕ'_m are the mobilized cohesion and friction angle, respectively.

The procedure for stability analysis generally involves the following steps.

1. Assess the refuse and soil properties.
2. Assess leachate levels.
3. Determine the type of analysis required.
   For waste facilities on soft, saturated clay or moderately over-consolidated clay, a short-term condition may be appropriate for analysis. In this case, the undrained shear strength of the deposit is

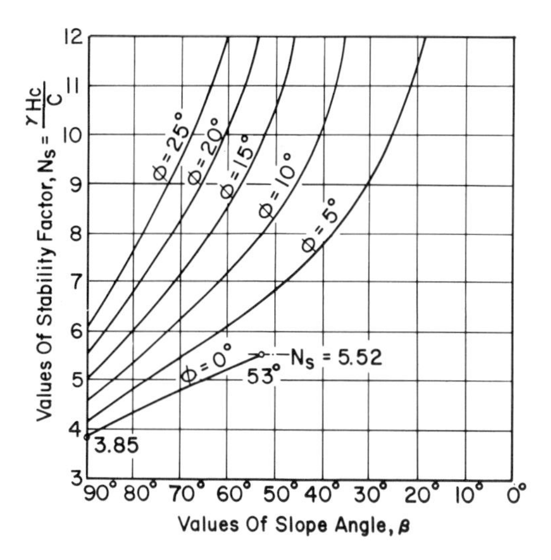

Fig. 11.5 Taylor stability chart for c, ϕ material (Terzaghi and Peck, 1967).

used and total stresses are considered. For waste facilities on heavily overconsolidated clay, or where cuts are made in such deposits, an effective stress analysis using long-term porewater pressures (assumed to be the same as ambient) may be more critical.

4. Select the geometry of the slip surface.
5. Compute the factors of safety for varied dimensions of the slip surface. The minimum value is the factor of safety for the case investigated.

The shape of the slip surface could be curved, planar or a combination of both. For homogeneous material an arc of a circle has been found to be satisfactory. A planar type slip surface is useful for analysis of soil covers where the slip surface is parallel to the slope or, in other cases, where planar surfaces are dictated by site conditions.

11.8.2 Stability charts

For simple slopes composed of homogenous materials and no seepage surface, the Taylor charts (Terzaghi and Peck, 1967) could be used for estimating the factor of safety. The chart in Fig. 11.5 is applicable for a constant (c, ϕ) material within the slope and the base. All the critical circles for friction angles more than about 3° are toe circles. The chart, therefore, could be used for refuse embankments on firm bases. The chart is based on full mobilization of the frictional component of strength and the cohesion value is the mobilized cohesion c_m. In order to apply the factor of safety to both c, and ϕ, an iterative procedure is used, as illustrated by the following example.

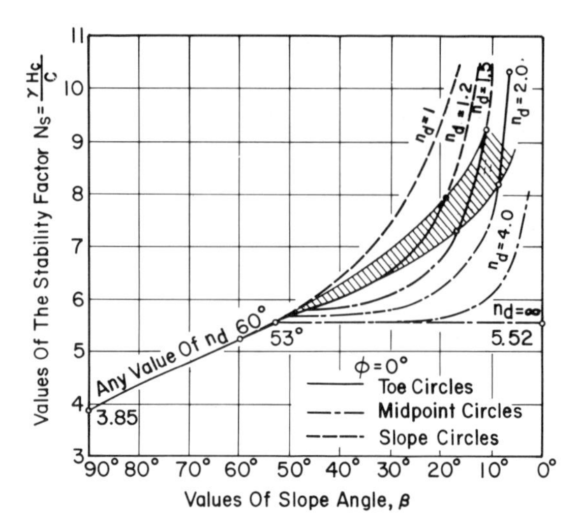

Fig. 11.6 Taylor stability chart for $\phi = 0$ analysis (Terzaghi and Peck, 1967).

Example 1

Estimate the factor of safety for a refuse embankment 40 feet high founded on stiff clay. The side slopes are 1 vertical on 1 horizontal, so the slope angle = $\tan^{-1}$ 1.0 = 45°. The unit weight is 7.85 kN/m^3 (50 pcf), c = 14.4 kPa (300 psf), and ϕ = 20°.

For trial 1 assume F = 2.0. Then using Eq. 11.15b

$$\phi_m = \tan^{-1} (\tan 20°/2) = 10.3°$$

From Fig. 11.5, N_s = 9.2 so c_m = (50)(40)/9.2 = 217.4 psf, and F = 300/217.4 = 1.38.

For trial 2 assume F = (1.38 + 2)/2 = 1.69. Then

$$\phi_m = \tan^{-1} (\tan 20°/1.69) = 12.15°$$

From Fig. 11.5, N_s = 10.55 so c_m = (50)(40)/10.55 = 189.6, and F = (300/189.5) = 1.58.

For trial 3 assume F = (1.58 + 1.69)/2 = 1.635. Then

$$\phi_m = \tan^{-1} (\tan 20°/1.635) = 12.6°$$

with N_s = 10.82, so that c_m = (50)(40)/10.82 = 184.8 psf, and F = 300/184.8 = 1.62.

The Taylor chart for ϕ = 0 is shown in Fig. 11.6 and is applicable when the slope and the base have uniform cohesion c. In order to use the chart for preliminary estimates, it is necessary to estimate an equivalent cohesion value along the critical circular arc. The position of the critical circle may be estimated using Fig. 11.7. The center of the

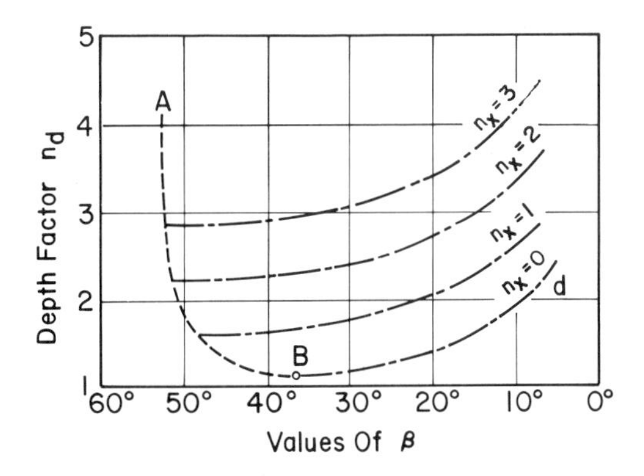

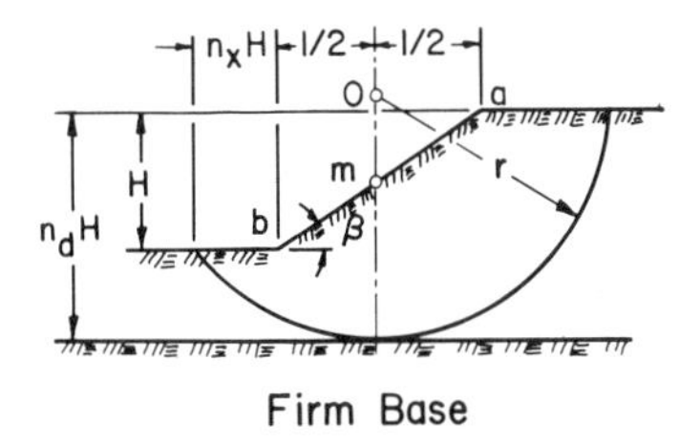

Fig. 11.7 Position of the critical failure surface for $\phi = 0$ material (Terzaghi and Peck, 1967).

circle could be assumed to be along a vertical line through the mid point of the slope. After the critical circle is drawn, the equivalent cohesion of the refuse and frictional layers may be estimated as follows.

1. Find the average inclination *a* of the slip surface through the layer of interest. For a layer having a friction angle ϕ, compute the equivalent cohesion, c_{eq}, as

$$c_{eq} = (W' \cos \alpha) \tan \phi / L \tag{11.16}$$

   where W' = effective weight of soil and refuse (i.e., buoyant weight under water) above length L, L = approximate length of the slip surface through the layer, and α = average inclination of the slip surface through the layer.
2. Compute the weighted average of cohesion c_{av} for the entire slip surface based on the cohesion and slip surface arc length in each layer.

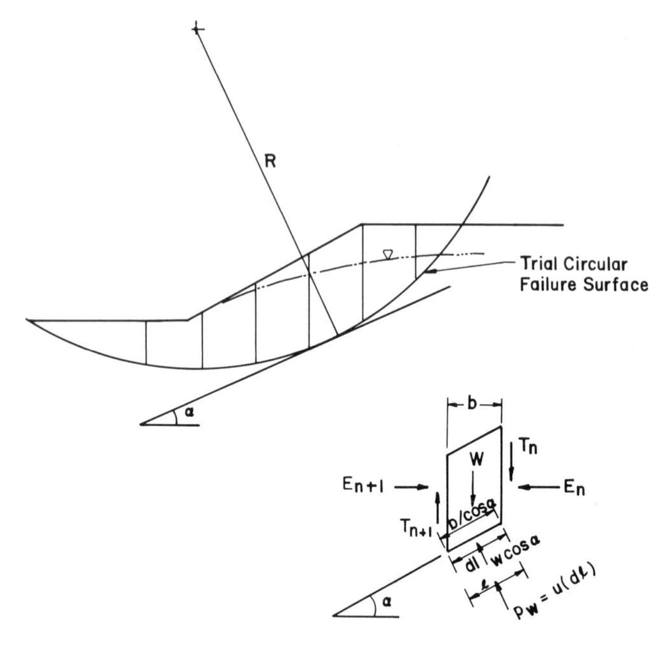

Fig. 11.8 The method of slices.

11.8.3 The method of slices

This designation is applied to a variety of methods where the potential sliding mass is divided into slices, as illustrated in Fig. 11.8. The equilibrium of each slice, with its base on an assumed failure surface, is considered. The differences between various methods are the shape of the failure surface and the number of equilibrium conditions satisfied. In the ordinary method of slices (also known as the Swedish method or the Fellenius method), the interslice forces are assumed zero. The factor of safety is defined as the ratio of resisting moment around the center of a circular arc to the driving moment.

$$F = \frac{\Sigma c(\delta l) + (W \cos \alpha - u(\delta l)) \tan \phi}{\Sigma W \sin \alpha} \tag{11.17}$$

where δl is the length of the base of the slice, W is the total weight of the slice, α is the inclination of the base of the slice, u is the average porewater pressure acting on the base of the slice, and Σ indicates summation over the total number of slices.

In the modified Bishop method, the factor of safety is defined as the ratio of the shear strength to the mobilized shear strength. In order to calculate the factor of safety, the vertical equilibrium of each slice is considered and side forces are set to zero. The resulting expressions are as follows.

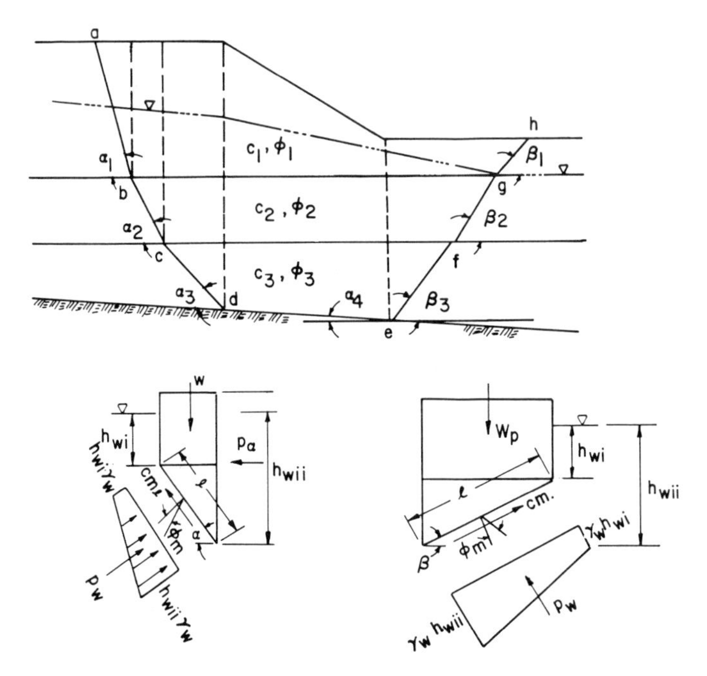

Fig. 11.9 The translational wedge method.

$$F = \frac{\Sigma c(\delta l) \cos \alpha + (W - u(\delta l) \cos \alpha) \tan \phi / m_\alpha}{\Sigma W \sin \alpha} \tag{11.18}$$

$$m_\alpha = \left(1 + \frac{\tan \alpha \tan \phi}{F}\right) \cos \alpha \tag{11.19}$$

The factor of safety appears on both sides of the equation for F. A trial and error procedure is used to achieve a solution. The angle α is positive when the inclination of the base of the slice is in the same quadrant as the ground surface.

A number of computer programs are available to obtain solutions with various methods of slices. The programs will compute the factor of safety for many failure surfaces and locate the surface that gives the lowest factor of safety.

11.8.4 The translational wedge method

This method is described in the Naval Facilities Engineering Command Design Manual DM7 (1982). The potential failure surface is approximated by planar segments, as shown in Fig. 11.9. The mass of soil or refuse above each segment forms a potential sliding wedge. At failure, the inter wedge force is assumed horizontal (i.e., smooth vertical planes) and equivalent to the active force calculated by considering the equilibrium of the wedge. In general, where the factor of safety is other than 1, the inter-wedge force is calculated using the mobilized cohesion c/F and mobilized friction ($\tan \phi/F$). This active condition prevails where the inclination of the failure planes is in the same quadrant as the slope (zone abcde in Fig. 11.9). For the α wedges, the failure plane through each layer is assumed to be at an angle $\alpha = (45 + \phi/2)$ unless the angle is dictated by stratigraphy (e.g. α_4 in Fig. 11.9). Where for a particular wedge α is small (e.g. α_4 in Fig. 11.9), the α force may have a negative sign indicating a net resisting force.

For planar surfaces at angles that are not in the same quadrant as the slope, the wedges sliding along those planes are passive wedges. At a factor of safety of 1.0, the inter-wedge forces are assumed horizontal and equivalent to the passive pressure force using c, and ϕ for the material. In general, this force (β force) is computed using the mobilized friction and cohesion. The angle β is assumed to be equal to $45 - \phi/2$ through each layer.

In order to perform the analysis, the equilibrium of the central wedge is considered. The α forces are produced by the sliding mass on each plane (α forces P_α). On the passive side, the β forces are computed. For the equilibrium of the central wedge, the following must be satisfied.

$$\Sigma P_\alpha = \Sigma P_\beta \tag{11.20}$$

For each active wedge

$$P_\alpha = (W - c_m L \sin \alpha - P_w \cos \alpha) F_\alpha - (c_m L \cos \alpha - P_w \sin \alpha) \tag{11.21}$$

For each passive wedge

$$P_\beta = (W + c_m L \sin \beta - P_w \cos \beta) F_\beta + (c_m L \cos \beta + P_w \sin \beta) \tag{11.22}$$

where

$$F_\alpha = \frac{\tan \alpha - [(\tan \phi)/F]}{1 + (\tan \alpha \tan \phi/F)} \tag{11.23}$$

$$F_\beta = \frac{\tan \beta + [(\tan \phi)/F]}{1 - (\tan \beta \tan \phi)/F} \tag{11.24}$$

The functions F_α and F_β are shown in Figures 11.10 and 11.11, respectively.

11.8.5 Stability of the cover

The stability of the cover could be assessed based on the translational wedge method. Figure 11.12 shows the elements of the method as applied to the cover system. The interface strength is characterized by

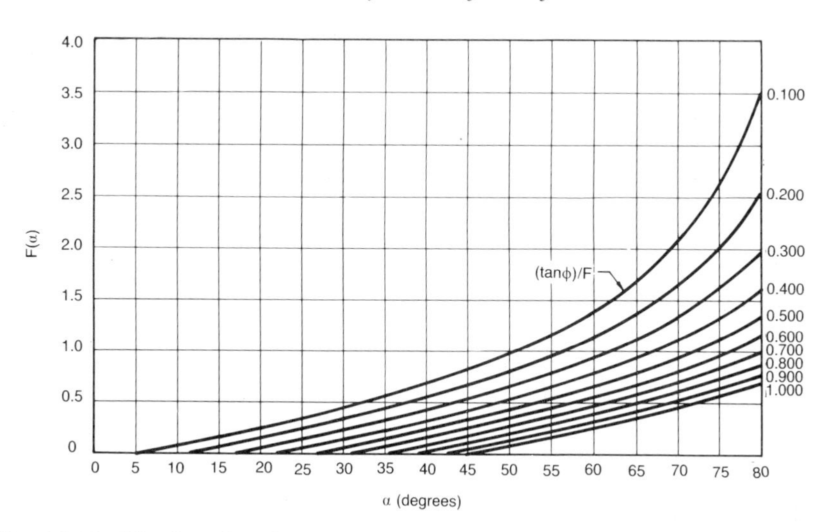

Fig. 11.10 The function F_{α}.

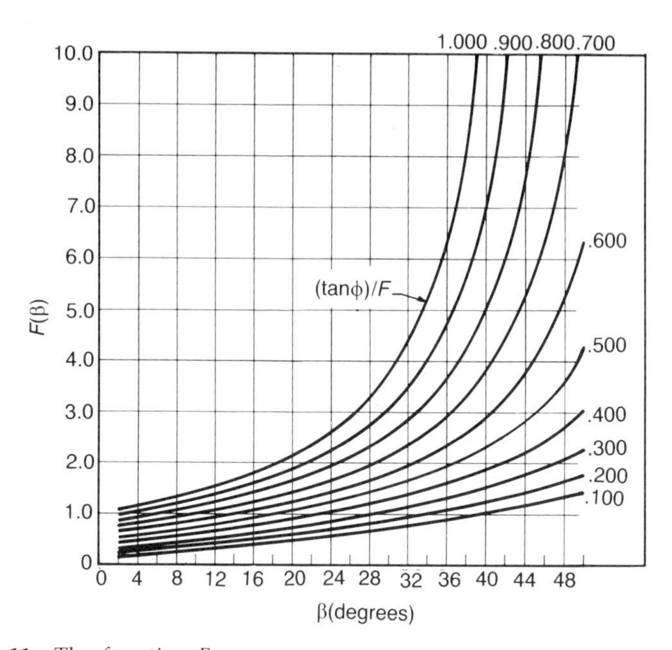

Fig. 11.11 The function F_{β}.

c_i and δ_i. Cover reinforcement per unit width is indicated by force T. The potential failure surface is abcd. Considering the horizontal equilibrium of the central wedge, the stability equation is written as

$$P_a + P_c = P_p \tag{11.24}$$

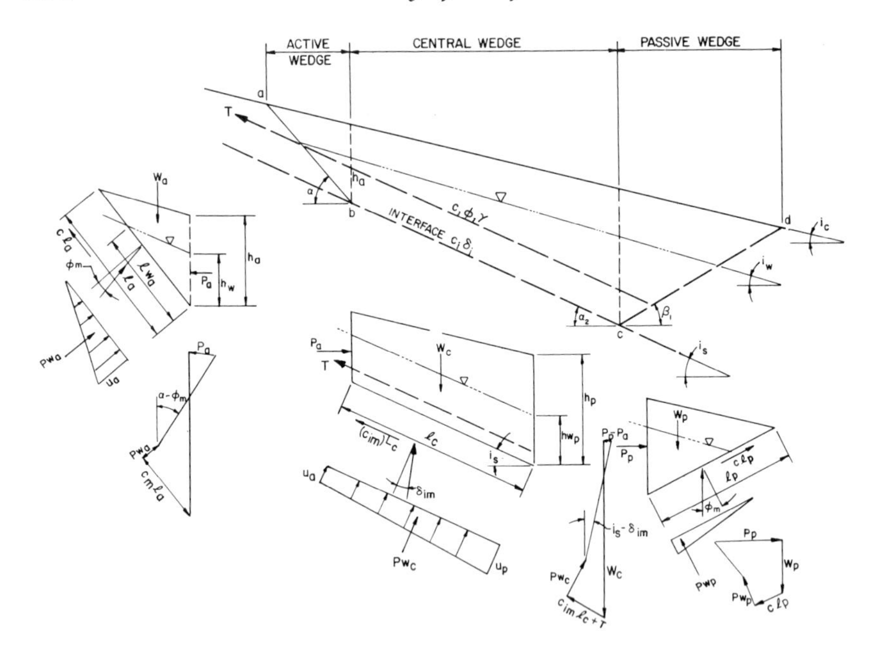

Fig. 11.12 Stability of the cover.

where

$$P_a = [W_a - (c/F)L_a \sin \alpha - P_{wa} \cos \alpha]F_\alpha - [(c/F)L_a \cos \alpha - P_{wa} \sin \alpha] \quad (11.25)$$

$$P_c = [(W_c - ((c_i/F)L_c + T) \sin i_s - P_{wc} \cos i_s)]F_c - [((ci/F)L_c + T) \cos i_s - P_{wc} \sin i_s] \quad (11.26)$$

$$P_p = [W_p + (c/F) \sin \beta - P_{wp} \cos \beta]F_\beta + [(c/F)L_p \cos \beta + P_{wp} \sin\beta] \quad (11.27)$$

where

$$F_\alpha = (F \tan \alpha - \tan \phi)/(F + \tan \alpha \tan \phi) \quad (11.28)$$

$$F_c = (F \tan i_s - \tan \delta_i)/(F + \tan i_s \tan \delta_i) \quad (11.29)$$

$$F_\beta = (F \tan \beta + \tan \phi)/(F - \tan \beta \tan \phi) \quad (11.30)$$

The angles α and β are chosen to maximize P_α and minimize P_β. The angles could be computed by considering the equilibrium of active and passive wedges and a factor of safety of 1.0. The resulting expressions are

$$\tan \alpha = \tan \phi + \sqrt{1 + \tan^2\phi - \tan i_c/\sin \phi \cos \phi} \quad (11.31a)$$

$$\tan \beta = -\tan \phi + \sqrt{1 + \tan^2 \phi - \tan i_c/\sin \phi \cos \phi} \quad (11.31b)$$

for $i_c = 0$, $\alpha = 45 + \phi/2$, and $\beta = 45 - \phi/2$.

For long slopes, P_α and P_β become relatively small and the equilibrium condition reduces to $P_c = 0.0$. For a cover of uniform thickness h, and a seepage surface parallel to the slope at depth h_w above the interface, the stability equation reduces to

$$F = (1/1 - t)\,[(c/\gamma h)\,(1/\sin i_s \cos i_s) + (\tan \delta_i/\tan i_s)\,(1 - \gamma_w h_w/\gamma h)] \quad (11.32)$$

where

$$t = (T/\gamma h L_c)\,(1/\cos i_s \sin i_s) = T/W_c \sin i_s \quad (11.33)$$

11.8.6 Seismic considerations

Ground motion induced by earthquakes add to the destabilizing force and hence reduce the factor of safety. During ground shaking, the factor of safety could be reduced to 1.0 for a very short time before a reversal in the direction of the seismic force or the end of the significant duration of ground motion. Newmark (1965) evaluated the seismic effects on embankments in terms of deformation. The elements of the Newmark method are further explained by Franklin and Griffin (1981), where the method was applied using earthquake records. Using simplifying assumptions and assuming downhill motion only, the slope movement could be approximated by (Newmark, 1965) as

$$U = \frac{V^2}{2gN}\frac{A}{N}\left(1 - \frac{N}{A}\right) \quad (11.34)$$

where U = displacement, V = ground velocity, A = ground acceleration, and N = acceleration required to produce a factor of safety of 1.0 (i.e., resistance factor).

The parameter N could be estimated by a pseudo-static analysis where a horizontal static force is substituted for the dynamic force (Sarma, 1975; Chang *et al.*, 1984). The ground velocity could be estimated from ground response analyses or available correlations (Krinitzsky and Chang, 1988; Krinitzsky *et al.*, 1988).

Surveys of landfills in California after earthquakes revealed cracks that may have been generated prior to earthquake occurrence (Siegel *et al.*, 1990). The OII landfill in Southern California, about 24 m high and founded on rock was monitored to assess earthquake effects (Siegel *et al.*, 1990). The Pasadena earthquake (Dec. 3, 1988, M = 5.0. epicenter located 15 km from the landfill) produced a horizontal acceleration of 0.22 g at the base and 0.1 g at the top of the landfill. Amplification occurs at frequencies less than 2 Hz, which constituted a small segment of the frequency content. Attenuation occurred for frequencies greater than 2 Hz. The Malibu earthquake (January 18, 1989, M = 5, epicenter located 50 km from the landfill) produced about 0.01 g at the base of the landfill and top of the landfill. Frequencies lower than 3 Hz amplified while attenuation occurred for the above 3 Hz frequency content. Considering a predominant frequency of 2 Hz, the average shear wave velocity of the landfill would be 195 m/s (640 ft/s) (uniform layer over rigid base). Sharma *et al.* (1990) reported a velocity of 200 m/s (650 ft/s) for a landfill in Richmond, California, from down-hole shear wave velocity measurements.

There are no case histories of failure or damage to landfills due to liquefaction of saturated refuse. Refuse in landfills with a leachate mound could conceivably experience a loss of strength. The hydraulic conductivity of typical municipal refuse is in the order of 10^{-3} cm/s (Oweis *et al.*, 1990) and under strong seismic shaking excess pore pressures may develop. Because the undrained shear strength for refuse is usually not known, it is prudent to apply a larger than usual factor of safety on the static strengths for stability assessment of refuse embankment with leachate mounds in seismically active areas.

REFERENCES

Bishop, A.W. (1971) Shear strength parameters for undisturbed and remolded soil specimens, in *Stress Strain Behavior of Soils* (ed. R.H.G. Parry), G.T. Foulis & Co., Ltd., Henley-on-Thames, Oxfordshire, pp. 3–58.

Chang, C.-J., Chen, W.F. and Yao, T.P. (1984) Seismic displacement in slopes by limit analysis, *J. Geotech. Eng.*, ASCE, **110**(7), 860–74.

Clausen, C.-J.F., Graham, J. and Wood, D.M. (1984) Yielding of soft clay at Mastemyr, Norway, *Geotechnique*, **34**(4), 581–600.

Collins, R.J. (1978) Highway construction use of incinerator residue, *Geotech. Pract. for Disposal of Solid Waste Materials*, ASCE, New York, 246–66.

Duplancic, N., Dayal, U. and Colella, J.C. (1987) Hazardous waste landfill cap system stability, *Geotech. Pract. for Waste Disposal '87, Geotechnical Special Publication 13* (ed Richard D. Woods), ASCE, New York, 432–46.

Fang, H.Y., Slutter, R.J. and Koerner, R.M. (1977) Load bearing capacity of compacted waste materials, in *Proc., Specialty Session on Geotech. Eng. and Environ. Control, 9th Int. Conf. on Soil Mechanics and Foundation Eng., Tokyo*, IV/2, 265–78.

Franklin, A.G. and Griffin, M.E.H. (1981) Dynamic analysis of embankment sections, Richard B. Russell dam, in *Earthquakes and Earthquake Engineering – Eastern United States* (ed. James E. Beavers), Ann Arbor Science, Ann Arbor, Michigan, pp. 623–42.

Gibbs, H.J. and Holtz, W.G. (1957) Research in determining the density of sands by spoon penetration testing, *Proc., 4th Int. Conf. on Soil Mechanics and Foundation Engineering*, London, England, 35–9.

Krinitzsky, E.L. and Chang, F.K. (1988) Intensity related earthquake ground motion, *Bulletin of the Assoc. of Eng. Geol.*, **XXV**(4), 425–35.

Krinitzsky, E.L., Chang, F.K. and Nuttli, O.W. (1988) Magnitude-related earthquake ground motion, *Bulletin of the Assoc. of Eng. Geol.*, **XXV**(4), 399–423.

Ladd, C.C., Foott, R., Ishihara, K., Schlosser, F. and Poulos, H.G. (1977) Stress-deformation and strength characteristics, *Proc., 9th Int. Conf. on Soil Mechanics and Foundation Eng.*, 421–94.

Lambe, T.W. and Whitman, R.V. (1969) *Soil Mechanics*, John Wiley & Sons, Inc., New York.

Landva, A.O. and Clark, U.J. (1986) Geotechnical testing of waste fill, *Proc. 39th Canadian Geotech. Conf.*, Ottawa, Canada, 371–85.

Landva, A.O. and Clark, J.I. (1990) *Geotechnics of waste fill – theory and practice*, ASTM STP 1070 (eds A.O. Landva and G.D. Knowles).

Leonards, G.A. (1962) Engineering properties of soils, *Foundation Eng.* (ed. G.A. Leonards), McGraw Hill, New York, p. 213.

Martin, J.P., Koerner, R.M. and Whitty, J.E. (1984) Experimental friction evaluation of slippage between geomembranes, geotextiles and soils, *Proc., Int. Conf. on Geomembranes*, Denver, CO, Industrial Fabrics Association International, St. Paul, Minnesota, 191–6.

Mayne, P.W. (1988) Determining OCR in clays from laboratory strength, *J. Geotech. Eng.*, **114**(1), 76–92.

National Solid Waste Management Association (1985) *Basic data: solid waste amounts, composition and management systems*, National Solid Waste Management Assoc. Technical Bulletin No. 85–6, Dublin, Ohio.

Naval Facilities Engineering Command (1982) Design Manual 7.1, *Soil Mechanics*, Department of the Navy, Alexandria, VA.

Newmark, N.M. (1965) Effects of earthquakes on dams and embankments, *Geotechnique*, **15**(2), 139–60.

Oweis, I. and Khera, R. (1986) Criteria for geotechnical construction on sanitary landfills, *Environ. Geotech.* (ed. H.Y. Fang), Envo Publishing Co., 205–22.

Oweis, I. and Khera, R. (1990) *Geotechnology of Waste Management*, Butterworth Publishers, London, England.

Oweis, I., Mills, W., Leung, A. and Scarino, J. (1985) Stability of sanitary landfills, *Geotechnical Aspects of Waste Management*, Foundation and Soil Mechanics Group, Metropolitan Section, ASCE, New York.

Oweis, I., Smith, D.A., Ellwood, R.B. and Green, D.S. (1990) Hydraulic characteristics of refuse, *J. Geotech. Eng.*, ASCE, **116**(4), 539–53.

Poran, C.J. and Ali, F.A. (1989) Properties of solid waste incinerator fly ash, *J. Geotech. Eng.*, ASCE, **115**(5), 1118–33.

Sarma, S.K. (1975) Seismic stability of earth dams and embankments, *Geotechnique*, **25**(4), 743–61.

Schoenberger, R.J. and Fungaroli, A.A. (1971) Incinerator residue – fill site investigation, *J. SMFD*, ASCE, **97**(10), 1431–43.

Seed, H.B. (1987) Design problems in soil liquefaction. *J. Geotech. Engin.*, ASCE, **113**(8), 827–45.

Seed, H.B. and De Alba, P. (1986) Use of SPT and CPT tests for evaluating the liquefaction resistance of sands, in *Use of In Situ Tests in Geotechnical Engineering* (ed Samuel P. Clemence), Geotechnical special publication no. 6, ASCE, New York, pp. 281–302.

Seed, R.B., Mitchell, J.K. and Seed, H.B. (1990) Kettleman Hills waste landfill slope failure II: stability analyses, *J. Geotech. Eng.*, **116**(4), 669–90.

Seed, H.B., Mitchell, J.K., and Chan, C.K. (1960) The strength of compacted cohesive soils, *Proc. Res. Conf. on Shear Strength of Cohesive Soils*, ASCE, New York, 897–964.

Sharma, D.H., Dukes, M.T. and Olsen, D.M. (1990) Field measurements of dynamic moduli and Poisson's ratios of refuse and underlying soils at a landfill site, *Geotechnics of Waste Fills – Theory and Practice,* ASTM STP 1070 (eds A.O. Landva and G.D. Knowles), Am. Soc. for Testing and Materials.

Siegel, A.R., Robertson, J.R. and Anderson, D.G. (1990) Slope stability investigation at a landfill in Southern California, *Geotechnics of Waste Fills – Theory and Practice,* ASTM STP 1070 (eds A.O. Landva and G.D. Knowles), Am. Soc. for Testing and Materials.

Skempton, A.W. (1954) The pore-pressure coefficients A and B. *Geotechnique,* **4**, 143–7.

Stoll, O.W. (1971) *Mechanical properties of milled refuse,* ASCE National Water Resources Engineering Meeting, Phoenix, Arizona, January 11–15.

Takasumi, D.L., Green, K.R. and Holtz, R.D. (1991) Soil geosynthetics interface strength characteristics: a review of state-of-the-art testing procedures, *Geosynthetics '91,* Industrial Fabrics Assoc. Int., St. Paul, Minnesota, **1**, 87–100.

Tchobanoglous, G., Theisen, H. and Eliassen, R. (1977) *Solid Wastes, Engineering Principles and Management Issues,* McGraw Hill, New York, p. 209.

Terzaghi, K. and Peck, R.B. (1967) *Soil Mechanics in Engineering Practice,* John Wiley and Sons, Inc., New York.

Williams, N.D. and Houlihan, M.F. (1987) Evaluation of interface friction properties between geosynthetics and soils, *Geosynthetics '87,* Industrial Fabrics Assoc. Int., St. Paul, Minnesota, **2**, 616–27.

CHAPTER 12

Mine waste disposal

Dirk Van Zyl

12.1 INTRODUCTION

Mine wastes are a broad group of waste materials resulting from the extraction of metals and non-metals. The wastes include solid, as well as liquid waste, and can be inert or can contain hazardous constituents. In general, mine waste consists of high volume, low toxicity wastes (EPA, 1985).

It was estimated in 1982 that about 500 and 30 million tonnes of waste were produced in the US from copper and molybdenum mines, respectively. The main reason for the high volumes of mine waste is the very low concentration of metals contained in ores. For example, gold mining is presently conducted using heap leach technology for ores containing as little as 0.5 g per tonne of gold. It is, therefore, clear that virtually every tonne of rock which is mined remains as waste after extraction of the gold. In open-pit mining, there is also a considerable volume of overburden and waste rock produced which contains no mineralization or levels of mineralization too low to make extraction economical. The term stripping ratio in open pit mining refers to the tonnes of waste stripped (discarded as waste rock) as a ratio of the tonnes of ore mined. Stripping ratios in the order of 2 to 6 are not unusual. Consider the case of an open-pit copper mine where 66 000 tonnes of copper ore is mined per day and the average grade of the ore is 1.5% (i.e., one tonne of ore contains 1.5 kg of copper) and the stripping ratio is 3. In this case, 200 000 tonnes of waste rock is discarded daily, while 65 000 tonnes of other waste (typically tailings) remain after extraction of the copper. In a year with 350 days of production, 69 000 000 tonnes of waste rock and 22 750 000 tonnes of tailings are produced.

Geotechnical Practice for Waste Disposal.
Edited by David E. Daniel.
Published in 1993 by Chapman & Hall, London. ISBN 0 412 35170 6

Mine wastes remain after the extraction of materials which are sold on the commodity markets. The markets are not controlled by the producers but are controlled by the free market system of supply and demand. The extra cost of mine waste disposal can, therefore, not be passed on to the consumer but must be absorbed within the economic constraints of the project. This aspect, combined with an appreciation of the high volume and low toxicity characteristics of the waste are important in considering the overall waste disposal strategies for mine waste.

The disposal of mine waste represents capital and operating cost to a mining operation. For this reason, mine waste has historically been disposed of at the lowest cost. This often resulted in considerable environmental impacts. For example, tailings containing hazardous constituents were directly discharged onto the land without containment, or in some cases, directly into rivers, while waste rock was dumped as close as possible to the mine adit or shaft without consideration of environmental impacts. Some of these and other historical impacts of mine waste on the environment have been surveyed by Smith (1987).

Some of the historical mine waste disposal practices led to catastrophic failures. While not all such failures have been well documented some of the recent failures have been evaluated. Three such failures which caused a considerable number of deaths were the Buffalo Creek, Aberfan and Chilean tailings impoundments (Dobry and Alvarez, 1967; Bishop *et al.*, 1969; Wahler and Assoc, 1973). These failures focused considerable attention on the geotechnical aspects of designing safe mine waste disposal facilities. This attention was the impetus for developing the science related to geotechnically sound mine waste disposal facilities over the last three decades (NBRI, 1959; U.S. Dept. of Interior, 1975; ASCE, 1979; Vick 1983; Van Zyl and Vick, 1988).

Attention has been focused over the last decade on the environmental considerations related to mine waste disposal. Activity in this area has recently increased many fold (Van Zyl, 1985; University of California at Berkeley, 1988; Lootens *et al.*, 1991). Much has been learned about the environmental impacts of mine waste; however, it is an area for potential fruitful future research contributions. In the early 1980s, cyanide use (mostly for the extraction of gold and silver) attracted considerable attention and the disposal of such wastes was considered a big environmental risk. Research and practical experience since that time have shown that cyanide waste can be disposed of safely and that cyanide can best be considered a 'transient pollutant'. Of much greater concern is the disposal and long-term effects of acid-generating wastes, also referred to as acid mine drainage (AMD) or acid rock drainage (ARD), (Steffen, Robertson and Kirsten, 1989). This area of research has attracted much attention and practical methods are available, while others are being developed, to safely dispose of acid-generating wastes or to treat effluents from such facilities.

Chemical characteristics of tailings solids and solutions are determined by the characteristics of the ore, as well as the chemicals used during extraction of the metals. It is beyond the scope of this book to discuss chemical characteristics of mine waste. However, it is an important topic, which in most cases, determines the final design of a mine waste facility.

This chapter presents a brief review of mine waste disposal practices for various types of mine waste. It is not intended to be exhaustive, as this topic could fill a multi-volume text. The references will concentrate on recently published summary documents instead of an extensive list of individual papers, except where such papers present specific information.

12.2 WASTE ROCK

Waste rock consists of unmineralized rock, as well as rock containing mineralization which is too low to extract economically with existing technologies. Materials overlying the orebody, often referred to as overburden, are also usually included in the waste rock category. It is obvious that because waste rock can contain some mineralization, today's waste may become tomorrow's ore. This fact has been proven many times and especially so more recently with the development of low cost hydrometallurgical processes for extraction of gold (such as heap leaching). Many waste rock dumps are presently being reworked as ore.

The characteristics of waste rock depend on the characteristics of the host rock, as well as the rock surrounding the ore body. Waste rock is very often competent, durable, hard rock consisting of particle sizes from 2 m diameter to silt and clay size. However, waste rock can also consist of very weak materials, such as weathered volcanics or sediments, which can produce considerable stability concerns. A recent summary of waste rock disposal practice was edited by McCarter (1985).

Waste rock and overburden may be disposed of in valley fills, side-hill dumps, or open piles. The type of waste rock disposal facility is a function of the volume of waste, site topography and drainage, and haul distance from the mine. Unless located in an arid climate, these waste management units will have flood runoff protection ponds for purposes of sediment control, and surface runoff diversion ditches to minimize natural runoff intruding on the pile. Even in arid climates, there may be a need to install flood diversions to deal with flash flood events.

Disposal of waste rock and overburden takes place by end-dumping from trucks, or stacking using conveyors. The material is dumped over the face of a lift. The lift may be the full height of the dump, i.e., up to 100 m or more, or 15- to 30-m lifts with intermediate benches. The working face of a lift is at the angle of repose of the material. As the dumped material cascades down the face, segregation of particles occurs. The large boulders roll to the bottom and the gravels, sands, and fines remain near the top. The higher the working face of a lift, the more prominent the segregation. As a result, the bottom of disposal facilities is frequently free-draining rock, while the upper elevations and working surfaces are less permeable.

The upper surfaces of the pile are usually flat. They are gently sloped away from the dump face toward the natural terrain to enhance dump truck safety during operation, and also to prevent surface runoff, during storm events, from cascading over the edge of the piles.

Valley fills usually are started at the upstream end of the valley. Dumping progresses downstream, with increasing lift thickness. Sidehill piles are formed by dumping along hillsides or valley slopes so that the toe of the dump does not cross the drainage. Dumping is continued along the natural contour and away from the slope. Surfaces are graded to keep runoff away from the exposed face of the dump. An open pile may be constructed over relatively flat topography, or as a ridgeline fill, and can be extended to form a sidehill dump.

Reclamation of waste rock dumps can be done through a number of approaches. Dump slopes can be armored with competent, durable hard rock which resembles scree slopes. In this case, these dumps will typically be left at their natural angle of repose which could be on the order of 37 degrees. The slopes of waste rock dumps can also be smoothed with dozers to shallower slope angles than the angle of repose, for example 2H:1V. There is considerable controversy about the reclamation of such slopes through the establishment of vegetation. One school of thought is that vegetation can not be established on slopes steeper than 3H:1V; however, examples exist of successful revegetation of slopes at 2H:1V. It is important to control the length of the slope to reduce the formation of gully erosion.

Major concerns with waste dump construction include operational stability, as well as long-term stability. Weathering of waste rock following deposition can lead to a change in rock shear strength characteristics and subsequent failure of the structure. Acid rock drainage (ARD) from waste dumps is a specific concern which must be addressed at the time of design. ARD can lead to significant environmental damage if not controlled. ARD from waste rock disposal facilities is by far the biggest environmental concern in mine waste disposal. This topic is outside the scope of this book. However, it must be given full attention during the design of a waste rock disposal facility.

The geotechnical characteristics of waste rock range from that of rockfill to that of soil. Uhle (1988) summarized and statistically evaluated most of the rockfill data published to date. The shear strength characteristics of rockfill are typical of granular materials. However, waste rock demonstrates a much more pronounced curved Mohr envelope than is typical of sands. The shear strength, under low normal stress, is therefore, higher than a linear failure envelope will indicate.

Waste rock disposal facilities for new mines are designed on the basis of information obtained during the geological exploration. Core samples of the ore and waste will be available for mineralogical, metallurgical and geotechnical evaluations. There are very often only small quantities of core available and it is therefore difficult to perform sufficient geotechnical testing to confirm the expected shear strength of the waste. The design engineer must rely on experience in the selection of design parameters. Engineering geological description of the core during the exploration program is a very useful tool in the characterization of the waste. Information such as rock hardness, mineralogy, rock quality designation (RQD), fracture spacing, and presence of weathered products are necessary to select appropriate design parameters.

It is difficult, if not impossible, to measure the shear strength of waste rock in the laboratory due to the large sizes of the particles disposed during operations. Direct shear and triaxial testing of smaller particles may lead to the selection of overly conservative shear strength values. Point load testing of intact rock particles, such as cores, can be used to estimate the shear strength of the waste rock. The following empirical relationship suggested by Hoek and Bray (1981) can be used to estimate the shear strength:

$$t = A\sigma_c(\sigma/\sigma_c - T)^B \tag{12.1}$$

where: A, and B are the constants defining the shape of the Mohr failure envelope; the uniaxial compressive strength, σ_c is given by $\sigma_c = 24I_s$, where I_s is the point load strength; and $T = \frac{1}{2}(m - \sqrt{m^2 + 4s})$, where m and s are dimensionless constants depending on particle shape.

Waste rock disposal facilities must be designed to be well drained. Coarse particles deposited at the bottom of the waste rock dump as a result of natural segregation can form an adequate drain in many instances. However, if springs are present on the site then specific underdrains must be installed to intercept these springs and drain them under controlled conditions. Underdrains consisting of large rock particles wrapped in geotextile are usually specified. These drains can also include perforated pipes, however it is good practice to design the rock drain with sufficient capacity to carry the expected flow.

Since waste rock disposal facilities are constructed by end dumping, they are, therefore, deposited in a relatively loose state. They are subject to settlement under selfweight, as well as settlement when structures are placed on them. There are examples of mining structures being constructed on waste rock disposal facilities. In this case the waste rock disposal facility must be constructed as an engineered fill. Placement of the material in 1- to 2-m-thick lifts compacted by multiple passes of construction traffic will usually result in an adequately compacted structure.

In summary, it can be stated that the geotechnical concerns of waste rock disposal facilities are related to characterization of the materials in terms of shear strength and the construction of these facilities for maximum economy and reliable short and long term stability. Experience plays a large role in the design of waste rock disposal facilities.

12.3 TAILINGS IMPOUNDMENTS

Tailings are the products remaining after the extraction of metals from ore by physical and chemical methods. The ore is finely-ground and slurried to allow the processing. Grain sizes of tailings are typically in the fine sand and silt range and can contain as much as 80% smaller than the US No. 200 sieve (0.074 mm). Tailings are typically deposited as a slurry into specially designed impoundments. Special depositional methods have been developed to enhance the stability of such impoundments and to allow for environmentally safe disposal of the tailings. Vick (1983) describes the production and disposal design of tailings impoundments in much detail. Van Zyl and Vick (1988) present papers describing the recent practice of tailings disposal. An extensive bibliography on tailings disposal is presented by ICOLD (1989).

Tailings are typically deposited in an impoundment formed by an embankment or dike. The embankment can be constructed of fill material or, in some instances, of the tailings themselves. Construction with tailings material involves controlled spigoting of the tailings, i.e., the discharge of tailings through several small pipes or spigots, and allowing the solutions to drain from the solids. In some instances, the tailings may be cycloned to separate the coarser fraction, which then is used for embankment construction.

An impoundment can be formed by constructing a single embankment across a valley or a three-sided embankment on the side of a valley or on sloping ground. On flat ground, an embankment or a 'ring' dike that completely surrounds the deposited tailings materials is required (Robertson and van Zyl, 1980).

Slurried deposition is the most common and practical means and can be accomplished by the following methods: subaqueous deposition;

managed deposition; conventional point discharge; and cycloned deposition. Each is now considered in turn.

Subaqueous deposition

This involves discharge under water in the impoundment. Deposition is uncontrolled, and the tailings settle as a soft bottom sediment, or are transported and dispersed over a large area. The resulting low density material is generally very soft and may require special construction techniques should a cover be required during closure.

Managed deposition

This method, which has also been referred to as subaerial deposition (see discussion below) is typically accomplished by discharging the slurry from one or more points around the perimeter of the impoundment area. As the slurry drains toward the low point in the impoundment, the solids spread out over the bottom to form beaches or deltas. The free water drains to a central location, or low point along the perimeter, where it either is recycled into the processing system, decanted to a separate containment facility for reuse or evaporation, or evaporated in place. The point or points of discharge are moved around the impoundment to allow the exposed solids to dry and increase in density. Subsequent layers are then deposited, and the cycle is repeated.

Conventional point discharge

This method is similar to managed deposition, except that no effort is made to add additional discharge points or to frequently move the discharge points. As a result, the tailings are usually not as dense or dry as with managed deposition. Also, with conventional point discharge, the finer portion of the solids tends to settle out in the same area and create a soft area that is difficult to access, and subject to additional settlement/consolidation over time.

Cycloned deposition

This involves separation of the coarser tailings fraction prior to deposition. This type of deposition creates a zone or pile of coarse solids relatively free of excess solutions, plus a pool of solutions and wet fine solids which may not drain readily and, therefore, consolidate. Operation of a cycloning system requires planning and a significant amount of management if the system is to function properly. It usually is done, at least in part, to provide coarse tailings for use as fill in raising the embankment.

The moisture content and density of deposited tailings are functions of the ore extraction process, method of deposition, grain size distribution, and specific gravity. Moisture content can vary from 20 to over 60%. For slurried deposition, the material is saturated and moisture contents typically vary from 40 to 60% and dry densities from 1 to 1.3 g/cm^3. At the end of deposition the density of a tailings deposit generally increases with depth. The lower tailings have usually consolidated, due to the weight of the upper tailings with time, while the upper tailings are still in the process of consolidating.

The surface of the tailings will settle after deposition ceases and solution will drain from the settling deposit. Finite strain consolidation theory is well suited for evaluating the consolidation of a tailings deposit during construction as well as after the end of deposition (Caldwell *et al.*, 1984).

Where managed, e.g., layered, deposition of tailings has occurred, the densities may be higher than the slurried deposition, i.e., typically 1.3 to 1.45 g/cm^3. As a result there is likely to be less settlement of the surface of the tailings after closure. With lower moisture contents, typically 20–40%, and no standing water, the surface of managed tailings becomes accessible for closure activities in a shorter period of time than slurried tailings, and usually without the need for special construction techniques.

It is also possible to 'dry' tailings so that deposition can take place using 'landfill' techniques. Not all tailings materials are amenable to this type of treatment as it is dependent on the material characteristics, especially the presence of clays in the ore. The presence of clay can make it practically infeasible to 'dry' the tailings, or can result in a tailings product which still has a high moisture content (Robertson *et al.*, 1982).

'Dewatering' or 'drying' of the tailings occurs in the metals recovery plant using unit process, such as thickening, pressure filtration, vacuum filtration, and a combination of these. The residual moisture content is usually at or just below saturation, and generally varies from 15 to 25%, however, for very fine-grained material, higher values can occur (Robertson *et al.*, 1982). The dry density of these materials will generally vary from 1.2 to 1.6 g/cm^3 depending on how much compaction occurs when the material is deposited.

Deposition of dried tailings is accomplished by either truck dumping and spreading with a bulldozer, or by conveyor transport and stacking. This type of deposition results in a pile with no pools of free-standing solution. Direct precipitation onto the pile is allowed to drain off and collect in a separate holding pond. Depending upon the degree of dewatering, residual seepage of solution can occur as a result of gravity drainage and consolidation of the tailings. Where the tailings contain a high proportion of silt- and clay-sized fractions, some form of containment may be necessary to prevent the tailings from spreading.

Tailings impoundments can be constructed by using tailings to raise the embankment or by using compacted earth fill as the containment structure. Tailings impoundments containing toxic materials, such as cyanided gold tailings, are often constructed as fully-lined impoundments.

Tailings impoundments raised with tailings can be constructed using the upstream, centerline, or downstream construction methods (Vick, 1983). Figure 12.1 presents schematics of these construction methods. For the upstream construction method, tailings deposition can take place through spigots, i.e., multiple open-ended discharges which will allow natural segregation of the particles along the beach, so that coarse particles are deposited close to the embankment, while finer particles are deposited further down the beach. A beach slope is also formed which is characteristic of the slurry material and the discharge rate. Cyclone separation of fine and coarse particles can also be used so that coarse particles are deposited for embankment construction, while the finer particles are deposited in the pool area.

Centerline and downstream construction are typically done through the use of cyclones, where the embankment raises are constructed with the coarse materials (cyclone underflow), and the fine materials (cyclone overflow) are deposited behind the embankment. Sufficiently large volumes of coarse material must be available following cyclone separation to allow for centerline and downstream construction. The schematics in Fig. 12.1 clearly show that large volumes of sands are required to construct centerline and downstream embankments, especially towards the end of the facility's life.

As described above carefully managed deposition can take place through the regular rotation of depositional areas, allowing for the deposition of thin layers, which are allowed to dry and increase in density before the next layer is deposited. Such thin-layer managed tailings deposition has been practiced in the South African gold mining industry since the early part of this century (Ruhmer, 1974). Most recently, a number of publications have referred to this method as 'subaerial deposition' implying that it is tailings deposition which is not done under water (Knight and Haile, 1983; Cincilla *et al.*, 1991). Although this terminology has found its way into the literature, it does not specifically refer to thin-layer managed deposition. Any depositional method which allows for deposition of tailings 'under air', such as cycloned tailings deposition, should also be considered as subaerial deposition. This author therefore prefers to refer to this method as thin-layer managed tailings deposition vs. the catch-all term 'subaerial' deposition.

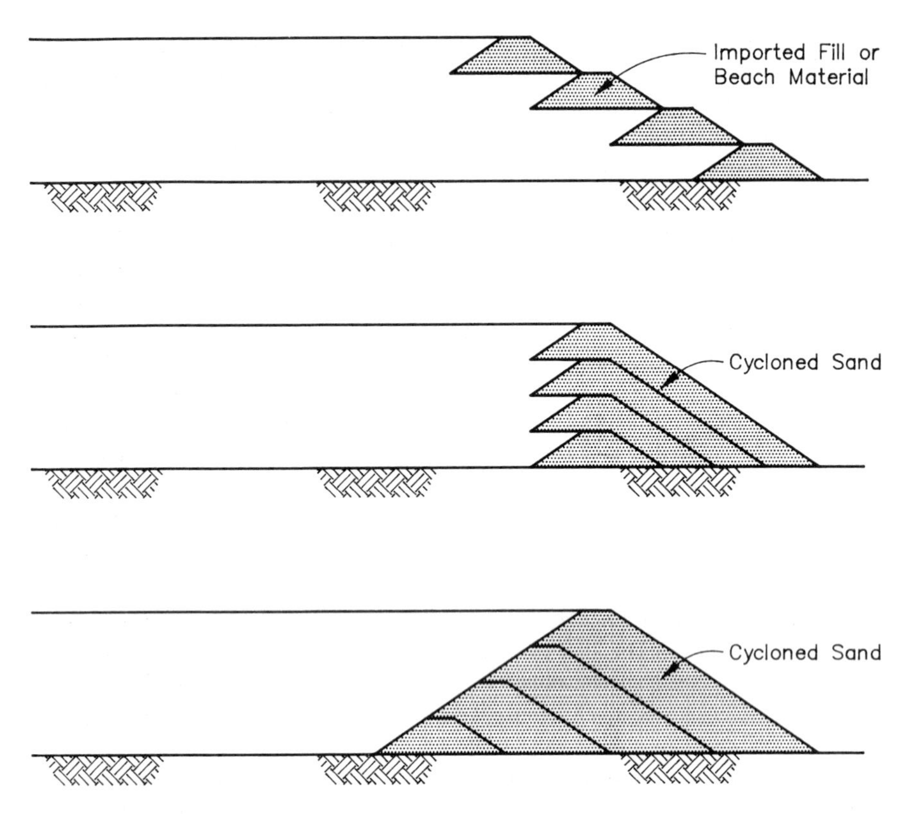

Fig. 12.1 Upstream, centerline and downstream construction of tailings impoundments.

Thin-layer managed tailings deposition has a number of advantages, however, it is important that the operator of such a tailings impoundment must make a firm commitment to managed deposition as it does require considerable attention. From an operating standpoint, it is, therefore, more expensive than less managed tailings deposition methods.

In order to reduce seepage losses to the environment from tailings impoundments containing toxic materials, a number of alternatives can be considered. Fully-lined impoundments can be constructed or the thin-layer managed technique can be combined with underdrained tailings impoundments to reduce the hydraulic head on the liner, and thereby reducing the potential for seepage losses.

During the initial stages of tailings deposition, the drain layer enhances the drainage of the deposited tailings. After the drain layer is covered with slimes, the tailings permeability above the drain layer reduces significantly and with time, effectively form a second liner. Hydraulic conductivities as low as 5×10^{-7} cm/sec can be obtained in tailings (depending on particle size and clay mineral content) when subjected to high overburden stresses. Seepage is removed from the drain layer during operations, thereby reducing the head on the synthetic liner to very low values. This approach has been successfully practiced at a number of newly constructed gold tailings impoundments in the US.

The sedimentation behavior of tailings along a beach following deposition has been studied by a number of researchers and has resulted in the definition of a beach profile as well as the particle size segregation along the beach. The deposition of particles from the slurry and, therefore, the segregation along the beach, as well as the beach profile, are functions of specific gravity of solids, percent solids in slurry, and discharge rate of the slurry. Melent'ev *et al.* (1973) propose a model for the development of a beach, as well as segregation of particles along the beach.

In 1979, this author, with the help of a translator, translated Chapter 2, Principles of Hydraulic Fill Theory, of the text by Melent'ev *et al.* (1973) in a specific project. Subsequently, a number of researchers have used this translation and have successfully shown that Melent'ev's theories are valid and can be applied successfully to the deposition of gold and platinum tailings (Blight, 1987). Considerable effort in this area, combined with seepage analyses, have also been presented by Abadjiev, (1985).

The profile of a beach is generated by the gravitational sorting of particle sizes as the slurry flows down the beach. A reduction in particle size occurs along the beach which results in the reduction of the hydraulic conductivity of the tailings as a function of distance from the depositional point (Blight, 1987). A master profile of the beach is developed which can be expressed as:

$$\frac{H}{Y} = \left(1 - \frac{H}{X}\right)^{n} \tag{12.2}$$

where H is the length of beach, Y, the elevation between point of deposition and pool, X, the distance along beach, and n, the dimensionless constant dependent on tailings characteristics.

Blight (1987) shows that this expression models the master beach profile for various tailings materials. Figure 12.2 presents the dimensionless beach profiles for various tailings materials.

Abadjiev (1976, 1985) has suggested the following relationship for the change of saturated hydraulic conductivity for deposited tailings as a result of the material segregation along the beach.

$$k = ae^{-bH} \tag{12.3}$$

where a and b are constants characteristic of the beach, and H is the distance along beach from deposition point.

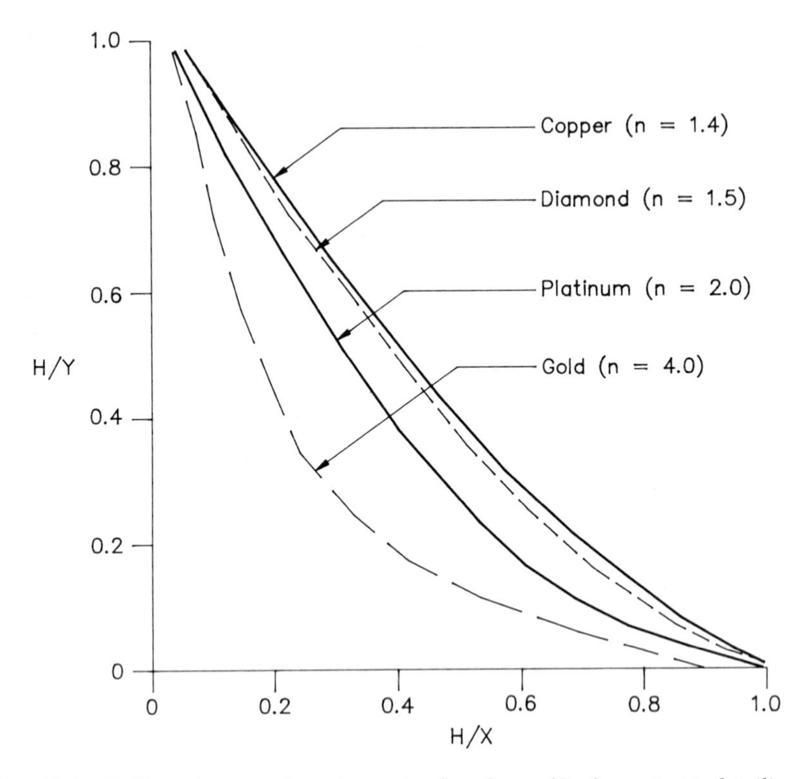

Fig. 12.2 Tailings impoundment master beach profile for spigotted tailings, after Blight (1987).

Further characterization of master beach profiles and depositional behavior is important for the more accurate modelling of tailings deposition and therefore the geotechnical characteristics of a tailings deposit.

Geotechnical concerns of tailings impoundment design center on static and especially dynamic stability, while the environmental concerns center on seepage of contaminated liquids into the environment, surface discharge of contaminated liquids, and dusting. Static stability is determined by the shear strength of the tailings and the presence and location of a phreatic surface. The potential of a liquefaction failure of a tailings impoundment must be considered in seismically active areas. Much has been published about this topic and the interested reader is referred to the literature (e.g., ICOLD, 1990).

The geotechnical characteristics of tailings depend largely on the amount of clay minerals in the tailings. If the tailings are free of clay minerals, then the tailings will behave as a frictional material. A specific concern during the construction of a tailings impoundment is the consolidation behavior of the finely-ground slurried materials. If a tailings impoundment is constructed rapidly (at a high rate of rise), then excess pore pressures do not have time to dissipate and can lead to liquefaction failures. Such failures can occur under static loading conditions, however, they are obviously more prevalent under dynamic loading conditions.

The angle of internal friction of non-clayey tailings material under effective strength conditions is typically in the order of 35 degrees, due to the angular nature of tailings particles. This value is maintained for coarse as well as fine tailings, i.e. sands and slimes. Direct shear as well as triaxial shear tests can be used to obtain shear strength characteristics of tailings. In most cases reconstituted samples must be used because of the sandy nature of tailings (Chen and Van Zyl, 1988). *In situ* testing of tailings impoundments can be done successfully using piezocone equipment. The results can be used to interpret shear strength, consolidation as well as permeability parameters (Rust *et al.*, 1984).

The geotechnical design of a tailings impoundment requires a thorough understanding of tailings production, depositional behavior, stability considerations, and environmental design requirements. Although not discussed in this section, liner design is becoming an important consideration in tailings impoundment design. In most cases a single composite liner provides sufficient environmental protection. Liner design details are discussed elsewhere in this text.

12.4 SPENT HEAP LEACH ORE

Heap leaching is a process where low-grade ores are stacked on a low-permeability base and irrigated with a lixiviant to extract the metal. In the case of gold, cyanide is used as a lixiviant and in the case of oxide copper, sulfuric acid is used as a lixiviant. This method of extraction was first developed on a commercial scale in the US for gold deposits in the late 1970s, although copper and uranium heap leaching had been practiced on a world-wide basis since the earlier part of the 20th century. Considerable advances have been made over the last decade in the design of heap leach facilities (Van Zyl *et al.*, 1988).

A large percentage of gold is presently extracted in the US using heap leach facilities. It is therefore clear that the design of heap leach facilities is receiving considerable attention. After extraction of the metal, the spent heap leach ore remains as a waste product. The long-term disposal of such waste must be carefully considered during the initial design of the heap leach facility.

Heap leach projects can be developed on permanent pads. For example, the ore is left on the pad during and after leaching and the spent ore, therefore, remains on a lined facility. Reusable pads can also be used where the spent ore is removed after metal extraction, and disposed of at a separate facility. Valley leach facilities are used in steep terrain to contain the ore. After leaching and rinsing, the ore remains in its containment and the liner is punctured to allow drainage of infiltration into the foundation.

Heap leach ores can consist of run-of-mine material, where a blasted ore is placed on the pad without any further preparation. The ore can also be prepared, for example, through crushing or crushing and agglomeration. In the latter cases, the ore particles are typically smaller than 75 mm and very often smaller than 20 mm. Excessive fines in the crushed ore can be agglomerated onto the coarse particles by using Portland cement as a binder and the tumbling action in a rotating drum (McClelland and Van Zyl, 1988). It is important to maintain uniform percolation through the heaps, so that the leachate is evenly distributed and contacts all the sources of the metal in the ore. Application rate of leachate is typically in the range of 0.1–0.2 L/min/m^2.

After leaching, the ore must be rinsed so that the heaps can be reclaimed. In the case of cyanide heap leaching, considerable attention is presently paid to rinsing of heaps and new methods and results from research and practice are expected in the next few years. The degradation of cyanide in heaps is related to the reactive nature of cyanide which leads to rapid volatilization and oxidation, as well as biological processes which take place over a longer period of time (Chatwin, 1989). Bacterial degradation of cyanide occurs readily in spent ore heaps, as well as natural soils.

The spent ore heap is usually considered rinsed when the effluent from the heap has a weak acid dissociable (WAD) cyanide content of 0.5 mg/l or lower. Although the cyanide level can be reduced, it is possible that leachate of metals could be of greater concern in the long-term after closure of a heap leach facility. It is, therefore, important to reduce long-term infiltration into a reclaimed heap. Final reclamation may consist of reshaping of the heap to allow placement of a growth medium and finally revegetation of the heap. In certain extreme climatic conditions, where high net precipitation is experienced, it may be necessary to construct a liner over the top of the reshaped heap.

Liner requirements for heap leach facilities differ between the various states in the US. Single composite liners provide sufficient environmental protection for permanent pads and reusable pads where low hydraulic heads are maintained. In the case of valley leach facilities a double composite liner may be required because of the higher hydraulic head being maintained in the heap. Similarly, for the ponds associated with heap leach facilities, it is recommended that double composite liners be used so that the hydraulic head on the lower composite liner can be reduced when the leak collection system is evacuated.

The slope stability of heaps constructed on clay or geomembrane liners must be evaluated. Site-specific geometric and material characteristics must be used. If it is assumed that the heap is constructed on top of an underlying pad, consisting of a geomembrane constructed onto compacted low permeability soil, then the interface between the low permeability soil and the geomembrane constitutes the weakest part of the structure. A block-type failure is the most critical. It can further be assumed that the heap material, as well as the interface shear strengths, can be characterized by an angle of internal friction.

The friction angle of a liner interface is dependent on the liner type as well as the material in contact with the liner. Site specific testing must always be conducted. However the following general statements can be made about interface friction angles:

1. more rigid liners, such as high density polyethylene (HDPE) have lower friction angles than flexible liners, such as very low density polyethylene (VLDPE) and polyvinylchloride (PVC);
2. textured (surface-roughened) geomembranes have higher friction angles against soil than smooth geomembranes;
3. the presence of clays against the geomembrane interface results in lower friction angles than granular materials; and
4. the friction angle of interfaces between two layers of synthetic materials can be very low, e.g., HDPE and a geotextile.

The design of heap leach facilities is well regulated in the US where such facilities are found. The geotechnical engineer must use site specific design criteria to develop an environmentally protective design. Leakage and stability considerations are foremost in the design of heap leach facilities for operational and post closure conditions.

12.5 LIQUID WASTE

Liquid waste streams at mines consist of tailings water, leachate from heap leach facilities, smaller leach streams containing higher concentrations of contaminants, and water produced through dewatering of pits or collection of surface runoff.

The design of liquid waste containments is dependent on the chemical characteristics of the contained liquid. Sludges often collect in liquid waste ponds and such sludges can become a concern at the time of reclamation of the waste.

Closure of liquid waste ponds is usually done through treatment and disposal of the liquids, such as chemical treatment or evaporation. Sludge remaining in the bottom of the pond may be removed and deposited in one of the other mine waste facilities on site, or may have to be solidified in place through the use of portland cement, or in some cases, may have to be removed and deposited in a hazardous waste landfill. The pond liners can be folded over and removed and placed in one of the other mine waste facilities or can be buried in place.

12.6 OTHER MINE WASTES

12.6.1 Coal mining

In the removal of coal, large volumes of overburden and interburden wastes must be removed. Because of the continuous nature of coal beds, such mining is done by draglines where the overburden is cast in the mined-out area. The characteristics of the coal waste are typical of sedimentary environments consisting of interbedded sandstones and shales. In coal strip mining, the mine spoil is recontoured and reclaimed as the mine develops.

12.6.2 Industrial minerals

Industrial minerals include sand and gravel, zeolites, vermiculite, and other similar minerals. The wastes from such mining operations are diverse in their characteristics. The mining methods can consist of mechanical separation of the minerals from waste, most often using dry processes. Each of these wastes must, therefore, be considered on a site-specific basis.

REFERENCES

Abadjiev, C.B. (1976) *Seepage Through Mill Tailings Dams*, Trans. of the 12th ICOLD, Mexico City, **1**, 381–93.

Abadjiev, C.B. (1985) Estimate of the physical characteristics of deposited tailings in the tailings dam of non ferrous metallurgy, *Proc., 11th Int. Conf. on Soil Mechanics and Foundation Eng.*, San Francisco, **3**, 1231–4.

ASCE (1979) *Current Geotechnical Practice in Mine Waste Disposal*, papers collected by the Committee on Embankment Dams and Slopes of the Geotech. Eng. Div.

Bishop, A.W., Hutchinson, J.N., Penman, A.D.M. and Evans, H.E. (1969) Geotechnical investigations into the causes and circumstances of the disaster of 21st October 1966, in *A Selection of Technical Reports Submitted to the Aberfan Tribunal*, Item 1, 1–80, H.M.S.O., London.

Blight, G.E. (1987) The concept of the master profile for tailings dam beaches, in *Proc. of the Int. Conf. on Mining and Ind. Waste Management*, (eds J.A. Wates and D. Brink) South African Inst. of Civ. Eng., Johannesburg, South Africa, pp. 95–100.

Caldwell, J.A., Ferguson, K., Schiffman, R.L. and Van Zyl, D. (1984)

Application of Finite Strain Consolidation Theory for Engineering Design and Environmental Planning of Mine Tailings Impoundments, *Sedimentation/Consolidation Models – Predictions and Validation*, (eds R.W. Yong and F.C. Townsend), ASCE, pp. 581–606.

Chatwin, T.D. (1989) *Cyanide Attenuation/Degradation in Soil*, Resource Recovery and Conservation Company, Salt Lake City, Utah.

Chen, H.W. and Van Zyl, D. (1988), Shear Strength and Volume Change Behavior of Copper tailings under Saturated Conditions, in *Hydraulic Fill Structures*, (eds D.J.A. van Zyl and S.G. Vick), ASCE.

Cincilla, W.A., Dye, R.A. and East, D.R. (1991), Nevada Goldfields Aurora project: a case history of subaerial tailings deposition vs. mechanical dewatering and disposal, paper presented at the 94th Nat. Western Mining Conf., Colorado Mining Assoc., Feb. 24–27.

Dobry, R. and Alvarez, L. (1967) Seismic failures of Chilean tailings dams, *J. Soil Mechanics and Foundations Div.*, A.S.C.E., **93**(SM6), 237–60.

EPA (1985), Report to Congress: Wastes from the extraction and beneficiation of metallic ores, phosphate rock, asbestos, overburden from uranium mining and oil shale, EPA/530-SW-85-033.

Hoek, E. and Bray, J. (1981), *Rock Slope Engineering*, revised 3rd edn, The Institution of Mining and Metallurgy, London.

International Commission on Large Dams (ICOLD) (1989) *Bibliography: mine and industrial tailings dams and dumps*, Bulletin 44a.

International Commission on Large Dams (ICOLD) (1990) *International Symposium on Safety and Rehabilitation of Tailings Dams*, Sydney, Australia, May 23.

Knight, R.B. and Haile, J.P. (1983), *Subaerial Tailings Deposition*, 7th Pan-American Conf. on SM & FE, Vancouver, B.C.

Lootens, D.J., Greenslade, W.M. and Barerk, J.M. (eds) (1991) *Environmental Management for the 1990s*, Soc. for Mining, Metallurgy, and Exploration, Inc., Littleton, CO.

McCarter, M.K. (ed.) (1985) *Design of Non-Impounding Mine Waste Dumps*, Soc. of Mining Eng., Littleton, CO.

McClelland, G.E. and Van Zyl, D. (1988), Ore preparation: crushing and agglomeration, Ch. 5. in *Introduction to Evaluation, Design and Operation of Precious Metal Heap Leaching Projects*, (eds D.J.A. Van Zyl, I.P.G. Hutchison and J.E. Kiel), Soc. of Mining Eng., Littleton, CO, pp. 68–91.

Melent'ev, V.A., Kolpashnikov, N.P. and Volnin, B.A. (1973) *Hydraulic Fill Structures*, Energy, Moscow.

National Building Research Institute (1959) *An investigation into the Stability of Slime Dams with Particular Reference to the Nature of the Material of their Construction and the Nature of their Foundation*, Pretoria, South Africa.

Robertson, A. MacG. and van Zyl, D.J.A. (1980) Design and construction options for surface uranium tailings impoundments, Proceedings of the First International Conference on Uranium Mine Waste Disposal, *Soc. of Mining Engrs*, AIME, Chapter 11, pp. 101–119.

Robertson, A. MacG., Fisher, J.W. and Van Zyl, D. (1982) Handling and disposal of dry uranium tailings, *Proc., 5th Symp. on Uranium Mill Tailings Management*, Colorado State Univ., Fort Collins, Colorado, pp. 55–69.

Ruhmer, W.T. (1974), Slimes-dam construction in the gold mines of the Anglo-American group. *J. South African Institute of Mining and Metallurgy*, **74**(7), 273–84.

Rust, E., Van Zyl, D. and Follin, S. (1984) Interpretation of piezometer cone testing of tailings, *Proc. of 6th Ann. Symp. on Management of Uranium*

Tailings, Low-Level Waste, and Haz. Waste, Colorado State Univ., Fort Collins, Colorado, pp. 627–38.
Smith, D.A. (1987) *Mining America: the industry and the environment, 1800–1980*, University Press of Kansas.
Steffen, Robertson and Kirsten (1989) *Draft Acid Rock Drainage Technical Guide*, BiTech Publishers, Vancouver.
Uhle, R.J. (1986) *A statistical analysis of rockfill data – shear strength deformation parameters with respect to particle size*. MS thesis, Colorado State University, 214pp.
University of California at Berkeley (1988) *Mining Waste Study Team, Mining Waste Study, Final Report*, California Water Resources Control Board.
US Department of the Interior (1975) *Engineering and Design Manual*, Coal Refuse Disposal Facilities, Mining Enforcement and Safety Administration.
Van Zyl, D. (ed.) (1985) *Cyanide and the Environment*, 2 vols, Geotechnical Engineering Program, Colorado State University.
Van Zyl, D.J.A., Hutchison, I.P.G. and Kiel, J.E. (eds) (1988) *Introduction to Evaluation, Design and Operation of Precious Metal Heap Leaching Projects*, Soc. of Mining Eng., Littleton, CO.
Van Zyl, D.J.A. and Vick, S.G. (eds) (1988) *Hydraulic Fill Structures*, ASCE, Geot. Special Publication No. 21.
Vick, S.G. (1983) *Planning, Design, and Analysis of Tailings Dams*, John Wiley and Sons.
Wahler, W.A. and Associates (1973) *Analysis of Coal Refuse Dam Failure, Middle Fork, Buffalo Creek, Saunders, West Virginia*. US Department of the Interior, Bureau of Mines, 2 vol.

PART THREE

Remediation Technologies

CHAPTER 13

Strategies for remediation

Larry A. Holm

13.1 INTRODUCTION

Within the last several years, waste remediation has become a significant area for the application of geotechnical engineering. In particular, waste remediation efforts at 'uncontrolled' waste sites have drawn increasing attention from the technical, regulatory, legal, and public communities. Recognition of potential human health and environmental risks has heightened the efforts on national and international agendas.

Waste remedial planning and implementation is a complex technical, regulatory, legal, and public environment in which the geotechnical engineer applies both general and specific knowledge and experience. In many ways, waste remedial planning and implementation offer significant challenge and opportunity for the geotechnical engineer. Capable application of specific geotechnical technologies (discussed in other chapters) is clearly needed. Also, the geotechnical engineer has much to contribute to an overall waste remediation approach by virtue of understanding decisionmaking under uncertainty. Through traditional training and experience with natural geologic conditions and materials, geotechnical engineers possess an understanding of the thought processes and techniques for dealing with uncertainty. Although waste remediation embodies many concerns and technologies that may be new to the geotechnical engineer, the basic understanding of uncertainty can provide a valuable contribution to the overall planning and implementation of remedial activities.

By definition, waste 'remediation' deals with corrective efforts on pre-existing problems. The remedial process for such problems is fraught with uncertainty about the source, cause, and condition at

Geotechnical Practice for Waste Disposal.
Edited by David E. Daniel.
Published in 1993 by Chapman & Hall, London. ISBN 0 412 35170 6

failure, the nature, rate of spread, and extent of the problem conditions, and the potential risks posed to humans and the environment. The application of new, some relatively unproven, technologies to these problems further adds to uncertainty. Remedial work, therefore, often requires a different perspective and technical approach than is demanded of many traditional engineering endeavors.

Waste remediation efforts are heavily influenced, and in the US often dictated, by statutory and regulatory compliance. It is increasingly common that regulatory standards and guidance provide very prescriptive procedural and technical requirements. Furthermore, waste and environmental regulations are constantly evolving, generally increasing in scope and stringency.

Waste remediation is forensic in character. The litigative potential of waste remediation activities, at least in the US, often emphasizes legal-evidential considerations. Such emphasis can expand efforts and divert attention from the primary goal of determining adequate technical solutions. In addition, public interest, awareness, and participation in many remedial planning and implementation efforts require that the engineer formulate, conduct, and present the rationale and results of these efforts in a manner that is understandable to an audience that varies widely in technical background. For success of a project, remedial strategies and approaches must be formulated with knowledgeable consideration of the many, sometimes conflicting, issues and goals involved.

13.2 APPROACH TO REMEDIAL PLANNING AND IMPLEMENTATION

In formulating a successful approach to waste remediation, the geotechnical engineer should recognize the fundamental differences between remedial efforts and design of new facilities. Central to the differences is the high degree of uncertainty associated with remedial efforts, particularly in terms of definition of the source of contamination, extent of contamination, and effectiveness of proposed remediation methodologies. A sound grasp of such uncertainties is essential to any logical approach to remediation and requires a thorough understanding of the nature of uncertainties involved.

13.2.1 Comparison to traditional design and failure evaluation

Geotechnical engineering and, indeed, most engineering and scientific disciplines function successfully based on accumulated experience; the applied methodologies are essentially the result of trial and error, however well-founded in scientific principles. In traditional endeavors, new designs proceed to successful completion through a process that draws upon, and translates, previous experience to the specific project at hand. While uncertainties do exist, they are managed by application of accepted techniques of data gathering (sampling and testing), data interpretation, analysis, and application of factors of safety. The techniques may vary, but each has its fundamental basis in previous experience. While the techniques of sampling, testing, interpreting, and analyzing (including statistical methods) can provide greater insight into the range or degree of uncertainty, they do not eliminate inherent uncertainty. Uncertainty is still managed through application of empirical judgment, often in the form of an appropriate conservatism, application of factors of safety, and/or adherence to accepted design standards.

When failures occur, back-analyses using traditional design methodologies are often found to be lacking. Simply setting the factor of safety equal to unity may neither predict the failure nor demonstrate the characteristics evidenced by the failure. The empiricism embodied in the design methodology often renders the methodology incapable of adequately modeling the failure. Furthermore, uncertainties in basic parameters of the analyses allow (and may even encourage) accepting or 'proving' a failure hypothesis initially perceived to be the correct one.

Approaches have been developed and applied in geotechnical engineering that significantly improve the ability to model and understand the causes, mechanisms, and results of failures. These approaches recognize and attempt to avoid potential pitfalls of such efforts. In any event, it is seldom possible to understand completely all details of the cause and result of a failure.

Remedial planning and implementation are most closely comparable to failure evaluation in the traditional sense; that is, the conditions to be remediated represent, or are the consequences of, a 'failure'. All of the limitations and potential pitfalls of traditional failure evaluation are present in remediation. Furthermore, many more potential uncertainties are present in waste remediation than in tradtional geotechnical engineering. Successful remediation demands that the nature and character of the uncertainty in remediation be understood.

13.2.2 Understanding uncertainties in waste remediation

Uncertainties inherent throughout remedial planning and implementation can be understood in the context of potential variability and variations of the conditions requiring remediation, the criteria that remediation is intended to achieve, and the performance of alternative or selected technologies.

Uncertainty of conditions

Any waste site involves both physical and chemical conditions that will influence remediation. In a general sense, these include: conditions of the source(s); contaminant release mechanism(s); fate, transport, and interactions in the subsurface; and conditions of actual or potential exposure to receptors. As illustrated in Fig. 13.1, these conditions and interactions, in even a simplified case, are complex. Although complete definition of site conditions is impossible, formulation of a conceptual model in this context can facilitate understanding and communication of critical conditions requiring remediation and can enhance the understanding of general, inherent uncertainties.

It is often useful, or even essential, that the conceptual model of the site provide distinction between 'onsite' and 'offsite' conditions. In the case of a localized, single source (such as a confined facility), the distinction may be straightforward. In cases of widespread, multi-source situations (such as regional groundwater contamination or distributed airborne contamination), the determination of the 'site' boundary may be problematic. Though boundary determination may ultimately be jurisdictional rather than technical, the definition of 'onsite' and 'offsite' can influence the technical approach to remedial planning and implementation and should be carefully considered.

General understanding of the source(s) of contamination and their release mechanism(s) may include estimating the location of the source and the type, total volume, and rate of release of contaminants. At some waste sites such information is readily available and reasonably complete. More commonly (particularly for large, old, multi-user sites),

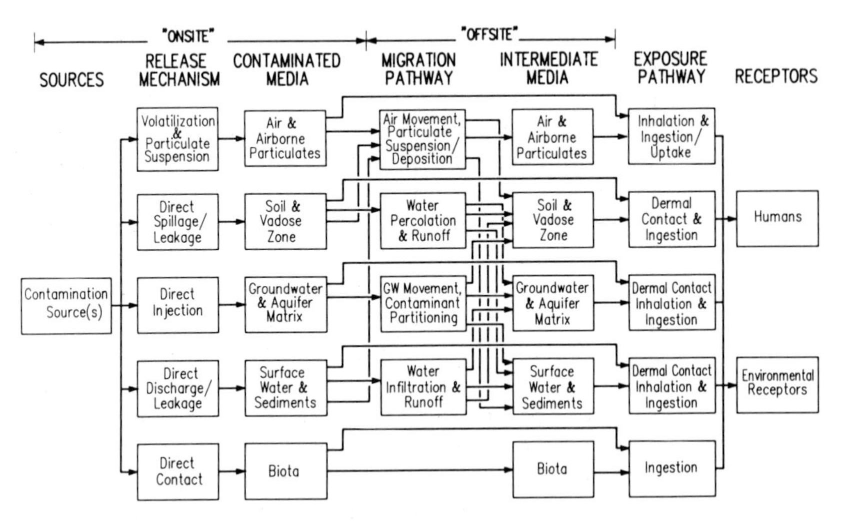

Fig. 13.1 Simplified source, media, pathway, receptor diagram.

such information is only partially available and may be of questionable accuracy. Although optimum use should be made of available information by reviewing records and interviewing people knowledgeable of the history of the site, often only rough estimates of the source and release conditions are possible.

The conditions of the contaminated media at a waste site usually receive considerable scrutiny during remedial planning. Commonly, these efforts are referred to as 'determining the nature and extent' of contamination. The efforts provide information on:

- the types of contaminated media present (e.g., air, soil and sediments, facilities, surface water, groundwater, and biota);
- the physical characteristics of each medium (e.g., temperature and wind direction for air, total organic carbon content, cation-exchange capacity, and index/engineering properties for soils and sediments, and temperature, pH, hardness, alkalinity, gradients, and flow rates and quantities for surface water and groundwater);
- the types (both general categories and specific compounds or elements), concentrations, and characteristics of contaminants present in each medium; and
- the lateral and vertical extent of each contaminated medium.

The interactions between contaminated media can be important. As illustrated in Fig. 13.1, interactions may include numerous, complex pathways for past, ongoing, or potential contaminant migration. The contaminant fate and transport mechanisms of many of the interactions may be only partially understood and defy scientific quantification, thereby allowing only qualitative assessment.

Assessment of background conditions, both physical and chemical, is often an important part of site characterization. Physical information, remote from the site, may be needed to estimate ambient and boundary conditions for air, surface water, and groundwater. Assessment of chemical background may include estimation of the presence and spatial distribution of chemical types and concentrations ambient to the site. For sites in relatively pristine ambient environments, estimation of background chemical conditions may be straightforward. More commonly, site locations in areas influenced by urban and industrial environments pose significant questions regarding the appropriate determination of background conditions.

Since the ultimate goal of any remedial effort is to mitigate impacts on humans and/or the environment, a reasonable understanding of exposure and receptor conditions is often critical to remedial planning and implementation. Exposure conditions include not only the physical/chemical mechanisms by which contaminants may be available to receptors, but also the assimilation mechanism (e.g., inhalation, ingestion, or dermal absorption) and the dose–response characteristics (e.g., type, concentration, quantity, and duration of contaminant contact). For contaminants with known or suspected health effects, methodologies for estimating exposure conditions are available (as discussed in the following section). Significant uncertainty may remain for individual contaminants for which demonstrated or suspected health effects are not available and for multiple-contaminant conditions where synergism may influence cumulative health effects. With respect to receptor populations, estimates may include the number and type of human and/or environmental receptors affected by current and future site-use scenarios. In particular, forecasting future site use may play a pivotal role in the overall remedial approach.

Uncertainty of criteria

Criteria that influence remedial planning and implementation (as impediment or impetus) include national, state or provincial, and/or local statutes, regulations, standards, advisories, or guidance (termed 'regulations' here). Pertinent regulations may include those specific to waste remediation but also those of more general or environmental nature. In addition, specific health-related (toxic and carcinogenic) criteria often supplement regulations in application to remedial activities.

Although comprehensive discussion of criteria is beyond the scope of this chapter, this section attempts to illustrate their importance in remedial activities and to describe, however generally, potential influences that may result from the multiplicity, complexity, and sometimes conflicting character of the waste remediation criteria. Constant evolution can introduce significant uncertainty as changes in remedial criteria occur both during and after remedial planning and implementation.

In general, regulations can be categorized as location-specific, contaminant-specific, or action/technology-specific. Location-specific regulations include those related to where the contamination and remediation occur. Contaminant-specific regulations include those relating to type, quantity, and/or character of the waste or contamination to be remediated. Action/technology-specific regulations include those pertaining to the means and methods of remediation. An important aspect of regulation categories relates to when in the remedial process each category of regulation can reasonably be considered. While location-specific regulations can be brought to bear on the remedial considerations relatively early (when site identification has been made), consideration of contaminant-specific regulations may await site characterization information. Likewise, action/technology-specific regulations can only be considered after the concepts, alternatives, and technologies of the remedial implementation are formulated.

As an example of the regulatory framework for remedial planning and implementation, the United States Environmental Protection

Agency (USEPA, August 1988) has articulated regulatory compliance criteria for hazardous waste remediation in terms of applicable or relevant and appropriate requirements (ARARs). In the terminology of the USEPA, 'applicable' refers to promulgated, legally enforceable laws and statutes that specifically address waste substances or pollution. 'Relevant and appropriate' (and it is important that it must be both) refers to promulgated laws and statutes that relate to situations sufficiently similar to the particular waste situation and that are well suited to the situation. In addition to ARARs, USEPA provides guidance for consideration of other criteria, phrased 'to be considered (TBC)', that are nonpromulgated, nonenforceable regulations that are not strictly 'applicable' nor 'relevant and appropriate' but that may warrant inclusion based on specific circumstances. While the intent is that remediation comply with ARARs, USEPA recognizes six conditions of waiver of ARARs: interim measures, greater risk to health and environment, technical impracticability, equivalent standard of performance, inconsistent application of State requirements, and fund balancing. The US experience clearly demonstrates the myriad of regulations that may be brought to bear on remedial planning and implementation.

In the US, public participation in remedial planning is mandated by law (CERCLA, 1980; SARA, 1986; USEPA, October 1988; USEPA, June 1988). The degree of public participation may vary between sites but, at a minimum, generally includes formal public notification, public review and comment on documents, public meeting, and written response to public comment. While public involvement may not dramatically alter a well-conducted, technically sound remedial effort, it does require that the engineer formulate, conduct, and present the rationale and results of these efforts in a manner that is understandable to an audience that varies widely in technical background.

In addition to specific regulations, criteria are considered in remedial planning and implementation through preparation of exposure and risk assessments; typically these include both environmental (ecosystem) and human health considerations. Exposure and risk assessments may influence the degree and methods of remediation, and indeed, may dictate whether remediation is required at all. They can also be useful in determining required remedial action monitoring (prior to, during, and after remedial implementation). The results of exposure and risk assessments may establish remediation targets that vary from, and are in addition to, criteria set forth in regulations.

The USEPA has developed guidance for the conduct of Exposure Assessments (USEPA, 1988) and risk assessments for both environmental receptors (USEPA, 1989) and human health (USEPA, March 1989; USEPA, September 1989). These guidance documents provide general methods and considerations to be applied in the qualitative and quantitative evaluation of exposure routes and actual or potential risk associated with exposure.

Uncertainty of technology performance

In concept, potential remedial actions and technologies can be considered in the context of Fig. 13.1. The overall goal of protection of human health and the environment can, depending on the conditions present, be accomplished by removing the sources and/or contaminated media, eliminating release, migration, and/or exposure pathways, or removing potential environmental and human receptors. Each has found application in remediation; often, combinations of actions are required. In any case, uncertainties in technology performance are inextricably linked to uncertainties in site conditions.

Many of the actions finding application to waste remediation include technologies such as removal and containment that are familiar in traditional geotechnical engineering. In addition, the emphasis on providing long-term reduction of the mobility, toxicity, and volume of wastes has increased interest in application of new and innovative technologies.

In application to waste remediation, traditional control technologies such as vertical cutoff walls, liners, and cover systems may be needed to perform under conditions and for purposes that are outside of traditional experience. For example, while the performance of soil liners as water barriers is relatively well understood and predictable, their application as vaporous or liquid waste barriers must satisfy additional issues and constraints related to performance and performance prediction.

In addition, traditional technologies may often be subject to constraints during implementation that are not present in nonwaste applications. For example, soil excavation, which may pose little problem under traditional circumstances, may be subject to significant concerns relating to vapor emissions and/or contaminated water control during implementation in waste remediation applications. Such 'secondary' concerns can represent significant challenge in application of traditional geotechnical technologies to waste remediation.

The consideration and application of new and innovative technologies in waste remediation offer considerable challenge to geotechnical engineers and scientists. Innovative technologies are usually considered in the context of their ability to modify the chemical or physical characteristics, or otherwise destroy the hazardous component of the wastes. Because of the potential for permanent remediation, such technologies warrant careful consideration in remedial planning and implementation. Uncertainties in the application of innovative technologies to specific waste situations may include:

- the ability of technology to treat the full range of contaminant types present, and potential adverse effects on the technology resulting from the presence of other contaminants;

- the character of residuals resulting from the application of the technology, including technical and/or regulatory requirements for final disposition;
- the types and character of any by-products of the technology application; and
- the performance of the technology in full-scale remedial application (i.e., scale-up considerations).

While some of the considerations in the application of innovative technologies go beyond the technical areas of many geotechnical engineers, most technologies require geotechnical involvement in assessing the geologic conditions and impacts (for *in situ* technologies) and/or handling, preprocessing, and post-processing requirements (for other technologies). In this context, most of the uncertainties inherent in traditional geotechnical technologies are present in innovative technology applications.

In the application of both familiar and innovative technologies, the ability to predict and monitor performance and the relative accuracy of such predictions and monitoring can have significant impacts on remedial planning and implementation.

13.2.3 Requirements of successful remedial approach

Geotechnical engineering has much to contribute to the overall approach of remedial planning and implementation. While many technical aspects of traditional geotechnical engineering translate directly to remedial activities, to be successful, a basic appreciation of many 'new' issues must be incorporated into remedial planning and implementation.

Any successful approach to remedial planning and implementation must address a multiplicity of technical and regulatory issues. Such complexities demand the detailed attention of a coordinated, multidisciplinary project team and integration of issues in a way perhaps unprecedented in other engineering and scientific endeavors.

Because remedial projects address 'failures', an objective and efficient approach is required for determination of site conditions. Leonards (1979), although referring to traditional endeavors, articulated an approach to failure situations, which is also appropriate in approaching remedial activities.

1. The first step is to conceive all potential mechanisms of failure, avoiding a priori assessments of which mechanism is more likely to be the correct one. The first step is important; if the initial listing of failure mechanisms does not include the true cause, the probability of its eventual discovery is greatly diminished.
2. Each mechanism is then examined in turn, using all available

evidence, seeking to *disprove* its validity. It is easier to disprove a hypothesis than to prove it to be correct; moreover, this approach minimizes the possibility of personal bias entering into the calculations . . . even 'fundamentals' can be interpreted in a way that encourages accepting a hypothesis initially perceived to be the correct one.

3. The hypothesis that resists all attempts at being disproved is reexamined with a view towards identifying other conditions that must prevail if it is to be the only correct one. If a hypothesis is correct, features of the existing data previously overlooked are now found to fit into the picture, or simple additional investigations suggest themselves as a crucial test of validity, or both.

Inherent uncertainties in conditions, criteria, and technology performance must be recognized in formulating a remedial approach. Regardless of the level of effort expended, uncertainty cannot be eliminated and can only be managed through an approach that recognizes this fact, strives to understand the certainties, and establishes and implements an adaptive methodology. The approach must provide a perspective for flexibility and incorporate methods that can foresee and logically respond to changes.

13.3 METHODOLOGY FOR REMEDIAL PLANNING AND IMPLEMENTATION

The process of remedial planning and implementation has evolved as a general methodology that parallels the study-design-build paradigm familiar in traditional geotechnical, and other, engineering and scientific efforts. As its basic steps the methodology includes remedial investigation (RI) and feasibility study (FS), remedial design (RD), and remedial action (RA).

Figure 13.2 schematically illustrates the typical sequence and flow of the process. The figure and the following text are intended only to illustrate the general technical components of the methodology. Specific technical activities and activities relating to regulatory compliance or approval are not illustrated; while obviously important, these can only be determined based on the regulatory situation of the project.

13.3.1 Pre-RI/FS activities

Pre-RI/FS activities include preliminary site inspection and assessment, compilation and review of existing data, and formulation and planning of the remedial approach for the site.

The preliminary site inspection and assessment usually involves

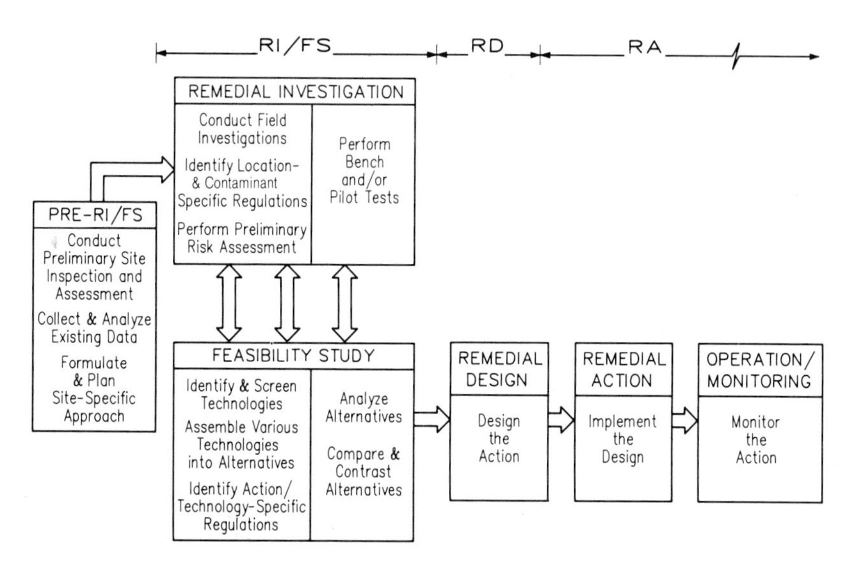

Fig. 13.2 Schematic of the remedial process, after US EPA (1988).

reconnaissance of the site and surrounding area, initial contact with site and regulatory personnel, and obtaining general information about the site conditions to allow intelligent planning of future activities.

Prior to beginning RI/FS activities, available information and data pertinent to the site and potential remedial activities should be gathered, compiled and analyzed, and summarized. Existing information and data may be site-specific, such as records of operations and previous investigations, and/or regional including surveys of geology, hydrology, surface soils, and meteorology. The magnitude of potential costs and effort required to gather new data (during field investigations) dictates that the best possible use of existing information be made. Often the amount of existing information far exceeds the amount of new data that can be obtained even in the most aggressive remedial investigation; at best, new investigations may only supplement existing data. While existing data may be of varying quality, thorough review prior to initiated investigations provides initial characterization of the site and area to allow efficient planning of investigations and preliminary identification of potential remedial responses.

The formulation of a site model, as illustrated in Fig. 13.1, can be useful at this stage. While it is seldom possible to define in detail the sources, media, pathways, and receptors based on existing information, formulation of the overall site situation in this context can assist in focusing investigations on critical conditions and improve the general understanding of inherent uncertainties. A remedial approach and plan can then be formulated, based on initial understanding of the site situation.

Situations that pose obvious, substantial, and immediate threat should be addressed by immediate action. Immediate actions are usually limited to actions such as removal of containerized concentrated wastes, and exposure-control actions such as site access control or provisions for temporary potable water supply. Immediate actions need not involve substantial data gathering or alternative evaluation; typically they proceed to design (often with only limited, confirmatory data) and directly to implementation.

For large complex sites, the project may be divided into manageable subprojects (sometimes termed operable units). An operable unit may be defined as a specific facility (such as a waste pile or lagoon) or may be media-specific (such as surface water, soils, or groundwater). Operable units should be defined and selected, based on the potential for beneficial early and/or separate action, taking into account the following factors.

- *Understanding of conditions* i.e., the ability to sufficiently define the conditions using existing or readily obtainable data and information. If lengthy data gathering efforts are required in the remedial planning process, the utility of the operable unit may be diminished.
- *Potential for separate action.* If potential actions for an operable unit cannot be planned and implemented separately within an overall remedial plan for the site, the operable unit may not be practical.
- *Scope of the operable unit.* The scope of the operable unit should not be so large (or the potential to dramatically expand so great) that decisions on early action are discouraged by funding or other considerations.

Operable units generally follow the same remedial process illustrated in Fig. 13.2. Properly selected and defined, operable units will allow the remedial planning and implementation to proceed more rapidly than may be possible for the site as a whole. When multiple operable units are employed, the overall remedial planning and implementation may take on a phased approach, which can provide substantial benefit, but also must include assessment and evaluation of the combined effects of the operable units.

Whether for an individual operable unit, or for the site as a whole, the scoping efforts in the pre-RI/FS include preliminary identification of potential response actions, preliminary identification of regulatory requirements, definition of data needs and data quality objectives, specific planning of field and laboratory investigation tasks, and, if appropriate, establishing community relations requirements.

Objective identification of the potential range of remedial response actions can ensure that investigation tasks provide the data necessary for later detailed evaluation of remedial technologies and alternatives. Preliminary identification of regulatory requirements ensures that field and laboratory investigations provide data appropriate for comparison to standards, criteria, advisories, and guidance that may be brought to bear on the remediation.

Data needs are identified to define data types, quantity, and quality necessary to assess general site conditions and regulatory issues, and to evaluate the applicability of remedial actions and technologies. Data quality objectives are developed that define specific requirements of the data, often expressed in terms of accuracy and precision, dictated by specific data uses (e.g., general site characterization, regulatory compliance, risk assessment, technical evaluation of technologies, and/or legal-evidential uses).

These early efforts in the remedial planning and implementation are essential to efficient and effective project execution. They form the basis for detailed planning and execution of remedial investigations, provide preliminary insight into the site model (sources, media, pathways, and receptors) and associated uncertainties, and establish essential overall direction to the remedial planning and implementation process based on concepts of potential action.

13.3.2 Remedial investigation (RI)

Remedial investigation activities include field and laboratory studies to gather site data and estimate the conditions present at the site. Investigations may include:

- inventory of site facilities and surface features;
- subsurface exploration of the vadose zone, soils and geologic stratigraphy, and hydrogeologic conditions;
- data gathering related to surface water and hydrology; and
- meteorological, demographic, and ecological data gathering.

For each activity, data on the physical and chemical conditions of each medium (soil, groundwater, surface water and sediments, and air) may be obtained.

While many of the RI techniques and their applicability, particularly in subsurface explorations, are familiar in traditional geotechnical engineering, remedial investigation applications pose additional concerns, such as:

- health and safety of field personnel;
- equipment decontamination (before, during, and after use);
- sample handling, shipment, and chain of custody;
- media cross-contamination; and
- disposal of exploration-derived wastes.

Remedial investigation includes the compilation and interpretation of field exploration data, and conduct of a preliminary risk assessment to better define the initial site model. Regulations, especially location- and contaminant-specific, are identified.

As shown in Fig. 13.2 and discussed further in the following section, remedial investigation activities should overlap with feasibility study activities. The overlap and implied interaction and iteration is important in that it focuses the investigation on obtaining those data that are necessary and sufficient for remedial planning.

Data gaps may be identified, during the RI, based on needs to better define site conditions or potential technology performance. While data gaps may warrant phasing of remedial investigations at complex sites, experienced judgment should be applied in determining sufficiency of the data for purposes of remedial planning and implementation. Improved definition of conditions may always be desirable but must be balanced against potential costs and delays.

13.3.3 Feasibility study (FS)

As shown in Fig. 13.2, the FS and RI overlap and interact to allow inclusion of investigation findings into the FS considerations, and feedback to the RI regarding data needs of specific technologies. The interaction is essential to avoid obtaining data which, although interesting in defining site conditions, may be of little or no specific relevance to potential action.

As a first step in the feasibility study, remedial objectives are formulated. Objectives typically include the broad requirement of protection of human health and the environment. More specific objectives, often based on preliminary identification of regulatory and risk criteria, must be considered in a range. While it may be desirable to firmly commit to specific objectives, these can seldom be decided prior to the effectiveness and cost evaluations resulting from the FS.

Feasibility study includes identification of general response actions and specific technologies and technology options, assembly of technologies and options into remedial alternatives, and the evaluation and comparison of assembled alternatives.

An example of the relationship among general response actions, technologies, and technology options is illustrated in Fig. 13.3. The heading 'general response actions' denotes general functions of actions; 'no action' and 'institutional controls' often warrant consideration along with more aggressive actions. Under the heading 'Technology' there are general types of action for each function, and in the third column options describe the more specific remedial processes or means to be employed. The figure does not attempt to list all potential technologies and options for soil remediation; more are available, some proprietary, and continually more are being developed. Thorough con-

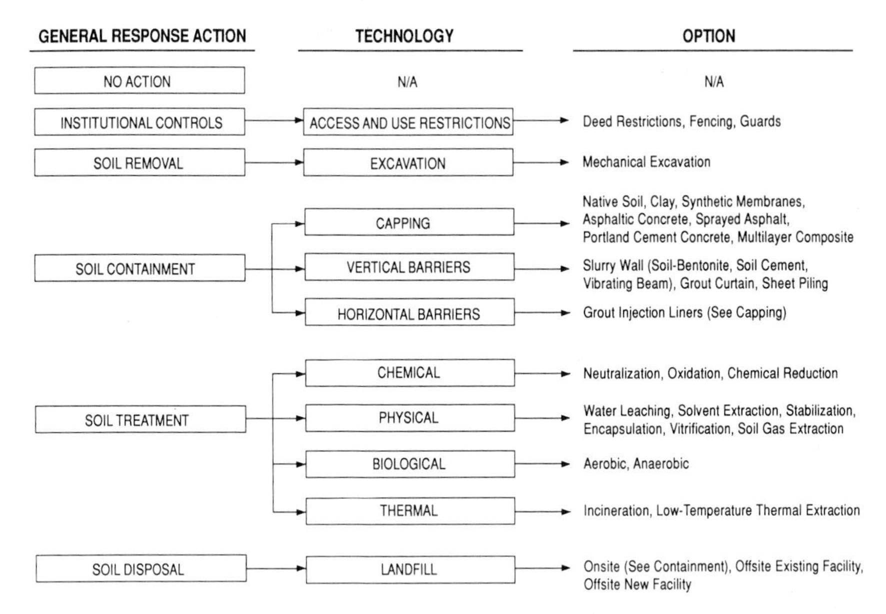

Fig. 13.3 Example general response actions, technologies and options – media soil.

sideration of remedial actions requires that the geotechnical engineer stay current with remedial technology developments.

While Fig. 13.3 addresses only soil, the complexity and number of potential alternative combinations demonstrate the need for careful screening prior to assembly of alternatives; the complexity is compounded when other media are involved. Technologies and options are screened based on general understanding of their applicability to conditions of the site, effectiveness, implementability, and (to a lesser degree, at this stage) cost. While the technology/option identification should include a universe of those available, only those technologies and options having the greatest potential are carried forward for assembly into alternatives.

Surviving technologies and options are developed and assembled into remedial action alternatives that address the full range of media present at the site and that achieve the range of remedial objectives. Action/technology-specific regulations are identified and considered in alternative development.

At this point, it is often necessary or desirable to conduct bench- or pilot-scale testing of technology options to assist in the detailed evaluations. Proprietary technology options may also require testing and may pose special challenges in maintaining vendor confidentiality while still providing objective, independent assessment of performance.

In detailed evaluation, the assembled remedial alternatives are analyzed and compared considering their effectiveness, implementability, and cost, using information on site conditions and technology performance obtained during the RI and regulatory and risk criteria identified during the RI and FS.

Effectiveness involves the degree to which overall protection of human health and the environment and compliance with regulations are achieved. Effectiveness should consider both short-term (during implementation) and long-term (during operation) effects and achievement. In addition, technical and regulatory preference for reduction in mobility, toxicity, and/or volume may require that these issues be specifically addressed in detailed evaluation of each assembled alternative.

Implementability includes both technical and administrative feasibility of constructing and operating and maintaining the alternative. For example, implementability considerations may include capacity of the site to support necessary construction efforts, availability of current and future disposal facilities, and/or likelihood of obtaining and renewing required permits.

Costs for each alternative are prepared to include capital, annual operation and maintenance (O&M), and present worth. While cost estimates may, with appropriate allowances and contingencies, generally be expected to be within −30% to +50% for the scope of alternative understood at the time of the estimate, uncertainties in site conditions, criteria, and action performance along with variations in market conditions, may result in wide disparity between FS-level cost estimates and actual costs. Such potential disparity may warrant inclusion of cost-sensitivity analyses and consideration of cost ranges in comparing alternatives.

The detailed evaluation of alternatives provides the basis for selection of a remedial alternative for the site. The process of actual selection, particularly when such a process involves regulatory approval, is beyond the scope of this chapter. Although alternative selection is not illustrated in Fig. 13.2, it usually occurs between FS and remedial design (RD) and may involve formal regulatory and public participation.

13.3.4 Remedial design and remedial action

Remedial design may involve preliminary design and various stages to completion of final design (USEPA, June 1986). The process of analyses and design leading to completion of contract documents (plans and specifications) that are suitable for construction of the remedy is well understood in traditional engineering. Preliminary design involves early summary of the engineering parameters (including regulatory criteria and monitoring) to be addressed by the remedy; draft plans and specifications should outline the design concepts and process-flow diagrams. Final design usually proceeds through a series of staged completion (at points of various percent complete) to allow review and, if necessary, mid-course corrections. While the process is similar to traditional engineering design, several issues warrant special consideration in remedial design, including the following.

- Design investigations may be required to confirm or refine previous estimates of site conditions.
- Bench-, pilot-, and/or demonstration-scale testing may be required to refine performance estimates or scale-up factors.
- Obtaining site access and normal permits (as well as specific environmental permits), because of the character and stigma of waste remediation, may present unique problems.
- Health and safety of personnel (including construction, operation and maintenance, and community) may influence both the configuration of the design and the methods of implementation.

A discussion of regulatory reviews during design is beyond the scope of this chapter. The process for interface and regulatory review, if required, should be established early in the remedial design.

Because of inherent uncertainties and the associated need for modifications during construction and implementation, contracts should clearly define procedures for change orders and claims. Careful consideration should be given to the use of performance versus method specification. Performance specification may often be preferable but must be consistent with equitable payment and assignment of liability established in the construction documents.

Procurement for construction services should consider both the technical requirements and other support needs (such as personnel health and safety). Highly specialized activities or proprietary technologies may warrant inclusion of prequalification stipulations. While openly advertised competitive bids (with contract award based on price) are common, contracts procured on a negotiated price basis may be appropriate.

Remedial implementation may include construction or implementation of the designed remedy, operation and maintenance of the remedy, and monitoring of remedy performance.

13.3.5 Potential shortcomings of the methodology

The remedial planning and implementation methodology, previously described, follows the study-design-build paradigm similarly applied in other engineering and scientific endeavors. While the methodology is basically sound and appropriate, practical difficulties and delays have been experienced in its application to waste remediation efforts. Often the difficulties result in failure to move forward in the methodology in the face of the inherent, unavoidable uncertainty associated with waste remediation projects. The consequence of stalemate can be evidenced beyond individual projects and the technical community (Wallace, 1987).

A shortcoming of the methodology is the implied separation of the decision-making process into specific steps (i.e., RI/FS, RD, RA) and implied finality of the decisions made at each step. In addition, the methodology may imply a remedial goal is 'reached' and the process complete, at (or soon after) completion of construction (Brown *et al.*, 1990).

The combined effect may manifest itself within waste remediation projects as:

- hesitancy to establish objectives for remediation;
- insistence that additional data be gathered – often this is coupled with the mistaken impression that additional data will eliminate uncertainty.

Projects that do move forward in the absence of thorough understanding of the inherent uncertainties may experience unanticipated problems and be without a course for expeditious response. Still more dire is the potential that unanticipated conditions may continue or worsen, yet remain unnoticed during the remedial effort.

Potential shortcomings of the methodology warrant recognition and understanding. The shortcomings can be minimized if supplemented with a previously established approach for decision-making uncertainty (Dunnicliff and Deere, 1984; Duplanic and Buckle, 1989; US National Research Council, 1989; Wallace and Lincoln, 1990; Brown *et al.*, 1990; D'Appolonia, 1990).

13.4 APPLICATION OF THE OBSERVATIONAL METHOD

Peck (1969) articulated an approach for dealing with decisionmaking under the uncertainties inherent in traditional geotechnical engineering. The approach, phrased the **observational method**, has been applied with success in geotechnical engineering in situations that are remarkably similar to waste remediation:

> Whenever . . . some unexpected development has occurred, or whenever a failure of accident threatens or has already taken place, an observational procedure may offer the only satisfactory way out of the difficulties. Under these circumstances perhaps most engineers would instinctively adopt such a procedure. The mere observation of events . . . often suggests remedial measures that prove successful. Yet the results are sometimes disappointing

and occasionally disastrous because the observations do not constitute part of a well-considered programme encompassing all of the applicable steps in the complete . . . procedure.

(Peck, 1969, p. 173)

13.4.1 General description of the observational method

Peck summarizes the key elements of the complete observational method in the following ingredients:

(a) Exploration sufficient to establish at least the general nature, pattern and properties of the deposits, but not necessarily in detail.
(b) Assessment of the most probable conditions and the most unfavorable conceivable deviations for these conditions. In this assessment, geology often plays a major role.
(c) Establishment of the design based on a working hypothesis of behavior anticipated under the most probable conditions.
(d) Selection of the quantities to be observed as construction proceeds and calculation of their anticipated values on the basis of the working hypothesis.
(e) Calculation of the values of the same quantities under the most unfavorable conditions compatible with the available data concerning the subsurface conditions.
(f) Selection in advance of a course of action or modification of design for every foreseeable significant deviation of the observational findings from those predicted on the basis of the working hypothesis.
(g) Measurement of quantities to be observed and evaluation of the actual conditions.
(h) Modification of design to suit actual conditions.

(Peck, 1969, p. 173)

Knowledgeable application offers great opportunity for savings of time and money, and may, in many situations encountered in waste remediation, offer the only rational way to proceed to confident remedy.

13.4.2 The observational method in waste remediation

Successful incorporation of the observational method into the methodology for waste remediation (section 13.3) requires that the process of remedial planning and implementation be recognized as a continuum. That is, the interpretations of conditions, criteria, and performance that are made in the process are merely working hypotheses. Reassessment of the hypotheses continues from pre-RI/FS through implementation;

throughout the process, additional data needs are assessed based on their potential to disprove, as suggested by Leonards (1979), the hypotheses. Decisions are finalized only after the remedy is in place and sufficient monitoring has been conducted to confirm the hypotheses of conditions and performance.

Figure 13.4 illustrates the remedial process incorporating the observational method. Compared to Fig. 13.3, the process is similar in its general sequence and flow. While incorporation of the observational method into the process maintains basic logic of the study-design-build paradigm, significant yet subtle differences should be noted.

In the remedial investigation, conditions and criteria are interpreted for both expected conditions and potential deviations. Models (whether conceptual or analytical) may provide essential insight into the potential range of conditions to be addressed by remediation. This range represents the basis for alternative development.

In the feasibility study, technologies and options as well as assembled alternatives are identified and evaluated based on their ability to address the range conditions and criteria. Models of alternative performance (whether conceptual or analytical) provide the working hypothesis for the remediation and, more specifically, provide the essential performance prediction necessary to guide the selection of parameters to be monitored. Mark *et al.* (1989) provide a case study of the application of the observational method through RI/FS.

Remedial design involves design of the remedy for both expected conditions and potential deviations, prediction of remedy performance

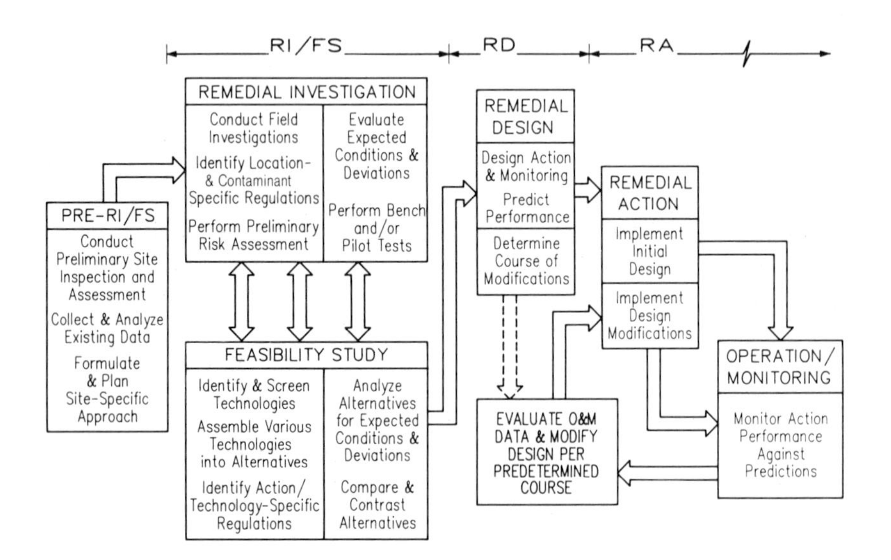

Fig. 13.4 Schematic of the observational method.

over the range of conditions and criteria, design of a monitoring system to measure specific conditions and performance, and predetermination of a course of action for each potential deviation.

Remedial action involves construction and implementation of the design for the expected conditions and implementation of the monitoring system. It is essential that the data obtained during construction and implementation be analyzed and compared to predictions in a timely fashion and that feedback evaluations (as illustrated in Fig. 13.4) provide the ability for expeditious modification of the action.

13.4.3 Cautions in application of the observational method

Application of the observational method requires that the remedial process be viewed as a continuum in which action is planned and implemented with prior consideration for the uncertainties involved. The complete process demands anticipation of specific unfavorable conditions, prediction of specific performance, specific monitoring against each, and predetermination of a course of action to be followed based on the results of the monitoring. Only with thorough and knowledgeable incorporation of these elements is the observational method complete; any lesser approach that implements and then reacts to the results is not the observational method. The observational method is only viable if implementation can be modified based on the results of monitoring. Success of the method requires that the right quantities be monitored. The potential for progressive failures must always be considered in that they may not allow sufficient time for modification based on observations.

While the observational method offers great opportunity for improving the approach and methodologies of waste remediation planning and implementation, it is not new. Indeed, the adaptation of the observational method to waste remediation illustrates that there is nothing new except the history that we never learned. The challenge is clear to geotechnical engineers who understand the method.

While the ingredients may appear to offer a proceduralized, stepwise process for project execution, the observational method actually represents a philosophy more than a methodology. Peck, in the introduction to his 1983 presentation of the Rankine Lecture (Dunnicliff and Deere, 1984), expressed his concern, which is applicable to the adaptation of the method to waste remediation: '. . . I still feel that my efforts to formalize the observational method were too contrived, too rigid.'

The observational method is not as difficult as it may appear when presented in procedures and methodologies intended to 'clarify' the method. Such efforts may be useful to communicate or illustrate the method, but they seldom 'clarify' it. We must be careful that the simple and elegant fundamentals of the method are not clouded or lost completely in the efforts to convert a 'way of thinking' to a 'way of doing'.

REFERENCES

Brown, S.M., Lincoln, D.R. and Wallace, W.A. (1990) Application of the observational method to hazardous waste engineering, *J. Management Eng.*, ASCE, **6**(4), 479–500.

CERCLA (1980) 96th US Congress *Comprehensive Environmental Response, Compensation, and Liability Act of 1980 (CERCLA)*, Public Law 96-510, December 11th.

D'Appolonia, E. (1990) Monitored decisions, twenty-fourth Karl Terzaghi lecture, *J. Geotech. Eng.*, ASCE, **116**(1), 4–34.

Dunnicliff, J. and Deere, D.U. (1984) *Judgement in Geotechnical Engineering – The Professional Legacy of Ralph B. Peck*, John Wiley & Sons.

Duplanic, N. and Buckle, G. (1989) Hazardous data explosion, *Civ. Eng.*, Dec. 68–70.

Leonards, G.A. (1979) Discussion of: foundation performance of Tower of Pisa, *J. Geotech. Eng. Div.*, ASCE, **105**(GT1), 95–105.

Mark, D.L. *et al.* (1989) Application of the observational method to an operable unit feasibility study – a case study, *Proc. of Superfund '89*, Hazardous Materials Control Research Institute, Silver Spring, Maryland, pp. 436–442.

Peck, R.B. (1969) Ninth Rankine lecture, advantages and limitations of the observational method in applied soil mechanics, *Geotechnique*, **19**(2), 171–87.

SARA (1986) 99th US Congress, *Superfund Amendments and Reauthorization Act of 1986 (SARA)*.

USEPA (June 1986) *Superfund Remedial Design and Remedial Action Guidance*, OSWER Directive 9355.0-4A.

USEPA (1988) *Superfund Exposure Assessment Manual*, OSWER Directive 9285. 5-1.

USEPA (June 1988) *Community Relations in Superfund*, Interim Version, OSWER Directive 9230.0-3B.

USEPA (August 8, 1988) CERCLA Compliance With Other Laws Manual, OSWER Directive 9234.1-01, Draft.

USEPA (October 1988) *Guidance for Conducting Remedial Investigations and Feasibility Studies Under CERCLA*, OSWER Directive 9335.3-01, Interim Final.

USEPA (March 1989) *Risk Assessment Guidance For Superfund – Environmental Evaluation Manual*, OSWER Directive 9285.7-01, Interim Final.

USEPA (September 1989) *Risk Assessment Guidance For Superfund Volume I: Human Health Evaluation Manual*, OSWER Directive 9285.701a, Interim Final.

US National Research Council (1989) Commission on Engineering and Technical Systems, The Geotechnical Board, *Geotechnology – Its Impact on Economic Growth, the Environment, and National Security*, National Academy Press.

Wallace, W.A. (1987) The cleanup of hazardous waste sites in the United States: positioning for least cost, *Proc. World Conf. on Haz. Waste*, Budapest, Hungary, Elsevier Science Publishers, B.V.

Wallace, W.A. and Lincoln, D.R. (1990) *How Scientists Make Decisions About Groundwater and Soil Remediation*, National Academy Press.

CHAPTER 14

Geophysical techniques for subsurface site characterization

Richard C. Benson

14.1 INTRODUCTION

Geophysical methods encompass a wide range of surface and downhole measurement techniques which provide a means of investigating subsurface hydrogeologic and geologic conditions. These methods have also been applied to detecting contaminant plumes and locating buried waste materials.

The primary factors which limit the accuracy of site characterization are the limitations of the number of sample points, and the lack of correct spatial sampling. Geophysical measurements can be made relatively quickly, thereby providing a means to increase sample density. Continuous data acquisition along a traverse line can be employed with certain techniques at speeds up to several kilometers per hour. Because of the greater sample density, anomalous conditions are more likely to be detected, resulting in a more accurate characterization of subsurface conditions.

All of the geophysical methods, like any other means of measurements, have advantages and limitations. There is no single, universally applicable geophysical method, and some methods are quite site specific in their performance. Thus, the user must select the method or methods carefully and understand how they are applied to specific site conditions and project requirements.

Unlike direct sampling and analysis, such as obtaining a soil or water sample and sending it to a laboratory, the geophysical methods provide non-destructive, *in situ* measurements of physical, electrical or geochemical properties of the natural or contaminated soil and rock.

Geotechnical Practice for Waste Disposal.
Edited by David E. Daniel.
Published in 1993 by Chapman & Hall, London. ISBN 0 412 35170 6

The success of a geophysical method depends on the existence of a contrast between the measured properties of the target and background conditions. If there is no measurable contrast in the measured property, the target will not be detected. Furthermore, if a layer is sufficiently thin, or if the size of a localized target is sufficiently small or if it is too deep, it may not be detected.

Geophysical methods are not new; they have been used for many decades in the exploration for deep oil and gas as well as mineral exploration. The methods have also been applied to geotechnical applications and regional water resources development for the last few decades. The application of the methods to hazardous waste site investigations are more recent, beginning in the mid to late 1970s. Applications to the engineering geology, geotechnical application and hazardous waste sites are somewhat different in their application than oil, gas and mineral exploration because they are used to obtain relatively shallow data (typically less than 30 m or so in depth) and often are required to produce higher resolution.

14.2 BACKGROUND

Often, traditional approaches to subsurface field investigations at waste disposal sites have been inadequate. Site investigations rely only upon the use of direct sampling methods such as:

- soil borings for soil and rock samples;
- monitoring wells for gathering hydrogeologic data and water samples;
- laboratory analysis of soil and water samples to provide a quantitative assessment of site conditions; and
- extensive interpolation and extrapolation from these points of data.

This approach has evolved over many years in the geotechnical areas and is commonly considered a standard in which to deal with field investigations. However, there are numerous pitfalls associated with this approach, which can result in an incomplete or even erroneous understanding of site conditions. These pitfalls have been the subject of numerous papers and conferences over the years (Lysyj, 1983; Hileman, 1984; Perazzo *et al.*, 1984; Walker and Allen, 1984; Dunbar *et al.*, 1985). They have also precipitated the tightening of groundwater monitoring regulations (USEPA, 1985).

The single most critical factor faced in site evaluation work is accurately characterizing the site's hydrogeology (Benson and Pasley, 1984). With an accurate understanding of a sites hydrogeology, predicting the location of movements of contaminants, or designing a clean-up operation would be reasonably straightforward. If all sites were simple, horizontally stratified geology with uniform properties, site characterization would be easy, since data from just one boring would be sufficient to characterize the site. However, in most geologic settings, this will not be the case. Even at sites whose geology appears to be uniform there are sometimes subtle variations which can cause significant errors in site characterization.

In the design of many soil and rock sampling programs and monitoring well networks, the placement of borings and wells has been done mainly by educated guesswork. The accuracy and effectiveness of such an approach is heavily dependent upon the assumption that subsurface conditions are uniform. This approach usually assumes that regional hydrogeology and groundwater flow (as obtained from literature) is valid for the site specific setting, and that data from only a few site specific borings or monitoring wells can be used to characterize the site. These assumptions are frequently invalid, resulting in non-representative locations for borings piezometers and monitoring wells and erroneous generalizations from limited information. To improve the accuracy of the site investigation, a large number of borings may be required.

14.2.1 Sample density

A soil or core sample obtained from drilling may be representative of only a limited area surrounding the hole. Therefore, fractures, cavities, bedrock channels, sand lenses, and local permeable zones can easily be missed by borehole programs.

An insight into the number of discrete samples or borings that are required for accurate site characterization can be obtained by considering detection probability (Benson and LaFountain, 1984). Fig. 14.1a shows a target area which is 1/10 of the total site area. This 'target area' (whose size and location are usually unknown) could be a waste burial site, a plume from a chemical spill, or an old sinkhole. Based upon probability calculations, the number of samples or borings required to achieve various detection probabilities are shown in Fig. 14.1b. For example, 10 uniformly spaced borings would produce a 10% probability of detection for a site-target of 10. For smaller targets, such as a narrow sand lens or fractures, the site-to-target area ratio will increase significantly, thus 100 to 1000 borings may be required to achieve a confidence level of 90% making the subsurface investigation like 'looking for a needle in a haystack'.

It becomes obvious that achieving a good statistical evaluation of complex hydrogeologic site conditions requires samples or borings to be placed in a close-order grid, which would reduce the site to 'Swiss cheese'. In many cases, direct sampling alone is not sufficient to characterize accurately site conditions. This is the primary reason for the application of the geophysical methods.

a

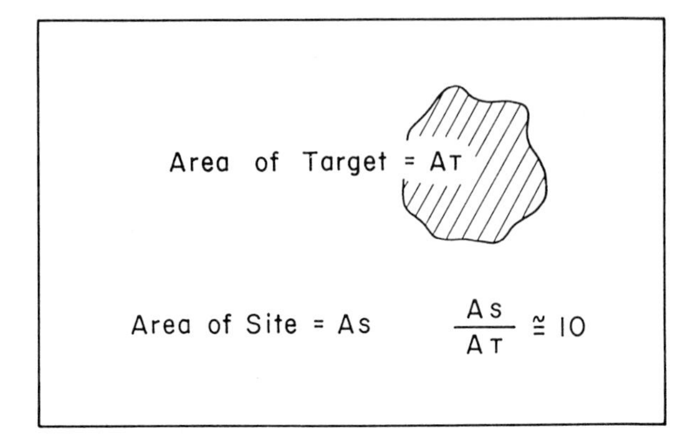

b

Probability of Detection	As/At = 10	As/At = 100	As/At = 1000
100	16	160	1600
98	13	130	1300
90	10	100	1000
75	8	80	800
50	5	50	500
40	4	40	400
30	3	30	300

Fig. 14.1 Spatial sampling requirements; (a) illustrates site to larger area ratio – As/At = 10 in this case; (b) number of sample points required for various As/At ratios and probability of detection – this table assumes uniform sample grid; if a random placement is used, the number of samples must be increased by a factor of 1.6. *Source*: Benson and LaFountain (1984).

14.2.2 How geophysical methods are used

To provide a greater volume of measurement

Data obtained from borings or monitoring wells come from a very localized area and are representative of material conditions at the borehole. Geophysical methods, on the other hand, usually measure a much larger volume of the subsurface material (Fig. 14.2).

As anomaly detectors

As a result of geophysical measurements being relatively rapid, a larger number of measurements can be taken, for a given budget. With a greater number of measurements plus the fact that the measurement

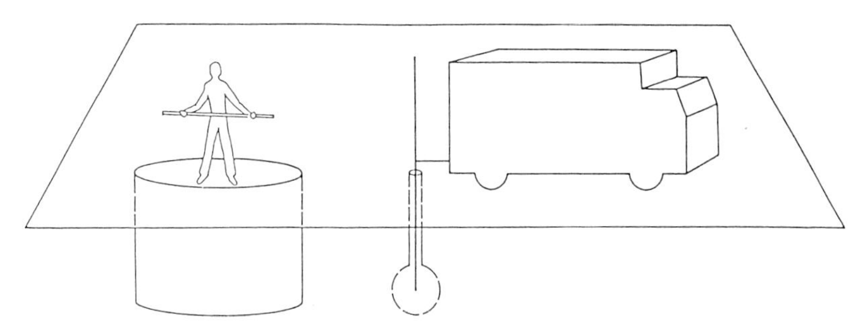

Fig. 14.2 Comparison of volumes sampled by geophysical method and a borehole: (left) a typical geophysical measurement integrates a larger volume of soil and rock while (right) the volume of soil and rock sampled by drilling is relatively small.

encompasses a larger volume of subsurface material, the geophysical methods can detect anomalous areas which may pose potential problems, and thus are essentially anomaly detectors.

Once an overall characterization of a site has been made using geophysical methods and anomalous zones identified, drilling and sampling programmes are made more effective by:

- locating boreholes and monitoring wells to provide samples that are representative of both anomalous and background conditions;
- minimizing the number of borings, samples, piezometers and monitoring wells required to characterize accurately a site;
- reducing field investigation time and cost; and
- significantly improving the accuracy of the overall investigation.

This approach yields a much greater confidence in the final results, with fewer borings or wells, and an overall cost savings. It makes good sense to minimize the number of monitoring wells at a site while optimizing the location of those installed. If the wells are located in the wrong position, they do not provide representative data and a large amount of relatively useless data would accrue.

Using the geophysical method in a systematic approach, drilling is no longer being used for hit-or-miss reconnaissance, but is being used to provide the specific quantitative assessment of subsurface conditions. Boreholes or wells located with this approach may be thought of as **smart holes** because they are scientifically placed, for a specific purpose, in a specific location, based on knowledge of site conditions from geophysical data (Benson and Pasley, 1984). While smart holes might sometimes be placed without the use of geophysical methods, they often can be placed more reliably if the geophysical methods are incorporated into the subsurface investigation program.

If piezometers and wells have already been installed at the site, the geophysical methods can still provide significant benefits. The location of existing borings and monitoring wells relative to background, down-gradient and anomalous site conditions can be assessed, thus providing a means of evaluating the representativeness of existing data. Then, if additional borings or wells are needed to fill gaps in the overall site coverage, they can be placed accurately as smart holes.

To aid in understanding the overall hydrogeologic setting

Assessment of site conditions will often require that an area larger than the site itself be considered. Contaminant transport by groundwater and the geohydrologic factors which control groundwater flow, do not stop at arbitrary site boundaries or property lines. Insight into the character of local hydrogeologic conditions must be derived from the knowledge of the broader understanding of the overall hydrogeologic setting. Omitting an adequate understanding of the overall hydrogeologic setting will often result in a number of critical gaps in information about the site itself. The geophysical methods provide a means of rapid reconnaissance over larger areas, and often can be employed to obtain the data to support an understanding of the overall setting, as well as to provide site specific data.

To provide continuous measurements when possible

Geophysical surveys often involve making measurements of subsurface properties at discrete points over a site. That is, the instrumentation is located at a station along a survey line or a grid and measurements are made one point at a time. The data is then analyzed and plotted to provide profile lines on contour maps. The degree of detail of such data is proportional to the number of measurements made. However, some geophysical methods can provide measurement of subsurface parameters continuously as the instrumentation moves along the survey line.

Continuous methods should be employed whenever possible to maximize the amount of data obtained, to achieve maximum resolution and to minimize project costs. This approach is particularly necessary when site conditions are suspected of being highly variable and small features such as fractures need to be identified.

Although the continuous surface geophysical methods referred to in this chapter (see section 14.4) are typically limited to a depth of 15 m or less, they are applicable to many site investigations. They can provide a continuity of subsurface information, not practically obtain-

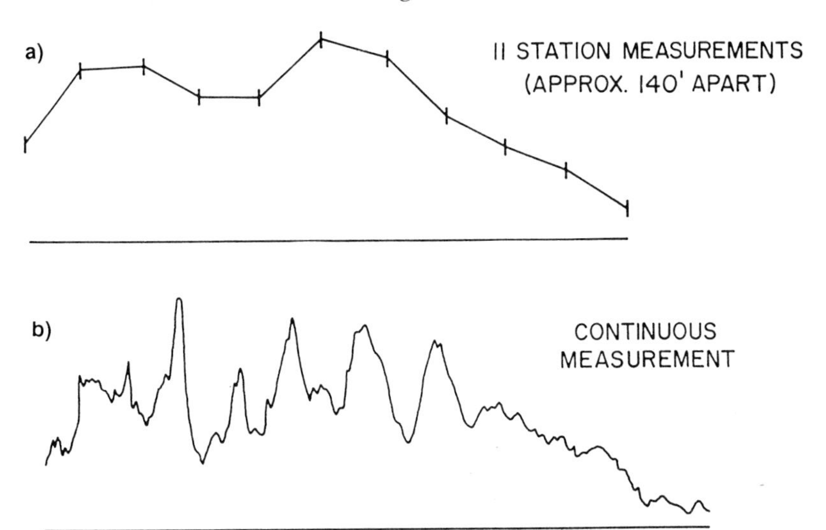

Fig. 14.3 Comparison of (a) station and (b) continuous measurements from the same site. Note: the data was obtained with a truck-mounted EM-34 with a 10-meter coil spacing. The higher values of electrical conductivity are caused by moisture within the underlying gypsum rock.

able from station measurements. Continuous surface geophysical methods can be applied at speeds of 1–5 km/hr or more, resulting in a cost-effective approach for relatively shallow survey work.

The benefits of continuous measurements are seen by comparing station measurements and continuous measurements shown in Fig. 14.3. The set of data in Fig. 14.3b reveals the highly variable nature of a site indicated by a continuous measurement technique, while those in Fig. 14.3a shows the loss of information that can result from making a limited number of station measurements at the same site and interpolating between sample points. Limiting the number of measurements results in a distorted set of data and can lead to errors in interpretation of site conditions especially when the target size is significantly smaller than station spacing.

The data in Fig. 14.3 was obtained by electromagnetic measurements of subsurface electrical conductivity. The higher conductivity values in Fig. 14.3b indicate fractures within underlying gypsum rock. These fractures show up because they are more electrical conductive due to water content and weathering of the gypsum rock. The fracture provides a pathway for groundwater.

By running closely-spaced parallel survey lines with continuous methods, subtle changes in subsurface parameters can often be mapped and total site (100%) coverage can be achieved if necessary.

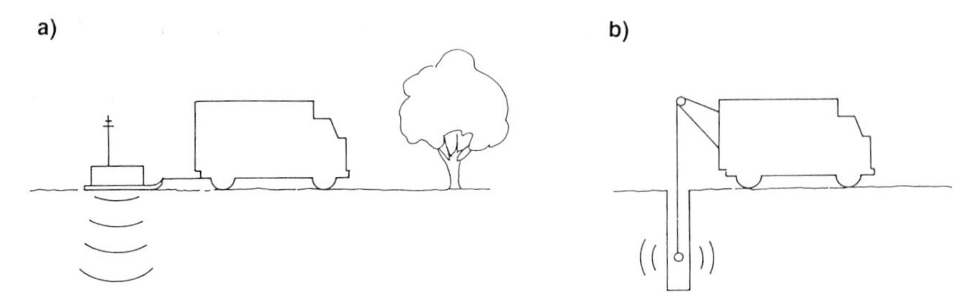

Fig. 14.4 Modes of geophysical measurements; (a) surface and (b) downhole geophysics.

Site investigation methods are scale dependent

All site investigation methods, including geophysics, are scale dependent. For example, aerial photography is an effective tool to be used in obtaining the big picture at a local site investigation, but, it will not provide any information about site-specific soil conditions at a depth of 3 m. Conversely, a boring will provide detailed information on soil or rock conditions versus depth, but information from the boring is only valid for a very limited extent immediately around the borehole. Surface geophysical measurements (Fig. 14.4a) can be used to determine detailed soil and rock conditions over a few tens of square meters or over many square kilometers, while geophysical logging measurements made down a borehole will extend the measurements from the borehole itself radially to a distance of several centimeters to 1–2 m depending upon the method used (Fig. 14.4).

Therefore, the site investigation methods employed must be selected to suit the particular scale of data needed to meet project requirements (Benson and Scaife, 1987). A complete subsurface investigation must include measurements of the big picture (aerial photography), intermediate picture (surface geophysical measurements), and the very local details (boring, logging and sampling data).

Summary

Geophysical methods:

- usually measure a larger volume than direct sampling;
- are very effective at detecting and mapping anomalies;
- are used to locate or assess representative locations of boreholes, piezometers, and monitoring wells;
- can be used for reconnaissance measurements to aid in characterizing the regional setting;
- can be used for gathering both reconnaissance and detailed site specific data;

- provide a means for obtaining continuous measurements, both on the surface and downhole; and
- aid in integrating the wide range of scale dependent measurements from aerial photography to core samples.

14.3 APPLICATIONS FOR GEOPHYSICAL MEASUREMENTS

There are three major areas for the application of geophysical methods at waste disposal sites. They are:

- assessing hydrogeologic conditions;
- detecting and mapping contaminant plumes; and
- locating and mapping buried wastes and utilities.

14.3.1 Assessing hydrogeologic conditions

Probably the most important task of a site investigation will be assessing the natural hydrogeologic conditions. A variety of geophysical methods can be used to determine depth to bedrock, degree of weathering, sand and clay lenses, fracture zones, and buried relic stream channels (Keys and MacCary, 1976; Benson and Glaccum, 1979; Benson *et al.*, 1982; Benson and Yuhr, 1987). Accurately understanding the hydrogeologic conditions and anomalies can make the difference between success and failure in site characterization, because these features will often control groundwater flow and contaminant transport.

14.3.2 Detection and mapping contaminant plumes

A major objective of many site investigations is the detection and mapping of contaminant plumes, and geophysical methods can be employed in two ways to solve this problem. Some methods can be used for the direct detection of contaminants, but in those cases where the contaminant cannot be detected directly, geophysical methods can be used to assess detailed hydrogeologic conditions which are controlling groundwater flow, such as fractures or a buried channel. Once the contaminant flow pathways have been identified, direct sampling methods can be used to further assess conditions (Cartwright and McComas, 1968; McNeill, 1980; Benson *et al.*, 1982, 1985; Greenhouse and Monier-Williams, 1985).

14.3.3 Locating and mapping buried wastes and utilities

Geophysical methods can also be used to locate and map the areal extent and sometimes the depth of buried wastes in trenches and landfills (Benson, *et al.*, 1982). There are also methods which can be employed to detect buried drums, tanks, and utility lines. In many cases the trenches associated with pipes and utilities will be of interest because they form a permeable pathway for contaminant migration.

14.4 SURFACE GEOPHYSICAL METHODS

The surface geophysical methods discussed in this section are

- ground penetrating radar
- electromagnetics
- resistivity
- seismic refraction
- seismic reflection
- micro-gravity
- metal detection
- magnetics.

While these methods are not the only ones which may be employed, they are presented as the primary methods because they are regularly used and have been proven effective for waste site assessments.

14.4.1 Ground penetrating radar

Ground penetrating radar uses high frequency electromagnetic waves from less than 100 Mhz to 1000 Mhz to acquire subsurface information. Energy is radiated downward into the ground from a transmitter and is reflected back to a receiving antenna. The reflected signals are recorded and produce a continuous cross-sectional picture or profile of shallow subsurface conditions.

Reflections of the radar wave occur whenever there is a change in dielectric constant or electrical conductivity between two materials. Changes in conductivity and in dielectric properties are associated with natural hydrogeologic conditions such as bedding, cementation, moisture, clay content, voids, and fractures. Therefore, an interface between two geologic layers which have a sufficient contrast in electric properties will show up in the radar profile (Benson and Glaccum, 1979; Benson *et al.*, 1982; Benson and Yuhr, 1987).

The radar record is similar to the view we would get if we observed the cross-section of soils in a trench or a cross-section of rock at a road cut. Figure 14.5 shows a radar record (cross-section) of a clean quartz sand over a clay loam. Note the level of detail that can be obtained in mapping the top of the clay loam. The numerous small hyperbola in the upper portion of the record, within the quartz sands, are due to tree roots or animal burrows.

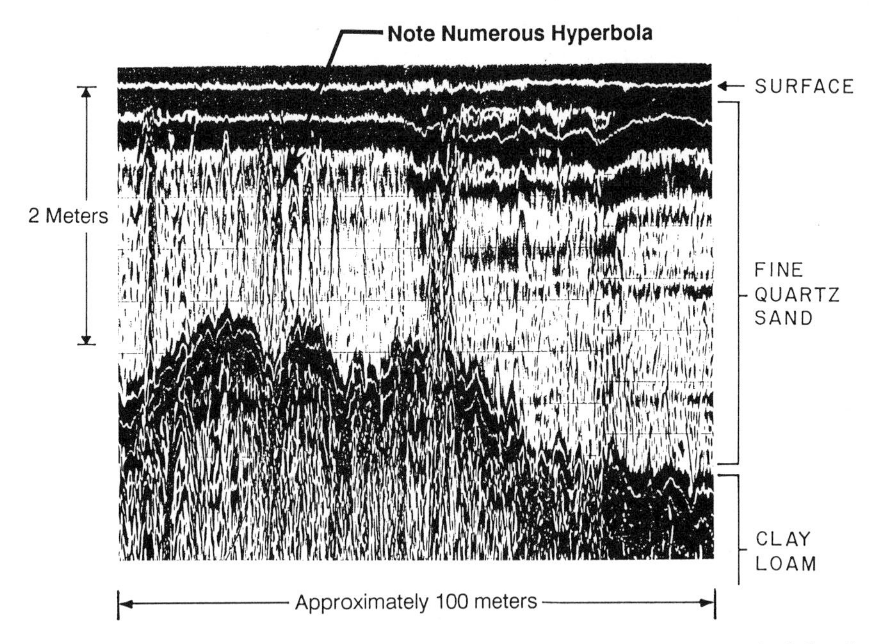

Fig. 14.5 Radar profile of quartz sand over clay loam. Note the level of detail that can be obtained.

The vertical scale of the radar profile is in units of time (nanoseconds, or 10^{-9} sec). The time it takes for an electromagnetic wave to move down to a reflector and back to the surface (two-way travel time) is relatively short since the waves are travelling at almost the speed of light. The time scale then is converted to depth by making measurements or assumptions about the velocity of the waves in the subsurface materials.

Depth of penetration of the radar wave is highly site-specific. The method is limited in depth by attenuation due to the higher electrical conductivity of subsurface materials or scattering. Generally, radar penetration is greater in coarser, dry, sandy or massive rock, and less in wet, fine grained clayey (conductive) soils. However, data can be obtained in saturated materials if the specific conductance of the pore fluid is sufficiently low. For example, radar has been applied to map the sediments in fresh water lakes and rivers. While radar penetration in soil and rock to more than 30 m has been reported, penetration of 5–10 m is more typical. In silts and clays, penetration may be limited to a meter or less. Yet, in some situations, useful results can be obtained in silts and clays (Benson and Yuhr, 1990). The water table can be detected in coarser grained materials but not in finer grained sediments with a large capillary boundary. Both metallic and nonmetallic buried pipes and drums can also be detected.

The continuous data produced by the radar method offers a number of advantages over some of the other geophysical methods. Continuous profiling permits data to be gathered much more rapidly, thereby providing a large amount of data for a given budget. In some cases, total site coverage of an area can be obtained. Radar data may be obtained at speeds up to 5–10 km/hr or more. Very high lateral resolution data can be obtained by towing the antenna by hand at much slower speeds (less than 1 km/hr).

Radar has the highest resolution of all of the surface geophysical methods. Vertical resolution of radar data can range from a few centimeters to about 1 m depending upon the depth and the frequency used. A variety of antennas can be selected to cover frequencies from less than 50 Mhz to 2000 Mhz. Lower frequencies provide greater depths of penetration with lower resolution and the higher frequencies provide less penetration with higher resolution.

Two types of radar systems are currently available. In the first system, a separate transmitter and receiver antenna are held in contact with the ground by either one or two persons at some given distance apart. The radar data are gathered and then the two antennas picked up and moved some increment of distance, put in contact with the ground and a new set of radar data acquired. Each set of data is recorded on station by station basis and stored in a lap-top computer. Once the field data has been acquired in this manner, the data is processed on a lap-top computer, similar to a seismic reflection data. This **station measurement radar system** is probably effective in mapping slightly deeper geologic horizons and horizons with less reflection coefficient in which the radar data can be improved by repeated sampling (stacking) and subsequent processing. It is also usable in areas with brush and trees where a continuous radar antenna could not be towed. In the second system, the **continuously towed radar system**, the antenna is moved continuously across the ground recording data (real time) producing a graphic record (cross-section). This system has much higher spatial resolution, both laterally and vertically since it acquires infinitively more data.

Preliminary field analysis of radar data is possible using the picture-like record. Despite its simple graphic format, there are many pitfalls in the interpretation of radar data. There are multiple bands within the data due to ringing, and these may obscure layers and cause confusion in interpretation by untrained persons. Overhead reflections from trees or power lines may also appear on the record when not using shielded antennas (although this is only a problem with lower frequency antennas). System and geologic noise can sometimes clutter up the record, making interpretation difficult. Although radar can be recorded on magnetic media and processed by computer the necessary information can usually be derived from the raw graphic records.

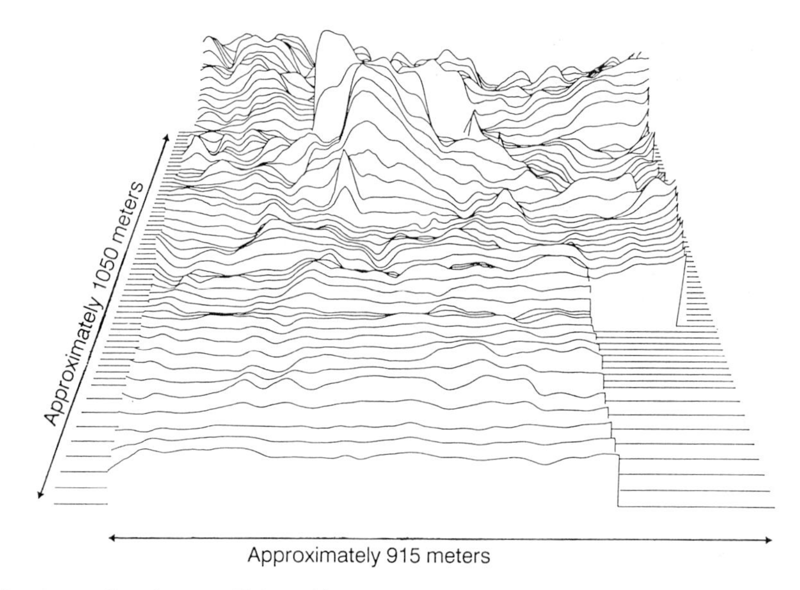

Fig. 14.6 Continuous EM profile measurements show a large inorganic plume (center rear) and considerable natural geological variation.

14.4.2 The electromagnetic and resistivity methods

The electromagnetic (EM) and resistivity methods are similar in the sense that they both measure the same parameter, but in different ways. Electrical conductivity values (millisiemens/meter or millimhos/meter) are the reciprocal of resistivity values (ohm/meter) although units of ohm-feet are also common in the US. Electrical conductivity (or resistivity) is a function of the type of soil and rock, its porosity and the fluids which fill the pore spaces. The specific conductance of the pore fluids often dominates the measurement. Both methods are applicable to the assessment of natural hydrogeologic conditions (Griffith and King, 1969; McNeill, 1980; Telford *et al.*, 1982).

Natural variations in subsurface conductivity, or resistivity, may be caused by changes in basic soil or rock types, thickness of soil and rock layers, moisture content, and depth to water table. Localized deposits of natural organics, clay, sand, gravel, or salt-rich zones will also affect these values, while structural features such as fractures or voids can also produce changes in them.

The absolute values of conductivity (or resistivity) for geologic materials are not necessarily diagnostic in themselves, but their spatial variations both laterally and with depth, can be significant. It is the identification of these spatial variations or anomalies which enable the electrical methods to find potential problem areas rapidly (Fig. 14.6).

Because the specific conductance of the fluids in the pore spaces can

a)

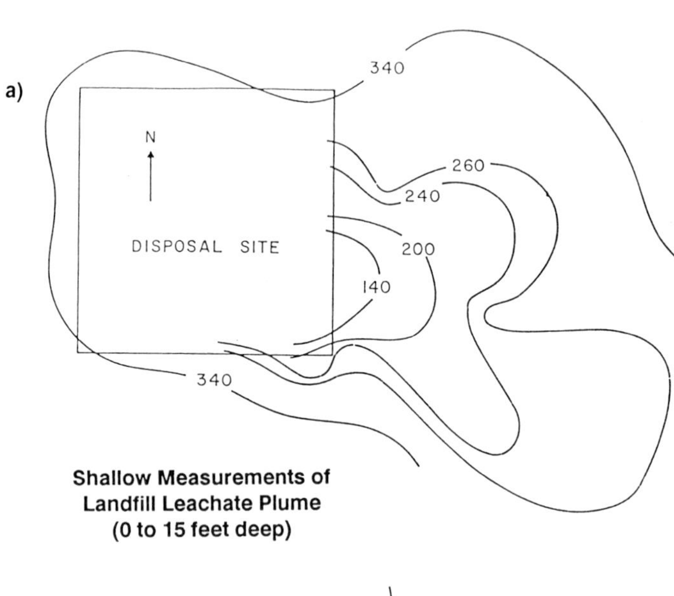

b)

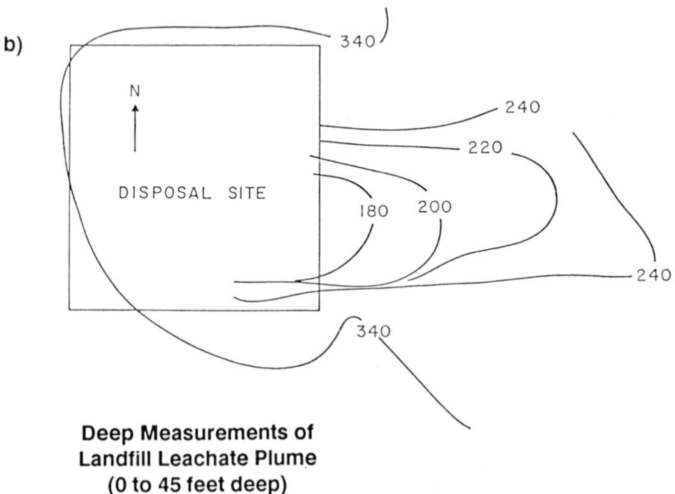

Fig. 14.7 Resistivity map of leachate plume from a landfill. Values are in ohm-feet; landfill is approx. 1 sq mile. (a) shallow measurements of landfill leachate plume (0–15 ft deep); (b) deep measurements of landfill leachate plume (0–45 ft deep).

dominate the measurements, detection and mapping of contaminant plumes can often be accomplished using the electrical methods. Because inorganics in sufficient concentrations are often more electrically conductive than groundwater, both the lateral and vertical extent of an inorganic plume can be mapped using the electrical methods. Correlation between groundwater chemistry data and results using electrical methods to map inorganics from landfills has been as good as

0.96 at the 95% confidence level (Benson *et al.*, 1985). Electrical methods provide a means of directly mapping the extent of the inorganic contaminants *in situ*, obtaining direction of flow, and estimating concentration gradients (Fig. 14.7). These measurements can also be used for time-series measurements to obtain plume dynamics, and thus provide vital information for the modeling of groundwater flow (Benson *et al.*, 1988).

If the contaminate plume consists of a mix of organics and inorganics, such as leachate from a landfill, a first approximation to the distribution of the organics can be made by using electrical methods to map the more electrically conductive inorganics (Fig 14.7). Correlation between groundwater chemistry data for total organic carbon in a landfill leachate and results using electrical methods has been as good as 0.85 at the 95% confidence level (Benson *et al.*, 1985).

In cases where pure organics such as trichloroethylene (TCE) exist, electrical as well as other geophysical methods can be used to define permeable pathways or buried channels through which the contaminants may migrate. Direct detection of hydrocarbons can sometimes be accomplished by looking for a conductivity low (resistivity high) associated with the organics. The possibility for such an anomaly exists where large amounts of hydrocarbons have been in place for a long period of time and there is a sufficient contrast in electrical values between the natural background values and the hydrocarbons. To date this approach has had limited success.

Both EM and resistivity methods may be used to obtain data by 'profiling' or 'sounding'. Profiling provides a means of mapping lateral changes in subsurface electrical conductivity (or resistivity) at a given depth (Figs. 14.6 and 14.7). Profiling measurements are made by obtaining data at a number of stations along a survey line, the spacing between the measurements depending upon the variability of the setting and upon the lateral resolution desired. At each station along the profile line, data may be obtained for one depth or a number of depths depending upon project requirements. It is useful to take at least two measurements, a shallow one and a deeper one, so that the influence of the highly variable shallow soils and cultural influences can be assessed (Fig. 14.7). Profiling is well suited to the delineation of hydrogeologic anomalies, mapping of contaminant plumes and location of buried material.

The sounding method provides a means of determining the vertical changes in electrical conductivity (or resistivity) correlating with soil and rock layers. In this case, the instrument is located at one location and measurements made at increasing depths. Interpretation of sounding data provides the depth, thickness and conductivity (or resistivity) of subsurface layers with different electrical conductivities (or resistivities); (Fig. 14.8).

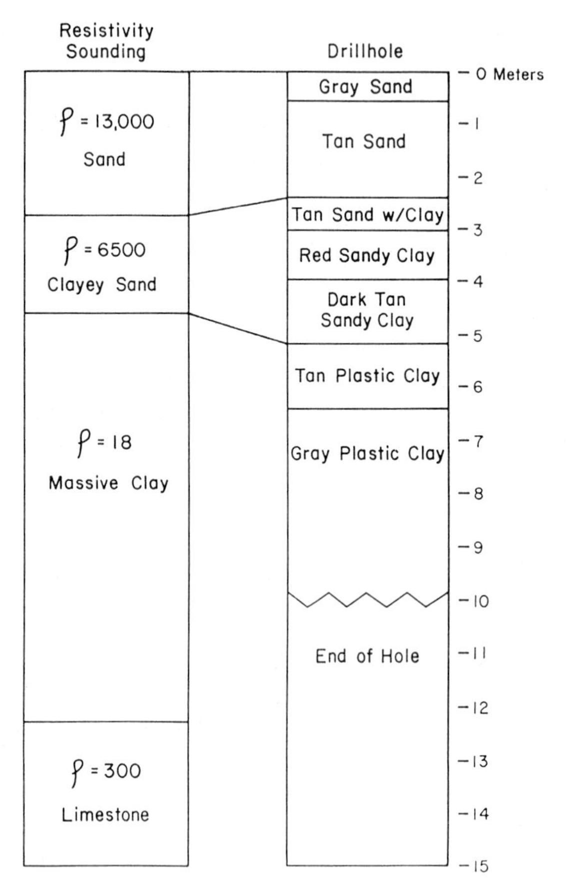

Fig. 14.8 Resistivity geoelectric section showing correlation with a driller's log. Resistivity values are in ohm-feet.

Electromagnetics

Two types of electromagnetic instrumentation both of which induce currents into the ground by electromagnetic induction, are in use: time-domain and frequency-domain. The most common is the frequency-domain system in which the transmitter is radiating energy at all times, measuring changes in magnitude of the currents induced within the ground (McNeill, 1980). The time-domain system, in which the transmitter is cycled on and off, measures changes in the induced currents within the ground as a function of time.

Because the electromagnetic instruments do not require electrical contact with the ground, measurements may be made quite rapidly. Lateral variations in conductivity can be detected and mapped by profiling. Using commonly available frequency-domain EM instruments, profiling station measurements may be made to depths ranging from 1 to 60 m.

Continuous EM profiling data can be obtained to a depth of 15 mm (Benson *et al.*, 1982). These continuous measurements significantly improve lateral resolution for mapping small hydrogeologic features (Fig. 14.3b). Data can be recorded on an analog strip chart recorder or a digital data acquisition system. The excellent lateral resolution obtained from continuous EM profiling data has been used to outline closely-spaced burial pits, to reveal the migration of contaminants into the surrounding soils (Fig. 14.6), or to delineate complex fracture patterns (Fig. 14.3b; Benson *et al.*, 1982).

In addition to evaluation of natural hydrogeologic conditions and mapping of contaminant plumes, some of the electromagnetic instrumentation can be used to locate trench boundaries, buried wastes and drums, and metallic utility lines. Frequency-domain EM instruments provide two outputs consisting of an in-phase component and an out-of-phase component. The out-of-phase component is used to measure electrical conductivity and can be used to locate pipes. The in-phase component is a measure of the magnetic susceptibility, but it can also be used to detect both ferrous and nonferrous metal. For example, using the in-phase component, a single 200 L steel drum can be detected at a depth of about 2 to 3 m.

Vertical variations in conductivity can be determined by sounding. The instrumentation is placed at one location and measurements are made at increasing depths by a changing coil spacing and/or coil orientation. Sounding data can be acquired at depths ranging from 0.75 to 60 m by combining data from a variety of commonly available frequency-domain EM instruments. The vertical resolution of frequency-domain EM soundings is relatively poor since only a limited number of measurements are made at a few depths. However, they do provide a quick means of obtaining limited vertical information. Time-domain transient EM systems on the other hand are capable of providing detailed sounding data from depths of 50 m or more than 300 m.

Resistivity

As with EM measurements, electrical resistivity measurements are a function of the type of soil or rock, its porosity, and the fluids which fill the pore spaces. The method may be used in many of the same applications as the EM method (Cartwright and McComas, 1968; Griffith and King, 1969; Zohdy *et al.*, 1974; Mooney, 1980; Benson *et al.*, 1982; Telford *et al.*, 1982).

The resistivity method requires that an electrical current be passed through the ground from a pair of surface electrodes. Both direct current and switched direct current power sources are used. The voltage resulting on the surface of the ground due to current is measured between a second pair of electrodes. This requires that metal stakes be driven into the ground or that non-polarizing (copper-copper sulfate) electrodes be used. The greater the electrode spacing, the greater the depth of the measurement. Usually the depth is less than the spacing between electrodes. There are a number of electrode geometries that can be used, including the Schlumberger, dipole-dipole and many more, but the simplest, in terms of geometry is the Wenner array, which consists of four equally spaced electrodes all in a line. The apparent resistivity of the soil and rock is calculated based on the electrode separation, the geometry of the electrode array, the applied current, and measured voltage.

The resistivity technique may be used for profiling or sounding, similar to EM measurements. Profiling provides a means of mapping lateral changes in subsurface electrical properties to a given depth, and is well suited to the delineation of hydrogeologic anomalies and mapping of inorganic contaminant plumes (Fig. 14.7).

Sounding measurements provide a means of determining the vertical changes in subsurface electrical properties. Interpretation of sounding data provides the depth, thickness and resistivity of subsurface layers. Data can be interpreted using master curves for 2–3 layers (Orellana and Mooney, 1966). Computer models are commonly used and may be used to handle more than 2–3 layers (Mooney, 1980). Sounding data are used to create a geoelectric section which illustrates changes in the vertical and lateral resistivity conditions at a site. Figure 14.8 shows a geoelectric section developed from a single resistivity sounding, along with a drillers log showing the correlation.

One drawback with resistivity sounding is that the electrode array requires considerable space. For example, a Wenner array sounding (with electrodes equally spaced) may require that the spacing between the electrodes be as much as 3–4 times the depth of interest. Therefore, a sounding to a depth of 30 m could require an overall array length (from current electrode to current electrode) of 300 to 400 m. At many sites this space may not be available.

Comparison of electromagnetic and resistivity measurements

The frequency-domain electromagnetic method is often preferred for making profiling measurements since it requires less space for a measurement to a given depth. Also because the electromagnetic method does not require that electrodes be driven into the ground, measurements can be made more rapidly and are not influenced by shallow geologic noise associated with the electrodes used in resistivity measurements. On the other hand, because resistivity methods provide better vertical resolution than the frequency-domain EM method, the resistivity method is commonly employed for sounding measurements. When space is limited and deep soundings are needed, there are advantages to using the time-domain electromagnetics system for soundings since it requires less space than long resistivity arrays.

Electromagnetic measurements may be affected by buried metal pipes, metal fences, nearby vehicles, buildings and power lines, etc. as are resistivity measurements. But resistivity measurements are often less sensitive to many of these problems, sometimes permitting resistivity measurements to be made near cultural metal where electromagnetic measurements cannot be made.

Electromagnetic and resistivity measurements from the same location may not agree, due to the difference in the volume of material being sampled and in the differences in current distribution inherent to the two methods. Measurements will agree if they are both made over a uniform media and the instruments have been properly calibrated.

14.4.3 Seismic refraction and reflection

Seismic refraction and reflection techniques are often used to determine the top of bedrock, to determine depth of water table, and to assess the continuity of geologic strata, faults and buried bedrock channels. The refraction method may be used to characterize the type of rock and degree of weathering based upon the seismic velocity of the rock. The seismic velocity in rock is related to its density and hardness. Therefore, characterization of the material on the basis of seismic velocity can indicate the degree of weathering, and rippability.

Seismic waves are transmitted into the subsurface by a source. These waves are refracted and reflected when they pass from one soil or rock type into another which has a different seismic velocity. An array of geophones placed on the surface measures the travel time of the seismic waves from the source to the geophones. The refraction and reflection techniques use the travel times of the waves and the geometry of the source-to-geophone wave paths to model subsurface conditions. The unit of time is milliseconds (10^{-3} seconds). For most refraction work, the first refracted compressional wave arrivals (P-waves) are used. For reflection work, the latter arriving reflected compressional waves are used.

A seismic source, geophones, and a seismograph are required to make the measurements. The seismic source may be a simple sledge hammer or other mechanical source with which to strike the ground. Explosives may be utilized for deeper applications which require greater energy. Geophones implanted in the surface of the ground translate the ground vibrations of seismic energy into an electrical signal. The electrical signal is displayed on the seismograph, permitting measurement of the arrival of time of the seismic wave and displaying the wave forms from a number of geophones. Geophone spacing can be varied from about 1 m to more than 100 m depending upon the depth of interest and the lateral resolution needed.

Since the seismic refraction and reflection methods measure small ground vibrations, they are inherently susceptible to vibration noise from a variety of natural (i.e., wind and waves) and cultural sources (i.e., walking, vehicles and machinery).

Seismic refraction

The refraction method is commonly applied to shallow investigations up to about 100 m (Griffith and King, 1969; Benson *et al.*, 1982; Telford *et al.*, 1982; Haeni, 1986). However, with sufficient energy, surveys to several hundred meters are possible. Up to three or four layers of soil and rock can normally be determined, if a sufficient velocity exists between adjacent layers. A typical refraction line for a shallow investigation might consist of 12 or 24 geophones set at equal spacings as close as 1.5–3 m. The refraction survey may require a maximum source-to-geophone distance 4–5 times the depth of investigation. Significantly greater source energy will be required as the depth of investigation increases. Two inherent limits to the refraction method are its inability to detect a lower velocity layer beneath a higher velocity layer and thin layers.

Seismic refraction work can be carried out in a number of ways. The simplest approach in terms of field and interpretation procedures can be carried out by creating two separate seismic impulses, one at each end of the geophone array. The results of travel time measurements are plotted in a time-distance plot as the first step to the analysis (Mooney, 1973). The results of this simple measurement provide two depths and thus, the dip of rock under the array of geophones.

A more detailed refraction survey can be carried out so that depths are obtained under every geophone (Fig. 14.9). This survey will produce a detailed profile of the top of rock. Lateral resolution will depend upon the geophone spacing which might range from 1.5 to 15 m (Redpath, 1973). The general reciprocal method (Palmer, 1980) will accommodate varying velocities within each layer, while calculating the depth under each geophone.

Seismic reflection

By comparison, a seismic reflection survey is capable of much deeper investigations with less energy than the refraction method. While reflections have been obtained from depths as shallow as 3 m, the

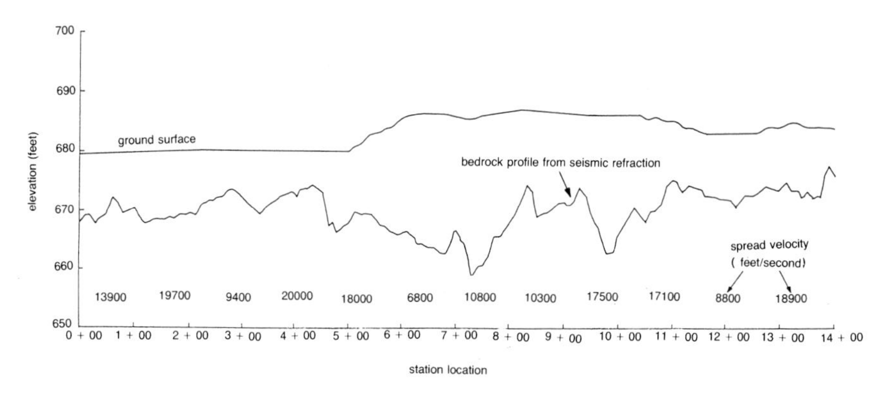

Fig. 14.9 Profile of top of bedrock from seismic refraction survey. (Depth to rock determined under each geophone. Geophone spacing is 3 m (10 ft).)

method is more commonly applied to depths of 15–30 m or greater. The reflection technique can be used effectively to depths of several hundred meters or more and can provide relatively detailed geologic sections (Fig. 14.10).

As in radar reflections, the vertical scale is measured in two-way travel-time i.e., the time it takes for a wave to travel down to an interface and back up to the surface again. The time scale must then be converted to depth making some assumptions regarding seismic velocity within the strata.

There are two approaches currently used to obtain shallow seismic reflection data, both of which have only evolved in the mid 1980s: the common offset method (Hunter *et al.*, 1982); and the common depth point (CDP) method adapted from the oil industry (Lankston and Lankston, 1983; Steeples, 1984). The common offset method uses low cost equipment and software but has some site specific limitations which are not inherent in the CDP method. The common depth point has fewer site specific limitations, but is more dependent upon sophisticated hardware and software capabilities. Hardware and software for the shallow CDP method have just recently become readily available in the late 1980s.

The shallow high resolution reflection methods discussed here attempt to utilize the highest frequencies possible (150–600 Hz) to improve vertical resolution, and relatively closely spaced geophones (0.3–6 m apart) to provide good lateral resolution. Because of the need for higher frequencies, attention must be given to selection of a seismic source and its optimum coupling to soil or rock, as well as geophone placement.

The reflection method is limited by its ability to transmit energy, particularly high frequency energy, into the soil and rock. Loose soil near the surface limits the ability of the soil system to transmit high

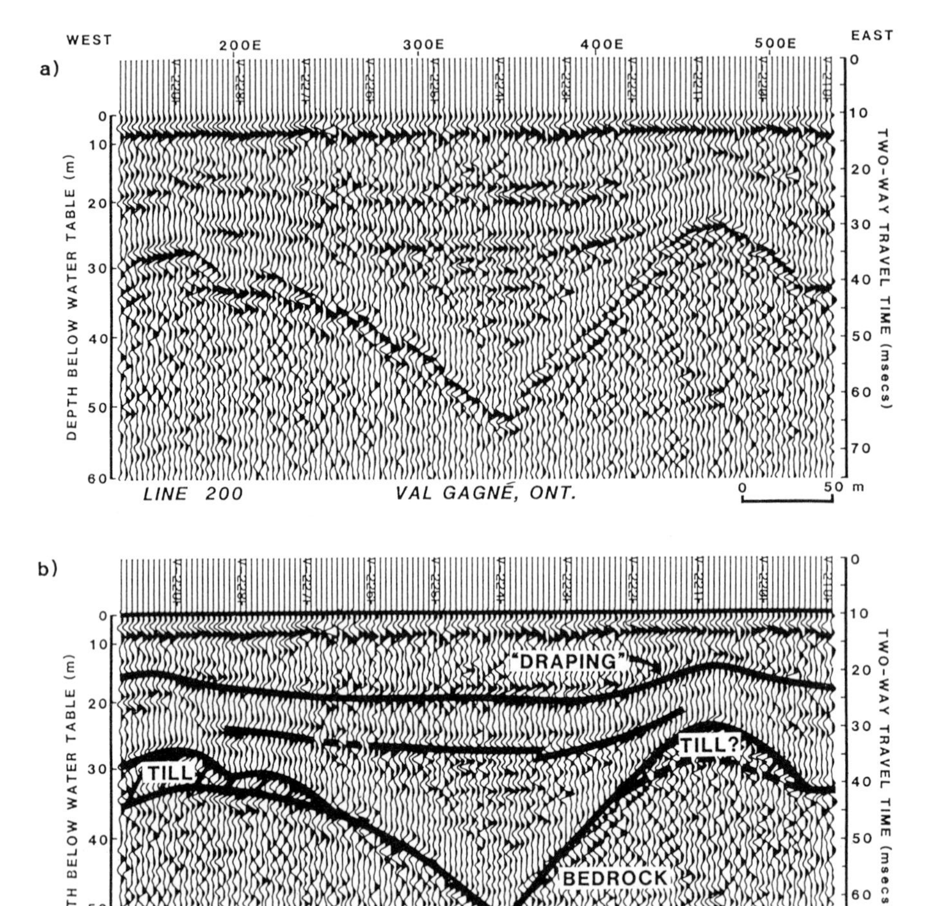

Fig. 14.10 Common offset seismic reflection data showing channel in bedrock. *Source*: Hunter *et al.* (1982).

frequency energy into and out of the rock, limiting the resolution which can be obtained.

14.4.4 Micro-gravity

Gravity measurements respond to changes in the earth's gravitational field caused by changes in the density of the soil and rock. By measuring the spatial changes in the gravitational field, variations in subsurface geologic conditions can be determined (Griffith and King, 1969; Telford *et al.*, 1982). There are two basic types of gravity survey: a regional gravity survey and a local micro-gravity survey. A regional

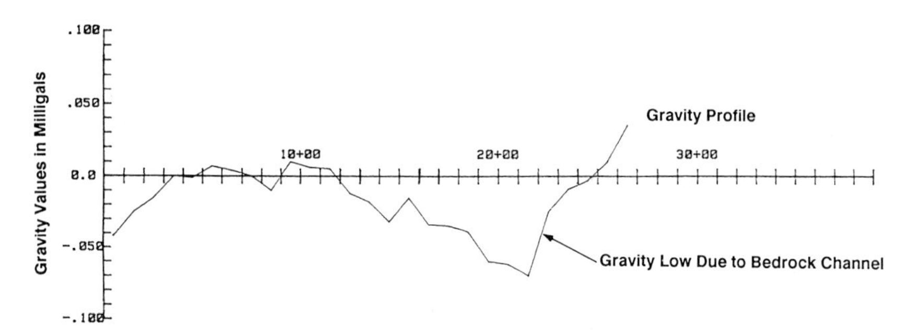

Fig. 14.11 Micro gravity profile showing bedrock channel. *Source*: Technos Inc.

gravity survey which employs widely spaced (a few hundred meters to a few kilometers) stations is carried out with a standard gravity meter. Micro-gravity surveys, with closely spaced stations of 1.5–6 m, are carried out with a very sensitive micro gravimeter. Regional surveys are used to assess major geologic conditions over many hundreds of square kilometers. Micro-gravity surveys are used to detect and map shallow, localized, geologic anomalies such as bedrock channels, fractures, and cavities.

The unit of acceleration used in gravity measurement is the gal. The earth's normal gravity is 980 gals. Micro-gravity measurements are sensitive to within a few micro gals (10^{-6} gals).

The micro-gravity survey results in a Bouguer anomaly which is the difference between the observed gravity values and theoretical gravity values. The Bouguer anomaly is made up of deep-seated effects (the regional Bouguer anomaly) and shallow effects (the local Bouguer anomaly). It is the local Bouguer anomaly that is of interest in micro-gravity work (Fig. 14.11).

A gravimeter is designed to measure extremely small differences in the gravitational field and is a very delicate instrument. The instrument is thermostatically controlled to minimize drift caused by temperature variations. Considerable care must be taken in shipment and general field use to avoid shock to the instrument. Gravity measurements may be affected by ground noise (similar to the seismic methods), winds, and temperature. To compensate for small instrument drift throughout the day, measurements must be made at a base station every hour or so, so that drift corrections can be applied to the data. Corrections must also be made for the constantly changing earth tides, changes in elevation (to the nearest millimeter), and topography. Gravity data may be presented as a profile or as a contour map, depending upon project needs.

14.4.5 Metal detection

Metal detectors are commonly used by utility and survey crews for locating buried pipes, cables, and property stakes. They are also useful for detecting buried drums and for delineating the boundaries of trenches containing metallic drums or trash (Fig. 14.12; Benson *et al.*, 1982). Metal detectors can detect both ferrous metals such as iron and steel, and non-ferrous metals such as aluminum and copper.

Metal detectors have a relatively short detection range, since the detector's response is proportional to the cross-section of the target and inversely proportional to the sixth power of the distance to the target. Small metal objects about 10–20 mm in size can be detected at a distance of approximately 0.5–1 m. Specialized metal detectors will detect larger objects like standard steel petroleum drums at depths of 1–3 m, and massive piles of drums may be detected at depths of up to about 3 m. The metal detector is a continuously-sensing instrument and is used with a sweeping motion while moving forward along a survey line. It may also be held steady while a traverse line is walked and the results recorded on a strip-chart recorder. The area of detection of a metal detector is approximately equal to its coil size or coil spacing (typically 0.3–1 m). Metal detectors can be affected by nearby metallic pipes, fences, cars, buildings and in some cases by changes in soil conditions.

14.4.6 Magnetometry

A magnetometer measures the intensity of the earth's magnetic field. As with gravity surveys, a magnetic survey can be used to map regional geologic conditions over large areas. In certain geologic environments, magnetics can be used to map depth to bedrock, channels, and fractures (Griffith and King, 1969; Breiner, 1973; Telford *et al.*, 1982). The primary application of magnetic measurements at hazardous waste sites is in detecting buried drums, tanks, and pipes (Breiner 1973; Benson *et al.*, 1982). A magnetometer will only respond to ferrous metals (iron and steel) and will not detect nonferrous metals. The presence of buried ferrous metals creates a local variation in the strength of the earth's magnetic field, permitting the detection and mapping of buried ferrous metal (Fig. 14.13).

Two types of magnetic measurements are commonly made: total field measurements and gradient measurements. A total field measurement responds to the total magnetic field of the earth, any changes caused by a target, natural magnetic, and cultural magnetic noise (ferrous pipe, fences, buildings and vehicles).

The effectiveness of total field magnetometers can be reduced or totally inhibited by noise or interference caused by time-variable

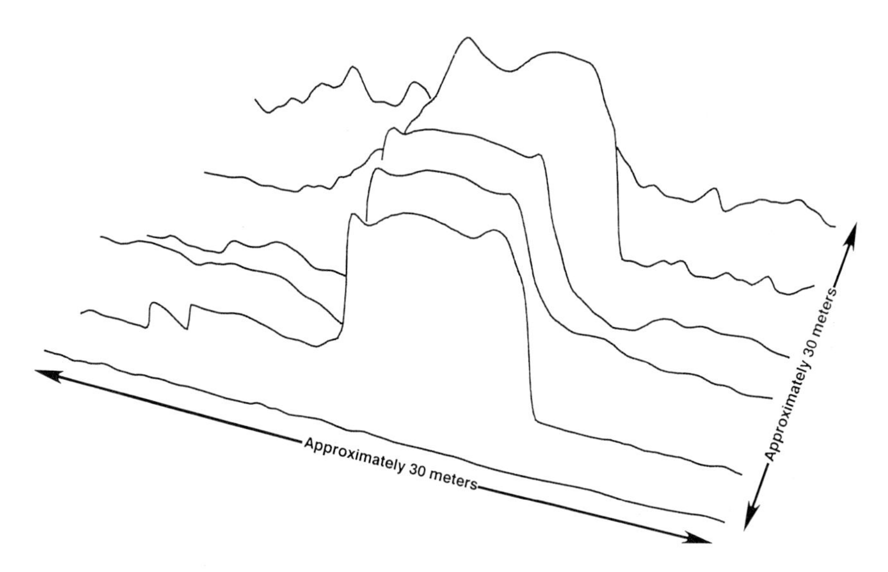

Fig. 14.12 Results of a metal detector survey to locate a burial trench.

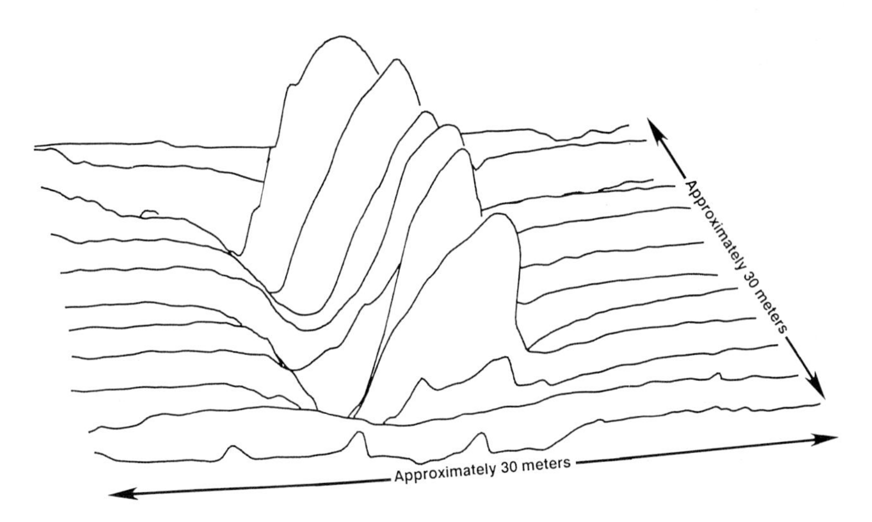

Fig. 14.13 Magnetic gradient over a trench with buried drums. The trench is approx. 20 m by 100 m.

changes in the earth's magnetic field, or spatial variations due to magnetic minerals in the soil, steel debris, pipes, fences, buildings, and passing vehicles.

A base station magnetometer can be used to reduce the effects of natural noise by subtracting the base station values from those of the search magnetometer. This can minimize any errors due to natural long period changes of the earth's field. Cultural noise, however, will remain a problem with total field measurements. Many of these problems can be avoided by use of gradiometer measurements and proper field techniques.

Gradient measurements are made by a gradiometer, which is simply two magnetic sensors separated vertically (or horizontally) by 1 to 2 m. Gradient measurements have some distinct advantages over total field measurements: they are insensitive to natural changes in the earth's magnetic field and minimize most cultural effects. Because the response of a gradiometer is the difference of two total field measurements it responds only to the local gradient. As a result, it is better able to locate a relatively small target, such as a buried drum. The disadvantage of a gradiometer is that it is slightly less sensitive than a total field instrument.

A total field magnetometer's response is proportional to the mass of the ferrous target and inversely proportional to the cube of the distance to the target (such as a drum). A gradiometer response is inversely proportional to the fourth power of the distance to the target (such as a drum), making it less sensitive than the total field measurement. While gradiometers are inherently less sensitive than total field instruments, they are also much less sensitive to many sources of noise. Under ideal conditions, a single drum can be detected at depths up to about 6 m with a total field magnetometer (about 3 m with a gradient magnetometer). Massive piles of drums can be detected at depths up to 15 m or more with a total field magnetometer (about 7 m with a gradient magnetometer).

A total field or gradient proton procession magnetometer normally requires the operator to stop to take a measurement, while fluxgate gradiometers permit the continuous acquisition of data as the magnetometer is moved across the site. Continuous coverage is much more suitable for detailed (high resolution) surveys to identify local targets, such as drums, and the mapping of areas in which complex anomalies are expected.

14.5 DOWNHOLE GEOPHYSICAL MEASUREMENTS

One of the most common subsurface investigation techniques is that of sampling soil and rock at discrete intervals (typically 1–2 m) as a boring is advanced. This method provides gross information on subsurface lithology but sand lenses, fractures, or other subtle changes in geology which can affect permeability, can easily go undetected. Though continuous sampling or coring can improve the description of geologic conditions, it is costly, time-consuming, and drillers or geologist's logs are somewhat subjective. Furthermore, 100% recovery of the sample is rarely achieved.

A number of downhole logging techniques are available for determining the characteristics of soil, rock, or fluid along the length of a boring, or monitoring well (Keys and MacCary, 1976). These methods provide continuous high resolution *in situ* measurements that are often more representative of hydrogeologic conditions than are the samples obtained from borings. An adequate assessment of conditions will often require that multiple logs be used since each log responds to a different property of the soil, rock, or fluid. Some of these techniques will provide measurements from inside plastic or steel casing, and some will allow measurements to be made in the unsaturated zone as well as the saturated zone.

Downhole logging measurements can be correlated to the known geologic strata in one hole and then can be used to identify and correlate geologic strata in other holes without soil or rock sampling. Thin layers and subtleties, not readily detected in soil or core samples, can often be resolved by logging. Logging can significantly improve one's ability to accurately characterize and correlate strata between borings by providing high-resolution data, independent of subjective interpretations of soil and rock type.

A number of soil and rock properties can also be measured *in situ*. Values for soil and rock porosity, density, seismic velocity and elastic moduli can be obtained to facilitate engineering design. Even more important, is the ability to identify the uniformity or lack of uniformity of subsurface conditions. Downhole measurements can be used to identify permeable zones such as sand lenses in glacial tills, weathered zones, and fractures or cavities in rock. The same measurements are also effective for identifying impermeable zones such as aquitards, and help to assess their continuity and integrity.

Monitoring wells which have been in place for years provide the basis for long-term chemical monitoring. For many of these wells, neither geologic logs nor installation records are available. Using downhole techniques, it is often possible to obtain geologic information and well construction details. In addition, logging may be used to determine whether a problem exists with well construction and what type of remedial work, if any, is necessary to correct it.

By running nuclear logs in existing holes with steel or PVC casing, geologic strata outside the casing can be characterized. Under some conditions, contaminants outside PVC casing can be characterized by running electromagnetic induction logs. A downhole television camera can be used within cased wells, to assess monitoring well conditions or can be used within an uncased borehole to assess the existence of fractures.

While each log is susceptible to both natural and cultural noise, borehole diameter and/or well construction will probably be of most concern. Most logs provide measurements within a radius of 150 to 300 mm from the hole. Therefore, as the hole diameter becomes larger, the measured results become more dominated by drilling and construction aspects.

A description of the commonly used logs is given below. Table 14.1 lists the conditions in which these logs can be used and some inherent limitations.

14.5.1 Nuclear logs

Natural gamma log

A natural gamma log records the amount of natural gamma radiation that is emitted by rocks and unconsolidated materials. The chief use of natural gamma logs is the identification of lithology and stratigraphic correlation in open or cased holes above and below the water table.

The gamma-emitting radioisotopes normally found in all rocks and unconsolidated materials are potassium-40 and daughter products of the uranium and thorium decay series. Because clays and shales concentrate these heavy radioactive elements through the processes of ion exchange and adsorption, the natural gamma activity of clay and shale-bearing sediments is much higher than that of quartz sands and carbonates. Therefore, the gamma log, which indicates an increase in clay or shale content by an increase in counts per second (Fig. 14.14), is useful for evaluating the presence, variability, and integrity of clays and shales.

The radius of investigation for the natural gamma log is from about 150 to 300 mm from the borehole wall (Keys and MacCary, 1976).

Gamma-gamma (density) log

A gamma-gamma log is used to determine the bulk density of the soil and rock as well as to identify lithology. The log can be used in open or cased holes above and below the water table (Fig. 14.15).

The gamma-gamma log is an active probe containing both a radiation source and a detector. This log provides a response in counts per second, that is averaged over the distance between the source and the detector. The count rate is inversely proportional to density and therefore, the count rate can be used to indicate relative changes in density. The count rate can be calibrated to provide actual density of the *in situ* material. The radius of investigation for the gamma-gamma log is relatively small (only about 150 mm; Keys and MacCary, 1976).

Table 14.1 General characteristics and use of downhole geophysical log

downhole log	*parameter measured (or calculated)*	*casing*			*saturated*	*unsaturated*	*radius of measurement*	*effect of hole diameter and mud*
		uncased	*PVC*	*steel*				
Natural gamma	natural gamma radiation	yes	yes	yes	yes	yes	150–300 mm	moderate
Gamma-gamma	density	yes	yes	yes	yes	yes	150 mm	significant
Neutron	porosity below water table–moisture content above water table	yes	yes	yes	yes	yes	150–300 mm	moderate
Induction	electrical conductivity	yes	yes	no	yes	yes	750 mm	negligible
Resistivity	electrical resistivity	yes	no	no	yes	no	300 mm to 1.5 m	significant to minimal depending upon probe used
Single point resistance	electrical resistance	yes	no	no	yes	no	near borehole surface	significant
Spontaneous potential (SP)	voltage–responds to dissimilar minerals and flow	yes	no	no	yes	no	near borehole surface	significant
Temperature	temperature	yes	no	no	yes	no	within borehole	NA
Fluid conductivity	electrical conductivity	yes	no	no	yes	no	within borehole	NA
Flow	fluid flow	yes	no	no	yes	no	within borehole	NA
Caliper	hole diameter cased hole diameter	yes	yes	yes	yes	yes	to limit of sensor typically 0.6–0.9 m	NA

NA: not applicable.

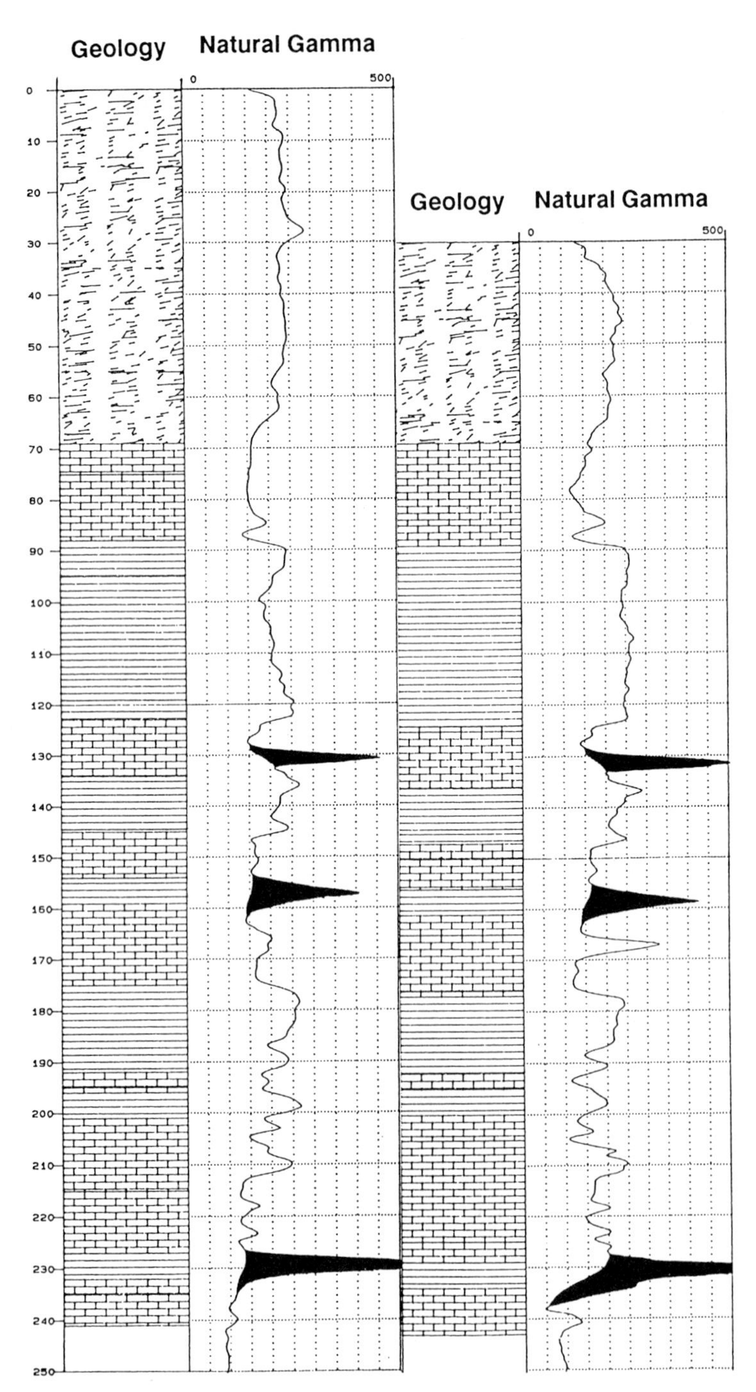

Fig. 14.14 Natural gamma logs from two nearby boreholes, 100 ft apart. Note the characterization and correlation of the shale and limestone units. *Source*: Technos Inc.

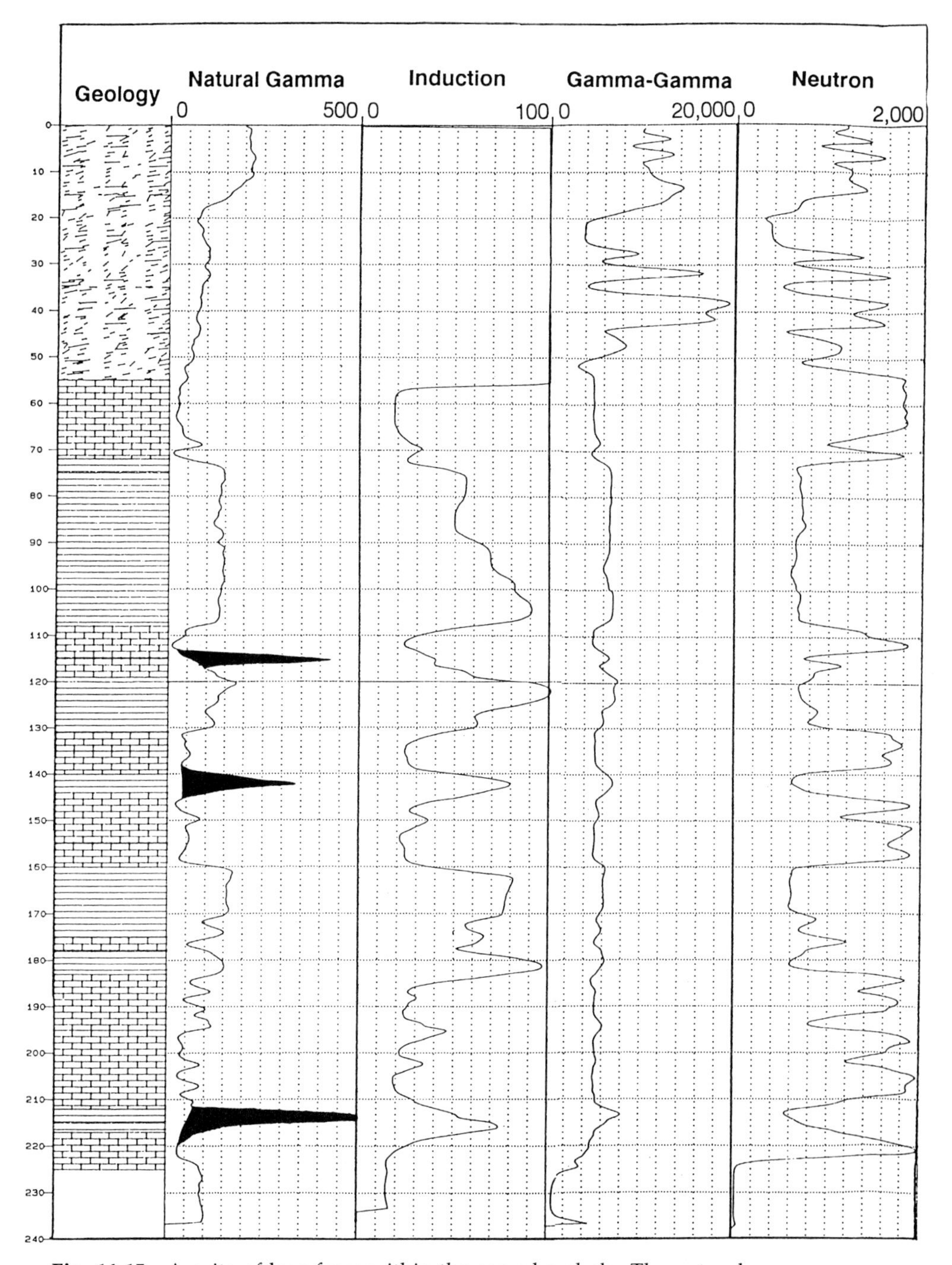

Fig. 14.15 A suite of logs from within the same borehole. The natural gamma log provides a means of characterizing the shale. The gamma-gamma log provides a measure of density and the neutron log provides a measure of porosity within the shale/limestone units. *Source*: Technos Inc.

Therefore, borehole diameter variations and well construction factors can affect this log more than other logs.

Neutron-neutron (porosity) log

A neutron-neutron log provides a measure of the moisture content above the water table and total porosity (based upon the water content of the rock) below the water table (Fig. 14.15). The log can be run in open or cased holes, above and below the water table. The neutron-neutron log is an active probe with both a radiation source and a detector. It provides a response, in counts per second, that is averaged over the distance between the source and the detector. The count rate is inversely proportional to water content and therefore, the count rate can be used to indicate relative changes in moisture content above the water table and porosity below the water table. The count rate can be calibrated to provide actual moisture content or porosity of the *in situ* material. Since clays and shales contain bound water, this log will indicate an increase in moisture or porosity in these materials, since the log cannot resolve the differences between bound and free water. Therefore, caution must be used in interpreting this log. The radius of investigation for the neutron-neutron probe is approximately 150 mm (up to 300 mm in very porous formations; Keys and MacCary, 1976). Borehole diameter variations and well construction factors can affect this log, but not as severely as the density log.

14.5.2 Non-nuclear logs

Induction log

The induction log is an electromagnetic induction tool for measuring the electrical conductivity of soil or rock in open or PVC-cased boreholes above or below the water table (similar to EM measurements made on the surface). The induction log can be used for identification of lithology and stratigraphic correlation. Electrical conductivity is a function of soil and rock type, porosity, permeability, and the fluids filling the pore spaces. Since the logs response (in millisiemens/meter or millimhos/meter) will be a function of the specific conductance of the pore fluids, it is an excellent indicator of inorganic contamination (Fig. 14.16) and in some cases organic contamination (when organics are mixed with inorganics, or when thick layers of hydrocarbons are present). Variations in conductivity may also indicate changes in clay content, permeability of a formation, or fractures. An induction log provides data similar to that provided by a resistivity log (because conductivity is the reciprocal of resistivity). However, the induction log can be run without electrical contact with the formation, and therefore,

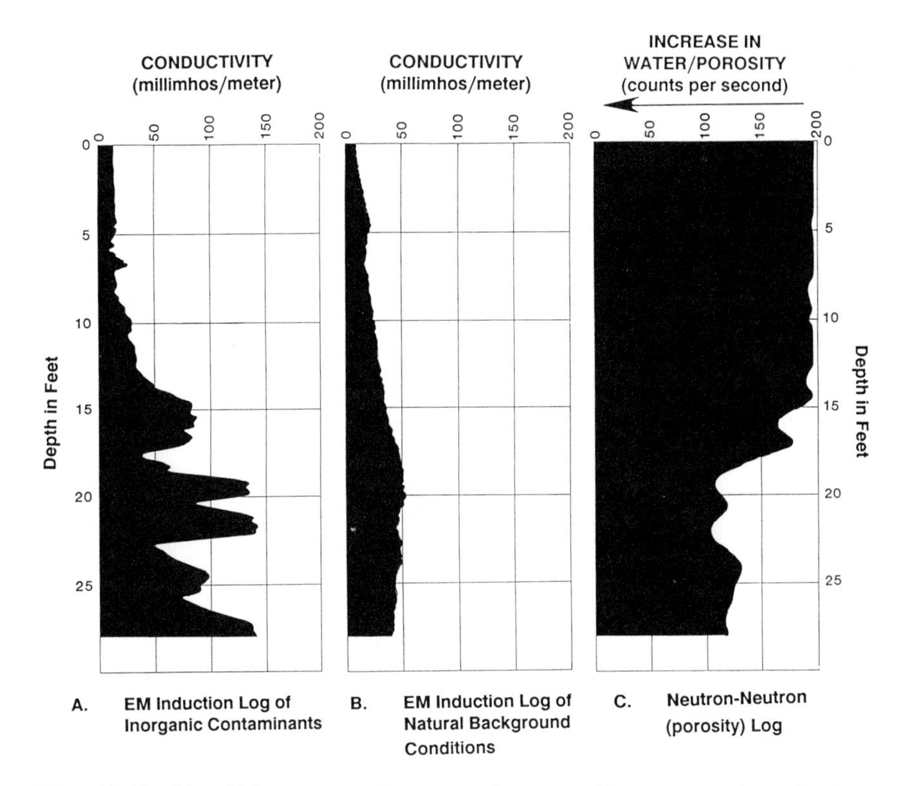

Fig. 14.16 Identifying contaminants and permeable zones using (a) EM induction log of inorganic contaminants, (b) EM induction log of natural background conditions and (c) neutron-neutron (porosity) log. *Source*: Technos Inc.

it can be used in both the vadose zone and the saturated zone and to log through PVC casing.

The radius of investigation for the induction log is approximately 0.75 m from the center of the well. Because this log has a much larger radius of investigation than other logs, it is almost totally insensitive to borehole and construction effects, and as such it is a good indicator of the overall soil and rock conditions surrounding the borehole.

Resistivity log

The resistivity log measures the apparent resistivity (ohm-feet or ohm-meters) of rock and soil within a borehole. Because resistivity is the reciprocal of conductivity, which is the property measured by an induction log, the resistivity log responds to and measures the same properties and features as the induction log. However, because of the need for electrical contact with the borehole wall, the resistivity log can only be run in an uncased hole filled with water or drilling fluid.

There are a number of electrode spacings or geometries which may be used for resistivity logs. 'Short normal' probes (typically an electrode spacing of 400 or 450 mm) give good vertical resolution and measures the apparent resistivity of the formation immediately around the hole. 'Long normal' probes (typically an electrode spacing of 1.7 m) have less vertical resolution but measure the apparent resistivity of undisturbed rock at a larger radius from the hole, similar to the induction log (Keys and MacCary, 1976).

Resistance log

A resistance log (sometimes referred to as single-point resistance) measures the resistance (in ohms) of the earth materials lying between a downhole electrode and a surface electrode. It can only be run in uncased holes in the saturated zone. The primary uses of resistance logs are geologic correlation and the identification of fractures or wash-out zones in resistive rocks. A resistance log should not be confused with the resistivity log, which provides a quantitative measure of the material resistivity.

The radius of investigation of the resistance log is quite small. It is in many cases as strongly affected by conductivity of the borehole fluid as it is affected by the resistance of the surrounding volume of rock (Keys and MacCary, 1976).

Spontaneous-potential log

The spontaneous-potential (SP) log measures the natural potential, in millivolts, developed between the borehole fluid and the surrounding rock materials. It can only be run in uncased holes within the saturated zone. The SP voltage consists of two components. The first component results from electrochemical potential caused by dissimilar minerals or fluids. The second component is the streaming potential caused by water moving through a permeable medium.

SP measurements are subject to considerable noise from the electrodes, hydrogeologic conditions and borehole fluids. Even though these measurements do not provide quantitative results they have a number of applications, including characterizing lithology, providing information on the geochemical oxidation reduction conditions, and providing an indication of fluid flow.

The radius of investigation of the SP log is highly variable (Keys and MacCary, 1976).

Temperature log

A temperature log provides a continuous record of the temperature of the borehole fluid immediately surrounding the sensor as it is lowered within an open borehole. It is usually run in uncased holes within the saturated zone. The temperature log will often indicate a zone of ground water flow within the uncased portion of a borehole. Flow is indicated when an increase or decrease in water temperature occurs. Changes in temperature can also be used to monitor leaks in casing where damage or corrosion has occurred. A temperature log may have a sensitivity of 0.5 °C or better.

Fluid flow

There are many ways of measuring fluid flow within a borehole (Keys and MacCary, 1976). The most common uses an impeller-type flow meter which provides counts per second. It can only be run in uncased holes within the saturated zone. The count rate can usually be calibrated to provide results in the desired units.

Fluid conductivity

A fluid conductivity log provides a measurement of the specific conductance of the borehole fluids (micromhos/centimeter). It can only be run in uncased holes within the saturated zone. If accurate values are needed, a temperature log must also be run so that corrections can be made.

Caliper log

A caliper log provides a record of the diameter of an open borehole or of the inside diameter of a well casing. The log can be run in open or cased holes, above or below the water table, although it is most commonly run in open holes. The caliper probe consists of spring-loaded arms which extend from the logging tool so that they follow the sides of the borehole or casing.

Caliper logs are utilized to measure hole diameter and to locate fractures or cavities in an open hole. The caliper log can also be used to determine well construction details and casing diameter. It can also be used to reveal any deterioration due to extreme corrosion or accumulation of minerals on the interior of the well casing.

14.6 APPLICATIONS

There is no simple way to select the geophysical method(s) required to solve a particular problem. Tables 14.2–4 are provided to illustrate how the geophysical methods may be used to carry out assessment of hydrogeologic conditions, detecting and mapping contaminants, and locating and mapping buried wastes and utilities. However, simple tables and rules of thumb often fail when considering specific project needs and site specific conditions, therefore, the tables presented here should be used only as a guide.

14.6.1 Assessing hydrogeologic conditions

The first and often the most important task of any site investigation is the evaluation of the natural hydrogeologic conditions. A description of the overall hydrogeologic conditions of the setting and site specific conditions along with any hydrogeologic anomalies is usually required. Knowledge of the natural anomalies, in relation to the setting and the site itself ensure that drilling and sampling is done at locations which will be most likely to yield useful information.

Table 14.2 lists possible applications of surface geophysical methods and some of their advantages and limitations in evaluating hydrogeologic conditions. Variations in the shallow natural setting are best evaluated with ground penetrating radar, which provides the highest resolution of all of the surface methods. However, depth of penetration of radar is highly site-specific, and is typically less than 16 m. When silts and clays are at the surface, penetration may be limited to only 1–2 m.

Even with these limitations, ground penetrating radar can often help solve problems deeper than its sensing range. For example, by looking for anomalies in shallow marker beds, or observing shallow soil piping, shallow radar data can be used to predict the presence of cavities and fractures much beyond its range. Investigation of such near-surface indicators (NSI) with radar and other methods to evaluate deeper conditions is a powerful technique (Benson and Yuhr, 1987).

High-resolution seismic reflection can be used with radar to provide a more complete depth profile. While this seismic reflection method has less resolution than radar, information can be acquired to depths of tens or even hundreds of meters. The reflection method is often found to be ineffective at depths shallower than 7–15 m where radar is most effective. Therefore, these two methods are quite complementary for developing detailed geologic profiles. It should be noted, however, that the cost of seismic work is considerably greater than the cost for a radar survey.

Seismic refraction and resistivity soundings provide good vertical information, such as depth to water table and depth to rock though they are not capable of achieving the lateral and vertical resolution of radar, or in some cases, the vertical resolution of seismic reflection.

The frequency-domain EM techniques have excellent lateral resolution in the continuous profiling mode to depths of about 15 m. In general, the frequency-domain EM are limited in their capability to

Table 14.2 Surface geophysical methods for evaluation of natural hydrogeologic conditions*

method	*general application*	*continuous measurements*	*depth of penetration*	*major limitations*
Radar	profiling and mapping; highest resolution of any method	yes	to 30 m (typically less than 10 m)	Penetration limited by soil conditions
EM (Frequency-domain)	profiling and mapping; very rapid measurements	yes (to 50 feet)	to 60 m	Affected by cultural features (metal fences, pipes, buildings, vehicles)
EM (time-domain)	soundings or profiling and mapping	no	to few hundred meters	Does not provide measurements shallower than 150 feet
Resistivity	soundings or profiling and mapping	no	No limit (commonly used to 100 to 300 m)	Requires good ground contact and long electrode arrays Integrates a large volume of subsurface Affected by cultural features (metal fences, pipes, buildings, vehicles)
Seismic refraction	profiling and mapping soil and rock	no	No limit (commonly used to 100 to 300 m)	Requires considerable energy for deeper surveys Sensitive to ground vibrations
Seismic reflection	profiling and mapping soil and rock	no	to few hundred meters	Shallow surveys, <100 feet are most critical Sensitive to ground vibrations
Micro gravity	profiling and mapping soil and rock	no	No limit (commonly used to 100 to 300 m)	Very slow, requires extensive data reduction Sensitive to ground vibrations
Magnetics	profiling and mapping soil and rock	yes	No limit (commonly used to 100 to 300 m)	Only applicable in certain rock environments Limited by cultural ferrous metal features

* Applications and comments should only be used as guidelines. In some applications, an alternate method may provide better results.

produce vertical detail (sounding data). Yet, these EM methods can often provide some relative sounding information (i.e., thick versus thin or shallow versus deep) very quickly and more cost effectively than resistivity or seismic refraction.

Probably the two best techniques to map lateral variations in soil and rock, from a speed and resolution point of view, are ground-penetrating radar and continuous EM measurements. While the radar performance is highly site specific, the EM technique can be applied in almost any environment and can often provide deeper information, but with much less vertical resolution than radar. Continuous EM profiling measurements provide high lateral resolution and can be run at speeds from 1–5 km/hr depending on the detail required. The rapid speed at which EM measurements can be obtained and the option of continuous profile measurements at depths of up to 15 m makes EM the best choice for profile work under most situations.

The resistivity method can also be used for profile measurements by moving the array in small increments to provide data at closely spaced intervals. This is a slow process relative to an EM survey, and resistivity data can be affected by near-surface geologic noise at the electrodes.

Sometimes, one method may work and another will fail under a given set of site conditions. For example, in some cases, resistivity measurements can be made relatively close to a chain link fence or a buried pipe line (when they are relatively isolated electrically) where EM measurements cannot.

In order for any geophysical method to work, there has to be a contrast in the parameter being measured. The best method is the one in which the parameters being measured have the greatest contrast and will be least influenced by site specific conditions and noise. The final decision must be made on a site-by-site basis.

Once the surface methods have defined the 2-dimensional or even 3-dimensional conditions reasonably well, boring locations can be selected. These locations should be selected to be representative of the normal background conditions at the site and to investigate any anomalies identified by geophysical or other methods identified by geophysical or other methods.

The drilling program should be designed to provide a means of accurately characterizing soil and rock conditions to the greatest extent possible within the budget available. If an adequate downhole logging program is used, most of the holes can be drilled without extensive sampling. However, it is always good practice to continuously sample (or core) and log at least one borehole and then geophysically log this hole as an on-site 'calibration' hole. Then the other boreholes can often be drilled and geophysically logged without sampling. This approach provides a reference for the logging data to compare to site-specific soil samples or rock core. Then the logs can be used to extrapolate soil and rock type and other conditions to nearby boreholes. For example, the natural gamma logs shown in Fig. 14.14 clearly show the limestone (low counts) and the shale units (high counts). In this case, correlation of the stratigraphy from natural gamma logs in adjacent holes is easily and accurately made.

When the appropriate suite of logs are combined, continuous *in situ* logging measurements can be obtained in both the unsaturated and saturated zones to characterize hydrogeologic conditions. Geologic formations can be identified and easily correlated from hole to hole. Relative estimates of clay content, density and porosity can be obtained. Permeable sand lenses and fractures can be identified, as can impermeable clay and shale zones. In addition, the continuity of impermeable zones can be assessed. The maximum amount of data should always be obtained from each hole because borings are few and costly.

Figure 14.15 shows a suite of natural gamma, density, and porosity logs from the same hole. The density log shows variable conditions in the overlying soil, but fairly uniform density within the shale and limestone units. In contrast the porosity log shows considerable variation throughout both the shale and limestone units. Without calibration these logs can be used to indicate relative changes in density and porosity. By calibrating these logs, quantitative results for density and porosity may be obtained in some situations.

14.6.2 Detecting and mapping of contaminant plumes

Table 14.3 illustrates how the surface geophysical methods can be applied to the mapping of contaminant plumes. The fundamental approach to evaluating the direction of groundwater flow and the possible extent of a contaminant plume is by determining the hydrogeologic characteristics of the site (that is, determining the presence of pathways such as buried channels, fractures, and permeable zones).

'Direct' detection of inorganics (or organics mixed with inorganics) can be accomplished by the electrical methods (EM and resistivity) as well as ground penetrating radar, as shown in Table 14.3. When inorganics are present in sufficient amounts, the higher specific conductance of the pore fluids acts as a tracer by which the plume can be mapped. In cases where inorganic plumes have a very low specific conductance, or plumes consist of dissolved organics, they will not be detectable by the electrical methods.

Where suitable penetration is possible, ground penetrating radar can provide a means for mapping the depth to the top of and lateral extent of shallow inorganic plumes. However, because of the site-specific behavior of radar, the EM or resistivity methods are most often used. Of the two methods, EM measurements are preferred for profile work,

Table 14.3 Surface geophysical methods for mapping of contaminant plumes*

Feature to be mapped	*Comment on approach*
Permeable pathways, bedrock channels, etc.	The fundamental approach to evaluating the direction of groundwater flow and the possible extent of a contaminant plume is by determining the hydrogeologic characteristics of the site (see Table 14.2 for evaluation of natural hydrogeologic conditions).
Inorganics or mixed inorganics and organics	When inorganics are present in sufficient concentrations above background or organics are part of such an inorganic plume, they can be detected by the electrical methods and sometimes radar. The higher specific conductance of the pore fluids acts as a tracer by which the plume can be mapped.
Hydrocarbons	When sufficient hydrocarbons have been present in the soil or floating on a shallow water table, for a sufficient period of time they may sometimes be detected by the electrical methods or by radar. Because of their low conductivities (high resistivity) they may sometimes be detected by the electrical methods. Due to changes in dielectric constant or suppression of the capillary zone they may sometimes by detected by radar (in some situations where degradation of hydrocarbons is occurring, conductivity may increase). These applications are limited and should be treated with caution. A more reliable approach is to map natural permeable pathways (see Table 14.2 for evaluation of natural hydrogeologic conditions and table 14.4 for mapping of cultural pathways).
Radar	Limited applications – may sometimes be used to detect shallow floaters (0–6 m) to map hydrocarbons in soil. May detect thickness in some cases.
EM	May be applicable to detect low conductivity at some sites.
Resistivity	May be applicable to detect high resistivity at some sites.

* Applications and comments should be used only as guidelines. In some applications, an alternate method may provide better results.

particularly where continuous sampling can be employed. Resistivity is generally preferred for sounding work.

Both resistivity and EM measurements can miss a contaminant plume if the measurements are in the wrong location. However, rapid EM profiling by either continuous or station measurements allows coverage of a site with closely spaced data. It is not unreasonable from a cost perspective to have overlapping measurements, therefore providing total site coverage using the EM profiling method.

Both resistivity and EM are capable of providing vertical sounding data. The frequency-domain EM method provides a depth of penetration that is limited to about 60 m and provides less resolution than the resistivity method since measurements are made only a few depths. The depth to which resistivity sounding data can be obtained is virtually unlimited. Depths of one to several hundred meters are obtainable. However, the long resistivity arrays necessary for deep measurement may not be practical in many areas due to space restrictions and cultural factors. Here, the time-domain EM transient systems, which have a smaller coil size, would be the choice for measurements to depths from 3 m to a few hundred meters.

In some cases organics can be mapped because they are mixed with inorganics. Figure 14.6 shows the inorganic plume from a chemical/ drum recycling center which also contains organics. Figure 14.7 shows the plume from a landfill that contains low levels of organics. The results in Fig. 14.17 show an excellent comparison of measurements by EM, resistivity, and organic vapor analyses (OVA) responses obtained from a mixed plume of organics and inorganics confined in a buried channel. Clearly, if inorganics are present, they should be used as an easily detectable tracer that will provide a first approximation of where the organics may be.

Some investigators have suggested that direct detection of shallow, major hydrocarbon spills can be accomplished by looking for EM conductivity lows (or resistivity highs) associated with the organics. Recent spills of petroleum products do not seem to yield a resistivity high or an EM conductivity low. However, the possibility for an anomaly exists where the product has been in the ground for some time, where there is a substantial amount of floating product, and where the conductivity of the natural soil conditions is high enough (or resistivity low enough) to provide a reasonable contrast between the hydrocarbons and the natural soil.

Once the spatial extent of a contaminant has been mapped by surface geophysics and after boreholes have been installed, continuous, downhole logging can be used to evaluate changes in the vertical hydraulic conductivity of soil and rock, as well as the distribution of contaminants. The vertical distribution and concentration of contaminants at a site can vary significantly as a result of local changes in hydraulic conductivity. Because hydraulic conductivity can change by more than an order of magnitude in less than a meter, it can have a significant impact on the test results obtained from a monitoring well. The chemical concentration in a well may be low, average, or high, depending upon screen length and location. Two downhole logging techniques particularly well suited for permeability evaluations are the

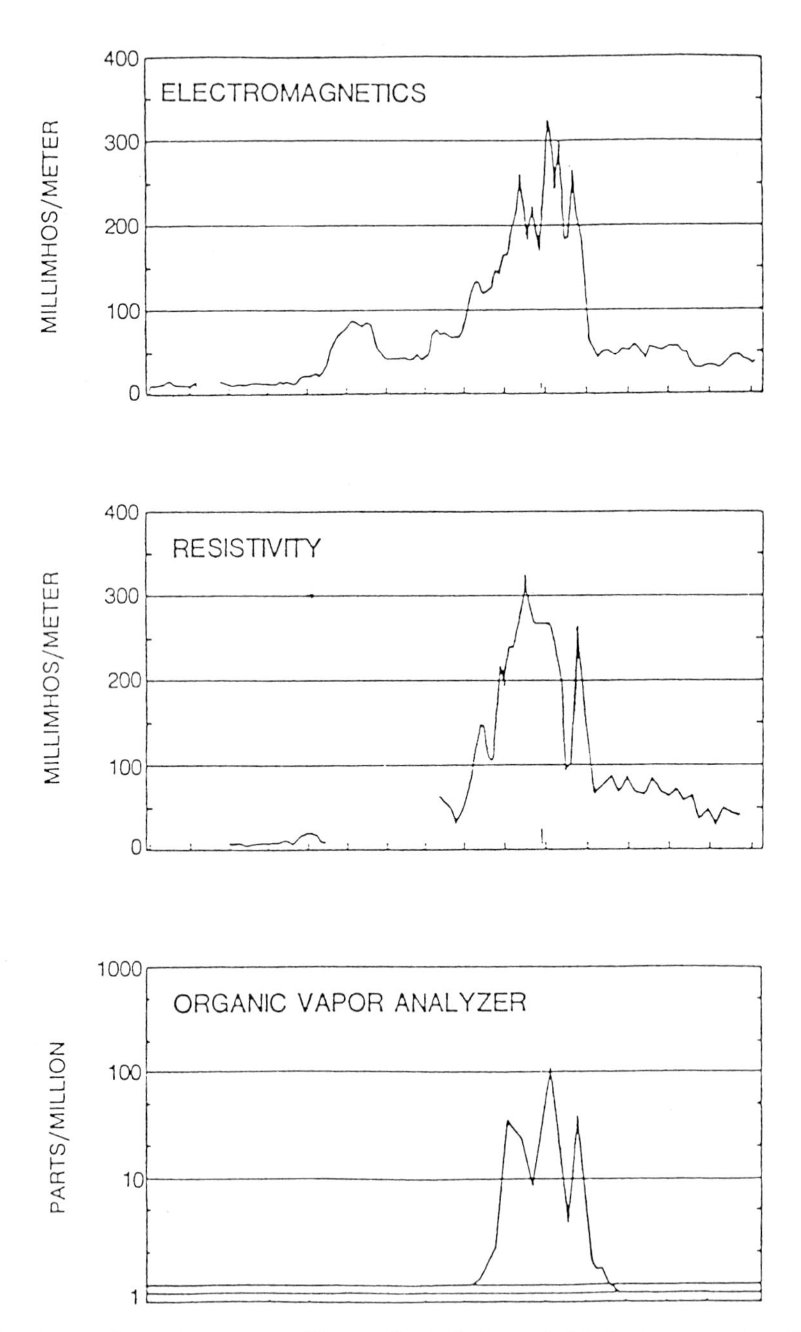

Fig. 14.17 Organic vapor profile over buried channel: (a) electromagnetics, (b) resistivity and (c) organic vapor analyzer. Note correlation between resistivity and EM measurements. This is an excellent example of a buried channel controlling flow and the level of correlation between organic and inorganic contaminants.

electromagnetic (EM) induction log (or resistivity logs – only in open hole) and the neutron (porosity) log. Both of these logs can be run in an open hole or within an existing PVC-cased well, above or below the water table (Fig. 14.16).

The EM induction log, shown in Fig. 14.16a, indicates the presence of inorganic contaminants which have preferentially migrated within five discrete zones of increased permeability in the limestone. These zones are indicated by higher conductivity values and range from 0.5 to 1.5 m in thickness. The presence of high hydraulic conductivity zones detected by the EM induction method was confirmed by using a downhole television camera which visually located small cavities and fractures in each zone. Figure 14.16b shows an EM induction log of natural conditions taken in a background well. Here the permeable zones are indicated by the log since there are no contaminants present. Figure 14.16c shows a neutron log taken in the same background well. In this log, zones of variable porosity are revealed whether contaminants are present or not. Conditions shown by this log are also representative of conditions at the contaminated well.

An adequate assessment of conditions in a hole often requires that more than one log be run. At this site, an EM induction log was used to identify the contaminated zones, and a neutron log was used to identify zones of increased porosity. Once conditions at a site are understood, a reliable and representative monitoring well system can be designed and data from existing monitoring well systems can be more accurately evaluated.

14.6.3 Location and mapping of buried wastes and utilities

Location and mapping of buried wastes, utilities, drums, and tanks is a common application of geophysical methods. Table 14.4 lists the surface geophysical methods applicable to this problem. Locating buried bulk wastes where no metal is present, can often be accomplished by ground-penetrating radar, if soil conditions are suitable. Often the shallow edges of trenches can be detected even in soil conditions which provide poor ground-penetrating radar penetration. Shallow EM tools are also effective for most location problems. When metals are present, EM conductivity, metal detectors, and magnetometers are the primary choices. Metal detectors and magnetometers are unaffected by most soil types or by the presence of contaminants. However, EM measurements are influenced by both variations in soils and the presence of contaminants.

To locate buried steel drums, the use of metal detectors, magnetometers, or the in-phase component of measurements EM are recommended. All three methods can be used to locate single

Table 14.4 Surface geophysical methods for location and mapping of buried wastes and utilities*

bulk wastes method	*bulk wastes w/o metals*	*steel drum w/metals*	*pipes and drums*	*tanks*
Radar	Very good if soil conditions are appropriate Sometimes effective to obtain shallow boundaries in poor soil conditions	Very good if soil conditions are appropriate Sometimes effective to obtain shallow boundaries in poor soil conditions	Good if soil conditions are appropriate (may provide depth)	Very good for metal and non-metal if soil conditions are appropriate (may provide depth)
EM	Excellent to depths less than 6 m	Excellent to depths less than 6 m	Very good (single drum to 2–3 m)	Very good for metal tanks
Resistivity	Good (sounding may provide depth)	Good (sounding may provide depth)	NA	NA
Seismic refraction	Fair (may provide depth)	Fair (may provide depth)	NA	NA
Micro gravity	Fair (may provide depth)	Fair (may provide depth)	NA	NA
Metal detector	NA	Very good (shallow)	Very good (shallow)	Very good (shallow)
Magnetometer	NA	Very good (ferrous only; deeper than metal detector)	Very good (ferrous only; deeper than metal detector)	Very good (ferrous only; deeper than metal detector)

* Applications and comments should be used only as guidelines. In some applications, an alternate method may provide better results.
NA: Not applicable

200 L steel drums as well as large piles of drums within their depth limitation.

Both the metal detector and the EM will respond to ferrous and non-ferrous metals, while a magnetometer will respond only to ferrous metals. Therefore, it is necessary to assess what metals may be present in order to select the appropriate instrument.

While ground-penetrating radar can be used to find drums, it will often be unable to detect a single drum if it is not oriented so that energy is reflected back to the antenna. Furthermore, many natural and man-made objects may have a radar response similar to that of a drum.

For small, discrete, critical targets such as a single steel drum, continuous measurements (on closely spaced lines of about 2 m) are required to provide a reasonable probability of detection. Ground-penetrating radar, shallow EM measurements, metal detectors, and certain magnetometers can provide these continuous measurements. However, there may be cases in which the proximity of other metal structures may limit the use of EM to locate drums or trenches, making ground-penetrating radar a clear choice.

Metal detectors and ground-penetrating radar both provide reasonably good spatial resolution to pinpoint the location of a target, while EM and magnetometers do not provide the same target resolution because the shape of their response curve is broader and often more complex.

Metal detectors, EM, and total field magnetometers are highly susceptible to interference from nearby metallic cultural features. Any of these features can produce an erroneous response that may be incorrectly interpreted as a subsurface target. Because metal detectors are relatively short-range devices, they can be operated closer to such sources of interface, than can most magnetometers.

Measurements made with a total field proton procession magnetometer are susceptible to interference from high magnetic gradients, natural changes in the earth's magnetic field, and nearby power lines, whereas fluxgate gradiometers do not suffer from these short-comings.

Seismic refraction, resistivity, magnetic, and gravity techniques may also be used to locate boundaries of larger trenches and landfills. These techniques are much slower and will provide less resolution than the previously described methods. However, they are often the only techniques which can be used to estimate the thickness of a landfill or trench. It should be noted that interpretation of such data should be done with caution by experienced personnel.

REFERENCES

Benson, R.C. and Glaccum, R.A. (1979) Radar surveys for geotechnical site assessment, in *Geophysical Methods in Geotechnical Engineering*, Specialty Session, ASCE, Atlanta, Georgia, pp. 161–78.

Benson, R.C., Glaccum, R.A. and Noel, M.R. (1982) *Geophysical Techniques for Sensing Buried Waste and Waste Migration*, EPA – Environmental Monitoring Systems Laboratory, Las Vegas, Nevada.

Benson, R.C. and LaFountain, L.J. (1984) Evaluation of subsidence or collapse potential due to subsurface cavities, in *Sinkholes, Their Geology, Engineering and Environmental Impact, Proc. of the 1st Multidisciplinary Conf. on Sinkholes*, Sinkhole, Orlando, Florida, pp. 201–15.

Benson, R.C. and Pasley, D.C. (1984) *Ground Water Monitoring: A Practical Approach for a Major Utility Company*, 4th Nat. Symp. and Exposition on Aquifer Restoration and Ground Water Monitoring, National Water Well Association.

Benson, R.C. and Scaife, J. (1987) Assessment of flow in fractured rock and karst environments, *Proc. 2nd Multidisciplinary Conf. on Sinkholes and the Environmental Impacts of Karst*, Orlando, Florida.

Benson, R.C., Turner, M., Turner, P. and Vogelson, W. (1988) *In situ*, time-series measurements for long-term ground-water monitoring, in *Ground Water Contamination: Field Methods*, ASTM STP-963 (eds A.G. Collins and A.I. Johnson), Am. Soc. for Testing and Materials, pp. 58–72.

Benson, R.C., Turner, M., Vogelson, W. and Turner, P. (1985) Correlation between field geophysical measurements and laboratory water sample analysis, *Proc. Nat. Water Well Association/EPA Conf. on Surface and Borehole Geophysical Methods in Ground Water Investigations*, National Water Well Association.

Benson, R.C. and Yuhr, L. (1987) Assessment and long-term monitoring of localized subsidence using ground-penetrating radar, *Proc. 2nd Multidisciplinary Conf. on Sinkholes and the Environmental Impact of Karst*, Orlando, Florida.

Benson, R.C. and Yuhr, L. (1990) Evaluation of fractures in silts and clay using ground penetrating radar, presented at the 4th Radar Conf., Denver, Colorado.

Breiner, S. (1973) *Applications Manual for Portable Magnetometers*, Geometrics, Sunnyvale, California.

Cartwright, K. and McComas, M. (1968) Geophysical surveys in the vicinity of sanitary landfills in northeastern Illinois, *Ground Water*, **6**, 23–30.

Dunbar, D., Tuchfeld, H., Siegel, R. and Sterbentz, R. (1985) Ground water quality anomalies encountered during well construction, sampling and analysis in the environs of a hazardous waste management facility, *Ground Water Monitoring Review*, **5**(3), 70–4.

Greenhouse, J.P. and Monier-Williams, M. (1985) Geophysical monitoring of ground water contamination around waste disposal sites, *Ground Water Monitoring Review*, **5**(4), 63–9.

Griffith, D.H. and King, R.F. (1969) *Applied Geophysics for Engineers and Geologists*, Pergamon Press.

Haeni, P. (1986) *Application of Seismic-Refraction Techniques to Hydrologic Studies*, US Geological Survey, Open File Report No. 84-746, Hartford, CT.

Hileman, B. (1984) Water quality uncertainties, *Environ. Sci. and Tech.*, **18**(4), 124–6.

Hunter, J.A., Burns, R.A., Good, R.L., MacAulay, H.A. and Cagne, R.M.

(1982) Optimum field techniques for bedrock reflection mapping with the multichannel engineering seismograph, in *Current Research, Part B, Geological Survey of Canada, Paper 82-1 Part B*, pp. 125–9.

Keys, W.S. and MacCary, L.M. (1976) Application of borehole geophysics to water-resources investigations, *Techniques of Water-Resources Investigations of the US Geological Survey*, ch E1.

Lankston, R.W. and Lankston, M.M. (1983) *An Introduction to the Utilization of the Shallow or Engineering Seismic Reflection Method*, Geo-Compu-Graph, Inc.

Lysyj, I. (1983) Indicator methods for post-closure monitoring of ground waters, in *Proc. Nat. Conf. on Management of Uncontrolled Haz. Waste Sites*, Hazardous Materials Control Research Institute, pp. 446–8.

McNeill, J.D. (1980) Electromagnetic resistivity mapping of contaminant plumes, in *Proc. Nat. Conf. on Management of Uncontrolled Haz. Waste Sites*, Washington, DC, pp. 1–6.

Mooney, H.M. (1973) Engineering seismology, in *Handbook of Engineering Geophysics, Vol. 1: Bison Instruments*, Minneapolis, Minnesota.

Mooney, H.M. (1980) Electrical resistivity, in *Handbook of Engineering Geophysics, Vol. 2: Bison Instruments*, Minneapolis, Minnesota.

Orellana, E. and Mooney, H.M. (1966) *Master Tables and Curves for Vertical Electrical Sounding Over Layered Structures*, Interciencia, Madrid, Spain.

Palmer, D. (1980) *The Generalized Reciprocal Method of Seismic Refraction Interpretation* (ed. K.B.S. Burke), Dpt. of Geology, Univ. of New Brunswick, Fredericton, N.B., Canada.

Perazzo, J.A., Dorrler, R.C. and Mack, J.P. (1984) Long-term confidence in ground water monitoring systems, *Ground Water Monitoring Review*, **4**(4), 119–23.

Redpath, B.B. (1973) *Seismic Refraction Exploration for Engineering Site Investigations*, Technical Report E-73-4, US Army Engineer Waterways Experiment Station Explosive Excavation Research Laboratory, Livermore, CA.

Steeples, Don W. (1984) High resolution seismic reflections at 200 Hz, *Oil and Gas J.*, 86–92, Dec 3.

Telford, W.M., Geldart, L.P., Sheriff, R.E. and Keys, D.A. (1982) *Applied Geophysics*, Cambridge University Press.

USEPA (1985) RCRA Ground-Water Monitoring Technical Enforcement Guidance Document.

Walker, S.E. and Allen, D.C. (1984) *Background Ground Water Quality Monitoring: Temporal Variations*, 4th Nat. Symp. and Exposition on Aquifer Restoration and Ground Water Monitoring, National Water Well Association.

Zohdy, A.A., Eaton, G.P. and Mabey, D.R. (1974) Application of surface geophysics to ground-water investigations, *Techniques of Water-Resources Investigations of the US Geological Survey, ch D1*.

CHAPTER 15

Soil exploration at contaminated sites

Charles O. Riggs

15.1 INTRODUCTION

Most of the drilling methods currently used for subsurface environmental exploration are used on a day-to-day basis for geotechnical or mineral exploration or for groundwater development and are readily available. It is important to understand that the most economical or the most readily available drilling method that would be used at a site for geotechnical sampling or testing, or for mineral exploration, or for water well drilling may not be the most technically appropriate method for exploration of contaminated sites. As an example, a particular site might be most economically drilled and sampled using the straight hydraulic (mud) rotary method to obtain geotechnical design data or to install a production water well; however, it might be better to use hollow-stem augers or casing advancement to sample contaminated materials, even if auger or casing advancement installation would be more time-consuming and more expensive.

When contaminated sites are investigated, sampling quality without cross-contamination is usually the most important consideration. It is usually the primary goal in subsurface environmental exploration to advance a borehole and obtain a soil or groundwater sample without altering the chemistry of the sampled material by the drilling process, by drilling fluid additives, or by the sampling device and the sampling mechanics. Environmental exploration of and field installation oversite

Geotechnical Practice for Waste Disposal.
Edited by David E. Daniel.
Published in 1993 by Chapman & Hall, London. ISBN 0 412 35170 6

personnel (inspectors) should have a thorough understanding of all generally available drilling methods and techniques, and particularly the possible effects of the drilling method and technique on sample quality.

15.2 AUGER DRILLING

Auger drilling with its many technical and economic advantages was only slowly accepted in geotechnical engineering practice. However, auger drilling has rapidly evolved as a primary method of subsurface exploration at contaminated or potentially contaminated sites.

15.2.1 Continuous flight 'solid stem' auger drilling

Continuous flight, 'solid stem', augers (Fig. 15.1) have been used in geotechnical exploration for many years. They are particularly efficient and technically effective when drilling relatively shallow exploration borings in fine grained soils above the groundwater level. However, it is practically impossible to use solid stem, continuous flight augers with intermittent soil sampling through a chemically contaminated ground surface without downwardly transporting contaminants. (The same problems, probably magnified, occur with the use of 'hand augers'.)

Continuous flight solid stem augers are usually fabricated with steel flights welded to a relatively small diameter steel tube. They usually have 'solid', hexagonal shank and socket, pinned connections (Fig. 15.1, Table 15.1). The flighting is usually 'pitched and timed' to be practically continuous through each hexagonal shank and socket connection of the auger column. A continuous flight auger column is advanced with rotary and axial motion from a drill rig. The articulated auger column is usually comprised of 1.52 m long sections. The auger column must be completely removed from the borehole to permit a sample of soil to be taken from beneath the bottom of the hole.

15.2.2 Hollow-stem auger drilling

The hollow-stem auger drilling method was first used for geotechnical exploration during the 1960s. A hollow-stem auger column has a continuously open axial stem. Hollow auger tools consist of outer components and inner components (Fig. 15.2, Table 15.2). When the borehole is advanced to the depth of sampling, the pilot assembly and center rods are removed. A sampling tube is inserted through the hollow axis of the auger column and then pushed or driven into soil located beneath the base of the string of augers. The sampling tube is

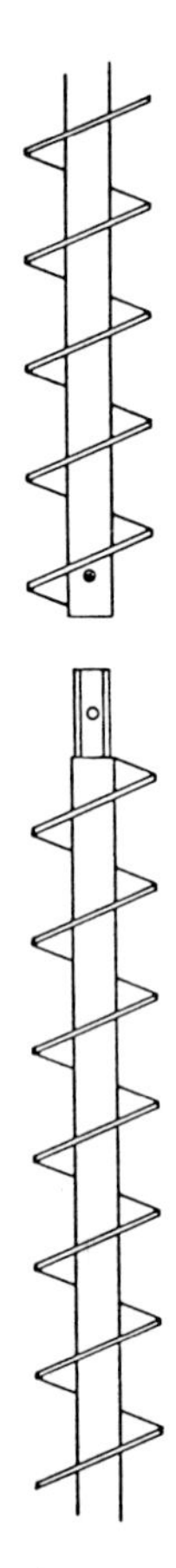

Fig. 15.1 Continuous flight solid stem auger.

then withdrawn from the hollow stem and the pilot assembly reinserted. If a monitoring well is being installed, the casing is inserted through the hollow stem.

Typical hollow auger sections are 1.52 m in length and are available for monitoring well installations with inside, hollow-stem diameters ranging from 83 mm to 207 mm (Table 15.2). Most hollow augers have keyed, box and pin connections for transfer of drilling torque through the coupling and for easy coupling and uncoupling (Fig. 15.3). Hollow augers with threaded connections are available, but lubricants are usually required for the threads. Lubricants may introduce contamination to the soil. If lubricants are used, materials that will not be misinterpreted as contamination must be used.

Hollow-stem augers have been used to advance soil exploration

Table 15.1 Typical continuous flight auger sizes

auger series	*hexagonal shank* *in.*	*(mm)*	*auger diameter* *in.*	*(mm)*	*auger head cutting diameter* *in.*	*(mm)*
1125	$\frac{13}{16}$	(21)	$2\frac{5}{8}$	(67)	3	(76)
1500	$1\frac{1}{8}$	(29)	$2\frac{1}{2}$	(64)	3	(76)
	$1\frac{1}{8}$	(29)	3	(76)	$3\frac{1}{8}$	(79)
	$1\frac{1}{8}$	(29)	4	(102)	$4\frac{1}{2}$	(114)
	$1\frac{1}{8}$	(29)	$4\frac{1}{2}$	(114)	5	(127)
	$1\frac{1}{8}$	(29)	$5\frac{1}{2}$	(140)	6	(152)
2000	$1\frac{5}{8}$	(41)	4	(102)	$4\frac{1}{2}$	(114)
	$1\frac{5}{8}$	(41)	$4\frac{1}{2}$	(112)	5	(127)
	$1\frac{5}{8}$	(41)	$5\frac{1}{2}$	(140)	6	(152)
	$1\frac{5}{8}$	(41)	6	(152)	$6\frac{3}{4}$	(171)
	$1\frac{5}{8}$	(41)	7	(178)	$8\frac{1}{4}$	(210)
2875	$1\frac{5}{8}$	(41)	$7\frac{7}{8}$	(200)	9	(229)
	$1\frac{5}{8}$	(41)	$8\frac{7}{8}$	(225)	10	(254)
	$1\frac{5}{8}$	(41)	11	(280)	12	(305)

Note: auger diameters are without hardsurfacing and should be considered minimum manufacturing dimensions.

borings to depths greater than 100 m but are usually used only to maximum depths of 20–45 m, depending on the density or hardness of the soil, the depth of the groundwater level, the power available at the drill spindle and the dead weight of the drill rig.

Groundwater levels and zones of perched water are usually immediately detected and can be accurately logged during hollow auger drilling. This important observation capability can not be accomplished during drilling with 'wet' drilling methods. When hollow augers are used to drill below the groundwater level, water levels are often controlled within the auger stem to prevent excess seepage stresses and a 'quick condition' from occurring below the auger head. Seepage stresses below the auger head can be controlled by injecting drilling fluids into the hollow augers during drilling. This procedure essentially provides a straight rotary drilling system requiring only a small amount of water to maintain the balancing head within the hollow auger. Under some circumstances the pilot assembly is replaced with a simple 'knock-off' plate on the bottom of the hollow auger head.

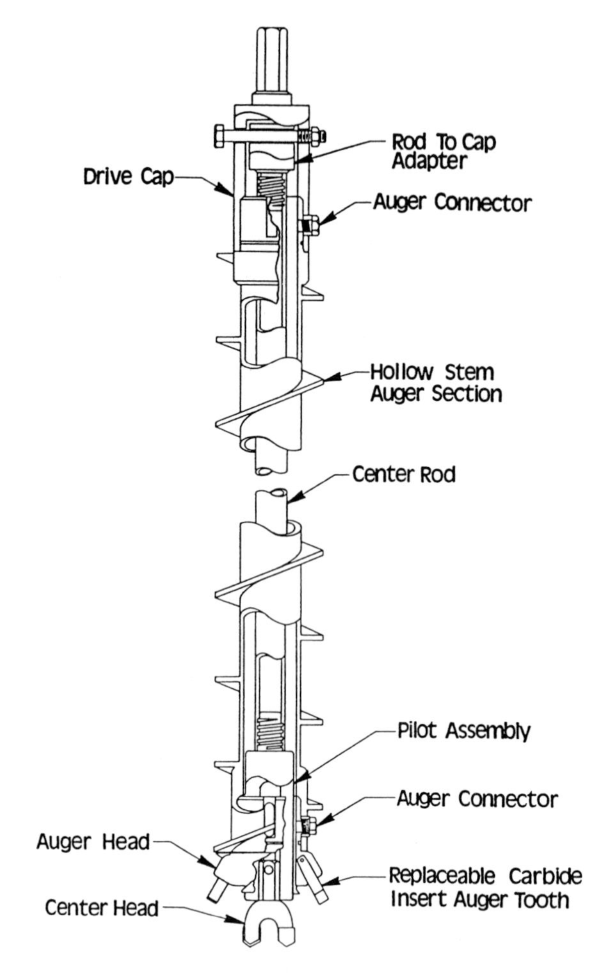

Fig. 15.2 Continuous flight hollow stem auger.

15.3 DIRECT ROTARY DRILLING

Direct (straight) rotary drilling is often called mud rotary drilling; however, the circulation medium of rotary drilling can be clear water, some liquid other than water, or air, with or without additives. Core drilling with liquid or air circulation is a form of direct rotary drilling.

When the 'direct' or 'straight' hydraulic rotary method is used for subsurface exploration, a bit on the bottom of a drill rod column is simultaneously rotated and advanced into the soil or rock. Soil or rock cuttings produced by the drilling process are removed from the borehole with a fluid that is pumped downward through the drill rod

Table 15.2 Typical hollow-stem auger sizes

hollow-stem inside diameter		*fighting diameter*		*auger head cutting diameter*	
in.	*(mm)*	*in.*	*(mm)*	*in.*	*(mm)*
*$2\frac{1}{4}$	(57)	$5\frac{5}{8}$	(143)	$6\frac{1}{4}$	(159)
*$2\frac{3}{4}$	(70)	$6\frac{1}{8}$	(156)	$6\frac{3}{4}$	(171)
$3\frac{1}{4}$	(83)	$6\frac{5}{8}$	(168)	$7\frac{1}{4}$	(184)
$3\frac{3}{4}$	(95)	$7\frac{1}{8}$	(181)	$7\frac{3}{4}$	(197)
$4\frac{1}{4}$	(108)	$7\frac{5}{8}$	(194)	$8\frac{1}{4}$	(210)
$6\frac{1}{4}$	(159)	$9\frac{5}{8}$	(244)	$10\frac{1}{4}$	(260)
$8\frac{1}{4}$	(207)	$11\frac{5}{8}$	(295)	$12\frac{1}{2}$	(318)

Note: auger fighting diameters should be considered minimum manufacturing dimensions.
* Seldom used for monitoring well installations.

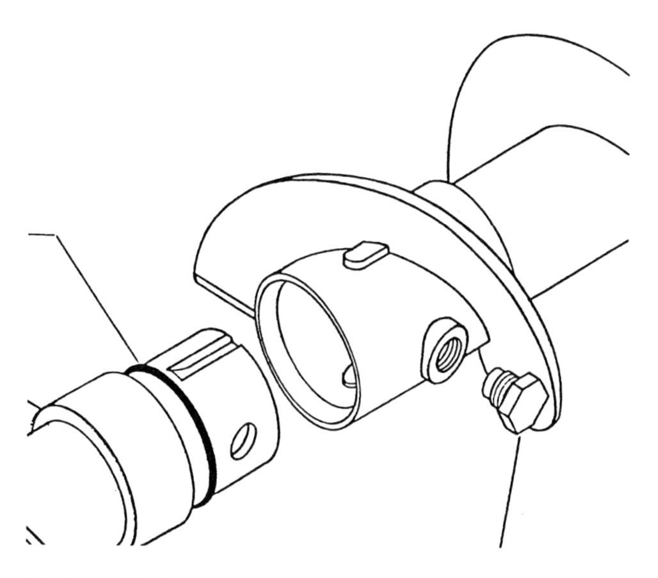

Fig. 15.3 Joining of hollow stem augers.

column, passing through the bit and then upward within the annulus of the borehole and the drill rods (Fig. 15.4).

In comparison to auger drilling the complete rotary drilling process and system is rather complex. The system typically consists of the drill rods and bit, a mud pit, a suction hose, a recirculation pump, a pressure hose to a swivel and a rotation spindle which is a part of the drill rig.

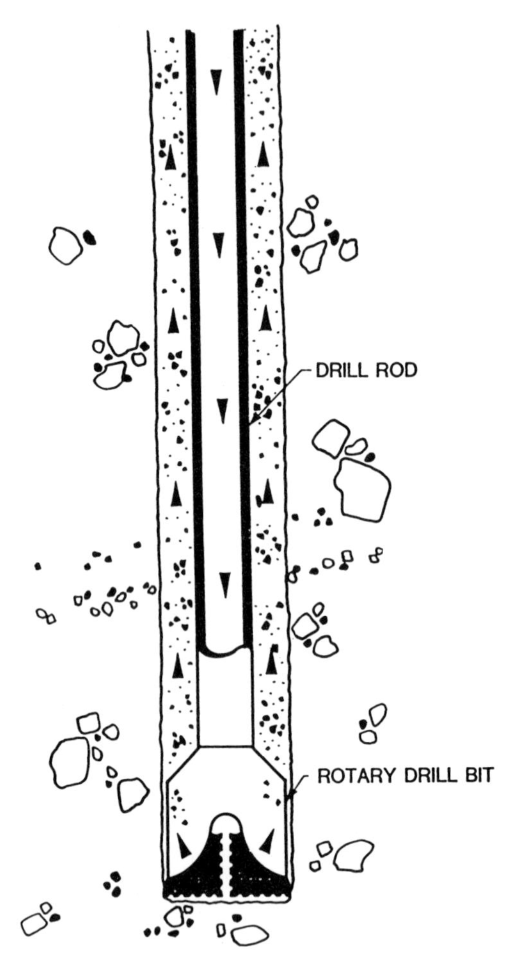

Fig. 15.4 Direct rotary drilling.

15.3.1 Drilling mud

Drilling mud is the liquid circulation medium of 'hydraulic' rotary drilling. It usually consists of water and one or more additives.

Commercially available, processed sodium montmorillonite clay powder (bentonite to a driller) is the most frequently used rotary drilling fluid additive. The higher quality bentonite additives consist of sodium montmorillonitic clay powder with sodium carbonate and polymer additives plus impurities consisting of various other naturally occurring mineral particles.

Other often used drilling fluid additives include standard bentonite (without polymers), cellulose polymers, organic polymers, anionic acrylimide polymers usually in combination with a plyacrylate, sodium carbonate (soda ash), potassium chloride (potash), barium sulfate (barite), and various lost circulation materials such as wood fibers and shredded paper. Extreme care must be taken to ensure that additives or leachable components of additives will not introduce any chemicals that might be misinterpreted as contamination.

A correctly 'designed' drilling fluid with additives can perform some combination of the following as required:

1. it can provide a viscous medium for improved transportation of cuttings from the borehole;
2. it can form a membrane seal to prevent excessive loss of water through the borehole wall;
3. it can react chemically with active clays or shale units to improve bit performance or to deter swelling of clay or shale units along the borehole wall; and
4. it can clean and cool the bit.

15.3.2 Viscosity for cuttings transport

Pure water without additives is the 'ideal' drilling fluid but can be used only under ideal circumstances. For example, when competent rock, other than a swelling shale is drilled with a core barrel, water without additives is usually a very adequate drilling fluid. Because the annulus of a cored hole is usually relatively small (if the drill rods are size-matched to the core barrel) and the cuttings generated by coring bits are usually small, upflow velocities adequate for cuttings transport are obtainable at relatively low fluid pumping rates. The upflow velocity within the borehole annulus can be computed by dividing the pumping rate by the section area of the annulus. When boreholes in rock are cored and when the drill rods are size matched to the core barrel, e.g. NW rods with NW barrel or when wireline rods are used, upflow velocities greater than 20 m/min are readily attainable and usually adequate for cuttings transport without using drilling fluid additives.

When the soil or rock is drilled for geotechnical exploration or monitoring well installations using the direct rotary method, both the borehole annulus and the cuttings are usually larger than with core drilling. To provide transport capability of the larger cuttings within the larger annulus without using a pumping rate that will erode the wall of the borehole, the viscosity of the drilling fluid is usually increased by adding bentonite powder or a polymer or possibly both.

A quick measure of the viscosity of drilling liquids can be obtained with a Marsh funnel. The time in seconds for one liter of liquid to flow by gravity through a standard orifice in the funnel provides a quick and easy measure of viscosity. Water has a funnel viscosity of about 26 s. Typical drilling fluids have Marsh funnel viscosities of 36–60 s, although 35–40 s is often considered ideal. In other words, typical drilling fluids are often too viscous, but sometimes with good purpose.

When drilling fluids are too viscous, cuttings do not adequately separate or 'settle out' in the mud pit but are recirculated through the pump, swivel and drill rods. Recirculation of cuttings can damage components of the pump and restrict, if not totally block, the recirculation process.

15.3.3 Borehole support

If a boring is drilled using water without additives through a soil or rock of low hydraulic conductivity, water will not be lost to the ground and the water in the borehole will exert a pressure on the wall of the borehole of approximately 2.9 kPa per meter of depth below the fluid surface. This water pressure, and almost nothing else, keeps the borehole open. If a boring is drilled into relatively permeable soils, the water will apply the same magnitude of pressure to the borehole wall; however, at the same time the water will seep into the soil. When circulation is stopped, or when the source of water is depleted, the water level will recede within the borehole, seepage will reverse from the ground to the borehole, and the borehole walls will usually collapse. However, a high quality bentonite or some other additive can be used to form a 'mud cake' on the borehole wall, impeding fluid loss, thus maintaining the liquid level and the supporting pressure on the borehole wall. The mud cake will also impede the flow of water through a well screen and filter pack placed in a borehole rotary drilled to the complete well depth.

The occurrence of artesian groundwater pressures at depth can cause the net supporting pressure on the borehole wall to approach zero or even cause out-flow at the top of the boring. A weighting additive such as barite (barium sulfate) can be added in controlled amounts to maintain borehole support or in severe cases prevent artesian flow. The addition of too much barite, however, can raise fluid densities excessively high and contribute to hydraulic fracturing of ground. Barite is usually added to bentonite base mud systems that have an initial funnel viscosity in excess of 50 seconds in order to maintain the barite in suspension.

15.3.4 Inhibiting clay activity

The same reaction of water with sodium montmorillonite that produces a borehole viscosifier or sealer from high quality 'bentonite' powder can cause a shale stratum along the borehole wall to swell against the core barrel or against the drill rods causing the barrel to 'lock up' in the hole or even 'twist off'. It can also cause a build-up of a ring of soil cuttings called a 'boot' on the outside of a core barrel or on drill rods. And, it can contribute to smearing or plugging of the bit. These undesirable occurrences are often 'inhibited' with the use of polymers or potash in the drilling fluid. Carboxymethylcelulose (CMC) base polymers can be used alone as an inhibitor or in combination with bentonite powder, in which case the bentonite must be the first additive mixed with the makeup water. Muriated potash can also be used alone as an inhibitor or in combination with a polyacrylamide viscosifier, in which case the potash should be first totally dissolved in the water before adding polymer.

15.3.5 Air rotary drilling

There are acknowledged advantages and disadvantages to the overall concept of drilling with air. The main advantages are that there is no cost for air and air is readily available at the drilling site. The main disadvantage involves the costs of pressurizing (pumping) the air before using it and cleaning the air before and after using it.

Air rotary drilling for geotechnical exploration is usually limited to depths of about 75 m and for drilling in soil or rock above or only a short distance below the groundwater level. Air rotary drilling is used to much greater depths using large compressors for other purposes such as groundwater development. An air rotary drilling system is similar to a hydraulic rotary system. The mud pit, suction hose and pump are replaced with an air compressor and a dust collector and usually some kind of air discharge cuttings removal system.

For air rotary drilling, air is the circulation medium. Air provides hole cleaning and cools the bit but cannot, even with foaming additives, provide borehole stability. When air is used for straight circulation, a minimum upflow air velocity of about 1200 m/min is often stipulated for adequate hole cleaning; however, upflow velocities of much less than 1200 m/min have been successfully used. The adequacy of a compressor to provide enough air is determined by dividing the capacity of the compressor by the theoretical, uneroded section area of the borehole annulus. Often the true annular area becomes oversized from air blast erosion. Larger air flows are then required to transport cuttings through the enlarged zone. The use of foaming agents can reduce the air requirements by as much as one-half.

To increase upflow velocity, the borehole annulus can be reduced by using larger drill rods but keeping the bit size the same; however, a minimum radial clearance of about 10 mm is usually maintained for passage of larger 'chips'.

A 4–9 bar compressor is usually adequate when NX, NW or larger drill rods are used for drilling typical small diameter geotechnical exploration borings above the groundwater level to depths of less than 75 m. External flush drill rods should be used to minimize turbulence and erosion of the borehole annulus.

15.4 CORE DRILLING OF ROCK AND SOIL

For many circumstances, core drilling is the preferred drilling method for exploration in rock. Core observation allows optimal evaluation of rock characteristics and placement of sampling devices.

A core barrel is rotated and simultaneously axially advanced with rotary and axial motion applied through a drill rod column from a drilling machine. A drilling fluid flushes and cools the bit, transporting the rock or soil cuttings upward and out of the drilled hole. If the procedure is successful, the rock or soil at the advancing end of the core barrel is cut or abraded by the bit into a cylinder that enters and is protected by the hollow axis of the core barrel (Table 15.3).

This description seems simple enough; however, there is little to be called simple about core drilling. There are a multitude of variables that severely complicate the process. The variables are usually more numerous when core drilling is performed to obtain geotechnical design parameters in shallow, often weathered or broken rock, than when core drilling is performed for other purposes such as deep mineral exploration. The geotechnical investigator or the monitoring device installer is usually interested in the structural continuity and

Table 15.3 Common coring bit sizes

size designation	*bit set O.D. mm*	*bit set I.D. mm*	*reaming shell O.D. mm*
EX	37.7	21.5	37.7
EWD3	37.7	21.2	37.7
AX	48.0	30.1	48.0
AWD4, AWD3	48.0	28.9	48.0
AXCWL, AWC3, AQWL	48.0	27.1	48.0
BX	59.9	42.0	59.9
BWD4, BWD3	59.0	41.0	59.9
BWBWL, BWCE, BQWL	59.9	36.4	59.9
NX	75.7	54.7	75.7
NWD4, NWD3	75.7	52.3	75.7
NXBWL, NWC3, NQWL	75.7	47.6	75.7
NC	92.7	69.5	92.7
HWD4, HWBWL, HWD3	92.7	61.1	92.7
HQWL	96.3	63.5	96.3
CP, PQWL	122.1	93.1	122.1

mass permeability of shallow rock units which are commonly fractured or jointed and difficult to core drill. At greater depths where rock is less frequently fractured or jointed, and becomes relatively easy to core drill, there is usually no further need for geotechnical or monitoring design information. There is a further complication in that geotechnical or water well drillers are often rarely required to core drill in rock.

15.4.1 Core drilling systems

Drill rigs that are designed and manufactured specifically for core drilling can often be used secondarily but usually less efficiently for other purposes such as auger drilling or straight rotary drilling of soil or rock. Other drill rigs, however, particularly multipurpose auger-core-rotary drills, that are designed and manufactured specifically to efficiently perform drilling other than core drilling are often used secondarily and sometimes less efficiently for core drilling. The capability to core a particular rock unit is related to the selection of the bit and six very important drilling system factors:

1. the range of 'vibration free' rotary velocity available at the spindle of the drill rig;
2. the capability to control the rate of axial feed of the spindle;
3. the force reaction characteristics of the drill rig and the capability to control and limit the rock force reaction, therefore, the force on the bit;
4. the design and condition of the core barrel;
5. the type, size, section length and condition of the drill rods; and
6. when drilling with liquids, the pumping characteristics of the circulation pump including the capability to control the flow of the drilling fluid – when drilling with air circulation, the performance characteristics of the air compressor are equally significant.

The combined characteristics of the complete system and the capability of the drill crew all contribute to the performance of the core barrel and the bit, and the quality of the recovered core.

15.4.2 Conventional core barrels

Several core barrel designs have been standardized by the Diamond Core Drill Manufacturers Association (DCDMA). However, the most recent core barrel designs, which usually provide the highest quality core and the most efficient coring capability, are not DCDMA standard barrels. It should be understood that unique improvements to any drilling tool system are usually patented. Such improvements are seldom 'standardized' until the patents expire, by which time, invariably, further patented improvements are made.

15.4.3 Wireline core barrel

Wireline core barrels are similar to conventional core barrels with the distinguishing exception that the inner tube assembly with core encased can be removed from the outer tube with an 'overshot' device lowered into the borehole on a wireline and hoisted to the surface through open coupling, wireline drill rods. The outer barrel assembly and the wireline drill rods remain in place in the hole during core removal. After reinserting the same or another inner tube assembly, coring can be resumed. A wireline core barrel is very similar to a conventional core barrel.

There are several proprietary wireline latching and overshot systems. Most work very well. A latching and overshot system should provide not only the basic capability of efficiently removing and reinserting the inner tube assembly but also the driller assurance that the inner tube assembly is positively latched in the correct position within the head and the outer tube when and if the inner tube is lowered into place with a lowering tool.

Drillers often use two inner tube assemblies so that an alternate inner tube can be lowered or 'dropped' into the core barrel and drilling resumed while the core is being inspected or removed from the other inner tube.

15.4.4 Split inner tubes

The solid inner tube of conventional and wireline core barrels can be replaced with a split inner tube that is held together by nylon reinforced tape wrapped around the barrel in full circle tape slots. After the split tube is removed from the outer barrel, the tape on the inner tube is cut and half of the tube is removed to allow inspection and logging of the core in the configuration that it entered the barrel.

Another type of split inner tube system has an intermediate inner tube which encases and holds the split tube together. Following a core run, the split tube and core is hydraulically extruded from the intermediate inner tube, usually with the same drilling fluid pump that is used to drill the hole. These barrels are called triple tube core barrels.

Both split inner tubes and triple tubes are available for use with both wireline and conventional barrels.

15.4.5 Coring bits

The coring bit is the tool that is attached to the bottom of a core barrel such that if a rotary action, an axial advance and a flushing fluid are properly applied through the drill rods, the rock (or soil) will be cut into a cylinder that enters the core barrel.

Bits are manufactured in many sizes and configurations. Accordingly, the nomenclature of coring bits is quite complex. Bits can be generally classified as:

1. diamond-surfaced set;
2. diamond-impregnated;
3. carbide insert; and
4. polycrystaline diamond insert.

Most geotechnical and particularly mineral exploration core drilling is now performed with impregnated diamond bits. However, surface-set diamond bits are often used and were the first type used for core drilling. 'Surface-set' means that individual diamonds are set on the face and sides of the crown of the bit. The nomenclature of coring bits evolved to describe the details of surface set bits.

For most rock coring situations the use of impregnated bits is much more economical than the use of surface-set bits, i.e. for drilling in most hard and abrasive igneous and metamorphic rock units and in the hardest of sedimentary rocks. The diamonds of impregnated bits are distributed within the matrix crown. These diamonds are usually the manufactured type and, therefore, are much smaller than the mined diamonds that are used for surface set bits.

Some bits have tungsten carbide cutters to core soil or relatively soft, sedimentary rock. These bits usually have wide kerfs to cut an 'oversized' hole for easier passage of large cuttings. Some harder sedimentary rocks can be efficiently cored with similar coring bits that have cutting inserts each of which consists of a thin sheet of manufactured (polycrystaline) diamond bonded to a carbide insert.

15.4.6 Soil sampling core barrels

Some rock core barrels can be used as soil sampling core barrels if the appropriate bits are used. In addition there are core barrels that are specifically designed and identified as soil sampling core barrels. The two most popular soil sampling barrels are the Denison type and the Pitcher type.

The Denison type barrel was first used by the Denison (Texas) District of the US Army Corps of Engineers. It has an inner tube with a shoe or inner tube extension that can be set approximately flush with the face of the bit or can be extended variable distances beyond the face of the bit. The variable inner tube extension was initially accomplished by varying the length of the shank of the bit. The extensions on recently manufactured Denison barrels are made by using different lengths of inner tube shoes. The shoe can be set about 75 mm ahead of the face of the bit for sampling the softest materials and flush with the face of the bit when the soil (or rock) is hard enough to require the coring action of the bit for penetration. Denison barrels are usually manufactured with either a 0.6 m or a 1.5 m nominal length barrel.

The Pitcher type barrel was initially manufactured by the Pitcher Drilling Company in California, and is very similar to the Denison barrel. The basic difference is that the Pitcher barrel has a coil spring between the inner tube and the swivel, such that a single length, thin wall inner tube is automatically advanced variable distances in front of the face of the bit according to the hardness of the soil being sampled. When relatively soft ground is encountered during coring, the force of the spring causes the inner tube to advance ahead of the face of the bit. When relatively hard ground is encountered, the spring is compressed and the leading edge of the sample tube is pushed back. It can be pushed back as far as the face of the bit. Pitcher type barrels are usually manufactured in 1.5 m or shorter lengths.

Soil sampling core barrels are usually operated at rotary velocities in the range of 50–150 RPM. This range of rotary velocity minimizes drilling vibrations and gives the driller better capability to control the rate of penetration and the circulation fluid pumping rate.

Some of the bits used on soil sampling core barrels can produce relatively large cuttings. Therefore, larger upflow velocities are required within the borehole annulus than are usually required for coring rock. An adequate upflow velocity may not be obtainable when using water without viscosifier additives because of the often limited capacity of the fluid pump that is available on an exploration drill rig. For these circumstances, drilling fluid additives must be used to increase the viscosity of the drilling fluid. The increased viscosity increases cuttings transport capability but decreases settling rates of cuttings in the recirculation pit. The consequent recirculation of cuttings can cause plugging of the core barrel.

The usual solution to the cuttings transportation problem is to use as large a drill rod as available and practical, to use a well designed fluid recirculation pit of adequate size and to control the drilling fluid viscosity to allow the settling of cuttings. For many ground conditions, drilling fluid additives may be required not only to remove the cuttings but to seal the borehole wall for stability.

15.4.7 Core drilling with air circulation

Core drilling with air is generally limited to coring of soil and sedimentary rocks that can be scratched with a steel knife blade.

When air is used as the principal bit coolant and circulation medium, large surface set diamonds or carbide type wide kerf bits are usually used. The use of wide kerf bits is often necessary to prevent a ring of cuttings, a 'boot', from building up on the outside of the core barrel or on the outside of the drill rocks.

When coring with air, the bit is usually rotated at a rotary velocity that will provide a circumferential bit velocity of about 10–20 m per minute. These slow velocities are about 1/4 to 1/10 of the velocity that would be used to core the same rock with liquid circulation. Nevertheless, greater penetration rates than can be attained with liquid circulation often are the result. It is believed that the absence of liquid from the surface of the rock eliminates a normal stress effect and allows the rock to be more easily sheared. An annulus upflow velocity of about 900–1500 m per minute is usually required for air coring.

15.5 ROTARY PERCUSSION DRILLING

There are two basic types of rotary percussion drills. Rotary percussion drilling involves:

1. variations of systems that utilize surface operated air-powered or hydraulically-powered rotary-impact rock drills; and
2. variations of systems that utilize down-the-hole air-operated rotary-impact hammers.

Air and hydraulic drifters can be used to drill holes in the size range of about 25–125 mm in diameter. Currently available down-the-hole air hammers can be used to drill holes in the size range of about 100 mm to greater than 400 mm in diameter. Down-the-hole air hammers can be used with most multipurpose or straight rotary drill rigs.

Both down-the-hole hammer systems and rotary impact drifters employ the same basic drilling mechanism. A bit on the end of an articulated hollow drill rod column is rotated relatively slowly while being axially oscillated at a relatively rapid rate. The action of the bit hammering on the rock face causes the rock to fracture in tensile rebound following each compressive impact loading. There are two general types of percussion bits:

1. those with 'X' shaped impact surface (called full face bits); and
2. those with spherical carbide inserts (called button bits).

Under most circumstances, bits with carbide buttons will advance in rock at a much faster rate than full faced bits of equal diameter.

15.6 PERCUSSION DRILLING

Percussion or cable-tool type drilling, once the principal method of the water well industry, is seldom used in the US for geotechnical exploration mainly because of economic factors. Cable-tool drills are sometimes used in geotechnical exploration and water well drilling practice to advance casing, particularly on environmental exploration projects where drilling fluids are not permitted. Sometimes a cable-tool drill and a drill with augering capability can be used in combination to advance and case a large diameter borehole for installation of a groundwater sampling or monitoring device.

A cable-tool drill rig consists of a horizontal walking beam that is operated off a rotating crank such that the walking beam with attached cable sheaves is used to raise and drop a heavy string of impact drilling tools. Each raising and dropping of the tool string provides one impact of a percussion drilling bit on the rock face at the bottom of the borehole (or one impact to a casing column).

When sufficient drilling progress has been made, the drilling tool string is withdrawn and the rock cuttings are removed with a bailer. The progress of drilling is relatively slow and inefficient with respect of rotary methods.

15.7 CASING ADVANCEMENT METHODS

For some ground conditions casing advancement (and retraction) methods may provide the most technically acceptable method of installing a monitoring well. There are various rotary and rotary percussion drilling systems for installing casing. Sometimes they work very well and sometimes the problems during drilling are insurmountable. Cable-tool drills can also be effectively used to drive and retract casing. Casing clean out is accomplished with dual tube reverse-air circulation, or by bailing the hole, or with a rotary or washing process. There are a very limited number of powerful, 'drill-through' casing drivers available in North America. Casing advancement methods are most applicable to the case of monitoring in very deep soils.

15.8 SOIL SAMPLING EQUIPMENT

Soil sampling devices consist of tubes that are pushed or driven into the soil beneath the bottom of the borehole. To sample with a hollow-stem auger, the sampling tube is lowered down the center of the hollow auger and then pushed or driven into the underlying soil. With other drilling methods, the drilling tool must be removed from the hole and the sampling tube lowered down the boring on a separate pipe or wire. In either case, there is a risk that the sampling tube will become contaminated from soils or liquids that overlie the sampling point.

Samples may be taken at various frequencies. In traditional geotechnical subsurface exploration, samples are obtained at an interval of approximately 1–2 m, or at a significant change in stratigraphy. For environmental exploration, continuous sampling or sampling at very close intervals is common because infrequent sampling can cause the exploration team to miss a thin contaminated zone or an important hydrogeologic anomaly.

Soil sampling devices may be grouped into two categories:

1. devices for obtaining high-quality, 'undisturbed' samples; and
2. devices for obtaining disturbed samples.

The degree of disturbance may be evaluated from the area ratio (A_R), which is defined as follows.

$$A_R = \frac{\text{cross-sectional area of sampler}}{\text{cross-sectional area of sample}} \times 100\% \qquad (15.1)$$

$$= \frac{\dfrac{\pi D_o^2}{4} - \dfrac{\pi D_i^2}{4}}{\dfrac{\pi D_i^2}{4}} \times 100\% = \frac{D_o^2 - D_i^2}{D_i^2} \times 100\% \qquad (15.2)$$

where D_i is the inside diameter of the sampling tube and D_o is the outside diameter. A sampling tube that provides 'undisturbed' samples should have an area ratio of $\leqslant 11\%$, e.g., as required in ASTM D1587. Sampling tubes with larger area ratios may cause significant disturbance of the soil and should not be regarded as undisturbed samplers. Samplers that do provide 'undisturbed' samples are often called **thin-walled tube samplers** or **Shelby tube samplers**.

15.8.1 Thin-walled tube samplers

In the US, thin-walled tube samplers are readily available in nominal diameters of 50, 76, and 127 mm, with corresponding wall thicknesses of 1.24, 1.65, and 3.05 mm. The usual length is nominally 750 mm. The sharpened sampling end of the tube is usually crimped to a slightly smaller inside diameter than the rest of the tube to provide inside clearance that reduces sliding resistance between the inside of the tube and the soil sample. The inside clearance ratio, C_R, is defined as follows.

$$C_R = \frac{D_i - D_e}{D_e} \times 100\% \qquad (15.3)$$

where D_i is the inside diameter of the sampling tube and D_e is the inside diameter between machined faces at the sampling end of the tube. If the sampling end of the tube has not been crimped at all, then

$D_i = D_e$ and $C_R = 0$. More typically, some inside clearance is provided and C_R is between 0.5 and 1.5%. The ASTM practice for undisturbed sampling (D1587) requires that $C_R = 1\%$.

Thin-walled sampling tubes are normally pushed into the subsoil to minimize disturbance. Sometimes the sampler is driven, but driving may increase disturbance. The **recovery ratio** is defined as the ratio of the length of the plug of soil sample retained in the sampling tube divided by the distance the sampler was pushed or driven into the soil. Under ideal circumstances, the recovery ratio will equal 1.0, more often than not the sliding resistance between the sampled material and the tubing wall will reduce the recovery ratio to much less than 1.0 (soil beneath the plugged tube is pushed away from the tube).

Once the thin-walled sampling tube has been returned to the surface, it may be handled in one of several ways. One technique is to extrude the sample in the field and then carefully wrap and seal the sample in a way that minimizes water loss or (for contaminated soil) loss of any volatile constituents. Another technique is to seal the ends of the tube with wax or special end caps to prevent moisture loss. If volatiles are present, the sample is normally sealed in a way that keeps air in the tube (termed **head space**) to an absolute minimum. The samples are then transported back to the laboratory. If the soil samples will be subjected to tests aimed at defining *in situ* properties (e.g., shear, consolidation, or hydraulic conductivity tests), the test specimens or sampling tubes containing the specimens must be carefully packaged and transported in a manner that causes minimal disturbance (see ASTM D4220).

15.8.2 Disturbed samples

Disturbed samples may be taken for many purposes, e.g., determination of major soil constituents, analysis of water content, or analysis of contamination. The advantage of obtaining a disturbed sample is that a thick-walled sampling tube can be used; the thick-walled tube is robust and can be used many times before replacement. Also, the tube can contain small, thin tubes (called **liners**) on the inside of the thick-walled sampling tube. The soil is contained in the liners, which are extremely convenient for storing the samples during transport back to the laboratory. The liners can be made of inert materials, e.g., stainless steel, that will not react chemically with the contaminants.

The most common type of sampling device for obtaining disturbed samples is the **split-barrel sampler**. A split-barrel sampler is split in half down the length, but for sampling the two halves are clamped together at both ends by a sampling **shoe** and sampling head that threads into the sampling rod. The split barrel sampler is convenient because when the sampler is removed from the borehole, the shoe and head are unscrewed and then the two halves of the sampling barrel are separated, revealing the soil sample.

Sampling tubes that obtain disturbed samples come in many lengths and diameters. One very commonly-used sampler is the standard split-spoon sampler, which has an inside and outside diameter of 34.9 and 50.8 mm, respectively. The standard split-spoon sampler may be driven into the soil with a 64 kg hammer to obtain the penetration resistance, which is known as the standard penetration **N value** (ASTM D1586). Larger-diameter samplers (typically about 50–75 mm in diameter) are commonly used, particularly for sampling with liners inside the sampling barrel. It is becoming common practice to use sampling tubes with a length of 1.5 m inside hollow-stem augers to sample while the auger is advanced.

15.8.3 Cleaning sampling equipment

The following guidelines are recommended for sampling contaminated, or potentially contaminated, soil. As a minimum, sampling equipment should be washed with a non-phosphate detergent solution and then rinsed with a control water. If inorganic contaminants are present, a series of acid-control water rinses may be needed, with the acid serving as a desorbing agent. If organic contaminants are present, an organic desorbing agent (such as isopropanol, acetone, or methanol) should be included in the rinse sequence. The reader may wish to consult ASTM D5088 for further information.

15.9 CONTAMINANT CONTROL

An important consideration in subsurface exploration is control of contaminant migration within the borehole. Contaminants can be transported vertically in the borehole, either upward or downward. Vertical migration of contaminants is undesirable if one is trying to delineate contaminated and uncontaminated zones.

A most common problem during any drilling process is the transport of contaminated materials downward in a borehole. It is practically impossible, without casing the upper part of the hole, to prevent cross contamination of materials when drilling through contaminated ground.

The suggested technique for drilling through contaminated materials into uncontaminated materials is as follows:

1. drill to a depth slightly greater than the depth of suspected contamination;

2. set casing to the depth of the base of the borehole; and
3. continue drilling a smaller diameter hole through the casing.

This technique can be slow and expensive, but it provides a reasonably effective way to minimize downward transport of contaminants. In some cases, more than one casing may be set, with each successive casing having a smaller diameter. In addition, one may need to drive casing while drilling or employ other non-standard techniques.

Contaminants may also be transported upward during drilling. It is possible to contaminate near-surface materials that might otherwise not be contaminated. Again, casing may be needed to prevent such contamination, particularly if the source of contamination is ground water that might seep laterally into an uncontaminated vadose zone.

15.10 HEALTH AND SAFETY CONSIDERATIONS

Drilling into contaminated materials represents a serious hazard to the drillers and others near the drilling operation. Health and safety considerations must be given the highest priority on all projects involving the possibility that contaminants will be encountered. In general, the procedure is to assume the worst until data are collected to indicate that less severe conditions exist. Under worst-case conditions, a separate air supply is maintained for all the drilling crew and anyone near the drilling operation, and the highest degree of skin protection is provided. If conditions are less severe, face masks with respirators may prove adequate for breathing. In any case, air quality is normally monitored continuously (e.g., for methane if drilling into old municipal solid waste or volatile organics if drilling into soil contaminated with chemical waste) in case an unanticipated and dangerous situation is encountered.

In the US and most other industrialized countries, extensive training courses are available and are required for drillers, borehole loggers, and observers. The problem is not so much in identifying appropriate health and safety practices as it is in making sure that workers are fully aware of those practices and are willing to take required precautions despite the often significant inconvenience and personal discomfort involved.

CHAPTER 16

Vapor analysis/extraction

Lyle R. Silka and David L. Jordan

16.1 INTRODUCTION

Interest in the fate, behavior, and remediation of volatile organic compounds (VOCs) in unsaturated soil, or the vadose zone, has increased rapidly over the past five years. The US Congress' Office of Technology Assessment (OTA, 1984) identified VOCs as being one of the more ubiquitous groups of hazardous chemicals present in contaminated ground water nationwide. A major reason for this is the widespread use of VOCs in the manufacture of pesticides, plastics, paints, pharmaceuticals, solvents, and textiles. Since volatile constituents of petroleum, as well as a variety of synthetic solvents, comprise a significant portion of the contamination cases encountered today, it is no wonder that such interest has been born.

Research and development of technologies directed towards the identification and clean-up of volatile organics in soil and groundwater has led to the application of a variety of principles and techniques to remediation, both off-the-shelf and innovative. Research has been conducted into the behavior of volatile compounds in the solid, liquid, and vapor phases of soil and groundwater systems. In addition, computer modeling has been applied in order to understand better the complex interrelationships of the many factors affecting the migration and fate of VOCs in the subsurface.

This chapter focuses on the basic principles governing the behavior and fate of VOCs in the vadose zone, methods for detection and identification of VOCs, and on remedial technologies relying on vapor-phase transport.

Geotechnical Practice for Waste Disposal.
Edited by David E. Daniel.
Published in 1993 by Chapman & Hall, London. ISBN 0 412 35170 6

16.2 VAPOR TRANSPORT IN THE VADOSE ZONE – THEORY

When a liquid VOC is spilled on the soil or leaks from a tank into the soil, the VOC partitions among the liquid and vapor phases and becomes dissolved in soil moisture and adsorbed onto the surfaces of soil minerals and organic matter. The degree of partitioning of the VOC among these four components will depend on the volatility and water solubility of the VOC, the soil moisture content, and the type and amount of soil solids, i.e., the minerals and organic matter.

16.2.1 Partitioning of VOC between pure liquid and soil gas

Partitioning between the pure liquid and soil gas is controlled by the vapor pressure of the VOC and the temperature (Thibodeaux, 1979). At equilibrium, the mole fraction of a VOC in the air space above the pure liquid VOC at a specified temperature is expressed as

$$y_a = \frac{p_a}{p_T} \tag{16.1}$$

where y_a is the mole fraction of chemical a, p_a is the vapor pressure of chemical a, and p_T is the total pressure in the air space.

Vapor pressures for many VOCs at ambient temperatures are available in the literature (Perry and Chilton, 1973; Callahan *et al.*, 1979; Verschueren, 1983; Montgomery and Welkom, 1990). Kerfoot (1991) has evaluated the effects of temperature on the partitioning which is discussed at the end of this section.

16.2.2 Partitioning of VOC between soil gas and soil moisture

Partitioning between the VOC vapor in the soil gas and VOC dissolved in soil moisture may be expressed as, K_H, the ratio of its concentration in each of the two phases.

$$K_H = \frac{C_G}{C_L} \tag{16.2}$$

where C_G is the concentration of the VOC in soil gas, and C_L is the dissolved concentration of the VOC in the water phase. At equilibrium, this ratio is constant for constant temperature and is referred to as Henry's Law constant (Thibodeaux, 1979).

Henry's Law constant may also be expressed as a function of the VOC vapor pressure, the concentration of the VOC in water, and temperature as (Thibodeaux, 1979).

$$K_H = \frac{16.04 p_a M_a}{T C_L} \tag{16.3}$$

where M_a is the gram molecular weight of the VOC, T is the temperature (in kelvin), and the other parameters are as previously defined.

Dilling (1977) reports values of Henry's Law constant for numerous chlorinated solvents. Values of K_H for selected VOCs are presented in Table 16.1. Empirically derived values of Henry's Law constants reported by Dilling (1977), Swallow and Gschwend (1983), and Lappala and Thompson (1983a) are in reasonable agreement with the calculated values of K_H, keeping in mind the temperature dependence of K_H.

Ong and Lion (1991) found through empirical tests that sorption of VOCs at low vapor concentration (<10 mg/l) into soil moisture can be accounted for by the linear partitioning process as described by Henry's Law. However, they found that at very low moisture contents (equivalent to less than eight monolayers of water), the sorption isotherm became nonlinear, with the observed partitioning into the water phase being higher than that predicted by Henry's Law. Ong and Lion (1991) theorized that the increased sorption at low moisture content may be due to sorption onto the bound water on mineral surfaces, as opposed to being solely dissolved. To expand on this theory, the observed increase in VOC sorption by soil moisture over that predicted by Henry's Law may be attributable to a greatly expanding surface area of water across which partitioning may occur as soil moisture decreases below a critical value, i.e., a nonlinear relationship

Table 16.1 Reported values of Henry's Law constant, vapor pressure, and solubility at 25 °C for selected chlorinated solvents

chemical	*solubility in water (ppm)*	*vapor pressure (mm Hg)*	*Henry's Law Constant (dimensionless)*	
			calculated	*found*
1,1,2,2-Tetrachloroethane	3000	6.5	0.019	NA
1,1,2-Trichloroethane	4420	23	0.038	NA
1,1-Dichloroethane	8700	82	0.050	0.04*
Tetrachloroethylene	140	18.6	1.2	0.50† 0.43*
Trichloroethylene	1100	74	0.49	0.33‡
trans-Dichloroethylene	6300	326	0.27	NA
cis-Dichloroethylene	3500	206	0.31	NA

NA: not available. *Source*: Dilling (1977).
* Empirical values reported by Dilling (1977).
† Empirical values will vary from calculated values due to differences in temperature.
‡ Empirical values reported by Lappala and Thompson (1983a).

may exist between soil water-surface area and soil moisture content. Mineral morphology may also play an important role in the observed increased sorption rates. The data of Ong and Lion (1991) indicate that the critical soil moisture content below which vapor-water partitioning becomes nonlinear varies according to soil type. While the critical soil moisture for clay minerals such as montmorillonite and kaolinite appears to be eight monolayers, that for alumina is much lower, as low as four monolayers of water molecules. However, these low moisture contents will be limited principally to arid regions and shallow soil during dry months in humid regions.

16.2.3 Partitioning of VOC between soil moisture and soil solids

Ong and Lion (1991) describe sorption of VOCs onto soil solids from soil vapor as a two-step linear process for soil moisture above a critical value. Above that soil moisture content, soil solids will be coated with layers of water molecules. VOC vapor will partition from the vapor phase into the liquid water phase. Once in the liquid water, some of the VOC will be adsorbed onto the soil mineral and organic matter. At equilibrium, the degree of partitioning between the soil solids and the soil moisture has been expressed as the linear isotherm (Roberts and Valocchi, 1981)

$$K_D = \frac{S}{C_L} \tag{16.4}$$

where K_D is the partition coefficient or distribution coefficient (with units of l^3/m), S is the mass of chemical adsorbed per unit dry mass of soil solids, and C_L is the concentration of the chemical in the soil moisture.

It has been observed that strongly hydrophobic organic chemicals tend to adsorb more strongly onto the soil solids (Roberts and Valocchi, 1981). Empirical studies by Karickhoff *et al.* (1979) and Karickhoff (1984) found that K_D was proportional to the organic carbon content of the soil as well as the octanol:water partition coefficient (K_{OW}) which is a measure of the hydrophobicity of an organic chemical. For the equilibrium condition, this relationship has been expressed by the following equation (Karickhoff *et al.*, 1979) and the essentially same relationship found by Rao *et al.* (1985).

$$K_D = 0.63 K_{OW} f_{OC} \tag{16.5}$$

where K_D is the distribution coefficient of Eq. (16.4), f_{OC} is the soil organic carbon content, and K_{OW} is the octanol:water partition coefficient.

The amount of carbonaceous matter in the soil is the dominant factor controlling the extent of adsorption of organic chemicals. Karickhoff

Table 16.2 Reported values of octanol: water partition coefficient and calculated values of partition coefficient for selected chlorinated solvents

chemical	*octanol: water partition coefficient**	*calculated partition coefficient†, fraction organic carbon*	
		0.001	*0.01*
1,1,2,2-Tetrachloroethane	2.56	0.23	2.3
1,1-Dichloroethane	1.79	0.04	0.4
Tetrachloroethylene	2.88	0.48	4.8
Trichloroethylene	2.29	0.12	1.2
trans-Dichloroethylene	1.48	0.02	0.2
cis-Dichloroethylene	1.48	0.02	0.2

* From Callahan *et al.* (1979).
† Calculated from $K_D = 0.63 K_{OW} f_{oc}$.

et al. (1979) also found that the particle size of the mineral fraction was important. For example, the partition coefficients for semivolatile compounds, such as pyrene and methoxychlor, for the sand-sized fraction are approximately 100 times less than their partition coefficients for the silt- and clay-sized fractions, due primarily to the lower organic carbon content of the sand (Karickhoff *et al.*, 1979). Table 16.2 presents data for K_{OW} and calculated values of K_D using Eq. (16.5) for selected VOCs. From the example calculations of partition coefficients in Table 16.2, these VOCs are not strongly adsorbed onto the soil solids due to their relatively low octanol:water partition coefficients. In comparison, compounds with low volatility and low solubility, such as pentachlorophenol with a log K_{OW} of 4.74, has a K_D of 35, i.e., pentachlorophenol will be preferentially adsorbed to the soil solids by a factor of 100 to 1000 times greater than the chlorinated solvents listed in Table 16.2.

16.2.4 Effects of temperature and soil moisture content on partitioning

Kerfoot (1991) evaluated the theoretical effects of temperature and pore-water content on the partitioning of VOCs between the soil gas, soil water, and soil solids, assuming thermodynamic equilibrium conditions. For compounds with a Henry's Law constant of 0.05 (for example, the chloroethanes), effects of temperature on partitioning are negligible up to about 40 °C, above which partitioning of the VOC to the soil-gas phase increases. For VOCs with a Henry's Law constant of 1.0 (such as tetrachloroethene), a relatively constant increase in partitioning to the gas-phase is predicted by Kerfoot (1991). Based on his sensitivity analysis, there should be no significant difference in the relative partitioning between the three phases with respect to temperature for a 25% water saturation versus a 75% water saturation under the normal temperature range.

Smith *et al.* (1990), based on data developed from a site investigation and laboratory experiments, found that TCE adsorption decreases as soil moisture content increases from zero to the saturation soil moisture content (the soil moisture content that is in equilibrium with 100% humidity). They also found that results of soil and soil-gas sample analyses indicated that the ratio of the concentration of TCE adsorbed on the vadose-zone soil to its concentration in the soil gas was one to three orders of magnitude greater than the ratio predicted by equilibrium models. They concluded that the apparent disequilibrium resulted from the slow desorption of TCE from the organic matter in the soil relative to the faster volatilization loss of TCE from the soil system. Therefore, the assumption of Henry's Law constant and equilibrium adsorption isotherms may not hold in the field.

16.2.5 Transport of VOC vapor through soil gas

Transport of VOC vapor through soil gas may be under density gradients, thermal gradients, pressure gradients, or concentration gradients.

Density gradient

Density-induced convective flow of VOC vapor may be important where concentrations of denser VOCs are high, especially near the source. Falta *et al.* (1989), Sleep and Sykes (1989), and Mendoza and Frind (1990a; 1990b) concluded that convective flow due to density gradients may be important. Sleep and Sykes (1989) concluded that transport due to a density gradient was important when the density contrast is more than several times the ambient soil-gas density. In conducting sensitivity analyses of density-dependent VOC vapor transport, Mendoza and Frind (1990a) concluded that density gradients would be important at relative densities greater than 1.15.

The importance of density gradients to the transport of VOC vapor is related to permeability. Falta *et al.* (1989) and Mendoza and Frind (1990a) concluded that density-driven transport can be neglected for permeabilities of less than $10^{-11}\,m^2$ (10.1 darcy), i.e., equivalent to a clean sand (Freeze and Cherry, 1979). Since the effective permeability of the material to gas is affected by the water content, density-driven transport should be expected to become less important as the moisture content increases.

Aside from consideration of permeability, density-driven transport of VOC vapor through the unsaturated zone will be important for regions close to a source of dense volatile compounds and will be important early in the history of a spill when free-phase liquids are present. As the source dissipates and as distance from the source increases, density-driven transport will become less important. Lin (1990) concluded that density contrast would be an unimportant factor in vapor transport from contaminated groundwater unless nonaqueous phase liquid (NAPL) were present above the water table.

Thermal gradient

Thermally-induced convective transport of VOC vapors is generally limited to the near surface due to the rapid attenuation of temperature fluctuation with depth. While no research on the thermally-driven transport of VOC vapors through soil has been reported, research on transport of water vapor has been. For example, Hillel (1972) stated that the warming of soil lowers the suction pressure and raises the vapor pressure of soil water. The resulting thermal gradient induces water vapor migration from warmer to cooler regions in the soil. Therefore, the effect of the warming of the soil surface during the day would be to decrease the upward flux of water vapor.

Thermal gradients in soil that may be significant to soil-gas transport are generally limited to the upper several feet of the soil column. Marshall and Holmes (1979) state that the amplitude of thermal fluctuations in soil decreases exponentially with depth below the surface. They present the following empirically derived equation for estimating the temperature variation as a function of depth and time.

$$T_{(z,t)} = T_a + A_0 e^{-\left(\frac{\pi}{\tau\kappa}\right)^{0.5z}} \sin\left(\frac{2\pi t}{\tau} - \left(\frac{\pi}{\tau\kappa}\right)^{0.5z}\right) \tag{16.6}$$

where T is soil temperature, T_a is the average soil temperature, A_0 is the amplitude of the soil surface temperature fluctuation, t is time (day or year), τ is the period of the cycle (day or year), and κ is the thermal diffusivity. The thermal diffusivity is assumed constant, although, in reality, it varies with moisture content.

Soil temperature fluctuates on diurnal and annual cycles. Based on Eq. (16.6), diurnal soil temperature fluctuations caused by the rising and setting of the sun are damped out within a short distance below the soil surface. Figure 16.1 presents the results of applying Eq. (16.6) for the following conditions:

midsummer with an average soil temperature of 30 °C;
a surface soil temperature amplitude of 20 °C; and
a constant average soil moisture content of 20–30% resulting in a constant thermal diffusivity of $2.0 \times 10^{-7}\,m^2/s$.

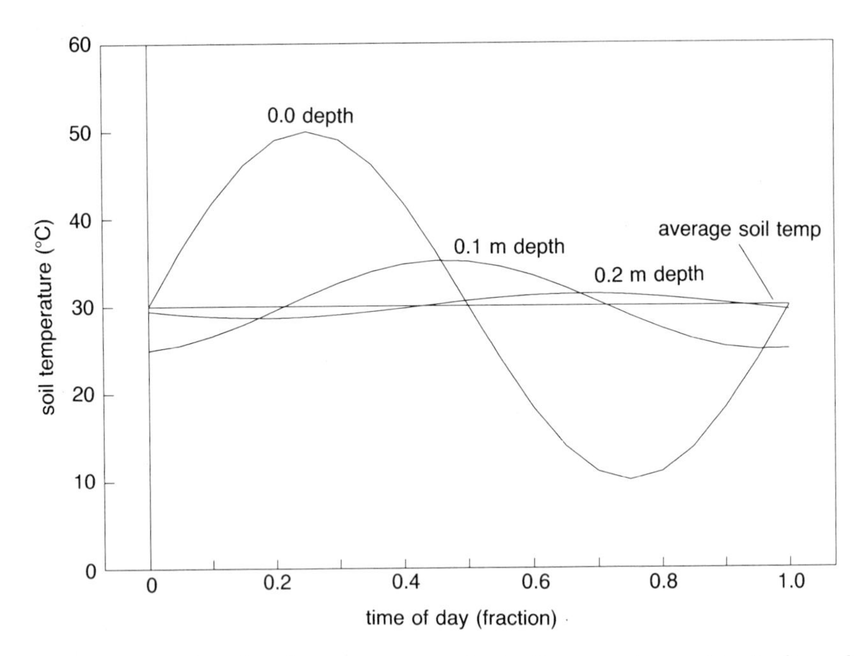

Fig. 16.1 Depth of soil temperature fluctuations under ambient diurnal surface temperature changes for an average midsummer soil temperature of 30 °C (77 °F) for three soil depths.

These conditions would be common to a large portion of humid temperate climatic regions with vegetated soil or partial shade of the soil surface (daytime maximum temperatures for either bare or drying and sparsely vegetated soils in direct sunlight would be expected to be higher). Figure 16.1 shows the diurnal soil temperature variation at three depths: 0.0, 0.1, and 0.2 m. At a depth of 0.2 m, the diurnal soil temperature variation has been damped out to less than 5% of the average soil temperature. Thus, in general, diurnal soil temperature fluctuations are not an important factor in VOC transport through the vadose zone.

For seasonal soil temperature changes, Fig. 16.2 shows the results for the soil depths of 0.0, 1.0, and 3.0 m for the same conditions as Fig. 16.1, except the average annual soil temperature is 17 °C and the amplitude is 13 °C. Over the year, the fluctuation in the average soil temperature is not damped out until much deeper, i.e., the largest fluctuation is less than 10% of the average soil temperature at a depth of 3.0 m.

Soil temperature fluctuations may be an important factor on the partitioning of the VOC between the liquid, gaseous, and adsorbed phases (Kerfoot, 1991). However, the vertical soil temperature profile below a depth of 0.2–1.0 m can be considered constant over the short term of several months. Laterally, where soil surface conditions may

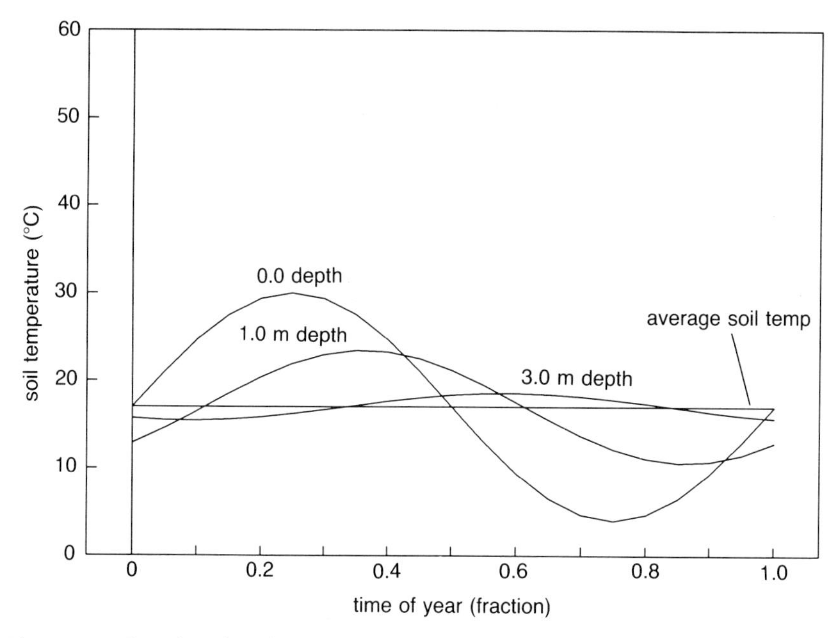

Fig. 16.2 Depth of soil temperature fluctuations under ambient seasonal surface temperature changes for an average annual soil temperature of 17 °C (58 °F) for three soil depths.

change from bare soil, to vegetated soil, to shaded soil, to pavement, localized lateral thermal gradients may exist at these lateral boundaries for short periods at shallow depths due to differences in the depth to which the thermal fluctuation penetrates the soil. These localized regions experiencing thermal gradients will produce insignificant effects on the soil-gas transport with respect to the typical scale involved.

Pressure gradients

Barometric pressure effects in the subsurface have been observed by Turk (1975), Freeze and Cherry (1979), and Nazaroff *et al.* (1987). Barometric pressure changes due to the weight of the atmospheric column overlying the ground surface at any instant will have spatially uniform effects over large areas, especially for open ground surfaces. Barometric pressure changes due to the passage of atmospheric highs and lows will act to depress and raise the water table slightly (Turk, 1975; Freeze and Cherry, 1979). This barometric pressure fluctuation will cause a slight compression of soil gas during the passage of a high atmospheric pressure system and a slight expansion of soil gas during the passage of a low atmospheric pressure system.

The insignificant effect of atmospheric pressure on the unsaturated zone was substantiated by Turk (1975) who found diurnal pressure fluctuations of up to 60 mm depth in soil. More recently, Kerfoot (1991) reported that pressure in the unsaturated zone remained stable over an 8-month period to a depth of 2.1 m and concluded that barometric pressure changes have a negligible effect on soil-gas concentrations of VOCs.

Significant pressure gradients can be established in the unsaturated zone due to the effects of manmade structures. Nazaroff *et al.* (1987) reported increased flow of gaseous contaminants, gasoline vapors, and radon, respectively, towards basements. Nazaroff *et al.* (1987) found that the artificial venting, or depressurization, of houses, such as for radon gas mitigation, can increase the pressure differentials by 10–20 times natural differentials, and the velocity of radon gas through the soil towards the basement can approach 1.0 m/h on a transient basis.

Concentration gradients

In general then, except for the region affected by density-induced transport of concentrated DNAPL vapors, the primary transport mechanism for VOCs in the unsaturated soil is by diffusion through the soil gas under concentration gradients. The distribution of VOC concentration in the soil gas can be modeled by Fick's second law which in one dimension is expressed by the following equation (Thibodeaux, 1979).

$$\frac{C}{t} = \frac{D^2C}{z^2} \tag{16.7}$$

where C is concentration of the VOC in air, D is the diffusion coefficient, and z is the distance traveled.

Assuming the outer boundary condition is zero concentration, Eq. (16.7) can also be expressed as (Thibodeaux, 1979)

$$\frac{C_{(z,t)}}{C_{(z=0,t=0)}} = \mathrm{erfc}\left[\frac{z}{4Dt^{0.5}}\right] \tag{16.8}$$

where $C_{(z,t)}$ is the concentration (as mole fraction) at a distance z and time t, $C_{(z=0,t=0)}$ is the initial concentration, and erfc is the complimentary error function.

Swallow and Gschwend (1983) conducted limited controlled laboratory experiments using a glass tank. Although their experimental design prevented a direct measurement of the concentration of VOCs in the unsaturated zone, Swallow and Gschwend (1983) concluded that volatilization can be adequately modeled by Fick's second law.

Lin (1990) also found reasonable agreement between observed VOC concentrations in soil gas and those predicted by a diffusion-based model developed by Silka (1986).

16.2.6 Diffusion coefficient in soil gas

The estimated diffusion coefficient for VOC vapor in air is $0.43\,\mathrm{m^2/d}$ (Jury *et al.*, 1983). This single value for D has been used for intermediate molecular weight VOCs based on the similar values obtained in measurements of the gas diffusion coefficient for numerous organic chemicals of intermediate molecular weight (Jury *et al.*, 1984; Lin, 1990).

However, the diffusion coefficient in soil gas has been found to be reduced from that in air by a tortuosity factor which accounts for decreased cross-sectional area for flow and increased length of the flow path through soil pores. Jury *et al.* (1983; 1984) concluded that the Millington–Quirk tortuosity formula adequately defines the soil-gas diffusion coefficient. Bruell and Hoag (1986) confirmed the validity of the Millington–Quirk model in column experiments. This formula is expressed by

$$D_G = \frac{Da^{\frac{10}{3}}}{n^2} \tag{16.9}$$

where D_G is the diffusion coefficient in soil gas, D is the diffusion coefficient in air, e.g., $0.43\,\mathrm{m^2/d}$, a is the volumetric air content of the soil, and n is the total soil porosity.

Since the VOC vapor may partition between the gas, liquid, and solid phases, an effective diffusion coefficient can be formulated that incorporates that partitioning. The removal of VOCs from the soil gas by partitioning into the soil moisture and soil organic matter results in a reduction in the apparent diffusion rate, and consequently, the apparent, or effective, diffusion coefficient. Jury *et al.* (1983) developed the following relationship between the diffusion coefficient in soil gas from Eq. (16.9) and the effective diffusion coefficient

$$D_E = \frac{D_G}{\dfrac{bK_D}{K_H} + \dfrac{w}{K_H} + a} \tag{16.10}$$

where D_E is the effective diffusion coefficient in soil gas corrected for effects of partitioning, D_G is the diffusion coefficient from Eq. (16.9), b is the bulk dry density of the soil, K_D is the soil partition coefficient from Eq. (16.5), K_H is Henry's Law constant from Eq. (16.3), w is the volumetric soil moisture content, and a is the volumetric air content (where n, total porosity, is equal to $a + w$). Jury *et al.* (1984) report that the model for the effective diffusion coefficient expressed in Eq. (16.10) gives results that are in good agreement with empirically derived values of D_E.

Research completed on the transport of VOC vapor in the unsaturated zone remains limited primarily to theoretical modeling studies and empirical observations from field data.

Although there has been considerable effort to characterize the transport and fate of pesticides in soil, the properties of the pesticides studied differ substantially from those of VOCs, resulting in a considerable difference in their behavior in the unsaturated zone. Research is still needed to improve the understanding of VOC vapor transport and fate in unsaturated soil in order to provide improved interpretation of soil-gas survey results and assist in designing more cost-effective clean-up programs.

16.3 SOIL-GAS SURVEYS

Soil-gas surveys are becoming more widely accepted as a tool for the preliminary determination of the extent of soil and ground water contamination by volatile organic compounds (VOCs). These surveys are being applied to the investigation of a variety of problems including groundwater plume tracking (Glaccum *et al.*, 1983; Lappala and Thompson, 1983a,b; Marrin, 1984; Voorhees *et al.*, 1984; Malley *et al.*, 1985; Silka, 1986; Kerfoot, 1987a,b; Marrin and Thompson, 1987; Silka and Spectre, 1991; Fusillo *et al.*, 1991), leaking tanks and pipelines (Mehran *et al.*, 1983; Feere and Silka, 1987; Stuart *et al.*, 1989), surface spills (Silka, 1986; Evans and Thompson, 1986; Marks *et al.*, 1989), potential releases from landfills, impoundments, and other sources (Los Angeles County, 1985; Smith *et al.*, 1990; Millison *et al.*, 1990; Thomsen and Joyner, 1990), and environmental audits and preconstruction assessments (Silka, 1986; Rizvi and Fleischacker, 1991; Viellenave and Hickey, 1990).

Since soil contamination by liquid VOCs typically causes vapor concentrations in soil gas in the immediate vicinity of the source to increase over several thousand to tens of thousand parts per million, outward diffusion of the VOC vapor through soil pores results in easily detectable concentrations of the VOC in soil gas up to tens of meters away from the source.

Due to the presence of VOCs at many sites of contamination, there has been increasing interest in the sampling and measurement of VOCs to determine their extent in soils and ground water. With the recent development of portable gas chromatographs that rely on a variety of detectors, quantitative, as well as semiquantitative, field analysis of VOCs in soil is now possible.

The portable gas chromatograph, a relatively new technology, has been shown to be especially applicable to the investigation of soil and ground water contamination through the analysis of shallow soil gas. Under diffusive transport, VOCs volatilize from ground water and move upward through the unsaturated zone, ultimately venting to the atmosphere. It has been shown that the concentration of VOCs in samples of shallow soil gas are related to the concentration of VOCs in ground water (Glaccum *et al.*, 1983; Lappala and Thompson, 1983a and 1984; Swallow and Gschwend, 1983; Marrin, 1984; Silka and Spectre, 1991). Soil-gas surveys are being recognized as a valuable tool, both alone and in conjunction with other techniques, that can provide data on the location and extent of soil and ground water contamination and aid in the design of more detailed ground water studies involving soil borings and monitoring well networks.

The application of soil-gas monitoring has been found to be useful not only in delineating soil and groundwater contamination at known contamination sites, but also as a valuable exploratory tool for identifying unknown sources of contamination. Soil-gas surveys have been successfully applied to environmental assessments for potential liabilities related to real estate transactions. In this mode, soil-gas surveys can be used to assess whether potential subsurface VOC contamination is present. For current owners of industrial plant sites, soil-gas surveys can be used as an integral part of environmental audits to identify VOC contamination from previous accidental spills and leaking tanks and pipelines.

The Los Angeles County Department of Water and Power, under a cooperative funding agreement with the US Environmental Protection Agency used soil-gas surveys for the identification of as yet unknown sources of VOC contamination in the San Fernando ground water basin.

The successful conduct of a soil-gas survey depends on several factors, including the size and age of the source, the moisture content and organic carbon content of the unsaturated zone, and the volatility and solubility of the VOC. Prior to the conduct of a soil-gas survey, the effects of these factors should be evaluated in order to optimize the design of the soil-gas survey. Through review of theory and the application of computer models, one can evaluate the operational limitations of soil-gas surveys. This evaluation provides additional value in allowing improved interpretation of data obtained from a soil-gas survey.

16.3.1 Summary of soil-gas sampling and analytical methodologies

A number of soil-gas sampling methods have been developed, including:

1. grab sampling of soil cores for headspace analysis;
2. surface flux chambers;
3. downhole flux chambers;
4. accumulator devices; and
5. suction ground probes.

Diverse analytical protocols also have been developed and used in soil-gas surveys. For example, the American Petroleum Institute (API, 1985), Silka (1986), Zdeb (1987), Marrin (1988) and Kerfoot (1988) evaluated soil-gas sampling methodologies, while Steinberg *et al*. (1990) compared the responses of the flame ionization detector (FID) versus photoionization detector (PID).

Headspace method

The grab sample method with headspace analysis by GC involves the collection of a core of soil, sealing of the sample with the exclusion of as much air as possible, and laboratory GC analysis by the headspace technique for VOCs coming out of the core. This method is simple, but the soil gas composition is subject to alteration during handling. Another problem with grab sampling for subsequent laboratory analysis is that additional time and expense are incurred when results are not obtained in the field and resampling is required.

Surface flux chamber

The surface flux chamber method involves the installation of an enclosure on the surface of the ground. Clean, dry air is added to the chamber at a known rate, and the concentration of VOCs is determined by a portable GC as the clean air exits the flux chamber. Knowing the concentration of VOCs in the exiting air and the rate of air flow, the flux or rate of VOCs emanating from the ground can be calculated. Although the technique allows rapid determination of the flux of VOCs through the ground surface, the detection limit of the technique is necessarily reduced by the dilution effect of the clean air that is introduced (API, 1985). Millison *et al.* (1990) reported on the use of a surface flux chamber to assess potential releases of VOCs to the air for purposes of deriving a hazard ranking score (HRS) for EPA's Superfund program. While they also employed the suction probe method, they did not compare the two methods.

Downhole flux chamber

The downhole flux chamber method follows the same principle as the surface flux chamber, except the chamber is driven or emplaced below the surface of the ground. This technique also has the basic drawbacks as the surface flux chamber method, although its detection limit is improved somewhat by the generally higher concentrations encountered with increasing depth below the surface. However, the technique is more labor intensive.

Surface accumulator method

The accumulator device technique also is referred to as the surface static collection method and involves the passive sampling of soil gas by the trapping of VOCs onto an adsorbent contained within a glass tube (Zdeb, 1987; Kerfoot and Mayer, 1986). A static trap is placed just below the soil surface and left in the ground for from several hours to as long as thirty days. The static trap consists of a ferromagnetic wire coated with activated charcoal that is contained in an inverted test tube. The trap passively collects diffusing VOCs by adsorption onto the activated charcoal. The trap is sealed and taken to the laboratory for analysis of VOCs by Curie point desorption mass spectrometry. The static collection method provides accurate, time-integrated determination of gas flux (Malley *et al.*, 1985). The sampling time can be adjusted to provide the necessary sensitivity and the technique is simple and low cost (API, 1985). However, the longer sampling time prolongs the field study. API (1985) reported uncertainty concerning the retention efficiency of the activated charcoal.

In addition, the method is limited to shallow soil depths, and vertical profiling is not feasible. Since the traps are returned to the laboratory for analysis, the accumulator method does not allow the sampling design to be altered in the field on the basis of previous results. Another complication in using the static accumulator traps is that they will trap any VOC that happens to impinge on the trap. Thus, extraneous atmospheric sources, such as VOCs from transient auto emissions, can be caught by the accumulator trap. To control for this potentially confounding source of VOCs, extra traps must be set and collected periodically during the course of the total accumulation period to assess potential effects of transient sources.

Viellenave and Hickey (1990) reported on the use of surface accumulator traps for an environmental assessment of an office building. After a one-week sampling period, the results indicated low level TCE and PCE contamination in the subsurface. Their results illustrate the potentially low detection limits possible with the accumulator traps.

Suction ground probe

Ground probe techniques for sampling soil gas involve the insertion of a tube into the ground to a desired depth, placing a vacuum on the tube, and analyzing the extracted air for VOCs, typically by field GC. API (1985) concluded that the technique is relatively sensitive and, because the sampling depth can be varied, surface interferences can be avoided, and sensitivity can be increased if needed. Although API (1985) considered the ground probe technique labor intensive and not suited to finer grained, clayey soils or where bedrock is near the surface, the technique has been successfully applied to numerous sites having a wide range of environmental conditions (e.g. Silka and Spectre, 1991 and Thomsen and Marrin, 1987).

Thomsen and Joyner (1990) used hand-hammered shallow probes to 1.0 m, hydraulically pressed probes to about 6.0 m, and hallow-stem auger drilling with a driven probe ahead of the auger down to 36 m to obtain vertical profiles of VOCs in soil gas. Hammered and hydraulically driven probes are not good in resistant soil such as coarse gravel or dense bouldery till. Hand-held electric rotary hammers fitted with small diameter augers and small hydraulically driven augers have been used frequently to depths of 6.0 m.

Selection of the sampling method depends on the study objectives, time constraints, and budget. The suction ground probe technique has been reported in the literature more often than other collection methods. The following points have been made concerning this method.

1. In combination with laboratory grade GCs, the method is sensitive to a wide range of VOCs and allows separation of component VOCs in soil gas.
2. The method provides rapid turn-around of VOC analyses so that sampling stations can be adjusted in the field to define better the extent of soil contamination.
3. In combination with the hand-held rotary hammer method of collection, the suction ground probe allows sampling in areas of restricted access.
4. The method allows sampling of soil gas through pavement as well as open ground.
5. The method is easily adapted to vertical profiling which often is important for the identification of the VOC source.

16.3.2 Analytical methods and quality control

Soil-gas surveys have been conducted with a full gamut of instruments from the portable organic vapor analyzer (OVA) equipped with a flame ionization detector (FID) and photoionization detector (PID), to portable gas chromatographs, to laboratory grade GC or GC/MS instruments. Steinberg *et al.* (1990) presented a comparison of the response of field PID versus FID and field FID versus lab GC-FID. The response factor for the OVA with FID changes with hydrocarbon composition and must be corrected for to make accurate interpretations of field data. Steinberg *et al.* (1990) found that the response factor for the FID decreases geometrically with carbon number. In their comparison of the field screening FID versus lab GC-FID, they found good agreement (correlation of 0.94) between the method with an almost one-to-one relationship (slope of 0.89).

Marks *et al.* (1989) made a comparison of PID versus FID for soil-gas surveys involving petroleum hydrocarbon. They found that the evaluation of a site based only on the concentrations of total hydrocarbons can result in losing valuable information. Indeed, the total hydrocarbon reading with an FID instrument cannot discriminate between the target compounds of concern and natural methane and light aliphatic hydrocarbons, which are abundant in nature. Marks *et al.* (1989) also concluded that, when calibrated with benzene and toluene standards, both FID and PID GCs will give comparable results, but the total hydrocarbon response between the two detectors can vary by five to ten times.

In general, the PID is more responsive to aromatic and ring compounds. The FID is a good general detector for hydrocarbons, which has led to its frequent application in soil-gas surveys, not to mention the past ready availability of FID instruments. However, more recently, the PID-equipped instruments have become the most commonly used due to their ease of use.

The electron capture detector (ECD) is the most sensitive detector for halogenated organics, but can be easily contaminated and difficult to operate in the field. The argon ionization detector (AID) has been overlooked for application to soil-gas surveys, although it has significant advantages in being very sensitive to a wide range of VOCs including both aromatic and hydrocarbon compounds.

Because of the poor detection limits and lack of qualitative chromatographic information available from field FID-OVAs and PID-TIPs, field GCs and transportable laboratory-grade GCs are considered state-of-the-art in soil-gas analyses.

Marks *et al.* (1990) discussed quality assurance and quality control concerns for soil-gas surveys. They found that soil-gas surveys have a 100–500% variability for calibrated constituents and up to 1000% for noncalibrated and total volatile hydrocarbons analyses. They believe that only a small amount of the variability is derived from the instrumentation, and that most of the variability is associated with sampling error and spatial/temporal variability. Marks *et al.* (1990) found that variability in sampling depth and variability in vacuum pressure during soil-gas collection are important factors that are difficult to come to grips with. In addition, variability can be introduced through short-circuiting of atmospheric air along the probe shaft.

Computer modeling of soil-gas transport

Although several investigators have developed models for the simulation of the transport of organic chemicals in the soil (Leistra, 1973; Jury *et al.*, 1983, 1984; Rao *et al.*, 1985) these models were limited in their application to the simulation of the diffusion of VOCs in soil gas. In general, the previous models were developed for application to the modeling of pesticide movement and fate in soils. The models are one-dimensional analytic solutions that do not allow for heterogeneous soil properties and initial conditions. Also, these models incorporate transport of the chemical in the liquid phase, as they are intended for the study of leaching and volatilization of pesticides from soils (Jury *et al.*, 1983). In the case of the model developed by Rao *et al.* (1985), transport by vapor diffusion was omitted.

Silka (1986) described a two-dimensional finite-difference vapor diffusion model and the results of sensitivity analyses of vapor diffusion through soil gas. Sensitivity analyses demonstrated the importance of soil moisture and organic matter content in controlling the migration of VOC vapor through the unsaturated zone.

Sleep and Sykes (1989) developed a two-dimensional finite-element model which incorporated transport of VOCs in variably saturated media. Their model included dissolved phase flow and transport, as well as density-dependent flow and transport of vapor. Mass transfer was controlled by mass transfer coefficients, rather than traditional partitioning coefficients. The mass transfer coefficients were varied to give reasonable results.

Mendoza and McAlary (1990), Mendoza and Frind (1990a,b) developed a two-dimensional finite-element model which incorporates diffusion, advection due to density gradients, and advection due to the vapor mass released by vaporization at the source. For their numerical experiments, they assume equilibrium partitioning between an oil, adsorbed, dissolved, and vapor phase.

As an example of a soil-gas model, the two-dimensional vapor diffusion model, 2D-DIFF, developed by Silka (1986) is discussed. This model is a finite-difference, forward difference approximation relative to time, and is based on Fickian diffusive transport. The model is based on the following assumptions.

1. Diffusion is described by Fick's Second Law.
2. Partitioning coefficients are linear and system is at equilibrium with respect to partitioning.
3. The Millington–Quirk tortuosity formula is valid.
4. The soil properties of bulk density and total porosity are homogeneous.
5. The diffusion coefficient in air is constant for all VOCs.
6. The soil gas is at constant atmospheric pressure.
7. The system is isothermal.

The model allows for heterogeneous initial concentrations with either constant concentration sources or instantaneous spike sources. The

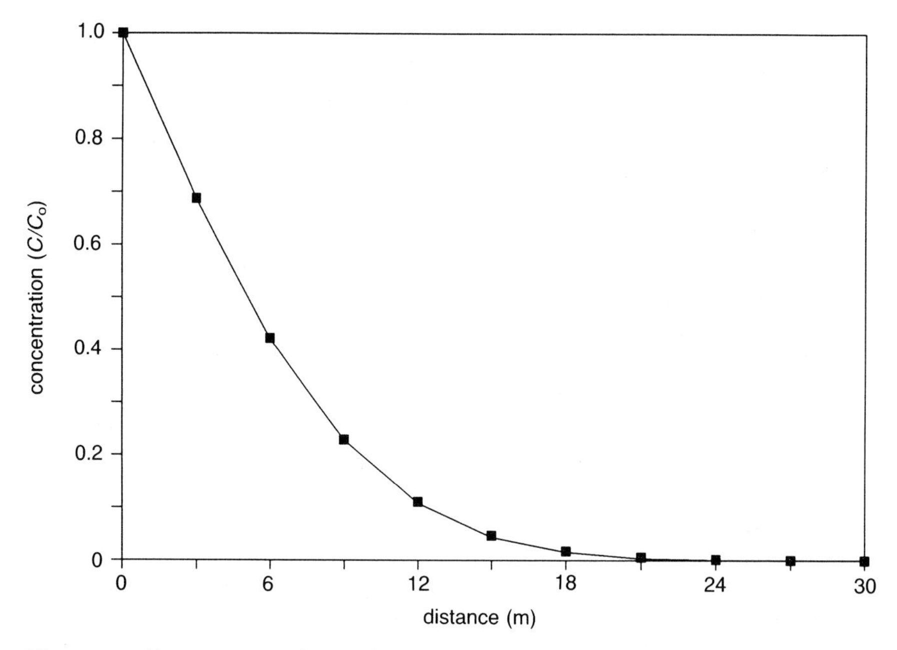

Fig. 16.3 Comparison of one-dimensional analytical solution of the Fickian diffusion equation (solid line) to the one-dimensional simulation using the soil–gas vapor diffusion model 2D-DIFF after 90 days from (solid squares) a constant concentration source, from Silka (1986).

diffusion coefficients may be varied over the finite-difference grid and in the x- and y-directions by weighting coefficients. The effects of partitioning are incorporated by using the effective diffusion coefficient as defined in Eq. (16.10).

16.3.4 Model verification

The verification of the computational algorithms for the vapor diffusion model 2D-DIFF has been done by comparisons with computed results from the one-dimensional analytical solution. The results of this comparison are presented in Fig. 16.3 for an elapsed time of 90 days. The analytical solution is represented by the solid line and the solution computed by 2D-DIFF is presented as solid squares.

16.3.5 Model validation

Preliminary validation of 2D-DIFF has been performed by Lin (1990) based on field data from Marrin and Thompson (1987) and Kerfoot (1987a,b). An adequate validation problem requires spatial information on the moisture content and organic matter content of the soil, the soil texture, as well as the source characteristics, whether a surface spill or contaminated ground water. In many instances this information is not gathered as part of a routine soil-gas survey. Further complications arise when one considers that field conditions are dynamic, i.e., always changing. This problem is especially acute for soil moisture content which, over the large scale, fluctuates seasonally. Also, field data are not available in sufficient detail to allow description of the spatial variability of soil conditions resulting in necessary oversimplification of the physical setting.

Jury (1986) lists several potential problems that must be considered to carry out successfully a field validation experiment of a vapor diffusion model:

1. lateral and vertical variability of transport and retention parameters that may introduce heterogeneities and anisotropy that are not included in the model;
2. macropores, cracks, plant root holes, and animal burrows that may create discontinuities difficult to account for in the model;
3. time-dependent boundary conditions, such as depth to water table and seasonally saturated zones that alter the system geometry from that modeled;
4. problems with validity of measurement techniques for characterizing field properties; and
5. scale effects that may be important in the field that are not accounted for in the model, such as localized departure from isothermal conditions or physical heterogeneities.

Striegl (1987) published an account of an apparently successful modeling of the diffusion of methane from a waste disposal trench in Illinois using a two-dimensional finite-difference solution of Fickian diffusion similar to the model described here.

The validation exercise completed by Lin (1990) using 2D-DIFF provides an example of these difficulties listed by Jury (1986). In the first validation case, Lin (1990) used data presented by Marrin and Thompson (1987) from a TCE spill site in Arizona, where the water table is approximately 30 m deep. Samples were collected through a probe which was driven ahead of a hollow-stem auger. Included in these data were soil texture, water content, and air porosity. The soil organic fraction was assumed to be 0.005 based on a sandy matrix. Initial vapor concentrations above the water table were calculated based on measured groundwater concentrations. On the first profile, the simulated data are, in general, lower than the actual data. A sharp drop in soil gas concentration across a sandy layer from 3 to 6 m was both observed in the field and predicted by the model. Results from a second profile were qualitatively similar where concentrations dropped across a clay layer, but in general the simulated results were about an order of magnitude different than the measured concentrations (Lin, 1990).

In the second case, Lin (1990) used data from Kerfoot (1987a,b) on a chloroform spill in Nevada, where the water table is 3–4 m below grade. The soil type is a sandy gravelly loam, with several percent clay. Samples were collected through a probe hammered into the ground to a depth of about 30 cm. Vertical profiling was also conducted at several points. Initial vapor concentrations were calculated using measured groundwater concentrations. In the analysis by Lin (1990), simulated results matched field observations more closely near the surface, while they were somewhat elevated above observed levels at deeper depths. The observed concentration profile with depth exhibited a linearity usually associated with a steady-state regime. However, the modeled results did not reach a steady state in a reasonable amount of time. In this case there may be errors in the input data or some environmental factor that is not being taken into account by the model (Lin, 1990). Several possibilities include advection effects and chemical or biological degradation. In addition, errors may have been introduced during soil-gas sampling.

Mendoza and Frind (1990a) compared their numerical results with laboratory experiments. The laboratory apparatus consisted of a 10 × 10 × 3 m sand-filled box instrumented with gas samplers, to which one liter of TCE was added over 30-cm square area at a depth of 45 cm. In general, the match between the observed and simulated results is good, but it is improved by the inclusion of density driven advection in the numerical model. This effect is more pronounced deeper in the soil.

16.3.6 Importance of advection

According to Mendoza and Frind (1990a), significant advection of subsurface vapors may be caused by density gradients due to high vapor pressures and high molecular weights of chlorinated solvents. Advection may also be caused by vaporization of free product at the source. Advection seems to be an important mechanism for dense, volatile compounds.

However, advection due to density gradients may only be important near the source. For example, TCE has an ideal gas phase vapor density of 5.64 kg/m^3, while air has a density of 1.26 kg/m^3. An air mixture saturated with TCE would have a density of 1.46 kg/m^3, producing an increase in density of only about 15% (Sleep and Sykes, 1989). Ignoring differences in gas phase density leads to underestimates of dispersion, and hence the spreading of the vapor plume. A saturated vapor phase will only exist in the vicinity of the free organic liquid, and thus density differences may only be important near pools of free product. Saturated vapor conditions could also exist in a high-porosity material with a significant amount of residually saturated product and a low organic matter and water content, such as sand residually saturated with TCE. A sand with a fairly even residual saturation of free product would serve as a significant reservoir for vaporization and would probably be affected by advection due to density gradients.

In numerical simulations by Sleep and Sykes (1989), incorporation of advection results in greater horizontal spreading of the vapor plume. In addition, gas-liquid partitioning reduces the concentration of dissolved TCE. The water acts as a sink for TCE vapor and the movement of the vapor is retarded by gas-liquid partitioning. Concentration gradients in the aqueous phase are diminished while concentration gradients in the vapor phase are increased. Simulations involving diffusion only showed reduced spreading of the vapor and dissolved plume. Decreased rates of dissolution and volatilization were also noted, since advection was not available to transport extra mass away from the source.

16.3.7 Sensitivity analyses

Silka (1988) conducted sensitivity analyses with 2D-DIFF to assist in designing soil-gas surveys. The objectives were to:

1. estimate the optimum grid spacing for soil-gas surveys;
2. determine what the optimum soil conditions are, especially moisture content, to maximize the probability of successfully detecting a surface source from a distance; and
3. determine the level of interference due to upward diffusion of VOC vapor from underlying contaminated ground water and determine if the ground water source of VOC vapor may interfere with the detection of a surface source and vice versa.

The parameters utilized in the sensitivity analyses are presented in Table 16.3.

Porosity was assumed constant in all runs at a value of 0.4. Divergence from this value in the field will affect the value of D_G computed in Eq. (16.9). A change from 0.4 to 0.3 for total porosity results in D_G increasing by a factor of 1.8, while decreasing porosity from 0.4 to 0.35 increases D_G by a factor of 1.3. Therefore, the value of porosity selected will not affect the results significantly when compared to uncertainties in other parameters.

Partitioning of the VOC between the vapor and the soil moisture and soil solids produces significant retardation of the migration of the vapor through the soil gas. For the purposes of this sensitivity analysis, the organic carbon content is assumed zero. However, inspection of

Table 16.3 Parameters used in modeling diffusion of 1,1,2,2-Tetrachloroethane and Trichloroethylene through soil–gas using 2D-DIFF.3

Unsaturated zone	
Porosity	0.4
Moisture content	0.0, 0.04, 0.1, 0.2
Fraction organic carbon content	0.0, 0.01, 0.03
Temperature	20 °C
Bulk density	1350 kg/m^3
1,1,2,2-Tetrachloroethane	
Molecular weight	168 gm/mole
Saturated vapor concentration	6 000 ppm
Diffusion coefficient in air	0.43 m^2/d
Henry's Law coefficient	0.02
Octanol: water partition coefficient	300
Trichloroethylene	
Molecular weight	131 gm/mole
Saturated vapor concentration	72 000 ppm
Diffusion coefficient in air	0.43 m^2/d
Henry's Law coefficient	0.4

Eqs. (16.5) and (16.10) shows that even the slightest amount of organic matter will greatly reduce the effective diffusion coefficient.

Soil moisture content, on the other hand, was varied over four values (0.0, 0.04, 0.1, and 0.2, equivalent to 0%, 10%, 25%, and 50% saturation, respectively) of the total pore space filled with water. This range in soil moisture is typical of those observed for sand (Marshall and Holmes, 1979).

An additional assumption utilized in the sensitivity analyses of Silka (1988) is that the soil moisture fluctuates seasonally, a common occurrence in humid temperate regions. The soil moisture in storage sustains a net decrease during the summer months when potential evapotranspiration (ET) exceeds available precipitation. No net downward flux of water occurs to wash VOCs out of the unsaturated zone. However, during the fall, winter, and spring months, the soil moisture in storage increases as the potential ET drops off. During these nine months, a net downward water flux occurs.

Although 2D-DIFF is not designed to simulate the transport of VOC in the dissolved form in downward percolating water, intuitively, the addition of clean recharge water to the soil will extract VOC vapor from the soil gas. The magnitude of the reduction in the concentration of VOC vapor in the soil gas by dissolution in the incoming recharge water will depend on the mass of recharging water and the value of the Henry's Law constant.

The net effect of the rainy-season recharge is assumed to essentially wash out all VOC vapor in the unsaturated zone. Therefore, the sensitivity analyses are 90 day simulations of the summer months. This assumption will not be conservative for the arid southwest and will result in the underestimation of diffusion distances. In the humid temperate regions, this assumption may overestimate the diffusion distances. Results of the sensitivity runs based on the parameters in Table 16.3 are presented in Figs. 16.4 through 16.6.

The sensitivity analyses of 2D-DIFF demonstrate the importance of soil moisture and organic matter content, as well as the VOC-specific properties of volatility and solubility, to the transport of VOC vapor in the soil gas. These results have been applied to the design of soil-gas surveys in order to optimize grid spacing.

Mendoza and Frind (1990b) performed extensive sensitivity analyses of their finite-element model. They also choose to neglect partitioning to the solid phase. A base case which includes the effects of both advection and diffusion shows a large plume with high vapor and dissolved concentrations developing rapidly. The density effect of their generic high molecular weight compound is evident in the sinking of the plume. Near the plume front, where concentrations in the vapor phase decrease, the density effects are much less apparent. In the diffusion-only case, the plume does not move out as fast from the source without the extra action of density-induced advective forces. In addition, the plume spreads more uniformly with depth, rather than sinking towards the water table. Since less mass is sinking through the soil matrix, more of the source mass is lost to the atmosphere.

Sensitivity analyses for permeability revealed that permeabilities below which advection becomes the dominant transport process correspond to those of a relatively dry clean coarse sand.

Diffusion will be the dominant transport process in medium sand and finer-grained matrices. Variations in source vapor density produce effects similar to those caused by varying the permeability. The advective component of transport increases with increasing density, and vice versa.

In the pure diffusion case, the mass vaporized as well as the net mass in the system increases with the source concentration, but is not affected by variations in permeability. The case is similar when advective mass flux due to vaporization is included, but increases in source concentration will cause a correspondingly greater increase in the mass vaporized and the net mass in the system.

With advection due to density gradients included, an increase in the source concentration causes very large increases in the mass vaporized and net mass in the system. Mass retention increases with permeability due to sinking vapors (Mendoza and Frind, 1990b).

16.3.8 Soil-gas survey grid spacing

Determination of the appropriate grid spacing for soil-gas surveys has been left primarily to best judgment of the investigator. Godoy and Naleid (1990) stated that grid spacings typically are two to four times the depth to the water table and that 15–30 m grid spacings are frequently employed for screening large areas. Under ideal, homogeneous, dry conditions, with no organic matter, soil gas would be expected to diffuse laterally to a distance equal to the thickness of the unsaturated zone. In coarse-grained soils, this may hold. In fine-grained soils, however, a significant capillary fringe may be present which could diminish the effective depth to the water table. Also, increasing soil heterogeneities, water content, and organic matter, as well as age of the source of VOC contamination will decrease the distance of VOC migration through soil gas.

The simulation of the diffusion of trichloroethylene (TCE) and 1,1,2,2-tetrachloroethane (TET) vapors through the soil away from a surface spill (Silka, 1986) indicate that a minimum grid spacing of 7–11 m is adequate to allow the detection of a spill in a low organic soil. For soil that has been dry for a period of several months, such as during summer droughts, the grid spacing may be increased to 13–16 m.

The results of the sensitivity analyses for diffusion of TCE and TET vapors under varying soil moisture conditions demonstrate the importance of soil moisture to successfully conducting the soil-gas survey. Dry soil represents the optimum soil moisture conditions for the soil-gas survey. Soil moisture is not as critical a factor for detection of a source of TCE or other more volatile VOC. However, soil moisture is a critical factor for successfully detecting a source of TET or other less volatile VOCs and VOCs with small Henry's Law constants.

The computer simulations reviewed indicate that, as the soil moisture content increases above 0.0%, the rate of diffusion rapidly decreases (as shown in Fig. 16.4). Thus, the grid spacing used in a soil-gas survey is critically dependent upon the soil moisture content. Under dry soil conditions, a grid spacing of up to 30 m may be sufficient to ensure detection of a source. However, under wet soil conditions, a grid spacing of less than 10 m may be required. These grid spacings would decrease for unsaturated zones that are less than 30 m.

16.3.9 Discriminating groundwater versus surface sources

A problem that arises in the course of soil-gas surveys for both detection of surface sources, but more so, for mapping ground water contamination, is the potential interference caused by VOC vapors from

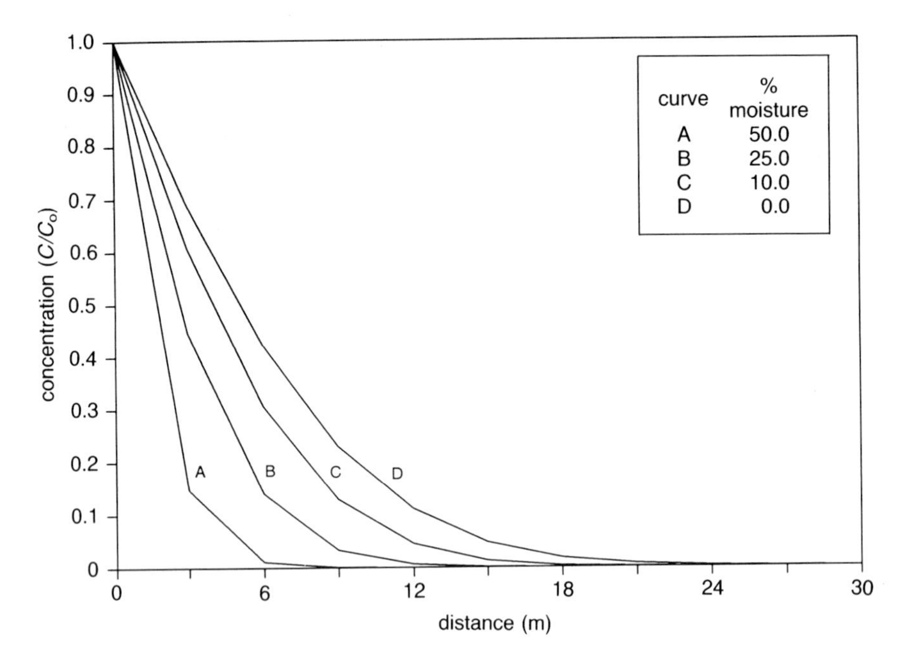

Fig. 16.4 Distance of diffusion of trichloroethylene under varying soil moisture content (no soil organic matter) after 90 days from a constant concentration source, from Silka (1986).

one source confusing the interpretation of the field data concerning the other source. The interferences are especially problematic in highly industrialized areas with multiple sources. Where the GC fingerprint differs, interpretation is made easier. However, where similar compounds are involved in multiple sources, discrimination between the VOC sources is made more difficult.

Computer simulations were completed to estimate the interference that may arise from the upward diffusion of VOC vapors emanating from contaminated ground water. For the modeling, it was assumed that the highest VOC concentration in ground water underlying the area of the soil-gas survey was 100 ppb. The depth to the water table was almost 10 m. This level of ground water contamination could result in soil-gas concentrations in the upper one meter of soil to be as high as 150 ppb. Thus, a soil-gas survey for surface sources would not be able to distinguish surface sources until the vapor concentration due to the source was above the 150 ppb level.

Conversely, the results of soil-gas surveys for mapping ground water contamination will be uninterpretable in regions near surface sources of VOC vapor. Inspection of Figs. 16.4 and 16.5, for instance, indicates that, out to a distance of about 30 m, the concentration of VOC in soil gas due to the surface source may be greater than 0.001 of the source

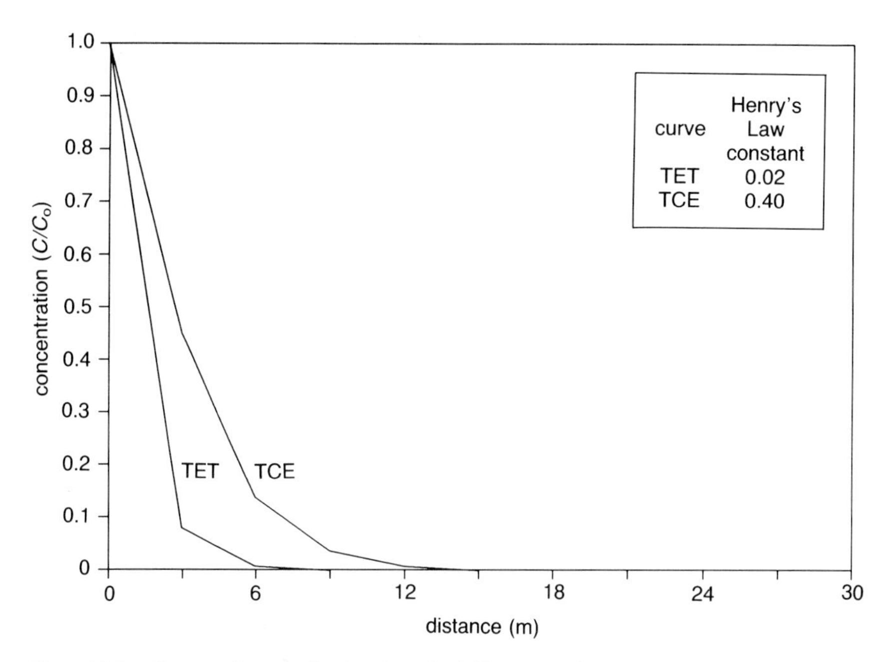

Fig. 16.5 Comparison of simulated diffusion of 1,1,2,2-tetrachloroethane (TET) and trichloroethylene (TCE) in soil with 25% moisture content and no organic matter, showing the effects of the difference in vapor/water partition coefficient represented by the Henry's Law constant, from Silka (1986).

concentration. When the saturated vapor concentration for TCE is 72 000 ppm (see Table 16.3), the concentration in soil gas at 30 m may be as high as 72 ppm. For TET, with a saturated vapor concentration of 6 000 ppm, the concentration in soil gas at a distance of 30 m may be as high as 6 ppm. The vapor concentration from surface sources can be sufficient to mask the detection of vapors coming from ground water contamination when the vapor concentration due to ground water contamination is of the order of less than 1 ppm – a problem reported by Marrin (1984).

16.3.10 Delineating surface contamination

Results of sensitivity analyses using the model 2D-DIFF have been presented to assist in designing and interpreting soil-gas surveys for contaminated soil (Silka, 1986) and leaking underground storage tanks (Ferre and Silka, 1987). With regard to the use of soil-gas surveys for identifying shallow contaminated soil (Silka, 1986), the optimum grid spacing for the soil-gas survey was found to be primarily dependent upon soil moisture and the value of Henry's Law constant, and, to a lesser degree, organic matter content. Figure 16.7 illustrates the

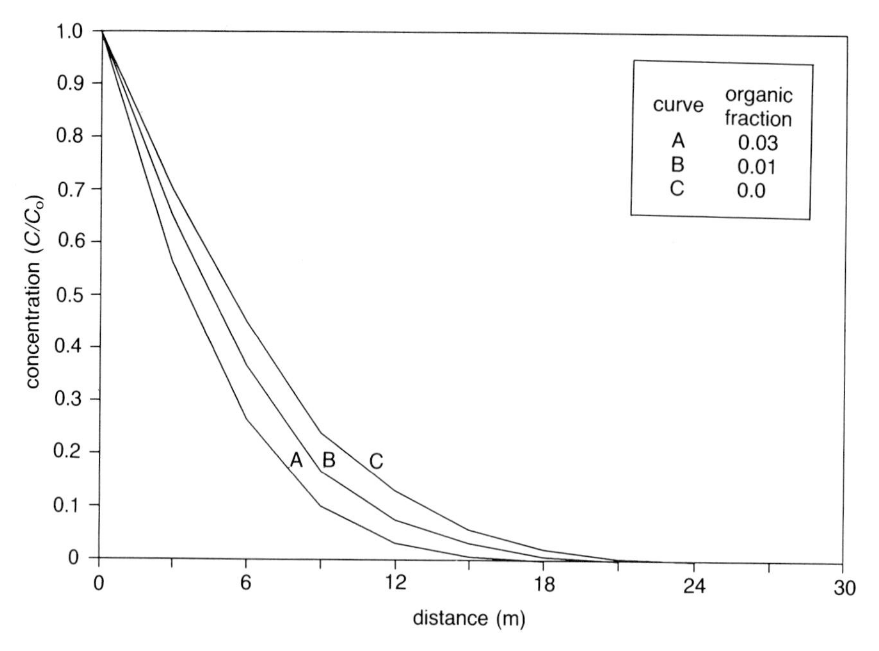

Fig. 16.6 Distance of diffusion of 1,1,2,2-tetrachloroethane (TET) for three soil organic matter contents in dry soil after 90 days from a constant concentration source, from Silka (1986).

dependence of the effective diffusion coefficient on K_H and fraction of pore volume occupied by water (w/n). The reduction in D_E is represented by the ratio of D_E/D_G, where D_G is the diffusion coefficient corrected by the Millington–Quirk tortuosity formula in Eq. (16.9). Therefore, in dry soil D_E/D_G is unity.

Optimum conditions for soil-gas surveys are obtained when dry soil conditions have prevailed prior to the survey. Since moist soil is the rule, though, especially in the more humid regions, the optimum grid spacing will generally be less than 30 m. For VOCs with even moderate values of K_H, for example trichloroethylene (TCE) with K_H of 0.4 and 1,1,2,2-tetrachloroethane (TET) with K_H of 0.02, the reduction in the effective diffusion coefficient is sufficient to reduce their distance of migration. Compared to an ideal VOC that does not partition into soil water, TCE migrates to only about 60% of the distance, while TET migrates to only about 20% of the distance of the unretarded ideal VOC within the same time frame. In homogeneous soils, the maximum extent of the migration of vapor through soil gas will be limited by the thickness of the unsaturated zone. However, many soils are heterogeneous and stratified. Stratified soils can cause greater lateral migration of the vapor if the upward diffusion is blocked by strata

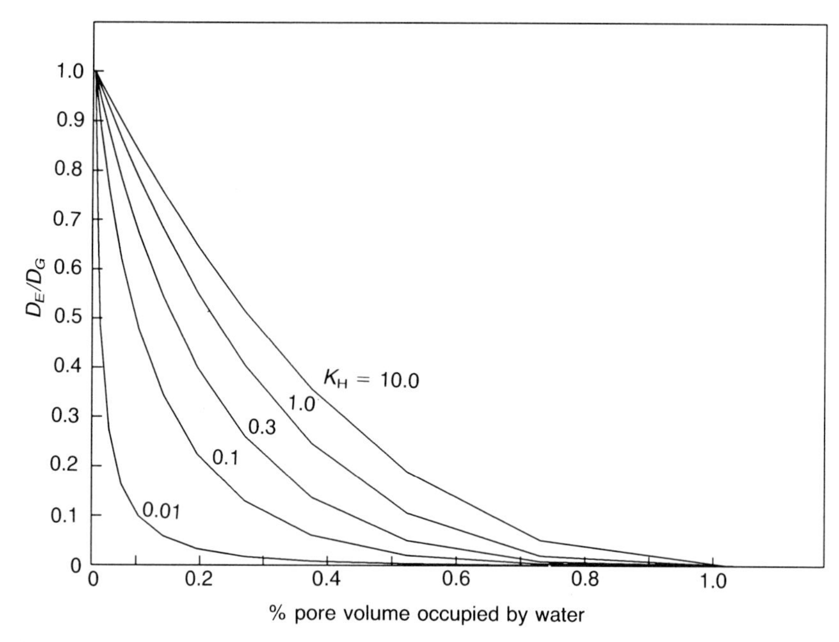

Fig. 16.7 Dependence of the effective diffusion coefficient, expressed as the ratio of D_E/D_G, on moisture content, expressed as the percentage of total porosity occupied by water, and Henry's Law constant, K_H, from Silka (1988).

containing finer grained soil having a higher moisture content or higher organic carbon content.

16.3.11 Mapping groundwater plumes

Several observations reported in the literature concerning the interpretation of soil-gas surveys for groundwater plume mapping have been investigated using the model 2D-Diff (Silka, 1988). It has been reported that concentrations of VOCs in soil gas decrease from the source at the water table to the surface by up to five orders of magnitude (Lappala and Thompson, 1983a). This field observation conforms to that expected for a diffusion dominated transport system. Since the soil-gas system is bounded below by an essentially constant concentration source and above by a constant zero-concentration boundary, i.e., the atmosphere, the vapor concentration will decrease logarithmically from the water table to the surface in an ideal, homogeneous soil.

Figures 16.8 and 16.9 show two cases for the distribution of vapor concentration with distance above the centerline of a plume of contaminated ground water where the water table is at a depth of 10 m. The concentration is presented in dimensionless units. Figure 16.8

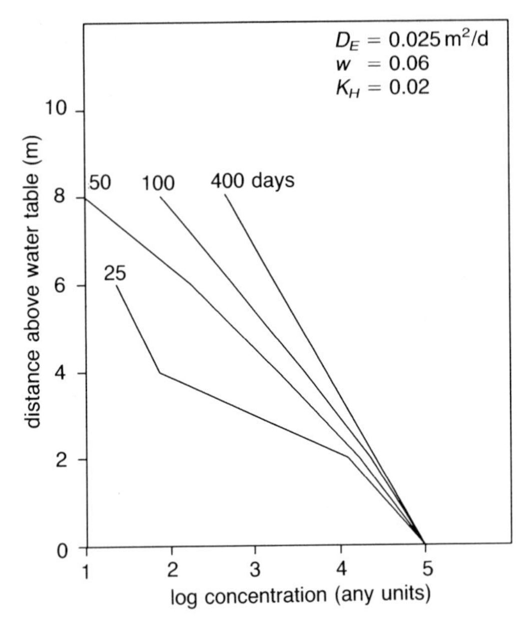

Fig. 16.8 VOC concentration in soil gas versus time and distance above the water table under nearly dry ($w = 0.06$) soil conditions, from Silka (1988).

shows the results for a relatively dry soil having only 20% water content ($a = 0.32$, $w = 0.08$) and a VOC with a K_H of 0.02. Figure 16.9 shows the results for a wetter soil with a 50% water content ($a = 0.2$, $w = 0.2$). In both cases, there is greater than three to five orders of magnitude change in the VOC concentration in soil gas from the water table to the surface, even when the concentration profile approaches steady state.

For the dryer soil (Fig. 16.8), the higher effective diffusion coefficient results in the concentration profile approaching steady-state faster than the wetter soil case (Fig. 16.9). Thus, the dryer soil conditions will produce a more responsive concentration profile as compared to the wetter soil. Shallow soil-gas measurements in the dryer setting will better reflect the distribution of the VOC in the ground water at that point in time.

Evans and Thompson (1986) concluded that aerobic biodegradation of hydrocarbon vapors was the cause of lower than expected concentrations, $<10\,\mu g/l$, in shallow soil gas at depths of less than 1.5 m when compared to the $>1000\,\mu g/l$ concentrations in the 1.8–4.3 m interval. However, they also reported at least one instance when the concentration gradient reversed and decreased with depth below the 1.8–4.3 m interval.

In order to substantiate the interpretation that biodegradation was

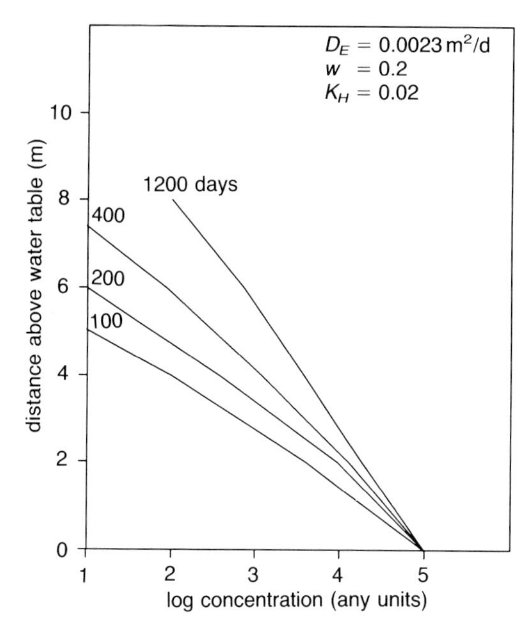

Fig. 16.9 VOC concentration in soil gas versus time and distance above the water table under wet ($w = 0.2$) soil conditions, from Silka (1988).

occurring, active microbial populations and degradation by-products should be confirmed in the soil column. It is more likely, though, that the lower-than-expected concentration in the uppermost 1.5 m was due to the normal decrease in concentration expected under the concentration gradient established by diffusion alone. Inspection of Figs. 16.8 and 16.9 shows that transient concentration profiles can produce a concentration decrease of greater than three orders of magnitude from the 1.8–4.3 m interval to the less than 1.8 m interval. Considering that the concentration decrease due to the diffusion gradient would be greater in wetter soils, the theory that biodegradation was the cause of the observed decrease can not be confirmed and is not necessary.

Evans and Thompson (1986) also reported the observation that vapor concentrations decrease rapidly, by two to three orders of magnitude just beyond the edge of the ground water contamination zone. Simulations were conducted using 2D-DIFF for an advancing front of a non-aqueous phase liquid (NAPL) floating on the water table. The model was set up with a constant saturated vapor concentration of 6 000 ppm at the NAPL-soil gas interface and an initial concentration of 6 000 ppm along the left side to represent the downward path of liquid VOC migration. Two variations were run, one with a vapor diffusion rate that was faster than the NAPL front velocity, and the second with a vapor diffusion rate that was slower than the NAPL front velocity.

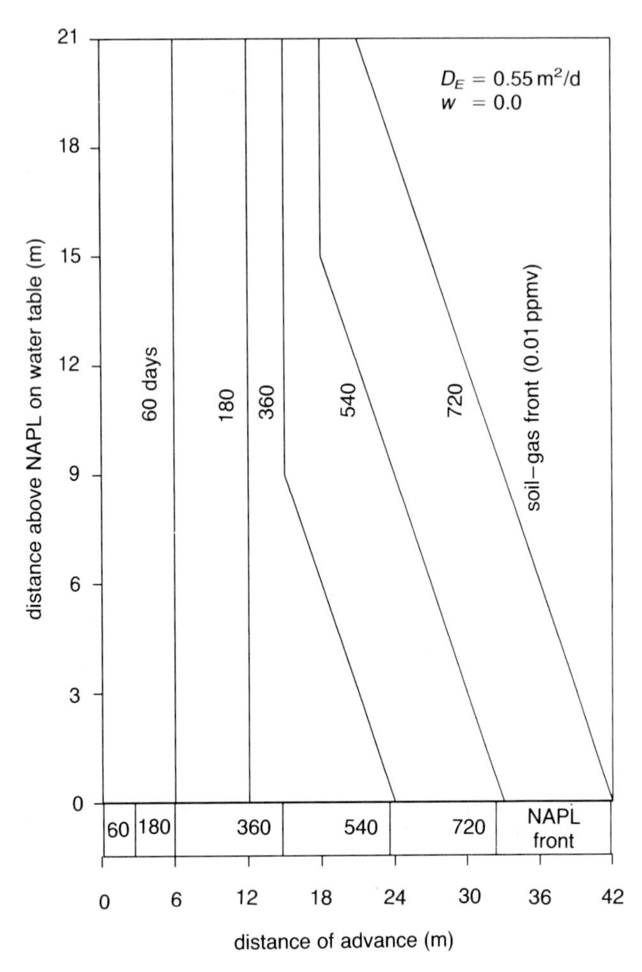

Fig. 16.10 Extent of migration of 0.01 ppm contour for VOC in soil gas from a vertical spill source and advancing NAPL front floating on the water table under dry soil conditions, from Silka (1988).

Figures 16.10 and 16.11 show the results in terms of the relative positions of the NAPL front and the 0.01 ppmv concentration contour. The results presented in these figures can be applied to the case of only dissolved VOC in a ground water plume by dividing 0.01 ppm by 6000 ppm, i.e., the contours in the figures would be equivalent to 1.6×10^{-6} times the concentration of the VOC in the soil gas just above the water table.

Figure 16.10 indicates that for a source front, i.e., NAPL or dissolved contaminant plume, that moves slower than the diffusion rate, the VOC does diffuse beyond the plume edge as observed by Evans and Thompson (1986), up to a height of 12 m above the water table. Above

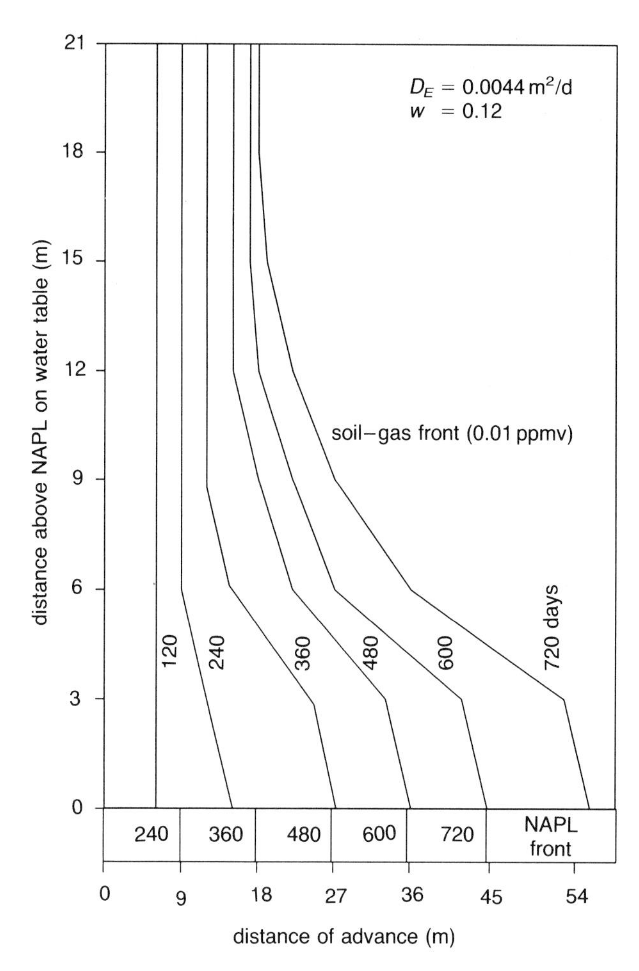

Fig. 16.11 Extent of migration of 0.01 ppm contour for VOC in soil–gas from a vertical spill source and advancing NAPL front floating on the water table under slightly moist soil conditions, from Silka (1988).

a height of 12 m from the water table, the vapor front falls behind the liquid front. However, Fig. 16.11 indicates that when the diffusion rate in soil gas is less than the velocity of the source front, the VOC distribution in the soil gas will lag behind the source front or edge of the plume at a much lower height above the water table. In the case illustrated in Fig. 16.11, the VOC diffusion in the soil gas begins to lag behind the front at a height of about 4.5 m above the water table. The lag increases to about 18 m at a height of 9 m above the water table. Therefore, a soil-gas survey may not always detect VOCs beyond the edge of the plume, resulting in the underestimation of the extent of the plume.

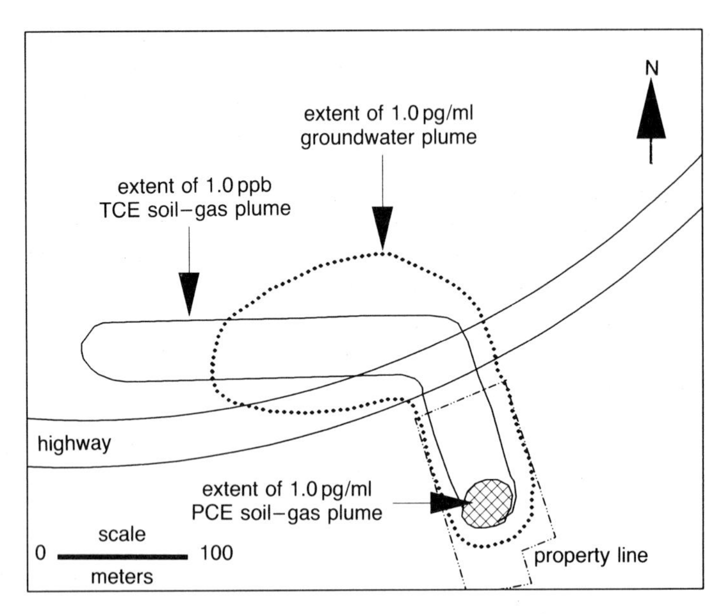

Fig. 16.12 Comparison of soil–gas survey results versus groundwater concentrations in total chlorinated species for southern California site, from Silka and Spectre (1991).

In fact, the lag of the soil-gas plume behind an advancing groundwater plume was observed in a study of a PCE contamination site reported by Silka and Spectre (1991). In this case, the site is in the semi-arid northern Los Angeles County of California. The geology consists of stratified alluvial silt and sand with some clay with the water table at a depth of approximately 10 m. Figure 16.12 shows the comparison of the soil-gas plume measured at a depth of approximately 1.0 m with the groundwater plume, illustrating the phenomenon of the lagging soil-gas plume behind the groundwater plume. Figure 16.12 also shows the soil-gas plume is wider than the groundwater plume, which is due to the fact that the groundwater plume is not growing sufficiently in the transverse direction to overcome the velocity of the vapor migration.

Discriminating between ground water and surface sources

A problem that arises in the course of soil-gas surveys for both detection of surface sources, but more so, for mapping ground water contamination, is the potential interference caused by VOC vapors from another source. These interferences are especially problematic in highly industrialized areas with multiple sources. The vapor diffusing laterally outward from a surface source may be sufficient to prevent the detection of vapors emanating from contaminated ground water when the vapor concentration due to ground water contamination is much less than that due to the diffusion from the surface source. This potential problem was recognized by Marrin (1984) and further evaluated by Silka (1986).

Silka (1986) applied the model 2D-DIFF to estimate the potential interference from the upward diffusion of VOC vapors emanating from contaminated ground water. For the modeling, it was assumed that the highest VOC concentration in ground water underlying the area of the soil-gas survey was 100 ppb. The depth to the water table was about 10 m. This level of ground water contamination could result in soil-gas concentrations in the upper 1 m of soil as high as 150 ppb.

In comparison, a surface source of TCE, with a saturated vapor concentration of 72 000 ppm, could cause a concentration in soil gas at a distance of 30 m of as high as 72 ppm. Even with a relatively low saturated vapor concentration, for example TET at 6000 ppm, the concentration in soil gas at a distance of 30 m could still be several parts per million. Vapor concentrations of less than a part per million due to diffusion from contaminated ground water would be completely masked by such surface sources. However, more advanced GC fingerprinting of the VOCs would be valuable in deciphering complex site contamination.

16.3.2 Conclusions for soil-gas surveys

VOC vapor migration through the unsaturated zone is controlled primarily by diffusion. Where vapor is denser than soil gas by more than 1.15 times, such as near a DNAPL front, and the vadose zone is coarse grained and relatively dry, density-driven transport can be important. Vapor diffusion is adequately described by Fick's second law, and the effects of partitioning between soil-gas and soil moisture can be adequately explained by Henry's Law. Adsorption of VOCs onto soil organic matter can be accounted for by the empirical equilibrium relationships between the octanol-water partition coefficient and the liquid-solid partition coefficient. However, desorption of VOCs from soil solids into soil water appears to be kinetically controlled and can produce disequilibrium conditions.

Design of soil-gas surveys should be developed with an understanding of the potential extent and distribution of contaminants in the subsurface. Preliminary modeling of the diffusive transport using a model such as 2D-DIFF can provide useful criteria for designing the survey and interpretation of subsequent results. Modeling results demonstrate the importance of soil moisture content for the design of the soil-gas survey. Optimum conditions for soil-gas surveys occur when lengthy, dry soil conditions have preceded the survey, which usually occur during July, August, and September over much of the United States.

Interpretation of soil-gas survey results are hampered by unknown or poorly defined parameters, such as soil porosity, moisture content, organic matter content, as well as source history and characteristics. In general, the variation in soil moisture will have the greatest influence on the rate of diffusion of VOC vapors through the unsaturated zone, especially for those VOCs with small values of K_H. Slight increases in soil moisture dramatically reduce the effective diffusion rate and increase the time required for concentrations in soil gas to approach steady-state values. Dry soils in arid regions may allow quasi-steady-state concentration profiles to be approached. However, steady-state conditions probably are never approached in the humid, temperate regions where frequent, episodic wet and dry periods occur.

16.4 VAPOR EXTRACTION SYSTEMS

16.4.1 Methods

Vapor extraction systems (VES) can be effective in remediating unsaturated and saturated soils contaminated by volatile organic compounds (VOCs). The operating principles of VES are straightforward. Contaminated soil is flushed with fresh air via a vacuum extraction well, drawing VOCs from the soil. As the contaminant is drawn off, more VOCs go into the vapor phase as equilibrium is shifted towards the gas phase, and are again drawn off by the vacuum. This action serves to remove vapors from soil gas, promote volatilization and mass reduction of any pure product phase present in the subsurface, force dissolution of VOCs from soil water, and cause desorption of VOCs from soil solids. In addition, the increased aeration of the soil may increase biodegradation of the heavier compounds, which are not as amenable to vapor extraction. Numerous case studies have shown that VES is effective in remediating soils contaminated by VOCs such as gasoline and solvents (Clister and Roberts, 1990). There are a number of methods of vapor extraction that have been utilized. The simplest of these is the installation of one or more vapor extraction wells in a contaminated area with the application of a vacuum to extract VOCs. There are a number of variations on this theme which may increase the efficiency of the system.

Plunkett and Simmons (1990) suggest excavating contaminated soils, placing them in a sealed, above-ground area, and remediating them using a system of horizontal vapor extraction 'wells' placed in the contaminated soil pile. Several authors (Pezzullo *et al.*, 1990; Hutton,

1990) describe the utilization of dual extraction wells for extraction of contaminated groundwater and soil vapors from the same well. The groundwater table is lowered due to pumping from the extraction well, hence creating an induced vadose zone. This induced vadose zone is then remediated by vacuum extraction.

Air injection has been used in some cases to increase airflow to the subsurface and volatilization from the groundwater table. Marley *et al.* (1990b) demonstrated the use of air injection wells (sparging) to enhance volatilization in the saturated zone. A pulsed injection system was used to prevent formation of short-circuiting pathways.

Walter *et al.* (1990) propose using horizontal 'vapor drains' installed under a clay cap. The system was designed to operate at low flow rates to minimize treatment costs for exhaust gases, and to operate over the life of the clay cap (about 30 years).

16.4.2 Design parameters

The design of a VES system depends on a number of factors. The physical properties of the contaminant, such as its Henry's Law constant and vapor pressure, are important. In complex mixtures, such as gasoline, only the more volatile constituents will be removed by vapor extraction. The semivolatile fractions will remain resistant to removal by vapor extraction.

A study of site conditions is important in evaluating the possible effectiveness of VES. According to modeling performed by Johnson *et al.* (1990a), the factors which most significantly affect the performance of a VES are vapor flowrate, contaminant composition, and vapor flowpath relative to location of the contaminant. They also present modelling results for estimation of the time required to achieve steady-state vapor flow, which depends on permeability, radial pressure distributions, vapor flow rate per unit length of well screen, radial linear velocity distribution, effects of soil type on vapor flowrates, and water upwelling caused by vacuum wells. A soil-gas survey is recommended at the site to define the extent of contamination and the concentrations present in the subsurface.

Site conditions which are most important include the aerial extent of VOCs, depth to groundwater, soil type, and soil heterogeneity. Important soil properties include soil structure, porosity, permeability, and soil moisture content. The presence and location of soil heterogeneities will play an important part in the efficiency of any VES. Dry, coarse soils, with low organic carbon content will be remediated faster than moist, fine soils, with high TOC. As the percent moisture increases, air permeability decreases. In addition, more water available in the subsurface will cause greater amounts of VOCs to partition into the water phase. Increased organic carbon content will cause more

VOCs to sorb onto the soil matrix, and hence, lengthening the time that vapor extraction is required.

Contaminant concentrations in the subsurface can be measured directly in the field using a portable gas chromatograph, or inferred from more precise laboratory methods performed on collected soil, water, or vapor samples. Johnson *et al.* (1990b) gives the following equation to estimate vapor concentrations for a multicomponent mixture (such as gasoline) in the subsurface based on compounds known to be present

$$C_{est} = \sum \frac{x_i P_i^v M_{w,i}}{RT} \tag{16.11}$$

where: C_{est} = vapor concentration estimate (mg/L); x_i = mole fraction of component i; P_i^v = pure component vapor pressure at temperature T; $M_{w,i}$ = molecular weight of component i (mg/mol); R = gas constant (0.0821 atm/mol K); and T = absolute temperature (K). From C_{est}, one can calculate the estimated removal rate of VOCs by

$$R_{est} = C_{est} Q \tag{16.12}$$

where R_{est} is the estimated removal rate per well and Q is the vapor flow rate per well. The estimated removal rate can then be compared to an acceptable removal rate, given by

$$R_{acceptable} = \frac{M_{est}}{\tau} \tag{16.13}$$

where M_{est} is the estimated mass of the spill, and τ is the maximum acceptable cleanup time (Johnson *et al.*, 1990b). The number of wells can then be estimated by dividing $R_{acceptable}$ by R_{est}.

In some cases it may be possible to infer air permeabilities from measurements of saturated hydraulic conductivity, but this does not take into account such factors as variable saturation and presence of a pure product phase. Vacuum pump tests are needed to define air permeability accurately. Walter *et al.* (1990) measured air permeabilities under field moisture conditions at several different scales both in the laboratory and in the field. Their results varied over several orders of magnitude, apparently as a function of the different scales at which the measurements were made.

Air permeabilities can be measured in the field via pumping tests where air is drawn from an extraction well, and pressure drawdowns are measured at several distances from the pumping well. Pressure drawdown data can be used to estimate air permeability using (Johnson *et al.*, 1990b)

$$P' = \frac{Q}{4\pi m \dfrac{k}{\mu}} \int_{\frac{r^2 \varepsilon \mu}{4kP_{atm}t}}^{\infty} \frac{e^{-x}}{x}\, dx \tag{16.14}$$

For $(r^2\varepsilon\mu/4kP_{atm}t) < 0.1$ Eq. 16.14 can be approximated by

$$P' = \frac{Q}{4\pi m \dfrac{k}{\mu}}\left[-0.5772 - \ln\left(\frac{r^2\varepsilon\mu}{4kP_{atm}}\right) + \ln(t)\right] \tag{16.15}$$

where: P' = gauge pressure measured at distance r and time t; m = stratum thickness; r = radial distance from the pumping well; k = soil permeability to air flow; μ = viscosity of air (1.8×10^{-4} g/cm sec); ε = air-filled soil void fraction; t = time; Q = volumetric vapor flow rate from pumping well; and P_{atm} = ambient atmospheric pressure (1.0 atm = 1.013×10^6 g/cm sec)

Using Eq. 16.15, it can be shown that a plot of P' versus $\ln(t)$ produces a straight line with a slope A and intercept B given by

$$A = \frac{Q}{4\pi m \dfrac{k}{\mu}}$$

$$B = \frac{Q}{4\pi m \dfrac{k}{\mu}}\left[-0.5772 - \ln\left(\frac{r^2\varepsilon\mu}{4kP_{atm}}\right)\right] \tag{16.16}$$

If both Q and m are known, then the equation for slope can be rearranged and solved for k. If Q or m are unknown, then k must be solved for using the equations for A and B (Eq. (16.16)).

$$k = \frac{r^2\varepsilon\mu}{4P_{atm}} \exp\left(\frac{B}{A} + 0.5772\right) \tag{16.17}$$

Q may be difficult to estimate, due to short-circuiting effects. The total flowrate from the extraction blowers may not be acting over the full range of the formation, and flow from the ground surface may act to short-circuit the system.

Marley *et al.* (1990a) developed analytical and numerical air flow models to evaluate soil properties based on results of pilot VES operations for the design of full-scale VES operations. Their numerical solution allows simulation of soil heterogeneities and flow to partially penetrating wells. In addition, the model can be used to determine vapor extraction well spacings, screen intervals, air flow rates, and system equipment requirements. A steady-state finite-difference air flow model is calibrated against field results from an air pumping test to solve for the air permeability.

Walter *et al.* (1990) developed analytical expressions for computing drain spacings required to circulate a specified number of pore volumes in a given time for a given pressure drop. Johnson *et al.* (1990b) present a flowchart/decision-tree approach to decide if vapor extraction is practical for a given scenario. The decision to use soil venting is based upon vapor concentrations, vapor flow conditions to achieve acceptable removal rates, feasible vapor flow rates, presence of any residual material after venting, and possible negative side effects of soil venting.

16.4.3 Radius of influence

The radius of influence of a soil venting well is important to consider in the design of any VES system. Mutch *et al.* (1989) found much larger radii of influence on the field than predicted by theoretical models, and attribute this to soil heterogeneities, rather than simple anisotropy. Kuo *et al.* (1990) define the radius of influence as 'the radial distance where there is sufficient air flow to reduce the contaminant concentrations below an acceptable level within a pre-specified time frame'. The radius of influence depends on the characteristics of the vacuum equipment, the physical and chemical characteristics of the contaminant, and the characteristics of the soils. Kuo *et al.* (1990) developed a numerical model to estimate radius of influence and study the effects of changing various parameters associated with it. They found that the radius of influence was greatest when the ratio k_r/k_z equaled infinity, where k_r is the permeability in the radial direction, and k_z is the permeability in the vertical direction. The radius of influence is proportional to the ratio k_r/k_z. Trowbridge and Malot (1990) present some empirical observations of radius of influence.

16.4.4 Exhaust treatment

There are several major types of treatment currently used to deal with VES emissions: dispersion, activated carbon, catalytic oxidation, condensation and liquid extraction, internal combustion and thermal incineration. The simplest of these is dispersion, where the vapors are diluted in a dispersion stack. However, this may be problematic in a regulatory environment. Activated carbon is commonly used where emissions of VOCs to the atmosphere need to be controlled. Activated carbon requires monitoring for breakthrough of VOCs, and replacement, which can increase costs. Trowbridge and Malot (1990) discuss the use of a catalytic oxidizer with a conversion efficiency of 99.8%, which was able to meet stringent local air quality regulations. Rippberger (1990) discusses an innovative technique whereby all VOC emissions are burned in an internal combustion engine fitted with a catalytic converter. In turn, the engine is used to drive the vacuum extraction equipment. This results in a significant cost savings over carbon canisters.

16.4.5 Fractionation of vapors

Various effects have been observed in the operation of VES where the composition of the contaminant is altered by the removal of the lighter fractions. Christenson *et al.* (1989) report that the vapor extraction seemed to remove more of the chlorinated compounds than the aromatic compounds. Pezzullo *et al.* (1990) found that their system enhanced volatilization of DNAPLs present in fractured bedrock in the subsurface. Johnson *et al.* (1990b) found a shift in composition to less volatile compounds in the first 20 days of vapor extraction, but relatively constant concentrations after that time. They also recommend using a boiling point curve distribution, inferred from GC data, to help identify the shift in composition of a complex mixture such as gasoline. Modeling results from Johnson *et al.* (1990a) indicate that vapor extraction removes compounds in the order of their volatilities. Their model also predicts the loss of soil moisture from vapor extraction, which may be significant enough to affect soil microbial activity.

16.4.6 Advection vs. diffusion

Walter *et al.* (1990) used tracer tests to attempt to quantify VES system performance. Field results indicated a retardation factor of 10–20, while laboratory column experiments and theoretical calculations indicated that the retardation factor should be closer to two or less. This discrepancy was probably due to low permeability layers present in the field, i.e., similar to the 'dual porosity' in the case of saturated solute transport.

Heterogeneous soil may contain isolated or dead-end pores and have variable saturation, and thus not be strongly affected by the VES. Since advective movement of the VOC in these stagnant areas is negligible, diffusion of the contaminant will dominate. Diffusion driven transport is much slower than advective transport, and thus will impede the efficiency of a VES.

In a soil where some regions allow transport of VOCs by advection and other regions limit VOC transport to diffusion, the advection dominated portions of the soil will be quickly flushed of VOCs, while flushing the diffusion dominated portions will require much longer. The VOC extraction rate from such a soil will be high in the early stages of operation, during the advection dominated portion of flushing. Once the advection dominated region has been flushed, the extraction rate will exhibit considerable tailing, as VOCs slowly diffuse out of the stagnant regions of the soil into the advection dominated region. Therefore the extraction rate of a VES in the early stages of operation should not be extrapolated to the entire flushing period, since that early observed rate may only reflect the rate of contaminant removal from the advection dominated region of the soil. Thus the VES process should be quantified with special concern for the potential importance of diffusion dominated regions to the operation of the VES.

16.4.7 Previous studies

Previous studies include a one-dimensional model developed by Baehr and Hoag (1987) which describes advection and diffusion of gasoline vapors in an unsaturated sand column. They found that the rate of volatilization of organics was high relative to the rate of air flushing in areas of fairly open porosity. However, they also mention that this may not hold true in areas where vapor extraction is diffusion limited. Their results also show that even in an advection dominated system, the extraction rate will exhibit exponential decay, due to a shift in the organic phase to less volatile components, as the more volatile components are flushed out first.

Bowman (1987) cites empirical results from a VES installed at the Twin Cities Army Ammunition plant in Minnesota, which displays an exponential decline in extraction rate (Fig. 16.13). From the figure, it is apparent that the extraction rate is quite high in the early stages of operation, but drops off quickly and approaches a steady state value

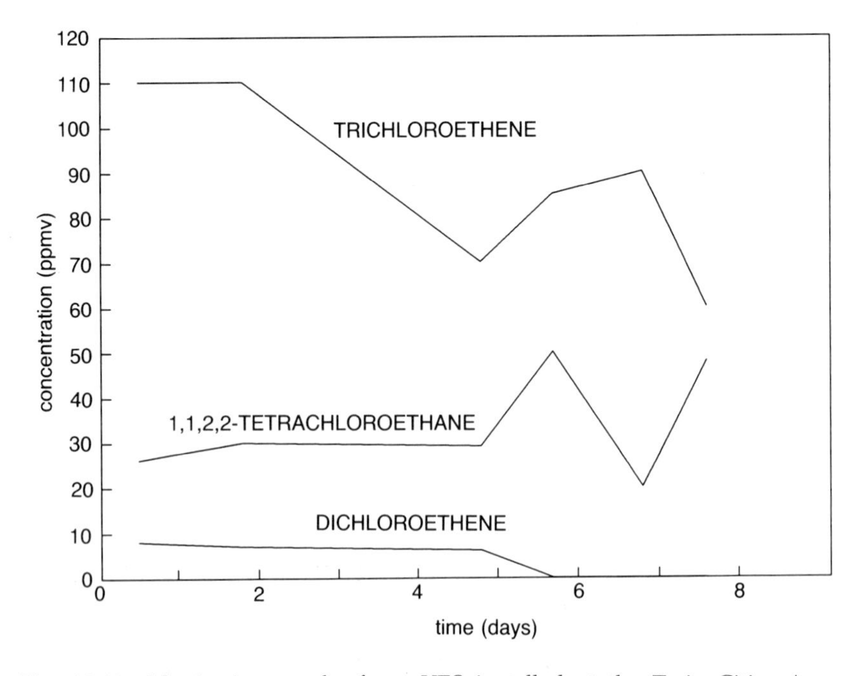

Fig. 16.13 Monitoring results for a VES installed at the Twin Cities Army Ammunition plant in Minnesota, from Bowman (1987).

asymptotically. Even after adding another blower at 120 days, the rate of extraction remained essentially constant at the asymptotically low value throughout the remainder of operation. The flattening of the curve is probably due to both fractionation of VOCs and diffusion dominated transport.

Woodward and Clyde (Woodward–Clyde Consultants, 1985) also present empirical data from a contaminated site in Tacoma, Washington (Fig. 16.14). Their data indicate that the extraction rate reached an equilibrium value within 90 days over a 260 day test. One possible explanation of their data is that the rate of air flushing is lower than the rate of volatilization from soil water and desorption from soil solids, thus the vapor being extracted was continually being replenished with VOCs. Therefore, the system remained in the advection dominated stage, and did not reach the stage where the rate of VOC removal would be limited by diffusion. The laboratory data for dichloroethylene (DCE) shows that the rate of extraction drops off significantly after an initial plateau, and remains fairly constant for three days. This may represent a transition from the advection dominated to the diffusion dominated regime.

Mutch *et al.* (1989) report that total VOC concentrations were in the range of 100–140 ppmv at startup and declined asymptotically to

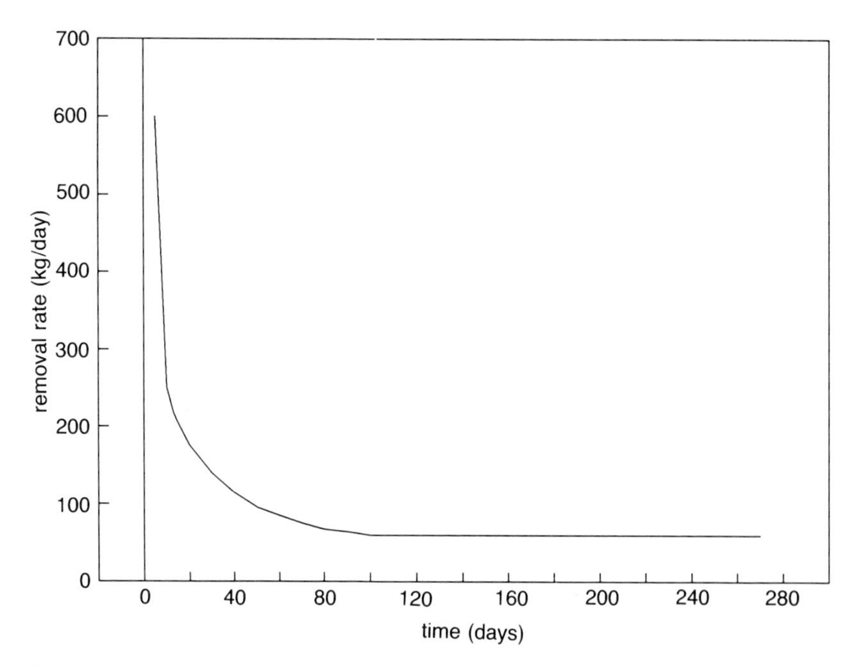

Fig. 16.14 Results of a vapor extraction test at the South Tacoma Channel-Well 12A Superfund Site showing total VOCs over time, from Woodward–Clyde Consultants (1985).

values of 60–70 ppmv. Crow *et al.* (1987) reported the most significant declines in concentrations of hydrocarbon vapors in the first few days of vapor extraction. Several case studies from Trowbridge and Malot (1990) show a diffusion-dominated regime at late times. Hutton (1990) compares analytical and pilot results. Pilot results showed an advection-dominated regime during the first 8–12 weeks, with diffusion dominating after that point. Their analytical solution fits well during early (advection-dominated), but not during later, diffusion-dominated time. They attempted to model the diffusion-dominated regime with an empirical boundary-layer solution, but with little success.

Numerical modeling results by Silka *et al.* (1989) of a vapor extraction system qualitatively agreed with results from several empirical studies. The proposed model consisted of two zones: an unsaturated, high porosity region which is advection dominated, and a saturated, low porosity zone which is diffusion dominated. The response of such a system to a VES has been shown to consist of an initial period of rapid vapor extraction, whereby the advective zone is flushed of soil gas. The advective period is followed by a sharp exponential decrease in the rate of extraction as the system becomes diffusion dominated.

Thus it is important to consider the diffusion stage when designing and pilot testing a VES, since predictions of extraction time for a given remedial action based on the early response of the VES often will be erroneous.

One other factor should also be considered in designing a VES. In a highly adsorptive soil, such as a soil with a high organic content, a large amount of contaminant will be resident in the solid matrix of the soil. This adsorbed contaminant mass will slowly desorb into the soil water and then into the soil gas, and keep the soil-gas concentrations from declining further. This will necessitate further soil-gas extraction. The amount of contamination in the different phases of the soil should be measured to gauge the magnitude of these effects. The combined effects of low-permeability zones (diffusion dominated) and delayed desorption can be assessed during VES operation by shutting the system down and monitoring the soil-gas concentrations. A soil with significant VOC mass contained in low-permeability zones and/or adsorbed to soil solids will exhibit an increased concentration in the VES discharge upon restart.

16.4.8 Boundary effects

The ground surface boundary condition is an important parameter to consider with VES. Numerical simulations by Mendoza and McAlary (1990) showed that loss of VOCs to the atmosphere could significantly reduce contaminant concentrations, especially at shallower depths.

Modeling results of Kuo *et al.* (1990) show that an impermeable boundary over an area being vacuum extracted has little effect. Field results from Diaz and Abidi (1990) show that air intake wells had little effect on the performance of the VES, and were subsequently closed.

Sensitivity analyses performed by Mendoza and McAlary (1990b) showed that the extent of subsurface contamination is sensitive to the ground surface boundary condition. For a diffusion-only case with an open ground surface which allowed mass to escape from the subsurface, the plume migrated more slowly than a plume beneath a partially or fully covered ground surface. The boundary effect is reversed when advection is included, with the plume under the open ground surface travelling the farthest.

Staes *et al.* (1992) report on the use of a three-dimensional groundwater flow model for assisting in the design of a vapor extraction system. While unavailable for review as of this writing, they probably converted the normal saturated water permeability to the gas-phase permeability, using the intrinsic permeability and density and viscosity of water versus air. The major assumption of the groundwater flow model being violated with this approach is that the model is based on the assumption that the fluid is not compressible. The potential impact of the violation of this assumption is that the model would tend to overpredict the mass transfer of gas.

16.4.9 Other VES design considerations

In addition to the considerations of subsurface conditions and their effects on the operation and performance of the VES, above ground system design considerations are important. Diehl (1992) reviews the minimization of emissions from soil vapor extraction (unavailable for review as of this writing). Depending on jurisdictions, GAC or other VOC removal may be required for the VES emissions. Innovative solutions, such as combustion, catalytic converters, and biodegradation have also been employed with varying degrees of success.

Fuerst and Underwood (1992) discuss vacuum extraction of volatile and semivolatile compounds at a Superfund site and Lawn (1992) discusses short-term vapor extraction system design (both unavailable for review as of this writing).

REFERENCES

API (American Petroleum Institute) (1985) *Detection of Hydrocarbons in Groundwater by Analysis of Shallow Soil Gas/Vapor*, API, Washington, DC.

Baehr, A.L. and Hoag, G.E. (1987) A modeling and experimental investigation of induced venting of gasoline-contaminated soils, in *Proc., Conf. on Environ. and Public Health Effects of Soils Contaminated with Petroleum Products*, 30–31 Oct 1987, Univ. of Massachusetts, Amherst, Massachusetts.

Bowman, R.S. (1987) Manipulation of the vadose zone to enhance toxic organic chemical removal, *Proc., 2nd Int. Workshop on Behavior of Pollutants in Porous Media*, 14–19 June. Bet Dagan, Israel.

Bruell, C.J. and Hoag, G.E. (1986) The diffusion of gasoline-range hydrocarbon vapors in porous media, experimental methodologies, in *Proc., NWWA/API Conf. on Petroleum Hydrocarbons and Organic Chemicals in Ground Water – Prevention, Detection and Restoration*, Nat. Water Well Assoc., pp. 420–43.

Callahan, M.A., Slimak, L.W., Gabel, N.W., May, I.P., Fowler, C.F. *et al.* (1979) *Water related environmental fate of 129 priority pollutants, vol. II*. EPA-440/4-79-029b, Environmental Protection Agency, Washington, DC.

Christenson, D., Loo, W., Relf, M. and Karst, G. (1989) *In situ* treatment of soil containing chlorinated solvents, ketones and aromatic organic compounds: 3 case histories, in *Proc. of Hazmacon 90*, Anaheim, California, Assoc. of Bay Area Governments, Oakland, CA., pp. 199–209.

Clister, W.E. and Roberts, B.D. (1990) Operating principles and case histories of soil vapor extraction systems, in *Proc. of the 11th Nat. Conf.*, Washington, DC, Haz. Mat. Control Res. Inst., pp. 646–57.

Crow, W.L., Anderson, E.P. and Minugh, E.M. (1987) Subsurface venting of vapors emanation from hydrocarbon product on ground water, *Ground Water Monitoring Rev.*, **7**(1), 51–7.

Diaz, G.M. and Abidi, S.F. (1990) Vapor extraction – a case history, in *Proc. of Hazmacon 90*. Anaheim, California, Assoc. of Bay Area Governments, Oakland, CA., pp. 247–55.

Diehl, K.M. (1992) Minimizing emission from superfund air strippers and soil vapor extractors, in *Proc. of HMC-South '92 Conf.*, HMCRI, Greenbelt, MD.

Dilling, W.L. (1977) Interphase transfer processes, II: Evaporation rates of chloro methanes, ethanes, ethylenes, propanes, and propylenes from dilute aqueous solutions; comparisons with theoretical predictions, *Envron. Sci. Tech.*, **11**(4), 405–9.

Evans, O.D. and Thompson, G.M. (1986) Field and interpretation techniques for delineating subsurface petroleum hydrocarbon spills using soil gas analysis, in *Proc., NWWA/API Conf. on Petroleum Hydrocarbons and Organic Chemicals in Ground Water – Prevention, Detection and Restoration*, Nat. Water Well Assoc., pp. 444–55.

Falta, R.W., Javandel, I., Pruess, K. and Witherspoon, P.A. (1989) Density-driven flow of gas in the unsaturated zone due to the evaporation of volatile organic compounds, *Water Resources Res.*, **25**(10), 2159–69.

Ferre, P.T. and Silka, L.R. (1987) Application of the soil-gas survey to underground storage tank monitoring and leak detection, in *Proc., Eastern Regional Ground Water Issues: A Conf.*, NWWA, Dublin, Ohio, pp. 187–206.

Freeze, R.A. and Cherry, J.A. (1979) *Groundwater*, Prentice-Hall, Inc., Englewood Cliffs, NJ.

Fuerst, D. and Underwood, B. (1992) Vacuum extraction of volatile and semivolatile compounds at a superfund site, in *Proc., HMC-South '92 Conf.*, HMCRI, Greenbelt, MD.

Fusillo, T.V., Keoughan, K.M. and Norwood, J.M. (1991) The use of soil gas investigations for delineating chlorinated solvent contamination in soils and ground water, *Ground Water Management*, Book 7, Nat. Water Well Assoc., Dublin, Ohio, 163–75.

Glaccum, R., Michael, N., Evans, R. and McMillion, L. (1983) Correlation of geophysical and organic vapor analyzer data over a conductive plume containing volatile organics, in *Proc., 3rd Nat. Symp. on Aquifer Restoration and Groundwater Monitoring*, May 25–27, Columbus, Ohio, Nat. Water Well Assoc., pp. 421–7.

Godoy, F.E. and Naleid, D.S. (1990) Optimizing the use of soil gas surveys, *Haz. Mat. Control*, **3**(5), 23–9.

Hillel, D. (1972) *Soil and Water, Physical Principles and Processes*, Academic Press, New York.

Hutton, J.H. (1990) Remediation of ground water contaminated with volatile organic chemicals by *in situ* aeration: a case study, in *Proc., Hazmacon 90*, Anaheim, California, Assoc. of Bay Area Governments, Oakland, CA., pp. 223–34.

Johnson, P.C., Kemblowski, M.W. and Colthart, J.D. (1990a) Quantitative analysis for the cleanup of hydrocarbon-contaminated soils by *in situ* soil venting, *Ground Water*, **28**(3), 413–29.

Johnson, P.C., Stanley, C.C., Kemblowski, M.W., Beyers, D.L. and Colthart, J.D. (1990b) A practical approach to the design, operation, and monitoring of *in situ* soil-venting systems, *Ground Water Monitoring Rev.*, **10**(2), 159–78.

Jury, W.A. (1986) Chemical movement through soil, in *Vadose Zone Modelling of Organic Pollutants*, (eds S.C. Hern and S.M. Melancon) Lewis Publishers, Inc. Chelsea, Michigan, pp. 135–58.

Jury, W.A., Spencer, W.F. and Farmer, W.J. (1983) Behavior assessment model for trace organics in soil: I. Model description, *J. Environ. Qual.*, **12**(4), 558–64.

Jury, W.A., Spencer, W.F. and Farmer, W.J. (1984) Behavior assessment model for trace organics in soil: IV. Review of experimental evidence, *J. Environ. Qual.*, **13**(4), 580–6.

Karickhoff, S.W. (1984) Organic pollutant sorption in aquatic systems, *J. Hydraulic Eng.*, **110**(6), 707–35.

Karickhoff, S.W., Brown, D.S. and Scott, T.A. (1979) Sorption of hydrophobic pollutants on natural sediments, *Water Res.*, **13**, 241–8.

Kerfoot, H.B. (1987a) Shallow-probe soil-gas sampling for indication of ground-water contamination by chloroform, *Int. J. Environ. Anal. Chem.*, **30**, 167–81.

Kerfoot, H.B. (1987b) Field evaluation of three methods of soil-gas measurement for delineation of groundwater contamination, in *Proc., 3rd Ann. Symp. on Solid Waste Testing and Quality Assurance*, **1**, Amer. Public Works Assoc., pp. 1–67 to 1–80.

Kerfoot, H.B. (1988) Is soil-gas analysis an effective means of tracking contaminant plumes in ground water? What are the limitations of the technology currently employed? *Ground Water Monitoring Rev.*, **8**(2), 54–7.

Kerfoot, H.B. (1991) Subsurface partitioning of volatile organic compounds: effects of temperature and pore-water content, *Ground Water*, **29**(5), 678–84.

Kerfoot, H.B. and Mayer, C.L. (1986) The use of industrial hygiene samplers for soil-gas surveying, *Ground Water Monitoring Rev.*, **1**(4), 74–8.

Kuo, J.F., Aieta, E.M. and Yang, P.H. (1990) A two-dimensional model for estimating radius of influence of a soil venting process, in *Proc., Hazmacon 90*, Anaheim, California. Assoc. of Bay Area Governments, Oakland, CA., pp. 235–46.

Lappala, E.G. and Thompson, G.M. (1983a) Detection of ground water contamination by shallow soil gas sampling in the vadose zone, in *Proc. of the Characterization and Monitoring of the Vadose (Unsaturated) Zone, Dec 8–10, 1983*, Las Vegas, Nevada, Nat. Water Well Assoc., pp. 659–79.

Lappala, E.G. and Thompson, G.M. (1983b) Detection of ground water contamination by shallow soil gas sampling in the vadose zone, in *Proc., Nat. Conf. on Management of Uncontrolled Haz. Waste Sites, Nov 7–9, 1984,* Washington, DC, Haz. Mat. Control Res. Inst., Silver Spring, MD, pp. 20–8.

Lawn, A.M. (1992) Vapor extraction system design for short-term remediation, in *Proc., HMC-South '92 Conf.*, HMCRI, Greenbelt, MD.

Leistra, M. (1973) Computation models for the transport of pesticides in soil, *Residue Rev.*, **49**, 87–130.

Lin, S. (1990) Soil gas analysis for the investigation of groundwater contamination by volatile organic chemicals, unpublished Ph.D. thesis, Univ. of California, Los Angeles.

Los Angeles County (1985) Remedial investigation of the San Fernando Valley groundwater basin, request for proposal for Consulting Eng. Services, RFP No. 739, Los Angeles County Dept. of Water and Power.

Malley, M.J., Bath, W.W. and Bongers, L.H. (1985) A case history: surface static collection and analysis of chlorinated hydrocarbons from contaminated ground water, in *Proc., Conf. on Petroleum Hydrocarbons and Organic Chemicals in Ground Water: Prevention, Detection, Restoration, Nov 13–15, 1985*, Nat. Water Well Assoc.

Marley, M.C., Nangeroni, P.E., Cliff, B.L. and Polonsky, J.D. (1990a) Air flow modeling for *in situ* evaluation of soil properties and engineered vapor extraction systems design, in *Proc., Hazmacon 90*, Anaheim, California, Assoc. of Bay Area Governments, Oakland, CA., pp. 208–22.

Marley, M.C., Walsh, M.T. and Nangeroni, P.E. (1990b) Case study on the application of air sparging as a complementary technology to vapor extraction at a gasoline spill site in Rhode Island, in *Proc., 11th Nat. Conf., Washington, DC*, Haz. Mat. Control Res. Inst., pp. 636–9.

Marks, B.J., Winsor, T.R. and Eames, M.A. (1989) Interpretation of soil gas survey results for petroleum hydrocarbons, in *Proc., Hazmacon 90*, Anaheim, California, Assoc. of Bay Area Governments, Oakland, CA., pp. 57–67.

Marks, B.J., Baltezore, M.L. and Radke, H. (1990) Quality assurance and quality control in soil-gas surveys, in *Proc., Hazmacon 90*, Anaheim, California, Assoc. of Bay Area Governments, Oakland, CA., pp. 31–41.

Marrin, D.L. (1984) Remote detection and preliminary hazard evaluation of volatile organic contaminants in ground water, PhD dissertation, Univ. of Arizona.

Marrin, D.L. (1988) Soil-gas sampling and misinterpretation, *Ground Water Monitoring Rev. J.*, **VIII**(2), 51–3.

Marrin, D.L. and Thompson, G.M. (1987) Gaseous behavior of TCE overlying a contaminated aquifer, *Groundwater*, **25**(1), 21–7.

Marshall, T.J. and Holmes, J.W. (1979) *Soil Physics*, Cambridge University Press, New York.

Mehran, M., Nimmons, M.J. and Sirota, E.B. (1983) Delineation of underground hydrocarbon leaks by organic vapor detection, in *Proc., Nat. Conf. on Management of Uncontrolled Haz. Waste Sites*, Oct 31–Nov 2, 1983, Washington, DC, Haz. Mat. Control Res. Inst., Silver Spring, MD, pp. 94–7.

Mendoza, C.A. and Frind, E.O. (1990a) Advective-dispersive transport of dense organic vapors in the unsaturated zone: 1, model development, *Water Resources Res.*, **26**(3), 379–87.

Mendoza, C.A. and Frind, E.O. (1990b) Advective-dispersive transport of dense organic vapors in the unsaturated zone: 2, sensitivity analysis, *Water Resources Res.*, **26**(3), 388–98.

Mendoza, C.A. and McAlary, T.A. (1990) Modeling of ground-water contamination caused by organic solvent vapors, *Ground Water*, **28**(2), 199–206.

Millison, D., Marcotte, B. and Harris, J. (1990) Applications and Comparison of Soil Gas, Flux Chamber and Ambient Air Sampling Results to Support Risk Assessment at a Hazardous Waste Site, in *Proc. 11th Nat. Conf.*, Washington, DC, Haz. Mat. Control Res. Inst., pp. 290–6.

Montgomery, J.H. and Welkom, L.M. (1990) *Groundwater Chemicals Desk Reference*, Lewis Publishers, Inc., Chelsea, Michigan.

Mutch, R.D., Clarke, A.N., Clarke, J.H. and Wilson, D.J. (1989) *In situ* vapor stripping: preliminary results of a field-scale US EPA/Industry Funded Research Project, in *Proc., 10th Nat. Conf.*, Washington, DC, Haz. Mat. Control Res. Inst., pp. 562–9.

Nazaroff, W.W., Lewis, S.R., Doyle, S.M., Moed, B.A. and Nero, A.V. (1987) Experiments on pollutant transport from soil into residential basements by pressure-driven airflow, *Environ. Sci. Tech.*, **21**(5), 459–75.

OTA (1984) *Protecting the Nation's Groundwater from Contamination*, **1**, OTA-O-233, US Government Printing Office, Washington, DC.

Perry, H.P. and Chilton, C.H. (eds) (1973) *Chemical Engineers Handbook*, 5th edn, McGraw-Hill Book Company, New York.

Pezzullo, J.A., Peterson, R.M. and Malot, J.J. (1990) Full-scale remediation at a superfund site using *in situ* vacuum extraction and on-site regeneration: case study – phase I, in *Proc., 11th Nat. Conf.*, Washington, DC, Haz. Mat. Control Res. Inst., pp. 624–7.

Plunkett, J.B. and Simmons, T.P. (1990) On-site vapor extraction – demonstrated effectiveness, in *Proc., 11th Nat. Conference*, Washington, DC, Haz. Mat. Control Res. Inst., pp. 641–5.

Rao, P.S.C., Hornsby, A.G., Kilcrease, D.P. and Nkedi-Kizza, P. (1985) Sorption and transport of hydrophobic organic chemicals in aqueous and mixed solvent systems: Model development and preliminary evaluation, *J. Environ. Qual.*, **14**(3), 376–83.

Rippberger, M.L. (1990) Remediation of gasoline-contaminated groundwater: spray aeration/internal combustion oxidation, in *Proc., 11th Nat. Conf.*, Washington, DC, Haz. Mat. Control Res. Inst., pp. 865–7.

Rizvi, S.S.H. and Fleischacker, S.J. (1991) Determination of VOCs in subsurface soils using gas sampling: a case study, in *Proc. Haz. Mat. Control/Superfund '91*, HMCRI, pp. 479–84.

Roberts, P.V. and Valocchi, A.J. (1981) Principles of Organic Contaminant Behavior during Artificial Recharge, in *Quality of Groundwater, Proc. Internat. Symp., Studies in Environ. Sci.*, Vol. 17 (eds van Duijvenbooden, W., Glasbergen, P. and van Lelyveld, H.). Elsevier Scientific Pub. Co., the Netherlands, pp. 439–50.

Siegrist, R.L. and Jenssen, P.D. (1990) Evaluation of sampling method effects on volatile organic compound measurements in contaminated soil, *Environ. Sc. Tech.*, **24**(9), 1387–400.

Silka, L.R. (1986) Simulation of the movement of volatile organic vapor through the unsaturated zone as it pertains to soil-gas surveys, in *Proc., NWWA/API Conf. on Petroleum Hydrocarbons and Organic Chemicals in Grand Water – Prevention, Detection and Restoration*, Nat. Water Well Assoc., pp. 204–24.

Silka, L.R. (1988) Simulation of vapor transport through the unsaturated zone –

interpretation of soil-gas surveys, *Ground Water Monitoring Review J.*, **VIII**(2), 115–23.
Silka, L.R. *et al.* (1989) Modeling applications to vapor extraction systems, presented at the Soil Vapor Extraction Technology Workshop, June 28 & 29, US Environmental Protection Agency Risk Reduction Engineering Laboratory, Releases Control Branch, Edison, NJ.
Silka, L.R. and Spectre, Y. (1991) Soil-gas survey for mapping chlorinated solvent plume in groundwater, *Ground Water Management,* (5), Nat. Ground Water Assoc., 519–31.
Sleep, B.E. and Sykes, J.F. (1989) Modeling the transport of volatile organics in variable saturated media, *Water Resources Res.*, **25**(1), 81–92.
Smith, J.A., Chiou, C.T., Kammer, J.A. and Kile, D.E. (1990) Effect of soil moisture on the sorption of trichloroethene vapor to vadose-zone soil at picatinny Arsenal, New Jersey, *Environ. Sc. Techn.*, **24**(5), 676–83.
Staes, E.G., Ratzlaff, S.A. and Stellar, J.R. (1992) Vapor extraction design using a three-dimensional groundwater flow model, in *Proc., HMC-South '92 Conf.*, HMCRI, Greenbelt, MD.
Steinberg, S.M., Countess, R.J. and Lin, C.C. (1990) Comparison of the response of VOC analyzers to total hydrocarbon measurements by GC-FID-PID, in *Proc., Hazmacon 90*, Anaheim, California, Assoc. of Bay Area Governments, Oakland, CA., pp. 42–51.
Striegl, R.G. (1987) Transport of methane in the unsaturated zone, in *Proc., Assoc. of Ground Water Scientists and Engineers Program, Focus: Contaminant Hydrogeology, NWWA Ann. Meeting and Expo*, September 14–16.
Stuart, J.D., Lacy, M.J. and Roe, V.D. (1989) Ground water and soil-gas monitoring at sites of subsurface gasoline contamination in the Northwest, in *Preprinted Extended Abstracts*, Div. of Environ. Chem., Am. Chem. Soc., pp. 489–90.
Swallow, J.A. and Gschwend, P.M. (1983) Volatilization of organic compounds from unconfined aquifers, in *Proc. 3rd Nat. Symp. on Aquifer Restoration and Groundwater Monitoring*, Nat. Water Well Assoc., pp. 327–333.
Thibodeaux, L.J. (1979) *Chemodynamics, Environmental Movement of Chemicals in Air, Water, and Soil*, John Wiley & Sons. New York.
Thomsen, K.O. and Joyner, S. (1990) Source Identification and Characterization Using Areal and Vertical Soil Gas Sampling Techniques, in *Proc. 11th Nat. Conf.*, Washington, DC, Haz. Mat. Control Res. Inst., pp. 277–84.
Thomson, G.M. and Marrin, D.L. (1987) Soil gas contaminant investigations: a dynamic approach, *Ground Water Monitoring Review*, **7**(3), 88–93.
Trowbridge, B.E. and Malot, J.J. (1990) Soil remediation and free product removal using *in situ* vacuum extraction with catalaytic oxidation, in *Proc. Hazmacon 90*, Anaheim, California, Assoc. of Bay Area Governments, Oakland, CA., pp. 235–46.
Turk, L.J. (1975) Diurnal fluctuations of water tables induced by atmospheric pressure changes, *J. of Hydrology*, **26**, 1–16.
Verschueren, K. (1983) *Handbook of Environmental Data on Organic Chemicals*, 2nd edn, Van Nostrand Reinhold, New York.
Viellenave, J.H. and Hickey, J.C. (1990) Use of High Resolution Passive Soil Gas Analyses to Characterize Sites Contaminated with Unknowns, Complex Mixtures, and Semi-Volatile Organic Compounds, in *Proc. 11th Nat. Conf.*, Washington, DC, Haz. Mat. Control Res. Inst., pp. 277–84.
Voorhees, K.J., Hickey, J.C. and Klusman, R.W. (1984) Analysis of

groundwater contamination by a new static surface trapping/mass spectrometry technique, *Anal. Chem.* **56**, 2604.

Walter, G.R., Yiannakakis, A., Bentley, H.W., Hauptmann, M. and Valkengurg, N. (1990) Gas in a hat, in *Proc. 11th Nat. Conf.*, Washington, DC, Haz. Mat. Control Res. Inst., pp. 557–64.

Woodward–Clyde Consultants (1985) *Performance evaluation, pilot scale installation and operation, soil-gas vapor extraction system, Time Oil Company site*, USEPA Work Assignment No. 74-ON14.1.

Zdeb, T.F. (1987) Multi-depth soil-gas analyses using passive and dynamic sampling techniques, in *Proc., Petroleum Hydrocarbons and Organic Chemicals in Ground Water: Prevention, Detection, and Restoration*, Nat. Water Well Assoc., Dublin, OH.

CHAPTER 17

Vertical cutoff walls

Jeffrey C. Evans

17.1 INTRODUCTION

Vertical cutoff walls are installed in the subsurface to control horizontal movement of groundwater and contaminants. This chapter presents the objectives and applications for vertical cutoff walls, describes the types of cutoffs available and presents detailed information to choose, design and construct the appropriate cutoff wall.

Use of vertical cutoff walls in the subsurface initially began with groundwater control and structural applications in Europe before 1950 (Xanthakos, 1979). In these applications, vertical cutoff walls were installed into the subsurface to act as a barrier to horizontal groundwater flow during excavation. In addition, they were often designed with reinforced concrete to provide structural stability for the excavation within the cutoff wall perimeter. Slurry trench cutoff walls have been used in environmental applications since the 1970s and have come into widespread usage since the 1980s as a component in the overall remedial system to control flow of groundwater (Spooner *et al.*, 1984).

17.2 OBJECTIVES

Depending upon the objective of the vertical cutoff wall, different criteria are considered in the design. For example, if the objective is to minimize the rate of contaminant transport off-site, it is necessary to consider contaminant transport through the wall, potential degradation of the wall, and consequences of inadequate cutoff wall performance. Contrast this objective to the use of a cutoff wall to minimize the rate of on-site migration of uncontaminated groundwater from the up-

Geotechnical Practice for Waste Disposal.
Edited by David E. Daniel.
Published in 1993 by Chapman & Hall, London. ISBN 0 412 35170 6

gradient area. In this case, there is a reduced concern about deficiencies in the cutoff wall and potential degradation due to the flux of uncontaminated groundwater through the wall.

In the majority of the situations employing vertical cutoff walls, environmental control systems include a combination of the techniques described in this book. If the objective, for example, is to minimize the rate of contaminant transport off-site, the environmental control system may include a vertical cutoff wall coupled with a low permeability cover, groundwater withdrawal system, and treatment systems for the pumpage.

In defining the vertical cutoff wall objectives, it is important to determine whether the barrier is to act as a ground water barrier with low hydraulic conductivity or a contaminant transport barrier. For example, the flux of groundwater through a wall with known hydraulic conductivity can easily be calculated. However, if the wall is to act as a contaminant barrier, it is necessary to consider modes of contaminant transport in addition to advective transport. As described in Chapter 3, contaminant transport processes include advection and dispersion (dispersion includes both mechanical dispersion and molecular diffusion). As a result of establishing different objectives, the type of vertical cutoff wall and the design criteria for the cutoff wall depend very much upon the defined objectives of that cutoff wall.

17.3 APPLICATIONS

Vertical cutoff walls are traditionally considered for environmental applications as circumferential barriers; that is, they typically surround the site (Fig. 17.1). In this system, an inboard subsurface drainage system and cover over the contaminated area are typically used in conjunction with the circumferential vertical cutoff wall to protect offsite groundwater.

Vertical cutoff walls need not be circumferential. They may be used to effect a vertical barrier for only a portion of the site. For instance, at the Rocky Mountain Arsenal, the cutoff wall was used down-gradient of the contamination to improve the effectiveness of a groundwater withdrawal and treatment system and to permit recharge of treated water further down-gradient (Campbell and Quintrell, 1985).

Vertical cutoff walls are also employed in up-gradient locations. For instance, Evans and Hill (1984) describe a vertical cutoff wall that serves to reduce the rate at which clean, uncontaminated groundwater flows onto the site from up-gradient locations.

Vertical cutoff walls frequently key into a stratum of naturally low hydraulic conductivity (Fig. 17.1). A key is not always necessary or cost effective when contaminated groundwater is being extracted or when

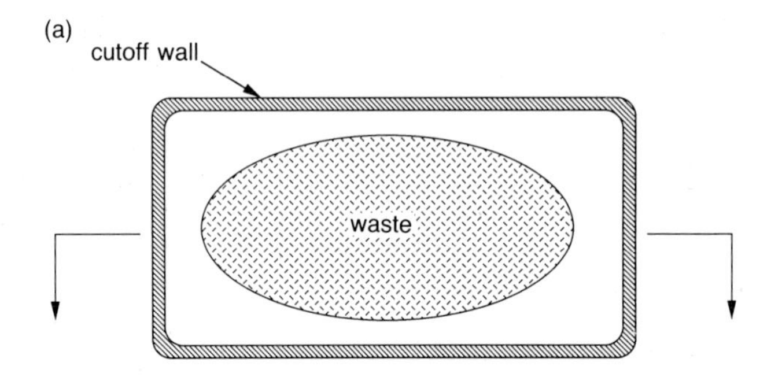

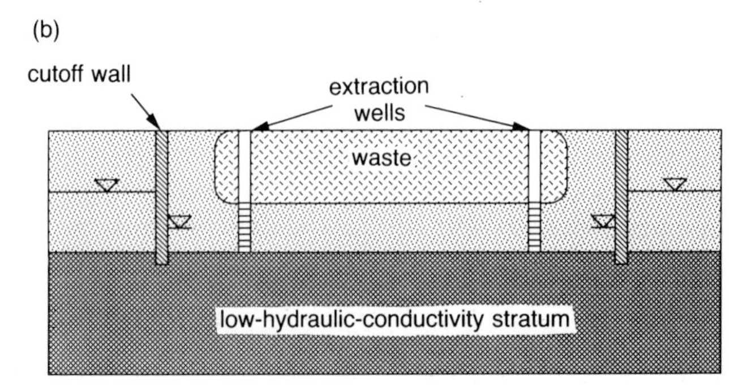

Fig. 17.1 Schematic diagram of vertical cutoff wall configuration for typical site remediation project: (a) plan view and (b) cross-section.

the contaminants are concentrated near the ground surface or floating on the water table.

17.4 TYPES OF VERTICAL BARRIERS

17.4.1 Vertical cutoff walls of low hydraulic conductivity

Traditionally low permeability vertical cutoff walls are constructed by installing a vertical barrier into the subsurface. The vertical barrier has a lower hydraulic conductivity than the associated formation. These vertical cutoff walls may be divided into several groups based on construction methods.

Several types of vertical cutoff walls may be constructed using the slurry trench method of excavation, and hence are called 'slurry trench cutoff walls'. In the slurry trench method of construction, a vertical trench is excavated into the subsurface utilizing a slurry, typically of bentonite and water, to maintain the trench stability. The completed excavation is then used to form the geometry of the cutoff wall. The completed cutoff wall may consist of soil-bentonite (Millet and Perez, 1981), cement-bentonite (Jefferis, 1981), plastic concrete (Evans *et al.*, 1987), or structurally reinforced concrete (Boyes, 1975).

In addition to the slurry trench methods of construction for vertical cutoff walls, there are grouting methods where materials of low hydraulic conductivity are injected into the subsurface. The method of injection varies and helps define two sub-classes of vertical cutoff walls. Grout curtains are vertical cutoff walls constructed using one of the traditional grouting techniques. Vibrating beam cutoff walls are constructed by inserting a steel beam into the subsurface using a vibratory pile driver. The steel beam is modified with grout injection nozzles and the void space is grouted as the beam is withdrawn (Leonards *et al.*, 1985).

In addition to grouting and slurry trench techniques, cutoffs can be constructed using other materials or other composite techniques. For example, steel sheet pile may be driven to provide a cutoff in the subsurface. Also, cutoff walls may be constructed with augering techniques, where soils are augered, materials injected through the augers, and the blend is mixed *in situ*.

17.4.2 Cutoff walls of high permeability

In addition to vertical cutoff walls which impede groundwater flow by virtue of their low hydraulic conductivity, a new technique has been developed to install a vertical cutoff by pumping from within a vertical drain of high hydraulic conductivity. This technique employs a biodegradable drilling mud to maintain the trench stability. The completed trench is backfilled with a material having a high hydraulic conductivity. Pumping from this highly permeable vertical zone can create a hydraulic barrier via control of flow patterns.

17.5 FEASIBILITY STUDIES

The first step in the design of a vertical trench cutoff wall is to identify the design objective(s). For example, the objective may be to provide a specific hydraulic conductivity to reduce the rate of groundwater flow. Upon establishment of the vertical cutoff wall objectives, it is then necessary to determine which types of vertical cutoff wall can meet these objectives. This can often be accomplished through the use of groundwater and contaminant transport models. All too often the cutoff wall is selected as a solution without first establishing the associated engineering of cutoff wall objective and analysis.

The next step is to evaluate the hydrogeologic and geotechnical aspects of the site. A vertical cutoff wall may be the best solution in terms of satisfying the project objectives; yet, if the cutoff wall is not constructible due to boulders in the subsurface or fractured rock, or unacceptable health or safety hazards posed by contaminated materials, then it is necessary to look at alternative solutions.

During the feasibility studies it is important to examine both short- and long-term performance of the cutoff wall. What will its properties be in terms of hydraulic conductivity, strength and compressibility? Are these properties satisfactory to meet the objectives of the project? In the long term, will the wall be durable?

During the design feasibility studies it is important to consider alternatives. Although soil-bentonite or cement-bentonite may appear to be the cutoff wall of choice, it is necessary to consider other alternative techniques to construct vertical barriers in the subsurface as described in this chapter. In fact, one alternative may be to not use a vertical cutoff wall, but to accomplish the objectives using other techniques, such as increased pumping rates.

Finally, it is necessary to consider cost. If construction conditions are so difficult and the potential for degradation is so real that the cost of the successful cutoff wall is not competitive with other techniques, then one needs to consider other techniques. Generally speaking, soil-bentonite slurry trench cutoff walls are among the least expensive techniques available in the US for vertical barriers in the subsurface. In Europe, cement-bentonite cutoff walls are more prevalent.

17.6 SOIL-BENTONITE SLURRY TRENCH CUTOFF WALLS

17.6.1 General information and technology

Soil-bentonite slurry trench cutoff walls are constructed utilizing the slurry trench method of excavation. In this method, a trench is excavated into the subsurface and a bentonite-water slurry is used to maintain trench stability, as shown in the typical section of Fig. 17.2. The excavation can be completed to depths of 15–20 m using a conventional backhoe with an extended stick. For greater depths a dragline or clamshell is usually employed. One contractor can excavate to 27 m with a specially modified backhoe (Case International, 1982). The slurry is typically a bentonite-water mixture consisting of 5% bentonite and 95% water by weight. Bentonite, a montmorillonitic clay, swells in the presence of water, imparting a viscous nature to the fluid and aiding in the formation of a filter cake along the walls of the trench. For trench stability, the slurry level is maintained at or near the top of the trench, typically within 1 m. Trench widths vary from 0.6 to 1.5 m, with 0.9 m trench widths being typical. Trench stability results from the

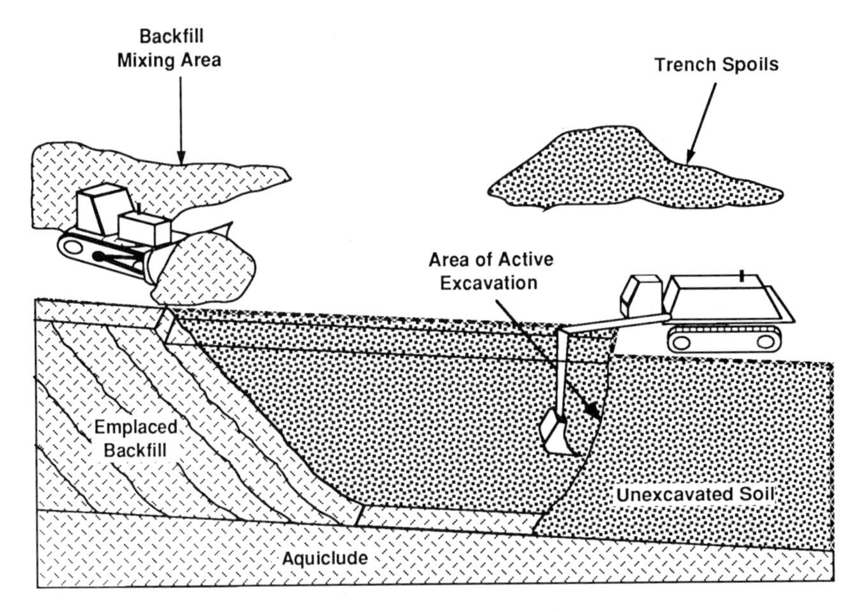

Fig. 17.2 Excavation of trench and placement of soil-bentonite backfill.

fluid pressures of the slurry exceeding the active earth pressures and the analysis of trench stability is well documented (Nash, 1974).

Once the trench is excavated to the desired depth, the bentonite-water slurry is replaced with a soil-bentonite backfill having a low hydraulic conductivity. The backfilled trench forms the completed vertical cutoff wall. The backfill optimally consists of a mixture of sand, silt, and clay, and bentonite-water slurry. The soil-bentonite backfill is placed in the trench at a consistency of high slump (100–150 mm) concrete. In order to achieve this consistency, the material needs to be 'fluidized' by adding bentonite-water slurry to soil.

17.6.2 Backfill placement

It is conventionally considered that the backfill sloughs forward and the emplaced backfill results in a homogeneous blend (D'Appolonia, 1980). Research incorporating analytical methods, laboratory model studies and field observations indicate that backfill 'sloughing forward' may not be the appropriate mechanism by which backfill displaces the slurry (Evans *et al.*, 1985b). Static equilibrium analysis using rotational equilibrium methods show that the critical failure surface is at a shallow depth. Backfill placed on top of the previously placed backfill causes shallow slope instability near the top of the trench. The mass in movement then slides down the top of the previously placed backfill as a

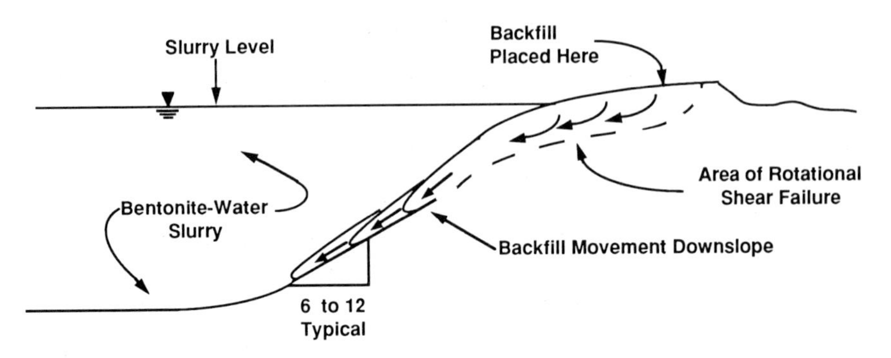

Fig. 17.3 Movement of soil-bentonite backfill into trench.

viscous fluid as shown on Fig. 17.3. Since backfill slides over previously placed backfill, entrapment of thin layers of soil-bentonite slurry between the layers is possible. It is believed that there is little or no mixing of one backfill batch with the next. Instead, backfill slides downslope on top of previously placed backfill.

17.6.3 Configuration of wall

During excavation of the trench for soil-bentonite slurry trench cutoff walls, the bentonite-water slurry within the trench will be level (horizontal). Thus, when excavating downslope or upslope, the depth of the slurry in the trench will vary with the site grade. If the surface of the slurry is not maintained at or near the ground surface, the risk of trench instability increases. As a result, there is a maximum grade for cutoff excavation. This limit is normally about 1% although it more precisely depends upon the results of stability calculations and the depth of excavation. A deeper excavation results in a longer open trench between the backfill and the excavation face. Final design studies must consider the site work necessary to adjust the grade of the cutoff wall including possible horizontal relocation of the vertical barrier cutoff wall.

The thickness of the vertical barrier wall depends upon the objectives, although walls with a width of 750–900 mm are typical. A wall that is too thin is difficult to excavate and backfill; a wall that is too thick may increase the costs unnecessarily.

The working surface next to the wall is about 15 m wide to provide for the backhoe and associated construction traffic. Wider surfaces allow for easier construction access. The working surface must provide a stable base to support the excavating equipment as well as any associated truck traffic. The working surface is normally constructed of on-site available materials because of the temporary nature of the use.

17.6.4 Excavation

Fluid levels are normally maintained within 1 m of the ground surface, although stability conditions may dictate stricter requirements. Fluid properties also are related to the stability needs, but fresh slurry has properties as shown on Table 17.1.

It is noted that excavations in cohesionless soils (sands and gravels) may experience some 'unraveling' of the slope at the top of the trench. In contrast, excavations in clay may become unstable due to pre-existing discontinuities which weaken the clay. In particular, vertical desiccation cracks may result in large chunks of clay falling into the excavated trench. The presence of tension cracks must be carefully controlled by limiting traffic along the sides of the trench and minimizing the time duration of the open trench.

The hydrated slurry filtrate loss of 25 cm^3 at a pressure of 689 kPa for 30 minutes indicate a high quality bentonite. A stabilized Marsh viscosity of 32–40 seconds provides a fluid material that is pumpable and thin enough to be displaced by the backfill, yet is adequately viscous to provide trench stability. The density of the slurry should be sufficient to maintain the trench stability, yet not too high to preclude backfill displacement of the slurry during placement.

17.6.5 Backfill

Final studies of the backfill mix design include bentonite content, natural fines content, grain size distribution of the base soil, slump, backfill hydraulic conductivity and permanence of the soil-bentonite backfill. Backfill can be prepared using soils excavated from the trench or using materials from a borrow source. The base soil is mixed with additional bentonite-water slurry into a homogeneous blend and then placed in the trench. As a result of the natural variability that may occur along the length of the trench, the grain size distribution of base materials from a borrow source may be easier to control and result in a more uniform soil-bentonite backfill. It is important to include natural fines in the backfill to reduce the hydraulic conductivity, although a strong relationship may not exist between fines content and hydraulic conductivity (Ryan, 1987).

As discussed earlier, the backfill is prepared at a consistency of high slump concrete, and the proper slump must be determined during the mix design studies. Laboratory studies are conducted to evaluate hydraulic conductivity and compatibility prior to construction, and it is important that these laboratory samples be prepared at the same slump (and water content) as required in the field. If slump is not controlled in the laboratory studies, the hydraulic conductivity and compatibility test data may not be representative of field conditions.

Table 17.1 Recommended properties of bentonite-water slurry

parameter	*bentonite slurry*		*cement-bentonite slurry*	
	as-mixed	*during excavation*	*as-mixed*	*during excavation*
density (g/cm^3)	1.01–1.04	<(β backfill −0.25)	1.10–1.24	1.10–1.4
viscosity, apparent (Seconds Marsh)	32–40	38–68	40–45	38–80
filtrate loss, ml	<25		100–300	
pH	7.5–12	10.5–12	12–13	
bentonite content, % by weight	4–7	~6	4–7	
other ingredients, % by weight	sand, <1	sand, <5	cement, 15–30	

Laboratory studies of the backfill during final design studies always include measurement of hydraulic conductivity, and for environmental applications, compatibility. Both organics and inorganics have been found to cause increases in bentonite permeability (Alther *et al.*, 1985). Chemical compatibility is typically evaluated by performing long-term permeability tests using the contaminants expected (Evans and Fang, 1988).

17.6.6 Technical specifications

In order to prepare the technical specifications for a soil-bentonite slurry trench cutoff wall, it is first necessary to establish the philosophy of specification. Some owners take the approach that the specification should provide guarantees for items over which the contractor may have little control. For example, the long-term performance of the cutoff wall may depend upon site and subsurface conditions that the contractor cannot control. The technical specifications should establish the reliability and responsibility of the contractor for those things that can be controlled. In this way, the vertical cutoff will be of high quality at a competitive cost.

The components of the technical specification for a vertical cutoff include contractor qualifications, site personnel qualifications, construction requirements, including materials, equipment and methods, quality control requirements, drawings and the basis of measurement and payment. The contractor qualifications section of the technical specifications often requires the bidder to establish experience in constructing vertical barriers for environmental applications.

The construction requirements established in the technical specification should clearly spell out which materials are and are not permitted, which equipment is and is not permitted, and what methods will and will not be permitted. The construction requirements should be as general as possible to permit contractor innovation, yet be as specific as possible to ensure that the selected construction method will not diminish the quality of the completed cutoff wall. The quality assurance/quality control (QA/QC) requirements of the technical specification must differentiate between that being provided by the contractor and that being provided by the owner through the design engineer. In general, the best philosophy is one which employs the design engineer's personnel to provide as much quality control as possible without alleviating the responsibility of the contractor for building a quality project.

The technical specifications must contain the basis for measurement and payment. For example, if there is excess slurry loss due to voids in the subsurface, is this the responsibility of the contractor to provide the extra bentonite-water slurry, or the owner? A well constructed technical specification will delineate slurry loss, which is expected, i.e., based upon the porosity of the subsurface materials, versus unexpected based upon information not provided to the contractor, i.e., the open pipe.

17.6.7 Construction considerations

Grades must be nearly level or discontinuities in grade (steps) must be considered for the vertical barrier construction. Corners must be carefully considered and space allocated for either criss-crossing or for a gentle radius to avoid backfill 'hang-up' at the corners and increased potential for backfill entrapment. The depth of the wall, often controlled by hydrogeologic considerations, must be reconsidered with respect to constructability. For example, shallow excavation during site preparation to permit a slightly lower grade may allow the entire project to be excavated utilizing only backhoe, without the need for specialized equipment required for greater depths. Platform width and traffic patterns impact the cost of construction. The easier it is for the contractor to build the job, the lower the price and the higher the quality.

As part of the final design studies, the continuity of the wall, the minimum depth of key, the slurry preparation and delivery to the trench, the disposition of unsuitable excavated materials and the stability of the trench must be evaluated. Cutoff wall continuity is typically assured by the method of excavation. The minimum key can be documented using soundings and sampling of excavated materials. Slurry mixing is accomplished using mixing ponds with overnight hydration or high speed colloidal shear mixers. Where space permits, the option should be provided to the contractor.

The homogeneity, slump, and the engineering properties of the backfill must be carefully considered so as to minimize slurry entrapment. Backfill mixing is accomplished along the side of the trench, or in a remote mixing area. The advantage of mixing along the side of the trench is cost, as no trucking of backfill is typically required for this alternative. The disadvantages include the need for a larger mixing area and the difficulty in adding natural fines (or sand and gravels) to augment the materials excavated from the trench for the base soil. The advantages of a remote mixing area include increased homogeneity, uniformity and control, reduced working pad widths needed along the trench, and finally the ease in which additional base soils can be brought into the area for mixing.

Slurry entrapment can be minimized by the contractor's method of construction including the need to ensure that no backfill is dropped through the slurry in the trench and that the soil-bentonite backfill is at the proper slump. Further, careful control of the backfill movement around corners must be made to ensure no hanging up at the corners.

The objective of the quality control program is to monitor the construction so that the vertical cutoff wall is built in accordance with the plans and specifications. It is equally important to verify that the site and subsurface conditions encountered during construction are consistent with those assumed during the design. It is common practice to include quality control duties with the contractor's responsibilities. In so doing, it is possible to address the first issue, which is to ensure that the wall is properly constructed. However, without a representative of the design engineer on site during construction, the second objective cannot be met, that is, assuring that the site and subsurface conditions are consistent with those assumed in the design. It is recommended that the quality control responsibilities be carried out by the design engineer responsible to the owner.

How do the responsibilities of the contractor and engineer differ on a slurry trench cutoff wall project? First, the contractor should be required to provide all of the supporting documentation for the quality of the materials. For example, the data sheets from the bentonite, grain size distribution analyses of any proposed borrow materials, specifications of the mixer, and the like. Second, the contractor should be responsible for providing all of the technical support necessary for the engineer to complete his quality control testing. For example, when the engineer needs to sample the mixed backfill, the bulldozer operator needs to momentarily halt the mixing process to provide the opportunity for the engineer to obtain his sample. Likewise, excavation must cease and the excavator must cooperate with the engineer in order to take soundings at the top of the aquatard and at the bottom of the key. The contractor must cooperate in other areas as well; in providing manpower assistance to take readings and to provide testing equipment where requested. It is the engineer's responsibility to conduct on-site tests for the control of the project. These tests must include measurements to document the excavation, the backfill and the materials used, including raw products and mixed products as discussed below. Thus, working as a team, not in an antagonistic relationship, the contractor and the engineer can work together to provide the owner with the best possible project at the lowest possible cost.

In order to provide control of the quality of the excavation, there are several tests and measurements for the engineer to make. First, soundings need to be obtained at the base of the trench excavation to determine final grade. Soundings are typically made at 3 m centers along the trench alignment. Further, for excavation into a key material, the engineer must be present to inspect visually the excavated material to assure that the excavation penetrates adequately into the key material. Testing includes soundings and sampling of the key material. Additional monitoring during excavation includes a one-time measurement of the backhoe bucket width to ensure that the trench width will meet the requirements and a measurement of the level of the working pad and excavation equipment to assure the verticality of the trench. Finally, testing of the bentonite-water slurry as-mixed and in the trench is required as discussed below.

Quality control testing of the backfill includes the measurement of slump to assure that the material is fluid enough to flow into the slurry trench cutoff wall, yet not so wet that it is too compressible for satisfactory long-term performance. The slope of the backfill below the slurry surface, optimally between 6H:1V and 12H:1V, is measured several times a day. In addition to samples for slump testing, samples are taken periodically for grain size distribution. Finally, samples are obtained and taken to document the laboratory hydraulic conductivity of field-mixed samples. It is noted that perfect homogeneity cannot be achieved under the field mixing conditions. For example, cohesive soils that are excavated from the underlying aquatard are typically broken down into clods 25 to 50 mm in diameter (a function of the tracks of the bulldozer used in blending backfill). Thus, adequate sampling, in quantity and quality, must be taken to ensure that the requirements of the project are being met despite this heterogeneity of the backfill. The methylene blue test may be used in the field to determine the bentonite content of the soil-bentonite backfill, provided a calibration curve is developed initially.

Quality control tests are required for the raw materials and the as-mixed products, specifically the bentonite and bentonite-water slurry. The bentonite should conform to the project specifications; usually a slurry grade bentonite with a yield between 90 and 150 barrels. In addition, the as-mixed bentonite-water slurry must have the proper Marsh viscosity, filtrate loss and mud density (API, 1984) characteristics to ensure slurry thin enough to be displaced by the backfill placement, yet thick enough (and of the appropriate density) to provide trench stability (Table 17.1). Finally, the unit weight of the as-mixed slurry is at least 1.02 g/cm^3 and is higher in the trench. A maximum density in the trench or a density differential may be specified. It is often specified that the bulk density of the slurry in the trench be a minimum of 0.25 g/cm^3 lighter than the backfill which is displacing the slurry.

17.6.8 Potential failure mechanisms

In order to design and construct properly a soil-bentonite slurry trench cutoff wall, it is useful to examine potential failure mechanisms. The failure mechanisms may be classified into two categories: (a) construction defects; and (b) post construction property changes. The following discussion addresses these potential failure mechanisms and identifies areas where the engineer and contractor can focus to minimize the potential for these failure mechanisms (Evans *et al.*, 1985a).

Improperly mixed backfill may contain lumps of unmixed granular soils and/or pockets of bentonite-water slurry. Either of these heterogeneities in the backfill result in zones of higher hydraulic conductivity within the cutoff wall. Improperly mixed backfill is particularly difficult to control when the backfill is mixed alongside the trench as compared to backfill mixed in a remote mix area.

Backfill inhomogeneity may result from slurry entrapment during placement. For example, if the backfill is too stiff (i.e., the slump is too low), the backfill may override the slurry and entrap pockets of slurry. Further, entrapped slurry may result if care is not taken during placement of the backfill and backfill is dropped through the slurry. Finally, the mechanism of backfill movement inherently risks the entrapment of thin layers of slurry, although it is expected that these consolidate to an inconsequential thickness.

Backfill inhomogeneity may also result from trench sediments covered by the backfill during placement. Trench sediments result from settlement of the more granular fraction from the slurry during periods on inactivity (i.e., overnight, weekends, holidays). Sediment accumulations need to be checked prior to the start of construction each day. The coarser fraction in the sediment results in an area of the trench with higher hydraulic conductivity.

Materials in the trench bottom also result from areas of trench instability visible from the surface but difficult for the contractor to remove. However, it is important to enforce the clean out of any cave-in from trench instability in order to provide a cutoff wall that does not contain windows. Finally, there may be spalling from the formation, which is not visible from the surface. Careful sounding of the trench bottom immediately prior to backfill placement can go a long way to ensure that backfill spalling does not occur.

Even with a properly constructed soil-bentonite slurry trench cut off wall without construction defects, changes in the properties of the soil-bentonite backfill which are time dependent can occur after construction. The time dependent property changes are discussed in the following paragraphs.

Increases in hydraulic conductivity can occur due to cycles of freezing and thawing, cycles of wetting and drying, the development of ice lenses within the backfill or desiccation. A properly designed final cover will prevent freezing and formation of ice lenses. Recent test results have shown increases in the hydraulic conductivity in the portion of the wall at and above the water table (Cooley, 1991).

Soil-bentonite slurry trench cutoff backfill is placed as a saturated cohesive material with high slump and, as a result, consolidates under the self-weight loads. Due to friction along the sidewalls of the trench, the consolidation settlement results in an unknown state of stress. If high hydraulic head conditions are imposed (such as beneath the dam), it is possible that hydraulic fracturing may occur. It has been found that hydraulic fracturing occurs when the hydrostatic stress exceeds the mean radial stress (Bjerrum *et al.*, 1972). It is unlikely that hydraulic fracturing will occur due to drawing down the water table on the down-gradient side of the cutoff wall.

The most important consideration for the long-term behavior of the soil-bentonite slurry trench cutoff wall is chemical incompatibility. Chemical incompatibility may be global as a result of contaminants dissolved in the ground water, causing hydraulic conductivity increases in the soil-bentonite. Local incompatibility may result from concentrated organics (non-aqueous phase liquids, or NAPLs) floating or sinking within the formation and attacking localized areas of the cutoff wall.

When using soil-bentonite slurry trench cutoff walls in waste disposal applications, it is necessary to control hydraulic conductivity increases due to chemical attack. Increases in hydraulic conductivity can be minimized by utilizing a well-graded base soil with a high natural fines content. In this way, the soil consists of a matrix of progressively smaller materials which are relatively inert (and not subject to changes in their diffuse ion layer as a result of changes in pore fluid chemistry). With a backfill designed in this manner, the hydraulic conductivity will remain quite low (i.e., about 10^{-6} cm/s) even with physiochemical attack and shrinkage of the bentonite.

It is also important to control settlement of soil-bentonite slurry trench cutoff walls so as to provide the maximum stress in the cutoff wall and reduce the potential for hydraulic fracturing. Fortunately, settlement can be controlled in the same way as control of increases in hydraulic conductivity through the use of a well-graded base soil with a high natural fines content.

17.7 CEMENT-BENTONITE SLURRY TRENCH CUTOFF WALLS

17.7.1 General information and technology

Cement-bentonite slurry trench cutoff walls are constructed using the slurry trench method of excavation. The difference between cement-bentonite and soil-bentonite vertical cutoff construction is that the slurry used in the excavation process is not replaced with a low permeability material as in the soil-bentonite slurry trench cutoff wall technique. Instead, the slurry, which contains bentonite, cement and water, is left to harden in place. The hydraulic conductivity of cement-bentonite slurry trench cutoff walls typically varies between 1×10^{-5} cm/sec and 1×10^{-6} cm/s, as measured in the laboratory on field mixed samples compared with 1×10^{-7} cm/s to 1×10^{-8} cm/s for soil bentonite.

17.7.2 Design feasibility studies

Since the materials excavated from the trench are not reused for cement-bentonite slurry trench cutoff walls, the ultimate disposal of these materials must be considered. The costs and logistics associated with disposal of the excavation quantities of hazardous waste must be considered in the feasibility analysis.

The shear strength of cement-bentonite increases with time and cement content. For typical mixes, the strength of cement-bentonite is similar to medium to stiff clays. Cement-bentonite backfill is much stronger and less compressible than soil-bentonite backfill but is more permeable. Where strength is a particularly important requirement, additional attention should be focused on this method of vertical cutoff wall construction.

Recently, progress has been made in developing construction techniques that permit the insertion of geomembrane or polymer sheet piling in a cement-bentonite wall. As the construction techniques are perfected and experience with them is gained, this type of design should see a large increase in use.

17.7.3 Final design studies

Final design studies incorporate most of the items described earlier for soil-bentonite slurry trench cutoff walls along with additional considerations unique to cement-bentonite. For example, it is important to study the hydraulic conductivity of cement-bentonite. Experience has shown that there is little relationship between hydraulic conductivity and cement content. Hydraulic conductivities $<10^{-7}$ cm/s can be achieved but only with difficulty. A hydraulic conductivity $\leqslant 10^{-6}$ cm/s is a typical design goal for cement-bentonite walls.

17.7.4 Technical specifications

Bentonite-water slurry used in the preparation of cement-bentonite slurry is normally prepared using a high speed colloidal shear mixer. In this mixer, a stabilized Marsh viscosity is obtained for the bentonite-water slurry in a matter of five minutes or less. The specifications must cite the requirements for the stabilized Marsh viscosity of the bentonite-water slurry. The cement is then added and the resulting cement-bentonite-water slurry is pumped to the trench.

As a result, at the end of each day, the trench is left filled with cement-bentonite and partially unexcavated. In the following construction day, it is important to fully remove the materials at the interface between the previously placed cement-bentonite and the unexcavated materials to assure a good key between the materials. This 'cold joint' is normally satisfactory, assuming a thorough removal of unexcavated materials from the joint area.

17.7.5 Construction considerations

The construction of a cement-bentonite slurry trench cutoff wall requires considerably less working area than the construction of a soil-bentonite slurry trench cutoff wall. There is no need to have large scale mixing ponds for the hydration of bentonite, nor is there a need to have a backfill mixing area. Thus the working platform can be relatively narrow, and the equipment used to mix the cement-bentonite can be very compact. The bentonite, the cement, and the mixers can all be efficiently located and the completed cement-bentonite slurry pumped to the trench. Further, concern about grades may be less significant for cement-bentonite. Since the trench is constructed incrementally (daily) and allowed to harden overnight, steps in the elevation of the trench can be made to permit construction along steeper site grades.

17.7.6 Quality control

The parameters measured for cement-bentonite quality control are similar to those for water-bentonite slurries. Note that filtrate loss is not normally measured for the cement-bentonite method of cutoff construction. This is because filtrate loss, a measurement of the quality of the bentonite, is greatly increased when the cement is added to the bentonite-water slurry.

It is important to obtain field mixed cylinders of cement-bentonite for subsequent laboratory permeability and strength testing. The cylinders should be cured at 100% humidity for the curing time (typically 28 days) prior to testing.

17.7.7 Failure mechanisms

The potential for construction defects due to backfill inhomogeneity and slurry entrapment is greatly reduced using cement-bentonite slurry trench method of construction. Because there is no backfill mixing or placement and the slurry is left to harden in place, the potential for construction defects is considerably reduced. The primary concern is an inadequately mixed batch of cement-bentonite-water slurry, or improper composition of the slurry. With proper construction control, however, the ingredients (bentonite, water and cement) can be added in the proper proportions and the possibility of construction defects considerably reduced.

Certain post-construction property changes are likewise of less concern for cement-bentonite than for soil-bentonite. Changes in long-term performance due to settlement and potential hydraulic fracturing are reduced due to the hydrated cement structure that develops. However, chemical incompatibility, both global and local must be considered. There is less data available on cement-bentonite than on soil-bentonite. Further, freeze-thaw, desiccation and wet-dry cycles, which could alter the performance of the wall as previously described, have similarly not been well studied.

Construction defects unique to cement-bentonite include accurate batching of mix components, 'first batch' effects, filtrate loss, and cold joints. The first batch effects result from the fact that the first batch of the day is prepared in a clean mixer, that is, clean water is added, bentonite hydrated, and then cement is added. This often results in a much thicker slurry than in subsequent batches, where a thin film of bentonite-cement slurry coats the mixer prior to the addition of the next batch of mix component. It is important to test the materials after the first batch is prepared to get an accurate representation of what typically appears in the field.

17.8 STEEL SHEET PILING

Steel sheet piling has found widespread application in conventional civil engineering construction for reduction of water flow in the subsurface. However, for environmental applications, steel sheet piling has rarely been employed. This is because steel sheet piling is subject to leakage through the interlocks, and as such, the leakage is higher than typically accepted for waste containment applications.

In order to reduce seepage through the interlocks, steel sheet piling has been used in conjunction with cement-bentonite on at least one occasion. In this project, a cement-bentonite slurry trench cutoff wall was excavated into the subsurface. Prior to the initial set of the cement-bentonite slurry, steel sheet piling was inserted into the cutoff wall. In this way, a composite wall was constructed which had the benefits of high resistance to hydraulic fracturing under the projected head conditions, plus the impedance of leakage through the joint by essentially grouting the joints with cement-bentonite. Although this solution was costly, the limited length of the required cutoff wall made it the most practical solution.

Until developments are made to reduce the leakage through the interlocks of steel sheet piling, it is likely that the use of steel sheet piling as vertical cutoffs in waste disposal applications will be limited. It is further noted that the relatively high cost of the steel sheet piling for a permanent installation provides further disincentive to its use.

17.9 PLASTIC CONCRETE

Recent developments in the study of materials for use in slurry trench cutoff walls had given rise to the use of plastic concrete in vertical cutoff wall applications. In a plastic concrete vertical cutoff, the trench is excavated in panels using the slurry trench method of construction, as in a soil-bentonite slurry trench cutoff wall. However, the low hydraulic conductivity of the completed wall is a result of replacement of the bentonite-water slurry with a plastic concrete material. This plastic concrete, often known as lean concrete, consists of a mixture of water, cement, aggregate, and bentonite. The resultant characteristics provide material that is flowable (as soil-bentonite backfill), but hardens in place (as cement-bentonite). The resultant plastic concrete wall is significantly stronger than soil-bentonite and has low hydraulic conductivities desirable for environmental applications.

The hydraulic conductivity of plastic concrete is typically less than 1×10^{-8} cm/sec. Further, the long-term hydraulic conductivity in response to permeation with concentrated organic fluids indicates no degradation in hydraulic conductivity. This is in significant contrast to the degradation which is demonstrated using soil-bentonite slurry trench cutoff walls. Thus, for environmental applications, plastic concrete appears to provide a combination of low initial hydraulic conductivity and excellent long-term endurance.

Plastic concrete has been used in several applications, but its use is limited compared with soil-bentonite because of high cost. The cost of a plastic concrete backfill is greater than that of a soil-bentonite backfill, particularly where the excavated material in a soil-bentonite trench can be reused. However, its durability and permanence in specific applications may outweigh the cost differential, particularly where the cutoff wall is used as a component in an overall groundwater remediation program.

17.10 DIAPHRAGM WALLS

Vertical cutoffs can be constructed in the subsurface as diaphragm walls. A diaphragm wall is typically constructed in panels, as shown on Fig. 17.4 using a bentonite-water slurry to maintain the excavation stability. The bentonite-water slurry is displaced by concrete using the tremie method of displacement. The diaphragm wall has historically been used for dewatering of structures where the completed cutoff wall is used in the soil retention system. The cutoff wall may then be incorporated into the overall structure. The structural integrity of the cutoff wall is then considered in the overall stability of the completed structure.

Trench excavation is completed utilizing one of several techniques, including clam shell. After completion of the trench excavation, the pre-assembled reinforcing cage is lowered into place, and concrete is tremied to displace the bentonite-water slurry. The strength and hydraulic conductivity of the final wall is similar to reinforced concrete. The panels are completed in an alternating sequence and special precautions are made to ensure a leakproof joint.

Diaphragm walls have been successfully employed on many projects and visual examination of the completed cutoff walls is possible as interior excavation proceeds. Problems occasionally develop where a window in the wall results from improper tremie technique. In most cases, diaphragm walls provide a positive vertical cutoff to groundwater flow in the subsurface.

Diaphragm walls have not been applied to environmental applications due to their high costs.

17.11 VIBRATING BEAM CUTOFF WALLS

Vertical cutoffs can be constructed in the subsurface utilizing the vibrating beam method of construction. This is not technically a slurry trench method of construction since a trench is not excavated in the subsurface utilizing a slurry to maintain the trench stability. In this application, a beam is modified with grout injection nozzles and connected to a vibratory pile driver. The beam is vibrated into the ground, creating a slot, which is then filled with grout as the beam is withdrawn. Grout is also injected as the beam is inserted to act as a lubricant for beam penetration. Successive penetrations of the beam are made, and the beam follows the previous footprint and a continuous wall is constructed. Several studies have been shown to demonstrate the viability of this technique (Leonards *et al.*, 1985). The grout material can consist of cement-bentonite material or alternative materials such as bitumen based asphaltic materials.

The principal advantage of this technology is that it is not necessary to excavate materials from the subsurface and handle or dispose of the excavated materials. Since there is no excavation process, health and safety considerations, and cost, are potentially reduced. The principal disadvantage comes in the assurance of the integrity of the cutoff wall. Where walls are to be constructed in shallow depths in relatively loose materials, the stiffness of the beam goes a long way to ensure the

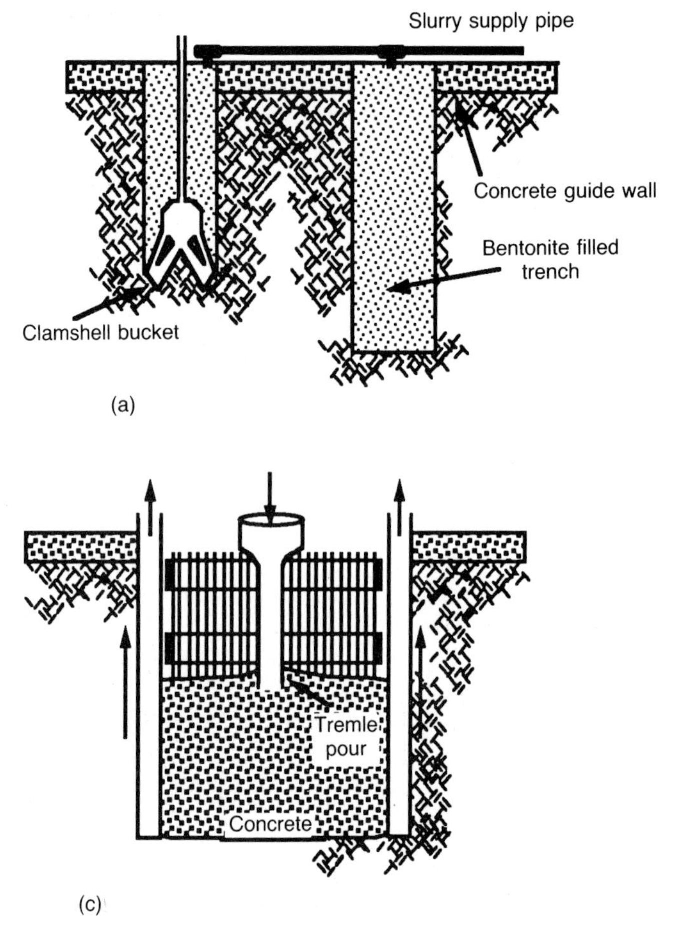

Fig. 17.4 Diaphragm cutoff wall: (a) excavate soil and replace with bentonite slurry; (b) place stop-end tubes and reinforcing steel into fully excavated panel; (c) pour tremie concrete to displace slurry, remove stop-end tubes, and (d) different construction phases.

continuity of the cutoff wall. However, where the barrier is very deep, or where the subsurface conditions vary, the stiffness of the beam can no longer ensure the continuity of the cutoff wall. Further, even in loose, granular materials which are initially easily penetrated with the vibrating beam, the vibration effects tend to densify the materials, making progressive penetrations successively harder. In this case, it may be necessary to pre-auger the materials (McLay, 1987).

The hydraulic conductivity of the vibrating beam cutoff wall will be

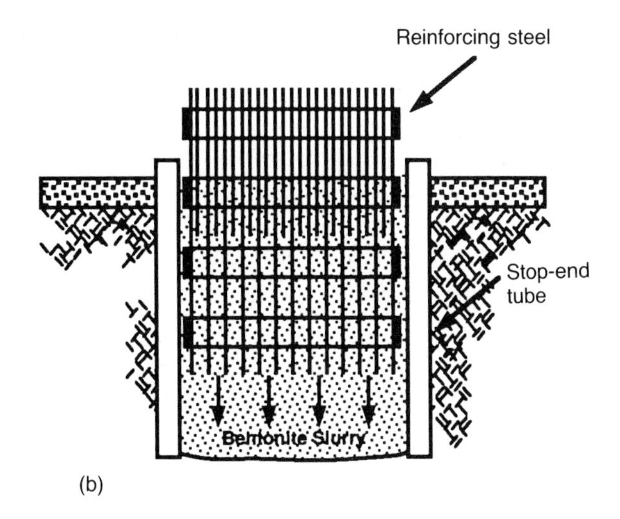

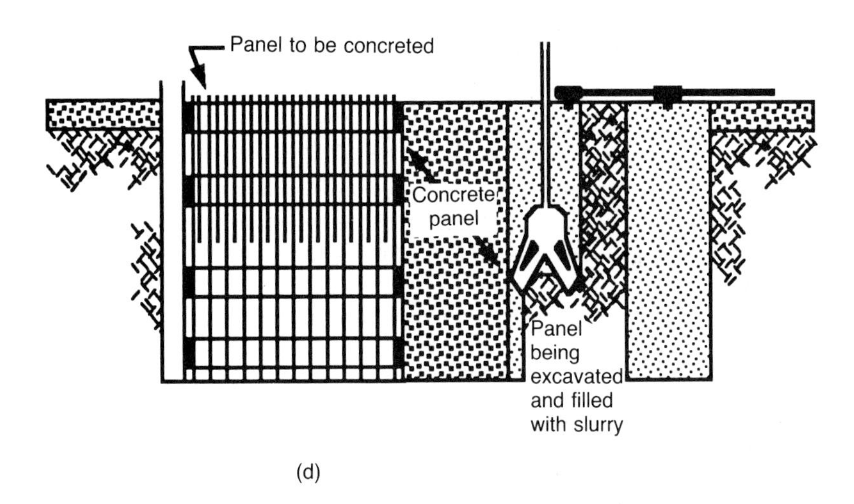

similar to that of cement-bentonite slurry trench cutoff walls and as a result may be higher than desirable. Further, since the wall thickness is related to the thickness of the beam, the thickness of these cutoffs are generally 150 mm or less.

The vibrating beam has found limited use in environmental applications. Design feasibility studies typically tip the scale toward this technique where the cost and hazards of excavating large quantities of hazardous materials impede the use of one of the more traditional techniques.

17.12 DEEP SOIL MIXED CUTOFF WALLS

A technique developed in Japan and recently used in the United States, known as deep soil mixing, can be employed to construct a vertical cutoff wall. A series of auger shafts, arranged in line, are used. The augers, which are modified to include mixing paddles, penetrate the ground while simultaneously allowing for addition of bentonite-water slurry. The result is a soil-bentonite cutoff wall mixed *in situ*. A continuous wall results from successive overlapping penetrations (Ryan, 1987).

The principal advantage is the reduced health and safety risk associated with *in situ* mixing. The principal disadvantage results from the engineer's lack of control over backfill properties.

17.13 VERTICAL DRAINS

Recent developments have resulted in the development of permeable vertical walls that serve as drains. Groundwater or leachate flows towards the vertical drain where the liquid is collected. From a theoretical standpoint, vertical drains can be analyzed as slots and flow to vertical permeable cutoffs is easily calculated.

The method of construction of the permeable cutoffs is similar to that for soil-bentonite slurry trench cutoff walls. A trench is first excavated under the head of a slurry. However, in the case of permeable cutoffs, the slurry is composed of a biodegradable drilling mud (a biopolymer) and water. In this way, a thickened mud is used during excavation, much the same as bentonite-water slurry. The slurry is then displaced with a permeable material such as sand and gravel. In time, the mud biodegrades, leaving an open formation. In this way, after the trench is backfilled and the mud biodegrades, a permeable slot remains in the subsurface from which groundwater can be pumped (Day, 1991).

The development of this technique provides tremendous advantages over previous construction techniques for the installation of drain tile collection systems. Previous to this large excavations with backslopes or trench supports were required, including the handwork of laborers within the trench. Using this technique, the trench support is provided by the biodegradable slurry, and the process proceeds much more rapidly, much more safely and at considerably lower cost.

17.14 CAPS FOR CUTOFF WALLS

The construction of a vertical cutoff in the subsurface usually necessitates the subsequent construction of a cutoff wall cap. Most of the barrier materials described are subject to desiccation and freeze-thaw degradation and may be relatively soft compared with the adjacent ground. The construction of a cap provides the necessary protection.

The final design studies must consider traffic loading, both parallel and transverse, and the development of design alternatives. Capping of cutoff walls can be easily accomplished using geosynthetics, geogrids, or high-strength geotextiles for reinforcement and geomembranes, compacted clay, or geosynthetic clay liners for control of water movement.

17.15 MONITORING PERFORMANCE

Since the cutoff wall is constructed beneath the head of bentonite-water slurry in the subsurface, as-built conditions are very difficult to ascertain. Once installed, the monitoring of vertical cutoff wall performance is also difficult. Methods to establish the cutoff wall performance include the establishment of background levels of groundwater and contaminant distribution. Wells can be strategically placed to monitor changes with time. These wells can monitor performance during implementation of the system and be used for field pump tests. In addition, geophysical approaches have been tried.

The most successful method to ensure adequate performance in the long term is to conduct thorough design studies described earlier and provide competent field control during construction.

REFERENCES

Alther, G.R., Evans, J.C., Witmer, K.A. and Fang, H.Y. (1985) Inorganic permeant effects upon bentonite, *Hydraulic Barriers in Soil and Rock*, ASTM STP 874, Am. Soc. for Testing and Materials, Philadelphia, pp. 64–74.

American Petroleum Institute (1984) *Standard Procedure for Field Testing Drilling Fluids, API Recommended Practice*, API RP 138, 10th edn, Dallas, Texas, June 1.

Bjerrum, L., Nash, J.K.T.L., Kennard, R.M. and Gibson, R.E. (1972) Hydraulic fracturing in field permeability testing, *Geotechnique*, **22**(2), 319–32.

Boyes, R.G.H. (1975) *Structural and Cut-Off Diaphragm Walls*, Applied Science Publishers Ltd., London, England.

Campbell, D.L. and Quintrell, W.N. (1985) Cleanup strategy for Rocky Mountain Arsenal, *Proc., 6th Nat. Conf. on the Management of Uncontrolled Haz. Waste Sites*, HMCRI, November 4–6, Washington, DC, pp. 36–42.

Case International Company (1982) *Case Slurry Wall Notebook*, Manufacturers Data, Case International Company, Houston, Texas.

Cooley, B.H. (1991) State of stress in soil bentonite slurry trench cutoff walls, Honors thesis, Bucknell Univ., Lewisburg, PA.

D'Appolonia, D.J. (1980) Soil-bentonite slurry trench cutoffs, *J. Geotech. Eng. Div.*, ASCE, **106**(4), 399–417.

Day, S. (1991) Extraction/interception trenches by the bio-polymer slurry drainage trench technique, *Haz. Mat. Control*, **4**(5), Sept/Oct, 27–31.

Evans, J.C. and Fang, H.Y. (1988) *Triaxial permeability and strength testing of contaminated soils*, Advanced Triaxial Testing of Soil and Rock, ASTM STP 977, Am. Soc. for Testing and Materials, Philadelphia, pp. 387–404.

Evans, J.C., Fang, H.Y. and Kugelman, I.J. (1985a) Containment of hazardous materials with soil-bentonite slurry walls, *Proc., 6th Nat. Conf. on the Management of Uncontrolled Haz. Waste Sites*, Washington, DC, Nov, pp. 249–52.

Evans, J.C. and Hill, G.H. (1984) *Phase I Remediation Studies, Necco Park, Niagara Falls, NY*, Woodward-Clyde Consultants, June 18.

Evans, J.C., Lennon, G.P. and Witmer, K.A. (1985b) Analysis of soil-bentonite backfill placement in slurry walls, *Proc., 6th Nat. Conf. on the Management of Uncontrolled Haz. Waste Sites*, Washington, DC, Nov, 357–61.

Evans, J.C., Stahl, E.D. and Droof, E. (1987) Plastic Concrete Cutoff Walls, *Geotechnical Practice for Waste Disposal '87*, ASCE Geotechnical special publication (13), June, pp. 462–72.

Jefferis, S.A. (1981) Bentonite-cement slurries for hydraulic cut offs, *Proc., 10th Int. Conf. on Soil Mechanics and Foundation Eng.*, Stockholm, June 15–19, A.A. Balkema, Rotterdam, 435–40.

Leonards, G.A., Schmednecht, F., Chameau, J.L. and Diamond, S. (1985) *Thin slurry cutoff walls installed by the vibrated beam method, hydraulic barriers in soil and rock*, ASTM STP 874, (eds A.I. Johnson, R.K. Frobel, N.J. Cavalli and C.B. Peterson) Am. Soc. for Testing and Materials, pp. 34–44.

McLay, D. (1987) Installation of a cement-bentonite slurry wall using the vibrating beam method – a case history, *Proc., 19th Mid-Atlantic Ind. Waste Conf.*, Bucknell Univ., Lewisburg, PA, June, 272–82.

Millet, R.A. and Perez, J.Y. (1981) Current USA practice: slurry wall specifications, *J. Geotech. Eng. Div.*, ASCE, **107**(GT8), Aug, 1041–56.

Nash, K.L. (1974) Stability of trenches filled with fluids, *J. Const. Div.*, ASCE **100**(4), 533–42.

Ryan, C.R. (1987) Soil-bentonite cutoff walls, *Geotechnical Practice for Waste Disposal '87*, (ed. R.D. Woods), ASCE, New York, pp. 182–204.

Spooner, P., *et al.* (1984) *Slurry Trench Construction for Pollution Migration Control*, EPA-540/2-84-001, US Environmental Protection Agency, Cincinnati, Ohio.

Xanthakos, P.P. (1979) *Slurry Walls*, McGraw-Hill Book Co., New York.

CHAPTER 18

Cover systems

David E. Daniel and Robert M. Koerner

18.1 INTRODUCTION

Final cover systems serve a variety of functions for both new waste disposal units as well as site remediation projects. The general principles of design of cover systems are the same for new disposal units as well as old waste disposal areas undergoing remediation.

Final cover systems are a critical component in the overall process of managing liquid and gas movement into and out of the underlying contaminated material. The main objectives in designing a final cover system are to separate the buried waste or contaminated material from the surface environment, to restrict infiltration of water into the waste, and (in some cases) to control release of gas from the waste. Because of the variety of waste and site characteristics, each cover design must be tailored to the specific requirements of each project. It is important for design engineers to have a thorough understanding of the basic principles involved in cover design because the final cover system profile cannot be generalized for all sites and all waste types.

old

18.2 BASIC CONCEPTS FOR COVER SYSTEMS

Cover systems may be called upon to perform one or more of the following functions:

- raise ground surface elevation in low-lying areas;
- minimize the amount of runoff of precipitation;
- promote controlled runoff of whatever precipitation is remaining;
- separate the waste from plants and animals;
- prevent migration of perched leachate out of waste on side slopes;

Geotechnical Practice for Waste Disposal.
Edited by David E. Daniel.
Published in 1993 by Chapman & Hall, London. ISBN 0 412 35170 6

- limit infiltration of precipitation into the waste; and
- control release of gas from the waste.

The US Environmental Protection Agency (EPA) considers keeping water out of the contaminated material to be the prime element in the final cover (USEPA, 1989).

The cover must perform the intended functions for extended periods of time. The 'design life' of a cover depends primarily upon the nature of the waste, the hydrogeology of the site, the length of time that maintenance of the cover will be provided, and the likelihood of funds being available for repair of the cover. Unfortunately, too many designers are forced to design a final cover system for an actively decomposing and deforming body of waste; a better approach for such wastes might be to construct a temporary cover and then wait until substantial decomposition of the waste occurs before attempting to construct a final cover system. Unfortunately, this latter strategy may require the passage of many years before most of the decomposition is complete; special financial arrangements would be required because the responsible party may not be viable when the time finally comes to construct the cover. Thus, final covers are usually constructed shortly after the last placement of waste in new landfills or concurrently with overall cleanup activities for sites undergoing remediation.

The design of the final cover is complicated by environmental factors that include:

- temperature extremes, possibly including freeze/thaw to significant depths;
- cyclic wetting and drying;
- penetration by plant roots, burrowing animals, worms, and insects;
- total and differential settlement caused by compression of the underlying waste or foundation soil;
- temporary (or permanent) surcharge loads, e.g., soil stockpiles or materials storage;
- down-slope slippage or creep;
- vehicular movements on haul roads traversing the cover;
- wind or water erosion;
- deformations caused by earthquakes in seismically-active areas,
- long-term moisture changes caused by water movement into or out of the underlying waste; and
- alterations caused by gas derived from volatile components of the waste or decomposition products.

Because of these site-specific environmental stresses and conditions, the design of a cover system can be very challenging. It is often more difficult to provide an effective hydraulic barrier layer in a cover system than in a liner system because the cover system is challenged by unknown and unquantifiable stresses that do not act on liner systems buried deep beneath the waste.

Most cover systems are composed of multiple components. As shown in Fig. 18.1, the components of a cover system can be grouped into five categories:

1. surface layer;
2. protection layer;
3. drainage layer;
4. barrier layer; and
5. gas collection layer.

Not all components are needed for all final covers. For example, a drainage layer may be needed at a site with large precipitation but not at an arid site. Similarly, a gas collection layer is a required component of covers placed over wastes that produce gas but would serve no useful purpose for waste that does not produce gas.

The range of materials used for the various components of a cover system are also listed in Fig. 18.1. The materials used to construct the components of a final cover system can include natural earth materials, geosynthetics, a blend of natural materials, or a blend of earth and geosynthetic materials, e.g., geosynthetic clay liners.

18.3 DETAILS OF COVER SYSTEM COMPONENTS

18.3.1 Surface layer

There are four alternative materials that can be used for the surface layer:

1. soil (Figs. 18.2a and 18.2b);
2. soil with a geosynthetic erosion control layer placed at the surface (Fig. 18.3);
3. cobbles (Figs. 18.4a and 18.4b); or
4. paving material (Fig. 18.5).

The advantages and disadvantages of the various materials are identified below.

Alternate A: uniform protective soil (Fig. 18.2a)

Advantages

1. Construction material is locally available.
2. Special topsoil is not needed.
3. Soil stores water that has infiltrated; plants return much of this water to the atmosphere via evapotranspiration.

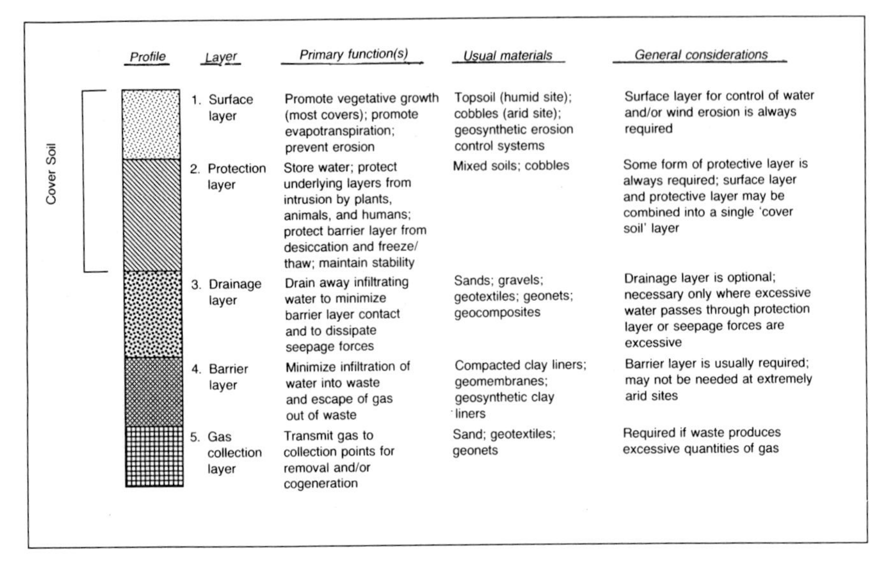

Fig. 18.1 Five possible components in a cover system.

4. Soil (if adequately thick) protects liner from freeze/thaw and wetting/drying.
5. Straightforward construction.

Disadvantages

1. Requires adequate precipitation to support growth of vegetation on soil.
2. Irrigation could be required during periods of drought to maintain vegetative cover.
3. Establishing vegetative cover may be difficult if weather conditions are unfavorable immediately after seeding.
4. Excessive wind or water erosion may occur prior to establishment of vegetative cover or if vegetation dies.
5. Vegetative cover may be difficult to establish (no topsoil).
6. Little resistance to intrusion by plant roots or burrowing animals.
7. Probably the thickest of all cross-sections adding to possible differential settlement.

Assessment

This design has been used where special topsoil is not needed to support growth of a suitable vegetative cover. This is the simplest of possible cover soil designs but provides little beyond physical separation between the ground surface and underlying layers and storage of water for evapotranspiration. The uniform protective soil layer should be relatively thick compared to other schemes.

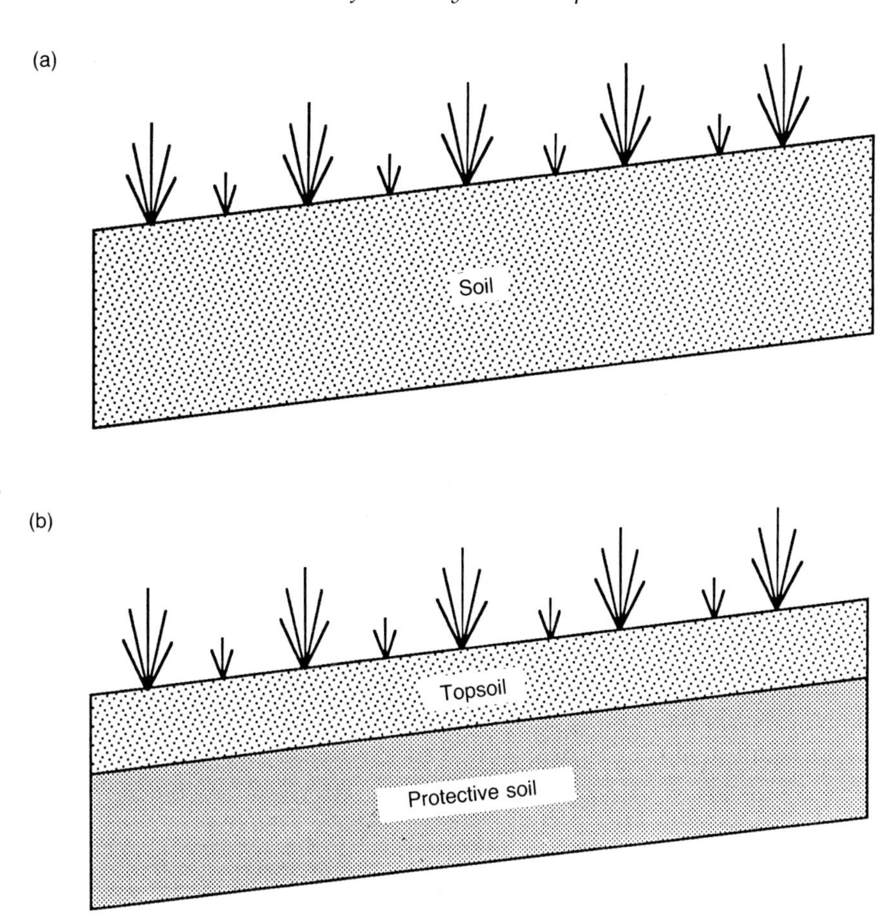

Fig. 18.2 (a) Uniform protective soil; (b) topsoil overlying protective soil.

Alternate B: topsoil overlying protective soil (Fig. 18.2b)

Advantages

1. Construction materials are usually available locally.
2. Topsoil and protective soil layer store water that has infiltrated; plants return much of this water to the atmosphere via evapotranspiration.
3. Plants not only return water to the atmosphere but also reduce soil erosion and stabilize the topsoil.
4. Thickness of protective soil layer can be adjusted as necessary (e.g., to prevent barrier soil from freezing) without affecting the thickness of topsoil.
5. Plant roots tend to be confined to nutrient-rich topsoil (deep penetration of roots is undesirable).

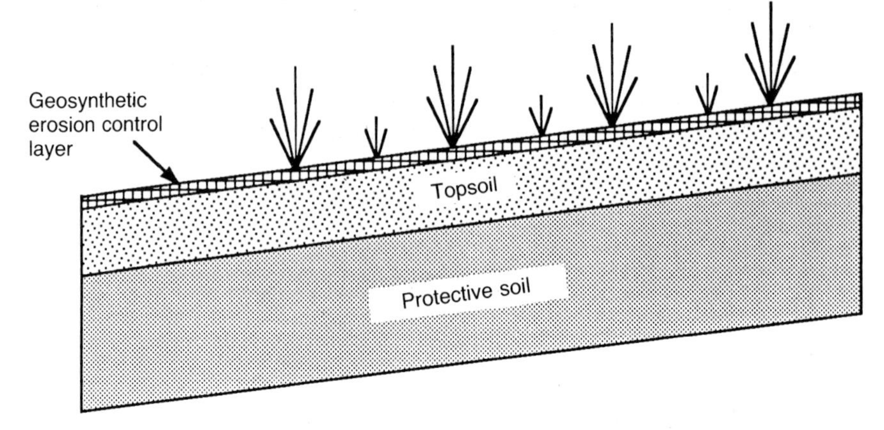

Fig. 18.3 Erosion control layer over topsoil over protective soil in a cover system.

Disadvantages

1. Requires adequate precipitation to support growth of vegetation on topsoil (design unsuitable for some arid sites).
2. Irrigation could be required during periods of drought to maintain vegetative cover.
3. Establishing growth of vegetation may be difficult if weather conditions are unfavorable immediately after seeding.
4. Excessive wind or water erosion may occur prior to establishment of vegetative cover or if vegetative cover becomes thin.
5. Little resistance to intrusion by burrowing animals.

Assessment

This is the most common soil cover design used in cover systems and is the minimum required cover soil design established by many regulatory agencies. The main problems experienced with this design are maintaining a vegetative growth during drought periods and formation of erosion gullies in the top soil during excessive precipitation periods. This design can be very effective at sites that receive adequate rainfall but generally requires maintenance.

Alternate C: erosion control layer over topsoil over protective soil (Fig. 18.3)

Advantages

1. Topsoil and protective soil layer store water that has infiltrated; plants return much of this water to the atmosphere via evapotranspiration.
2. Plants plus erosion protection layer minimize soil erosion; erosion

protection layer minimizes erosion while plant growth is established and erosion rills and gullies afterward.
3. Thickness of protective soil layer can be adjusted as necessary without affecting the thickness of topsoil.
4. Plant roots tend to be confined to nutrient-rich topsoil (deep penetration of roots is undesirable).
5. Erosion layer may offer resistance to burrowing animals.
6. Erosion control geosynthetic can be anchored or reinforced to add stability to steep side slopes.
7. The geosynthetic erosion control material can be preseeded.

Disadvantages
1. Requires adequate precipitation to support growth of vegetation on topsoil (design unsuitable for some arid sites).
2. Irrigation could be required during periods of drought to maintain vegetative cover.
3. Establishing growth of vegetation may be difficult if weather conditions are unfavorable immediately after seeding.

Assessment
This design is not nearly as common as the same design without the geosynthetic erosion control layer. However, many erosion problems have been observed on covers; the erosion control layer is recommended for many (if not most) sites.

Alternate D: cobbles over protective soil (Fig. 18.4a)

Advantages
1. No need to establish or maintain vegetative growth (particularly advantageous for arid sites or sites subject to periodic long droughts).
2. Cobbles provide protection from wind or water erosion; effectiveness of cobbles not dependent upon maintenance of vegetative growth.
3. Cobbles may minimize penetration from burrowing animals if rocks are of adequate size.
4. Cobbles minimize growth of plants with deep roots that could penetrate underlying barrier layers.

Disadvantages
1. Cobbles must be adequately large to resist washout; materials may not be readily available locally.
2. Minimal evaporation of water from underlying soil (cobbles allow water to percolate downward but insulate soil from surface evaporation).

(a)

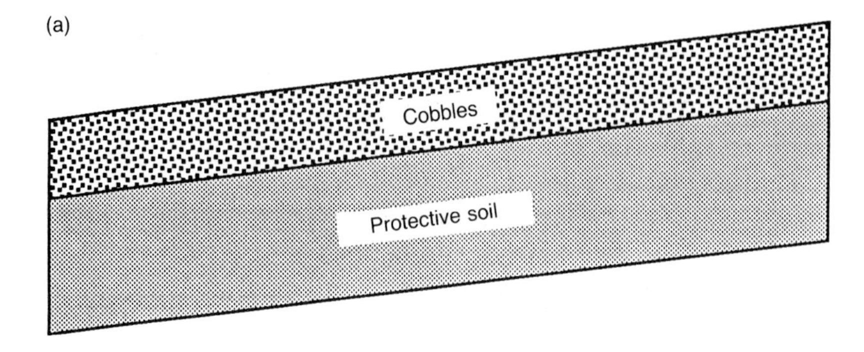

(b)

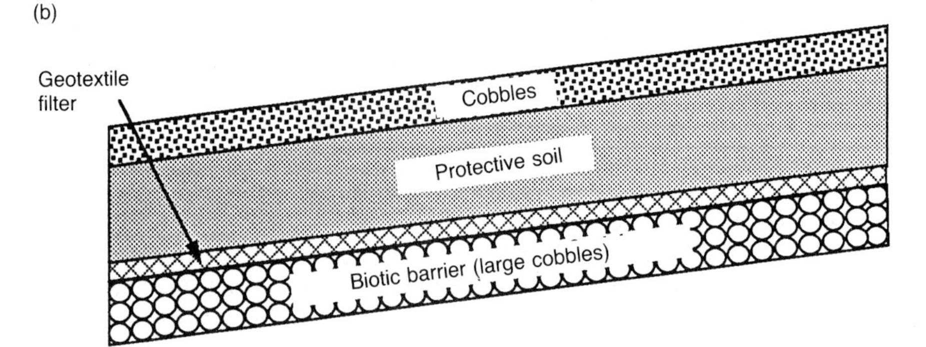

Fig. 18.4 (a) Cobbles over protective soil; (b) cobbles at surface and at depth.

3. Wind-blown materials may fill voids between cobbles; deeply-rooted plant growth may eventually become established and require maintenance.
4. Stability concerns for steep slopes.

Assessment
This design is usually considered only for arid sites. Provided problems with burrowing animals and lack of water evaporation from protective soil layer are not too serious, this design may be the most appropriate one for arid sites.

Alternate E: cobbles at surface and at depth (Fig. 18.4b)

Advantages
1. No need to establish or maintain vegetative growth (particularly advantageous for arid sites or sites subject to periodic major drought).
2. Cobbles at surface protect from wind or water erosion; effectiveness of cobbles not dependent upon maintenance of vegetative growth.

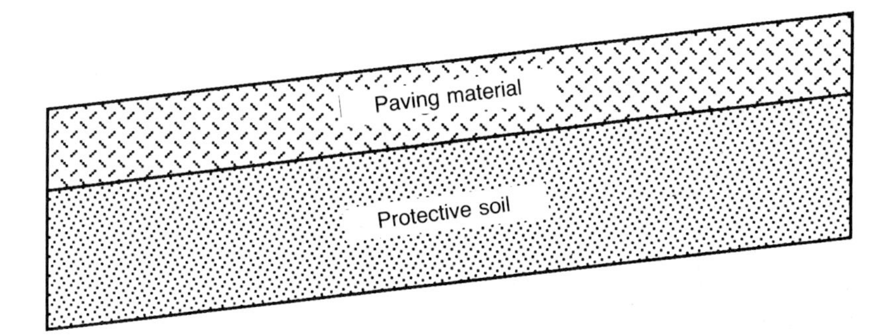

Fig. 18.5 Paving overlying protective soil in a cover system.

3. Cobbles buried in the cover (biotic barrier) minimize penetration from burrowing animals and plant roots if rocks are of adequate size.

Disadvantages
1. Cobbles at surface must be adequately large to resist washout and cobbles at depth must be very large to stop burrowing animals; materials may not be readily available locally.
2. Minimal evaporation of water from underlying soil (cobbles at surface allow water to percolate downward but insulate soil from surface evaporation).
3. Stability concerns for steep slopes.

Assessment
This design is usually considered only for arid sites requiring maximum security. A biotic barrier has rarely been used, except for low-level radioactive waste buried at arid sites. Provided problems with minimal water evaporation from protective soil layer are not too serious, this is a viable design for arid sites at which the main concern is uptake of buried waste constituents by plant roots or burrowing animals. This cross-section promises to be extremely costly.

Alternate F: paving overlying protective soil (Fig. 18.5)

Advantages
1. Construction materials are usually available locally.
2. May be able to use cover for parking of vehicles.
3. Surface layer also serves at least partly as hydraulic barrier.

Disadvantages
1. Paving will require maintenance.
2. No evaporation or evapotranspiration of water from protective soil.

Assessment

This design is only considered for site remediation projects where the subsurface materials provide stable foundation support and the paving material can be maintained.

Of the four main alternatives for the surface layer, soil has been by far the most commonly used material. Soil is generally available locally at nominal cost, and suitable soil will promote the growth of vegetation. Roots of plants reinforce the cover soil, restrict the rate of soil erosion, decrease runoff velocity, and remove water from the soil via evapotranspiration. Vegetation also creates a leaf cover above the soil to reduce rainfall impact and decrease wind velocity (Lutton *et al.*, 1979). Procedures for vegetating cover soils are reviewed by Gilman *et al.* (1985).

Guidelines for soil depths are given by USEPA (1989). The main problems experienced with soil are excessive erosion (especially on newly-seeded covers or steeply sloping covers) and a need to irrigate the soil during periods of drought in order to maintain growth of vegetation. A geosynthetic erosion control layer can be very beneficial in limiting problems with soil erosion in the period following installation and can even help to establish vegetative growth with seeds (even on a time-release basis) embedded in the layer.

Swope (1975) studied 24 landfill covers in Pennsylvania and found that 33% had slight erosion, 40% moderate erosion, and more than 20% had severe erosion. Johnson and Urie (1985) report that erosion is made more severe from the installation of hydraulic barriers (which, without a drainage layer, can produce soaked soil with large amounts of runoff after heavy rains) and vehicles driving on the surface of the cover. Many designers underestimate the problem of erosion of the surface layer prior to establishment of a mature vegetative cover.

Additionally, the problem of stability of the cover soil must be addressed. A number of failures of covers were reported by Boschuk (1991). Most of these failures were attributed to fundamental errors made in assessing the stability of the cover, or in several instances, lack of an analysis of the stability of the cover. Geogrid or geotextile reinforcement layers can be incorporated in the cover to increase factors of safety against slope instability to acceptable levels. Particularly critical interfaces along which slippage would be most likely will be described later.

In areas where the amount of precipitation is inadequate to support growth of a vegetative layer, an alternative is to armor the surface with cobbles. However, cobbles serve as a one-way window for infiltrating water, i.e., water can infiltrate downward through the cobbles, but due to lack of plant roots, little water is returned to the surface via evaporation.

Asphaltic concrete surfaces have been used on occasion for the surface layer (Repa *et al.*, 1987) but only for special remedial applications.

18.3.2 Protection layer

A protection layer (which will generally be some type of local soil) may serve several functions:

- to store water that has infiltrated into the cover until the water is later removed by evapotranspiration;
- to physically separate the waste from burrowing animals or plant roots;
- to minimize the possibility of human intrusion into the waste;
- to protect underlying layers in the cover system from excessive wetting/drying (which could cause cracking); and
- to protect underlying layers in the cover system from freezing (which could also cause cracking).

Figures 18.2–5 illustrate the alternative materials for the protective layer. Local soil is the most commonly used material. Soil can be very effective in storing water for later evapotranspiration and provides protection against freezing of underlying materials if the thickness of the protective layer is sufficiently large. Maps of maximum frost penetration depths are given in various publications, e.g., USEPA (1989).

Cobbles (Fig. 18.4) have occasionally been considered as a barrier to plant roots and burrowing animals (Cline *et al.*, 1982) but are not commonly used.

If the protective layer is placed directly on a barrier layer, a plane of potential slippage exists at the interface between the protective layer and the barrier layer. The risk of instability will be particularly acute following periods of heavy rain if no drainage layer is provided beneath the protective layer. The engineer must ensure that an adequate factor of safety exists at this and all other interfaces in the cover system. If the originally-designed cover system does not have an adequate factor of safety, the steps that are usually taken to increase the factor of safety include use of different materials (stronger soils or textured geomembranes), addition of a drainage layer, flattening of slopes, benches within the slope, or reinforcement of cover soils with geogrids or high-strength geotextiles.

18.3.3 Drainage layer

A drainage layer is often placed below the protective layer and above the barrier layer (Fig. 18.1). There are three reasons why a drainage layer might be desirable:

1. to reduce the head of water on the barrier layer, which minimizes infiltration;
2. to drain the overlying protective layer, which increases the water-storage capacity of the protective layer; and
3. to reduce pore water pressures in the cover soils, which improves slope stability.

There are four alternative materials that can be used for the drainage layer:

1. sand or gravel with either a soil filter (Fig. 18.6a) or a geotextile filter (Fig. 18.6b);
2. a thick geotextile (Fig. 18.7), which serves as both a drain and a filter;
3. a geonet with a geotextile filter/separator (Fig. 18.8); and
4. a geocomposite drain (Fig. 18.9), which consists of a polymeric drainage core and an overlapping geotextile filter/separator.

Selection of the material type is usually based on economics; all the materials listed above will function adequately if properly designed. If air space has an economic value (e.g., for a new waste containment unit), thin, geosynthetic materials rather than a thicker layer of sand or gravel will usually prove to be the most economical material.

The designer is cautioned to be careful about two details. First, water must discharge freely from the drainage layer at the base of the cover. If the outlet plugs or is of inadequate capacity, the toe of the slope will become saturated, develop excess pore water pressure, and potentially become unstable. Drainage pipes that commonly surround a site at its low elevations must have adequate capacity and cannot be allowed to plug. Conversely, the designer must beware of the possibility that the drainage layer may drain overlying soils so well that vegetation will not grow during parts of the year without irrigation. This problem is minimized by selecting surface and protective soils with adequate water-holding capacity to maintain vegetative growth.

The advantages and disadvantages of the various alternates are now considered in some detail.

Alternate A: sand or gravel drain with soil filter (Fig. 18.6a)

Advantages

1. Thickness of earthen materials helps to protect underlying layers from intrusion, puncture, and temperature extremes such as freeze/thaw.
2. Long history of use of soil filters and sand or gravel drains in engineering applications.

(a)

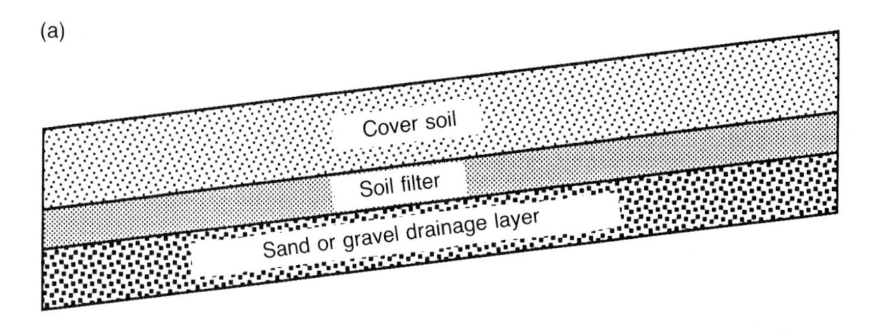

(b)

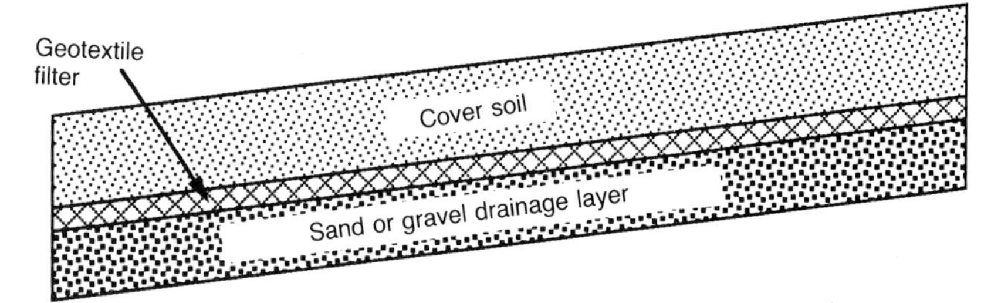

Fig. 18.6 Sand or gravel drain with (a) soil filter; or (b) geotextile filter.

Disadvantages

1. Suitable sand or gravel materials may not be readily available locally.
2. Gravel drainage layer may tend to puncture underlying geomembrane or geosynthetic clay liner; a cushioning layer may be needed or restrictions on maximum particle size of gravel may be required.
3. Fines in sand or gravel tend to migrate downslope, which can lead to buildup of unreleased seepage pressures and slope instability.

Assessment

This conventional design is technically sound, but its applicability can be limited by lack of suitable construction materials.

Alternate B: sand or gravel drain with geotextile filter (Fig. 18.6b)

Advantages

1. Thick sand or gravel drainage layer helps to protect underlying layers from temperature extremes such as freeze/thaw and from puncture.
2. Long history of use of sand or gravel drains in engineering works.

3. Extensive history of use of geotextile filters to prevent clogging of granular drainage layers.
4. Geotextile filters are readily available and can be shipped to any location.
5. Geotextile filter should prevent fines from entering sand or gravel layer.

Disadvantages

1. Suitable drainage materials may not be available locally.
2. Gravel drainage layer may tend to puncture underlying geomembrane or geosynthetic clay liner; a cushioning layer may be needed or restrictions on maximum particle size of gravel may be required.

Assessment

This conventional design is technically sound. Applicability may be limited by lack of suitable drainage materials in the local area.

Alternate C: thick geotextile drain (Fig. 18.7)

Advantages

1. Thick geotextiles are readily available and can be shipped to any location.
2. Specialty geotextiles can be fabricated for site-specific needs, e.g., high-denier fibers and composite geotextiles needlepunched together.
3. Lightweight installation equipment can be used; risk of damaging underlying liner during installation of drainage layer is minimal.
4. Thick, nonwoven geotextile provides good puncture protection for underlying barrier layer.
5. Few concerns with stability of drainage layer on steep slopes.
6. Geotextiles can be anchored, providing stability for covers on steep slopes.

Disadvantages

1. Thin drainage layer does not help very much to protect underlying materials from freeze/thaw effects.
2. May have inadequate transmissivity to provide drainage in high precipitation climates.

Assessment

This is a conventional design for a drainage layer that is placed directly on a geomembrane liner. If suitable drainage materials are not available locally, this is usually one of the best design alternatives for a drainage layer. The design can also be used for a drainage layer placed directly on a compacted clay liner.

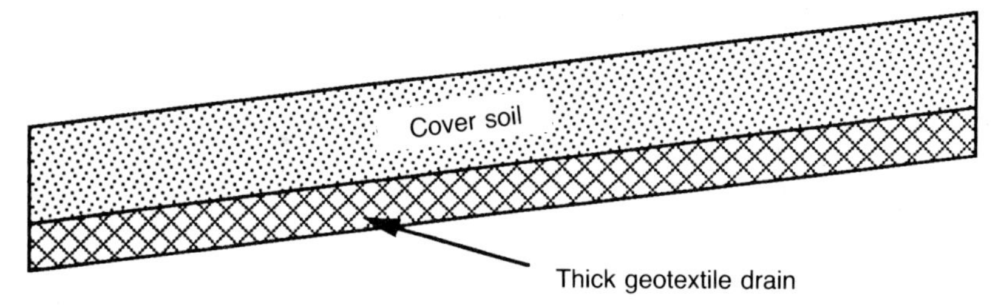

Fig. 18.7 Thick geotextile drain in a cover system.

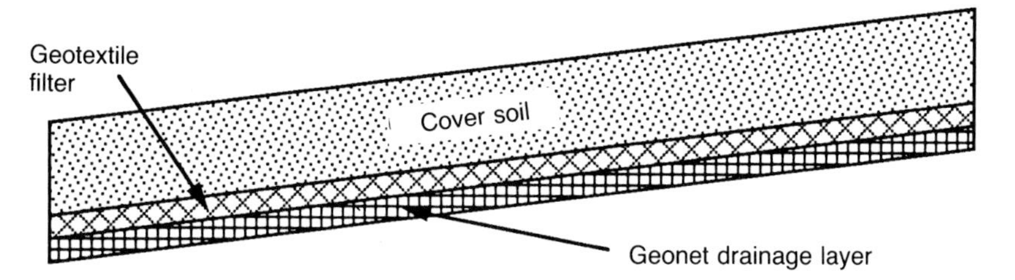

Fig. 18.8 Geonet drain with geotextile filter in a cover system.

Alternate D: geonet drain with geotextile filter (Fig. 18.8)

Advantages
1. Geotextiles and geonets are readily available and can be shipped to any location.
2. Lightweight installation equipment can be used; risk of damaging underlying liner during installation of drainage layer is minimal.

Disadvantages
1. Thin drainage layer does not help very much to protect underlying materials from freeze/thaw effects.
2. Potential slippage along geotextile/geonet interface; bonded geotextiles to geonets are preferred but long-term stability of bond must be assured. Geotextiles (bonded to geonet) can be anchored to improve stability.
3. Potential slippage along geonet/geomembrane interface, if underlying liner contains a geomembrane on the upper surface.
4. Seaming of geonets on long slopes is not straightforward and the intended connection must be assessed via laboratory tests.

Assessment
This is a conventional design for a drainage layer that is placed directly on a geomembrane liner. If suitable drainage materials are not available

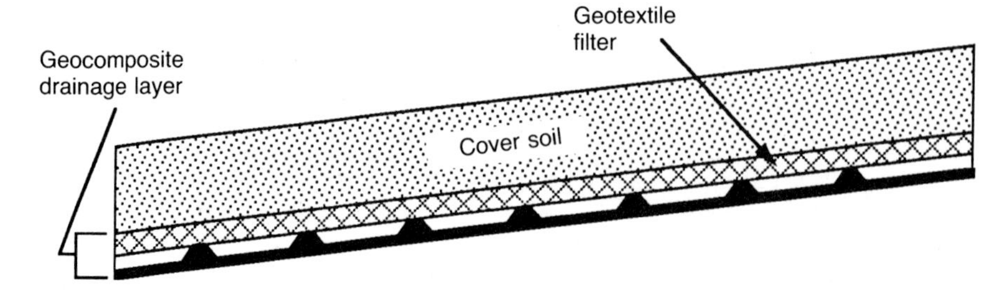

Fig. 18.9 Geocomposite drain in a cover system.

locally, this is often the best design alternative for a drainage layer. The design can also be used for a drainage layer placed directly on a compacted clay liner, but a sand or geotextile filter would have to separate the geonet from the compacted clay liner.

Alternate E: geocomposite drain (Fig. 18.9)

Advantages

1. Geocomposite drainage materials are readily available and can be shipped to any location.
2. Lightweight installation equipment can be used; risk of damaging underlying liner during installation of drainage layer is minimized.
3. Can be used on top of a compacted clay liner without need for an additional separation layer.
4. Connection of geocomposite cores is generally via a male/female interlock, which is very straightforward.

Disadvantages

1. Thin drainage layer does not help very much to protect underlying materials from freeze/thaw effects.
2. Potential slippage along geotextile/geocomposite interface. Bonded geotextiles to geocomposites are preferred, but long-term stability of bond must be assured.
3. Potential slippage along geocomposite/geomembrane interface, if underlying liner contains a geomembrane on the upper surface. Geotextile (bonded to geocomposite) can be anchored to assist stability.

Assessment

This design has not been used as frequently as a geonet drainage layer and geotextile filter. Technically, however, the geocomposite drainage layer can have much greater transmissivity, if required, than the geonet/geotextile combination.

18.3.4 Barrier layer

The barrier layer is often viewed as the most critical engineered component of the final cover system. The barrier layer minimizes percolation of water through the cover system directly by impeding infiltration through it and indirectly by promoting storage or drainage of water in the overlying layers and eventual removal of water by runoff, evapotranspiration, or internal drainage.

Historically, a compacted clay liner with a thickness of approximately 0.6 m and a design hydraulic conductivity of 1×10^{-7} cm/s has been the most commonly used barrier layer. This usage probably reflects the historic use of clayey materials for civil engineering projects requiring a layer that impedes water infiltration. Typically, the compacted clay liner has been used without a geomembrane liner and has been covered with a relatively thin (150–600 mm) layer of cover soil. For modern hazardous waste disposal facilities, however, a composite liner consisting of geomembrane/compacted clay is common in the US, but otherwise (for MSW facilities) a single compacted clay liner (CCL) is the dominant barrier material used in cover systems.

It is unfortunate that compacted clay has been so widely and indiscriminately used in cover systems because there are several problems which make the long-term performance of a compacted clay liner questionable as a barrier layer in many cover systems. The problems with clay liners used in cover systems include:

- clay liners are difficult to compact properly on a gas collection layer or a soft foundation (e.g., waste);
- compacted clay will tend to desiccate from above and/or below and crack unless protected adequately;
- compacted clay is vulnerable to damage from freezing and must be protected from freezing by a suitably thick layer of cover soil;
- differential settlement of underlying compressible waste will crack compacted clay if tensile strains in the clay become excessive; and
- compacted clay liners are difficult to repair if they are damaged.

In addition, a compacted clay liner is not as effective as a composite liner in impeding movement of liquid through the liner (Chapter 5).

Some designers have assumed that a compacted clay liner can be protected from desiccation with a thin layer of cover soil. Montgomery and Parsons (1989) describe field experiments in which three test plots were constructed and monitored for three years. A 1.2-m thick layer of compacted clay was covered with either 150 mm or 450 mm of topsoil at a temperate site in Wisconsin. After three years, excavations were made into the compacted clay. The condition of the two plots with 150 or 450 mm of top soil was about the same. The upper 200–250 mm of clay was weathered and blocky, cracks up to 12 mm wide extended up to 1 m into the clay, roots penetrated up to 250 mm into the clay in a continuous mat, and some roots extended up to 750 mm into the clay. Clearly, neither 150 mm nor 450 mm of topsoil was enough to protect the clay adequately.

Corser and Cranston (1991) describe test plots in which a layer of compacted clay was covered with:

1. a 600 mm layer of topsoil;
2. an unprotected geomembrane; or
3. a geomembrane overlain by 450 mm of topsoil.

The site was located in a relatively arid part of California. In less than a year, significant drying and cracking occurred in the plots with soil cover alone and the geomembrane cover alone, but no significant desiccation occurred in the clay covered with both a geomembrane and soil. However, the tests were short-term tests, and long-term drying may have eventually occurred even with the geomembrane and soil overburden. Nevertheless, the studies of Montgomery and Parsons (1989) and Corser and Cranston (1991), taken collectively, illustrate that the best and perhaps only practical way to protect a compacted clay liner from desiccation from the surface is to cover the clay liner with both a geomembrane and a layer of cover soil. To provide less protection is inappropriate if the designer's intention is for the compacted clay liner to remain intact and not crack. A geomembrane may not be necessary if an extremely thick layer of cover soil is used, but the layer would have to be so thick that for most projects soil alone would be impractical. Extremely thick layers of protective soil are only practical for unusual wastes, e.g., radioactive wastes.

Although designers have sometimes considered desiccation from above in the design of clay liners in cover systems, it appears that few designers have given careful thought to the potential for the clay liner to dry from below. Compacted clay liners (CCLs) are usually placed at water contents wet of optimum and with soil water potentials of 0 to −1 bar. The equilibrium relative humidity of the air phase in soils with water potentials >−1 bar is in excess of 99.5%. If the relative humidity of the gas phase in the underlying material (e.g., soil, final daily cover, gas collection layer, or the waste itself) is less than 95%, the clay liner will eventually dry and, depending on the amount of drying, possibly crack. Even landfill gas may possess the capacity to dry a CCL, and dry soils or waste materials certainly possess that capability. A geomembrane placed beneath the CCL is probably needed if there is potential for drying from below. However, placement and compaction of CCLs on top of a geomembrane is tricky; one cannot use extremely heavy compaction equipment because the geomembrane could be damaged.

Yet another problem with CCLs is potential damage from differential settlement. It is convenient to define **distortion** as the differential settlement Δ that occurs over a distance *L* (see Fig. 18.10). Distortion in a cover stretches the liner and, as a result, produces tensile strains in the cover components. Murphy and Gilbert (1985) computed the relationship between distortion and tensile strain for cover components, and this relationship is shown in Fig. 18.11. Murphy and Gilbert also summarized available data on the maximum tensile strain that compacted clays can sustain and found that compacted clays cannot sustain tensile strains larger than 0.1–1%. From Fig. 18.11, the maximum distortion Δ/L associated with a tensile strain of 0.1–1% is 0.05–0.1.

Jessberger and Stone (1991) performed centrifuge tests on 35- to 45-mm-thick prototype clay liners that would be used for cover systems.

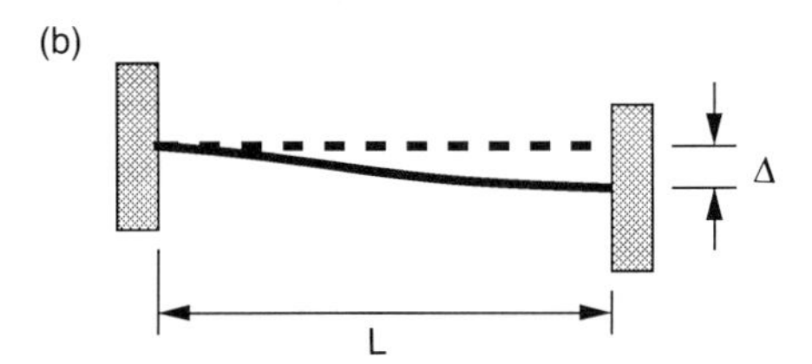

Fig. 18.10 Definition of distortion = Δ/L: (a) before settlement and (b) after settlement.

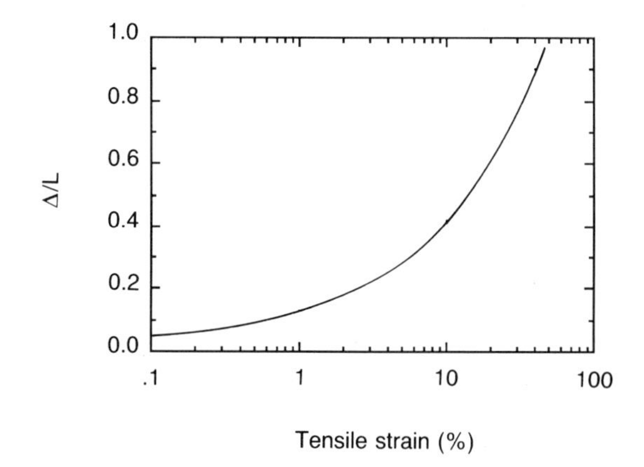

Fig. 18.11 Relationship between Δ/L distortion and tensile strain in a cover system, from Murphy and Gilbert (1985).

A 'trap door' beneath the liner was deformed to produce settlement like that which would be expected in a landfill. Flow rates through the soil were measured as a function of the distortion in the liner. It was found that the flow rate through the clay remained low until the distortion reached $\Delta/L = 0.1$, at which point the flow rate increased dramatically. Tension cracks were noted in later examinations of the material. Thus, the experiments reported by Jessberger and Stone support the findings from Murphy and Gilbert (1985), i.e., CCLs in cover systems cannot withstand Δ/L distortions greater than approximately 0.05–0.1 without cracking.

What does this level of distortion mean in practical terms? Suppose that one observes a circular crater in a landfill cover with a diameter of 6 m. What is the maximum settlement at the center of the crater before significant cracking would be expected in the liner? The horizontal distance from the edge to center of the crater, L, is 3 m, and 0.05–0.1 times L yields a maximum allowable settlement (Δ) of 0.15–0.3 m. It is the authors' experience that many, if not all, covers for municipal solid waste landfills have areas with distortions of this magnitude or larger. In such cases, the authors would argue that a CCL is an inappropriate material for a cover system. In defense of CCLs for hazardous waste disposal, cover systems for most modern hazardous waste landfills in the US probably experience Δ/L distortions less than 0.05 and differential settlement is not necessarily a problem. Also, for final covers placed over contaminated soil or other relatively incompressible material, differential settlement may not be a problem.

In contrast to the sensitivity of CCLs to cracking caused by differential settlement, geomembranes can withstand comparatively large tensile strains, even when stressed three-dimensionally. Indeed, a geomembrane suffers from few of the problems of a CCL in a cover system; the geomembrane can withstand large differential settlement and is not vulnerable to damage from desiccation or freeze-thaw. A single geomembrane liner is a lot more sensible liner component for most municipal solid waste cover systems than a single compacted clay liner.

There are actually seven options for the barrier layer component of a cover system that a designer should consider. They consist of CCLs, GMs, GCLs, or combinations thereof:

- a single compacted clay liner, or CCL (Fig. 18.12a);
- a single geomembrane, or GM, liner (Fig. 18.12b);
- a single geosynthetic clay liner, or GCL (Fig. 18.12c);
- a 2-component composite GM/CCL (Fig. 18.12d);
- a 2-component composite GM/GCL (Fig. 18.12e);
- a 3-component composite liner consisting of GM/GCL/GM (Fig. 18.12f); and

- a 3-component composite liner consisting of GM/GCL/GM (Fig. 18.12g).

The advantages and disadvantages of each of the seven options are described below and an assessment of each is given, as well. Further comments about the most desirable of these seven options are given later in this chapter.

Alternate A: single compacted clay liner (Fig. 18.12a)

Advantages
1. Convenient construction if clay is available locally.
2. If clays are not locally available, local soils can be blended with imported, processed clay, e.g., bentonite.
3. Conventional method and material with long history of use.

Disadvantages
1. Clay may dry out from below and crack (dry waste or gas).
2. Clay may dry out from above and crack (arid climate, drought conditions, or thin cover soil).
3. Tension cracks may form as a result of differential settlement.
4. Freeze/thaw may damage liner (cold regions, thin soil cover).
5. Limited hydraulic effectiveness due to lack of composite action.
6. Clay may be difficult to compact above compressible waste.
7. Clay may not be available locally.
8. Liner is difficult to repair if cracked.
9. Stability concerns for steep slopes.

Assessment
The main problems with this liner are that it almost certainly will be damaged by wet/dry cycles and (depending on site conditions) may also be damaged by settlement and/or freeze/thaw. The only way to prevent wet/dry and freeze/thaw damage is to use an extremely (and often impractically) thick layer of cover soil. Problems with settlement-induced cracking and drying from below often make this design unacceptable no matter how thick the cover soil is. A single compacted clay liner is not recommended for most cover systems and should only be considered where there will be a very thick cover soil, no significant differential settlement, and no drying from below.

Alternate B: single geomembrane liner (Fig. 18.12b)

Advantages
1. Straightforward and rapid installation.
2. Materials are readily available and can be shipped to any location.

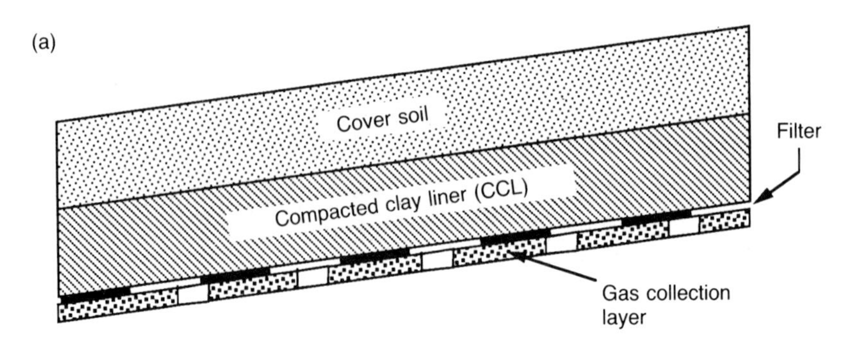
(a)
Cover soil
Compacted clay liner (CCL)
Filter
Gas collection layer

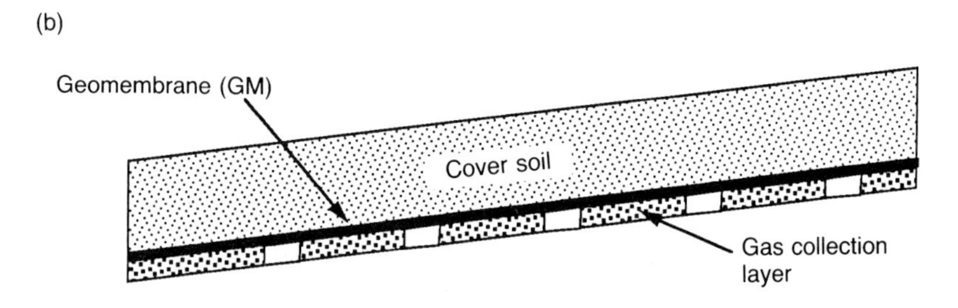
(b)
Geomembrane (GM)
Cover soil
Gas collection layer

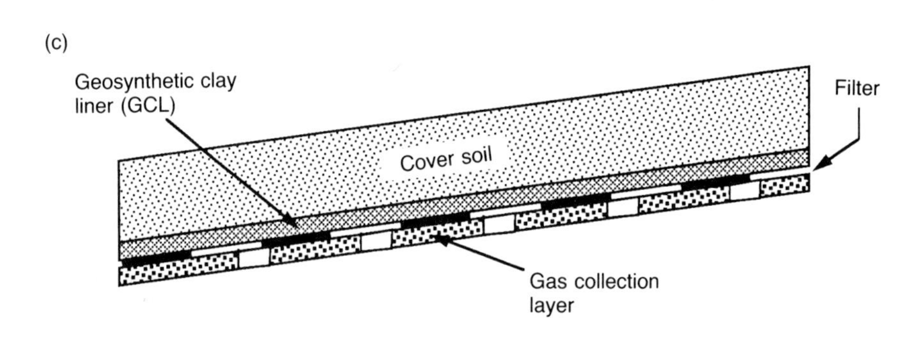
(c)
Geosynthetic clay liner (GCL)
Cover soil
Filter
Gas collection layer

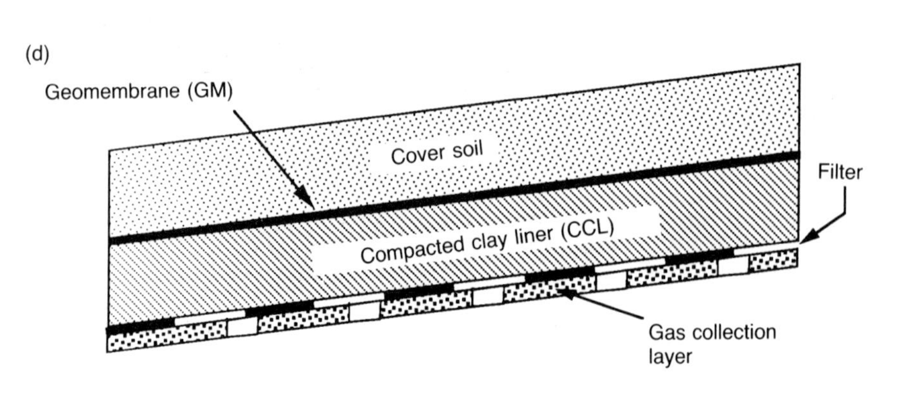
(d)
Geomembrane (GM)
Cover soil
Filter
Compacted clay liner (CCL)
Gas collection layer

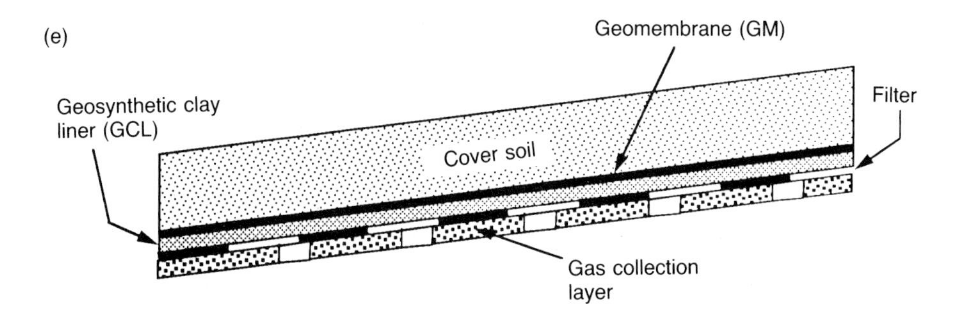

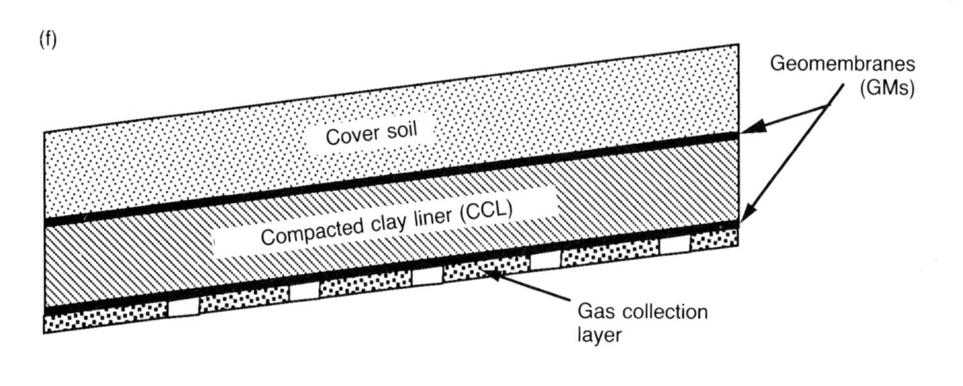

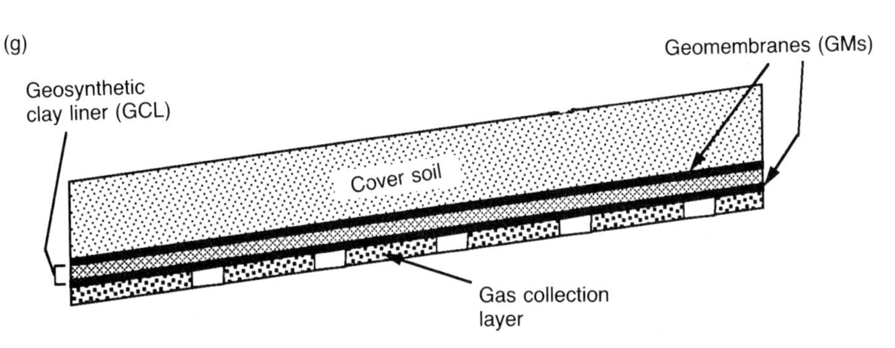

Fig. 18.12 Options for the barrier layer component: (a) single compacted clay liner; (b) single geomembrane liner; (c) single geosynthetic clay liner; (d) composite geomembrane/compacted clay liner; (e) composite geomembrane/geosynthetic clay liner; (f) composite geomembrane/compacted clay liner/geomembrane; and (g) composite geomembrane/geosynthetic clay liner/geomembrane, in a cover system.

3. Liner not damaged by wet/dry cycles; does not require a thick cover soil to prevent desiccation of liner.
4. Geomembrane and their seams do no appear to be affected by freeze/thaw cycling.
5. Some geomembranes can withstand very large differential settlement.
6. Easy to repair.
7. Gas collection penetrations are easy to construct (connection with a pipe boot).
8. Geomembrane can be anchored if stability is of concern.

Disadvantages
1. Potential slippage at interface between geomembrane and cover soil.
2. Tensile stresses may be mobilized in anchored geomembrane.
3. Vulnerable to puncture.
4. Limited hydraulic effectiveness due to lack of composite action.
5. Type and placement of cover soils is critical to minimize punctures.

Assessment
The main problem with this liner is that a geomembrane can be punctured during construction and later from accidental intrusion. This liner is only recommended when no other liner is practical and where some leakage through the liner can be tolerated.

Alternate C: single geosynthetic clay liner (Fig. 18.12c)

Advantages
1. Straightforward and rapid installation.
2. Materials are readily available and can be shipped to any location.
3. Geosynthetic clay liner possesses some self-healing capability from minor punctures, desiccation, and freeze/thaw (better in this regard than a compacted clay liner).
4. Gas collection penetrations are easy to construct.
5. Material is dry when placed and can be constructed in dry conditions without desiccation damage.
6. Geosynthetic clay liner can be easily repaired by patching if damaged.
7. Geosynthetic clay liner can be anchored if stability is of concern.
8. Geosynthetic clay liner can be reinforced if tensile stresses are of concern.

Disadvantages
1. Low shear strength of hydrated clay at mid-plane of some geosynthetic clay liners.
2. Potential slippage at interface between GCL and underlying filter.
3. Vulnerable to puncture.

4. Limited hydraulic effectiveness due to lack of composite action.
5. Placement of cover soil is critical to minimize punctures.
6. Dry GCL will not serve as a barrier to gas migration.
7. Differential settlement can cause clay shifting that could jeopardize low permeability of liner.
8. Alternating wet/dry cycles may cause shrinkage of GCL, resulting in leakage until rehydration and swelling reoccurs.

Assessment
Permeability of dry GCL to gas makes this liner alternative unacceptable for wastes that produce gas, unless one can be assured that the clay will be wetted soon after installation and will remain wet thereafter. The geosynthetic clay liner may be punctured during construction or after installation from accidental intrusion. This liner is only recommended for special cases where some leakage through the liner can be tolerated.

Alternate D: composite geomembrane/compacted clay liner (Fig. 18.12d)

Advantages
1. Composite liner very effective as a hydraulic barrier.
2. Minimum liner required by some regulatory agencies.
3. Geomembrane protects clay liner from desiccation if cover soil drys; does not require a thick cover soil to prevent desiccation of clay.
4. Thick liner cannot be punctured.
5. Geomembrane component can tolerate large differential settlement.
6. Geomembrane can be anchored to provide stability.

Disadvantages
1. Potential slippage along both soil/geomembrane interfaces.
2. Anchored geomembrane can cause tensile stresses to develop in the geomembrane.
3. Clay may dry from below and crack (dry waste or gas).
4. Tension cracks may form in clay as a result of differential settlement.
5. Freeze/thaw may damage clay liner (cold regions, thin soil cover).
6. Clay may be difficult to compact above compressible waste.
7. Clay may not be available locally.
8. Clay may not be easily repaired if damaged.

Assessment
A geomembrane/compacted clay liner is a common minimum requirement by various regulatory agencies for certain categories of waste. The primary concerns for this type of liner are with the long-term performance of the compacted clay component. Large differential

settlement and drying from below can cause the compacted clay to crack and increase its permeability. However, if the geomembrane remains largely intact and the permeability of the compacted clay does not increase excessively, the composite liner would still be expected to be reasonably effective.

Alternate E: composite geomembrane/geosynthetic clay liner (Fig. 18.12e)

Advantages

1. Relatively straightforward and rapid construction.
2. Materials are readily available and can be shipped to any location.
3. GCL possess some self-healing capability from minor punctures, desiccation, and freeze/thaw.
4. GCL is dry when placed and can be constructed in dry conditions without desiccation damage.
5. Liner can be repaired fairly easily.
6. Composite liner very effective as a hydraulic barrier.
7. Geomembrane component can tolerate large differential settlement.
8. GCL can be anchored if stability is a concern.
9. GCL can be reinforced if tensile stresses are of concern.

Disadvantages

1. Composite action questionable for most GCLs with upper geotextile.
2. Potential slippage along upper and lower surface of geomembrane and between some types of GCL and filter.
3. Low shear strength of hydrated clay at mid-plane of GCL for some products.
4. Vulnerable to puncture.
5. Placement of cover soil is critical to minimize punctures.
6. Differential settlement can cause clay deformation that could jeopardize low permeability of GCL in certain locations.

Assessment

This liner system offers redundancy and many practical benefits in terms of availability of materials and ease of construction. Questions can be raised about effectiveness of composite action between geomembrane and those GCLs with an upper geotextile – this is an area of research at this time. Provided that issues concerning stability and puncture are resolved, this system probably offers the best overall benefit/cost ratio of any liner type for cover systems.

Alternate F: composite geomembrane/compacted clay liner/geomembrane (Fig. 18.12f)

Advantages

1. Composite behavior of GM/CCL/GM extremely effective as a hydraulic barrier; with two GMs, some cracking of clay from differential settlement can be tolerated without hurting the overall hydraulic performance.
2. Geomembranes protect clay liner from desiccation from above or below; does not require a thick soil cover to prevent desiccation of the clay.
3. Thick clay liner cannot be punctured.
4. Geomembranes can tolerate large differential settlement.

Disadvantages

1. Difficult construction (especially compaction of clay on GM and separate anchorage systems for geomembranes).
2. Potential slippage along several interfaces or within CCL (for which excess pore water pressure cannot dissipate).
3. Freeze/thaw may damage clay liner (cold regions, thin soil cover).
4. Clay may not be available locally.
5. Clay not easily repaired if damaged.
6. Very complicated gas collection penetrations.

Assessment

This is a very conservative design which offers excellent redundancy of components and, more than any other design involving compacted clay, provides layers that protect the clay from damage due to desiccation. By keeping the clay wet, the protective geomembranes indirectly help to improve the resistance of the clay to settlement-induced cracking. However, this design is rarely (if ever) used mainly because construction is extremely complex and no regulatory agency has required that a geomembrane be placed below the compacted clay. This design is only recommended for extremely sophisticated waste containment facilities where an extraordinary degree of redundancy in the design is required.

Alternate G: composite geomembrane/geosynthetic clay liner/geomembrane (Fig. 18.12g)

Advantages

1. Relatively straightforward and rapid construction.
2. Materials are readily available and can be shipped to any location.
3. Composite behavior fairly good, even with imperfect GM/clay

contact or some settlement-induced clay deformation, because two geomembranes are used.

4. Geomembranes protect clay from wet/dry cycles.
5. Geomembranes can tolerate large differential settlement.
6. GCL is dry when placed and will not dry and crack during installation.
7. Liner can be repaired fairly easily.
8. Anchorage of all components is possible if stability concerns are present.

Disadvantages

1. Potential slippage along several interfaces and through mid-plane of some GCLs if GCL becomes fully hydrated.
2. Vulnerable to puncture.
3. Placement of cover soil is critical to minimize punctures.
4. Very complicated gas collection penetrations.

Assessment

This is a very conservative design that offers excellent redundancy of components and practical advantages such as availability and overall ease of construction (except gas collection penetrations). Stability can be assured by anchorage of components and use of high friction surface on top of upper geomembrane. Provided potential problems with puncture can be resolved, this liner probably offers the best combination of extremely low percolation of water through the liner and practical constructability.

18.3.5 Gas collection layer

The purpose of a gas collection layer is to collect gas for processing or discharge. Gas flows from the gas collection layer to **vents**. Flow of gas may occur passively under the pressure gradient generated naturally or actively through assistance from a vacuum pump at the surface. A gas collection layer is only necessary for capping wastes that can produce gas or volatiles that must be released in a controlled manner.

The gas collection layer must have a high in-plane gas flow rate and must not become plugged with fine-grained materials. The usual materials are:

- Sand or gravel with a soil filter (Fig. 18.13a);
- A thick geotextile which serves as both drain and filter (Fig. 18.13b);
- Geonet drain and a geotextile filter (Fig. 18.13c).

Any of these materials will function well if properly designed, and the advantages/disadvantages of each is considered below.

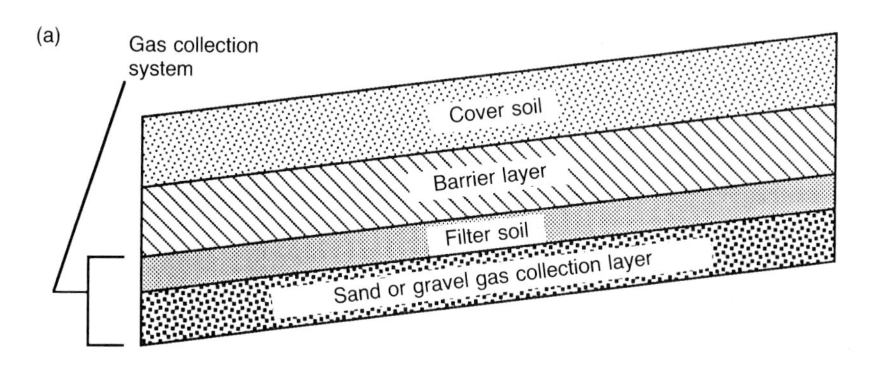

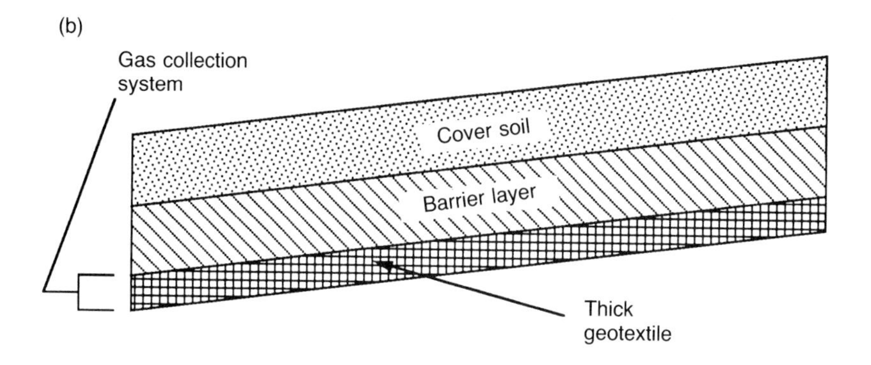

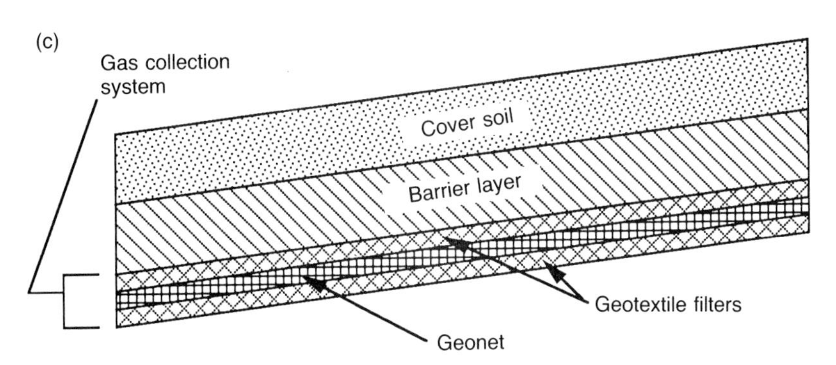

Fig. 18.13 Materials used in the gas collection layer: (a) sand or gravel drain with soil filter; (b) geotextile drain and filter; and (c) geonet drain and geotextile filters, in a cover system.

Alternate A: sand or gravel drain with soil filter (Fig. 18.13a)

Advantages

1. Long history of use of soil filters and sand or gravel drains in engineering works.

2. Easy to install perforated collection pipes in sand or gravel gas collection layer.
3. Base of vent pipes are easy to locate between filter and bottom of gas collection layer.

Disadvantages
1. Suitable materials may not be readily available locally.
2. Thickness of both components decreases air space within landfill.

Assessment
This is a technically sound design and probably the most common design when suitable construction materials are available locally. The major drawback is the loss of air space in the landfill facility.

Alternate B: geotextile drain and filter (Fig. 18.13b)

Advantages
1. Geotextiles are readily available and can be shipped to any location.
2. Virtually no airspace in landfill facility taken by gas collection layer.
3. Provides clean working surface if geosynthetics are used as lower portion of barrier layer.
4. Rapid, straightforward installation.
5. Specialty geotextiles can be fabricated for site-specific designs, e.g., high denier fibers and composite geotextiles needled together.

Disadvantages
Potential slippage between geotextile and overlying geomembrane (if the barrier layer has a geomembrane on the lower surface of the layer).

Assessment
This design is considered technically equal to designs that employ sand filter and gravel drainage materials. Selection of natural materials or geosynthetics is often based upon local availability and cost of natural materials and construction logistics. In many cases, this would be the least costly of the gas collection layer alternatives.

Alternate C: geonet drain and geotextile filters (Fig. 18.13c)

Advantages
1. Geotextiles and geonet are readily available and can be shipped to any location.
2. Geonet has extremely high in-plane gas transmissivity.
3. Geotextiles can be bonded to the geonet using a variety of fabrication methods.
4. Relatively no loss of airspace in the landfill facility.

5. Provides clean working surface if geosynthetics are used as lower portion of barrier layer.

Disadvantages
1. Potential slippage between geotextile and overlying geomembrane (if the barrier layer has a geomembrane on the lower surface of the layer) and between geonet and geotextiles.
2. Continuity of flow within the geonet must be assured across seams and edges.
3. Somewhat more complex connection to vent pipes than a single, thick geotextile drain and filter.

Assessment
This design is considered technically equal to designs that employ sand filter and gravel drainage materials. Selection of natural materials or geosynthetics is often based upon local availability and cost of natural materials and construction logistics. The geonet has a higher in-plane gas transmissivity than a geotextile, which could be advantageous if gas production rates are extremely large or vent pipes are more widely spaced than usual.

18.4 FACTORS IN SELECTION OF BARRIER LAYER ALTERNATIVES

The critical factors that affect selection of a barrier layer design are climate, the amount of differential settlement to which the cover will be subjected, the vulnerability of the cover soil to erosion or puncture, the amount of water percolation through the cover system that can be tolerated, the need for collection of waste-generated gas, and the steepness of the slope.

The seven barrier layer alternatives discussed previously and shown in Fig. 18.12 have been assessed by the authors in light of each of these factors. For each factor and the condition considered, each liner was assigned a factor ranging from 1 (not acceptable) to 5 (recommended). For example, a compacted clay liner was given a 1 (not acceptable) for cases involving large differential settlement whereas a geomembrane was assigned a factor of 5 (recommended) because the geomembrane can withstand large differential settlement. The assessments are summarized in Tables 18.1–6. The authors suggest that designers consider these assessments carefully for the unique conditions that exist at a particular site. There may be extenuating circumstances or special design precautions that have to be taken to mitigate problems that would normally be present; for this reason, the assessments are for general guidance only and should not be taken too literally.

Table 18.1 Influence of climate on selection of barrier for cover system

alternate	*liner component*	*general description of climate at the site*		
		arid	*cyclic (wet/dry seasons)*	*humid*
A	Compacted clay liner (CCL)	1 desiccation cracks	1 desiccation cracks	3 may get some desiccation cracks
B	Geomembrane liner (GM)	5	4	4
C	Geosynthetic clay liner (GCL)	3 unacceptable if waste produces gas	3 unacceptable if waste produces gas	4
D	Composite GM/CCL	2 long-term desiccation cracking	3 possible desiccation cracking	4
E	Composite GM/GCL	5	4	5
F	Composite GM/CCL/GM	4	4	5
G	Composite GM/GCL/GM	5	5	5

Scoring system: 1 = not acceptable; 2 = marginal; 3 = possibly OK; 4 = acceptable; and 5 = recommended.

Table 18.2 Influence of differential settlement on selection of barrier for cover system

alternate	*liner component*	*amount of differential settlement*		
		major	*moderate*	*nominal*
A	Compacted clay liner (CCL)	1 cracking of clay	1 cracking of clay	3 depends on site specific factors
B	Geomembrane liner (GM)	4	5	5
C	Geosynthetic clay liner (GCL)	2 lean areas of clay may develop	3 depends on severity of settlement	4
D	Composite GM/CCL	2 cracking of clay damage to GM	3 depends on severity of settlement	4
E	Composite GM/GCL	3 depends on severity of settlement	4	5
F	Composite GM/CCL/GM	3 cracking of clay	4	5
G	Composite GM/GCL/GM	4	5	5

Scoring system: 1 = not acceptable; 2 = marginal; 3 = possibly OK; 4 = acceptable; and 5 = recommended.

Table 18.3 Influence of cover soil on selection of barrier for cover system

alternate	*liner component*	*vulnerablity of cover soil to damage*		
		can be eroded or punctured	*erosion possible but not likely*	*can not be eroded or punctured*
A	Compacted clay liner (CCL)	1 desiccation of clay	2 desiccation of clay	3 may get some desiccation cracks
B	Geomembrane liner (GM)	1 UV degradation and puncture	1 UV degradation and puncture	3 depends on other site specific factors
C	Geosynthetic clay liner (GCL)	1 puncture	1 puncture	3 depends on other site specific factors
D	Composite GM/CCL	3 possible damage to GM	4	4
E	Composite GM/GCL	2 puncture	3 depends on site specific factors	4
F	Composite GM/CCL/GM	4	5	5
G	Composite GM/GCL/GM	4	5	5

Scoring system: 1 = not acceptable; 2 = marginal; 3 = possibly OK; 4 = acceptable; and 5 = recommended.

Table 18.4 Influence of percolation limits on selection of barrier for cover system

alternate	*liner component*	*allowable water percolation through barrier*		
		essentially none	*very little*	*moderate amount*
A	Compacted clay liner (CCL)	1 not a composite	2	3 depends on site specific factors
B	Geomembrane liner (GM)	1 not a composite	3 depends on site specific factors	5
C	Geosynthetic clay liner (GCL)	1 not a composite	2	3
D	Composite GM/CCL	3 depends on site specific factors	4	4
E	Composite GM/GCL	3 depends on site specific factors	4	5
F	Composite GM/CCL/GM	5	5	5
G	Composite GM/GCL/GM	5	5	5

Scoring system: 1 = not acceptable; 2 = marginal; 3 = possibly OK; 4 = acceptable; and 5 = recommended.

Table 18.5 Influence of gas generation on selection of barrier for cover system

alternate	*liner component*	*need for a gas collection system*	
		is required	*not required*
A	Compacted clay liner (CCL)	1 clay drying	1 cracking of clay
B	Geomembrane liner (GM)	4	5
C	Geosynthetic clay liner (GCL)	1 gas escapes when dry	5
D	Composite GM/CCL	3 depends on potential for clay drying	5
E	Composite GM/GCL	4	5
F	Composite GM/CCL/GM	4 vent connection detail complicated	5
G	Composite GM/GCL/GM	4 vent connection detail complicated	5

Scoring system: 1 = not acceptable; 2 = marginal; 3 = possibly OK; 4 = acceptable; and 5 = recommended.

18.5 GENERAL ASSESSMENT AND COMPARATIVE RANKING

The previous section presented charts of a number of important landfill cover design factors insofar as the liner, or barrier, system is concerned. While the rankings of the different liner sections within each chart are admittedly subjective, it is of interest to now extend the numerical assessments into a summary and then to contrast the resulting values to a general cost estimate. A generalized benefit/cost ratio will be computed for each of the seven alternate liner schemes. Critically important in such a ranking is the cost of the various schemes. The following estimated costs will be used.

- Compacted clay liners (CCLs). This is the most difficult element for a generalized cost estimation. In-place costs of CCLs typically vary from \$5.40 to \$21.50 per square meter depending on thickness, availability, size and type of facility. In extreme conditions (e.g., lack of clay locally), the cost can be much higher. However, for a 0.6 m thick compacted clay liner of 1×10^{-7} cm/sec hydraulic conductivity, \$7.50 per square meter will be the estimated installed cost.
- Geosynthetic clay liners (GCLs). Of the three commercially available

alternate	*liner component*	*steepness of slope*		
		flatter than 6:1 (<9°)	*6:1–3:1* (9–18°)	*steeper than 3:1* (>18°)
A	Compacted clay liner (CCL)	5	4	3 depends on clay properties and slope angle
B	Geomembrane liner (GM)	5	5	3 depends on material properties and slope angle
C	Geosynthetic clay liner (GCL)	4	3 depends on specific GCL material	3 depends on GCL material and slope angle
D	Composite GM/CCL	5	3 interfacial shear	2 interfacial shear; possible internal shear of CCL
E	Composite GM/GCL	5	3 depends on specific GCL material	2 interfacial shear; internal shear for some GCLs
F	Composite GM/CCL/GM	5	2 interfacial shear; depends on GM and CCL properties	1 no dissipation of pore water pressure in clay
G	Composite GM/GCL/GM	5	3 depends on specific GCL material	2 interfacial shear; internal shear for some GCLs

Scoring system: 1 = not acceptable; 2 = marginal; 3 = possibly OK; 4 = acceptable; and 5 = recommended.

Table 18.7 Summary of influence of individual factors on barrier layer for covers

alternate	*liner component*	*climate*			*settlement*			*cover erosion/ puncture vulnerability*			*allowable percolation*			*gas collection*				*slope inclination*
		arid	*cyclic*	*humid*	*major*	*mod.*	*nominal*	*major*	*mod.*	*low*	*ess. none*	*v. little*	*mod.*	*gas*	*no gas*	*<9°*	*9–18°*	*>18°*
A	CCL	1	1	3	1	1	3	1	2	3	1	2	3	1	1	5	4	3
B	GM	5	4	4	4	5	5	1	1	3	1	3	5	4	5	5	5	3
C	GCL	3	3	4	2	3	4	1	1	3	1	2	3	1	5	4	3	3
D	GM/CCL	2	3	4	2	3	4	3	4	4	3	4	4	3	5	5	3	2
E	GM/GCL	5	4	5	3	4	5	2	3	4	3	4	5	4	5	5	3	2
F	GM/CCL/GM	4	4	5	3	4	5	4	5	5	5	5	5	4	5	5	2	1
G	GM/GCL/GM	5	5	5	4	5	5	4	5	5	5	5	5	4	5	5	3	2

Scoring system: 1 = not acceptable; 2 = marginal; 3 = possibly OK (depends on site or product-specific conditions); 4 = acceptable; and 5 = recommended.

GCLs in the US, installed costs vary from $5.40 to 10.80 per square meter depending on site conditions. A cost of $7.50 per square meter is assumed.

- Geomembranes (GMs). The range of installed costs for geomembranes of the type generally used in cover systems is relatively narrow in comparison to the other liner materials. For geomembranes with good out-of-plane deformation properties and a thickness of approximately 1.0 mm, the installed cost probably averages about $7.50 per square meter.

Thus, it is seen that each of the three liner components is assumed to cost approximately $7.50 per square meter installed. Those liner schemes with two liner elements within them will cost twice those of a single liner element. Those with three components will cost three times that of a single component liner.

The benefit/cost ratio was computed by the authors as follows. The numerical rankings given in Tables 18.1–6 were tabulated into a single chart (Table 18.7). For each of the seven liner alternates, the numerical ratings in each horizontal row were summed to obtain an overall numerical 'benefit' for each alternative. Those liners with the highest overall ratings for all of the conditions considered scored highest. The 'benefit' number calculated by summing numbers in horizontal rows in Table 18.7 was then divided by the cost (in dollars per square meter) to determine the benefit/cost ratio. It is realized that this procedure is arbitrary, but it should give a good overall picture of which liner alternatives provide the best performance over a broad range of site conditions.

Table 18.8 lists results of the benefit/cost computations. The liner alternatives are grouped into one-, two- or three-layer systems. As a reminder of the systems: one-layer systems consist of a GM, CCL, or GCL; two-layer systems consist of GM/CCL and GM/GCL composites; and three-layer systems consist of GM/CCL/GM and GM/GCL/GM systems.

From Table 18.8, it can be seen that among the single-layer systems, the single geomembrane (GM) far outperforms the geosynthetic clay liner (GCL) in terms of benefit/cost ratio, and that both of these geosynthetic materials outperform a compacted clay liner (CCL) in terms of benefit/cost ratio. The compacted clay liner is the poorest overall technical choice of any single layer system and is also the scheme with the poorest benefit/cost ratio that one can utilize as a cover system's hydraulic barrier. Paradoxically, the CCL is the most widely used and permitted single liner material in use for current cover systems!

Of the two-layer (composite) systems, the GM/GCL outperforms the GM/CCL technically and on a benefit/cost ratio basis. This finding is also important since a composite liner is usually considered to be

Table 18.8 Benefit/cost assessment of various liner cross-sections

no of barrier layers	*design alternate*	*description*	*overall benefit**	*est. cost \$/m^2*	*benefit/cost ratio*	*ranking in group*
One layer	A	CCL	34	0.065	49	3
	B	GM	63	0.065	90	1
	C	GCL	46	0.065	66	2
Two layers	D	GM/CCL	58	0.13	41	2
	E	GM/GCL	66	0.13	47	1
Three layers	F	GM/CCL/GM	72	0.195	34	2
	G	GM/GCL/GM	77	0.195	37	1

*Determined by summing horizontal rows in Table 18.7.

a GM/CCL combination. The finding of this analysis is that, for cover systems, a GM/GCL is preferable to GM/CCL in terms of overall assessment.

Considering the triple layer systems, the GM/GCL/GM slightly outperforms a GM/CCL/GM system. However, the difference is relatively small.

In summary, it is obvious that designers must rethink their current cover system liner strategy. Compacted clay liners (CCLs) are *not* the general answer for the hydraulic barrier in the closure of a solid waste landfill or site remediation project. With desiccation from above the liner in arid or seasonally dry climates, drying of the liner from beneath, and cracking from differential settlement, CCLs *cannot* be expected to maintain their initially low hydraulic conductivity.

Conversely, the newer barrier systems, namely geomembranes (GMs) and geosynthetic clay liners (GCLs), should be seriously considered for cover systems. Geomembranes, in particular, can afford long-term barrier protection and undergo large differential settlement if required. Geosynthetic clay liners offer many of these same attractive features. Furthermore, the combination of a GM/GCL system results in high technical marks and provides a backup system which is generally assumed to be a conservative design strategy.

Lastly, if the authors were pressed to recommend the preferred barrier system currently available in a 'typical' solid waste landfill cover (for either a new landfill or an old landfill requiring closure or remediation), the authors would undoubtedly respond: a single geomembrane (GM) liner, or if a composite liner is desired, a geomembrane over a geosynthetic clay liner, i.e., a GM/GCL.

18.6 CONCLUSIONS

This chapter has summarized the basic principles employed for design of final cover systems for new waste containment units and site remediation projects. The most important conclusions relate to the barrier layer within the final cover. The main conclusions are as follows:

- compacted clay liners by themselves are *not* the general barrier system of choice;
- geomembranes or geosynthetic clay liners are better overall choices technically and on the basis of benefit/cost ratio than compacted clay liners; and
- the barrier system of choice for most municipal solid waste landfill covers should be a geomembrane or a geomembrane over a geosynthetic clay liner (GM/GCL).

Compacted clay may have a place in cover systems, but only for those limited situations where very small differential settlements are expected and adequate protection from desiccation and freeze/thaw is provided.

The conclusions listed above are based upon technical, rather than regulatory, considerations. The authors would be remiss, however, if regulatory issues were avoided completely. In this regard, RCRA covers for hazardous wastes and (by inference) municipal solid waste require a GM/CCL barrier in the cover. Even for site remediation projects, RCRA requirements may be applicable or relevant and appropriate. For an equivalency evaluation of the typical RCRA cross section, at least three issues must be addressed.

1. The hydraulic equivalency of the regulation-suggested CCL with respect to a replacement GCL must be demonstrated. This can be accomplished using Darcy's formula, but such an equivalency calculation does have assumptions, primarily that of no thin areas nor punctures in the GCL.
2. Some of the GCLs have an upper geotextile with high in-plane transmissivity which may compromise the 'intimate contact' of the clay and geomembrane. Work is actively ongoing by GCL manufacturers to decrease lateral flow within the upper geotextile to values similar to the clay itself.
3. Stability issues of geomembranes and hydrated GCLs must be addressed using proper geotechnical engineering principles.

With proper design and testing, however, the authors feel that regulatory concerns can be overcome with the net result of a superior long-term barrier system for solid-waste landfills and site remediation projects than are currently designed and constructed.

While it must be emphasized that considerable engineering design and performance testing is required, the construction quality control and quality assurance (CQC/CQA) issues are also essential. The best of design schemes and manufactured materials can be completely negated with poor CQC and the attendant CQA activities. Along with the necessity of proper placement, positioning and seaming of the GM and GCL materials, there are also barrier penetrations (such as gas vents that are required). Prefabricated geomembrane pipe boots (with proper connections and seams and carefully placed clay for CCLs and GCLs) are mandatory for the overall success of cover systems as proposed herein. With proper care by all parties involved, however, we feel that excellent long-term performance of the geosynthetic barrier systems of the type recommended in this paper will be realized.

REFERENCES

Boschuk, J. (1991) Landfill covers – an engineering perspective, *Geotech. Fabrics Rep.*, **9**(4), 23–34.

Cline, J.F., Cataldo, D.A., Burton, F.G. and Skiens, W.E. (1982) Biobarriers used in shallow burial ground stabilization, *Nuclear Tech.*, **58**, 150–3.

Corser, P. and Cranston, M. (1991) Observations on long-term performance of composite clay liners and covers, *Proc., Geosynthetics Design and Performance*, Vancouver Geotech. Soc., Vancouver, British Columbia.

Gilman, E.F., Flower, F.B. and Leone, I.D. (1985) Standardized procedures for planting vegetation on completed sanitary landfills, *Waste Management and Research*, **3**, 65–80.

Lutton, R.J., Gegan, G.L. and Jones, L.W. (1979) *Design and construction of covers for solid waste landfills*, US Environmental Protection Agency, EPA600/2-79-165, Cincinnati, Ohio.

Jessberger, H.L. and Stone, K.J.L. (1991) Subsidence effects on clay barriers, *Geotechnique*, **41**(2), 185–194.

Johnson, D.I. and Urie, D.H. (1985) Landfill caps: long-term investments in need of attention, *Waste Management and Research*, **3**, 143–8.

Montgomery, R.J. and Parsons, L.J. (1989) The Omega Hills final cover test plot study: three year data summary, presented at the 1989 Ann. Meeting of the Nat. Solid Waste Management Assoc., Washington, DC.

Murphy, W.L. and Gilbert, P.A. (1985) *Settlement and cover subsidence of hazardous waste landfills*, US Environmental Protection Agency, EPA/600/2-85/035, Cincinnati, Ohio.

Repa, E.W., Herrmann, J.G., Tokarski, F. and Eades, R.R. (1987) Evaluating asphalt cap effectiveness at a superfund site, *J. Environ. Eng.*, **113**(3), 649–53.

Swope, G.L. (1975) Revegetation of landfill sites, MS thesis, Pennsylvania State Univ., State Park, Pennsylvania.

US Environmental Protection Agency (1989), *Technical Guidance Document, Final Covers on Hazardous Waste Landfills and Surface Impoundments*, EPA/530-SW-89-047, Washington, DC.

CHAPTER 19

Recovery well systems

Bob Kent and Perry Mann

19.1 INTRODUCTION

Plato (*c*.427–347 BC) developed a sophisticated set of groundwater laws. In addition to laws on water conservation and water rights, Plato developed water pollution regulations, as quoted below.

> If anyone deliberately spoils someone else's water supply, whether spring or reservoir, by poisons or excavations or theft, the injured party should take his case to the City Wardens and submit his estimate of the damage in writing. Anyone convicted of fouling water by magic poisons should, in addition to his fine, purify the spring or reservoir using whatever methods of purification the regulations of the expounders prescribe as appropriate to the circumstances and the individual involved.
>
> (Saunders, 1970)

Although we have recognized the need for protection and remediation of groundwater resources for over 2000 years, major advances in remediation technology have only been developed in the last 10 years. Recovery wells are often used to 'capture' a contaminant plume. Once the contaminated groundwater is recovered, it is treated and released in an appropriate manner. Methods for recovering and treating contaminated groundwater are reviewed in this chapter.

19.2 SOURCES OF GROUNDWATER CONTAMINATION

Contaminants that pose a risk to groundwater originate from waste disposal practices, such as leachates from landfills and seepage from waste lagoons. A second category of contaminant is chemicals that

Geotechnical Practice for Waste Disposal.
Edited by David E. Daniel.
Published in 1993 by Chapman & Hall, London. ISBN 0 412 35170 6

are generally considered beneficial but when improperly managed or misapplied, can result in contamination of groundwater supplies. Gasoline from leaking underground storage tanks and fertilizers or pesticides applied to lawns and crops are examples of this latter category.

Industrial plants often are major sources of groundwater contamination. Leakage from a variety of sources within an individual facility results in the subsurface mixing of different contaminants and increases the difficulty of designing an efficient recovery and treatment system.

19.3 AQUIFER CHARACTERIZATION

Aquifer characterization is critical for determining the number, spacing, and pumping rate for recovery wells. Groundwater is in constant motion, flowing from areas of higher head to lower head. The natural velocity of groundwater flow ranges from <1 m/yr to >100 m/day. Determination of the direction and rate of groundwater flow in each contaminated aquifer is critical when designing groundwater recovery systems. If this is neglected, the spacing or pumping rate of wells may not be sufficient to intercept all of the contaminant plume.

The maximum rate of contaminated groundwater removal depends on the transmissivity of an aquifer, whereas the volume that can be removed by a well is dependent in part, on the storativity of the aquifer. These hydrologic parameters are the major controlling factors of the shape and size of the cone of depression, which is used to determine the 'radius of influence', of a pumping well. Transmissivity and storativity typically are determined by conducting aquifer pumping tests in the field. Although other methods exist for estimating these parameters, actual empirical determinations from field pumping tests are preferred for the design of recovery well systems. The authors are familiar with several inadequately designed recovery well systems that were based on transmissivity values derived from short-term slug tests. A pumping test of 24 or 48 hours is normally recommended for design consideration. It should be noted that, for high-yield aquifers that are grossly contaminated, the cost of storage and treatment of the water recovered during the pumping test may be prohibitive.

19.4 FLOW TO WELLS

Pumping groundwater from a well results in the lowering of the water level in the well (or 'drawdown') and a decrease in the groundwater head of the surrounding water-bearing formation. The greatest amount of drawdown occurs closest to the pumping well and decreases logarithmically with radial distance from the well. The radius of influence for a well completed in a confined aquifer typically is larger than that in an unconfined (water table) aquifer with equal transmissivities. Water pumped from a confined aquifer is released from storage by depressurization and slight expansion of the water and aquifer matrix (shrinkage of pore space) instantaneously and flows horizontally to the well. The water pumped from an unconfined aquifer, however, moves vertically by gravity drainage and results in a much slower release from storage, emptying of pores, and replacement of water with air. When a single well is pumped, a cone of depression is formed as the radius of influence caused by the discharging well expands. Theoretically, under ideal aquifer conditions, this radius expands to infinity. Equilibrium is sometimes reached over an extended period of time, resulting form recharge and/or discharge boundaries within the aquifer that prevent further expansion of the cone. The extent of the radius of influence for a single pumping well is typically between 10 and 1000 m, depending on the specific hydrologic conditions.

Well interference occurs when the cones of depression (i.e., the radii of influence) of two or more wells overlap. Interference between pumping recovery wells, although usually desirable, must be taken into account in order to maximize the drawdown of the aquifer in a given area. This interference, if significant, may result in lower pumping yields for individual wells within a multiple well system. As such, interference of pumping wells is a critical factor to be considered in the design of a recovery well system. Through careful design and planning, the efficiency of a recovery well system for collecting contaminants in groundwater can be optimized.

19.5 MODELING THE WELL FIELD

Computer models help simplify the process of determining an optimum number and spacing of recovery wells. Models can be used to simulate the distribution of drawdown within different arrays of wells completed within a given aquifer. These models are typically based on the Theis (1935) nonequilibrium equation, which describes drawdown of water levels in an ideal confined aquifer around a single pumping well. The effects of multiple pumping wells are then summed to determine the drawdown at any point in the aquifer at any time.

The Theis solution to the problem of radial flow is based on the following assumptions:

1. the aquifer is homogeneous, isotropic, of uniform thickness, and of infinite areal extent;

2. the discharging well is of infinitesimal diameter and completely penetrates the aquifer;
3. release of water from storage is instantaneous with the reduction in pressure due to drawdown;
4. flow to the well is radial and horizontal; and
5. the aquifer is confined (i.e., bound by impermeable strata both above and below).

As with most cases where models are used, many of the assumptions listed above will not be true for the actual aquifer. Hydraulic characteristics of an aquifer are not the same at every location in the aquifer and the drawdown predictions made by such models are not exact; that is, the actual drawdowns observed in the field may differ from those predicted by the model. In addition, the Theis solution is applicable to artesian aquifers, and most contamination occurs in semi-confined or water-table aquifers. As such, the solution derived normally is too optimistic for water-table conditions.

The optimum number of pumping wells and the optimum pumping rates may be determined by systematically changing the well spacings to find a spacing and number of wells that can achieve significant water-level drawdowns, and hence, increase the hydraulic gradient of the groundwater toward the recovery well throughout a plume without causing excessive drawdown at any one well. There are numerous commercially available models to assist an investigator in finding an optimum number and spacing of recovery wells.

Figure 19.1 is a conceptual diagram of a recovery well system that was installed around a leaky lagoon near the leading edge of a contaminant plume. If the recovery system were located a significant distance upgradient from the leading edge of the plume, the natural flow velocity would need to be included in the computer model to ensure the capture of the plume. Thomas Prickett's random-walk model is frequently used to determine the influence of natural flow on the ability of a recovery well to capture a plume.

An error that is often made in using computer programs to predict drawdowns and well spacings is to assume that the well is 100% efficient (no head loss at the well) and that maximum drawdown can be maintained. In thin water-table aquifers, the saturated thickness will be relatively small. Because the drawdown at a recovery well cannot be greater than the saturated thickness, the radius of capture possible for a shallow pumping well decreases with saturated thickness. The pumping rate predicted to maintain the maximum drawdown cannot always be maintained, because wells are seldom fully efficient and efficiency may decrease with time as a result of screen plugging. Thus, while the computer model may predict that a well can pump at a certain rate, the well may have significantly greater drawdown due to

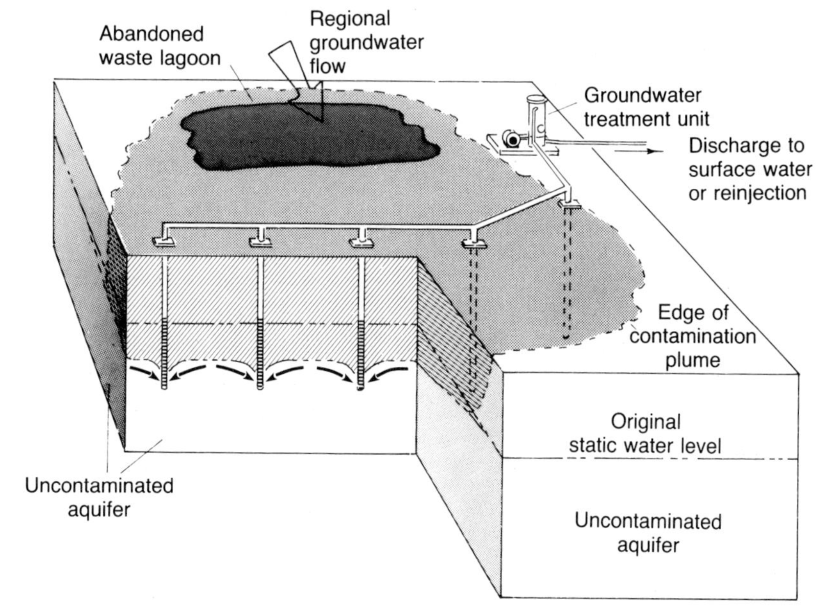

Fig. 19.1 Typical well recovery system.

head loss than predicted and reduced pumping rates may be required to prevent the water level in the recovery well from falling to the pump intake. With a lower pumping rate, the capture radius will not be as great as predicted by the models. The result will be less production than predicted and possibly no capture of portions of the contaminant plume, resulting in possible off-site migration of contaminants.

Another important factor that is often overlooked in designing well spacing is the hydraulic gradient. Given the same transmissivity and storage coefficient, an aquifer with a steeper hydraulic gradient requires closer spacing of recovery wells to prevent off-site plume migration.

19.6 MASS AND DISTRIBUTION OF CONTAMINANTS TO BE REMOVED

Before treatment methods can be evaluated, the total mass of contaminants to be treated should be known. The mass of contaminants in groundwater can be estimated by mapping concentrations determined by chemical analysis. Figure 19.2 indicates the concentration of volatile organic compounds in groundwater at a site in the northeastern United States. In this example, the following assumptions were made to estimate the mass of contaminants requiring treatment:

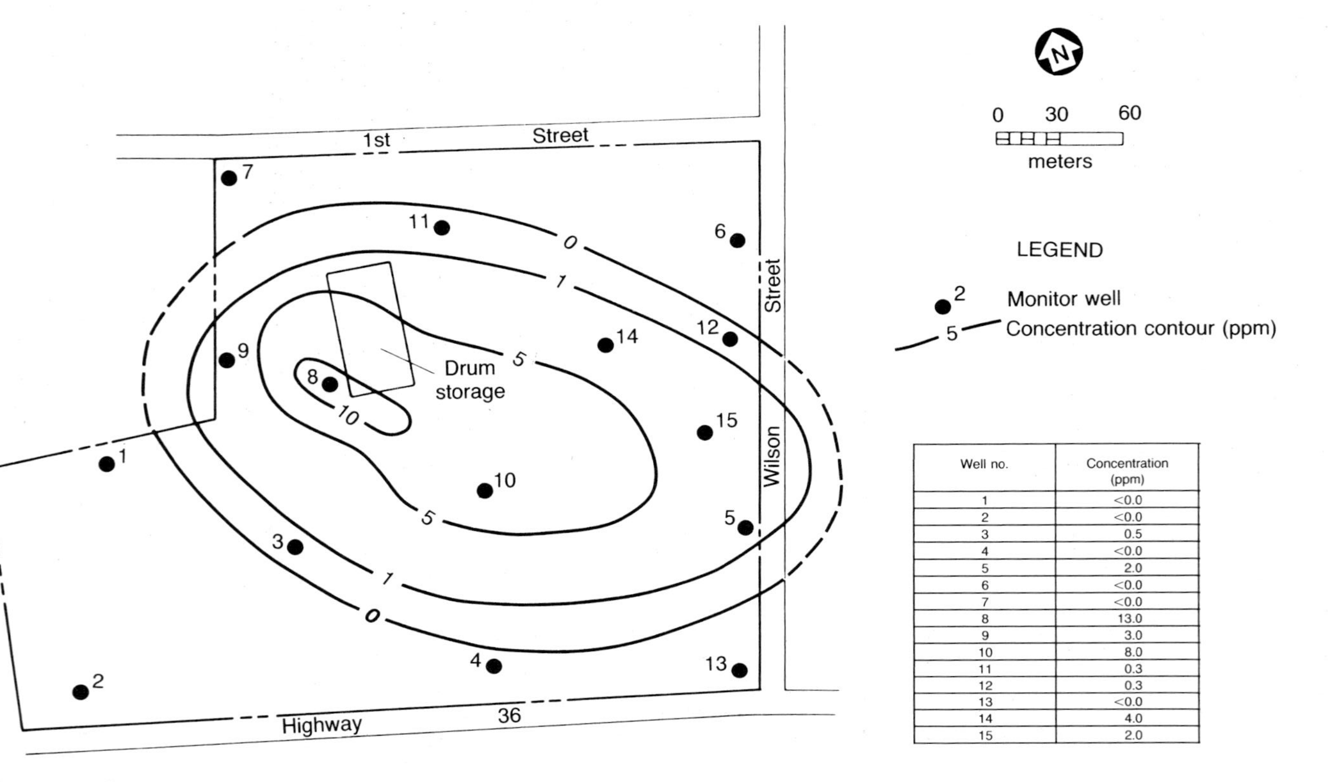

Well no.	Concentration (ppm)
1	<0.0
2	<0.0
3	0.5
4	<0.0
5	2.0
6	<0.0
7	<0.0
8	13.0
9	3.0
10	8.0
11	0.3
12	0.3
13	<0.0
14	4.0
15	2.0

Fig. 19.2 Concentration contour of compounds (ppm).

- the concentration of volatile organic compounds in the aquifer for a particular area of the site is equal to the average of adjacent contours (e.g., the concentration in groundwater for the region between the 5- and 10-mg/L contours is assumed to be 7.5 mg/L);
- the aquifer has a uniform concentration of volatile organic compounds with depth equal to the average of adjacent contours (e.g., all groundwater beneath the region bounded by the 5- and 10-mg/L concentration contour would be assumed to have a concentration of 7.5 mg/L for the full depth of the aquifer);
- the porosity of the aquifer is uniform; and
- the saturated thickness is uniform.

The following equation can be used to estimate the mass of contaminant in groundwater

$$M_t = \sum_{i=1}^{L} M_i = A_i B n C_i \tag{19.1}$$

where: M_i = mass of contaminant in groundwater for area segment i; A_i = area between concentration contours, area segment i; B = saturated thickness; n = porosity; C_i = average of bounding concentration contours for area segment i; L = number of area segments to be summed; and M_t = total of mass within all area segments.

In addition to determining the mass of contaminants dissolved in the aquifer, the mass of contamination in the vadose zone should also be determined, as well as any free phase that may exist. Ultimate recovered mass can be compared with predicted mass to assess the effectiveness of the recovery system.

Concurrent with calculating the mass of contaminants present, the horizontal and vertical limits of contamination must be determined. Frequently overlooked in contaminant studies is the vertical component of flow. Although most investigations do an adequate job of determining the horizontal extent of migration, the vertical extent is not as well defined. Cluster wells or multiple-depth wells are required to define the plume completely with the deeper wells being screened below the deepest zone of contamination.

19.7 CLEANUP TIME

Although it may be possible to estimate cleanup times by using computer programs, the length of many cleanup operations is much greater than predicted. For example, a recovery well system has recovered more than 140 000 cubic meters of contaminated water at a site in southern Louisiana for which the original plume volume was calculated to be only 14 000 cubic meters. Figure 19.3 demonstrates the

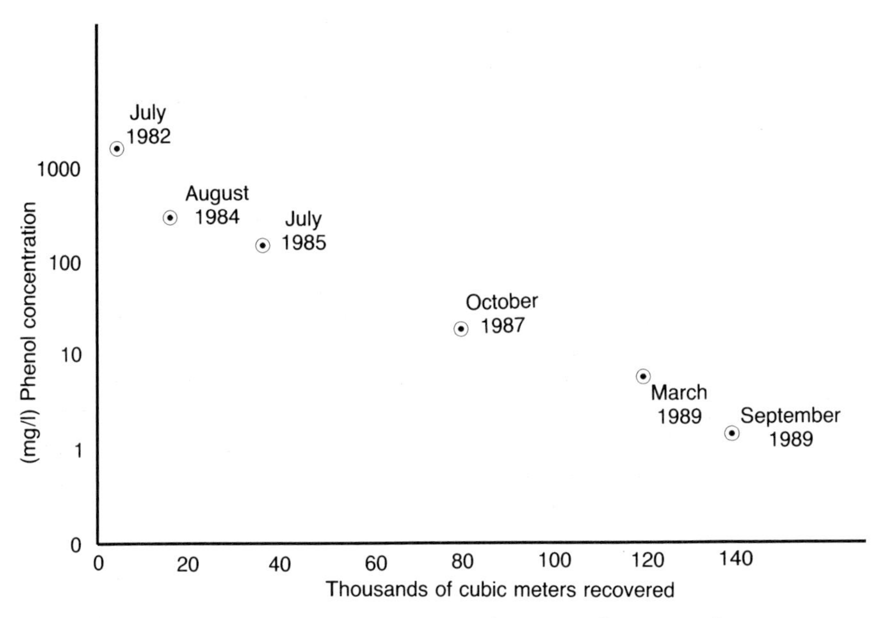

Fig. 19.3 Contaminant concentration (logscale) vs. total recovered water.

decreased concentration of contaminants in the recovered groundwater versus total volume pumped; although there has been a decrease in concentration, the aquifer is still contaminated. There is very little documented history of complete cleanup of a contaminated aquifer. Many recovery operations have been terminated, but most have not recovered all of the contaminant mass. The recovery systems have ceased operation because contaminant concentrations have been reduced to levels acceptable to regulatory authorities. A review of several case histories indicates that removing 10–25 plume volumes may be necessary to reduce contaminants to acceptable levels. Many professionals are beginning to believe that recovery-well systems may never reduce concentrations of organic contaminants to background levels. However, because many regulatory authorities have developed non-degradation policies for groundwater, conflicts are sure to develop when industries petition to terminate recovery systems. Health-based risk assessments have been suggested as one method to determine when to halt recovery operations.

In some recovery systems, contaminant concentrations have been observed to increase in a recovery well after an extended period of shutdown. In the case of organic contaminants, this has been attributed to the desorption of contaminants, predominantly from the clay matrix in an aquifer, into the water column. This has led to the concept that recovery wells should be operated in a cyclic manner to maximize recovery of contaminants. Some dewatering of the aquifer has occurred in some water-table aquifers after significant pumping. However, water and contaminants remain in the dewatered cone of depression. Cyclic pumping may also be of some benefit in restoring the dewatered zone in such cases.

19.8 RECOVERY WELL DESIGN

The design of recovery wells depends on the type of aquifer that has been contaminated and the recovery rate that is required. The recovery rate must be determined before a recovery well can be properly designed because the size and type of the pump will be major factors in determining the diameter of the casing and screen. Recovery wells range from 40-mm-diameter wells that use pneumatic displacement pumps that recover about 10 l/day to large-diameter wells equipped with high-production-rate pumps and large electric motors that can recover about 10 000 l/min. Examples of typical well completions for consolidated and unconsolidated aquifers are shown in Fig. 19.4.

19.8.1 Drilling methods

Recovery wells are drilled by the same general drilling techniques that are used to drill monitoring wells. Shallow recovery wells less than 100 mm in diameter are frequently installed through a hollow-stem auger (like a monitoring well). In unconsolidated sediments where the recovery well is greater than 150 mm diameter, rotary drilling methods are usually employed. When mud rotary is used to complete wells in unconsolidated sediments, the mud invades the formation and causes plugging. It is difficult to develop such a well to its full potential. Although organic drilling fluids have been discouraged for use in the construction of monitoring wells, they have an advantage over bentonite-based drilling muds when used for recovery wells because these organic drilling fluids break down and do not leave residuals that plug the aquifer pores. In consolidated formations such as sandstones and limestones, air-rotary drilling is a preferred method of drilling, especially in formations that have a history of lost circulation.

19.8.2 Materials of construction

Recovery wells are constructed of a variety of materials such as polyvinyl chloride (PVC), carbon steel, galvanized steel, stainless steel, and fiberglass. Monitoring well casing and screen are selected to preclude interference with the results of chemical analysis. Recovery well systems are installed in contaminated zones and precision analytical results are not usually a concern. For recovery wells, a more important

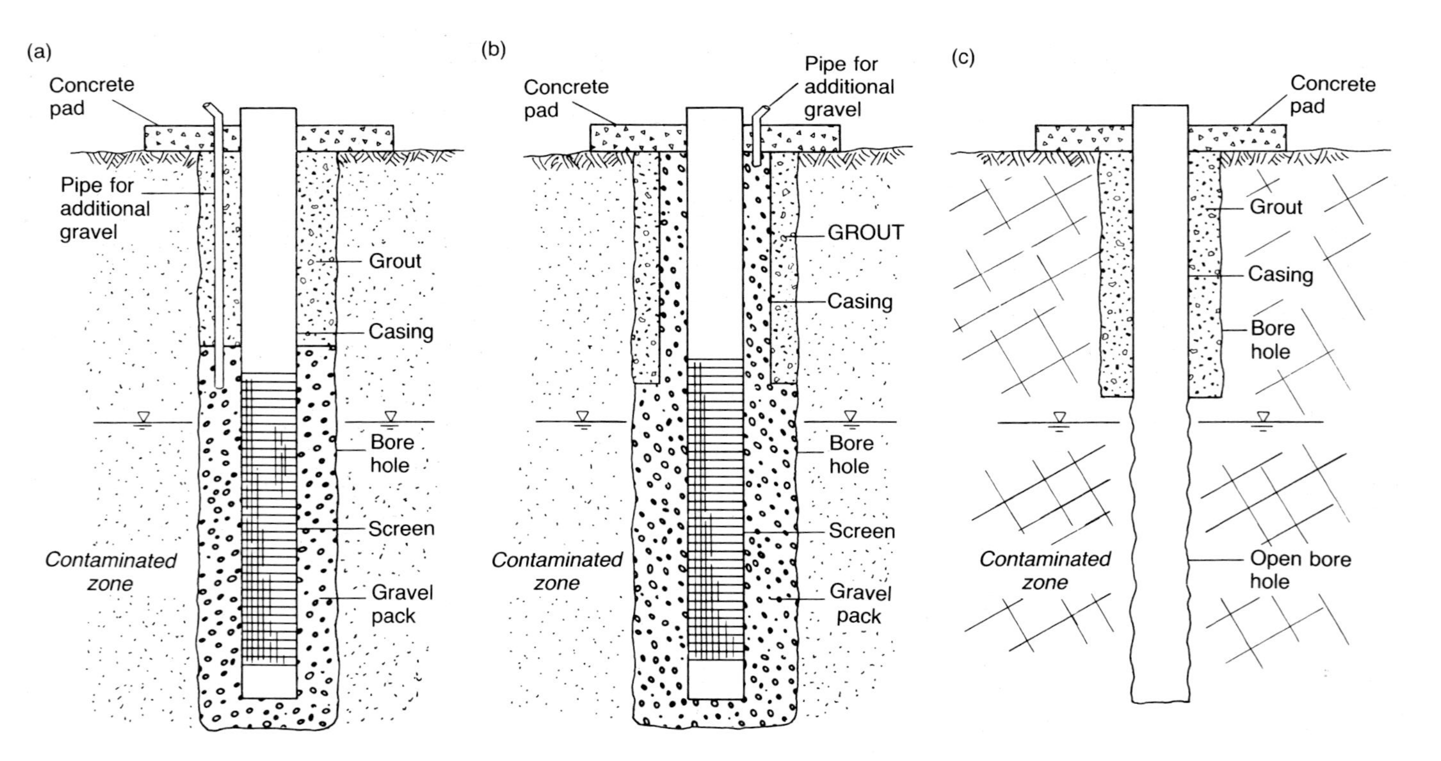

Fig. 19.4 Typical recovery well completion: unconsolidated aquifer with (a) straight casing and (b) telescoped completion, and (c) consolidated aquifer with open hole completion.

concern is the reaction of contaminated groundwater with the materials of construction resulting in decreased casing or screen life. PVC casing and screens may lose their structural integrity if they are placed in contact with high concentrations of organic contaminants. Wells for the recovery of free-phase hydrocarbons frequently are constructed of steel or fiberglass.

19.8.3 Filter pack design

Wells that are screened in unconsolidated aquifers are filter packed between the screen and the borehole. There are two types of filter packs: natural pack and artificial pack. A natural pack is produced by pumping and surging the well and removing the clay and silt matrix of the aquifer through the screen, leaving the coarser sands as an envelope around the screen. An aquifer is generally considered suitable for use of a natural pack if the aquifer is coarse grained and poorly sorted (well graded).

Artificial packs are used in recovery wells when the aquifer matrix is very fine grained or where sand, silt, and gravel are interbedded. The main advantage of an artificial pack is that the pack material can be coarser than the formation matrix and, therefore, larger screen slot sizes can be used. These larger slot sizes provide less opportunity for blockage or encrustation, either by chemicals or bacteria growth, which results in decreased well yield.

There is considerable debate over the design of filter packs. Thickness is an important consideration because it dictates the borehole size. The larger the borehole size, the more cost involved in construction of the recovery well. Under theoretical conditions filter pack of only a few millimeters is all that is required. However, because the screen is not exactly centered, a slightly thicker pack is normally recommended to prevent formation fines from migrating into the pumping well. Most recovery-well systems use a filter pack 50–150 mm thick. This requires drilling a borehole with a diameter 100–300 mm larger than the screen. If filter pack is too thick, it may be difficult to develop the well properly, especially if the well was drilled with mud-rotary techniques. Filter packs in excess of 150 mm are not usually recommended.

Filter packs can be installed using a tremie pipe or they can be poured into the casing-wellbore annulus if additional filter pack is required. Although it is not common practice, the filter pack can consist of different grain sizes placed opposite interbedded strata of different grain sizes. Most filter packs are not designed to be laminated, yet many shallow fluvial aquifers consist of interbedded lenticular strata, resulting in less than perfect gravel packs. For this type of aquifer, the authors have normally selected a filter-pack size on the basis of screen slot size rather than the more conventional grain-size curves of the aquifer matrix.

19.8.4 Well development

During recovery well drilling, some reduction in hydraulic conductivity will occur near the well bore and normally cannot be prevented. This damage to the formation can result in reduced production. Surging and backwashing will remove some of the damage to the wellbore that occurred during drilling and will increase the hydraulic conductivity of the material adjacent to the well bore. The greatest amount of damage to formations takes place during drilling with mud-rotary rigs because a mud cake builds up on the formation face during the drilling process. This mud cake supports the formation and minimizes the loss of drilling fluids into the formation. However, if this mud cake is not removed after drilling, the well will produce less water than anticipated. Development of the well and removal of the mud cake can become difficult if the filter pack is too thick or if the aquifer has low hydraulic conductivity.

19.9 REHABILITATION AND MAINTENANCE OF WELLS

Recovery wells require continued maintenance. Wells completed in consolidated or fractured aquifers without screens require less rehabilitation and maintenance than wells completed in unconsolidated sand formations. A major reason for maintenance is plugging of the gravel pack or screen resulting in a reduced well yield. Pump failure is also common in recovery-well systems.

Recovery wells are subject to the same corrosion and plugging phenomena that occur in water-supply wells. The most common plugging problems are caused by precipitation of calcium or magnesium carbonates and iron compounds. Recovery wells can also encounter maintenance problems in aquifers that are grossly contaminated. Many organic compounds can undergo chemical reactions that result in the precipitation or growth of organic sludges. These sludges grow in the screen or filter pack and are difficult to remove. Generation of carbon dioxide (CO_2) has been observed in several locations where the organic contaminant provides a food source to bacteria. This aerobic biological activity takes place in wells and gravel packs that have been dewatered and exposed to the atmosphere. CO_2 can cause problems in several ways: outgassing of CO_2 causes the pH to change and frequently results in precipitation of carbonates where as the generations of large quantities of CO_2 can cause cavitation in the pumps.

Several rehabilitation methods that have been developed for the water-well industry can be used to restore production to a contaminant recovery well. Acid is most commonly used to clean screens clogged with calcium carbonate. Hydrochloric acid can be pumped into the well to dissolve the scale. When acid reacts with carbonate material, CO_2 is generated. If acid is used to clean shallow recovery wells, the CO_2 may cause an eruption in the filter pack surrounding the recovery well or it may result in acid bubbling back to the ground surface. Acid workover of shallow recovery wells is generally not recommended. If the aquifer contains a significant amount of clay, acid can result in the breakdown of clays and subsequent formation plugging. In these cases, acidification of the well may result in the loss of production. In some instances, production can be re-established in a recovery well using mechanical screen scrapers. However, in many instances, the cost of attempted restoration of a recovery well may be more than the installation of a new well. It is not uncommon for recovery wells in grossly contaminated aquifers to be replaced several times before the recovery operation is complete.

Many recovery wells are completed in unconsolidated sediments of low hydraulic conductivity, such as silty sand or sandy silt. It is extremely difficult, if not impossible, to complete a well that does not produce some sediments. In some cases, the silt and clay are pumped through the treatment system and decrease the efficiency of the system or cause it to fail. Silt and sand grains are abrasive and can damage well screens, pumps, flow meters, and other components of the system. In some cases, the well becomes full of sediment and must be cleaned out. The most frequent method of cleaning is to pull the pump from the well, circulate clean water down the well bore through a drop pipe, and flush the sediment out.

19.10 PUMPING SYSTEM DESIGN

Two general types of pumping systems are used in the recovery of contaminated groundwater: double-pump and single-pump systems. Double-pump systems are used for the recovery of organic compounds that are not miscible in water, such as straight-chain hydrocarbons. Most petroleum products fall into this category. It should be noted that many components of hydrocarbons are soluble in water and, not only will there be a separate phase of a floating layer of hydrocarbons, but the water underlying the less-dense phase will contain soluble organic compounds such as benzene, ethylbenzene, toluene, and xylene.

19.10.1 Double-pump system

Figure 19.5 illustrates a typical double-pump recovery system used for the recovery of hydrocarbons that are less dense than water, such as gasoline. In this system, the water table is drawn down by the groundwater extraction pump and the floating hydrocarbons accumulate in the cone of depression. The hydrocarbons are removed

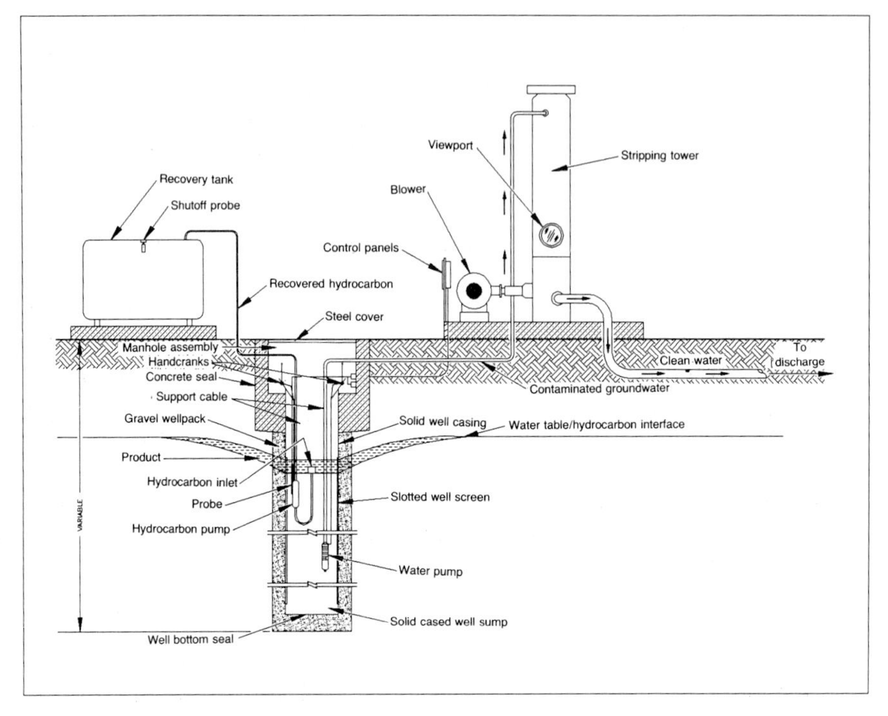

Fig. 19.5 Typical double pump recovery and air stripper for aquifer contaminated with floating and dissolved hydrocarbons.

via the hydrocarbon extraction pump and routed to a hydrocarbon recovery tank. The water that is pumped out by the groundwater extraction pump contains soluble contaminants such as benzene, ethylbenzene, toluene, and xylenes that are removed by a water treatment system.

19.10.2 Single-pump system

The double-pump recovery system was developed to separate the free-phase hydrocarbon and the groundwater in the recovery well. The single-pump system recovers both hydrocarbons and water and delivers a mixture of the two fluids through a single pipe to an above-ground hydrocarbon-water separator. The separator is used to separate the hydrocarbon phase, which can be recycled, from the water phase, which will require treatment. Both electric submersible pumps and gas-driven systems have been used.

Although leakage of hydrocarbons from underground storage tanks is a major problem, most instances of major groundwater contamination have resulted from the disposal of wastewater in evaporation ponds or treatment lagoons. This contaminated water is usually miscible with groundwater, and only single-phase-fluid recovery is necessary. Most recovery wells for miscible contaminants are equipped with either electric submersible pumps in medium-to high-yield aquifers or small gas or air pumps for low-yield aquifers. Figure 19.6 is a completion diagram of a recovery well with a submersible pump.

19.11 GROUNDWATER TREATMENT SYSTEMS

After the recovery-well system has been designed and the type and mass of contaminant(s) have been determined, consideration of the groundwater treatment facilities can begin. However, it is important to recognize that the design cannot be completed until key design parameters have been defined. These parameters are inlet concentration, desired effluent concentration, flow rate, fate of the residuals generated by treating the water, and options for disposing of the treated effluent. Several of the more important considerations are discussed below.

19.11.1 Inlet concentration

Development of a realistic estimate of the inlet concentration is important. Underestimating the inlet concentration can lead to a system that does not effectively treat the groundwater. Overestimating the inlet concentration can lead to the installation of a system that is more involved and expensive than it needs to be. A common error that should be avoided is the 'conservative' estimate that the inlet concentration to the treatment system will be the highest concentration for each constituent on the site. This implies that the entire site becomes equivalent to the hot spots for each constituent – a situation that grossly overstates the treatment requirements.

The average inlet concentration for the treatment system should be estimated from a material balance based on the flow rates and concentrations for each individual recovery well. The flow rates will have been determined during the development of the recovery well system. The initial inlet concentrations can be estimated at the location of each pumping well from the concentration contour, such as that shown in Figure 19.2. In some cases, it may be economically attractive to design and operate more than one treatment facility, one for the most concentrated area and one for less contaminated areas.

19.11.2 Outlet concentration

The outlet concentration should be established before designing the treatment system because the outlet concentration often influences the

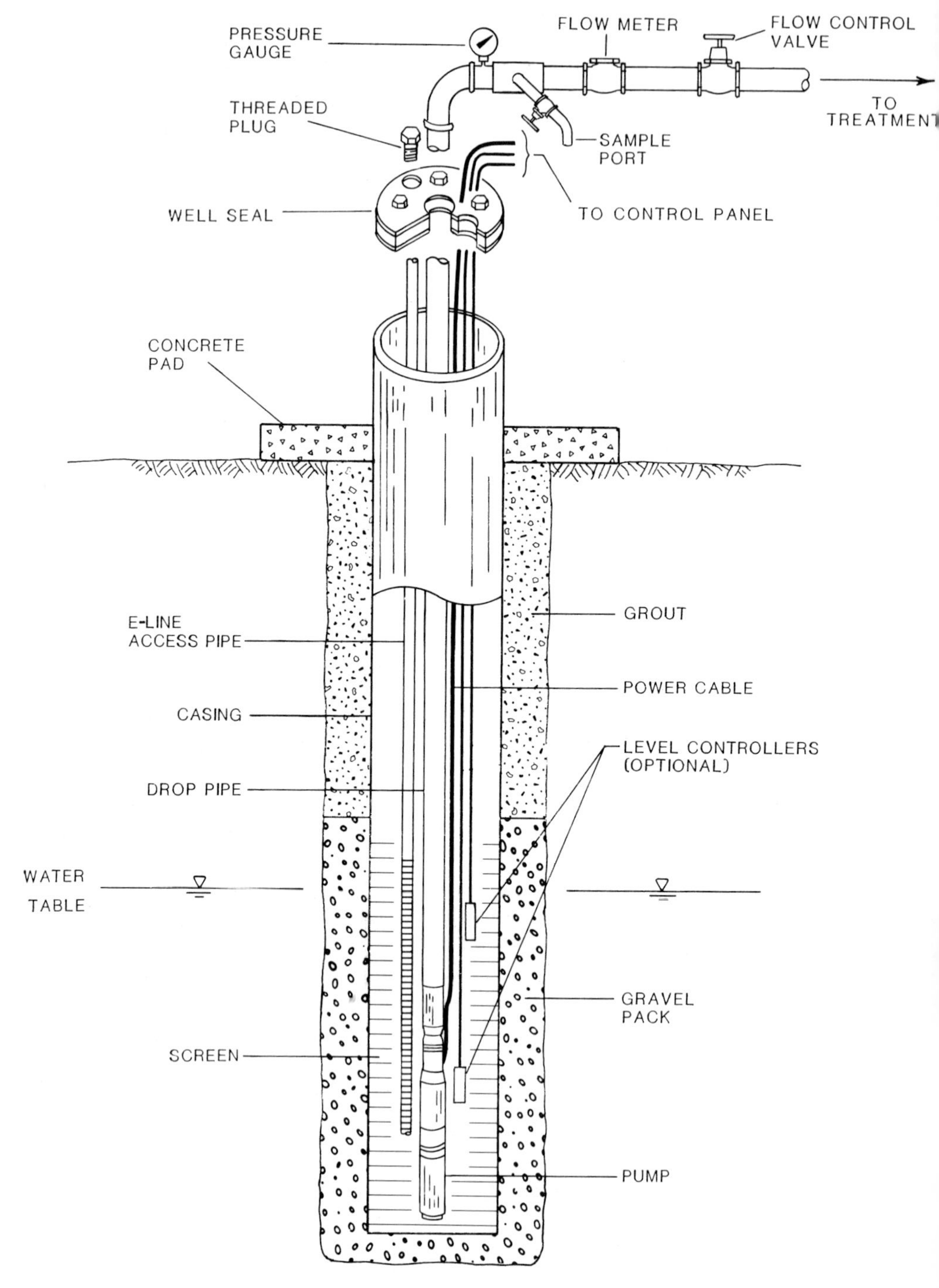

Fig. 19.6 Typical submersible pump recovery well system.

selection of the treatment technology. The outlet concentration will usually be heavily influenced by where the effluent will be discharged and by the potential impact of the individual chemical constituents on the receiving waters and potential receptors. Health-risk assessments are sometimes required in order to develop criteria for the outlet concentrations; however, most effluent limits are established by regulatory authorities.

19.11.3 Flow rate

The flow rate should be determined by carefully examining the pump scheme and the hydrogeology of the site. A realistic estimate for the flow rate should be developed prior to designing the treatment system because the flow rate determines the size of the treatment equipment, which, in turn, heavily influences the capital and operating costs. It is often tempting to specify a wide range for the flow rate; however, this should be avoided wherever possible because the hardware and instrumentation needed to effectively treat a highly variable flow rate involve higher capital and operating costs.

19.11.4 Treatment technologies

Although numerous systems are available for treatment of contaminated groundwater, air stripping and carbon adsorption are the most common treatment for organic contaminants. Carbon and air stripping are frequently used in series when both volatile and nonvolatile organic contaminants are present or when a very high removal efficiency is required.

A detailed assessment of treatment of contaminated groundwater is beyond the scope of this chapter. However, when designing recovery well systems, one must have some knowledge of the general treatment processes available. Selection of treatment technologies is influenced by the nature of the contaminants present. Groundwater contaminated by organic compounds can be treated by physical separation or oxidation techniques. Inorganic compounds cannot be oxidized to CO_2 and H_2O and, thus, must be physically separated from the groundwater.

pH adjustment may be used as pretreatment for groundwater contaminated with either organic or inorganic compounds. For organic compounds, pH adjustment can either optimize the conditions for treatment or prevent buildup of scale or bacteria in equipment such as strippers and carbon absorbers. For inorganic compounds, pH adjustment is used to change a physical or chemical property to enhance treatment, most often for removal of metals via precipitation.

Air stripping

The most-common air stripping process used to treat contaminated water is the packed column or packed tower (Fig. 19.5). The packed column method of treating groundwater uses a counter current flow system. Contaminated water is pumped to the top of the stripping tower and moves downward through the packing under the force of gravity. Air is forced upward through the column by a blower placed at the bottom of the tower. Volatile organic compounds are transformed from a liquid phase to a vapor phase and the volatilized contaminants then move up and out through an air outlet at the top of the column. The treated water collects in the bottom and is discharged or reinjected.

In some locations, the air discharged through the top of the tower is vented directly into the atmosphere. However, many regulatory authorities prohibit the direct discharge of organic contaminants to the atmosphere unless the total mass of emissions is very low.

Most manufacturers of air stripper equipment have developed computer programs to assist the design engineer. Most of these programs assume that the contaminant concentration remains constant. In many cases, this is not true and the quality of water from a recovery well varies with time. If the concentration at the recovery well exceeds the design input quality, the treatment system may not meet required discharge standards.

Packed-tower air strippers are not maintenance free. Periodic checks are required to maintain performance of the system. Frequent problems include scaling and clogging of the packing material by bacterial or algal biomass, calcium carbonate, or iron scale similar to that experienced with well screens. For design of a treatment system, the recovered water should be analyzed for scaling tendencies in addition to contaminant concentrations. Pretreatment may be required to prevent scaling and/or biological fouling.

Air strippers can be designed to be more than 90% efficient at removing volatile organic compounds. However, more typical removal efficiencies are indicated in Table 19.1 which contains the analytical results before and after treatment of groundwater contaminated by a fuel spill. Although contaminant concentrations were significantly reduced, additional treatment may be required in order to meet discharge standards. The primary contaminant of concern is benzene, which normally has a very low allowable discharge concentration limit. This limit can be attained by connecting an additional stripper in series or by carbon adsorption of the water after initial air stripping.

Carbon adsorption

Activated-carbon systems are widely used for the treatment of groundwater contaminated with organic compounds. Recovered water is

Table 19.1 Air stripper treatment

date	*influent concentration (ppb)*				*effluent concentration (ppb)*			
	benzene	*ethyl-benzene*	*toluene*	*xylenes*	*benzene*	*ethyl-benzene*	*toluene*	*xylenes*
01/19/89	4800	2100	14 000	13 000	121	<100	300	890
01/25/89	8000	2800	21 000	17 000	80	49	250	510
02/02/89	3400	2200	15 000	16 000	100	66	390	630

Flow rate: 6 GPM; tower size: 35.5 cm × 5.5 meter; % removal: Benzene, 97–98%; ethylbenzene, 97–98%; toluene, 97–99%; xylenes, 93–97%.

pumped through granular activated-carbon canisters, where the organic compounds are adsorbed. Adsorption is a surface phenomenon wherein the organic molecules are held in the internal pores of carbon granules. Activated carbon can only adsorb limited amounts of contaminants, after which time 'breakthrough' of the contaminants occurs. As an example, Table 19.2 lists the concentrations of contaminant detected in samples of groundwater before and after treatment by carbon adsorption. It demonstrates that 'breakthrough' times vary with different contaminants, when the groundwater has a mixture of contaminants.

Carbon consumption rates have been developed for many compounds and, if the groundwater contains only a few organic contaminants at low concentrations, then the anticipated life of the carbon can usually be estimated (frequently estimated as 10–15% by weight). However, where numerous organic contaminants are present or the plume concentrations are not uniform, more intensive evaluations are necessary to determine consumption rates. Batch test, column test, or in some cases, pilot tests are suggested. Numerous case histories exist in which the carbon usage rate was grossly underestimated on the basis of adsorption capacity data provided by the carbon suppliers.

19.11.5 Fate of residuals from treatment

Treatment of contaminated water usually entails physically separating the contaminants from the groundwater. The design of the treatment system is not complete until the fate of the residual contaminants is defined. In fact, the selection of the treatment technology often is directed by site-specific restraints or regulatory allowances for the disposal of the residual contamination. Some of the possible fates of residuals from air stripping and aqueous-phase activated-carbon systems are discussed in the following sections.

Table 19.2 Carbon treatment

date	*influent concentration (ppb)*				*effluent concentration (ppb)*			
	benzene	*ethyl-benzene*	*toluene*	*xylenes*	*benzene*	*ethyl-benzene*	*toluene*	*xylene*
08/28/87	9900	–	14 000	13 000	<100	–	<100	<100
11/05/87					<100	–	<100	<100
12/10/87					<100	–	<100	<100
12/14/87					<100	–	<100	<100
07/05/88	4500	1100	4 200	6 800	500	<100	<100	<100
07/12/88					2100	<100	1500	700
07/19/88					3300	<100	3500	600

19.11.6 Air emission options

Three potential options are available for controlling vapor emissions from an air stripper: direct discharge to the atmosphere, incineration/combustion, and carbon adsorption. Direct discharge is the most economical alternative. When the level of contaminants in the groundwater is low, direct discharge frequently is approved by regulatory authorities. However, when the contaminant level is high, vapor treatment may be required. Carbon adsorption of vapors is similar in concept to adsorption of contaminants from water. The greatest problem with carbon adsorption systems is the disposal or regeneration of the carbon. Off-gases from air strippers normally contain low concentrations of organic contaminants and cannot support a flame. If treatment is to be by combustion, supplemental fuel is usually required. Several companies market fume incinerators for use as emissions treatment. If the project is expected to be a short-term operation, then carbon adsorption may be advantageous, particularly because it involves a lower capital cost.

19.11.7 Carbon disposal

Several of the treatment options discussed use carbon for either primary treatment of water or secondary treatment for vapor control. Once the carbon has reached saturation and no additional adsorption is possible, the carbon and the adsorbed organic compounds pose a disposal problem. Disposal options may be limited by regulations depending on the source and type of contaminant. Carbon frequently can be returned to the supplier for thermal regeneration and reuse.

Other options are off-site incineration, disposal in a landfill, and possibly on-site regeneration using steam or other methods.

19.11.8 Effluent disposal options

Five major options are available for disposal of treated effluent. Each method has advantages and limitations. The final option that is selected frequently is selected on the basis of regulatory considerations rather than technical merit.

Reinjection

Treated water can be returned to the aquifer after treatment either via injection wells if the contaminated aquifer is deep or via infiltration trenches if the aquifer is shallow. If the aquifer is very permeable, then injection should be explored. However, if the hydraulic conductivity is low, or if the water table is high, reinjection may result in surface discharge of the injected fluids and may not be practical. Under those circumstances, other options may be more advantageous.

Sanitary sewer

Most industrial or urban areas are served by a municipal or regional sanitary sewer. Most cities or sewer authorities have ordinances that determine what is suitable for discharge to the sanitary sewer. If discharge is allowed, the sewer authority will specify the allowable quality and quantity of the effluent and any analytical and reporting requirements. Sewer tap fees frequently are required and monthly charges are assessed on the basis of discharge volume. High discharge fees often make this option less attractive than others. However, it is the most trouble-free and practical solution for low discharge volumes in urban areas.

Surface discharge or evaporation

Treated effluent may sometimes be discharged to a stream, storm sewer, or drainageway. Permits for this type of discharge are usually required by regulatory authorities. In many cities, municipal approval is also required. Permit applications can be quite lengthy and public hearings frequently are required. Other discharge options include irrigation, especially in rural areas where land is available. However, if the treated effluent contains low concentrations of hazardous compounds, securing regulatory approval for irrigation may be impossible. In these cases, lined evaporation lagoons may be an acceptable option.

Reuse

If the treatment plant is located at an industrial facility, some water reuse may be possible. Treated water could be used for plant process water or cooling water. However, this option generally is not viable for small industrial facilities.

19.12 RECORD KEEPING

In order to monitor the progress of a recovery well system, good record-keeping procedures must be developed. The original static water level, pumping rate, pumping level and contaminant concentrations should be determined and recorded for each well. During initial start-up operations, pumping rates and water levels should be measured daily. The time between measurements can gradually be increased to weekly and later, as the system stabilizes, to monthly or quarterly. Total volume pumped per well and for the total system should be recorded. A review of these records can detect early problems before complete component or system failure occurs. Without adequate records, it is almost impossible to determine the reasons for decreases in well performance without removing the pump and performing downhole investigations. Chemical analyses of the treatment system influent and effluent should be performed routinely to determine the mass of contaminants removed from the groundwater and the efficiency of the selected treatment option. In addition, monitoring well samples should be collected and analyzed periodically to measure the results of the cleanup operation. Without adequate analysis, updated predictions of cleanup times or treatment cost cannot be developed.

19.13 CONCLUSIONS

Although groundwater contamination has been recognized as undesirable for centuries, heightened environmental awareness in the last 25 years has significantly increased the demand for effective recovery and treatment systems and technological improvements. In the last 10 years, thousands of groundwater recovery wells have been installed in the United States. These systems range from small, single-well systems pumping at low rates to large, multiple-well systems pumping hundreds to thousands of liters per minute. Many recovery/treatment systems have experienced major operating problems and/or have not realized expectations. The following elements are responsible for many of the problems observed:

- inadequate determination of aquifer properties resulting in improper spacing of recovery wells and incorrect predictions of pumping rates;
- inadequate definition of vertical and horizontal limits of contamination;
- failure to anticipate chemical and biological reactions that lead to screen plugging or pump failure;
- failure to recognize the loss of efficiency in recovery wells resulting in lower well yields and smaller capture radius;
- inadequately defined groundwater chemistry resulting in poor selection of treatment systems and fouling of air strippers and carbon beds;
- inadequately defined design parameters resulting in undersized/oversized treatment systems;
- inadequate budgeting for maintenance of the recovery system;
- failure to maintain records of pumping rates, water levels, and water quality, resulting in incorrect conclusions concerning system performance; and
- failure to communicate to the client and regulatory agencies the limitations of recovery systems, resulting in overally-optimistic predications of system success.

The lessons learned during the last 10 years of operating recovery systems indicate that pump-and-treat recovery well systems alone are generally not sufficient to clean up most aquifers, particularly those contaminated with organic compounds. However, a recovery well system can immobilize a plume and prevent further migration of contaminants. In addition, the bulk of the total mass of contaminants is usually removed via the recovery system, thus lessening the risk to human health and the environment. Although recovery well systems will always be used, additional methods of remediation such as bioreclamation soil washing, or soil venting may be needed to reduce contaminants to acceptable levels.

REFERENCES

Saunders, T.J. (1970) *Plato; The Laws*, Penguin Books LTD, New York, p. 348.

Theis, C.V. (1935) The relation between the lowering of the piezometric surface and the rate and duration of discharge of a well using ground-water storage, *Trans. of the Am. Geophysical Union*, Washington, DC, 518–24.

CHAPTER 20

Bioremediation of soils

Raymond C. Loehr

20.1 BACKGROUND

Complex mixtures of chemicals, many defined as hazardous and toxic, increasingly enter surface soils through spills and unregulated land disposal. The regulatory and technical challenge is to use cost-effective control technologies that can treat complex chemical mixtures in contaminated soils and thereby reduce the resultant threat to human health and the environment. Bioremediation can be such a technology.

Bioremediation is a managed, demonstrated active treatment process that uses microorganisms to degrade and transform organic chemicals in contaminated soil, sludges and residues. Bioremediation also reduces the toxicity of the organics and the migration potential of hazardous constituents in the soils, sludges and residues being treated. As such, bioremediation processes can be considered as source control, pollution prevention, and risk reduction processes that can reduce or eliminate groundwater contamination and thereby reduce the need for costly and long-term groundwater treatment processes.

Bioremediation processes have resulted from the application of knowledge from microbiology, biochemistry, environmental engineering and chemical engineering. The fundamentals of biological treatment, as applied to contaminated soil conditions, are the key to the proper design and operation of the bioremediation processes. In such processes, established scientific and engineering principles are used to maintain satisfactory conditions for microbial degradation and loss of organics.

The advantages of contaminated soil bioremediation processes are that the processes:

- are used where the problem is located;

Geotechnical Practice for Waste Disposal.
Edited by David E. Daniel.
Published in 1993 by Chapman & Hall, London. ISBN 0 412 35170 6

- do not require transporting large quantities of contaminated material off-site;
- eliminate the problem rather than moving it somewhere else;
- minimize long-term liability;
- are ecologically sound and an extension of natural processes; and
- generally are cost-effective and competitive with other decontamination technologies for organics.

Bioremediation commonly is part of a total system for a site and can be used with other technologies to remediate a site and lower the overall cost of site cleanup.

Many bioremediation processes exist. Choosing the appropriate process is a function of the remediation goals to be achieved, the physical and chemical characteristics of the material to be treated, the environmental conditions that are created, the materials handling and equipment requirements, and the overall economics. The potential processes include:

- Solid phase processes, such as composting and land treatment; and
- Slurry phase processes, such as liquid solids systems in impoundments or bioreactors.

Bioremediation processes have been used in the United States at Superfund sites. An evaluation of the records of decisions (RODs) that have been made for Superfund site remediations through 1989 indicates that bioremediation has been recommended for source control at over twenty sites. The types of bioremediation processes that are used at these sites are noted in Table 20.1. The chemicals treated at these sites primarily have been polyaromatic hydrocarbons (PAH), volatile organics (benzene, toluene, ethylbenzene and xylene which collectively are noted as BTEX), pentachlorophenol, and phenols. Other chlorinated and non-chlorinated organics also were present at some sites.

Table 20.1 Bioremediation processes recommended for contaminated soils and sludges at Superfund sites*

type of process	*number of times recommended*
Excavation followed by land treatment	11
Excavation with on-site treatment	3
In situ treatment	4
Lagoon aeration	3
To be determined and other	2

*from review of Superfund RODs issued through 1989 (USEPA, 1990); some RODs specify multiple remedies.

Both solid phase and slurry phase processes, as discussed in this chapter, are applicable to surface soils that are contaminated as a result of spills; pits, ponds and lagoons that have leaked; or improper chemical disposal practices. The contaminated soils commonly are unsaturated and exist in the vadose zone above the groundwater. *In situ* bioremediation processes also can be considered for the treatment of contaminated groundwater and deep saturated soils. These latter processes are discussed in Chapter 21.

20.2 BIOREMEDIATION PERFORMANCE CRITERIA

The performance to be achieved is the key to the feasibility of any bioremediation process. The selection of performance criteria ('cleanup' standards) continues to be a subject for discussion and negotiation at the state and federal level. Soil cleanup criteria are established to assure protection of human health and the environment. Such protection is achieved by remediation methods that reduce the mobility, quantity and toxicity of chemicals in the contaminated soil such that human and environmental receptors are not threatened by any pathway.

In practice, soil cleanup criteria satisfy the following two goals. The first is that the concentrations of chemicals left in the soil should be at levels which protect people from risk due to

- ingestion,
- dermal contact,
- inhalation of fugitive dust, and
- inhalation of gaseous constituents volatilized from the soil.

The second is that the residual chemicals are in a form and at concentrations that will not leach and contaminate groundwater.

Acceptable contaminated soil cleanup concentration limits are established by the regulatory agencies on a site-specific basis prior to remediation. Such criteria should take into account site-specific factors such as contaminant sources, physicochemical conditions, environmental sensitivity, potential for human exposure, and subsequent land and/or water uses desired.

Risk assessments increasingly are used to identify appropriate cleanup criteria. An assessment of the risks at a site to be remediated can determine a target for the remediation effort. Such a risk assessment typically has the components identified in Table 20.2.

The results of a risk assessment help identify:

- the degree of control that is needed;
- the cleanup goals that should be achieved to minimize relative risk from this site; and

Table 20.2 Major components in a risk assessment to determine soil cleanup goals

Hazard identification	Determine the relative risks resulting from chemicals at a site: • select compounds to include; • identify exposed populations; • identify sensitive receptors; and • identify exposure pathways.
Exposure assessment	To how much of a compound are people and the environment exposed? For exposure to occur, the following must happen. • release • contact • transport • absorption or intake Together, these form an exposure pathway. There are many pathways, such as air, groundwater, surface water, crop and ingestion.
Toxicity assessment	What adverse health effects in humans are potentially caused by the compounds in question? Review the threshold and non-threshold effects potentially caused by the compounds at the environmental concentration levels.
Risk characterization	At the exposures estimated in the exposure assessment, is there potential for adverse health effects to occur and, if so, what kind of adverse effects and to what extent? This step develops a hazard index for threshold effects, estimates the excess lifetime cancer risk for carcinogens and estimates risk for other chemicals.

- the type of treatment or remediation that might accomplish the desired cleanup.

Cleanup goals are established as numerical standards or concentration limits. Because of the site and risk specific nature of these goals, a variety of standards or limits have been established at specific sites. Examples are (USEPA, 1988):

- 10 000 ppb total petroleum hydrocarbons, 67 ppb benzene, 200 ppm toluene, 44 ppm xylene;
- <1 ppm benzene, toluene, ethylbenzene, and xylene (BTEX);
- risk assessment performed at each site;
- <500 ppb total hydrocarbons (THC), <100 ppb total aromatics;
- to background levels if feasible;
- THC below 100 ppm;
- total BTEX 50–100 ppm or lower;
- non-detectable levels of volatiles; and
- <500 ppm BTEX and non-ignitables.

Until a comprehensive database is developed for soil cleanup criteria, numerical values established as criteria should not be considered as absolute standards, but rather as guidelines.

20.3 PROCESS FUNDAMENTALS

20.3.1 Overview

The microbial degradation of organic compounds has been recognized for centuries as a highly efficient and relatively inexpensive method to treat organic compounds. Biotreatment processes such as composting for sludges and organic refuse, the activated sludge and trickling filter processes for wastewater, and anaerobic digestion for manures and organic sludges have been used for many decades. However, the application of biological processes to degrade and detoxify industrial organics in soils is not as common. This is due primarily to a lack of understanding of:

- the scientific and technical principles involved; and
- how those principles can be used to design and operate bioremediation processes.

This section presents the pertinent principles and factors and discusses how they can be used and applied to such processes.

Bioremediation of a soil contaminated with organic wastes is accomplished by microbial degradation of specific organic compounds. Organisms in the soil require energy for food and growth. Microorganisms obtain this energy through the metabolic degradation of organic compounds. Thus, bioremediation involves the metabolic breakdown of organic chemicals as a food and energy source. Generally, microorganisms that currently exist in soil and water can degrade the organics in contaminated soils, assuming that the environmental conditions are suitable and nontoxic conditions exist. Microorganisms have evolved catalytic systems (enzymes) that degrade naturally occurring compounds present in the biosphere. However, contaminated soils can contain man-made organics that are difficult to degrade. With such chemicals, the natural microbial enzyme systems may have to adapt and acclimate to the chemical before degradation can occur. Acclimation results in an increase in the biodegradation rate of a chemical after the microbial community is exposed to the chemical for some period of time. The basic concepts that relate to bioremediation are:

- bioremediation is a source control, risk reduction and pollution prevention process;

- bioremediation processes reduce the toxicity and migration potential of organic compounds;
- biodegradation occurs in a wide variety of environments and both solid phase and slurry phase processes can be used for bioremediation;
- organic compounds are microbially converted to simpler compounds;
- microorganisms obtain the energy requirements for growth and maintenance from the compounds they degrade;
- microbial enzymes evolved for the degradation of naturally occurring organics can be acclimated for the bioremediation of many organics in contaminated soils; and
- suitable environmental conditions are necessary for the successful bioremediation of contaminated soils.

The factors that affect the performance of bioremediation processes are noted in Table 20.3.

The microbiological degradation of organics transforms carbon from the organic to the inorganic state. The transformation of organic carbon to inorganic carbon (CO_2) is accomplished through enzymatic oxidation, with molecular oxygen involved as a terminal electron acceptor (aerobic metabolism). This also can occur if the final electron acceptor is something other than molecular oxygen such as sulfate (SO_4) or nitrate (NO_3) (anaerobic metabolism). Aerobic degradation results in the production of CO_2; anaerobic degradation results in both methane (CH_4) and CO_2 production. Biological treatment produces innocuous end products – usually CO_2 and microbial biomass.

In the soil, a constituent may not be completely degraded, but transformed to intermediate product(s). The goal of bioremediation processes is the detoxification of a parent compound to a product or product(s) that are no longer hazardous. Thus, degradation may result in detoxification without complete mineralization.

If bioremediation can occur and is effective, why do some organics, such as those in contaminated soils, persist in the environment? This is a reasonable question that is answered by identifying the factors that affect microorganisms and therefore bioremediation processes. Even biodegradable organics may persist if adverse factors and non-optimum conditions exist. The factors that may prevent microbial degradation and bioremediation include:

- chemical concentrations that are toxic to microorganisms;
- inadequate type or numbers of microorganisms, such as due to toxic conditions;
- conditions too acid or alkaline;
- lack of nutrients such as nitrogen, phosphorus, potassium, sulfur or trace elements (many organic chemicals, for example, are not nutritionally balanced);

Table 20.3 Factors affecting bioremediation processes*

factor	*comment*
Microorganisms	Natural organisms are satisfactory, acclimation may be necessary, suitable environmental conditions need to be provided
Toxicity	Non-toxic conditions are needed
Available soil water	25–85% of water holding capacity desirable for solid-phase systems
Oxygen (O_2)	Aerobic conditions desired
Electron acceptors	Under aerobic conditions, O_2 is the terminal electron acceptor; when O_2 is not available, NO_3^-, Fe^{3+}, Mn^{2+} and SO_4^{2-} can act as electron acceptors
Redox potential (oxidative and reduced conditions)	Aerobes and facultative aerobes predominate at greater than 50 millivolts; anaerobes at less than 50 millivots
pH	5.5–8.5 for optimum degradation
Nutrients	Nitrogen, phosphorus and other nutrients are needed for microbial growth
Temperature	Degradation rates are affected by temperature
Water solubility	The water solubility of compounds in contaminated soil can affect degradation
Sorption	Many organic compounds are strongly sorbed to the organic matter in soil
Volatilization	Chemicals can be lost by volatilization
Loss rates	Total loss rates commonly are reported; first order rates usually used to describe losses

* from Huddleston *et al.* (1986) and Sims *et al.* (1989).

- unfavorable moisture conditions (too wet or too dry); and
- lack of oxygen or other electron acceptors.

Bioremediation processes are biological treatment processes that improve or stimulate the metabolic capabilities of microbial populations to degrade organic residues. Therefore, it is important to understand those conditions and reactions so that bioremediation processes can be successful. With such knowledge, it is possible to modify the non-optimum conditions so that microbial degradation can occur. For example, if there is inadequate nitrogen or phosphorus, such nutrients can be added to assure satisfactory microbial degradation. If the residues are too toxic, addition of other chemicals or uncontaminated soil may reduce the toxicity to the point that microbial degradation can occur. If inadequate type or numbers of microorganisms are present, acclimated organisms can be added.

The factors that affect bioremediation processes were identified in Table 20.3. The following sections describe the factors that affect the performance of bioremediation processes.

20.3.2 Microorganisms

Surface soils contain large number of microorganisms that include aerobic and anaerobic bacteria, fungi, actinomycetes and protozoa as well as earthworms and higher forms of life. Over one million organisms can be in one gram of agricultural soil. These organisms are capable of degrading most natural and synthetic organics that are in a soil. However, they can accomplish the degradation only if:

- non-toxic conditions exist;
- the organisms have or can develop the enzyme systems capable of degrading the organic compound; and
- other environmental conditions such as pH, nutrients, oxygen, temperature, and water are adequate.

The actual degradation that occurs in a bioremediation process is a result of the mixed microbial population that exists. One group of microorganisms may partially metabolize a compound and furnish a suitable substrate for another group of microorganisms. If an organic is biodegradable and environmental conditions are suitable, the natural organisms in the soil can adapt to degrade the organic compound. Specially developed microbial cultures have not been observed to be needed or successful in bioremediation systems.

20.3.3 Toxicity

Many individuals who consider remediation options for contaminated soil incorrectly assume that bioremediation is not feasible because toxic chemicals or hazardous wastes are in the soil. They infer that because such materials are toxic to living organisms, biological treatment is not a viable alternative. However, this assumption and inference is not correct. Many organic chemicals indicated as toxic or in hazardous wastes can be biodegraded if the contaminated soil is not toxic to the microorganisms and if other environmental factors, such as pH, nutrients and oxygen, are suitable.

In considering the feasibility of bioremediation processes, the relative toxicity of the contaminated soil should be determined. If the soil is found to be toxic to the soil microorganisms, either steps should be taken to reduce the toxicity or another remediation process should be considered.

The usual procedures to quantify toxicity of a chemical are toxicity assays which measure the effect of the chemical under specified test conditions. The toxicity of a chemical is proportional to the severity of the chemical on the monitored response of the test organism(s). Toxicity assays utilize test species that include rats, fish, invertebrates, microbes and seeds. The assays may use single or multiple species of test organisms. Although no single bioassay procedure can provide a comprehensive toxicity evaluation of a chemical, a valid toxicity screening test can provide information about the relative toxicity of a compound and can help predict non-inhibitory chemical application rates.

Toxicity assays using bacteria as the test organism are rapid, are easy to use, are cost effective, and use a statistically significant number of test organisms. One such bacterial toxicity assay method is the Microtox© assay. This method is relatively simple, rapid and inexpensive. The use of the Microtox© procedure to screen and predict the treatability potential of waste in soil and of contaminated soil has been evaluated and found to be satisfactory (Matthews and Bulich, 1984; Matthews and Hastings, 1987).

The Microtox© system is a standardized toxicity test which utilizes marine luminescent bacteria (*Photobacterium phosphoreum*) as indicator organisms. Bioluminescence of this test organism depends on a complex chain of biochemical reactions involving the luciferin-luciferase system. Chemical inhibition of any of the involved biochemical reactions causes a reduction in bacterial luminescence. The Microtox© toxicity assessment considers the physiological effect of a toxicant, and not just mortality.

This method utilizes an instrumental approach in which the indicator organisms are handled as chemical reagents. Suspensions of about one million bioluminescent organisms are 'challenged' by addition of serial dilutions of an aqueous sample. A temperature controlled photometric device quantifies the light output in each suspension before and after sample addition. Reduction of light output reflects physiological inhibition which indicates presence of toxic constituents in the sample. Such tests do not provide information on toxicity from a human health or an environmental standpoint. Rather, they are used as a relative toxicity screening method for contaminated soil and to identify the relative toxicity reduction that occurs when chemicals and waste are managed by bioremediation processes.

20.3.4 Water

The presence of water is essential for microbial activity. At low water concentrations, non-spore forming microorganisms will die or their concentrations will be reduced greatly. At high water concentrations, such as at or near saturation, the pores of the soil are filled with water and diffusion of oxygen from the atmosphere is restricted. Under the latter situation, anaerobic rather than aerobic conditions will occur unless oxygen is mechanically added, such as in an aerated slurry reactor. The water content of soil typically ranges from 15 to 35 volume percent. At 35%, most soils are water saturated. The water content can drop below 15% under arid conditions. Soil water content is commonly expressed as a percentage of soil water holding capacity. A soil water holding capacity range of 25–100% is the equivalent to a range of about 7–28 volume percent.

Over the range of about 30–90% of the water holding capacity of the soil, the moisture content has little effect on biodegradation rates (Dibble and Bartha, 1979). In solid phase bioremediation systems, moisture control may be necessary to achieve optimum biodegradation rates.

20.3.5 Oxygen

Adequate oxygen and aerobic conditions in bioremediation systems are important to:

- avoid odors produced by anaerobic conditions; and
- produce the most oxidized end-products.

Anaerobic conditions and degradation can occur in bioremediation systems but should be avoided since anaerobic biodegradation is slower and less complete and under reduced conditions, most metals are more water soluble.

For solid phase systems, aerobic conditions are maintained by:

- mechanically mixing the material, such as in composting, or tilling the mixture, as in land treatment;
- avoiding saturating the mixture with water; and
- maintaining the quantity of degradable material in the mixture such that the oxygen demand does not exceed the rate that oxygen is transferred from the atmosphere to the mixture.

For slurry phase bioremediation systems, oxygen is added by mechanical mixers and aeration systems. These systems also mix the liquid slurry to keep the particles in suspension and enhance oxygen transfer. Oxygen transfer in slurry systems usually is less than in dilute or clean water systems and the aeration equipment should be sized to transfer the necessary amount of oxygen under operating and not clean water conditions. To avoid oxygen limiting conditions, the dissolved oxygen concentrations in slurry reactors should be maintained above about 0.5 mg/l.

20.3.6 Electron acceptors and redox potential

In a bioremediation system, microorganisms metabolize organic compounds to obtain biological energy for microbial growth and maintenance. In this process, electrons from incompletely oxidized (reduced) compounds are transferred along respiratory electron transport chains, energy is captured by the microorganisms, and oxidized end-products (such as carbon dioxide, CO_2) result. Under aerobic conditions, oxygen (O_2) is the terminal electron acceptor. When O_2 is not available, nitrate (NO_3^-), iron (Fe^{3+}), manganese (Mn^{2+}), and sulfate (SO_4^{2-}) can act as electron acceptors if the organisms have the appropriate enzyme systems. The oxidation-reduction potential (redox potential) of a soil provides a measurement of the electron density of the system. The redox potential is expressed in millivolts.

Microbial degradation in bioremediation systems removes oxygen from the soil atmosphere in solid phase systems. If oxygen is not replenished, the systems become depleted in oxygen, reduced conditions result, and other substances are used as terminal electron acceptors. To maintain aerobic conditions, oxygen must be added to the bioremediation systems by the methods noted in section 20.3.5.

The loss of oxygen from the system also can cause a change in the microbial population. Facultative bacteria, which do use electron acceptors such as nitrate, sulfate, or oxygen, and anaerobic organisms become the dominant microbial populations. In bioremediation systems, the vast majority of the microorganisms are facultative.

20.3.7 pH

The optimum pH for microbial degradation is around neutral, generally in the range 6–8. However, biodegradation can occur outside this range although at reduced rates. In bioremediation systems, pH control rarely is needed unless very acid or alkaline conditions are encountered. pH control also can be needed if the biological activity causes a marked change in the pH.

20.3.8 Nutrients

Microbial metabolism and growth requires adequate macro and micronutrients. The soil normally supplies adequate micronutrients such as trace metals and minerals. However, it cannot be assumed that all soils have adequate macronutrients such as nitrogen (N), phosphorus (P) or potassium (K). Industrial organics in contaminated soils can be high in carbonaceous content but low in N and P.

The need for additional N and P is controlled by:

- the amount of N and P in the contaminated soil and the rate that they are available; and
- the amount and rate at which organic carbon is degraded in the mixture.

By estimating or measuring these factors, the need for additional N and P can be determined. A carbon to nitrogen (C/N) ratio of greater than 35–40 generally indicates inadequate nitrogen. A C/N/P ratio of about 100:10:1 will provide adequate nutrients in a bioremediation system. If inadequate nitrogen exists, the deficiency can be remedied by adding a chemical fertilizer such as ammonium sulfate or ammonium diphosphate. Over time, the need for additional nutrients can decrease since most of the initially added nutrients remain in the system for reuse.

20.3.9 Temperature

Biodegradation rates are affected by temperature, generally changing by a factor of two for a change of 10 °C in the temperature range of 5–30 °C. Low temperature is not lethal to microorganisms but will drastically reduce biodegradation rates. Biodegradation is essentially zero at freezing temperatures. Insulated reactors can be used to keep the reactor temperature 5–10 °C above ambient air temperatures.

In bioremediation systems, temperature control rarely is practiced. Different systems will operate at different temperatures, however. For example, solid phase composting type systems may operate at temperatures that are above ambient. The temperature in bioremediation systems normally are near ambient temperatures and will change slowly as ambient temperatures change. Abrupt temperature changes are unlikely.

20.3.10 Water solubility

Microbial degradation is carried out by microbial enzymes. In bioremediation systems, the enzymes are not released by the microbial cells and the substances to be degraded must contact or be transported to and into the microbial cell. Thus, the water solubility of the chemical is important since only the water soluble fraction of a chemical is readily degradable.

Many of the chemicals in contaminated soil are not very water soluble. This low solubility can reduce the availability of the chemical for degradation and will be a controlling parameter in bioremediation. The aqueous solubilities of chemicals commonly found in contaminated soils are noted in Table 20.4. The aqueous solubility decreases as the chemical becomes more complex.

Chemicals that have low water solubility generally are soluble in organic solvents. A commonly used measure of the relative organic solvent solubility is the octanol-water partition coefficient, K_{ow}, which is defined as the ratio between the concentrations of the compound in equal parts of octanol and water. A value greater than 1 indicates that the compound is more readily soluble in octanol than in water. Octanol models the behavior of other hydrocarbons and is used as a model nonpolar organic hydrocarbon solvent. K_{ow} values for many chemicals also are noted in Table 20.4.

The octanol-water partition coefficient is a key parameter in studies of the environmental fate of chemicals. K_{ow} has been found to be

Table 20.4 Solubility and octanol-water partition cofficients of organic chemicals*

compound	*pure compound aqueous solubility (mg/L)*	K_{ow}
Monocyclic aromatics		
Benzene	1787	135
Toluene	515	540
o-Xylene	213	1 320
p-Xylene	185	1 410
m-Xylene	146	1 585
Ethylbenzene	110	1 410
Other halogenated compounds		
Trichlorethene	1100	260
Pentachlorophenol	14	132 000
Polyaromatic hydrocarbons (PAH)		
Naphthalene	31	2 000
Acenaphthylene	3.93	5 500
Acenaphthene	3.42	8 300
Fluorene	1.98	15 100
Phenanthrene	1.29	28 800
Fluoranthene	0.26	79 400
Pyrene	0.135	75 900
Anthracene	0.066	28 200
Benz[a]anthracene	0.014	407 000
Benzo[a]pyrene	0.0038	933 000
Chrysene	0.002	407 000
Benzo[b]fluoranthene	0.001	3 720 000
Benzo[k]fluoranthene	0.0003	6 920 000
Benzo[ghi]perylene	0.00026	17 000 000
Benz[a]anthracene	0.014	407 000

* adapted from Tetra Tech (1989).

related to water solubility, soil and sediment adsorption coefficients and bioconcentration factors (BCF) (Lyman *et al.*, 1982).

Chemicals with low K_{ow} values, i.e., less than about 10, have high water solubilities, small adsorption coefficients and small BCF values. Chemicals with high K_{ow} values, i.e., greater than about 1000, are very hydrophobic with low water solubilities and high sorption coefficients.

The aqueous solubility (S) of a chemical can be estimated from K_{ow} values. A number of equations have been developed to correlate these two parameters. The equations generally take the form of

$$\log S = a \log K_{ow} + b \qquad (20.1)$$

where a and b are empirical regression equation constants. Examples of such relationships are noted in Table 20.5.

20.3.11 Sorption

Sorption is another parameter that affects organic chemical degradation. This occurs because the greater extent a chemical is sorbed to a soil, the longer it is retained in the soil and the greater amount of time that is available for degradation. Thus even chemicals that have slow degradation rates (long half-lives) can undergo satisfactory degradation in contaminated soil bioremediation systems.

The mobility of a chemical in a soil system can be expressed in terms of a retardation factor which, as noted in Chapter 3, can be calculated as

$$R = 1 + \frac{K_p \times \rho_b}{\theta} \qquad (20.2)$$

Table 20.5 Examples of empirical relationships between aqueous solubility and K_{ow} values for different chemicals*

equation	*units of S*	*chemical classes represented*
$\log S = -1.37 \log K_{ow} + 7.26$	µmol/L†	mixed classes; aromatics and chlorinated hydrocarbons well represented
$\log S = -0.922 \log K_{ow} + 4.184$	mg/L	mixed classes; pesticides well represented
$\log S = -1.49 \log K_{ow} + 7.46$	µmol/L	mixed classes; several pesticides
$\log(1/S) = 1.294 \log K_{ow} - 0.248$	mol/L	alkenes
$\log (1/S) = 0.996 \log K_{ow} - 0.339$	mol/L	benzene and benzene derivatives
$\log(1/S) = 1.237 \log K_{ow} + 0.248$	mol/L	alkanes

* adapted from Lyman *et al.* (1982).
† mol/L = moles per 1000 grams of water, i.e., molar solubility.

where R = retardation factor (unitless), K_p = soil-water partition coefficient (ml/g), ρ_b = dry bulk density of permeable material (g/ml), and θ = volumetric moisture content (decimal fraction). The partition coefficient, K_p, also is known as the sorption coefficient of a chemical to soil, and, as discussed in Chapter 3, the distribution coefficient, K_d. If a chemical does not interact with the soil, K_p = 0 and the retardation factor equals 1. Examples of such chemicals are chlorides and nitrates. Chemicals with high sorption coefficients are more tightly bound to the soil, are less mobile, and have lower degradation rates.

Adsorption is a surface phenomenon in which matter is extracted from one phase and concentrated at the surface of the second. Adsorption of a chemical from solution onto a solid occurs as the result of the lyophobic (solvent-disliking) character of the chemical relative to the particular solvent, or of a high affinity of the chemical for the solid.

Adsorption is modeled by an isotherm, which is a means of describing changes in adsorption at constant temperature (see Chapters 2 and 3). Adsorption of many chemicals found in contaminated soils commonly can be described by a Freundlich isotherm

$$q = KC^{\frac{1}{n}} \tag{20.3}$$

where q = mass of chemical sorbed per mass of soil at equilibrium, K = adsorption coefficient, C = concentration of chemical in the liquid, and n = an empirical constant representing adsorption intensity.

The adsorption coefficient, K, can be normalized to the organic carbon content of the mixture, K_{oc}

$$K = K_{oc} f_{oc} \tag{20.4}$$

where f_{oc} is the fraction of organic carbon content of the soil. This relationship works well for soils with organic carbon contents above about 0.5%.

Thus, it is possible to express the tendency of a chemical to be adsorbed in terms of K_{oc}, which is largely independent of the properties of the soil. K_{oc} may be thought of as the ratio of the amount of chemical adsorbed per unit weight of organic carbon (oc) in the soil to the concentration of the chemical in solution at equilibrium.

$$K_{oc} = \frac{\mu\text{g adsorbed/g organic carbon}}{\mu\text{g/mL solution}} \tag{20.5}$$

K_{oc} is a constant only if the adsorption isotherms are linear. Generally, linearity occurs when the equilibrium aqueous phase organic chemical concentration is below one half of the pure compound aqueous solubility of the chemical.

K_{oc} also has been related to the aqueous solubility, S, and the K_{ow} of a chemical. The relationships generally take the form of

$$\log K_{oc} = c \log S + d \quad (20.6)$$

or

$$\log K_{oc} = c \log K_{ow} + d$$

where c and d are empirical regression constants. Examples of such relationships are noted in Table 20.6.

20.3.12 Volatilization

Volatile constituents can be in many soils. Such constituents will be released to the atmosphere during bioremediation processes. Controls on volatile losses may be needed as part of the contaminated soil bioremediation system to meet regulatory requirements.

The occurrence of volatilization in such systems is a function of the vapor pressure of the chemical and the contact between the chemical in the soil and the gaseous phase or atmosphere. Bioremediation systems are mixed and/or aerated to provide adequate oxygen input and aerobic conditions. As a result, such systems offer excellent opportunities for volatile compounds to be released.

20.3.13 Chemical loss rates

Results from field systems and laboratory and pilot plant bioremediation systems indicate that the losses of organic compounds can be described by first-order rate reactions in which the rate of loss of a chemical is proportional to the chemical concentration.

$$\frac{dc}{dt} = -kc \quad (20.7)$$

Table 20.6 Examples of empirical relationships between K_{oc} values and other parameters for different chemicals*

equation	*chemical classes represented*
$\log K_{oc} = -0.55 \log S^{\dagger} + 3.64$	wide variety, mostly pesticides
$\log K_{oc} = -0.557 \log S^{\ddagger} + 4.277$	chlorinated hydrocarbons
$\log K_{oc} = 0.544 \log K_{ow} + 1.377$	mostly pesticides
$\log K_{oc} = 0.937 \log K_{ow} - 0.006$	aromatics, polynuclear aromatics, triazines and dinitroaniline herbicides
$\log K_{oc} = 1.00 \log K_{ow} - 0.21$	mostly aromatic or polynuclear aromatics

* adapted from Lyman *et al.* (1982).
† solubility (S) in mg/L.
‡ solubility (S) in micro-moles/L.

where c = chemical concentration (mass/mass), t = time, k = first-order rate constant (time^{-1}).

Chemical loss rates usually are discussed in terms of half-life ($t_{\frac{1}{2}}$), i.e., the time required to degrade or lose one-half of the chemical concentration. Mathematically, half-life can be calculated from the first-order rate constant

$$t_{\frac{1}{2}} = \frac{0.693}{k} \tag{20.8}$$

Examples of half-life values of chemicals in contaminated soil bioremediation systems are indicated in Table 20.7.

Most chemical loss rates that are reported represent total chemical loss and rarely are specific loss mechanisms identified. The losses can be due to biodegradation, chemical degradation, hydrolysis, photolysis and volatilization. While it commonly is assumed that microbial degradation is the major loss mechanism, as noted in section 20.3.12, volatilization also can be a loss mechanism for chemicals with a high vapor pressure.

Since remediation cleanup goals usually are related to concentration criteria (section 20.2), total loss rates are useful in determining the feasibility of bioremediation processes.

Table 20.7 Chemical loss half-life values in contaminated soil bioremediation systems*

organic	*half-life values ($t_{\frac{1}{2}}$) (days)*
benzene	0.1–1.0
ethylbenzene	6.1
toluene	6.4
o-xylene	93.9
m,p-xylene	8.5–14.7
anthracene	9–53
benzo(a)anthracene	41–231
chrysene	5.5–116
1-methylnaphthalene	12.6
naphthalene	8–30
phenanthrene	23–69
pyrene	10–100

*Data from situations where relatively fresh refinery wastes were being land treated and evaluated. Loss rates of consitituents in residuals at closure will be slower than constituents in fresh wastes.

20.3.14 Summary

Bioremediation processes can treat contaminated soils successfully. The microorganisms in such systems will grow and metabolize the organics present if environmental conditions are suitable. Aerobic conditions are desirable and are accomplished by providing mixing and mechanical aeration. Nutrients, such as nitrogen and phosphorus, may be needed. Toxic conditions should not exist and pH and temperature should be controlled as needed.

Many of the organics in contaminated soil have low water solubility, which can limit degradation, and are tightly sorbed to the organic matter in soil. The sorption constants and water solubilities of chemicals can be estimated from known partition coefficients such as K_{ow} and K_{oc}.

Loss rates, rather than biodegradation rates, are commonly reported. Such loss rates can be used to determine whether performance goals can be achieved and the overall feasibility of a bioremediation process.

With an understanding of how the fundamentals described in this section can relate to and affect the performance of bioremediation processes, such processes can be technically and economically feasible.

20.4 TOXICITY REDUCTION

When an organic chemical is degraded in a bioremediation process, it is transformed into other products through biological reactions with or without complete detoxification and immobilization. Measuring the loss of the parent compound does not assure that complete detoxification and immobilization occurs. Intermediate degradation products, which may be more mobile and/or toxic than the parent compound, may be generated as the parent compound degrades. Additional information on the transformation and/or detoxification of a chemical is necessary to establish that the loss of the parent compound leads to the complete detoxification of the chemical or waste. Such information can be obtained using either chemical or bioassay analyses.

Chemical analysis of detoxification products may yield information about biochemical degradation pathways, but it is time-consuming and expensive. Bioassays have been used successfully to demonstrate detoxification of the applied waste in the soil (Matthews and Bulich, 1984; Matthews and Hastings, 1987), and are less expensive and time-consuming. Such bioassays also have been used as a screening tool to evaluate the soil treatment potential of a chemical or waste.

The following describes the reduction in toxicity and in chemicals that occurred in a simulated contaminated soil biotreatment process (Loehr, 1989; Desappa and Loehr, 1991). The study was done in laboratory soil microcosms using phenol and eight chlorophenols.

The reduction of toxicity that occurred was evaluated by determining the toxicity of the water soluble fraction (WSF) of the chemical/soil mixture at the same sampling intervals used to obtain the degradation data. The chemical compounds that can be extracted with water represent the potentially leachable fraction of the chemical or any intermediate chemical detoxification products. The WSF of the chemical poses the greatest threat to groundwater contamination. Hence, evaluating the loss of the potentially leachable fraction of a chemical is important.

The concentration of the parent chemical in the WSF also was determined. This concentration was expressed in terms of quantity of chemical that was water extractable per kg of the soil. The toxicity of the WSF was determined using the Microtox© method (Beckman Instruments, 1982). To put the toxicity reduction data in perspective, the WSF toxicity reductions, the WSF chemical concentration reductions and the soil chemical concentration reductions were compared.

The loss of the chemical in the soil and in the WSF and the WSF toxicity reduction data were evaluated using the first-order kinetic data (half-lives) that resulted from the microcosm degradation studies. Figure 20.1 compares the half-life of the chemical in the soil and the WSF chemical loss for all of the nine chemicals tested. The correlation

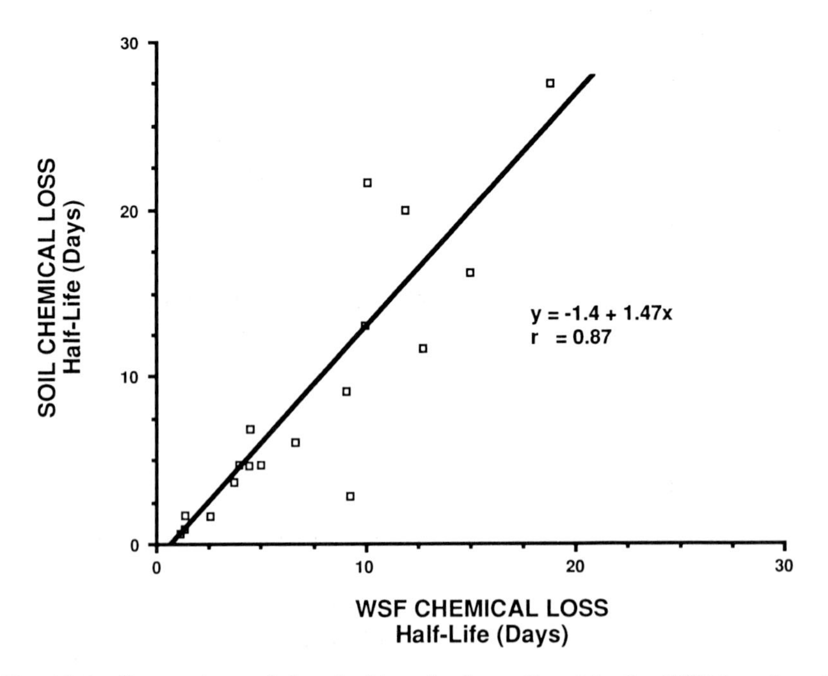

Fig. 20.1 Comparison of chemical loss in the soil and in the WSF for phenol and 8 chlorophenols, after Dasapa and Loehr (1991).

shows that, in general, the soil chemical half-life was about 1.5 times greater than the WSF chemical half-life. For these chemicals, this indicates that the loss of the chemical in the WSF was faster than the loss of the chemical in the soil. The correlation also indicated that no enhanced mobilization of applied chemical occurred as the degradation and detoxification took place.

Figure 20.2 compares the WSF chemical loss and the WSF toxicity reduction for all nine tested chemicals. The high correlation suggests that the WSF toxicity can be attributed to the target chemical concentration in the WSF and that no water extractable toxic intermediate products were formed. Thus, these chemicals were detoxified in the soil.

These results indicate that contaminated soil bioremediation processes not only reduce the concentration of organic chemicals originally present, they also reduce the loss of chemicals in the water soluble fraction in the soil and, based on one bioassay method, reduce the apparent toxicity of the water soluble fraction. Thus, bioremediation processes are chemical degradation and toxicity reduction processes.

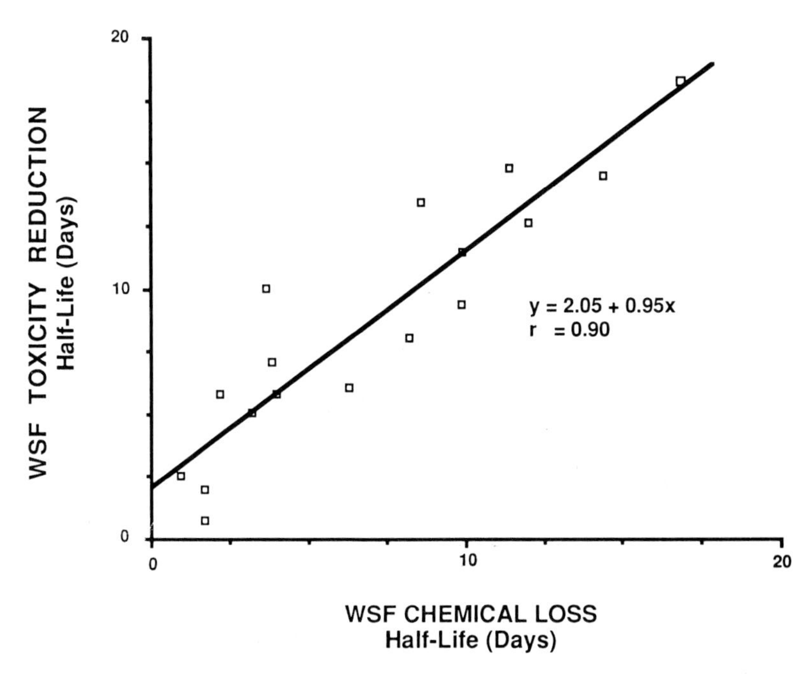

Fig. 20.2 Relationship between WSF chemical loss and toxicity reduction in the WSF, after Dasapa and Loehr (1991).

20.5 PROCESS DESCRIPTIONS

As noted in section 20.1, bioremediation processes can be categorized as solid and slurry phase processes. The solid phase processes include:

- surface soil and on-site modifications of land treatment; and
- a modification of the basic composting process.

Slurry phase processes include liquid-solids systems that can be used in existing impoundments or in tanks. The following describes each process and potential applications.

20.5.1 Solid phase land treatment

Land treatment is a managed technology that involves controlled application of a waste on the soil surface and/or the incorporation of the waste or contaminated soil into the upper soil zone. It is not the indiscriminate dumping of waste on land, and it is not landfilling. Land treatment technology relies on the dynamic physical, chemical and biological processes occurring in the soil. As a result, the constituents in the applied wastes are degraded, immobilized or transformed to environmentally acceptable components.

The design and operation of a land treatment facility is based on sound scientific and engineering principles as well as on extensive practical field experience. A land treatment site is designed and operated:

- to maximize waste degradation and immobilization;
- to minimize release of dust and volatile compounds as well as percolation of water-soluble waste compounds; and
- to control surface water runoff.

Land treatment can be a viable management practice for the treatment and disposal of hazardous and non-hazardous wastes as well as for bioremediation. Land treatment has been successfully practiced in all major climatic regions of the United States, Europe, and Canada under a wide range of hydrogeologic conditions. In the US, approximately 200 industrial land treatment systems have been in operation, including over 100 at petroleum refining facilities.

Both surface soil and on-site land treatment processes are among the more widely used bioremediation technologies.

Surface soil land treatment bioremediation

This involves:

- keeping contaminated surface soils in place;

- if needed, adding nutrients to assure adequate biodegradation and adjusting the pH toward neutral conditions;
- tilling the soil to increase the availability of oxygen and nutrients to the soil microorganisms; and
- possibly irrigating to assure adequate moisture for microbial degradation.

The organisms involved in the degradation are indigenous organisms unless none that can degrade the organics exist in the soil due to prior toxicity. In such cases, the toxicity needs to be reduced by:

- adding uncontaminated soil; or
- adjusting the environmental conditions and adding acclimated organisms.

The rate and extent that the organics in the soil are lost will depend on the combined effect of volatilization, sorption and biological degradation.

Both degradation and immobilization occur in a land treatment unit with most of the residual organics and metals remaining in the upper layers of the soil, essentially in or near the zone mixed by tilling. Figure 20.3 illustrates typical patterns of organics and metals in the soil from a land treatment unit that had treated petroleum refining wastes and residues for several decades.

Land treatment units also minimize the leaching of constituents applied to the unit or in the contaminated soil being bioremediated. Figure 20.4 illustrates the potential mobility of metals in soils, as measured by the toxicity characteristic leaching procedure (TCLP), at various depths at an industry land treatment unit. At well managed land treatment units, the pattern indicated in Fig. 20.4 is typical, i.e., the applied wastes are immobilized and zero to very low concentrations of the constituents in the surface soils are mobile and leach.

Site runoff must be contained and managed at all waste treatment operations. The runoff water which is collected may be allowed to evaporate on site, or it may be discharged to a wastewater treatment system, or may be reapplied to the land treatment site.

Note also that surface soil land treatment systems do not have a liner.

On-site land treatment systems

An on-site land treatment system does have a liner and is a constructed unit which contains contaminated soil or clean soil and wastes being bioremediated and has walls, a drainage system and a leachate collection system. It generally is above ground and allows more complete control of the process. An example of such a system is presented in Fig. 20.5.

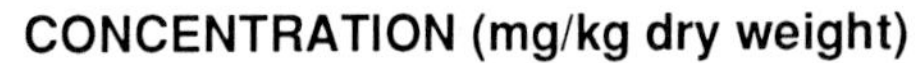

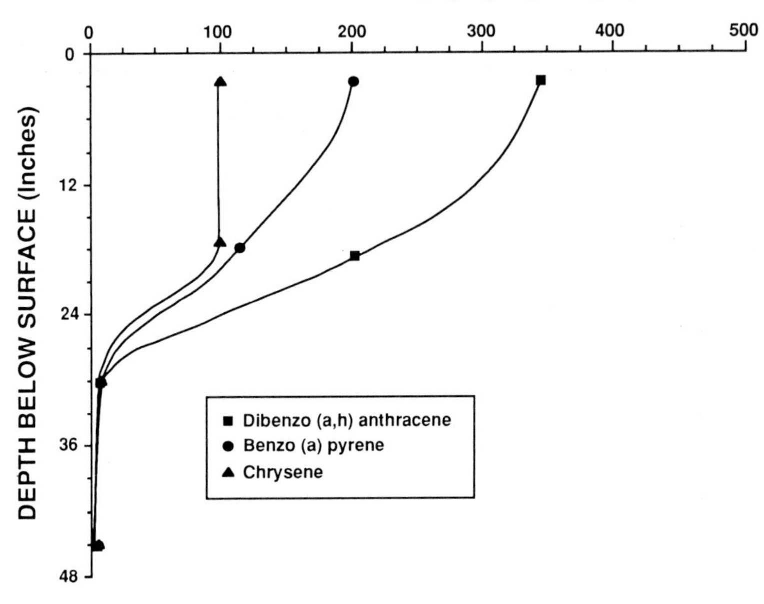

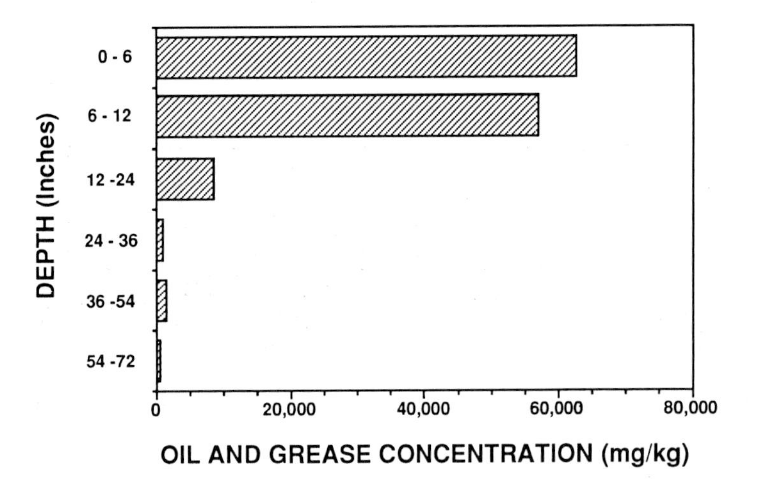

Fig. 20.3 Soil cores at an active hazardous waste land treatment site: (a) average PAH concentrations as a function of depth; (b) oil and grease (mg/kg) as a function of depth; after Loehr *et al.* (1990).

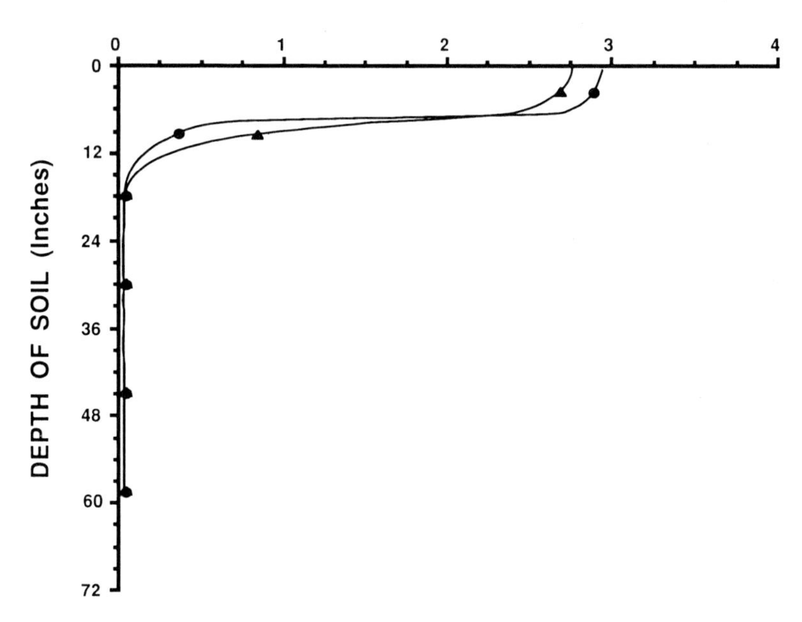

Fig. 20.4 Soil cores at a hazardous waste land treatment site: zinc concentration in the TCLP extracts; after Loehr *et al.* (1990).

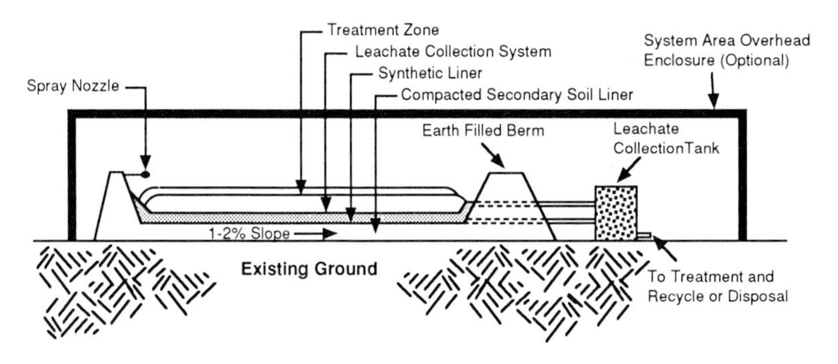

Fig. 20.5 Schematic of an above ground solid phase bioremediation unit.

An on-site system usually is constructed adjacent to the soil or wastes requiring bioremediation to minimize transportation costs and to provide better technical and managerial control of the process. Except for the fact that the on-site unit is constructed above ground, the fundamentals and operation of the on-site unit are the same as for the surface soil unit.

Both surface soil and on-site land treatment systems have been used successfully to bioremediate:

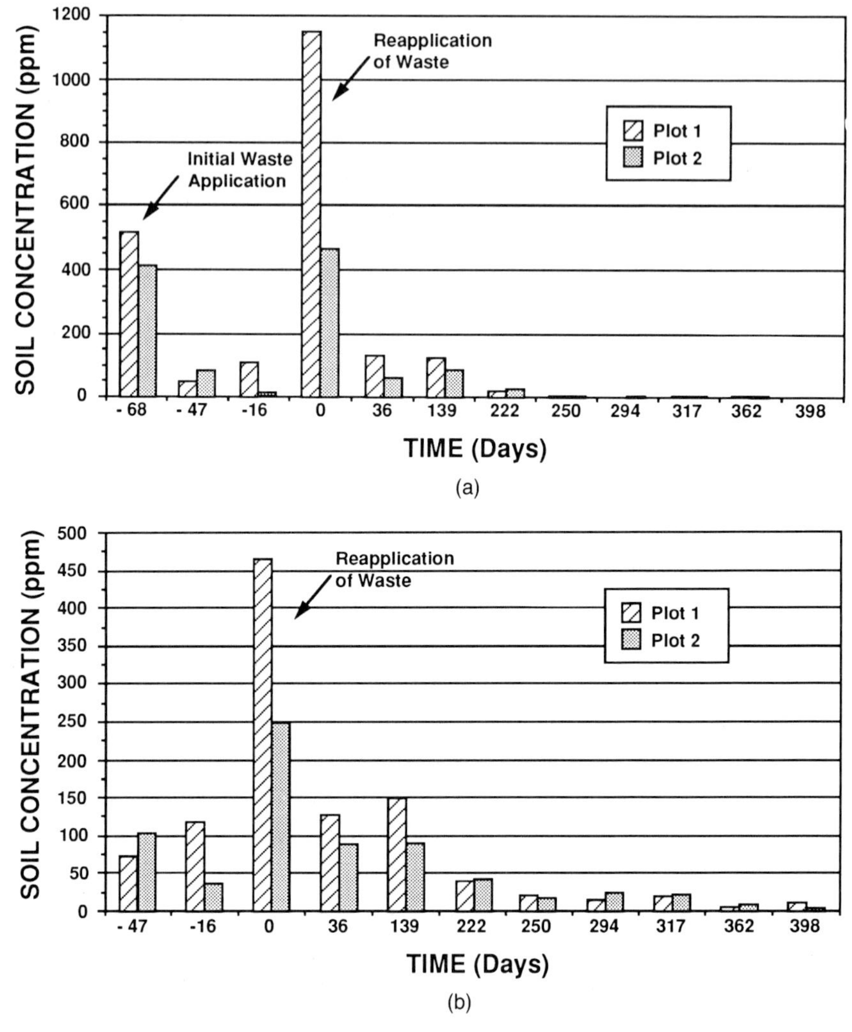

Fig. 20.6 Loss of (a) anthracence and (b) benzo(a)antracene, in a field land treatment soil bioremediation unit.

- contaminated soils at spill sites;
- industrial wastes and residues; and
- soils at surface impoundments and lagoons that are being closed.

Land treatment has been an effective bioremediation process for contaminated soils. Figure 20.6 illustrates field data that resulted from the land treatment bioremediation of wood preserving waste contaminated soil and sludge. At time zero in these plots, a second application of the sludge was made and the loss of PAH compounds was monitored over time. The loss rates ($t_{\frac{1}{2}}$) of the PAH compounds in these

Table 20.8 Loss rates of PAH compounds from the bioremediation of wood preserving waste contaminated soil and sludges using land treatment

compound	*half-life ($t_{\frac{1}{2}}$ days)*	*compound*	*half-life ($t_{\frac{1}{2}}$ days)*
naphthalene	95	chrysene	95
acenaphthylene	105	benzo(b)fluoranthene	290
acenaphthene	40	benzo(k)fluoranthene	140
phenanthrene	25	benzo(a)pyrene	100
anthracene	35	dibenzo(a,h)anthracene	260
fluoranthene	40	benzo(g,h,i)perylene	360
pyrene	40	ideno(1,2,3-cd)pyrene	600
benzo(a)anthracene	50		

plots are summarized in Table 20.8. The results are indicative of the type of losses that occur under field situations.

20.5.2 Composting

Composting as a bioremediation process for contaminated soils and industrial wastes and residues is similar to the process used for composting of leaves, garbage and food processing residues. The main difference is that high temperatures rarely are achieved in contaminated soil bioremediation composting and the purpose is degradation and loss of specific organic compounds rather than stabilization to produce a mulch or soil conditioner. At the completion of a bioremediation composting process, the treated material must be disposed of in an environmentally sound manner. However, both degradation and immobilization have occurred in this bioremediation process and the composted material should not cause surface or groundwater problems at its ultimate disposal site.

The essential elements of composting are the same as any bioremediation process:

- moisture,
- aeration,
- acclimated organisms,
- satisfactory carbon-nitrogen-phosphorus balance, and
- non-toxic conditions.

The characteristics of contaminated soils or residues considered suitable for composting are:

- constituents able to be lost by volatilization or degradation, or immobilized in the system;

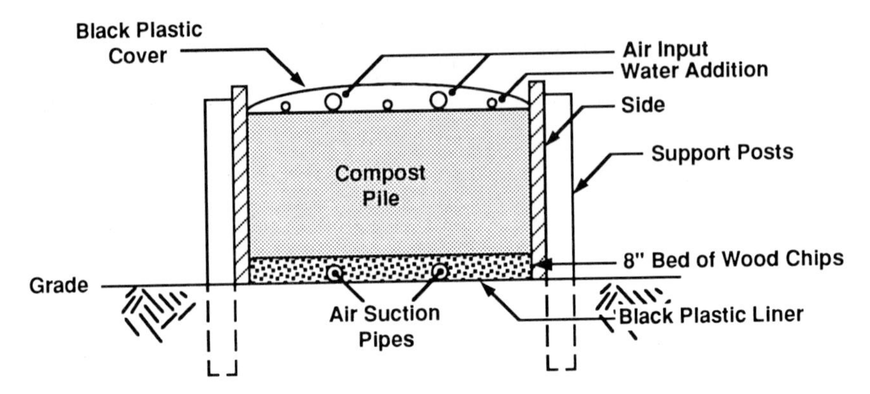

Fig. 20.7 Schematic cross-section of a compost type solid phase soil bioremediation unit.

- a low amount of free liquid so that aerobic conditions can be maintained;
- a high ratio of inert solids to biodegradable organics; and
- a mixture that can be easily broken up by mechanical turning and/or is porous to allow air to move through the composting solids.

Typical composting systems that can be used for bioremediation are the windrow, the Beltsville and the in-vessel systems. The windrow system is an open system, with periodic turning of the compost mix pile and no forced air. The Beltsville system is an open pile with an air distribution system under the pile. Air is sucked through the pile from the atmosphere and exhausted through a blower generally to an air pollution control system. The in-vessel system occurs in a closed or unclosed vessel in which mechanical mixing may occur and/or air is forced through the mixture by blowers. Bulking agents commonly are added to increase the porosity and assist the flow of air to maintain aerobic conditions.

For bioremediation composting systems, adequate operational control as well as control of all emissions, such as leachate and off-gases, is desirable. In-vessel systems provide such control. An example of such a system is presented in Fig. 20.7. If gaseous emission is not a concern, then a windrow or Beltsville system with positive leachate control can be satisfactory.

Bioremediation composting systems commonly are constructed and operated on-site so as to minimize transportation costs. The systems can be operated as continuous or batch units. Contaminated soil or residues are excavated from the site or wastes and residues are added periodically for treatment and remediation. After meeting specified clean-up criteria, the remediated mixture may be returned to the site or disposed of in a landfill or other acceptable location.

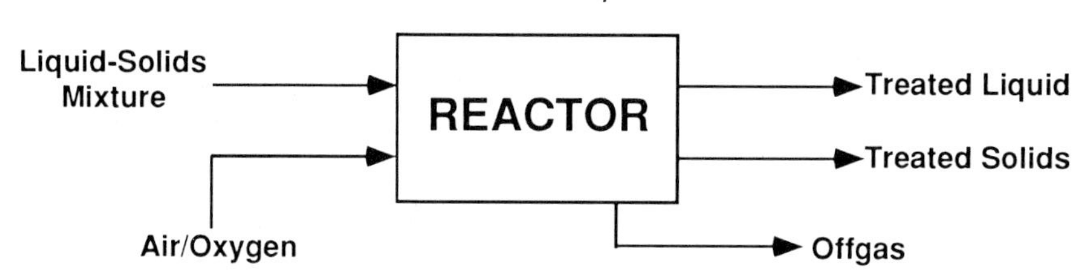

Fig. 20.8 Schematic process flow diagram of a single reactor liquid-solids bioremediation unit.

Composting has been used to remediate soils contaminated with diesel fuel and similar petroleum products.

20.5.3 Slurry phase processes

Liquid-solids treatment systems are slurry phase bioremediation systems operated to maximize mass transfer rates and contact between contaminants and microorganisms capable of degrading the contaminants. Because solids can be treated rapidly in contained reactors, much less area is needed for this on-site remediation process than for land treatment. An advantage of liquid-solids systems using tanks is that regulatory concerns related to the land disposal of hazardous wastes or to the contamination of groundwater may be eliminated.

Liquid-solids contact treatment is analogous to conventional biological suspended growth treatment (e.g., activated sludge). These units are designed to relieve the factors commonly limiting microbial growth and activity in soil, principally, the availability of carbon sources, inorganic nutrients and oxygen. To achieve this goal, the wastes are suspended in a slurry form and mixed. The mixing and aeration also prevent oxygen transfer limitations. Mixing can be provided by aeration alone or by aeration and mechanical mixing. Aeration is provided by floating or submerged aerators, or by compressors and sparges. Chemicals added to liquid slurry reactors can include nutrients and neutralizing agents to relieve limitations to microbial activity.

Liquid-solids systems are a relatively new approach for the remediation of contaminated organic materials. There have been several applications of these systems for the treatment of wood treating, coal tar and petroleum wastes. The most frequently used process configuration is a simple batch reactor system (Fig. 20.8). The reactor can be an above-ground tank or an existing pit or lagoon.

After the bioremediation clean-up goals are achieved in a batch reactor, the aeration and mixing are discontinued and the solids are allowed to settle. The treated liquid is decanted and discharged while the treated solids can be further treated by conventional land treatment or disposed of on-site. If the process is used in a lagoon or surface impoundment, it may be possible to decant the liquid and leave the treated solids in place.

Liquid-solids systems are suitable for contaminated slurries, such as may be in pits, ponds and lagoons, and contaminated soils which are too heavily contaminated for direct treatment by conventional land treatment. The technology will treat polyaromatic hydrocarbons, naphthalene, phenols, benzene, toluene, xylene and ethylbenzene. The degree of treatment of any one chemical is a direct function of its solubility in water and its rate of biodegradation.

Because of the mixing and aeration in these systems, volatilization is a significant factor in the removal of organics. Monitoring and management of the gaseous emissions may be necessary at locations where air quality concerns exist.

The phenomena which govern the process are mass transfer of organics from the solid phase to the aqueous phase and biodegradation of the aqueous phase organics. The design considerations important to properly apply these systems are:

- physical characteristics of the liquid-solids feed, particularly the fraction of the mix which is organic, the distribution of organics within the mixed slurry, and the viscosity and surface tension of the hydrocarbon phase;
- the nutrient, temperature and oxygen requirements to achieve the optimal biodegradation;
- solids and hydraulic residence times to perform adequate treatment and meet the clean-up goals;
- the characteristics of the offgas; and
- the degree of mixing (energy) required to keep the solids in suspension.

A liquid-solids reactor is one part of a slurry phase treatment system. Consideration must be given to how the soil or sludge will be removed from the source and transmitted to the reactor, if pretreatment such as thickening is desirable, the emission controls that may be needed, and what steps may be needed to handle and dispose of the treated slurry. The process options are indicated in Table 20.9.

20.6 SUMMARY

There are several bioremediation systems that can be used for contaminated soils and slurries. The benefits of such systems are:

- degradation and immobilization of organics;
- toxicity reduction;

Table 20.9 Soil slurry treatment process options

process step	*process options*
Excavation and Transport	dredge, earth moving equipment
Pretreatment	thicken or dilute
Emission Controls	carbon adsorption, biofiltration, combustion, recirculation of off-gas to slurry reactor
Slurry Treatment	batch, semicontinuous, continuous
Post-Treatment	land treatment, coagulate/flocculate, thicken, filter press, solidify, landfill

- source control and pollution prevention; and
- risk reduction.

Both solid phase and slurry phase processes are available and are being used with such materials. These processes can be used successfully if the fundamentals are understood and incorporated into design and operation. The costs of these systems generally are lower than alternative remediation methods. Because of the above benefits and the lower costs, bioremediation systems are an important technology for remediation and clean-up of specific sites.

REFERENCES

Beckman Instruments, Inc. (1982) *Microtox©* System Operating Manual, Beckman Instruments, Inc., Carlsbad, California.

Dasappa, S.M. and Loehr, R.C. (1991) Toxicity reduction in contaminated soil bioremediation processes, *Water Research*, **25**, 1121–30.

Dibble, J.T. and Bartha, R. (1979) Effect of environmental parameters on the biodegradation of oil sludge, *App. and Environ. Microbial.*, **37**, 729–39.

Huddleston, R.L., Bleckmann, C.A. and Wolfe, J.R. (1986) Land treatment biological degradation processes, in *Land Treatment: A Hazardous Waste Management Alternative*, (eds R.C. Loehr and J.F. Malina, Jr.) Water Resources Symposium No. 13, Center for Research in Water Resources, The Univ. of Texas at Austin, Austin, Texas.

Loehr, R.C. (1989) *Treatability Potential for EPA Listed Hazardous Wastes in Soil*, Robert S. Kerr Environmental Research Laboratory, USEPA, Ada, Oklahoma, PB 89-166-581AS, National Technical Information Service, Springfield, VA.

Lyman, W.J., Reehl, W.F. and Rosenblatt, D.H. (1982) *Handbook of Chemical Property Estimation Methods*, McGraw-Hill Book Company.

Matthews, J.E. and Bulich, A.A. (1984) A toxicity reduction test system to assist in predicting land treatability of hazardous organic wastes, in *Hazardous and Industrial Solid Waste Testing: 4th Symp.*, (eds J.K. Petros, Jr., W.J. Lacy and R.A. Conway) Philadelphia, ASTM/STP 886.

Matthews, J.E. and Hastings, L. (1987) Evaluation of toxicity test procedure for screening treatability potential of waste in soil, *Toxicity Assessment* **2**, 265–81.

Sims, J.L., Sims, R.C. and Matthews, J.E. (1989) *Bioremediation of Contaminated Surface Soils*, Robert S. Kerr Environmental Research Laboratory, USEPA, EPA/600/9-89/073, Ada, Oklahoma.

Tetra Tech, Inc. (1989) *MYGRT code version 2.0: an IBM code for simulating migration of organic and inorganic chemicals in groundwater*, EPRI EN-6531, Final Report, Project 2879–2, Electric Power Research Institute, Palo Alto, CA.

USEPA (1988) *Survey of State Programs Pertaining to Contaminated Soils*, Office of Underground Storage Tanks, Washington, DC.

USEPA (1990) *Selected Data on Innovative Treatment Technologies for Superfund Source Control and Ground Water Remediation*, Technology Innovation Office, Washington, DC, August.

CHAPTER 21

In situ bioremediation of groundwater

Gaylen R. Brubaker

21.1 INTRODUCTION

On October 18, 1989 the United States Environmental Protection Agency (USEPA) issued a memorandum which stated that 'pump-and-treat' technology is incapable of achieving the agency's groundwater remediation objectives. The reasons cited for this limitation included:

1. heterogeneity of the contaminated regions;
2. failure to control continued contamination from source zones;
3. inadequate system design; and
4. fundamental limitations of the process to remove occluded and adsorbed-phase material (Superfund Report, 1989).

As the limitations of the traditional 'pump-and-treat' groundwater treatment processes have become more broadly recognized (MacKay and Cherry, 1989; Keely, 1989), scientists and engineers have sought enhancements and alternatives which would provide more rapid and reliable remediation of contaminated aquifers. *In situ* bioremediation is an increasingly popular groundwater remediation technique, designed to remove occluded and adsorbed-phase contaminants and, thereby, provide more aggressive groundwater remediation.

The most common organic contaminants found in groundwater are chlorinated hydrocarbons (trichloroethene, trichloroethane, dichloroethene), benzene and its derivatives (toluene, ethylbenzene, and xylene), and polyaromatic hydrocarbons (naphthalene, phenanthrene, etc.). Since all of these compounds have very low aqueous solubilities and prefer to be sorbed onto the natural organics in soil, rather than be

Geotechnical Practice for Waste Disposal.
Edited by David E. Daniel.
Published in 1993 by Chapman & Hall, London. ISBN 0 412 35170 6

dissolved in water, it is very difficult to extract these contaminants from a soil/groundwater matrix. The aqueous solubilities, octanol-water partition coefficients, and typical retardation factors for a few of the more common groundwater contaminants are summarized in Table 21.1.

As the data in this table illustrate, the aqueous solubility of many common organic contaminants are less than 0.1% (1000 mg/l) and many are below 1 mg/l. In addition, these chemicals prefer to bind to

Table 21.1 Physical/chemical properties of selected groundwater contaminants

compound	K_{ow}	K_{oc}	K_d	R	*solubility (mg/l)*
Volatile aromatic hydrocarbons					
Benzene	135	85	0.9	5.3	1780
Ethylbenzene	490	309	3.1	16.4	515
Toluene	1412	890	8.9	45.5	140
o-Xylene	589	371	3.7	19.6	175
Non-volatile aromatics hydrocarbons					
Acenaphthene	10000	6300	63.0	316	3.42
Acenaphthylene	5000	3150	31.5	159	3.93
Anthracene	28100	17703	177.0	886	0.045
Di-n-butylphthalate	158500	99855	998.6	4993.8	400
Chlorobenzene	692	436	4.4	22.8	500
Chrysene	407000	256410	2564.1	12822	0.0018
p-Cresol	83	52	0.5	3.6	24000
p-Dichlorobenzene	2450	1544	15.4	78	49
Fluoranthene	79400	50022	500.2	2502	0.206
Fluorene	15800	9954	99.5	499	1.69
Naphthalene	1778	1120	11.2	57	31.7
Phenanthrene	28800	18144	181.4	908	1
Pyrene	75800	47754	477.5	2389	0.132
Volatile chlorinated hydrocarbons					
Carbon tetrachloride	436	275	2.7	14.7	800
Chloroform	93	59	0.6	3.9	8000
1,1-DCA	62	39	0.4	3.0	5500
1,2-DCA	30	19	0.2	1.9	8690
trans-DCE	30	19	0.2	1.9	800
PCE	398	251	2.5	13.5	150
1,1,1-TCA	150	95	0.9	5.7	4400
1,1,1-TCE	195	123	1.2	7.1	1100
Non-chlorinated solvents					
Acetone	0.58	0	0.0	1.0	miscible
Methylethylketone	1.82	1	0.0	1.1	miscible
Ethanol	0.18	0	0.0	1.0	miscible
Ethylene glycol	0.01	0	0.0	1.0	miscible

K_{oc} = organic carbon partition coefficient; K_d = soil water partition coefficient (assumes f_{oc} = 1%); and R = retardation coefficient (assumes f_{oc} = 1%).

the natural organics within soils rather than to dissolve in water. This preference is illustrated by K_{oc} and K_d. As a result of these properties, even after small portions of these molecules dissolve in groundwater, they move through the matrix at a small fraction of the rate of groundwater ($1/R$) because of continuous retardation by the soil. This results in a very inefficient remediation process for many organic contaminants using a pump-and-treat process.

The *in situ* bioremediation process has been developed to enhance the performance of the pump-and-treat process by injecting water soluble nutrients which will stimulate the natural biodegradation of organic contaminants within the aquifer matrix.

In situ bioremediation involves the introduction of microbial nutrients (typically ammonia and orthophosphate) and an oxygen source into a contaminated aquifer to promote the biodegradation of organic materials within this 'bioactive zone'. Groundwater is withdrawn from the aquifer to establish hydraulic containment and to increase the hydraulic gradient, thus increasing the rate at which groundwater moves through the formation. The captured water is typically treated to remove contaminants, and then a major portion of the water is amended with nutrients and reinjected. The basic process is illustrated in Fig. 21.1.

In order to design successfully an *in situ* bioremediation process at a particular site, it is helpful to understand how the general principles of

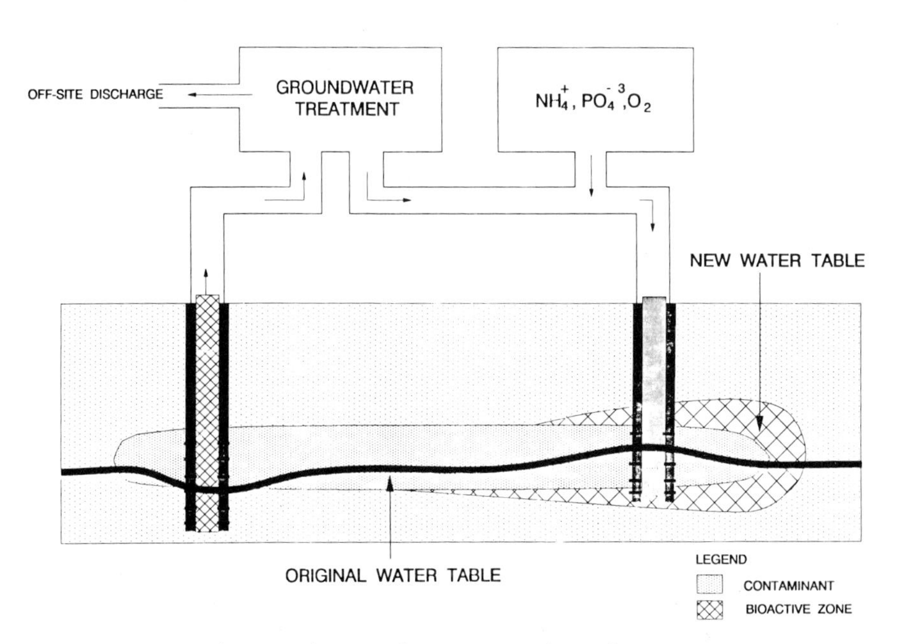

Fig. 21.1 *In situ* bioremediation of contaminated aquifer.

microbiology, hydrogeology and chemistry apply to the process. For example:

- the site must contain bacteria which are capable of degrading the contaminants of interest;
- the hydrogeology of the site must permit the movement of nutrient-enriched water through the contaminated region and allow for the controlled capture of this water; and
- the soil and groundwater must be chemically compatible with the nutrient sources provided.

The significance of each of these issues will be discussed further in the following sections. Since metals are not biodegradable, this discussion will be limited to organic contaminants.

A comprehensive literature review (Lee *et al.*, 1988) and several review articles (Thomas and Ward, 1989; Wilson *et al.*, 1986) have been published which provide summaries of *in situ* bioremediation technology. However, most of the case histories and more design-oriented articles have appeared in conference proceedings or other publications with limited distribution. As a result, it has been difficult to assemble sufficient information to gain an overall perspective on this emerging technology. This chapter is intended to fill that gap.

Section 21.2 is a review of early *in situ* bioremediation projects, presented as an introduction to the technology and to place the earlier literature in historical perspective. Advances in oxygen delivery are discussed in section 21.3. Section 21.4 presents a very brief overview of some of the relevant principles of microbiology. This is followed, in section 21.5, by a discussion of the microbial, hydrogeological, and chemical site characteristics which effect the technical feasibility and design of the process. This section also includes a discussion of laboratory treatability techniques, as well as design principles used as a project moves through the design phase. The final portion of the chapter looks to the future, with a very brief summary of six areas of research and development that are likely to expand dramatically the use of *in situ* bioremediation.

21.2 EARLY *IN SITU* BIOREMEDIATION PROJECTS

21.2.1 Ambler, Pennsylvania

The first field implementation of an *in situ* bioremediation process was initiated in 1972 at a site in Ambler, Pennsylvania (Raymond *et al.*, 1976). An oil pipeline owned by Sun Oil Company had leaked, spilling over 490 000 L of high-octane gasoline within the pumping zone of Ambler's municipal well. Over the next eight months, numerous monitoring and recovery wells were installed and almost two-thirds of the spilled gasoline was recovered from the fractured dolomite formation. However, further recovery efforts proved unproductive, leaving an estimated 150 000 L of gasoline in the formation.

Since the groundwater pumped from the site contained an average of 5 ppm of dissolved gasoline, remediation through groundwater pumping would have required almost 100 years of treatment, even if the city pumped more than a million liters of water per day. A group of scientists at Sun's Marcus Hook Pennsylvania Research Center were called in for consultation. They had been studying the factors which controlled the microbial degradation of petroleum products under various conditions and were convinced that the gasoline could be degraded within an aquifer, if a proper blend of microbial nutrients could be provided.

A concentrated solution of ammonium sulfate, disodium phosphate, and monosodium phosphate was periodically introduced into injection wells at the site. The blend was designed to provide the desired quantities of ammonia and phosphate, and to buffer the water at a pH of about 7. In addition, air was introduced into 10 monitoring wells by using air compressors and gas diffusers. No water was injected in this first process, but the three recovery wells did draw the nutrients through the contaminated area. Although there is little direct performance data from this site which conclusively demonstrates the biodegradation of gasoline, the process did appear to accelerate the reduction in gasoline concentrations. In addition, two sets of observations were made which suggested enhanced biodegradative processes. First, the aeration process increased the average dissolved oxygen content in the recovered water by 2–3 ppm, and second, the average population of gasoline-degrading bacteria increased by a factor of 100 in those areas where nutrients were added.

21.2.2 Other early projects

Following the encouraging results of the Ambler project, this process was used at approximately 20 other sites over the next 10 years, typically service stations near drinking water aquifers where rapid and complete cleanup was required. Although it is difficult to gauge the 'success rate' of these projects using today's standards, it appears that, in most instances, the concentrations of dissolved contamination and/or hydrocarbon vapors were reduced faster than would have been expected using traditional groundwater capture strategies (Du Pont, 1990; Lee *et al.*, 1988).

By the early 1980s, several other groups had implemented similar processes for *in situ* bioremediation. Groundwater contamination from methylene chloride, acetone, n-butyl alcohol and dimethylaniline was the motivation for a combined flushing/*in situ* strategy in Waldwick, New Jersey. The contaminated aquifer, located 2–5 m below grade, was flushed with water, and then the recovered water was treated in a surface bioreactor. Two trenches were used for water injection and nine wells for air injection while the recovered water was treated in aerated bioreactors on the surface, prior to reinjection. The nutrients from the surface bioreactor, plus the *in situ* aeration, thus provided the conditions for subsurface, as well as surface biodegradation (Jhaveria and Mazzacca, 1985).

The literature also contains other case histories in which surface biological processes were combined with *in situ* soil flushing processes. One example of this strategy was employed at an ethylene glycol spill at the Naval Air Engineering Center in Lakehurst, New Jersey (Flatham and Caplan, 1986). Approximately 15 000 L of cooling water containing 25% ethylene glycol leaked from a lined surface lagoon. An area measuring 14 m × 55 m was identified that had groundwater contamination in excess of 1400 ppm. Bench-scale studies showed that these concentrations were not toxic and that indigenous bacteria from the site could degrade the contamination, under proper pH and nutrient conditions.

A series of recovery wells was used to withdraw contaminated groundwater from the site. This water was treated in an activated sludge bioreactor which was mobilized on-site. The treated water then passed through a settling chamber and was reinjected, through five reinjection wells, to flush more contamination from the site and provide nutrients and dissolved oxygen for subsurface biodegradation. This strategy has also been used for gasoline, formaldehyde and mixed solvent spills, though in most of these projects, no special effort was made to provide additional oxygen to the subsurface environment. Although oxygen addition is not important when flushing contaminants such as alcohols or ketones, which are easily extracted from soil, it is a critical component of the *in situ* biodegradation process when the contaminants of concern are mostly sorbed within the soil matrix. As a result, this author will reserve the term '*in situ* bioremediation' to processes where the rate of biodegradation is being directly stimulated in the subsurface through the addition of an oxygen source (or alternative electron acceptor), as well as nutrients and/or bacteria.

21.3 ADVANCES IN OXYGEN DELIVERY

21.3.1 Laboratory studies

In 1982, the American Petroleum Institute (API) funded a study to enhance the oxygen delivery for *in situ* bioremediation processes (Texas

Research Institute, 1982). The first phase of this project, a literature review and analysis, suggested that hydrogen peroxide was the preferred alternative. Hydrogen peroxide is used as an oxidizing agent in many industrial applications and is often used as a backup source of oxygen for industrial waste treatment systems. Since one kilogram of hydrogen peroxide decomposes to form nearly one-half kilogram of oxygen, and hydrogen peroxide is totally miscible with water, it is an environmentally attractive alternative. These benefits seemed to outweigh its potential toxicity to bacteria, the need to store and handle peroxide in specialized equipment, and its cost.

A second phase of the study examined the toxicity of various concentrations of peroxide to gasoline-degrading bacteria. These studies showed that, although small populations of bacteria were sensitive to as little as 100 ppm of peroxide, more mature cultures could tolerate over 1000 ppm.

FMC Corporation, a producer of hydrogen peroxide, became interested in this new potential application of peroxide in 1983 (Brown *et al.*, 1984). FMC performed a series of column studies to examine the degradation of gasoline in the presence of various concentrations and sources of dissolved oxygen (DO). The test conditions included water saturated with air (8 ppm DO) and oxygen (40 ppm DO), as well as water amended with 250 ppm (118 ppm DO equivalents) and 400 ppm (200 ppm DO equivalents) of hydrogen peroxide. Several important observations were reported.

- Oxygen consumption was rapid under all test conditions.
- After two weeks, the number of total and gasoline-degrading bacteria increased as the oxygen delivery increased. The 200 ppm DO columns had greater than 100 times more bacteria than those which received 8 ppm.
- The loss in gasoline, accounted for through biodegradation, was proportional to the mass of oxygen supplied.

FMC then proceeded to inject a solution of hydrogen peroxide (the concentration was not reported) into a monitoring well at an active bioremediation site for 22 hours. Observations were made of the changes in dissolved oxygen in a second monitoring well 8 m downgradient. Significant increases in DO were observed within 10 hours, increasing to 15 ppm over the next 72 hours. Although they did not measure microbial activity in this instance, transport of dissolved oxygen from peroxide was demonstrated.

21.3.2 API field demonstration

Based on the increasing evidence that hydrogen peroxide would provide a significant advantage over simple aeration of aquifers, the

American Petroleum Institute (API) funded a field demonstration of the technology to improve documentation of basic design and performance issues (API, 1987; Brown and Norris, 1986; IT Corporation, 1986).

The site of interest was a petroleum products terminal in northern Indiana. In 1979, a major leak resulted in percolation of unleaded gasoline through the base of the bermed tank enclosure and down through sandy soil, until it reached the water table 7 m below grade. Three recovery wells were installed, soon after the spill, and the phase-separated product was collected. However, the soil below the original zone of contamination, and near the water table downgradient, contained significant petroleum contamination. The groundwater varied seasonally between about 7 and 8 m below grade and had a relatively flat surface.

The test area was designed to consist of two concentric triangular injection galleries, centered around an existing recovery well. The inner gallery was designed to receive water enriched with nutrients and hydrogen peroxide, while the outer gallery was designed to receive clean water only, thus isolating the test area hydraulically from the surrounding contamination. The interior and exterior galleries were constructed about 1.5 m deep and measured approximately 14 m and 27 m on a side, respectively. The pumping wells in the center of the treatment area were capable of yielding over 280 l/min (Fig. 21.2).

The site soils consisted of glacially-deposited sands and gravels. Although the site was initially identified as 'relatively homogeneous fine to medium grained sands', this assessment was amended, as initial coring data was gathered to generate baseline data for the biodegradation process. These borings indicated significant variability in the texture, ranging from tight, fine sands to loose, coarse gravels. In addition, several discontinuous layers of fine sands containing higher silt fractions were identified in the upper portions of the aquifer, roughly 1.5–2 m below grade.

The presence of the silt zone directly below the infiltration galleries had not been anticipated in the original design. As a result, the rate of groundwater injection was limited to 40–60 L/min, as opposed to the 200–300 L/min range originally planned. In addition, precipitation of minerals (iron and calcium phosphates) appeared to further reduce the rate of groundwater injection. After several months of operation, the injection design was altered to increase the amount of water (and thus nutrients) that could be added to the site, thus increasing the rate of biodegradation that would be observed during the test. First, nutrients were added to both the outer and the inner injection galleries, and then six injection wells were installed (two along the side of each outer triangle). As a result of these modifications, a total of 2.9 million liters of water containing 7700 kg of hydrogen peroxide, 4700 kg of am-

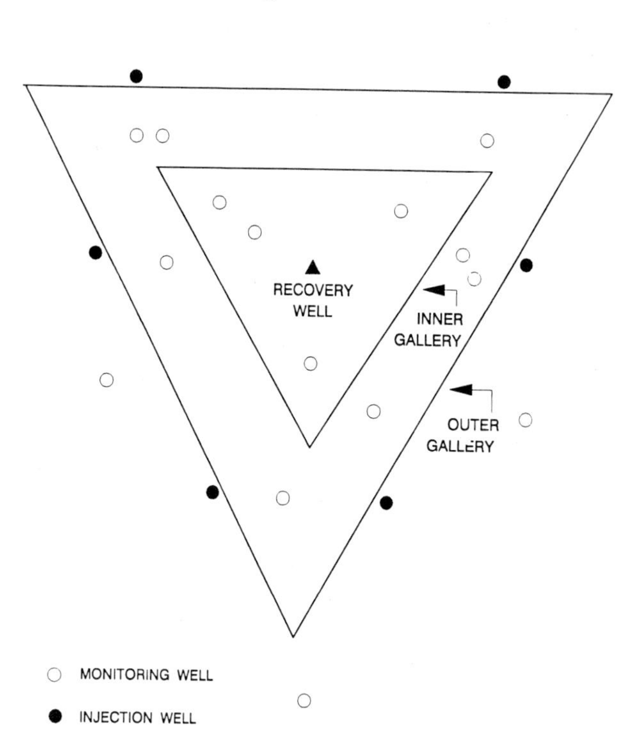

Fig. 21.2 Plan view of API biodegradation field demonstration.

monium chloride and 5500 kg of sodium phosphates and a variety of micronutrients were injected over the 24-week test.

The hydrogen peroxide was added continuously, initially at 100 ppm and then slowly increasing to 500 ppm, while the ammonium salts and phosphates were added in concentrated batches, two to three times per week. Throughout the test, groundwater from 15 observation wells within the site was analyzed weekly for nutrients, total and gasoline-degrading bacteria, dissolved oxygen, and pH. In addition, extensive soil coring was performed.

Soil samples were collected at 13 locations at the beginning of the test, 6 locations at week 5, 8 locations at week 16, and 19 locations at the end of the study. In the first three corings, samples were collected at three separate depths; a fourth depth was added for the final round of sampling. Samples were analyzed for petroleum hydrocarbon, total heterotrophic bacteria, gasoline-degrading bacteria, and nutrients. The conditions within the inner gallery, prior to and during the test, are summarized below. These data shown in Table 21.2, representative of a large mass of data from this project, provided the most convincing, documented support for *in situ* biodegradation to date.

Table 21.2 Conditions within inner gallery at treatment site

	before treatment	*post treatment*
ammonia (ppm)	<1	100–250
orthophosphate (ppm)	<1	100–250
heterotrophic bacteria (cfu/g*)	1.2×10^3	1.4×10^7
gasoline-degrading bacteria (cfu/g*)	2.0×10^2	1.1×10^6
petroleum hydrocarbon (kgs)	1900 ± 380	540 ± 230

*Colony forming units: a standard method of counting bacteria per gram of soil.

The existence of areas which had received only nutrients during a portion of the test, and the existence of areas of low hydraulic conductivity in the southern portion of the site, generated a large amount of data allowing for a number of other important observations.

- Areas which received the most nutrients (and oxygen) showed the highest increase in bacterial populations and the greatest decrease in petroleum hydrocarbon.
- Even though degradation was documented throughout the site, the concentration of dissolved oxygen in the monitoring wells remained low (<2 ppm) throughout the project.
- Average concentrations of petroleum hydrocarbon in the inner gallery decreased from 5373 to 1546 ppm over the 24-week project.
- Over 5400 kg of gasoline were degraded throughout the treatment area during the test. This suggests that each kilogram of degraded gasoline required 1.5 kilograms of oxygen.

Because of budget limitations, the API was not able to continue the test to determine what levels of remediation could be achieved, what quantity of dissolved oxygen would be required to complete the biodegradation process, or whether the rate of biodegradation would slow at lower levels of contamination. However, this study did demonstrate the tremendous potential of hydrogen peroxide as a means to stimulate aerobic biodegradation in contaminated aquifers.

21.4 PRINCIPLES OF MICROBIAL PROCESSES

Although microbial processes have long been recognized as important in subsurface environments (Back, 1989), microbiologists have only recently recognized the abundance and versatility of subsurface microbes in microbial degradation. This section provides a brief summary of the general nature of subsurface bacteria which are important in bioremediation processes. Many other types of microbial species exist in the subsurface, but discussion of these species is beyond the scope of this chapter.

Bacteria can adapt to survive under a wide range of environmental conditions, but they are most active in moist soils with a pH in the range 6.5–8. Although microbial activity is accelerated at warmer temperatures, metabolic rates in soil are usually limited by the availability of carbon or oxygen, and temperature is rarely a significant factor in controlling subsurface biodegradation. In spite of these 'preferred conditions' for microbial activity, viable bacteria, both aerobic and anaerobic, have been detected more than 100 m below land surface (Fliermans and Balkwill, 1989).

21.4.1 Characteristics of aerobic bacteria

Aerobic bacteria are single cell organisms, about the size of clay particles (0.5–3 μm), which exist in a variety of shapes, including rods, spheres, and chains. Aerobic bacteria grow by converting organic compounds into cell mass and carbon dioxide, and by converting oxygen to water vapor. Since the cells are made up of a variety of proteins, lipids, nucleic acids, and other complex constituents. Bacteria also need nitrogen, phosphorus, sulfur, and trace amounts of metals to grow. The non-water portion of a bacteria typically contains about 53% carbon, 12% nitrogen, 3% phosphorus, 1% sulfur, 20% oxygen, 7% hydrogen, and 7% ash, for a total mass of 10^{-12} per cell.

Due to the scarcity of available carbon and nutrient sources in most subsurface environments, these bacteria appear to live in a somewhat dormant state most of the time. However, when a suitable carbon source becomes available (as after a gasoline spill), the bacteria have the ability to multiply very rapidly, often doubling every 45 minutes. Pristine soils in shallow unconfined aquifers commonly contain 100–1000 aerobic bacteria per gram of soil, but this level may increase to 10^5 or more within one week of a contamination event. When this occurs, microbial activity is generally limited by the availability of dissolved oxygen. This increased oxygen demand can be observed by monitoring the dissolved oxygen levels in a contaminated aquifer. It is common for pristine, shallow groundwater to contain 4–6 ppm of dissolved oxygen (DO), while water contaminated by a biodegradable material will generally contain less than 0.5 ppm. (Commercial test kits are now available which allow rapid measurement of the dissolved oxygen levels in water.) Depressed levels near a suspected contaminant plume is evidence that the natural bacteria present at the site are degrading the material but that the rate is oxygen limited.

The bacteria present within an aquifer are a consortium of many different genera and species, each with slightly different biodegradation capabilities. Jamison and co-workers isolated 32 distinct cultures of bacteria from one site contaminated with gasoline, all of which could grow on gasoline as a sole carbon source (Jamison *et al.*, 1976). These included ten different species of the *Nocardia* genus, four *Acinetobacter* and eight *Pseudomonas* species, as well as several other bacteria. These workers then tested the ability of the various culture isolates to degrade various constituents of gasoline. They found that, although the complete consortium of bacteria could degrade all of the components of gasoline, there were several constituents which were not degraded by any of the pure isolates and no individual isolate degraded all constituents. In general, however, they found that *Nocardia* were primarily responsible for paraffin degradation and that the *Pseudomonas* cultures were responsible for aromatic degradation.

21.4.2 The nature of biodegradation

Essentially all commercial bioremediation processes involve accelerating aerobic biodegradation of a contaminant, where the contaminant acts as a carbon source and a source of energy for the bacteria. Bacteria convert the carbon source (benzene for example) to either cell mass or to carbon dioxide and water. In the process, the bacteria gain energy, which allows the cell to function. The overall biodegradation process can be expressed through simple chemical formulae, such as those shown below. Equation (21.1) shows the conversion of benzene and oxygen to cell mass and water, while Eq. (21.1) shows the final products of carbon dioxide and water.

Oxidation to cell mass

$$C_6H_6 + 2.5O_2 + NH_3 \rightarrow C_5H_7O_2N + CO_2 + H_2O \qquad (21.1)$$

Complete mineralization

$$C_5H_7O_2 + 5O_2 \rightarrow 5CO_2 + 2H_2O + NH_3 \qquad (21.2)$$

These equations illustrate that aerobic oxidation of hydrocarbons requires large quantities of oxygen, in this instance 3.1 kilogram of oxygen per kilogram of contaminant. Of course, the process is much more complex, in that bacteria actually perform a series of smaller reactions, each of which involves a particular microbial enzyme.

The enzymatic processes are well understood for petroleum hydrocarbons and most common non-chlorinated chemicals. Hydrocarbons such as octane, for example, are oxidized first to octanol, then to octylaldehyde and octanoic acid (Fig. 21.3) (Suflita, 1985). Octanoic acid is then converted to hexanoic acid, and a two-carbon segment becomes part of the tricarboxylic cycle of the bacteria. The process is then repeated, until the compound is completely degraded (Fig. 21.4). Branched and cyclic compounds tend to degrade more slowly than unbranched compounds because they do not 'fit' as easily into several

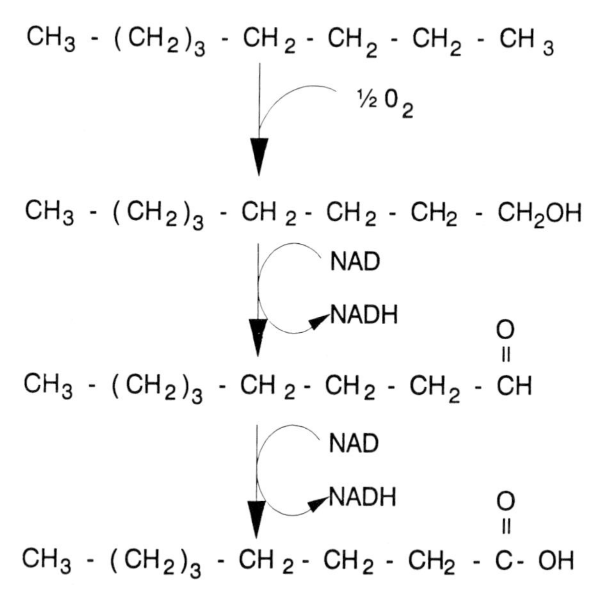

Fig. 21.3 Terminal oxidation of n-octane.

of the enzymes responsible for initial oxidation. The octane example also illustrates steps in the pathway where NAD (nicotinamide adenine dinucleotide) is converted to NADH (nicotinamide adenine dinucleotide dihydride) or where FAD is converted to FADH. These compounds are one means for the cell to store energy, which can then be used to perform other cell functions, at some later time, and return to the NAD or FAD form.

The biodegradation pathway for benzene in illustrated in Fig. 21.5. Similar enzymatic pathways are used for many substituted aromatic compounds, though the presence of chlorine, nitro, and some other substituents can deactivate the ring or prevent certain enzymatic reactions, reducing the rates of degradation.

Some of the enzymes which are responsible for degradation reactions are quite specific and will only react with chemicals of specific geometry. The enzyme responsible for the initial oxidation of octane to octanol, for example, prefers to react with terminal carbons (those at the end of a chain). As a result, cyclic hydrocarbons, which contain no terminal carbons, are degraded much more slowly than straight chain hydrocarbons of similar length.

Since the relative specificity of enzymes is frequently not known, it is very difficult to predict the relative rates of degradation of mixtures of compounds, even when the rates for the pure compounds are known. In one instance, an easily degraded compound will stimulate the production of an enzyme system needed to degrade a 'difficult'

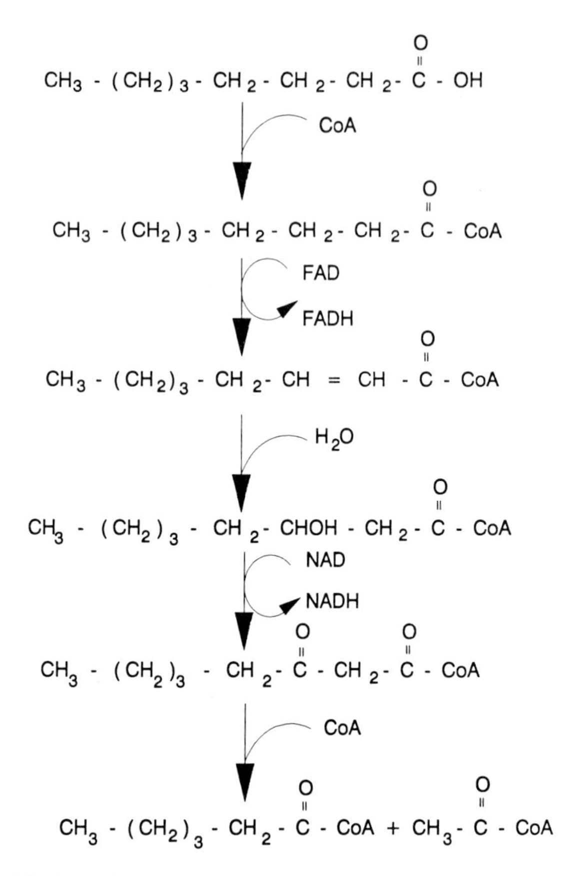

Fig. 21.4 β-oxidation of n-octane.

compound, thus accelerating the degradation of this more recalcitrant compound. In another instance, the recalcitrant compound will retard all microbial activity, while in a third instance, the bacteria will degrade only the easier material until it is depleted and then, gradually, begin to degrade the second carbon source. As a result of the unpredictability of this process, some type of treatability study is usually required during the design phase of any biodegradation project, which involves an unfamiliar combination of constituents.

Two general characteristics of the aerobic biodegradation process are especially important to note, when treating potentially hazardous compounds.

- The first oxidation step is usually the rate limiting step. After a compound becomes incorporated into the cell, it is usually degraded to its final end products quickly, with little accumulation of intermediate products.

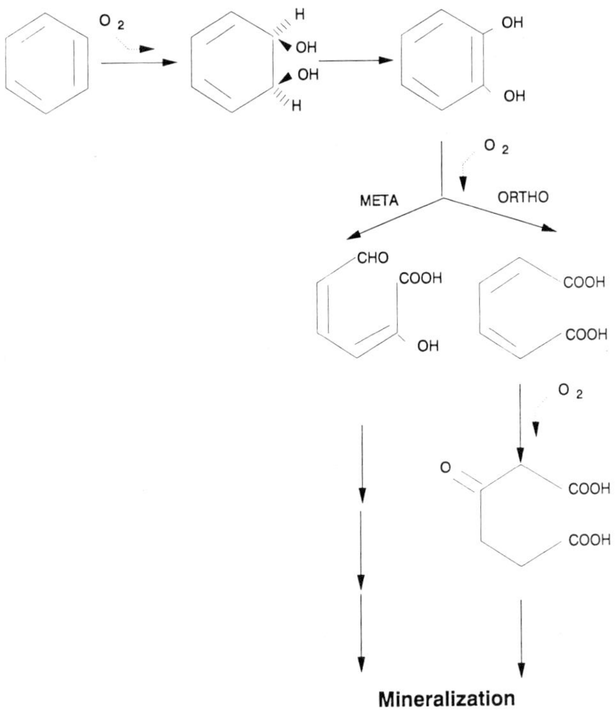

Fig. 21.5 Biodegradation of benzene.

- Contaminants are quickly transformed into the same intermediates which result from the degradation of 'natural' microbial substrates.

In those cases where the bacteria are able to make only limited changes to the substrate, the process is referred to as **biotransformation**. In those instances when bacteria convert the substrate to carbon dioxide and water (and other inorganic materials), the process is called **mineralization**. Although both can be considered biodegradation, mineralization of the contaminant is the desired process.

Some aerobic bacteria have the ability to use nitrate as an electron acceptor when molecular oxygen is not available. The process is called **denitrification**, and bacteria with this ability are called denitrifiers. In areas where neither oxygen nor nitrate is available, other types of bacteria may dominate. These bacteria can use sulfate (sulfate reducers), acetate (fermentative), or carbon dioxide (methanogenic) as electron acceptors. However, these bacteria are only active when the soil matrix is in a very reduced condition.

Recent studies have suggested that many of the contaminants of concern are degradable anaerobically through either denitrification or methanogenesis. Since these degradation pathways are not well understood and are not yet practiced commercially, their discussion will be deferred to a later section. However, two cautions are worth emphasizing for anaerobic processes. Under some conditions, anaerobic processes will create sulfide or methane as by-products of degradation. Secondly, anaerobic processes tend to be much slower than aerobic processes, such that intermediates can accumulate. These degradation intermediates for some anaerobic processes are more toxic than the starting compound. The degradation of trichloroethylene to dichloroethylene and vinyl chloride is the most commonly cited example of this phenomena. These properties of anaerobic processes can lead to major operational problems, if they have not been considered in the process design.

21.5 EVALUATION, DESIGN, AND IMPLEMENTATION

Although the design, predictability, and versatility of *in situ* bioremediation processes have improved significantly in the past 15 years, the majority of commercial projects are still petroleum hydrocarbons (gasoline, kerosene, fuel oil, diesel fuel, and jet fuel) in permeable aquifers (Yaniga *et al.*, 1985; Yaniga and Smith, 1986; Bell and Byrd, 1988; Brown *et al.*, 1988). This section will provide an overview of the contaminant, hydrogeological, and nutrient interaction issues which are typically considered in evaluating the suitability of a site for *in situ* bioremediation and in designing a remediation process.

21.5.1 Contaminant characteristics

Although the microbial environment affects the rate of biodegradation of a contaminant in a particular location, it is clear that certain chemical constituents are inherently more accessible to microbial degradation than others. The aqueous solubility, molecular size, and types of chemical bonds within a particular chemical structure affect the ability of a bacteria to metabolize the molecule and obtain energy from the process. Several literature reviews have been published which summarize the general 'biodegradability' of the types of compounds which are most frequently of concern (Alexander, 1981; Babea and Vaishnav, 1987). In addition, at least one reference provides a guide to the concentrations at which various contaminants may also be toxic to organisms (Verschueren, 1983). Collectively this literature can be used to provide the reader with an indication of the biodegradability of a particular compound. A brief summary of general trends of aerobic degradability is provided below (Brubaker, 1989). For the purposes of this analysis, an easily degradable material is one for which common bacteria can mineralize the compound (or mixture) of interest, with relatively short acclimation times, and can use the material as a sole carbon source.

- Simple hydrocarbons and light petroleum distillates, such as gasoline, kerosene, diesel, jet fuel, and light mineral oils are generally degradable. Their rate of degradation decreases with increasing molecular weight and decreasing solubility. Increased branching and cyclic structures also slow the degradation process.
- Aromatic hydrocarbons with up to two rings (including benzene, toluene, xylene, ethylbenzene, and naphthalene) are readily degradable. The rate of degradation of larger polyaromatic hydrocarbons (greater than three rings) decreases as size increases and solubility decreases.
- Alcohols, amines, esters, carboxylic acids, and nitriles are usually degradable, but these compounds are often toxic to unacclimated bacteria at high concentrations. Nitrobenzenes and ethers are usually more slowly degraded.
- Chlorinated hydrocarbons (both straight chain and aromatic) become increasingly difficult to degrade as the degree of chlorine substitution increases. Polychlorinated biphenyls (PCBs) and other polychlorinated hydrocarbons (chloroform, carbon tetrachloride, tetrachloroethylene, trichloroethylene, and dichloroethylenes) are not readily degraded aerobically.
- Pesticides are another very complex set of environmentally persistent compounds. In general, those which are found at hazardous waste sites (DDT, Lindane, Aldrin, Chlordane etc.) are not readily biodegraded.

The degradation of viscous organics materials, like number 6 fuel oil, creosote, and refinery wastes is often controlled by their physical condition in the soil/water matrix. When they exist as small droplets of oil, occluded within the pores of a soil, there is very little exposed surface area for degradation, and the biodegradation process will be inhibited.

Although it is difficult to quantify the importance of contaminant distribution to project feasibility, project success requires movement of nutrient-enriched water through those areas of the site which contain the highest concentrations of contamination. Sites which contain a few point sources of contamination, whether a lagoon or a leaking tank, can generally be treated fairly reliably with an *in situ* treatment method. However, *in situ* bioremediation strategies for sites which contain multiple and undefined sources of contamination become much more difficult to design and operate in a predictable fashion. The probability of successful remediation is definitely increased by a thorough understanding of the sources and transport mechanisms for the contaminants.

21.5.2 Hydrogeology

In most *in situ* bioremediation projects, mass transport of oxygen and nutrients is the largest single factor affecting the duration and efficiency of the project. Although the nature of the oxygen and nutrient sources are significant issues in this design, the design and operation of the groundwater recovery and injection system is of primary importance. Since the demand for oxygen far exceeds that of other nutrients, and microbial activity is essentially instantaneous relative to groundwater movement, the delivery of dissolved oxygen is the rate-limiting step. The objectives of the groundwater system are to establish hydraulic containment of dissolved phase contamination and to transport nutrients to the area of nutrient demand. (For very insoluble organics, such as four and five ring polyaromatic hydrocarbons, dissolution and microbial degradation may actually also be controlling factors in the process rate.)

Since both the soils and the contaminant distribution at a site are typically non-homogeneous, the initial determination of feasibility and cost will usually depend on an estimate of the total volume of degradable material, the oxygen demand of this area, the concentration of DO which can be delivered in groundwater, and the rate at which groundwater can be injected into the area of concern. These estimates are illustrated in the following subsection.

As in all groundwater treatment processes, hydraulic conductivity, saturated thickness and the natural groundwater gradient are important parameters. These parameters must be defined in order to determine the rate at which groundwater can be withdrawn from a well, the average groundwater velocity, and the capture zone defined by a specific pumping rate (Keely and Tsang, 1983). In a similar fashion, the thickness and hydraulic conductivity of the unsaturated zone will affect the amount of water which can be injected into a soil, as well as the sphere of influence of this injection point. The influence of injection points must overlap to provide nutrients to the entire contaminated zone, in the same manner that the capture zones must overlap for complete containment.

Successful design of this groundwater recovery and reinjection system requires a detailed understanding of the hydrogeology of the site. Although preliminary estimates of hydraulic conductivity can be made based on soil descriptions, grain size analysis, or slug tests, a pump test (preferably in combination with reinjection) should always be performed whilst the final design is being established.

A three-dimensional understanding of contaminant distribution is also required, so that the oxygen delivery is concentrated in the areas of highest oxygen demand. General indicator parameters, such as total organic carbon (TOC), chemical oxygen demand (COD), or total petroleum hydrocarbon constituents (TPHC), are much preferred over compound-specific analyses for these calculations. It is also important to be aware of other sources of oxygen demand (buried plant or animal wastes, septic or sewer systems, non-priority pollutant constituents) which will also be degraded during a bioremediation process, consuming oxygen in the process.

Since the concentration of contaminants often decreases exponentially with distance from a point source, it is often beneficial to concentrate injection point(s) near areas of high oxygen demand. This strategy will often allow the injected groundwater to follow the same migration paths as the contaminants. In larger sites, it may be necessary to divide the site into treatment cells, to allow for more efficient nutrient transport. In some instances this will result in cleaning up 80% of the site very rapidly and then concentrating on the primary source area for a longer treatment time. This possibility should be considered when designing the water treatment system and other process components, as it may create opportunities to stagger treatment in areas of low contamination and, thus, lower the overall project cost.

When designing a simple groundwater capture or pump-and-treat process, hydrogeologists are typically not concerned with the time required for the contaminant to migrate through the soil to the recovery point. However, for *in situ* bioremediation projects, the process efficiency is highly dependent on how long it takes for the nutrients to reach the areas where they are required by the bacteria. Although the sensitivity to this issue depends on the ability of the nutrients to move through the soil, peroxide stability, and oxygen demand, in general, it is often desirable to create a groundwater injection/recovery system which will deliver nutrients to all areas of the contaminated area within 60–90 days of the time of injection.

Of course, how far groundwater can move in 60 days is dependent on the permeability and porosity of the aquifer and the amount of hydraulic gradient created by the injection and recovery system. In a medium sand aquifer with a hydraulic conductivity of 5×10^{-4} m/s, an effective porosity of 0.35 and an induced gradient of 0.04, the average fluid velocity is 5 m per day between the point of injection and the point of recovery or treatment. As a result, groundwater remediation processes are very difficult to implement for soils with hydraulic conductivities less than about 10^{-6} m/s.

In addition, unconfined aquifers which are very thin allow very little drawdown, thus limiting the aquifer yield and the area of influence of a well. This can make a design difficult, unless a recovery trench can be used. Likewise, aquifers which are very close to the surface limit the amount of groundwater injection that can occur without flooding. This too, limits the hydraulic head that can be induced and the influence of a particular injection well.

Computer modeling is a very powerful tool in the design of the injection and recovery strategy for an *in situ* bioremediation process. Computer simulations help to evaluate whether the aquifer characteristics measured are internally consistent, and are useful in predicting the pumping rates, flow velocities, and capture zones which will result from various injection/recovery strategies. It is important to investigate the sensitivity of the system to various assumptions and parameters before finalizing the design, and then to be conservative in designing the injection wells. Since injection wells are notorious for losing their efficiency with time, the system should deliver nutrients to all areas of the site, even when the efficiency of a well is only 70%. Conservative design will also help to compensate for the non-homogeneity within a site. Maintenance of wells should also be planned on a routine basis.

In evaluating the computer simulations and examining design options, the drawdown at recovery wells should be minimized to the extent possible to avoid drawing the nutrients below the zone of highest contamination. The desire to minimize drawdown must be balanced against the conflicting objective of optimizing the hydraulic gradient and groundwater flow rates. The use of recovery trenches or a series of small recovery wells, in place of a few larger wells, will aid in this objective. Likewise, injection trenches or multiple injection wells are often used as a point of groundwater injection. In general, it should be assumed that the amount of groundwater that can be injected into an aquifer will be about 75% of the amount which can be withdrawn while maintaining good hydrogeologic control. The remaining 25% will need to be discharged off-site.

21.5.3 Oxygen demand and oxygen supply

The concentration of oxygen in injected water and the total volume of water that is injected become important factors in predicting the time required for the biodegradation of an easily degraded constituent. For example, if an aquifer were contaminated with diesel fuel over an area of 10 m × 20 m to a depth of 2 m at an average concentration of 200 ppm, (milligram of contaminant per kilogram of soil), this aquifer would contain 141 kg of contamination (assuming a bulk density of $1760\,kg/m^3$). Since about three kilograms of oxygen are required to mineralize one kilogram of hydrocarbon, this fuel creates an artificial oxygen demand of about 423 kg.

If this aquifer were a medium sand (hydraulic conductivity of $10^{-4}\,m/s$, and effective porosity of 0.35) and the natural gradient across the contaminated region were 1%, then the average velocity of water into the contaminated zone (10 m wide by 2 m deep) would be 0.0247 m/day. The daily volume of water moving into the area is the flux times the area of flow (10 m × 2 m) or, 1.73 m^3 of water per day. Since shallow groundwater typically contains about 5 ppm of dissolved oxygen, and water weighs about 1000 kg/m^3, this natural flow of groundwater will deliver 0.0087 kg of DO per day. At this rate, nature will deliver the oxygen required to mineralize the hydrocarbons present in about 130 years.

If a groundwater recovery trench were installed, clean water reinjected upgradient to increase the hydraulic gradient to 4%, and the reinjected water saturated with dissolved oxygen (8 ppm), the system would now deliver 0.055 kg of dissolved oxygen per day, reducing the treatment time to 21 years. (Of course, some of the constituents would be extracted from the site during this period.) By adding hydrogen peroxide as an oxygen source at 300 ppm (the equivalent of 150 ppm of DO), it would be possible to deliver 1.04 kg of DO per day and reduce the project time at this contaminated region to less than two years. (The actual calculation suggests seven months, but, in practice, one would not operate at the full 300 ppm for the entire treatment period.) This type of simple calculation provides a reasonable estimate of the time required for *in situ* bioremediation of a readily degraded contaminant.

21.5.4 Soil and groundwater chemistry

Microbial degradation processes are controlled by the presence of appropriate bacteria, the suitability of the environment for microbial degradation and the general chemical structure of the contaminants of interest. Unfortunately, very little information about the first two areas is typically obtained during initial site investigations. However, some information can often be obtained by examining the general groundwater parameters.

An examination of ammonia, chloride, phosphate, dissolved oxygen, conductivity, and pH data can provide considerable insight into the microbial environment. Groundwater pH is perhaps the most basic characteristic, since soil bacteria prefer pH in the range of 6.5–8. Elevated concentrations of ammonia, chloride, or sulfide can be an indication of septic or sanitary contamination or a highly anaerobic zone. This could indicate a large oxygen demand from other readily degradable materials, which would significantly impact cost and timing. There are several other chemistry issues which do not affect microbial health but may affect feasibility of the process by affecting nutrient transport.

Although simple in concept, the process of transporting an electron acceptor (typically hydrogen peroxide) through soils can present challenging problems at some sites. In 'ideal sites', hydrogen peroxide can provide a very effective means of delivering 150–250 ppm of dissolved oxygen in injected water. Hydrogen peroxide is infinitely soluble in water and decomposes in the presence of soil, such that 300 ppm of hydrogen peroxide produces about 150 ppm of dissolved oxygen. The main design concern in using peroxide is controlling the rate of this catalytic decomposition.

In those soils which contain high concentrations of natural organics, iron, nickel, and copper, the catalytic decomposition of peroxide can be so rapid that the groundwater quickly becomes supersaturated with oxygen. This can create gas blockage of the formation, loss of oxygen, and poor transport of DO to remote areas of the site. Although methods exist for conditioning soils to minimize these effects, this phenomenon can limit the concentration at which the peroxide can be used from 500 ppm, a common concentration in clean sands, to as little as 100 ppm.

Nitrogen and phosphorus are usually required in small amounts to support increased bacterial growth, while trace minerals are readily available in soil. These requirements are usually satisfied through the periodic addition of nutrient formulations containing a blend of ammonia and phosphate maintaining 2–10 ppm of each in the groundwater. However, ammonia and phosphate transport can be difficult at some sites, as these materials can be retarded and consumed while passing through soils. Ammonia can be lost through ion exchange with other cations in clays, and soils and phosphate can interact with cations, especially calcium and magnesium. Although these processes do not make the nutrients unavailable, they do limit the distance that nutrients can be moved effectively through the formation and/or impact the formulation of nutrients that are most effective. In addition, the general introduction of these nutrients, with associated ions, might lead to dispersion or swelling of clay particles.

For the purposes of an initial feasibility assessment, the interactions of phosphate with calcium and magnesium (both of which form insoluble precipitates with orthophosphate) are probably the only transport issues which can be assessed. Both cations inhibit phosphate transport at high concentrations. Heavy metals, such as cadmium, chromium, arsenic, and mercury, are toxic to bacteria in certain forms and concentrations. The presence of these materials, especially if they extend beyond a localized area, should be identified as an area for further analysis when evaluating any microbial-based process.

21.5.5 Laboratory treatability and design studies

The extent and complexity of laboratory treatability and design studies is very site specific. When working with common mixtures of contaminants (for example, gasoline) where the biodegradability is well known, it is probably not necessary to confirm biodegradation, but it may be desirable to evaluate the benefit of adding various levels of nutrients or of adjusting the pH. Since the bacteria responsible for biodegradation have adapted to the environmental conditions in the site soil, and the soil is also a source of nutrients, biodegradation studies should be performed in the presence of soil from the site. This is commonly achieved by creating a series of replicate samples, containing a slurry of soil and groundwater, and then varying the treatment conditions. Conditions tested usually include a microbial inhibited control (poisoned with mercuric chloride or sodium azide), a nutrient control (no added nutrients), plus several sets with various nutrients or additives. In some cases an anoxic control (no oxygen) is also included in this series. For the experiment to be meaningful it is critical to estimate the maximum possible oxygen demand of the system. The oxygen addition schedule must then be designed to satisfy this need.

Samples are sacrificed, after some appropriate period of time, and the entire contents of the vessel are extracted and analyzed for the contaminants of concern. The size and type of sample bottles and the time between samples is heavily dependent on the nature of the systems being tested. For gasoline or diesel fuel, a series of 'volatile organic analysis' bottles might be used to contain volatiles. Analyses would occur at regularly spaced intervals for 2–4 weeks. In this case, peroxide or oxygen would need to be injected regularly into the bottle to maintain the dissolved oxygen level. If, however, the contamination consisted of non-volatile compounds, like pesticides or polynuclear aromatic compounds, open mouth bottles might be used and the required incubation may be up to three months.

In general, biodegradation studies are helpful in predicting the relative rate at which various constituents will degrade and the maximum level of remediation which can be predicted. The actual rates in the field are likely to be limited by oxygen transport, desorption, and other mechanical factors, which are difficult to simulate in the laboratory. Many researchers in the area have attempted to use column studies to mimic field conditions. However, because of the variability inherent in soil studies and the inability to sample the soil matrix of a column without sacrificing the entire column, these studies require multiple sets of replicate columns in order to effect the desired reproducibility. As a result, most commercial design/treatability studies are performed in either static or slurried jars, rather than in columns.

Peroxide stability has become a topic of increasing concern in the past few years, since the rate at which peroxide decomposes upon contact with soil can be a major design factor for an *in situ* bioremediation project. In principle, a peroxide stability test consists of making up a slurry of hydrogen peroxide (typically 500 ppm) and soil, and then analyzing aliquots of the solution after 1, 2, 4, 8, and 24 hours to determine the concentration of remaining peroxide (Brubaker and Crockett, 1986). One recent presentation (Lawes, 1990) suggests that the results of this experiment are very sensitive to the details of how the experiment is performed (soil-to-water ratio, size of flask, rate of stirring, etc.). Since the test is generally useful only as a means of comparing the stability of various soils within a site, or between sites with which the engineer or scientist has operating experience, there is probably no 'best' method, as long as consistency is maintained.

Nutrient interaction with soil and/or groundwater is often evaluated as a means of determining the ease of transport through the soil matrix. These tests are often fairly simple. One example is exposing a solution of a nutrient-amended groundwater to site soil and then analyzing the water for the nutrients, to measure adsorption (Brubaker and Crockett, 1986). Some firms prefer to pass a solution of the nutrient through a column of soil and then analyze the effluent for breakthrough. Either method will provide an indication of the amount of ion exchange and/or adsorption of the nutrients with the soil. The column study is a little more elaborate and provides more detailed information.

21.5.6 Full-scale design and implementation

There are many site specific features which affect the ultimate design and implementation of a full-scale *in situ* bioremediation project. Many of these relate directly to the physical constraints of the site which limit the location of wells and/or trenches, groundwater treatment systems, and associated plumbing. In addition, the remediation objectives for sites vary, both in terms of the level of remediation required, but also relative to the rate of remediation. (It seldom makes economic sense, for example, to use *in situ* bioremediation to remediate an area until the source zones have been removed.)

Besides the design features outlined below, issues of water balance and disposal are often important. Since roughly 75% of the groundwater recovered from a site will typically be reinjected, it is generally advisable to remove the dissolved contamination prior to reinjection, even if the water which is discharged offsite does not require treatment. Reinjection of dissolved constituents can increase the possibility of microbial growth at the injection wells and it increases the oxygen demand in the aquifer. Any of the standard groundwater treatments can be used, but it is important to recognize that air strippers, carbon beds, and recovery wells may be more prone to biological fouling than normal, because of the nutrients added. (Fouling in injection wells is actually fairly uncommon when peroxide is used as an oxygen source.)

Since hydrogeologic control is a critical aspect of any *in situ* bioremediation project, it is common to operate the injection and recovery system for several days to weeks, prior to nutrient injection, to stabilize the water-flow rates. The addition of nutrients provides an opportunity to use chloride (present in the ammonium chloride) as a conservative tracer to confirm nutrient movement, flow rates, and capture. Hydrogen peroxide addition is typically initiated 3–10 days after the nutrients, starting at 100 ppm and then increasing to 300–600 ppm. Hydrogen peroxide is typically delivered onsite as either a 35% or 50% solution and is metered into the groundwater for dilution. As a concentrated solution, peroxide is corrosive to skin and can decompose violently, if contaminated with metals. Therefore, this material should only be handled according to manufacturer's directions and in equipment designed and designated for its use.

The amount of monitoring required at a site is dependent on the size of the site and how rapidly concentrations are likely to change. Groundwater monitoring is usually performed weekly to monthly for pH, DO, nutrients, bacteria, and contaminants. Most of these parameters can be monitored easily on-site using handy field test kits or meters. However, since the concentration of dissolved contamination usually increases during the initial phases of nutrient addition, the best means of verifying process performance is through soil coring. Since contaminant distribution is irregular, the decrease in concentration is usually not detectable until the process is about one-third complete, and even then, the most interpretable data will be collected if the cores used for monitoring are grouped to allow comparison of samples collected in the 'same location', before and after treatment.

As the process proceeds, the observed concentration of constituents in the groundwater often remain high until the remediation is nearly complete, and the mass of sorbed contamination is almost depleted. During this time the nitrogen and phosphate will also have moved further and faster than the dissolved oxygen. As the DO monitored downgradient of the contaminated area increases, the peroxide level can be decreased until the point where the oxygen demand of the aquifer has been satisfied. At the point where the natural oxygen supply to the aquifer is sufficient to sustain the continued biological degradation of the contaminants present, the addition of more nutrients or oxygen has no apparent benefit. Although this point may not necessarily reflect 'site closure', in most cases continued containment and/or monitoring will provide as much benefit as continued active nutrient addition.

21.6 ONGOING RESEARCH AND DEVELOPMENT

During the past five years, researchers and practitioners from a wide range of disciplines have contributed dramatically to our understanding of *in situ* bioremediation, asking new questions and providing new insight regarding the versatility, reliability, and optimization of this process (McCarty, 1987; Wilson and Ward, 1987; Thomas and Ward, 1989). This section contains a brief summary of a few of the more promising and innovative areas of development.

21.6.1 Biodegradation using denitrification

As recently as seven years ago, it was generally accepted that aromatic rings which lacked oxygen substitution were not biodegradable by denitrifying bacteria (Bouwer and McCarty, 1983). Recent research is challenging those assumptions and examining the possibility of using nitrate as an electron acceptor to enhance the biodegradation of organic contaminants. Although nitrate is often itself a groundwater contaminant, it is much cheaper than hydrogen peroxide and is more easily transported through soils. The ability to use this material as an oxygen source could significantly reduce the cost and complexity of the process. Denitrification also has advantages over other anaerobic processes in that this process uses the same enzymatic pathways as the aerobic process. This is important because these pathways are well-known and there are no undesirable by-products.

In the mid-1980s, Swiss workers (Kuhn *et al.*, 1985; Zeyer *et al.*, 1986; Kuhn *et al.*, 1988) reported the biodegradation of m-xylene by denitrifying bacteria which had been isolated from denitrifying river sediment and fed xylene continuously for several months. In a series of carefully controlled column studies, including experiments with ^{14}C labelled material, these workers observed the degradation of m-xylene and several other substituted benzenes. However, these bacteria were unable to degrade benzene, naphthalene, methylcyclohexane or 1,3-dimethylcyclohexane. Related research by a Canadian group (Major *et al.*, 1988) demonstrated the biotransformation of benzene, toluene and xylene in laboratory studies using a sandy aquifer material, and a recent study (Mihelic, 1988) describes the biodegradation of naphthalene, naphthol, and acenaphthalene under denitrifying conditions.

Field activities have been limited to one controlled field study, which failed to confirm degradation rates in excess of those expected from aerobic processes (Berry-Spark and Barker, 1987), and two attempts at field-scale remediation. The first involved a sandy aquifer contaminated with mineral oil. Water containing 300 ppm of nitrate was introduced into this contaminated zone for approximately 18 months, resulting in dramatic decrease in dissolved levels of both aromatic and aliphatic constituents (Geldner, 1987; Werner, 1985). The second project involved the injection of 45 ppm nitrate to an aquifer containing gasoline. Although data collected to date indicate a decrease in the levels of petroleum constituents at this site, it is difficult to gain any useful insight into the process from the limited information available (Sheehan *et al.*, 1988). The US EPA is in the process of performing much more elaborate evaluation of this technology at the Traverse City site (Hutchins, 1991). Laboratory data from this project confirm the presence of biodegradation of alkylated benzenes by denitrifying bacteria but benzene was not degraded under the conditions tested.

21.6.2 Methanogenic biodegradation

Although it has long been recognized that many aromatic compounds could be biodegraded under methanogenic conditions (Kuhn and Suflita, 1989), it is only very recently that laboratory studies under controlled conditions (Grbic-Galic and Vogel, 1987) have confirmed this observation. In addition, anecdotal field observations (Hayman *et al.*, 1988) suggest that there may be techniques of enhancing the anaerobic degradation of many organic constituents. This phenomenon is being examined in more detail at the EPA Traverse City project where zones of methanogenesis and of aerobic biodegradation were detected (Wilson and Ward, 1987).

21.6.3 Modelling the *in situ* bioremediation process

There have been increasing efforts to model both natural biodegradation processes and active bioremediation in the subsurface. These models are helpful in identifying the critical parameters which affect microbial activity and may some day be used in combination with risk assessments to define 'acceptable' levels of remediation at sites near sensitive receptors. BIOPLUME II was developed for the EPA, based on several years of study at the United Creosote site in Conroe, Texas, and it is now being applied to an aviation gasoline spill in Traverse City, Michigan (Rifai *et al.*, 1987; Rifai *et al.*, 1989). The current version has been 'packaged' in a manner which makes it fairly 'user friendly', though it is not yet in wide-spread use. Most of the other models being developed are more research oriented and not intended for commercial use (Widdowson *et al.*, 1988; Doyle and Piotrowski, 1989; Tim and Mostaghimi, 1989).

There are several research groups (Cunningham *et al.*, 1988; Rittmann *et al.*, 1988; Alexander and Scow, 1989; Chiang *et al.*, 1989; Swindoll *et al.*, 1988) looking at the kinetics and detailed mechanisms of biodegradation on a more molecular level, examining the role of biofilms under the types of environments found in aquifers. This research has the potential to lead to very important insights into the factors which control the biodegradation process.

21.6.4 Chlorinated hydrocarbons

Chlorinated solvents such as trichloroethylene, trichloroethane, dichloroethylene, and associated compounds are almost ubiquitous at industrial sites in many urban areas. Their presence and persistence results from widespread use as degreasers and cleaning solvents, their moderate solubility and volatility, their density (since they are denser than water they migrate down through aquifers), and their chemical and microbial stability (Vogel *et al.*, 1987). The carbon chlorine bond is a very 'stable' bond, and it appears that microorganisms do not generate much usable energy from the oxidation of this bond. As a result, even those bacteria able to biologically oxidize these compounds require another food source for energy.

During the past ten years, many researchers have been actively studying the chemical and microbial transformations of this class of compounds in subsurface environments (Fogel *et al.*, 1986; Wilson *et al.*, 1987; Nelson *et al.*, 1988). Methanogenic and aerobic processes have been studied under a broad range of laboratory conditions, and an extensive EPA-funded field study has been conducted to develop and evaluate techniques for implementing in situ bioremediation of these materials in a real aquifer (Semprini *et al.*, 1988). The results of this work can be summarized as follows.

- Halogenated aliphatic compounds undergo a limited number of chemical and biological reactions in natural settings. These include hydrolysis, dehydrohalogenation, and oxidation/reduction reactions.
- The rate of microbial oxidation decreases as the number of chlorine atoms increases. The rate of reductive dehalogenation increases as the number of chlorine atoms increases.
- Aerobic *in situ* processes are likely to require the addition of a readily degradable carbon source, which will stimulate the microbial population of interest (for example, methane or toluene), and sufficient oxygen to degrade both the original contaminants and the added carbon. The ratio of added carbon to contaminants is likely to be in the range of 30:1.

Based on the complexity of this process, both from a microbial and from an engineering perspective, it is doubtful that the *in situ* bioremediation of highly chlorinated solvents will become 'standard' technology in the next 3–5 years.

21.6.5 PAHs and other less soluble contaminants

Over 90% of all of the commercial experience with *in situ* bioremediation involves the biodegradation of contaminants which are soluble at the part per million level and whose remediation levels are defined in terms of part per billion of dissolved constituents. As projects progress into groundwater corrective action, an increasing number of sites will be examined to determine how to remediate soluble polyaromatics hydrocarbons (PAHs) to low part per billion or even part per trillion levels. *In situ* bioremediation of contaminants which have very low solubility may evolve into a very different process than that which exists today. Two of the unique aspects of this problem are summarized below.

- The rate of biodegradation of these compounds appear limited by its rate of dissolution even in well mixed systems. The rate of degradation of compounds which partition strongly to soil and desorb slowly may be dependent on these physical-chemical properties, in addition to nutrient transport.
- Current research suggests that the degradation of PAHs usually slows or ceases at levels of 10–50 ppb. The factors which control this phenomena are not well understood, nor are ways to continue remediation beyond this level.

Although a variety of groups (Scow *et al.*, 1986; Swindoll *et al.*, 1988; Borden *et al.*, 1989; Doyle and Piotrowski, 1989; Smith *et al.*, 1989; Wang *et al.*, 1990) are beginning to probe these and related issues, much more experience is required with these compounds before remediation of 'heavy' organics to sub-part per million levels will be routine.

21.6.6 *In situ* bioremediation in the unsaturated zone

Even though this chapter has focused on the bioremediation within the saturated portion of contaminated sites, it is important to recognize that contamination above the water table, in the unsaturated zone, is also of concern. Residual contamination in these areas can continue to contaminate groundwater as precipitation percolates through this region or as the water table fluctuates. Although unsaturated soils are often easier to excavate and treat than saturated soils, the inaccessibility of soils at many sites has created a need for a better understanding of biodegradation in unsaturated soils.

The introduction of vapor extraction techniques has led to an increased awareness of factors which control the fate and transport of organic vapors in unsaturated soils. One of the important insights from this process has been the increased awareness that aerobic biodegradation within the unsaturated zone can lead to oxygen depletion and carbon dioxide inhibition, even in shallow soils. Monitoring at the EPA Traverse City site indicated oxygen concentrations of less than 3% and carbon dioxide levels in excess of 10% at depths greater than 2 m (Kampbell, 1990).

The critical factors for vadose zone biodegradation appear to be the same as those involved in land treatment (see Chapter 20) and the *in situ* bioremediation of aquifers, with some subtle differences. Oxygen exchange is very important and is usually achieved by using the techniques that are developing in the vapor extraction area, drawing air through the unsaturated zone (Dineen *et al.*, 1990; Hinchie and Miller, 1992). Moisture is probably the next most important factor and must be maintained within a level that keeps the soils moist without seriously inhibiting air flow. Ammonia-nitrogen and phosphate are probably the next factors to be considered. Percolation of nutrient-enriched water, either through closely spaced injection points or through surface irrigation have both been used, as has gaseous ammonia. Although the amount of detailed study in this area still trails that of saturated zone treatment and surficial land treatment, this technique will certainly be used increasingly in the future.

REFERENCES

Alexander, M. (1981) Biodegradation of chemicals of environmental concern, *Science* **211**, 132–8.

Alexander, M. and Scow, K.W. (1989) Kinetics of biodegradation in soil, *Reactions and Movement of Organic Chemicals in Soils*, Madison, WI, 243–69.

American Petroleum Institute (API) (1987) *Field study of enhanced subsurface biodegradation of hydrocarbons using hydrogen peroxide as an oxygen source*, Publication No. 4448, Washington.

Babea, L. and Vaishnav, D.D. (1987) Prediction of biodegradability of selected organic compounds, *J. Indust. Microbiol.* **2**, 107–15.

Back, W. (1989) Early concepts of the role of microorganisms in hydrogeology, *Ground Water* **27**, 618–22.

Bell, R.A. and Byrd, R.E. (1988) Case history: gasoline spill in fractured bedrock addressed with bioreclamation, *Proc., HazTech Int. Conf.*, Cleveland, HazTech Int., Bellevue, WA, 2C-122–52.

Berry-Spark, K. and Barker, J.F. (1987) Nitrate remediation of gasoline contaminated groundwater: results of a controlled field experiment, in *Proc., NWWA/API Conf. on Petroleum Hydrocarbons and Organic Chemicals in Groundwater: Prevention, Detection and Restoration*, NWWA, Dublin, OH, pp. 127–44.

Borden, R.C., Lee, M.D., Thomas, J.M., Bedient, P.B. and Ward, H.C. (1989) *In situ* measurement and numerical simulation of oxygen limited biotransformation, *Ground Water Monitoring Rev.*, Winter, 83–91.

Bouwer, E.J. and McCarty, P.L. (1983) Transformation of halogenated organic compounds under denitrification conditions, *Appl. Environ. Microbiol.*, **45**, 1295–9.

Brown, R.A., Matson, C. and Benazon, N. (1988) The treatment of groundwater contaminated with petroleum hydrocarbons using *in situ* bioreclamation, in *Proc., HazTech Int. Conf.*, Cleveland, Haztech International, Bellevue, WA, pp. 2C-153–79.

Brown, R.A. and Norris, R.D. (1986) Field demonstration of enhanced bioreclamation, in *Proc., 6th Nat. Symp. on Aquifer Restoration and Groundwater Monitoring*, NWWA, Dublin, OH, p. 421.

Brown, R.A., Norris, R.D. and Raymond, R.L. (1984) Oxygen transport in contaminated aquifers, in *Proc., Petroleum Hydrocarbons and Organic Chemicals in Ground Water*, NWWA, Dublin, OH, pp. 421–41.

Brubaker, G.R. (1989) Screening criteria for *in situ* bioreclamation of contaminated aquifers, in *Proc., Haz. Wastes and Haz. Mat.*, HMCRI, New Orleans: pp. 319–21.

Brubaker, G.R. and Crockett, L.L. (1986) *In situ* aquifer remediation using enhanced bioreclamation, in *Proc., Hazmat 86*, Atlantic City, Tower Conf. Management, Glen Ellyn, IL.

Chiang, C.Y., Salanitro, J.P., Chai, E.Y., Colthart, J.D. and Klein, C.L. (1989) Aerobic biodegradation of benzene, toluene, and xylene in a sandy aquifer – data analysis, and computer modeling, *Ground Water*, **27**, 823–34.

Cunningham, A.B., Characklis, W.G. and Bouwer, E.F. (1988) Influence of microbial transformation on the *in situ* Biodegradation of groundwater contaminants, in *Proc., Petroleum Hydrocarbons and Organic Chemicals in Ground Water*, NWWA, Dublin, OH, pp. 669–84.

Dineen, D., Slater, J.P., Hicks, P. and Holland, L.D. (1990) *In situ* biological remediation of petroleum hydrocarbons in unsaturated soils, in *Petroleum Contaminated Soils, Volume 3* (eds P.T. Kostecki and E.J. Calabrese), Lewis Publishers, Chelsea, MI, pp. 177–93.

Dupont Biosystems (1990) Marketing Literature.

Doyle, R. and Piotrowski, M. (1989) *In situ* bioremediation of wood-treating compounds in groundwater, in *Proc., 6th Nat. Conf. on Haz. Wastes and Haz. Materials*, HMCRI, Silver Spring, MD, pp. 308–14.

Flathman, P.E. and Caplan, J.A. (1986) Cleanup of contaminated soils and groundwater using biological techniques, in *Proc., 3rd Nat. Conf. on Haz. Wastes and Haz. Materials*, HMCRI, Silver Spring, MD.

Fliermans, C.B. and Balkwill, D.L. (1989) Microbial life in deep terrestrial subsurfaces, *Bioscience*, **39**(6), 370–7.

Fogel, M.M., Taddeo, A.R. and Fogel, S. (1986) Biodegradation of chlorinated ethenes by a methane-utilizing mixed culture, *Appl. Environ. Microbiol.*, **51**, 720–4.

Geldner, P. (1987) Stimulated *in situ* biodegradation of aromatic hydrocarbons, in *Proc., 2nd Int. Conf. on New Frontiers for Haz. Waste Management*, Pittsburgh, PA, pp. 247–57.

Grbic-Galic, D. and Vogel, T.M. (1987) Transformation of toluene and benzene by mixed methanogenic cultures, *Appl. Environ. Microbiol.*, **53**, 254–60.

Hayman, J.W., Adams, R.B. and McNally, J.H. (1988) Anaerobic biodegradation of hydrocarbons in confined soils beneath busy places – a unique method of methane control, in *Proc., Petroleum Hydrocarbons and Organic Chemicals in Ground Water*, NWWA, Dublin, OH, pp. 383–96.

Hinchie, R.E. and Miller, R.N. (1992) Bioventing for In Situ Remediation of Petroleum Hydrocarbons, in *Bioventing and Vapor Extraction: Uses and Applications in Remediation Operations*, Air and Waste Management Association, pp. 39–48.

Hutchins, S.R. and Wilson, J.T. (1991) Laboratory and field studies in a fuel-contaminated aquifer under denitrifying conditions in, *In Situ Bioreclamation*, Butterworth-Heineman, London, pp. 157–72.

IT Corporation (1986) *In Situ Bioreclamation: A Case History – Gasoline Contamination in Northern Indiana*.

Jamison, V.W., Raymond, R.L. and Hudson, J.O. (1976) Biodegradation of high octane gasoline, in *Proc., 3rd Int. Biodegradation Symp.*, Applied Science Publishers, Englewood, NJ, pp. 187–96.

Jhaveria, V. and Mazzacca, A.J. (1985) Bio-reclamation of ground and ground water by *in situ* biodegradation: a case history, in *Proc., 6th Nat. Conf. on Uncontrolled Haz. Sites*, Washington, DC, pp. 242–52.

Kampbell, D. (1990) Simplified soil gas sensing techniques for plume mapping and remediation monitoring, in *Petroleum Contaminated Soils*, Lewis Publishers, Chelsea, MI, **3**.

Keely, J.F. (1989) *Performance evaluations of pump-and-treat remediations*, EPA/540/489/006.

Keely, J.F. and Tsang, C.F. (1983) Velocity plots and capture zones of pumping centers for groundwater investigations, *Ground Water*, Nov/Dec, 701–13.

Kuhn, E.P., Colberg, P.J., Schnoor, J.L., Wanner, O., Zehnder, A.J.B. and Schwarzenbach, R.P. (1985) Microbial transformation of substituted benzenes during infiltration of river water to groundwater and laboratory column studies, *Environ. Sci. Technol.*, **19**, 961–8.

Kuhn, E.P., Zeyer, J., Eicher, P. and Schwarzenbach, R.P. (1988) Anaerobic degradation of alkylated benzenes in denitrifying laboratory aquifer columns, *Appl. Environ. Microbiol.*, **54**, 490–6.

Kuhn, E.P. and Suflita, J.M. (1989) Dehalogenation of pesticides by anaerobic microorganisms in soil and groundwater – a review, in *Reactions and Movement of Organic Chemicals in Soils*, Soil Sc. Soc. of Amer., Special Publication No. 22, Madison, Wisconsin. pp. 111–80.

Lawes, B. (1990) Soil induced hydrogen peroxide decomposition, in *Petroleum Contaminated Soils*, Lewis Publishers, Chelsea, MI, **3**, ch 19.

Lee, M.D., Thomas, J.M., Borden, R.C., Bedient, P.B., Ward C.H. and Wilson, J.T. (1988) Biorestoration of aquifers contaminated with organic compounds, *CRC Critical Reviews in Environ. Control*, **18**, 29–89.

MacKay, D.M. and Cherry, J.A. (1989) Groundwater contamination: pump-and-treat remediation, *Environ. Sci. Technol.*, **23**, 630–6.

Major, D.W., Mayfield, C.I. and Barker, J.F. (1988) Biotransformation of benzene by denitrification in aquifer sand, *Ground Water*, **26**, 16–22.

McCarty, P.L. (1987) Bioengineering issues related to *in situ* remediation of contaminated soils and groundwater, presented at *Reducing Risks from Environmental Chemicals through Biotechnology*, Univ. of Washington, Seattle, Washington.

Mihelic, J.R. (1988) Microbial Degradation of Polycyclic Aromatic Hydrocarbons Under Denitrification Conditions in Soil-Water Suspensions. PhD thesis, Carnegie Mellon University, Pittsburgh.

Nelson, M.J.K., Montgomery, S.O. and Pritchard, P.H. (1988) Trichloroethylene metabolism by microorganisms that degrade aromatic compounds, *Appl. Environ. Microbiol.*, **54**, 604–6.

Raymond, R.L., Jamison, V.W. and Hudson, J.O. (1976) Biodegradation of high octane gasoline, in *AIChE Symp. Series*, **73**, 390–404.

Rifai, H.S., Bedient, P.B. and Wilson, J.T. (1989) *BIOPLUME model for contaminant transport affected by oxygen limited biodegradation, Environ. Res. Brief*, EPA/600/M019.

Rifai, H.S., Bedient, P.B., Borden, R.C. and Haasbeek, J.F. (1987) *BIOPLUME II – Computer Model of Two-Dimensional Transport under the Influence of Oxygen Limited Biodegradation in Ground Water, User's Manual, Version 1.0*, Rice Univ., Houston, TX.

Rittmann, B.E., Valochcchi, A.J., Odencrantz, J.E. and Bae, W. (1988) *In Situ Bioreclamation of Contaminated Groundwater*, HWRIC Report 031, Illinois State Water Survey, Savoy, IL.

Scow, K.M., Simkins, S. and Alexander, M. (1986) Kinetics of mineralization of organic compounds at low concentrations in soil, *Appl. Environ. Microbiol.*, **51**, 1028–35.

Semprini, L., Roberts, P.V., Hopkins, G.D. and McKay, D.M. (1988) *A Field Evaluation of In-Situ Biodegradation for Aquifer Restoration*, EPA Project Summary 600/S2-87/096.

Sheehan, P.J., Schneiter, R.W., Mohr, T.K.G. and Gersberg, R.M. (1988) Bioreclamation of gasoline contaminated groundwater without oxygen addition, in *Proceed., 2nd Nat. Outdoor Action Conf. on Aquifer Restoration, Ground Water Monitoring and Geophysical Methods*, Las Vegas, pp. 183–99.

Smith, J.R., Nakles, D.V., Sherman, D.F., Neuhauser, E.F. and Loehr, R.C. (1989) Environmental Fate Mechanisms Influencing Biological Degradation of Coal-Tar Derived Polynuclear Aromatic Hydrocarbons in Soil Systems. Presented at *The 3rd Int. Conf. on New Frontiers for Haz. Waste Management*, Pittsburgh, PA.

Suflita, J.M. (1985) Microbial principles of *in situ* remediation of aquifers, in *Proc. of HazPro 85*, Pudvan Publishers, Northbrook, IL.

Swindoll, C.M., Aelion, C.M., Dobbins, D.C., Jiang, O., Long, S.L. and Pfaender, F.K. (1988) Aerobic biodegradation on natural and xenobiotic organic compounds by subsurface microbial communities, *Environ. Toxicol. Chem.*, **7**, 291–9.

Texas Research Institute (1982) Feasibility Studies on the Use of Hydrogen Peroxide to Enhance Microbial Degradation of Gasoline, prepared for API, Washington, DC.

Thomas, J.M. and Ward, C.H. (1989) *In situ* biorestoration of organic contaminants in the subsurface, *Environ. Sci. Technol.*, **23**, 760–6.

Tim, U.S. and Mostaghimi, S. (1989) Modeling Transport of a Degradable Chemical and its Metabolites in the Unsaturated Zone. *Ground Water*, **27**, 672–81.

Verschueren, K. (ed.) (1983) *Handbook of Environmental Data on Organic Chemicals*, 2nd edn, Van Nostrand Reinhold, NY.

Vogel, T.M., Criddle, C.S. and McCarty, P.L. (1987) Transformations of halogenated aliphatic compounds, *Environ. Sci. Technol.*, **21**, 722–36.

Wang, Y.S., Subba-Rao, R.V. and DuPont, R.R. (1990) Personal communication.

Werner, P. (1985) A new way for the decontamination of polluted aquifers by biodegradation, *Water Supply*, **3**, 41–7.

Widdowson, M.A., Molz, F.J. and Benefield, L.D. (1988) Modeling multiple organic contaminant transport and biotransformation under aerobic and anaerobic (denitrifying) conditions in the subsurface, in *Proc. of Petroleum Hydrocarbons and Organic Chemicals in Ground Water*, NWWA, Dublin, OH, pp. 397–416.

Wilson, J.T., Leach, L.E., Henson, M. and Jones, J.N. (1986) *In situ* biorestoration as a ground water remediation technique, *Ground Water Monitoring Rev.*, Fall, 56–64.

Wilson, J.T., Fogel, S. and Roberts, P.V. (1987) Biological treatment of trichloroethylene *in situ*, in *Detection Control and Renovation of Contaminated Ground Water*, ASCE, New York.

Wilson, J.T. and Ward, C.H. (1987) Opportunities for Bioreclamation of Aquifers Contaminated with Petroleum Hydrocarbons. *Develop. Indust. Microbiol.*, **27**, 109–116.

Yaniga, P.M., Matson, C. and Demko, D.J. (1985) Restoration of water quality in a multiaquifer system via *in situ* biodegradation of the organic contaminants, in *Proc., 5th Nat. Symp. on Aquifer Restoration and Groundwater Monitoring*, NWWA, Dublin, OH, pp. 510–8.

Yaniga, P.M. and Smith, W. (1986) Aquifer restoration via accelerated *in situ* biodegradation of organic contaminants, in *Proc.: Management of Uncontrolled Haz. Waste Sites*, Washington, pp. 333–8.

Zeyer, J., Kuhn, E.P. and Schwarzenbach, R.P. (1986) Rapid microbial mineralization of toluene and 1,3-dimethylbenzene in the absence of molecular oxygen, *Appl. Environ. Microbiol.*, **52**, 944–7.

CHAPTER 22

Soil washing

Paul B. Trost

22.1 INTRODUCTION

Increasing costs, long-term liabilities, and bans on land disposal makes disposal in hazardous waste landfills increasingly unattractive. The high cost per tonne of incineration is further stimulating the rapid development and application of new and innovative on-site detoxification/volume reduction techniques.

Soil washing is a technique applicable to volume reduction/on-site detoxification, and can be defined as the removal and concentration of organic and/or inorganic contaminants from excavated soils. Typically, soil washing has developed along three parallel avenues of:

1. particle size fractionation;
2. aqueous-based systems using some form of mechanical and/or chemical method to liberate and concentrate the contaminants; and
3. counter current decantation (CCD) employing solvents for organic contaminants and acids/bases or chelating agents for inorganic contaminants.

22.1.1 Particle size fractionation

Particle size fractionation is based on the premise that contaminant distribution preferentially occurs in the fine fraction, e.g., <74 μm or −200 Mesh. In this small size range, clays and finely divided organics predominantly occur. To apply this method, tumblers, trommels, high pressure washing, sonification, hydrocyclones, or vibratory screens – in conjunction with sieve screens – are used to separate the fine fractions from the coarse fractions and thus achieve a volume reduction and concentration of contaminants.

Geotechnical Practice for Waste Disposal.
Edited by David E. Daniel.
Published in 1993 by Chapman & Hall, London. ISBN 0 412 35170 6

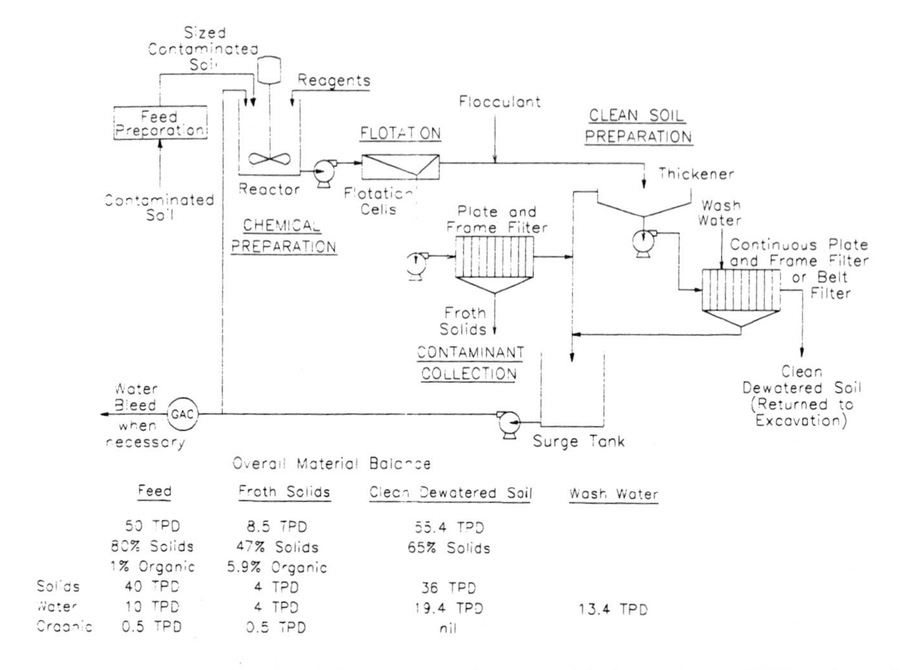

Fig. 22.1 Generalised flowsheet and material balance for 50 TPD soil wash plant. *Source*: MTARRI Golden, Colorado.

22.1.2 Aqueous-based soil wash systems

Aqueous-based soil wash systems may or may not employ particle sizing techniques but generally utilize methods such as froth flotation* to concentrate the contaminants in only 5–15% of the original volume. This concentration is achieved by the use of surfactants, alkaline agents, and/or minor amounts of solvents introduced into an aqueous slurry containing the sized or crushed soil. The slurry is 'conditioned' in a pre-mix step to liberate the contaminants from the soil mineral particles then pumped to the froth flotation cells to achieve separation/ concentration of contaminants into the froth. The cleaned soil is dewatered and recycled to its original excavation site. Please see the flowsheet in Fig. 22.1.

22.1.3 Counter current decantation (CCD)

CCD is a commonly applied technique in the chemical and minerals processing business. Basically, a series (1–7) of thickeners is utilized

*Froth flotation is utilized by the mineral processing industry and should not be confused with dissolved air flotation (DAF).

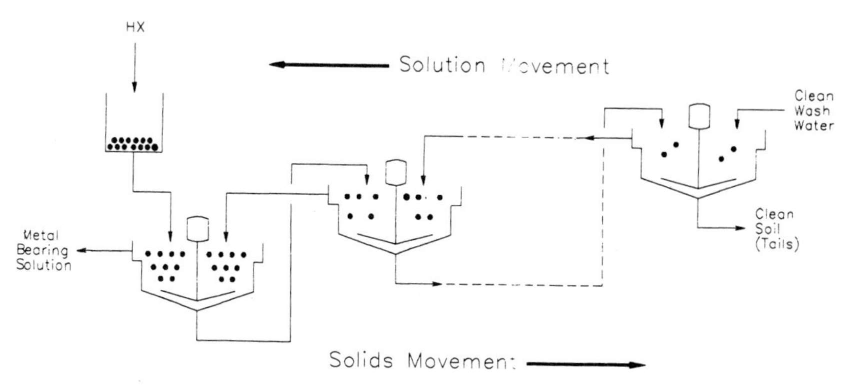

Fig. 22.2 Schematic flowsheet: metal removal by counter current decantation. *Source*: MTARRI Golden, Colorado.

with the contaminated soil introduced into tank 1 allowed to settle and then pumped from tank 1 to tank 2, and on to tank 3, etc. The solvent, acid/base, or chelating agent is pumped into the last tank (e.g., tank 3), allowed to mix with the soil then the liquid overflow containing contaminant and solvent is pumped to tank 2 and that overflow to tank 1 (Fig. 22.2). Thus, the equilibrium constant is shifted, driving the contaminants out of the soil and into the liquid phase.

A modification of this CCD system is a solvent extraction process similar to a small oil refinery where the contaminants are removed in vessels/towers utilizing unique solvents and/or critical fluids.

22.2 HISTORY/STATUS OF SOIL WASHING

Soil washing in the US has been investigated by the EPA at their Edison, New Jersey, facility since approximately 1982. Their operation consisted of high pressure water jets to achieve soil matrix breakup followed by screening and aqueous-based counter current decantation (CCD). Various companies in the US have conducted extensive bench-scale, and/or pilot-scale tests on different soil wash systems. Although no full-scale cleanup has been achieved as of late 1989, at least one is anticipated to be initiated in 1990. Numerous companies are simultaneously pursuing the particle sizing, aqueous-based systems, CCD systems and/or solvent extraction methods.

In Europe, full-scale cleanups using soil washing are presently occurring. Typically, these soil wash systems are fixed or semi-permanent installations capable of treating 200–600 tonnes/day at costs of \$100–200/tonne. Particle sizing is generally an integral part of their systems together with aqueous-based washing employing either froth flotation, ultrasonic vibration, low frequency vibration, solid-liquid separation, hydrocyclones and/or flocculation.

22.3 REGULATORY AND COMMUNITY ACCEPTANCE

The USEPA is actively promoting the continued development of soil washing technologies, and recent consent orders which specify cleanup levels and methods have included soil washing. A trend appears to be developing which, under these negotiated consent orders, the cleanup levels are slightly rising. Thus, if this trend continues, the market for soil wash may increase. If cleanup levels required a 99.99% removal rate of the contaminant, the market for soil washing would be limited. However, if contaminant removal rates are lowered to the 92–99% range, the market for soil washing will expand significantly. Soil washing operates as a volume reduction system with the contaminants going off site for final disposal/incineration.

22.4 FACTORS INHIBITING SOIL WASH TREATMENT

One main problem exists for soil wash treatment of contaminated soils, i.e., the wide variability of conditions encountered at most contaminated sites. Variability occurs both in the type of contaminants and the nature of the soil mineralogy. This wide variability requires that significant bench-scale testing be completed prior to ascertaining if a particular soil wash technology is applicable and can achieve the cleanup level goals at a competitive price. Because of this needed bench-scale testing combined with the relatively high contaminant removal rates required, soil washing may be more applicable to larger volume contaminated sites as opposed to smaller volume sites; thus, allowing more resources (time and money) for bench-scale testing to achieve these low cleanup levels. A likely breakpoint between soil washing vs. haul and dispose and/or incinerate appears to be about 2500 m^3.

22.4.1 Contaminant variability

Contaminant variability at most sites requiring remediation appears to be the norm rather than the exception. For example, at one Superfund site the main contaminant was polychlorinated biphenols (PCBs), both in the soil and in the groundwater. However, since transformers were filled with a mixture of PCBs and mineral oil, there is far more mass of mineral oil than PCBs. Also, this same site was previously a coal tar gas manufacturing site; thus, the PCB-rich mineral oil has migrated

Table 22.1 Organic contaminants at superfund sites

compound	*average concentration (mg/kg)*	*order of occurrence*
TCE	103	highest
toluene	1121	
benzene	17	
chloroform	<1	
PCBs	NA	
1,1,1-TCA	NA	
PCE	540	
phenol	511	
xylene	8400	
ethylbenzene	3200	
PAH	NA	lowest

Notes: concentrations may range from 1–100 000 mg/kg; NA = not available. *Source*: Esposito (1988).

through soils previously contaminated with polynuclear aromatic hydrocarbons (PAHs) such as chrysene and benzo(a)pyrene, and probably cyanide (CN). Cleanup must include all these chemicals.

Soil washing, if applied, must therefore simultaneously meet clean-up criteria for all of these contaminants, which have very different chemical activities.

Table 22.1 shows the order of occurrence and average concentration of the most commonly occurring organic contaminants at Superfund sites in the US. Typically, more than one of these organic contaminants is present at every site.

Although low in ranking, a significant cleanup target has been PAHs associated with past wood-treating operations (e.g., creosote) and coal gasification sites. Semi-volatile organic contaminants of significant concern are benzo(a)pyrene, dibenzo(a,h)anthracene, benzo(a)-anthracene, chrysene, anthracene, and naphthalene. Cleanup levels for these compounds may be 1 mg/kg or less with their total not to exceed 10 mg/kg.

Table 22.2 shows the average concentration of inorganic contaminants in Superfund-type soils and their relative order of occurrence. Not only is soil washing hampered by the wide variability of contaminants, but also by the mixture of both organic and inorganic contaminants. This mixture is shown in Table 22.3.

Soil washing of organic and inorganic contaminated soils usually requires combining two operational processes. Organic contaminant removal may be accomplished by froth flotation or a solvent extraction process; however, inorganic contaminant removal will require a CCD

Table 22.2 Inorganic contaminants at superfund sites

elements	*average concentration (mg/kg)*	*order of occurrence*
Pb	3100	highest
Cr	370	|
Zn	5000	|
Cd	180	|
As	90	|
Cu	2100	|
Hg	NA	|
CN	54	|
Ni	200	lowest

Notes: concentrations may range from <1–100 000 mg/kg; NA = not available. *Source*: Esposito (1988).

Table 22.3 Sites with mixed contaminants

historical usage of contaminated sites	*organic contaminant*	*inorganic contaminant*
Wood treating sites	PAH, PCP	Cu, Cr, As
Coal gas manufacturing sites	PAH	CN
Pesticide/herbicide manufacturing	DDT, Endrin, Aldrin Dieldrin, etc., Arsenious acid	As
Electronic manufacturing	TCE, PCE	Cu
Oil refinery sites	BETX*, PAHs, Crude oil	Cr, Pb, V

* BETX = benzene, ethylbenzene, toluene, xylene.

circuit or other leach circuit having acids/bases/chelating agents to solubilize the metals followed by precipitation of the removed metals or a significant volume of finer rejection.

The need for two operational processes will therefore increase the cost and may seriously impact any economic advantage of soil washing. This economic impact will be most seriously felt for on-site treatment at smaller sites, e.g., <10 000 tonnes where bench-scale testing, process optimization, and mob/demob costs will have a proportionately larger cost/tonne than for larger sites.

22.4.2 Mineralogic variability

Esposito (1988) cited PEI Associates' 1987 study of 151 contaminated sites in the US to show that contaminated soil is more likely to be composed of clay to sandy-clay soil as shown in Table 22.4.

Table 22.4 Soil types at superfund sites

soil type	*% of occurrence*
Sandy-clay	30
Clay	10
Sandy	20
No information	40

Source: Esposito (1988).

Thus, the soil wash system should be capable of handling relatively high percentages of fine-grained to clay-sized fractions, otherwise the system may have limited applicability and thus limited economic advantage due to high disposal costs of the contaminant-rich soil fraction.

22.5 DISTRIBUTION OF CONTAMINANTS IN SOILS

Generally contaminants are more concentrated in the finer fractions of the soil particles than in the coarser fraction. However, a significant amount of contaminants may still be present in the coarse particles, which also must be removed to achieve the cleanup goals. For example, Table 22.5 shows the distribution of pentachlorophenol (PCP), an organic contaminant, as a function of particle size.

Removal of the $-4.5\,\mu m$ size fraction by screening would result in a weight loss of only 6.5% of the total feed weight but would achieve removal of 65.5% of the contained PCP and, secondarily, would allow higher throughput of the remaining soil through a soil washing process by decreasing the settling and thickening times of the residual washed soil. Thus, screening for this soil is a very attractive alternative based on the projected economics of total treatment.

Table 22.6 shows distribution of selected inorganic contaminants as a function of soil particle size for an example site. If the soil in Table 22.6 were screened with the $-10\,\mu m$ size fraction removed, the weight to be disposed of would be 31.2% of the original feed weight and would contain 68.6% of the total arsenic present. Unfortunately, the Pb removal would not be as effective as the As and, thus, additional treatment would be required.

Contrarily, Table 22.7 shows that the distribution of PCBs in a quartz-rich but fine-grained soil are not particularly concentrated in the finer soil fraction and therefore very little advantage would be gained by screening and removal of the fine fraction since removal of 51.4% of the total soil weight would only achieve 73.8% removal of the PCBs.

For the clay-size fractions, sorption of inorganic cations is well documented (Chapters 2 and 3); clays also will sorb many organic

Table 22.5 Example of distribution of pentachlorophenol as a function of soil size fractions

size fraction (μm)	*wt% of fraction*	*% distribution PCP*
+45	89.5	28.5
−45 to +4.5	4.0	6.0
−4.5	6.5	65.5

Table 22.6 Example of distribution of inorganic contaminants as a function of grain size

size fraction	*wt% fraction*	*As*	*wt% dist.*	*Cd*	*wt% dist.*	*Fe*	*wt% dist.*	*Pb*	*wt% dist.*
+30 Mesh	8.9	800	6	60	7.2	27 000	9.7	500	6.7
−30 to +325	41.5	340	11.7	33	17.8	8 900	14.4	220	13.3
−325 to +20μ	12.1	670	6.9	54	8.7	12 000	5.8	680	12.3
−20 to +10	7.3	1100	6.8	110	10.7	18 000	5.2	1300	14.1
−10 to +5	4.2	2000	7.1	160	9.0	34 000	5.7	1500	9.4
−5	27.0	2700	61.5	130	46.7	55 000	59.2	1100	44.2

Table 22.7 Distribution of PCBs in soil fractions

size fraction (μm)	*wt% of fraction*	*% distribution of PCB*
+44 (coarse silt)	48.6	26.2
+20 to −44	40.8	55.0
+10 to −20	7.5	12.2
−10	3.1	6.6

compounds. For the larger sized particles, contaminant sorption may be partially explained by any or all of the following mechanisms:

1. sorption of cations by iron or manganese oxides coating individual mineral grains;
2. sorption of organic contaminants by iron and/or manganese oxides which have previously sorbed naturally occurring humates;
3. sorption of organic contaminants by naturally occurring humic acids which are coating individual mineral grains;
4. flocculation/precipitation; and
5. combination of any or all of the above.

Removal of contaminants from these larger sized particles is thus a prime requirement of soil washing techniques. For example, the

MTARRI process utilizes an aqueous system with froth flotation. Generally, higher removal rates of organic contaminants have been observed in a basic system as opposed to a neutral or acid system. One possible explanation would be the increased solubility of humic acids in higher pH systems, thus breaking the humate–contaminant bond and/or solubilizing the humate and contaminant off the mineral grain.

22.6 APPLICATION OF SOIL WASH PROCESSES

As previously stated, the biggest problem with application of soil washing processes is the diversity of the contaminants combined with the diversity of the mineralogy of the soil. Thus, prudent application of any soil washing system will generally be in four stages: bench-scale testing; process engineering; field pilot testing; and full-scale operation.

Phase I – bench-scale testing

Objective:
To conduct bench-scale testing on 100–1000 gm of contaminated soil to achieve the maximum removal of contaminants and/or meet cleanup levels while simultaneously producing the minimum quantities of reject material – containing the contaminant – which requires final disposal by incineration or burial.

Time:
2–12 weeks

Phase II – process engineering

Objective:
Based on the technical success of the bench-scale testing, conduct process engineering/optimization studies and tests to identify changes or modifications required by a particular soil wash process to treat a mineralogically unique soil.

Time:
1–24 weeks

Phase III – field pilot testing

Objective:
For larger size cleanups, e.g., >10 000 m^3, conduct pilot testing on site to determine operating parameters, final cleanup costs, and establish that cleanup levels can be achieved in an environmentally safe manner,

e.g., with minimal air emissions, spill control, process water treatment, and at an acceptable price.

Time:
1–4 weeks

Phase IV – full-scale operation

Objective:
Conduct the full-scale remediation and replace the cleaned soil back into the original excavation site.

Time:
1–4 years

The above four phases would typically be applied to projects having greater than 10 000 m^3 and with diverse contaminants present. However, for smaller size projects where the contaminants can be easily removed, e.g., volatile organic contaminants, the bench-scale testing, and process engineering steps should require minimal time and cost and a pilot test may not be required. This assumes the soil washing process to be applied has been sufficiently field tested so general operating parameters are already defined.

Europeans in Germany and the Netherlands have installed fixed soil washing facilities. Thus, no pilot test is needed and minimal process engineering testing/design is required for fixed time facilities since their operating parameters have been well defined.

22.7 DESCRIPTION OF PROCESSES

22.7.1 Operations using sizing, high pressure water washing and additives

EPA mobile extraction system

Process description
The USEPA has conducted both bench-scale and field testing of their soil washing process. Generally, their soil washing process relied on screening and removal of the >25 mm sized material followed by rotary drum washing of the <25 mm material. The <2 mm size fraction was subjected to counter current washing with water and in some cases additives. If cleaned, the <2 mm was filterpressed and deemed capable of being returned to the excavation site; if contaminant removal was insufficient, the soil was disposed of by landfilling and/or incineration. The >2 mm soil particles were subjected to high pressure water wash-

ing in the rotary drum with a 'water knife' and then discharged as clean.

Waste streams are process water and generally the <2 mm sized fractions. Process water was passed through activated carbon prior to discharge, whereas the <2 mm fraction generally had to be subjected to subsequent treatment, e.g., solidification or incineration.

Application/results

Testing was conducted on both organic and inorganic contaminated soils. According to a paper published by Scholz and Milanowski (1983), for large particles (2–12.7 mm) removal rates were: phenol – 99.1%, As_2O_3 – 52.1%, and PCB – 28.6% after 120 minutes of washing. For the <2 mm sized fractions, the counter current extraction produced removal rates of: phenol – 77.8–98.6%, As_2O_3 – 85% (at pH 1), and PCB – 37.5% (using Tween 80 surfactant).

Advantages/disadvantages

High processing rates can be achieved by using the rotary drum with high pressure washing. Unfortunately, residual contaminant levels in both the coarse and fine-grained size fractions appear to be relatively high and thus may not meet cleanup standards for many sites. The system was designed to be mobile and occupied three standard size flatbed trailers.

Stage of development

Bench-scale tests and pilot-scale field operations have been conducted. No full-scale cleanup was attempted.

Heijmans Millieutchniek BV (Nunno et al.*, 1989)*

Process description

This soil washing process is based primarily on particle sizing under the assumption the majority of the contaminants are concentrated in the finer fractions. All <63 µm sized particles are screened out and rejected, whereas the >63 µm particles are washed with detergents and oxidants, the contaminants removed by precipitation and/or flocculation.

Areas of application

Cyanides, heavy metals, mineral oil, and kerosine where the fines (<63 µm) are less than 20–30% by weight.

Results

Table 22.8 shows contaminant removal rates as reported by Nunno *et al.* (1989).

Table 22.8 Soil washing results for the Heijmans Millieutchniek BV process

soil type	*contaminant*	*before (mg/kg)*	*after (mg/kg)*
Silt/sand	total CN	250–500	10–15
	Cr	43–45	11–15
	Ni	250–890	40–70
	Zn	460–720	140–200
Coarse sand	kerosine	5000–7000	80–120
Coarse sand	PNA	250–400	0.5–10
Silt/fine sand	mineral oil	3000–8000	90–120

Source: Nunno *et al.* (1989).

Advantages/disadvantages
May have a high reject rate for finer-grained soils. Since the reject material must be further treated or disposed of, treatment costs per total tonne treated will far exceed the reported operating costs.

Stage of development
Operating full-scale as a fixed-base facility.

HWZ Bodemsanering, Amersfont, Netherlands

Process description
According to Nunno *et al.* (1989) the HWZ process is very similar to the Heijmans process described above.

Areas of application
As described above.

Results
See Table 22.9.

Advantages/disadvantages
Treatment of fine-grained soils appears to be a major problem; also, removal rates for both organic and inorganic contaminants may not be high enough to meet US standards.

Stage of development
Operating full-scale.

Harbaurer GmbH. Berlin, West Germany

Process description
Although this process also uses particle sizing, it rejects only the <15 μm clay-sized particles. Reportedly (Nunno *et al.*, 1989), low

Table 22.9 Soil washing results for HWZ process

contaminant	*before (mg/kg)*	*after (mg/kg)*
PNA	100–150	15–20
CN	100–200	5–15
Chlorinated hydrocarbons	20–30	<1
Pb	300	75–125

Table 22.10 Soil wash results Harbaurer process

contaminant	*before (mg/kg)*	*after (mg/kg)*
Total organics	5403	201
Phenol	115	7
Chlorinated organics	90.3	ND
PCBs	3.2	0.5

frequency vibration with detergents and/or other extractants significantly aids in desorption of contaminants from the remaining finer-grained particles. Water is treated by flotation, air stripping, ion exchange, and finally activated carbon.

Areas of application
Organic contaminants, phenols, PNA, PCBs.

Results
See Table 22.10.

Advantages/disadvantages
Lower overall operating costs are achieved by only rejecting the <15 μm material. This soil washing process was the only one reported by Nunno *et al.* (1989) that showed PCB removal rates, and hence may be applicable for PCB contaminated soils.

Stage of development
Full-scale operation at a fixed facility.

22.7.2 Operations using water washing and froth flotation

MTARRI process

Description
Solid feed from the site of excavation is classified by a screening, crushing or other means, and then fed to a mixing tank. Typically, all

Table 22.11 Waste-Tech bench-scale test results

contaminant	*before (mg/kg)*	*after (mg/kg)*
Gasoline	54 400	29
Benzene + toluene	38	ND
Chlorinated hydrocarbons e.g., TCE, PCE	40–100	ND
PNA	6–170	0.04–1.0
PCB	470	6.1
Cr	2 590	104
Zn	3 540	60
Hg	157	<0.8
As	12 100	<5.4

fines are retained. At the mixing tank chemical reagents of caustic and possibly surfactants are added together with recycled water. The slurry, being approximately 25% solids is then pumped to the froth flotation cells.

Residence time in the flotation cells is dependent on the nature of the contaminant and the soil type. At the flotation cells additional reagents are added as needed and the froth, containing the organic and a portion of the inorganic contaminants, is skimmed off. Solids progress continuously through the system with residence time per cell adjusted by means of weir gates. Solids are then fed to a thickener and if desired a belt filter press. Separated water is recycled to the mixing tank with bleed off water amenable to standard water treatment processes. If the metal content is sufficiently high, a subsequent counter current decantation (CCD) process can be employed to remove the residual inorganic contaminants.

Applications

The process has been bench-scale tested on sludges, creosote contaminated soils, gasoline and diesel contaminated soils, and soils having both volatile and semi-volatile contaminants, PCBs, As, Cr, and Zn. Recent field testing of a mobile 50-tonne-per-day soil washing unit yielded removal rates from kerosene-contaminated soils that exceeded anticipated results from bench-scale tests.

Results

See Table 22.11.

Advantages/disadvantages

By retaining most of the feed material and thus minimizing rejecting the fines, overall operating costs can be significantly reduced. If the

Table 22.12 Heidemij Uitvoering operating results

contaminant	*before (mg/kg)*	*after (mg/kg)*
PNA	19	0.34
Chlorinated hydrocarbons	5.3	0.4
Pesticides	650	14.4
Oils, hydrocarbons	>1 000	65
Toluene, benzene	3 000–18 000	20
Hg	11 900	110
As	67	1.5

cleanup levels achieved during bench-scale testing can be consistently achieved in the field, many of the US cleanup standards can be met. Disposal of the contaminant-rich froth equal to 5–15% of the original feed weight will be the other major expense. Typically, process water is recycled, thus disposal costs for water treatment is relatively low. Flotation cells were purchased with covers thus allowing air emission standards to be met. Treatment of high clay soils has been successfully demonstrated at the bench-scale.

Stage of development
Bench-scale testing and minor pilot-scale testing.

Heidemij Uitvoering, BV's Hertogenbosch, The Netherlands

Process description
Soils are initially screened to remove the <50 μm size fraction and hopefully a significant amount of contaminants. After screening, froth flotation is utilized to achieve separation and concentration of the contaminants from the mineral particles. The slurry is then settled, and washed according to Nunno *et al.* (1989).

Areas of application
CN, heavy metals, oils, pesticides. The system was not applicable to PCB removal.

Results
See Table 22.12.

Advantages/disadvantages
As with other froth flotation soil washing systems, quite high contaminant removal rates may be achieved. The main disadvantage is the apparent lack of certain PCB and PNA removal capabilities.

Table 22.13 Soil washing results for Biotrol process (bench scale)

soil type	*contaminant*	*before (mg/kg)*	*after (mg/kg)*
Silty sands	Pentachlorophenol	218	22
	Phenanthrene	102	1
	Fluoranthrene	16	0.5
	Benzo(a)pyrene	2.7	0.002
	Pyrene	16.7	0.36

Stage of development
Field operational.

Biotrol

Process description
Biotrol screens the feed soils to obtain a feed for soil washing. Screening rejects >12 mm material and the clay-sized material. The >12 mm material is treated by aqueous scrubbing, whereas the clay fraction is rejected to be disposed of by incineration or other treatment process. The −12 mm down to clay-size material is then soil washed with surfactants in a counter current system using froth flotation. Concentrated contaminants produced in the froth are then also disposed of by further treatment, e.g., incineration.

Areas of application
Pentachlorophenol and creosote-contaminated soils have been treated by this process.

Results
See Table 22.13.

Advantages/disadvantages
May have a high reject rate for finer-grained soils, thus total treatment costs may significantly increase. Cleanup levels appear to be low enough to meet many US standards.

Stage of development
A 250 kg per hour demonstration unit has been built and field tested.

22.7.3 Operations using solvents and/or solvent extraction

BEST process

A unique solvent extraction process (originally developed for organic extraction of a sewage sludge) has been modified, developed, and field tested for hazardous waste. This solvent extraction process uses triethylamine which is soluble in oil and water below 18 °C and only in oil above 54 °C. A two phase separation is then achieved and the triethylamine recovered for recycling by distillation. Solids are separated in a centrifuge and then dried. The process can operate at 100–1000 tonne per day. A 100 tonne per day skid-mounted plant is transportable to the site with four truck loads.

Applications
This process is amenable to separator sludges, other toxic organic sludges, and wood treating sludges from creosote. Additionally, the process has been tested on PCB-contaminated soils and sediments.

Results
According to Best, the process will remove $99^{+}\%$ of the organic contaminants and has been bench scale tested to remove 99% of PCB oil from a soil. Field testing has been completed on refinery sludges and the processed soils passed required leachability tests.

Advantages/disadvantages
The process is mobile but resembles a mini-refinery and therefore has limited mobility. However, for large remediation projects, e.g., 20 000 m^3, the system may be applicable. No metal removal is achieved, therefore requiring possible further treatment.

Stage of development
A mobile, field-operational unit has been employed in cleanups.

CF systems

Removal of unwanted chemical compounds by extraction with critical fluids is commonly practiced in the food and chemical industry. For example, the removal of caffeine from coffee beans is currently accomplished by using carbon dioxide as a critical fluid. This technology has recently been transferred to the hazardous waste business.

Critical fluids commonly employed in hazardous waste treatment are CO_2, propane or other light hydrocarbon gases. Their advantages are low cost, low toxicity and ease of separation from the extracted contaminants.

Description
Sludges are slurried to a conditioner tank where a homogeneous slurry mixture is produced and then fed to the solvent extraction unit. Recycled solvent also enters the extractor. From the extractor the slurry is pumped to a separator where the contaminant and critical fluid is

initially separated from the solids and water. A second solid/liquid separator, involving settling and filtration, produces a solids stream and a water stream. Solvent can be recovered from the water at this stage according to CFS. Solvents and organic contaminants are fed to a solvent recovery still where the solvents are recovered, compressed and recycled to the extractor with the heat from the recompression used to vaporize the solvent. The organic contaminant is then recovered.

Areas of application

The process has apparently been tested on separator sludges and oil contaminated soil. Presumably the technique would be applicable to both volatile and semi-volatile contaminants.

Results

According to CFS literature, removal efficiencies of organic contaminants from the solids range from 87–98% (as a percentage of feed). However, the metal concentrations (as a percentage of feed) were found to increase due to weight removal of water and organic contaminant. Purgeable organics (benzene, toluene, xylene) showed a 95–99^{+}% removal and the solids passed the TCLP test. Semi-volatiles such as naphthalene and phenanthrene showed a 99^{+}% removal and also passed the TCLP test. Recent testing on PCB-contaminated soils was reportedly encouraging.

Advantages/disadvantages

High contaminant removal rates may meet US cleanup level requirements for many sites. This technology may be especially suitable for emulsion treatment.

Results

As with the BEST system, removal of inorganic contaminants is not generally accomplished simultaneously with removal of the organic contaminants, hence a second step to remove metal (inorganic) contaminants is required, such as fixation.

Stage of development

A pilot-scale unit is currently being field tested.

ENSR process

Description

Soil is sized to <12 mm and then fed to a mixer where solvent and other chemical reagents are added to form a 33–50% (by weight) slurry. This slurry is pumped to a series of 3–5 counter current extrac-

Table 22.14 ENSR bench scale results

contaminant	*concentration before treatment (mg/kg)*	*concentration after treatment (mg/kg)*
PCB	1944	1
PCB	3300	non-detectable
PCB	1147	22

tion cells to achieve contaminant removal. Cleaned soil is dried and residual solvent removed. The contaminated solvent is distilled to recover and recycle the solvent, whereas the still bottoms – rich in contaminant – are sent for off-site disposal/destruction.

Application

Applicable to most organic contaminants and PCBs.

Results

See Table 22.14.

Advantages/disadvantages

High moisture content in soils may affect processing rates and/or cleanup levels attainable. However, that problem could be solved by predrying of the soil feed. Cleanup levels for PCBs are well within the standards set by the USEPA and therefore this process may be both technically and economically viable for PCB removal. As with other solvent extraction processes, the method does not address metal contaminant removal if a mixed organic/inorganic waste is present.

Stage of development

Bench-scale.

REFERENCES

Esposito, P. (1988) Characterization of RCRA/CERCLA sites with contaminated soil, presented at EPA Workshop on Extractive Treatment of Excavated Soil, Dec 1–2, Edison, NJ.

Nunno, T.J., Hyman, J.A. and Pheiffer, T. (1989) Part 7: European approaches to site remediation, *HMC Magazine*, Sept/Oct, 38–46.

Scholz, R. and Milanowski, J. (1983) *Mobile system for extracting spilled hazardous materials from excavated soils*. EPA-600/S2-83-100, US Environmental Protection Agency.

PART FOUR

Monitoring

CHAPTER 23

Monitoring wells

Bob Kent and Mark P. Hemingway

23.1 INTRODUCTION

23.1.1 Background

The installation of ground-water monitoring wells has developed over the past decade into a highly specialized activity, quite distinct from the fields of water supply well and geotechnical drilling from which it was derived. Stringent regulatory demands, sensitive analytical techniques, and exotic construction materials all require the monitoring well designer to comprehend a broad spectrum of technologies and practices.

The purpose of a monitoring well is, ultimately, to obtain a representative sample or measurement from the ground water within a particular hydrogeologic stratum or group of strata. This sample or measurement may be used to ascertain the presence or absence of introduced contaminants, as well as the concentrations of particular constituents. While this objective is conceptually simple, difficulty arises from the fact that, as with Heisenberg's electron, the act of observing alters that which is being observed.

The installation of a well is a process which is traumatic from the perspective of the aquifer. Traces of hydrocarbon fuels and lubricants may be introduced into the subsurface from the drilling rig hydraulics, packings, and exhaust, even when no leaks from these sources are visible. Drilling fluids mix with and alter the natural waters in the subsurface, and also produce a cake of foreign material on the borehole wall. Fragments of metal from the drill bit or auger flights are eroded and remain in the borehole. Any materials used to construct the well must interact to some degree with the water entering the well from the

Geotechnical Practice for Waste Disposal.
Edited by David E. Daniel.
Published in 1993 by Chapman & Hall, London. ISBN 0 412 35170 6

surrounding soils. Under the existing state of the science and engineering of well installation, these influences are unavoidable – they can be minimized, but not entirely eliminated.

From a realistic standpoint, then, the question is not whether the act of ground-water monitoring will affect the ground water, but rather the permissible extent of these effects. Given this, the objective of any designer of monitoring wells must be to determine the true purpose of the proposed well or well system, and to select practical and economical installation practices and protocols which address that purpose. This should include as large a contingency as possible for difficulties and problems which may arise during the installation process.

Design under such a philosophy is more than simply a rote obedience to a set of guidelines. It requires some pragmatic awareness of the potential for problems and inaccuracies in the field practices utilized.

Installing a monitoring well is typically a difficult and expensive process. There may be no second chance to accomplish a given task following an initial failure. Sound monitoring well design should therefore use methods which are more than simply possible; they must have an intrinsically high probability of success. The purpose of this chapter is to acquaint the reader with the practices of monitoring well design, installation, and sampling, with an emphasis on the realities involved in these activities.

23.1.2 Components of a monitoring well

Although well designs will vary widely depending upon purpose and installation setting, most wells have broadly similar configurations and components. The arrangement and nomenclature of these components for an idealized well are illustrated in Fig. 23.1a. The terminology used for particular components is subject to wide personal and regional variation, so the terminology used here may not correspond perfectly with that used by others.

A typical well consists most fundamentally of a length of pipe extending from the ground surface into the water-bearing stratum of interest. This penetration is usually accomplished by means of a borehole, which may be advanced using a variety of drilling methods. Within the targeted stratum, the pipe is perforated in some manner to permit the entry of ground water for sampling. The perforated section of the pipe is termed the well screen, and the unperforated section extending to the surface is termed the riser. Some wells are equipped with a short length of unperforated pipe below the screen called a sediment sump, for the collection of sediment which enters the well. The sump, screen, and riser may be collectively termed the well assembly or well string, as distinguished from the term 'well', which

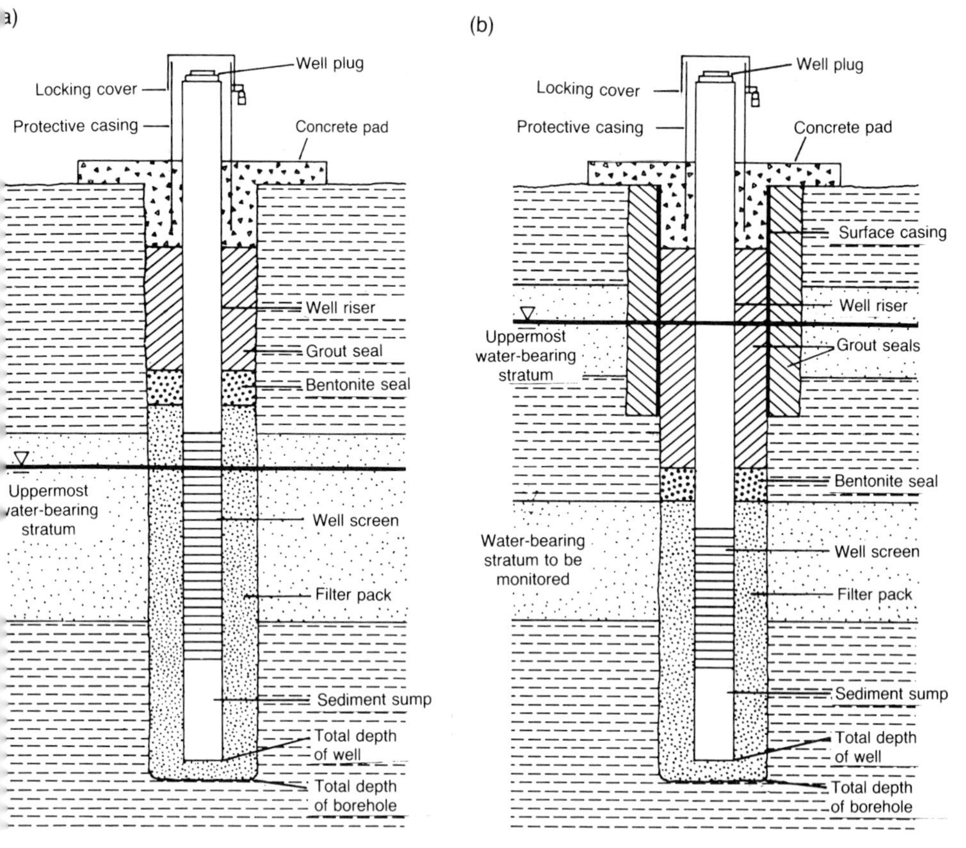

Fig. 23.1 Components of an idealized monitoring well: (a) single-cased and (b) multiple-cased.

includes both the well assembly, the borehole, and annular materials, as well as surficial structures.

The annulus surrounding the screen is typically backfilled with granular material, such as silica sand, to form a filter pack. The primary purposes of this filter pack are to minimize the amount of fine sediment which enters the well with the ground-water, and to minimize collapse of the borehole surrounding the screen. The annulus between the filter pack and the surface is sealed using a variety of low permeability materials, such as bentonite clay and Portland cement. This reduces the entry of surficial waters or waters from non-targeted water-bearing strata into the well screen.

At the surface, the riser may be protected against vandalism and vehicular damage by a metal protective casing or a manhole. A surface seal of concrete is typically placed around this protective structure to prevent surface drainage into the borehole.

For wells with multiple casings (illustrated in Fig. 23.1b), the well installation process is preceded by the installation of a non-perforated pipe string, which is sealed in place. This pipe string, termed a casing, allows the isolation of shallow water-bearing zones which are undesirable due to contamination or other factors prior to well installation. After the casing is installed, the borehole for the well is advanced and the well installed through the casing. If only one casing is utilized, it is typically termed the surface casing.

23.2 MONITORING WELL PLACEMENT

Most guidance suggests a minimum of four monitoring wells per potential source of ground-water contaminants. Three of these wells would be placed downgradient of the potential source, and one would be placed upgradient (USEPA, 1986). The upgradient well permits the assessment of the background character of the site's ground water, and the downgradient wells permit the detection of any contaminant plumes emanating from the source. This configuration would be the simplest possible, and may be adequate where hydrologic conditions and contaminant movement are very straightforward.

The proper placement of wells at most sites cannot be dealt with in such a simple manner. The depth to water, characteristics of flow, and pattern of aquifers and aquitards in both the saturated and unsaturated zone all affect the movement of contaminants. Aquifers of high hydraulic conductivity, for example, tend to promote the formation of narrow, elongated contaminant plumes, while broader plumes tend to develop within aquifers of lower hydraulic conductivity. The optimal placement of monitoring wells requires, therefore, both a basic understanding of the site hydrogeology, and of the possible migration paths of the suspected contaminant plume. Information on the latter topic is provided in examinations of the movement of immiscible and dissolved phase constituents through the subsurface performed by Schwille (1984, 1989), Mackay *et al.* (1985), Hinchee and Reisinger (1987), and Mercer and Cohen (1990).

One example of the influence of site-specific hydrologic conditions on the movement of contaminants was documented by Haug *et al.* (1989). The subject site was underlain predominantly by low hydraulic conductivity soils, but also included several narrow, elongated sand bodies. These sand strata behaved as preferential pathways for contaminant movement due to their higher hydraulic conductivity, but could have easily been missed by a cursory subsurface investigation due to their low width. An effective ground-water monitoring program at the site had to account for the presence of these features.

A second example would result from the presence of a low hydraulic

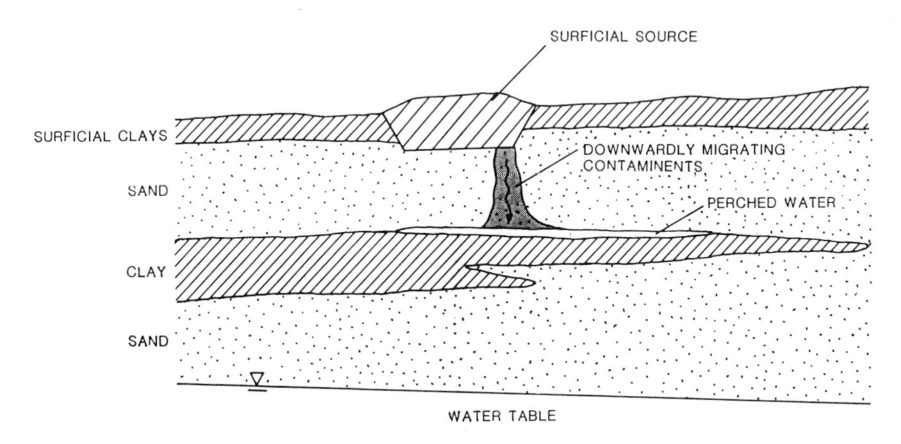

Fig. 23.2 Conceptual view of the potential effect of a shallow low hydraulic conductivity layer on the downward migration of contaminants through the unsaturated zone. Impact to the water table aquifer directly below the contaminant source is diverted to another area.

conductivity layer above the water table. Such a layer could alter the downward movement of contaminants from a surficial source (Fig. 23.2). Perched water above this confining layer might experience substantial impact, with little or no contamination detectable in the ground water from below the water table.

Given the number of factors which can play a controlling role in the movement of hydrocarbons through the subsurface, the selection of proper well locations can be a complex process, especially for sites within complicated hydrogeologic settings. For such a setting, many of the issues affecting well location cannot be resolved until a part of the investigative process is complete. For such settings, a phased approach to subsurface investigations is most appropriate. Such an approach requires an initial well installation phase, which establishes the basic hydrogeologic parameters of the site. The data from this phase is then utilized to select additional well locations, and to refine the monitoring well design.

23.3 MONITORING WELL DESIGN

23.3.1 Monitoring well materials

Monitoring well riser and screen are available in a host of materials. These include stainless and carbon steel, fluoropolymers such as polytetrafluoroethene (PTFE), fiberglass, thermoplastics such as polyvinyl chloride (PVC) and polypropylene, and aluminum. The relative advantages of these materials are summarized in Table 23.1. PVC is perhaps the most commonly used monitoring well material, due to its economy, strength, availability, and general inertness.

Most of these materials utilize threaded connections, or joints, to join the lengths of pipe into a continuous well assembly. This is a departure from early monitoring well installation practices, when wells were often constructed of glued-joint PVC. This practice was almost universally discontinued when analyses from several investigations indicated that the organic constituents of the cement used were bleeding into the water in the well, diminishing the usefulness of the wells (Fig. 23.3). Although glued-joint PVC may still be appropriate for use when monitoring for constituents such as metals or inorganic salts, the early problems with this type of joint have created an aversion to its use.

The thread design on most brands of riser and screen varies considerably from manufacturer to manufacturer. As a result, one brand of monitoring well pipe is typically not compatible with another. The quality of the threading and pipe is also highly variable from one manufacturer to another, which reduces the value of many low cost brands. The economy disappears when the joints fit together poorly, resulting in a difficult installation and possible losses is well integrity. The designer should use caution, therefore, when purchasing from an unfamiliar manufacturer.

Two factors should govern the material of construction used. These are material strength and chemical reactivity. The first parameter, structural strength, is of greatest significance when designing deep monitoring wells (>30 m). Under such circumstances, stresses upon the well material become considerable. The crush and bursting resistance, tensile strength, and joint strength of the material selected must all be sufficient to endure these stresses, or well failure will occur during or after installation. The importance of material strength became obvious when PTFE well materials first appeared. The low tensile strength and low coefficient of friction led to failures of the threaded joints, as threads flexed and slid across one another during installation.

A more common structural problem is faced when using PVC riser on deep (>30 m) wells in conjunction with annular seals consisting of cement or bentonite cement grouts. The hydration of the Portland cement in these grouts can generate considerable heat, which can soften the PVC, reducing its crush resistance. This, in turn can permit deformation of the well riser from the differential pressure on the outside riser. To moderate both the heat generation and the differential pressure, cement grouts should be emplaced around PVC riser in increments of approximately 20–25 m, allowing each increment to cure before emplacing the next. When necessary, the riser may also be kept full of potable water, which absorbs heat and helps balance the

Table 23.1 Relative advantage of the commonly utilized monitoring well pipe construction materials

material	*polyvinyl chloride*	*stainless steel*	*carbon steel*	*polytetrafluoroethylene*
Chemical resistance	Generally good, susceptible to attack by some separate-phase hydrocarbons, especially ketones, esters, and aromatics	Very good	Very good, except for corrosion	Very good, susceptible to attack only by certain exotic fluorinated compounds
Inertness	Moderately good, some sorption effects	Very good	Moderately good, although corrosion enhances sorption effects	Moderately good, some sorption effects
Corrosion resistance	Very good	Very good, although may corrode and leach chromium in highly acidic waters	Susceptible to attack in aggressive soils and waters	Very good
Strength	Weaker, less rigid than steel materials, soft at high temperatures, brittle at low	High strength at all temperatures	High strength at all temperatures	High impact strength, but low tensile strength and wear resistance compared to other plastics used for well materials. Screen slots may compress closed with time
Cost	Low	Moderate for riser High for screens	Low to moderate	High
Other	Quality highly variable Availability good	Much heavier than plastics	Much heavier than plastics	Poor availability

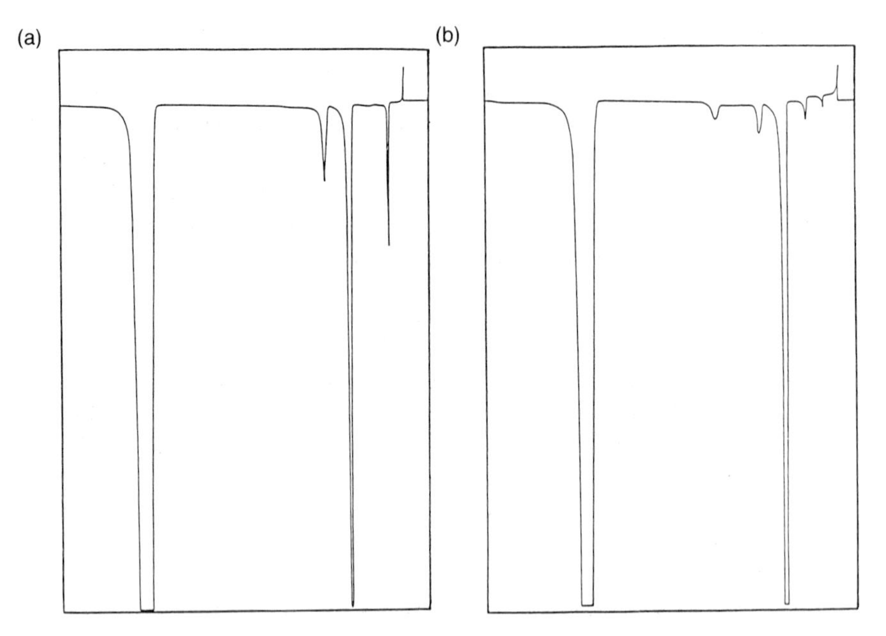

Fig. 23.3 Gas chromatographic scans of water samples from: (a) groundwater from a monitoring well constructed from glued-joint PVC pipe; and (b) distilled water in which a sample of glued PVC was immersed for several hours. The major peaks on the scans are 2-butanone, toluene, and tetrahydrofuran.

pressure differential. The disadvantage of the latter technique is that the water lost to the aquifer through the well must typically be recovered, lengthening the development effort.

The second characteristic of importance, chemical reactivity, reflects the potential for interaction between the materials of well construction and the ground water, natural dissolved constituents, and dissolved and separate-phase contaminants. Specific concerns are:

- alteration of natural or contaminant chemistry by the well materials, modifying the results of ground-water analyses;
- degradation of the well materials by the ground-water, such as accelerated corrosion of carbon steels under aggressive ground-water conditions; and
- degradation of the well materials by free-phase or high concentration dissolved phase organic constituents, such as the dissolution of PVC by some organic liquids.

The first of these concerns, the alteration of ground-water sample chemistry by the well material, typically receives the greatest attention. A study by Parker *et al.* (1990) indicates that, for organic contaminants, stainless steel affected constituent concentration the least. Interestingly, PVC performed better in this regard than PTFE. The critical point, however, is that PVC did not modify the concentrations of any of the tested constituents by more than 8% in a 24 hour period. Since the first step in properly sampling ground water is the purging of the well, the sampled water should be brought into first contact with the well materials shortly prior to sampling. While the sorption/desorption of constituents in wells within actual aquifers is substantially more complex than the tested laboratory microcosms, these findings do suggest PVC is adequate for ground-water monitoring purposes.

23.3.2 Monitoring well diameter

Although well materials exist in a wide array of sizes, from 10 mm to in excess of 600 mm in diameter, practicality and standardization of practice have narrowed the selection for most monitoring wells to either 50-mm or 100-mm nominal diameters. Larger sizes are less frequently used, since the increased production resulting from their greater open surface screened area is not usually of critical importance for ground-water sampling.

Narrowing the selection to these two sizes, however, still leaves the monitoring system designer with a series of decisions. The selection of a 50-mm diameter well lowers material costs and increases the ease of well installation, but the lower open surface area and smaller size render well development in poor aquifers difficult relative to the development of 100-mm diameter wells. Gass (1988) discusses the preferability of 100-mm wells in fine-grained strata to minimize sample turbidity. In addition, if a pump is to be used for well development, purging, sampling, or for conversion of the well from monitoring to ground-water recovery, the use of 50-mm wells restricts the pump options available. Very few types of electrical submersibles are sufficiently small to be used in wells of that diameter.

Well string materials (i.e., riser and screen) for 100-mm wells, on the other hand, are typically 2–4 times more expensive per unit length than 50-mm diameter materials. If relatively costly materials of construction such as PTFE or stainless steel are required, this factor can substantially inflate the cost of installation. In addition, for hollow-stem auger drilling and installation, 100-mm well strings require the use of an oversized auger for through-the-stem installation. As will be discussed later, the augers most commonly used in the environmental drilling industry have stem inner diameters (IDs) of approximately 75–100 mm. The installation of a 100-mm diameter well (usually 113-mm outer diameter (OD)) requires the use of an auger with a stem ID of at least 165 mm. Both drilling and sampling are more difficult with these larger augers, slowing production and further increasing installation costs.

For most purposes, however, the selection of 100-mm diameter well materials accrues advantages that offset these increased difficulties and costs. The larger surface area of these materials permits much more rapid and effective well development, especially in poor aquifers such as silty sands, silts, and clays, than with the use of 50-mm diameter wells. If dedicated electrical submersible pumps are required, readily available and economical units can be utilized, rather than specialized units constructed for the smaller diameter wells.

23.3.3 Screen length and placement

The screen length of a monitoring well should be governed by two factors: the thickness of the stratum of interest, and the potential vertical distribution of contaminants within that stratum. It is not reasonable, in most cases, to expect any uniformity in this vertical distribution, given the number of factors which influence the vertical movement of contaminants. A dense, high total dissolved solids (TDS) leachate plume may sink readily through an aquifer (Fig. 23.4a). In contrast, dissolved phase plumes in high flow velocity aquifers may stream along the upper surface for hundreds of meters with little vertical mixing (23.4b). Inhomogeneities in hydraulic conductivity caused by horizontal stratification could also restrict downward movement of contaminants. As with well placement, some degree of predictive capability may be necessary to properly place monitoring well screens.

After assessing these factors, the monitoring system designer must select a screen length which will minimize the vertical extent of saturated soils which will be monitored. Many hydrogeologic settings may be addressed by screening the entire extent of saturation within the targeted stratum. Prime examples include:

- the targeted stratum is relatively thin (<3 m);
- there is reason to suspect a generally uniform distribution of contaminants within the stratum; and
- the purpose of the well is to determine presence versus absence of contaminants, rather than precise concentrations.

The down side of this method, and the fact which restricts its application, is an averaging effect which occurs during sampling. The mixture of water from uncontaminated portions of the aquifer with impacted water will dilute constituent concentrations in samples from the monitoring well. This effect will be enhanced if the more highly productive (in terms of yield) portions of the targeted stratum are not contaminated. The dilution in such cases could entirely mask low levels of contamination.

One commonly utilized alternative to screening the entire targeted stratum is the use of monitoring well clusters. These consist of two or

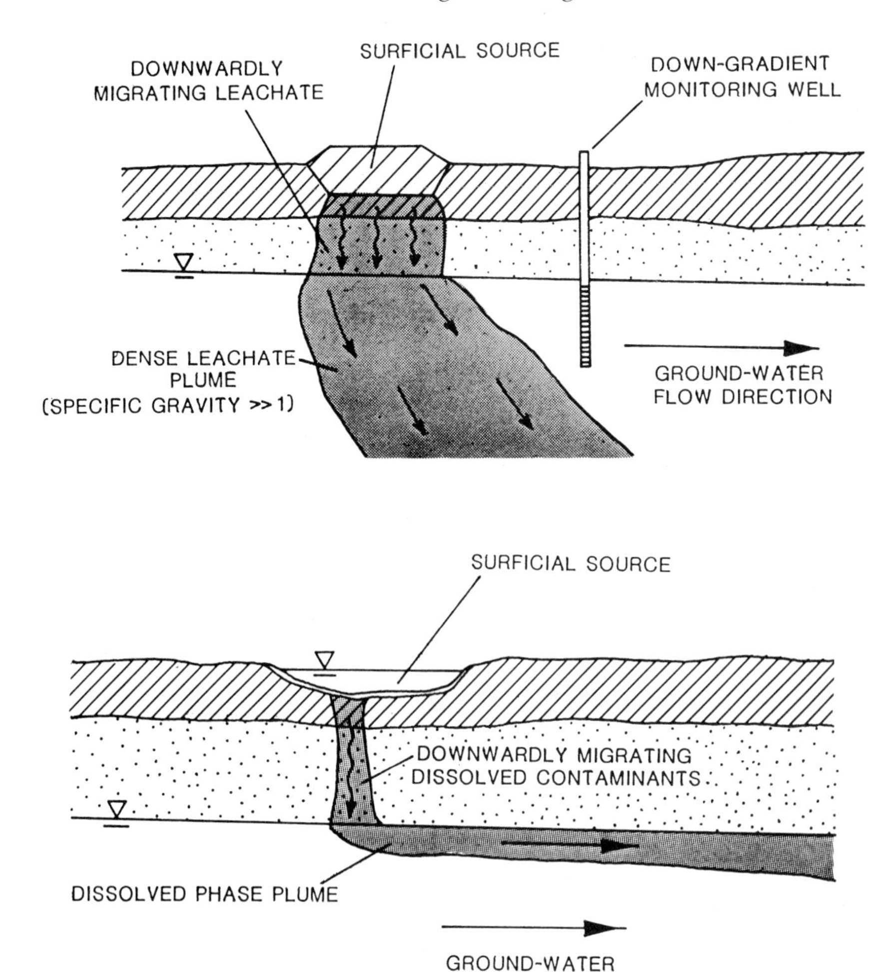

Fig. 23.4 Conceptual view of the potential movement: (a) of a dense, high-TDS plume through the saturated zone; and (b) of a dissolved-phase hydrocarbon plume in an aquifer of high ground-water flow velocity aquifer. Note that in (a) the illustrated downgradient well would not detect the presence of the plume (scenario based on an actual occurrence).

more wells at a single location, with screened intervals set at staggered depths (Fig. 23.5). This permits the assessment of relatively discrete portions of the targeted stratum during sampling. Although this type of installation has been performed with multiple well assemblies within a single borehole, the potential for a poor installation using this method is high. Multiple boreholes, with single well assemblies within each are currently more commonly utilized for cluster installation. The

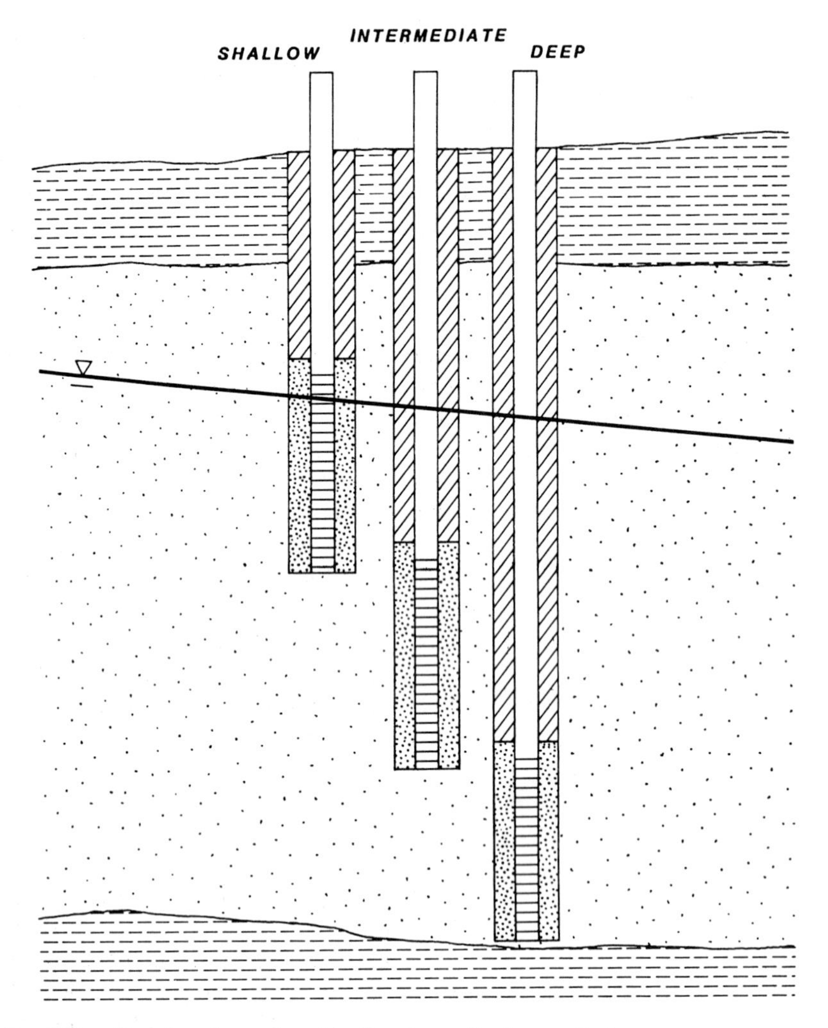

Fig. 23.5 Construction of an idealized monitoring well cluster.

primary disadvantages to the use of monitoring well clusters include the following:

- increased installation, sampling, and analytical costs, due to the increased number of wells and samples necessary; and
- for some locations within crowded facilities, it may be difficult to place multiple wells within a single locale.

If the potential detection ot free phase hydrocarbons is one purpose of the monitoring well, this will also affect the placement of the well screen. As discussed in section 23.2, free-phase hydrocarbons tend to accumulate at the upper surface of the capillary fringe. This is true both for those hydrocarbons with specific gravities less than that of water (referred to as light non-aqueous phase hydrocarbons (LNAPLs), or floaters), and, in many cases, for those more dense than water (dense non-aqueous phase hydrocarbons (DNAPLs), or sinkers). A screen placement across the upper surface of the capillary fringe is required for detection of these phases. For cases where DNAPLs have overcome the surface tension effects resisting their entry to water-saturated soils, placement of the screen across the base of the lowest portion of the targeted stratum is appropriate.

Although both these approaches seem relatively straightforward from a conceptual standpoint, the realities involved are not as co-operative. For hydrocarbon phases above the capillary fringe, it is first necessary to locate the upper surface of this fringe. This may not be possible during drilling, especially if mud rotary techniques are utilized, and so must often be estimated from water levels in nearby wells. In addition, once the capillary fringe is located, it is subject to temporal fluctuations related to seasonal water level changes. A well functional for floating hydrocarbon detection in December may therefore be useless for this purpose in June. As with well placement, it is necessary to have some understanding of site hydrologic conditions, and the possible temporal changes in those conditions, before selecting screen placements. As discussed, this may require the use of a phased approach to monitoring well installation.

23.3.4 Screen type

Well screens have been manufactured using a number of methods and opening styles, including torch- and saw-cut slots, louvered slots, down-hole perforated. For monitoring applications, however, two styles have emerged as predominant. The first is the mill-slotted, which consists of a section of pipe with controlled-width slots cut through the pipe wall. These slots are typically oriented perpendicular to the long axis of the pipe, and are cut using banks of precision saws in a milling operation. This type of screen, shown in Fig. 23.6a, is commonly available in PVC, and is also manufactured in stainless steel, polyethylene, PTFE, and other materials.

The second commonly used screen type is continuously-slotted (also referred to as wire-wrapped). This construction consists of a wire with a triangular cross-section, which is wound around a cylindrical base of rods such that an open space of controlled width is maintained between windings (see Fig. 23.6b). As shown in the figure, the triangular wire permits the entry without plugging of sediments too fine for the screen to exclude. These screens are manufactured in stainless steel, PVC, and fiberglass.

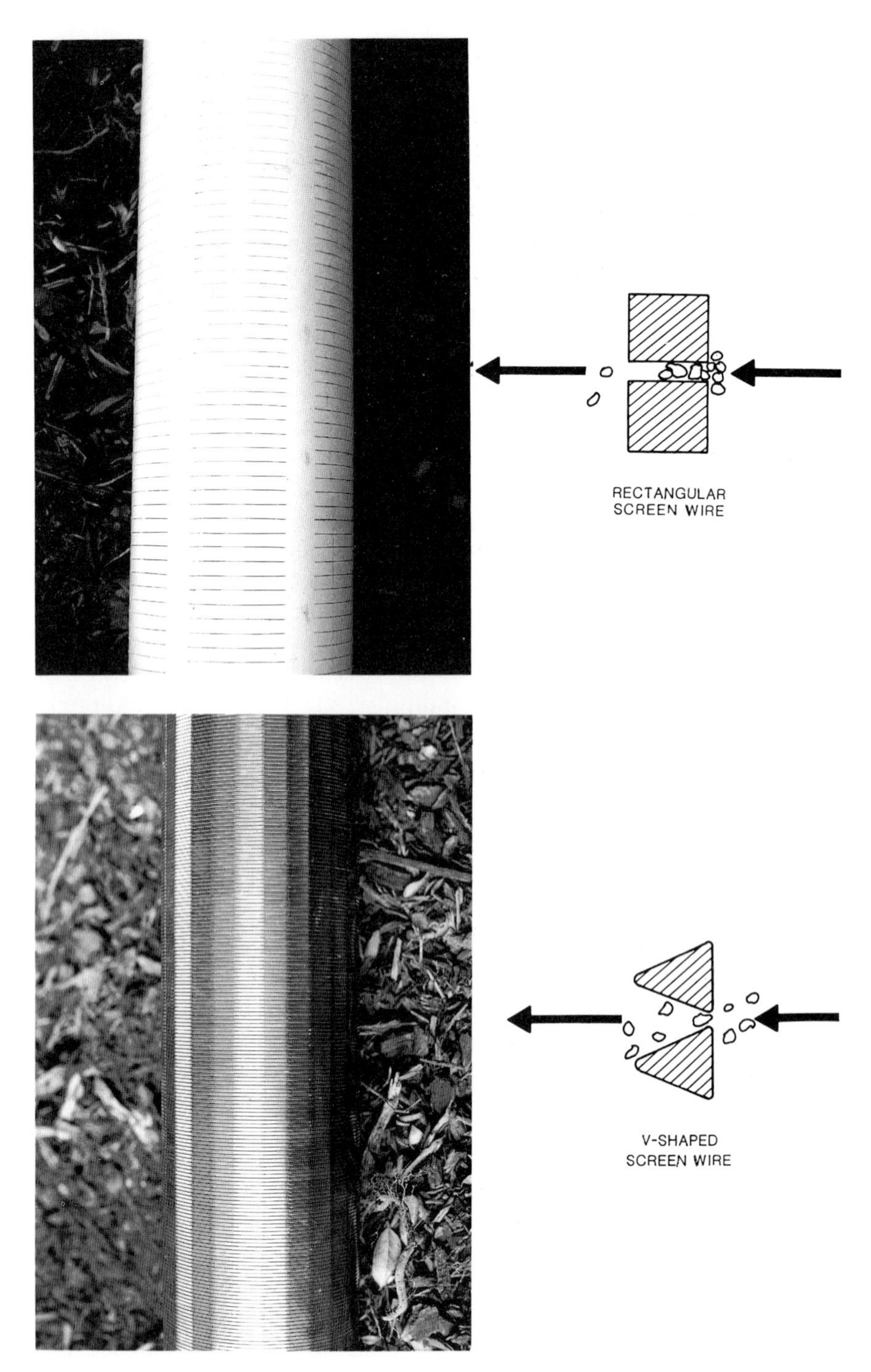

Fig. 23.6 (a) Mill-slotted well screen. (b) Continuously-slotted (wire-wrapped screen). The cross-sectional detail of sediment grains passing through the screen illustrates the effect of the triangular profile of the screen wires. Any sediment fine enough to pass through the screen slot will enter the screen freely without clogging. A screen constructed of rectangular profile wire, in contrast, can be readily plugged by grains of elongate or irregular shape.

Although more costly than the mill-slotted version, continuously-slotted screens are often preferred due to their larger open area per unit length, as well as their high tensile strength. Although the larger surface area is typically considered to enhance well yield, it should be considered that this will only occur when the screen open area is less than that of the surrounding soils or the filter pack. Once the open area of a well screen has reached that of these soils or the pack, the factor limiting flow into the well is the open area of the soils or pack, and not that of the screen. Most poorly graded (uniform) sand filter packs will have open areas in the range of 10–15%. If this can be attained by a mill-slotted screen, the use of a continuously-slotted type may not improve well performance (Clark, 1988). Additionally, if the purpose of the well does not require maximum well yield, mill-slotted screens may be adequate.

The slot width of the screen must be based upon the grain size characteristics of the targeted strata alone, for naturally packed wells, and upon the characteristics of both the stratum and the filter pack, for artificially filter packed wells. These concepts, and their bearing on screen slot size, will be discussed in more detail in section 23.3.6.

23.3.5 Sediment sumps

Many designers are now incorporating a sediment sump, also called a silt trap or sediment trap, into monitoring well constructions. This feature is a modification of a common water well practice, and consists simply of a blank (unperforated) section of pipe placed below the base of the screen (Fig. 23.7). Its purpose is to provide a catchtrap for fine sand and silt which bypasses the filter pack and screen and settles out within the well, a phenomenon known as 'silting up'. This sediment collects within the sump rather than within the screen, and therefore does not reduce the functional screened length of the well. This minimizes the need for periodic cleanouts of the screen to maintain the well's effectiveness.

Sumps may be a useful addition when it is anticipated that a large volume of fine sediment will be entering the well. This would include wells set in silt or clay strata, or installations for which difficulty in properly emplacing an artificial filter pack is anticipated. Gass (1988) recommends the use of a sump in fine-grained aquifers as a means of avoiding the great difficulty in producing low turbidity water from such low-producing strata.

In general, however, sumps are unnecessary except for situations where very turbid waters may be expected for the life of the well. If a filter pack can be successfully emplaced, therefore, a sump may be superfluous. Since a sump adds length to the well string below the

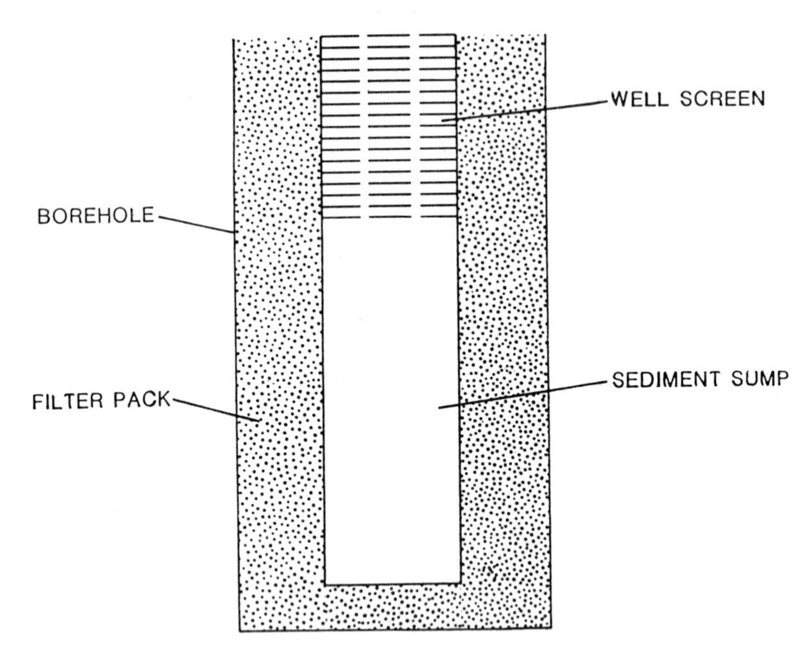

Fig. 23.7 Construction of a sediment sump.

base of the screen, it is necessary to add additional length to the borehole (or 'overdrill') in order to place the screen at the desired interval. For at least one commonly encountered hydrogeologic scenario, where the screen is placed at the base of an aquifer underlain by a thin confining layer, overdrilling for sump placement can potentially breach the lower confining layer.

If a sump is required, lengths of 0.6–1.6 m are adequate for most applications. Since most monitoring wells are pumped no more than four or five times per year, this length should be sufficient for several years' accumulation of sediment. Some wells which have longer screens (typically greater than 6 m) and more frequent usage may require 3-m sump lengths. Sumps on PVC wells may be field-fabricated using a piece of blank pipe sealed at the lower end with a slip-fit cap, which is secured with rivets or screws. Steel sumps are typically sealed at the lower end by a circular plate welded in place. Alternatively, sumps may now be obtained pre-fabricated from many monitoring well materials suppliers.

If a dedicated electric submersible pump is to be placed in a well equipped with a sediment sump, it is best to avoid placing the pump within the sump. This precludes the flow of water around the motor on the base of the pump, which results in insufficient motor cooling and shortened pump life.

23.3.6 Filter packs

Most of the discussion within this section will be pertinent to what are termed artificial filter packs; that is, filter packs comprised of material introduced into the borehole. These comprise the majority of filter packs utilized in monitoring wells today. An alternative worthy of mention, but which is often forgotten or neglected, is the natural filter pack. For this approach, the hydrogeologic unit being screened is permitted to collapse around the well screen. This type of filter pack is not appropriate when this targeted stratum consists largely of fine sand or smaller grain sizes. When this stratum is comprised of sand and gravel, however, with little or no fines to enter the screen and create turbidity, there are distinct advantages to the use of a natural filter pack. First, it eliminates one material which must be introduced into the well, which helps to minimize the impact of the well upon the subsurface environment, as well as the cost and effort of well installation. Second, for mud rotary installations, the collapse of the borehole during natural packing will disrupt the sheath of drilling mud which forms during drilling on the borehole's inner surface. This simplifies and enhances well development.

Although natural filter packs are appropriate in certain circumstances, they do not currently find wide application. The following discussions of filter pack selection and installation will therefore reflect practices related to artificial filter packs.

Gradation

Many monitoring wells are installed in strata which would not normally be considered aquifers; that is, in sand units which have high fines contents and/or abundant layers of fine sediments, or in silt or clay strata. This has become especially prevalent in the wake of regulatory guidelines or requirements which mandate monitoring of the uppermost water-bearing zones beneath the site of investigation. In such cases, a filter pack (less correctly referred to as a sand or gravel pack) is the preferred method of controlling the entry of fine sand, silt, and/or clay into the well screen. As discussed in section 23.1.2, such a filter pack is constructed by backfilling the annular space surrounding the screen with a granular, relatively inert material. In most cases, this consists of a clean, silica sand.

Filter pack selection techniques developed for water supply well applications are often suggested for monitoring well design. The difference in purpose for which these two types of wells are intended makes this practice inappropriate. For water supply wells, the filter pack must be carefully selected and emplaced to minimize resistance to water flow into the well. This may entail the analysis of grain size trends within the targeted stratum during the drilling process, and the selection of a variety of sand sizes for the filter pack. These filter pack sands are then emplaced to match the layering or other grain size variations present within the stratum. This technique was developed to maximize well efficiency and yield in the highly productive sands in which many water supply wells are installed.

For monitoring wells, in contrast, yield is not the primary consideration in well performance. The function of the monitoring well filter pack, therefore, is not identical to that of the water supply well filter pack. Monitoring well filter packs are intended primarily to:

- control the entry of fines into the well, so as to minimize the turbidity of ground-water samples; and
- provide structural support for the borehole in which the well is installed, and for the overlying annular sealing materials.

Neither of these purposes requires perfect compatibility between the grain-size gradation of the filter pack and that of the aquifer.

Despite this, monitoring well specifications or regulations often require the selection of a filter pack based upon detailed grain size analyses of the aquifer. As stated, the purpose of monitoring well filter packs makes such a degree of compatibility unnecessary. In addition, it is not practicable to attain this degree of compatibility, given the nature of the typical monitoring well installation process. Most monitoring well installations are performed as rapid, short-term operations, with materials purchased in advance and ready at the site. This expedites the installation of multiple well systems. The use of water supply well filter pack selection techniques would require a more deliberate approach, with filter pack material purchase delayed until the completion of the monitoring well borehole. This approach would dramatically increase both the cost and difficulty of well installation, without any significant resulting benefit.

It is more efficient in terms of material and installation practice to select a material which broadly suits the general characteristics of the aquifer, and is compatible with the screen slot selection. Several guidelines for this selection process are present in the literature, e.g. Driscoll (1986), and Gass (1988). In contrast to the well-graded filter packs used for water wells, these publications recommend the selection of a relatively uniform monitoring well filter pack, with the uniformity coefficient (C_u) for the filter pack ranging from 1 to 3. The uniformity coefficient, C_u, is equivalent to the ratio of the 60% finer grain size (D_{60}) of the material to the 10% finer grain size (D_{10}). Driscoll recommends selection of a filter pack material with a 50% finer (D_{50}) two times larger than that of the finest portion of surrounding soils. This produces a uniform filter pack only slightly coarser than the finest portion of the surrounding native material. Gass prefers a greater contrast between the natural soils and the filter pack, recommending that the D_{30} of the filter pack be 3–4 times larger than that of the finest portion of the surrounding soils. The slot size of the well screen should then be selected such that no more than 10% of the filter pack material can pass through the slots.

In addition to this portion of the filter pack, some designers recommend the placement of approximately 0.3 m of much finer sand (e.g., 'flour sand') at the upper portion of the pack. This secondary filter pack is intended to supplement the bentonite seal (this seal is discussed in detail in section 23.3.7) in precluding the entry of grout-related fluids into the screened interval of the well.

Filter pack materials and nomenclature

Although 'filter sands' are commercially available, ordinary silica blast sand is generally suitable for monitoring well filter pack construction, provided it is selected with some discrimination. Quality of blast sand may vary considerably from supplier to supplier, but is generally consistent for any given supplier through time. As with other monitoring well materials, then, it is best, when possible, to rely on a familiar supplier who has met specifications consistently in the past.

The main variable between brands is the mineral composition of the sand, since this is naturally dependent on the deposit from which the sand is quarried. It is best to identify one or two brands within a region which are composed of at least 90% quartz grains. This will assure the well installer of a filter pack composition in excess of 95% SiO_2. Such sands contain a minimum of the more soluble metals and other cations present in feldspars and accessory minerals such as micas, magnetite, and hornblende.

In addition to a minimum of potentially unstable minerals, it is important to have confidence in the cleanliness of the filter pack materials prior to their installation. Most sands are water-washed and thoroughly dried prior to sale, minimizing the potential for importing chemical contaminants. If there are specific reasons to suspect the filter pack materials of impurity, it may be advisable to collect a sample of the sand for leachate analysis. In most instances, however, this is probably an excess of caution.

Sands are normally given a product number or name by the supplier or manufacturer based upon their mean grain size and grading (sorting). There is no consistency in the product numbering systems utilized by different producers – they are specific to the company producing the sand. These numbers are, therefore, only minimally descriptive, and their usage should be avoided in well specifications and construction details. Instead, sands should be described on the basis of a mesh number, or sieve number range. This is a number which delineates the

Table 23.2 Opening sizes for US sieves

US standard sieve numbers	*opening size (mm)*	*opening size (microns)*
4	4.699	4699
6	3.327	3327
8	2.362	2362
10	2.000	2000
12	1.651	1651
14	1.400	1400
16	1.180	1180
18	1.000	1000
20	0.850	850
25	0.710	710
30	0.600	600
35	0.500	500
40	0.425	425
50	0.295	295
60	0.250	250
70	0.210	210
80	0.175	175
100	0.147	147
140	0.105	105
200	0.074	74

two US sieve numbers which bound the upper and lower grain size ranges of the blast sand. For example, describing a sand as a 16–40 Mesh indicates that approximately 95 weight percent of that sand falls between US sieve numbers 16 and 40 (see Table 23.2 for a listing of opening sizes for US sieves). These mesh numbers may be obtained by contacting the sand manufacturer and requesting details on the gradation of the subject sand.

Filter pack emplacement

Two methods are commonly used for filter pack emplacement. The most commonly used is to pour the sand into the annular space of the borehole at the surface, allowing it to settle into the space surrounding the screen. The sand is properly poured both slowly and steadily, with frequent soundings of the sand level with a weighted measure. These measures will minimize the possibility of bridging, which is the term used for sand and other materials lodging in the annular space above the depth intended (Fig. 23.8). Bridging can seriously impair proper filter pack emplacement, even to the extent of compromising the well.

This method may be effectively utilized in conjunction with hollow stem auger drilling techniques. This is accomplished by pouring the sand into the stem of the augers as the augers are removed from the

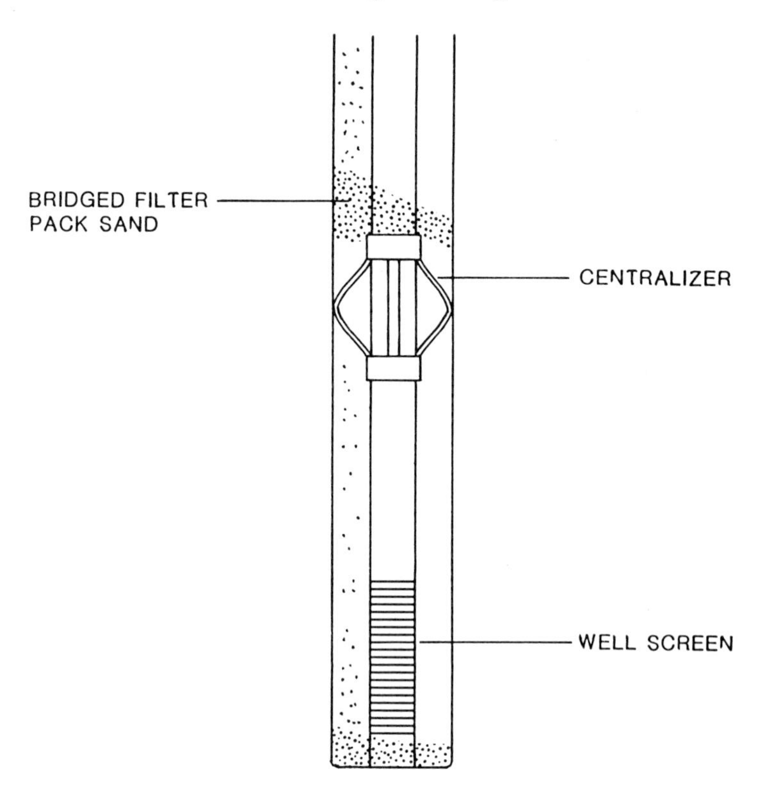

Fig. 23.8 An illustration of bridging, which occurs when a material such as sand or bentonite pellets lodge in the borehole above the depth for which they were targeted. If the bridge cannot be dislodged, the monitoring well construction may have to be aborted.

borehole in increments of less than 300 cm. This type of installation is very useful in situations when auger removal would result in excessive borehole collapse. Care must be utilized, however, to avoid accumulations of sand within the auger stem of more than approximately 300 cm, since this can wedge the well riser or screen inside the stem. In addition, the relatively small diameter of the auger stem reduces the working area available for sand passage, which in turn enhances the possibility of bridging.

To use this method for installations within mud rotary drilled boreholes, it is usually necessary to thin the viscous drilling mud within the borehole by mixing potable water. This is performed in order to induce the filter pack materials to settle through the annulus at a rate which is not prohibitively slow. Filter pack emplacement should commence as quickly as possible after this thinning is performed, however, since the thinning of the mud enhances the possibility of borehole collapse. The volume of water used for the thinning procedure should be kept to a minimum, to minimize both disturbance of the borehole ('wash outs'), and fluid losses to the surrounding soils which will extend development time.

The second option for the installation of filter pack materials in a mud rotary drilled borehole is emplacement through tremie pipe. For this method, a string of pipe is lowered into the annulus to the desired filter pack depth interval. The drilling fluid in the borehole is thinned using potable water until a slurry of filter pack sand and water can be induced to flow down the tremie pipe. As this slurry of sand and water exits the lower end of the tremie pipe, the sand settles out into the bottom of the borehole, forming the filter pack. This method is often regarded as necessary to counteract the tendency of the silica sands to form 'fining upward' sequences in the filter pack as they settle through the long column of fluid in the borehole.

For a properly selected monitoring well filter pack material, this method is rarely necessary. As mentioned earlier in this section, filter packs for monitoring wells should be uniform in gradation, that is, they should consist of a narrow range of grain sizes. This uniformity will minimize the potential impact of any fining upward sequences formed during filter pack emplacement. Further, given the purposes for a monitoring well filter pack – precluding the entry of fines and supporting the borehole and annular seal – any fining upward sequences which do occur will not impair its function. Since the tremie method results in the introduction of additional water to the borehole, which must then be recovered during development, and also requires additional time and effort to perform during installation, its use is not recommended.

Other options

In addition to the emplacement of granular filter packs into the annulus of the monitoring well, there are filtering mechanisms which may be installed as a part of the well assembly. These include the use of:

- well screens which have been wrapped in filter fabric; and
- prepacks, which are well screens which are manufactured of two nested screened pipes, with a silica sand filter pack between the screens. This entire assembly is installed as a single piece.

Several problems exist with the use of filter fabrics. They tend to clog easily, and as a result can severely restrict the entry of ground water into the well. In addition, many fabrics may restrict the passage of separate-phase hydrocarbons.

Prepacks are typically a more effective approach, but also have shortcomings. Because they are constructed of two nested screens, the outer diameter (OD) of the prepack is larger than that of a single screen of identical inner diameter (ID). This larger OD may complicate installations through an auger stem, or in other restricted settings. The thickness of the sand in a prepack is much lower than that within an annular filter pack, and may be less effective in the removal of some types of turbidity. The cost of prepacks is typically higher than that of a single screen and filter pack material purchased separately. If difficulties in filter pack installation are anticipated, however, prepacks are an alternative worth evaluating.

23.3.7 Annular seals

After the filter pack has been emplaced, it is necessary to fill the remaining annular space with a material of low hydraulic conductivity to serve as an annular seal. This seal is typically emplaced as a liquid, referred to as grout, which is usually comprised of either Portland cement, a slurry of bentonite clay, or Portland cement with an admixture of bentonite. The purpose of the annular seal is to prevent the exchange of fluids between the screened interval and higher strata and/or the surface. Without this precaution, cross contamination between aquifers could occur, limiting the possibility of obtaining a representative sample from the stratum or interval of interest. An improperly sealed well can actually enhance the spread of contaminants, by providing ground water from a contaminated aquifer access to less impacted strata.

The annular seal should be continuous with a seal at the surface, which is often referred to as a sanitary seal or surface seal. The surface seal should be designed to prevent the pooling of surface waters at the wellhead (the surface expression of the well). It should, more importantly, preclude the entry of these surface waters into the well and borehole. In addition, it should be durable enough to withstand stresses associated with weather and incidental impacts. Most surface seals are constructed of concrete.

Bentonite seals

To prevent the grout and grout-derived fluids from entering the filter pack of the monitoring well, it is standard practice to place a seal of bentonite directly upon the upper surface of the filter pack. Since cement-based grouts are a strong source of hydroxyl ions and free lime, this helps to minimize the impact of annular grouts on the ground-water chemistry of the targeted stratum. This seal should be emplaced as a solid, rather than a slurry, and should have a minimum thickness of 0.5–2 m.

Where stratigraphy permits, this seal may be placed to extend up into the nearest overlying aquitard. This helps to prevent the migration

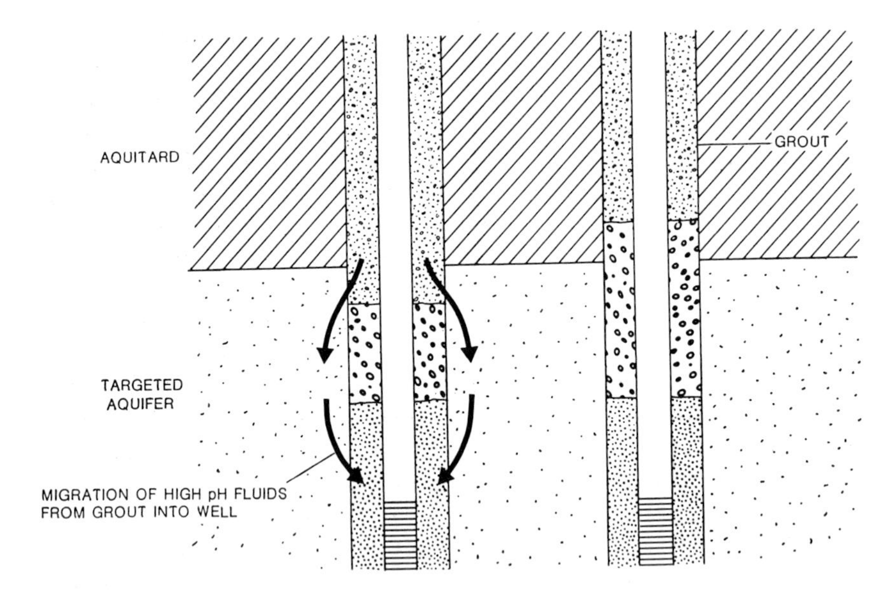

Fig. 23.9 Bentonite seal placement. The seal placement on the left permits high-pH fluids from the Portland cement grout to bypass the seal through the aquifer, and enter the well. By extending the seal upward another meter, right, the grout fluids may be better isolated from the well screen.

of high pH fluids from the grout around the seal via the aquifer soils (Fig. 23.9). This design modification is recommended in low hydraulic conductivity aquifers, or when 50-mm diameter wells are being installed. Under these two circumstances, it is often difficult to flush enough water through the filter pack and well to remove any grout-derived caustic fluids. An extension of this technique sometimes used when installing shallow wells is to entirely eliminate the use of liquid grouts, and extend the solid bentonite seal from the filter pack to the surface.

The most commonly utilized form of bentonite used for these seals is bentonite pellets. These are highly compressed spheres of finely-ground bentonite clay (a high-grade sodium or calcium montmorillonite), which are supplied in several sizes up to 12-mm diameter. Pellets of 'pure' bentonite, without admixtures, should be utilized. Upon contact with water, these hard dry pellets hydrate very rapidly, swelling and becoming cohesive. Within a few minutes, they can form a water-resistant seal. While this rapid hydration and cohesive character makes them very effective as a sealant material, it can also render their emplacement more difficult. After entering ground water or drilling fluid standing in a borehole, the pellets have a tendency to adhere to any obstacle they encounter, as well as to clump together. As a result, pellets tend to bridge (Fig. 23.8) at any constriction within the annular space.

This effect is quite important when design specifications call for emplacement of pellets through the stem of hollow stem augers. When the diameter of the well is relatively close to that of the auger stem, attempts to emplace pellets through the auger stem have a low success rate. The working area between the well and the inside of the stem in these cases is simply not sufficient to prevent the bridging of the sticky wet pellets. When this occurs, the well and/or augers must usually be removed to eliminate the bridge, and the well installation procedure returns to its starting point.

This effect can be directly addressed through the use of larger stem diameter augers, since this increases the open area available for pellet passage. One example would be the use of 165-mm stem diameter augers for the installation of a 50-mm well, rather than the 90-mm stem diameter augers which are more commonly used for this purpose. This practice will, incidentally, also decrease the risk of bridging during filter pack emplacement.

Another possible solution is the use of 'timed release' pellets, which have a slow-dissolving coating to delay the initial hydration period. This is not a complete solution to the problem, however, since the coating only reduces, and does not eliminate, the pellets' tendency to bridge within the augers. If this type of pellet is used, the manufacturer should be contacted to provide a chemical characterization of the coating, in order to ensure the absence of interferences with suspected ground-water constituents.

The question of hydration time for pellets is still a point of contention. While field testing of pellets (not the 'timed release' type) by the authors has indicated the formation of satisfactory seal within only a few moments of contact with water, many specifications for the installation of monitoring wells mandate hydration periods of one to several hours. While this period may be required for complete pellet hydration, such complete hydration is not necessary to produce a coherent seal. Once the points of contact between pellets have expanded sufficiently to eliminate the interconnection of pores between the pellets, the seal is sufficient to preclude the passage of grout-related fluids.

In addition to pellets, bentonite chips also find frequent application in bentonite seals. These chips are simply gravel-sized fragments of unground and untreated bentonite, which are graded based upon their size. Bentonite chips are compositionally identical to bentonite pellets, and are typically obtainable at approximately 25% of pellet cost. Since the chips have not been ground to a powder, they hydrate more slowly than pellets, reducing the likelihood of bridging in problem installations (such as through auger stems). The downside of this slower hydration and sealing is the greater time required between the emplacement of the chips and that of the grout. An absolute minimum of 30 minutes should be allowed for seal formation; even longer periods are preferable. This is in contrast to the 10–15 minutes required for pellet hydration.

Grouts

A grout typically consists of a slurry of cement and/or clays in water, although other materials are used for special applications. The requirements of a grout material usually include its hydration and solidification following emplacement, and the formation of a relatively permanent, low hydraulic conductivity seal.

Portland cement-based grouts

Neat cement grouts consist of Portland cement in water, with no admixtures of bentonite or other additives. Admixtures of sand or pea gravel to these cement grouts result in formation of a type of concrete, adding some structural strength to the grout, but having little effect on temperature, hydration time, or shrinkage.

More commonly used in the monitoring well installation practice are cement grouts with admixtures of 2–5% powdered bentonite by weight. This weight percentage is expressed as

$$\text{Percent bentonite} = \frac{\text{weight dry bentonite}}{\text{weight dry cement}} \times 100\%$$

Bentonite admixtures slow the development of compressive strength in the grout, and reduce the final strength of the solid grout (Smith, 1976; Dowell Schlumberger, 1984). Additional effects include an increase in viscosity, a decrease in thickening time and grout density, and reductions in fluid losses from the emplaced grout. The higher viscosity results in a better displacement of drilling mud from the well's annulus, resulting in a more complete final seal. The reduction in fluid losses may help reduce shrinkage of the grout after emplacement (Driscoll, 1986; Lehr *et al.*, 1988), as well as minimizing the loss of constituents which might affect ground-water chemistry. The reduction in shrinkage occurs only for bentonite mixtures up to 6%. Above this percentage, shrinkage may actually be exacerbated (Driscoll, 1986).

The mixture of bentonite with Portland cement should begin with the hydration of the bentonite alone, without the Portland cement. This permits relatively complete hydration of the bentonite. This hydration is not as complete in the presence of Portland cement, because it is inhibited to some degree by Ca^{++} ions released into the slurry by the cement (Dowell Schlumberger, 1984). After the bentonite is thoroughly mixed, the addition of Portland cement may proceed.

Most monitoring well specifications call for the mixture of 22–28 liters of water per 43-kg sack of Portland cement, with approximately 4 liters per sack additional for each percent bentonite added to the grout. These specifications are typically based on the standard practices of specialty cementing companies, such as oil well cementing operations. While these water-to-cement ratios are feasible using the mixing equipment available to this type of firm, the simpler, smaller scale equipment utilized by most monitoring well installers is typically not capable of meeting this mix specification. The grout simply becomes too viscous to pump during the mixing process. In deference to the realities of field installation, then, the designer should be prepared to consider a less stringent water-to-cement ratio.

Bentonite grouts

The use of bentonite grouts for monitoring well installation has gained some degree of acceptance during the past decade. Such grouts, which consist of high percentages of bentonite in a water slurry, do offer some distinct advantages. The thixotropic nature of bentonite-water slurries permits the pumping of the grouts for extended periods after mixing, which reduces the potential penalty of delays in grouting. The bentonite grouts generate far fewer calcium and hydroxyl ions than Portland cement, reducing the potential for high pH problems from grout leachates. In addition, the potential for problems with PVC pipe softening due to Portland cement heat of hydration is eliminated with bentonite grouts. Finally, the field mixing of these grouts is typically faster and easier than that of the Portland cement-based variety.

These grouts are not appropriate for many installations, however. The plastic and thixotropic nature of bentonite which eases and extends grout pumping also eliminates most of the mechanical strength of the bentonite grouts. They cannot, therefore, be used in well designs requiring a strong, non-plastic solid grout column. Since the water is less strongly bound in the chemical structure of bentonite slurries than in that of Portland cements, fluid loss and attendant shrinkage is always a possibility. This effect is probably of concern mainly in arid-zone settings, especially in strong capillarity soils such as clays and silts.

If bentonite grouts are utilized, the design specifications should delineate the permissibility of additives. Many of these grouts utilize organic polymers to improve pumpability, or metal oxides to improve yield. If these constituents could result in interferences with monitored constituents, their use should be barred in the well installation specifications.

Grout emplacement

Grouts are typically emplaced through a string of pipe lowered into the annular space, referred to as a tremie pipe. The tremie, which is typically 25- to 50-mm diameter, is suspended in the annular space

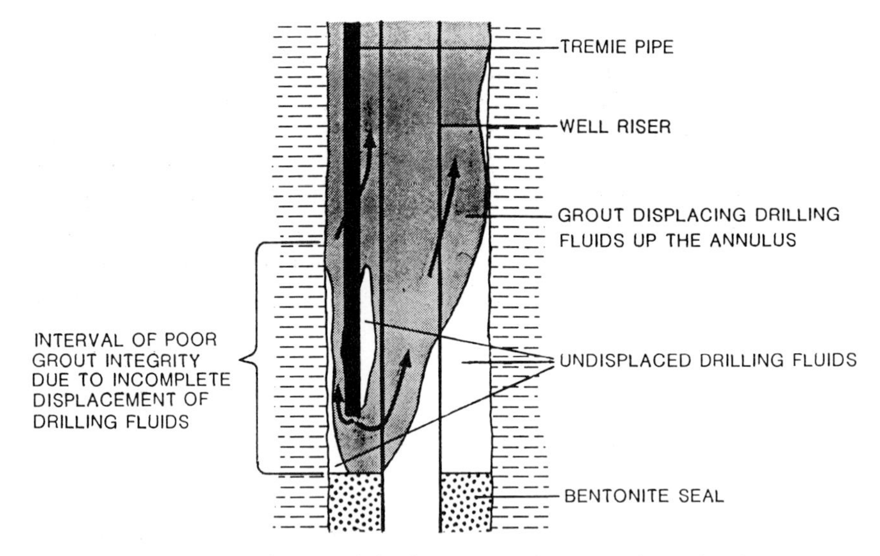

Fig. 23.10 Conceptual view of the formation of a zone of possible low annular seal integrity as a result of incomplete displacement of annular fluids by a pumped grout.

with its lower end approximately 1–2 m above the top of annular bentonite seal. The 1–2 m gap minimizes the possibility that the grout stream will wash out and displace part of the bentonite seal. Grout is pumped into the tremie at the surface, to discharge into lower portion of the annulus. In theory, the grout entirely fills the lower portion of the annular space, moving upward and displacing drilling fluid or natural waters in a plug.

While this method is certainly capable of providing a generally adequate seal, its potential short-comings should also be recognized. Cementing research in oil field drilling and well installation has long recognized the difficulty in emplacing a complete cement seal within a fluid-filled borehole. The main problem arises from the incomplete displacement of the fluid in the annulus by the incoming grout. If this fluid has a viscosity substantially greater than that of water, such as is the case for most drilling fluids, or for a mixture of auger cuttings and ground water, grouts will tend to stream or channel rather than plug flow. This effect would be strongest nearest the point of discharge from the tremie, so the grout seal would tend to be least complete near its base (Fig. 23.10).

While the scope of this problem has never been determined for monitoring wells, the implication is clear: the lower portion of an annular seal should not be relied upon as a perfect barrier to fluid migration. No solution to the potential problem is presently available, given the small scale of most monitoring well installations. It may be possible to reduce this channeling effect by using as large a tremie pipe diameter and as high a pumping rate as possible. The best approach, however, is to emplace the annular bentonite seal properly, so that minor gaps in the grout seal will not impair the function of the well.

Monitoring well installations often require the installation of grout into dry annular volumes, i.e. those which contain no standing liquids, such as ground water or drilling mud. In these situations, there is no disadvantage to introducing the grout into the annulus at the surface, without the use of tremie. Being liquid, the grout will effectively fill the annular volume.

23.4 DRILLING METHODS

Several drilling methods have been used for the installation of monitoring wells. These methods were discussed in detail in Chapter 15 and are reviewed here in terms of monitoring well installation. The most common methods of well installation are air rotary, direct mud rotary, reverse mud rotary, hollow stem augers, solid stem augers, and cable tool. The drilling method selected for well installation at a particular site must be suitable for the subsurface conditions likely to be encountered. In addition, it must be amenable to the particulars of well installation included in the design. One of the best methods of evaluating these factors is to contact consultants, drilling companies, and other agencies with experience in drilling within the subject area. These persons can often indicate which methods have historically been utilized with the greatest rate of successful installations.

In the following discussion, the focus will be placed upon the two methods most commonly used for monitoring well installations: hollow stem auger and direct mud rotary.

23.4.1 Hollow stem auger drilling

This method has attained a favored status in environmental drilling, especially among most of the regulatory agencies involved in groundwater monitoring, for several reasons. First, the absence of drilling fluids minimizes the potential impact of the drilling operation on the aquifer hydrogeochemistry. Second, the hollow stem of the auger typically permits convenient sampling as the borehole is advanced. The use of soil coring barrels, also called continuous samplers, even permits the collection of a relatively continuous soil sample as the augers are advanced. The hollow stem also permits the augers to serve as a temporary casing during well installation. In poorly consolidated formations which would collapse if unsupported, the stem may therefore permit the installation of the well pipe, the filter pack, and the annular seal before the augers are withdrawn.

Sampling using hollow stem augers may be complicated by the presence in the subsurface of saturated, unconsolidated sands (also known as running, flowing, or heaving sands). These sands typically enter and fill the lower part of the auger stem, often to an extent of 1 m or more. Once within the stem, these sands can be difficult to remove, and their presence can preclude both sampling and well installation. If these sands are known to be a subsurface condition, the use of alternative drilling methods may be preferable.

Although advances in drilling rig torque capacities and drilling technology are extending the range of depths and conditions within auger capabilities, auger drilling methods are still not appropriate for every setting. Case histories of augers being advanced to 100-m depths, for example, are not unknown, but the number of practitioners and rigs capable of performing these feats are still severely limited. As suggested in the introduction to this chapter, it is critical to recognize the difference between practices which are possible, and those with a high probability of success. Augers, as with other methods, should be utilized for situations well within their capabilities in order to minimize the risk of a failed or poor installation.

23.4.2 Direct mud rotary

Well installation is conceptually straightforward within boreholes advanced using mud rotary methods. When the total depth of the borehole is attained, the mud density is adjusted to ensure that the borehole will remain stable upon removal of the bit. The drill string is then removed, and the riser and screen of the well assembly lowered into the hole to the desired depth. Depending on the density of the drilling fluid in the borehole and the material used for well construction, the well assembly may have to be filled with clean water to counteract its buoyancy, and permit it to sink through the drilling mud. This is typically necessary when non-metallic well materials, such as PVC or PTFE, are utilized. The filter pack and annular seal are then introduced into the annulus of the borehole between the well and the borehole.

As with hollow stem augers, there are real, pragmatic limitations to the capabilities of this drilling method. Specifications which fail to recognize these limitations decrease the probability of successfully installing a well.

One example of a commonly applied design specification which may not recognize the physical limitations of the drilling method is the elimination of additives to the drilling mud. These specifications typically permit only potable water, with no bentonite, polymers, etc.

admixtures. The motivation is to remove any possible interferences between these additives and the suspected ground-water contaminants, as well as other impacts to the aquifer and ground-water character which are attendant to the use of these additives. In addition, the formation of a bentonite 'mud cake' on the borehole wall is hypothetically eliminated.

While the approach seems logical, its value decreases if it limits the capabilities of the drilling method so severely that drilling or well installation is compromised. Many aquifer materials, such as coarse sands or gravels, may not be drillable using neat water muds, since the lifting capability of the fluid returning up the hole is strongly dependent upon its viscosity. In addition, borehole stability in unconsolidated materials is typically tenuous in boreholes drilled in this fashion. Finally, one function of these additives is to minimize fluid losses from the borehole into the targeted stratum. Without this effect, the volume of potable water introduced into the subsurface, and the dilution effects which accompany this, may increase dramatically. One side effect of these fluid losses is the formation of an even thicker mud cake, consisting of natural clays, than might have been formed if higher quality clays such as bentonites had been used.

As with other methods, the well designer is forced to weigh both sides of the issue. In this case, this would involve balancing the possibility of chemical interference or hydraulic conductivity reduction from the bentonite against the possibility of a borehole which has collapsed around the well string, rendering proper filter packing and sealing impossible, or of large fluid losses into the aquifer.

23.4.3 Direct mud rotary vs. hollow stem augers

The relative advantages of each method for monitoring well installation are summarized and compared in Table 23.3. Briefly, hollow stem augers are favored because of their minimal impact on the aquifer and natural waters. Shallow soil sampling and well installation are relatively easy using this method, especially in the absence of flowing sands. The absence of introduced fluids permits a relatively accurate assessment of the depth to water.

Mud rotary is usable in a wider variety of geologic settings and depth ranges than augers. In addition, well installation is often much simpler, since the installer has the entire borehole for placement of annular materials, rather than just the auger stem. Mud rotary does require the introduction of some foriegn liquids into the aquifer, which may be undesirable for many investigations. Some types of geophysical logging can assist in the determination of depth to water, but the degree of certainty is much lower than that possible with auger drilling.

Table 23.3 Relative advantages of mud rotary vs. hollow stem auger drilling methods for monitoring well installation

criteria	*hollow stem auger*	*direct mud rotary*
Materials which can be drilled	Unconsolidated to poorly consolidated materials	Unconsolidated to consolidated materials
Drilling depth capabilities	Generally limited to less than 50 m	No strict depth limitations
Soil sampling	Rapid and simple through the auger stem Possible to continuously core while drilling	Slower for most sysems than with augers Typically requires bit removal to collect sample
Ground water level measurement	Can be measured during drilling	Cannot be readily measured until well installed
Ground water sampling	Can be sampled while drilling	Cannot be readily measured until well installed
Geophysical logging	Certain logs (e.g. natural gamma) can be run through the auger stem	Almost all logs can be run in borehole
Well installation	Can be performed through the augers for collapsible formations	Difficult in collapsible formations, but borehole more accessible than for installations through augers
Rapidity, mobility, access, footprint	Rapid setup and teardown, good access to remote locations, generally smaller footprint than comparably-sized mud rotary rigs	Setup and teardown generally slower, access not as good, and footprint larger than for comparably-sized auger rigs
Aquifer interferences	Minimization of fluid addition minimizes impact to aquifer geochemistry	Fluid interactions result in greater impacts than augering If managed, impact is typically still within acceptable limits
Development time	Typically short, less than for direct mud rotary	Short to long, dependent upon fluid losses, aquifer productivity, etc.
Cost per unit drilled	Typically low, less than for direct mud rotary, when applied appropriately	Variable, dependent upon site specific conditions Generally higher than auger costs in comparable settings

Table 23.3 *Continued*

criteria	*hollow stem auger*	*direct mud rotary*
Equipment decontamination	Typically easy and rapid, less difficult than for direct mud rotary	Typically moderate level of effort, more complex than for auger drilling
Cuttings control	Cuttings mostly solids, control straightforward	Cuttings mixture of liquids and solids, control more complex

Note: Comparisons are based on typical equipment and conditions. Individual site and drilling conditions may modify the favorability of one method versus another based on any of the above criteria.

23.5 MONITORING WELL DEVELOPMENT

As mentioned in this chapter's introduction, the drilling and well installation process is unavoidably traumatic from the perspective of the aquifer. The shearing and churning of soils, the suspension of natural and introduced clays and silts, and the introduction of non-aquifer fluids, all combine to damage the natural structure of the soils in the vicinity of the well. This damage is often, to some degree, irreparable. While the use of sound drilling procedures and reasonable caution can minimize these effects, well installation should always include post-installation efforts to repair the subsurface impacts of drilling. These efforts are termed well development.

23.5.1 Primary purposes of monitoring well development

Drilling fluid removal

During mud rotary installations, the borehole is kept full of a water-based mud to a level at or above surrounding grade. As a result, the head within the borehole is almost always greater than that within the subsurface soils, resulting in drilling fluid movement into the aquifer. These introduced fluids alter the ground-water chemistry surrounding the borehole by displacing and mixing with natural waters. This effect is also present, although to a lesser degree, when water must be added during an installation using hollow-stem augers. This is sometimes necessary to counteract the effect of flowing sands (section 23.4.1) or to dislodge an annular bridge (section 23.3.6).

One purpose of development in such cases is to remove the introduced water from the aquifer, so that subsequent sampling is more representative of the natural waters than of the drilling fluids. A logical starting point for this effort is to calculate the volume of fluid lost to the subsurface during drilling. This is done by maintaining a fluid balance during the drilling process. This balance tracks the volume of fluid used to advance the borehole and install the well, less any storage within the borehole and at the surface. In most cases, only losses to the targeted stratum are of interest, and not those which occur within overlying strata. This may not be true if the overlying strata are monitored by other monitoring wells at the same location (i.e., shallower wells within a monitoring well cluster).

The volume of fluid lost to the targeted stratum is the minimum volume of fluid which typically should be removed during development. Dilution of this lost fluid within natural waters, however, usually requires the removal of a larger volume to properly address the impact of the lost fluid. This volume may be several times that of actual fluid loss, and is dependent upon a range of site-specific characteristics.

Pragmatically, then, there is no way to accurately predict how much water the development process must remove to compensate for these fluid losses. This uncertainty is usually addressed by the field measurement of some basic physical and chemical characteristics of the waters removed during development. Drilling fluids tend to have elevated pH (typically 9 or above) and specific conductances. These parameters, as well as temperature, should be measured at intervals during development. Development is typically accepted as sufficient when these parameters stabilize to less than 10% variance for several measurements.

Mud cake removal

Drilling fluids are utilized to reduce fluid loss from a borehole by forming a thin, low hydraulic conductivity coating, typically called a mud or filter cake, on the borehole surface. This cake remains completely or partially intact following the well installation process. Once installation is complete, such a cake forms a low hydraulic conductivity sheath around the filter pack, and so may dramatically reduce the yield of the well.

Hollow-stem auger installations are not necessarily immune to the deleterious effects of a mud cake. If sufficient clays are encountered during the drilling process, these cohesive soils may be smeared into the equivalent of a mud cake by the rotation and downward motion of the augers (Fig. 23.11).

Although yield is typically not a primary consideration in monitoring well installation, large losses in well efficiency may impair its usefulness. If a strongly affected well is to be used for aquifer testing, for example, such as slug and pump tests, it may not accurately represent the hydrologic character of the tested strata. For a well placed within a low hydraulic conductivity stratum, the presence of a mud cake can greatly slow recovery by the well to the pre-sampling purge. This can

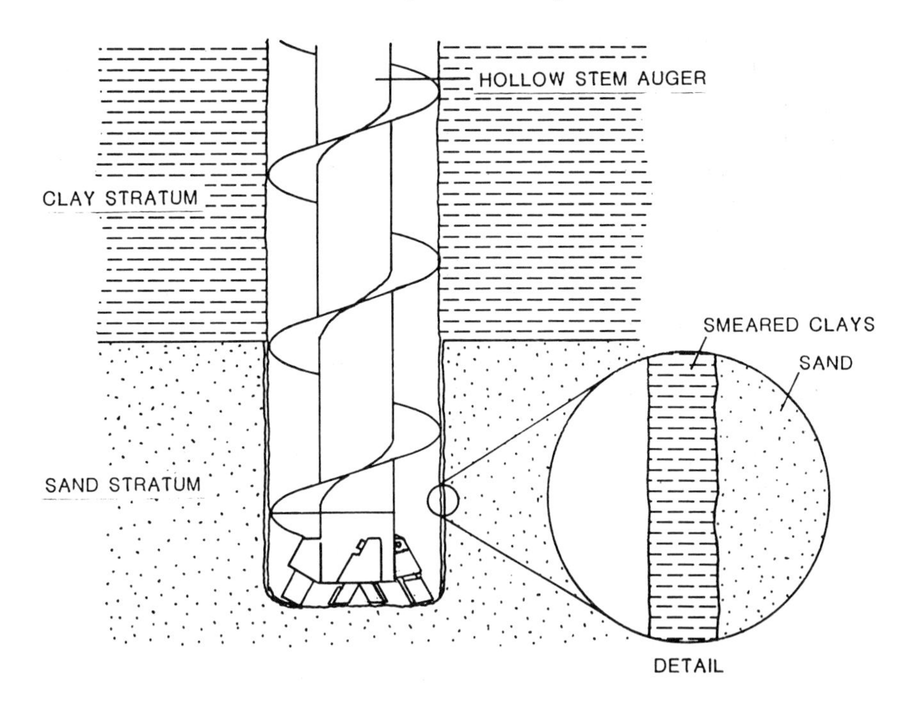

Fig. 23.11 Conceptual view of the formation of a 'mud cake' during auger drilling. Clays from the upper layer can be smeared across the surface of the lower portion of the borehole, creating a low permeability sheath around the borehole.

extend the time required to collect groundwater samples, in some cases by days.

Mud cake problems should be addressed promptly following the installation of the well. Hydrated clays, including those which comprise the bentonite used in most drilling fluids, tend to form an increasingly organized and coherent structure with increasing time. For this reason, the difficulty in removing a mud cake increases if the cake is left undisturbed following the drilling process. If left in place for several days or weeks, therefore, a mud cake can cause irreparable damage to the well's capabilities.

Turbidity reduction

As mentioned in section 23.3.6, filter pack installation is often imperfect due to bridging or borehole instability. If these problems occur, the physical result may be gaps in the filter pack, which will be occupied by natural sediments. If the aquifer contains significant percentages of silt or very fine sand, these fines will enter the well with the groundwater through these 'breaches' in the filter pack, resulting in high levels of turbidity. Due to interferences between these suspended sediments and certain analytical methods, this may result in water samples from the subject well being unsuitable for testing. Although turbidity can be eliminated after the fact by sample filtration, the practice of filtration is not universally considered acceptable. It is preferable, therefore, to minimize or eliminate the turbidity at the point of entry of ground water into the well. This may be accomplished, in some cases, by development of the well.

One approach is to remove sufficient water from the well to flush all of the fines capable of passing the screen into the well, where they can be collected and removed. This produces, in effect, a natural filter pack where the breach in the artificial filter pack exists. For situations where fines contents exceed 5–10%, however, or where these breaches are numerous or extensive, it may not be possible to implement this approach.

Under such circumstances, it is advisable to try an alternative approach. The turbid water within the well should be removed using a method which agitates the well and its surroundings as little as possible. Submersible pumping (preferably with a continuous, rather than a surging action) or double-tube airlifting are two possibilities. Once the turbidity levels drop, the well can be purged and sampled. The clarity of the water is not stable, however, since any agitation of the well will result in renewed entry of fines, and therefore turbid water, into the well.

23.5.2 Methods of development for monitoring wells

A large number of development options exist, with the required hardware ranging from sophisticated prefabricated tools to implements which may be constructed from readily available materials. In spite of this diversity, most can be classified as causing groundwater flow which is either unidirectional (water withdrawal) or bidirectional (surging) relative to the well.

Unidirectional methods, such as submersible pumping and airlifting, move water from the surrounding stratum into the well from the aquifer. These methods transfer a minimum of energy to the filter pack and surrounding soils, and so are not very useful for disruption of the mud cake. These methods are useful in removing large volumes of water from the well, in order to recover lost drilling fluids or flush out fine sediment which will pass through the slots in the well screen.

Bidirectional methods surge the well, moving water into the well from the surrounding soils, and back into those soils from the well in a reciprocating manner. Commonly used bidirectional methods include surging with a bailer or surge block, and the modification of unidirectional methods such as pumping to permit periods of backwash. One example of the latter would be the use of airlifting techniques to lift an air-water column almost to the surface, then shutting off the air feed to the system and allowing the column to fall back into the well. Repeated use of this technique can produce an effective surging action. The effectiveness of bidirectional methods is typically enhanced when performed on 100-mm diameter wells, since the larger open area of these wells permits more of the energy of the surge to reach the filter pack and surrounding soils.

23.6 DEDICATED PURGING AND SAMPLING SYSTEMS

Piror to the sampling of a monitoring well, the water standing in the well must be removed, permitting 'fresh' water from the aquifer to enter the well. This procedure, called 'purging', is intended to remove water which may have been subjected to chemical change due to extended contact with the unnatural conditions and materials within the well. Purge quantities are typically expressed in terms of casing volumes, with a casing volume consisting of the volume of water present within the riser and screen at the time of sampling. Purge quantities of 3–5 casing volumes are commonly specified, although many specifications now also call for the measurement of pH, specific conductance and temperature of purged waters at regular increments. Under these conditions, the purge is considered to be complete only when these parameters stabilize.

In order to minimize the need for equipment transportation and decontamination during sampling, purging systems which are dedicated to a single monitoring well are widely used. Such a system remains within the well continuously, and is used only for that well. Some types may be used for sample collection, as well as purging. Such dedicated purging systems are available in a range of types and degree of complexity. Some of the more commonly utilized varieties are discussed in the following sections.

23.6.1 Bailers

Bailers have long been the workhorses of groundwater sampling, due to their simplicity, low cost, and the wide range of bailer materials available. A bailer consists fundamentally of a length of pipe or tubing with a one-way valve fixed to the lower end. Bailers are available in most well construction materials, including stainless steel, PVC, and PTFE. When lowered on a rope or cable into the water within a well, the valve unseals, permitting the bailer to fill. As the bailer is raised from the water, however, the valve again seats and seals, retaining the water within. The bailer is raised to the surface and dumped, and the process repeated. The rope used is typically a relatively inert synthetic fiber, such as nylon or polypropylene.

While very convenient for shallow wells, they are not recommended for use in purging and sampling wells in which the depth to water is in excess of approximately 30 m. Beyond that depth, the mass of rope becomes unwieldy, the purging rate slows due to the time required for each interaction, and any small leaks in the bailer valve (a common phenomenon) result in the loss of most of the water load during the upward trip.

23.6.2 Pneumatic pumps

Pneumatic pumps are operated by air or gas pressure, which is used to displace a volume of fluid upward through a discharge line. Being air-driven, these pumps require a source of pressurized gas or air at the well-head. Samples are collected at the surface point of discharge. This type of pump includes bladder, piston, and other styles, with some variability in performance.

There are some advantages to the use of these systems for monitoring. They can be constructed of materials such as stainless steel and PTFE, which are widely considered to be relatively non-reactive with groundwater (section 23.3.1). They are widely accepted for sample collection for volatile organic chemical analysis, and may be used in 50-mm diameter wells. Finally, they can be left relatively unattended for several hours at a time for slow purging wells. Such systems are, however, relatively expensive, and pumping rates for most styles are generally low. Depending upon their construction, some types are limited by air pressure requirements to depths less than approximately 75 m.

23.6.3 Submersible electric pumps

These pumps consist of an electric motor and pump assembly, which are suspended within the well at the lower end of a string of discharge pipe or tubing. A source of electricity is required at the wellhead. This type of pump has the broadest range of applicability of the systems commonly used within monitoring wells. Models are available for flow rates ranging from a few liters per minute to several hundred liters per minute, and for depths to several hundred meters. Submersible pumps for well diameters as small as 50 mm are currently available. Costs are typically lower than for pneumatic systems.

Submersible pumps are currently not widely accepted for the collection of groundwater samples for volatile organic chemical analysis. This is due to concern regarding the off-gassing of these volatile constituents which may occur as the water passes through the pump's

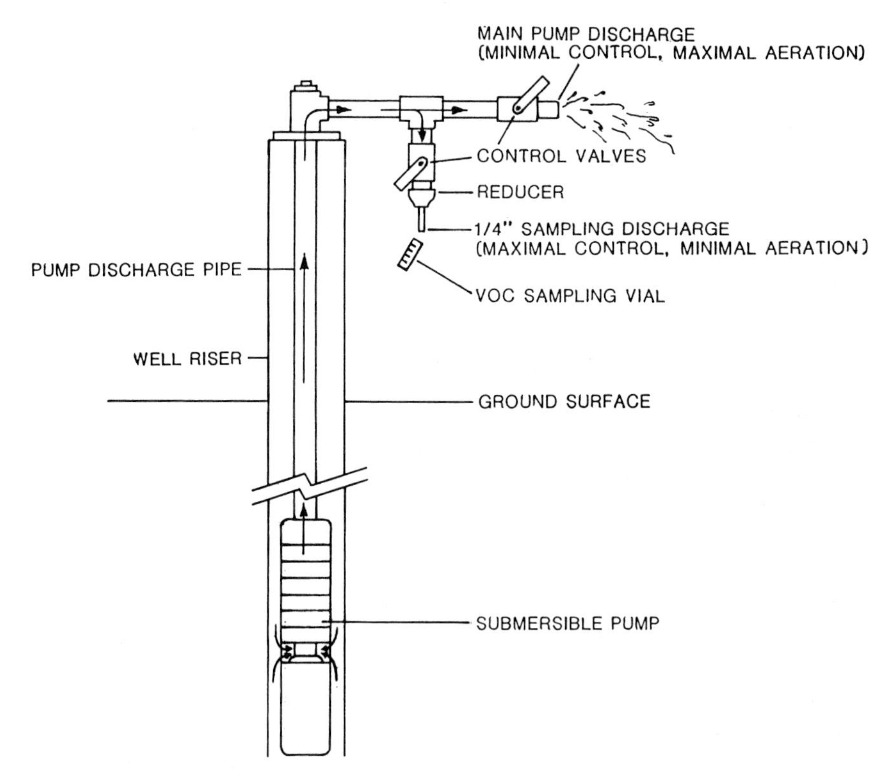

Fig. 23.12 Construction of a wellhead sampling tap to permit the collection of samples for VOC analysis from a submersible pump discharge.

rapidly rotating impeller system. It is probable, however, that the loss of volatiles attributed to the passage through the pump is actually due to aeration at the point of surface discharge. Since pumping rates from submersible pumps tend to be relatively high, the collection of a non-aerated sample can be quite difficult. This effect, however, can be counteracted by the proper construction of a wellhead sampling port for the collection of such samples, as illustrated in Fig. 23.12. Such a sampling port should be of small diameter, and be placed upstream of the main discharge. This permits sample collection prior to the aeration which occurs at the pump discharge. If sampling is being performed in order to satisfy regulatory requirements, however, the acceptability of this method to the overseeing regulatory body should be confirmed in advance of system installation.

23.6.4 Manual pumps

These are dedicated systems which are operated manually, and so require no electrical or air supply. One type of manual system uses two strings of PVC pipe, one nested within the other, and both placed within the monitoring well. When the inner pipe is moved up and down, the action of a plunger and valve system at its base moves water up the annular space between the two pump pipes. The water is then discharged from a spout on the larger pipe. Flow rates up to 20 l/min are feasible with these units. These pumps are not universally accepted for sample collection for volatile organic chemical analysis, due to concerns regarding aeration during the rise through the pump, and within the pump near the point of discharge.

A second style of manual pump consists simply of a foot valve on a length of flexible tubing extending into the well. These tubing or inertial pumps are operated by rapidly moving the tubing up and down within the well in a reciprocating manner. Since water can move into the tubing only on the downstroke, due to the action of the foot valve, the energy of the reciprocating movement gradually impels the water up the tubing. Discharge occurs at the top of the tubing. Flow rates of up to 16 l/min are possible, and operable depths extend to approximately 45 m for manual operation. The depth range may be extended to greater than 60 m by the use of a motorized assist to provide the reciprocating motion. Materials of construction include a variety of plastics, as well as PTFE. These pumps are probably more widely accepted for volatile organic chemical sample collection than the nested pipe pump discussed above, since the opportunity for water contact with air is limited within the small diameter tubing of the inertial pump.

23.7 MONITORING WELL SURFACE COMPLETIONS

After the well has been installed, a protective structure must be placed around the riser at the ground surface. This structure, termed a surface completion, is to protect the well against accidental impacts, tampering, or vandalism. Surface completions typically fall into one of two types.

- *Flush completions* This type is utilized in active, high traffic areas, where a high-profile completion would be an obstruction to site use. These completions are low in physical profile, usually extending no more than several centimeters above surrounding grade.
- *Above grade completion* This type is utilized in less active areas, or in areas where a more visible completion is helpful, such as in dense vegetation. These completions extend well above grade, typically by 0.6–1.2 m.

Flush completions consist of a sturdy steel or concrete manhole or meter box set in concrete around the top of the riser. The riser is cut off

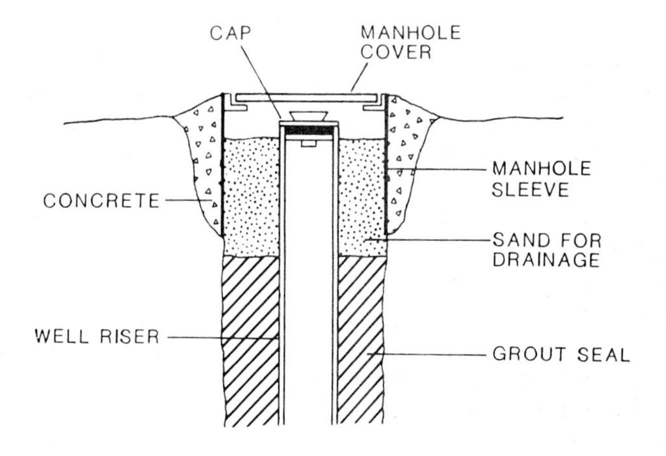

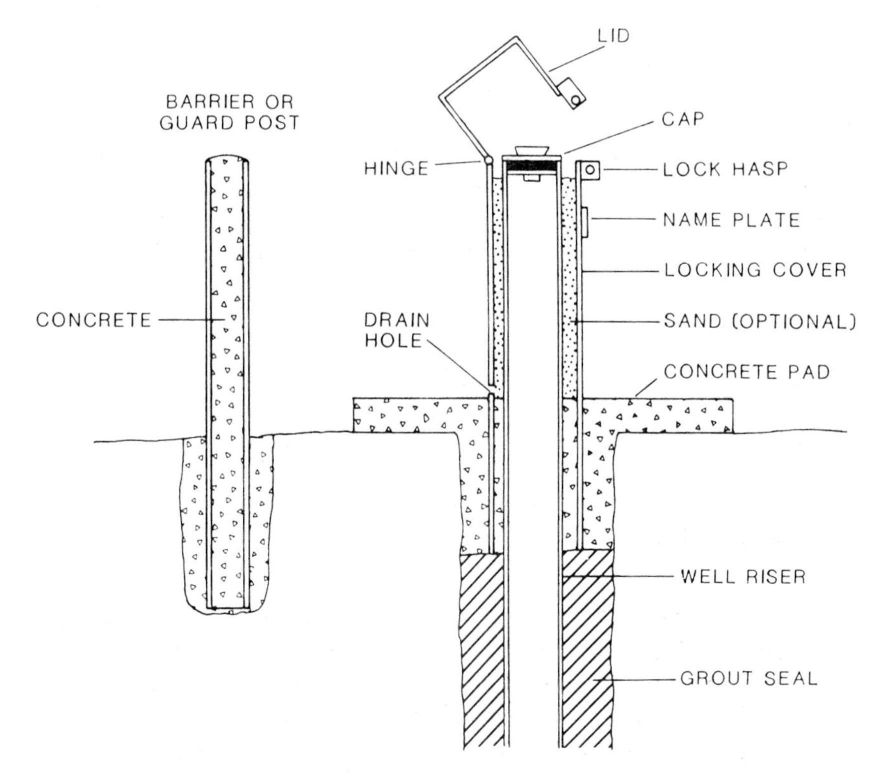

Fig. 23.13 Construction of a surface completion: (a) flush, using a manhole, and (b) above ground, using a locking cover and protective casing.

just below grade prior to setting the manhole, and the manhole itself is set slightly above grade to promote surface drainage away from the well (Fig. 23.13a). When constructing this type of completion, the manhole cover should never be in contact with the upper end of the well riser or any cap on the riser. If the lid is resting on the riser when a vehicle passes over the completion, damage to both the lid and well may result.

Flush completions are not recommended for use in areas where substantial surface water ponding may occur. They are also not well-suited for grassy or brushy areas, where plant growth may make the wells difficult to locate.

For above-ground completions, the riser of the well is terminated well above grade. A protective casing, usually constructed of steel, is placed over the riser and concreted in place. This casing is equipped with a lockable cap or lid to prevent tampering or vandalism. A square pad of concrete is constructed around the base of the protective casing (Fig. 23.13b). If the risk of damage from site activity is high, additional protection is afforded by the placement of steel barrier posts at regular intervals around the well. These surface completions are typically painted in bright colors to enhance their visibility.

Regardless of which style of surface completion is utilized, the protection of the well should always include a cap or plug at the upper end of the riser. This precludes the entry of condensation, rust, and dirt which may collect within the manhole or protective casing. Available styles of monitoring well caps or plugs include the following:

- slip caps which slide over the top of the riser and are held in place by friction;
- compression plugs which fit into the upper end of the riser (a wing nut on the upper side of the plug permits the compression of the plug, which forces a rubber bushing against the inner wall of the riser – this seals the well and holds the plug firmly in place); and
- threaded caps or plugs which thread directly onto a threaded fitting at the top of the riser – this type is useful only if the riser terminates in a threaded fitting at the desired completion type, or can be custom threaded in the field.

23.8 EQUIPMENT DECONTAMINATION

In association with ground-water monitoring installation practice, decontamination, or 'decon', of equipment is performed for three reasons:

- equipment entering a site may be deconned to remove contaminants acquired at previously visited sites, or during travel;

- equipment may be deconned between each utilization, to prevent the transport of contaminants between wells or boreholes, or between depths within a single borehole; or
- equipment exiting a site may be deconned to prevent the transport of contaminants off-site.

The most commonly utilized decontamination technique for large equipment consists of a high pressure (greater than 5 MPa), high temperature (greater than 90 °C) hot water or steam wash. This method, if performed in a thorough manner, will remove residues of oils and greases, caked mud, and dust and other loose particulates. For smaller equipment, such as that used for soil sampling, this method may be supplemented or replaced with solvent and/or detergent washes and deionized water rinses. Clean equipment should be covered with plastic sheeting for transport or storage.

Although a valid practice, decon is often carried to levels at which the additional protection accrued does not justify the effort and cost expended. Attempts to remove trace amounts of grease during decontamination, for example, may be fruitless if those same oils and greases are known to be generated by the operation of the drilling rig. Wrapping the well and riser in a sealing material for transport to the installation site will not protect against airborne contaminants if those same contaminants are present when the pipe is unwrapped at the installation site.

As mentioned during the introduction to this chapter, the installation of a monitoring well must be accompanied by some disruption of the subsurface, both physical and chemical. Rather than relying on an extreme level of decontamination for 'perfect' protection from contaminant introduction for installations of high protocol, it is more productive to utilize reasonable caution and sound installation techniques. Even if minute amounts of foreign constituents are introduced during installation, a properly installed and sampled well will be purged several times during development and the pre-sample purge. These purges should further reduce the probability of impact from any introduction of trace constituents.

REFERENCES

Clark, L. (1988) *The Field Guide to Water Wells and Boreholes*, Open University Press, John Wiley and Sons, New York, New York.

Dowell Schlumberger (1984) *Cementing Technology*, Nova Communications, Ltd., London, England.

Driscoll, F.G. (ed.) (1986) *Groundwater and Wells*, 2nd edn, Johnson Division, St. Paul, Minnesota.

Gass, T. (1988) Monitoring Well Filter Pack and Screen Slot Selection: A Reassessment of Design Parameters. *Water Well J.*, June, pp. 30–2.

Haug, A., Petrini, R.H., Grisak, G.E. and Klahsen, K. (1989) Application of geostatistical methods to assess positions and spacings of groundwater monitoring wells, in *Proc., Conf. on Petroleum Hydrocarbons and Organic Chemicals in Ground Water: Prevention, Detection and Restoration*, NWWA, Dublin, Ohio, pp. 159–74.

Hinchee, R.E. and Reisinger, H.J. (1987) A practical application of multiphase transport theory to groundwater contamination problems, *Ground Water Monitoring Rev.*, Winter 1987.

Lehr, J., Hurlburt, S., Gallagher, B. and Voytek, J. (1988) *Design and Construction of Water Wells*, NWWA, Van Nostrand Reinhold Company, New York, New York.

Mackay D.M., Roberts, P.V. and Cherry, J.A. (1985) Transport of Organic Contaminants in Groundwater, *Environ. Sc. and Tech.* **19**(5), 384–92.

Mercer, J.W. and Cohen, R.M. (1990) Transport of immiscible fluid in the subsurface, *J. Contaminant Hydrology*, **6**, 107–63.

Parker, L.V., Hewitt, A.D. and Jenkins, T.F. (1990) Influence of Casing Materials on Trace-Level Chemicals in Well Water. *Ground Water Monitoring Rev.*, Spring, pp. 146–56.

Schwille, F. (1984) Migration of organic fluids immiscible with water in the unsaturated zone, in *Pollutants in Porous Media: The Unsaturated Zone Between the Soil Surface and Groundwater*, (eds B. Yaron, G. Dagon, and J. Goldschmid) Springer-Verlag, New York, pp. 27–48.

Schwille, F. (1989) *Dense Chlorinated Solvents in Porous and Fractured Media: Model Experiments*, Lewis Publishers, Ann Arbor, Michigan.

Smith, D.W. (1976) *Cementing*, Amer. Inst. of Mining, Metallurgical, and Petroleum Eng., Inc., Dallas, Texas.

USEPA (1986) *RCRA Ground-water Monitoring Technical Enforcement Guidance Document*, OSWER-9950.1, US Government Printing Office, Washington, DC.

CHAPTER 24

Vadose zone monitoring

Lorne G. Everett

24.1 INTRODUCTION

Dr. Oscar E. Meinzer, in his tome entitled, *Hydrology*, which was first presented by the National Research Council in 1942, referred to the vadose zone as 'no man's land'. From 1942 through the formation of the US Environmental Protection Agency (EPA) in 1970, the vadose zone was fundamentally ignored. The first major national groundwater monitoring contract was awarded by EPA in 1972, on a sole-source basis, to General Electric's 'TEMPO' in Santa Barbara, California. In the years 1974–76, it became clear to the research team that groundwater investigations could not 'black box' the vadose zone. This initial effort resulted in a 15-step groundwater monitoring methodology published by Todd *et al.* (1976) and Everett (1980). Fundamental to this national strategy was an appreciation of contaminant transport through various parts of the vadose zone.

With the passage of the Resource Conservation and Reclamation Act (RCRA) of 1976, the first federal requirement for vadose zone monitoring was introduced. Specifically, soil core monitoring and soil pore-liquid monitoring were required as a part of an unsaturated zone monitoring strategy for hazardous waste land treatment units. Although RCRA passed in 1976, the first guidance document did not appear until ten years later (USEPA, 1986). Current state and federal legislation in the US, especially related to solid waste disposal sites and underground storage tank sites, are now heavily directed towards vadose zone monitoring.

Philosophically and practically, our approach to groundwater monitoring of hazardous waste sites has been flawed. In effect, designing a groundwater monitoring strategy to detect potential contamination

Geotechnical Practice for Waste Disposal.
Edited by David E. Daniel.
Published in 1993 by Chapman & Hall, London. ISBN 0 412 35170 6

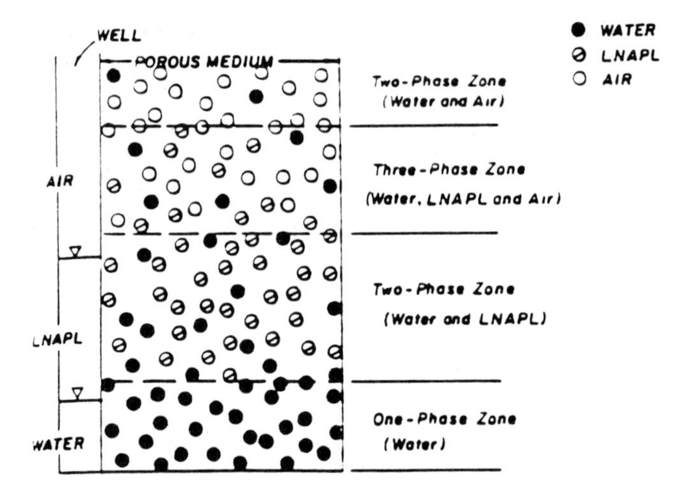

Fig. 24.1 Conceptual representation of a distribution of LNAPL, water, and air in a porous medium and monitoring well.

from a hazardous waste site is akin to a health monitoring strategy which tells you when the patient is dead. For the most part, when contamination, such as a light non-aqueous phase liquid (LNAPL) (Fig. 24.1) or a dense non-aqueous phase liquid (DNAPL), has got through a thick vadose zone, the remediation costs can be staggering. The philosophy of monitoring closed hazardous waste sites exclusively through the use of monitoring wells is an even further indication of a defective monitoring strategy. Currently, the EPA is evaluating amendments to RCRA and CERCLA which will introduce early alert or vadose zone monitoring strategies as a part of a closure program.

Prior to a discussion of vadose zone monitoring techniques, it is imperative to realize a fundamental appreciation of storage and flux in the vadose zone. The physical properties of the vadose zone associated with storage of water include:

1. total thickness;
2. porosity;
3. bulk density;
4. water content;
5. soil/water characteristics;
6. field capacity (specific retention);
7. specific yield; and
8. fillable porosity.

Technical information on each of these properties may be found in reference works by Davis and de Wiest (1966); Childs (1969); Hillel (1971); Colley *et al.* (1972); Bouwer (1978); Brakensiek *et al.* (1979); Freeze and Cherry (1979); Everett *et al.* (1984).

The flux of water through the soil and lower vadose zone underlying a source of contamination is staggeringly complex. Elements contributing to the complexity of flow include variations in the state of water saturation and spacial variations in the physical and hydraulic properties of the vadose zone. The difficulties in attempting to describe water movement in the vadose zone preclude presenting exact techniques for estimating travel time of water and water borne pollutants through this region. A number of indirect methods have been described by Everett *et al.* (1984) covering:

1. infiltration at the land surface;
2. unsaturated flow in the vadose zone; and
3. flow in the saturated regions of the vadose zone.

This chapter on vadose zone monitoring is directed towards developing a geotechnical appreciation of the vadose zone through a discussion of soil moisture/tension relationships, soil pore-liquid monitoring, and some indirect soil moisture monitoring techniques.

24.2 VADOSE ZONE DESCRIPTION

The hydrogeologic profile extending from the ground surface to the upper surface of the principal water-bearing formation is called the 'vadose zone'. As pointed out by Davis and de Wiest (1966), Bouwer (1978), Driscoll (1986), USEPA (1986), and Everett *et al.* (1988), the term 'vadose zone' is preferable to the often-used term 'unsaturated zone' because saturated regions or a wide distribution of tension-saturated regions are frequently present in the vadose zone. Davis and DeWiest (1966) subdivided the vadose zone into three regions, designated as: the soil zone, the intermediate vadose zone, and the capillary fringe (Fig. 24.2).

Soil zone

The term 'soil' is often used incorrectly to refer to the loose and discontinuous blanket of decayed rock debris overlying solid bedrock. This generic material is more appropriately referred to as the 'regolith'. Soil, on the other hand, is only the very uppermost part of the regolith where chemical and physical weathering are the most active. The movement of water in the soil zone occurs mainly as unsaturated flow caused by infiltration, percolation, redistribution, and evaporation (Klute, 1965). The depth of the soil water zone typically varies from 1–10 m. Within these soils, horizons of heavy clay may develop which can create shallow perched water tables (Everett, 1980; Everett *et al.* 1988). The water is held in soils by adhesive and cohesive forces in

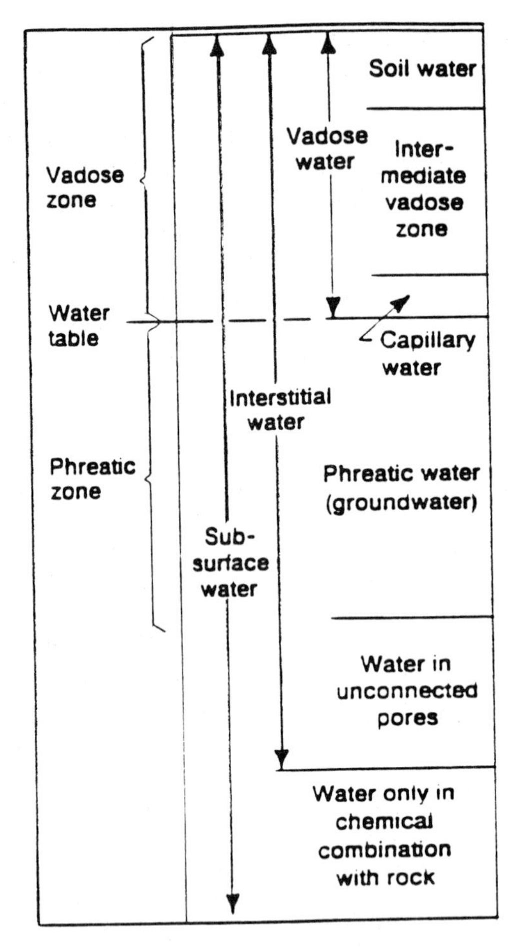

Fig. 24.2 Classification of subsurface water (Davis and de Wiest, 1966).

addition to body forces which are often called the 'force of gravity'. This attraction of the water molecules for each other and the attraction of the water molecule to the soil surface results in a thin film along the surface of each soil particle. When the water-holding capacity of the capillary forces is exceeded, water begins to percolate downward under the force of gravity.

Intermediate vadose zone

As the weathered materials of the regolith merge with the underlying deposits, which are often unweathered, a transition is made to the intermediate vadose zone. In some regions, this zone may be practically non-existent with the soil zone merging directly into the bedrock.

The intermediate vadose zone in alluvial basins, however, may be hundreds of meters thick, in which case the zone is unlikely to be laid down in a uniform pattern. In fact, it is more likely that these sedimentary basins may contain micro- or macro-lenses of silts and clays which can be interbedded with gravels. Water in the intermediate vadose zone may exist primarily in the unsaturated state and some regions may receive very little inflow from above. Occasionally, perched groundwater may develop in the interfacial deposits of regions containing varying textures. Such perching may be hydraulically connected to ephemeral or perennial stream channels, wherein temporary or permanent perched water tables may develop.

Capillary fringe

The capillary fringe is the base of the unsaturated zone that merges with underlying saturated deposits of the principal water bearing formation. This zone is not characterized as much by the nature of geologic materials as by the presence of water under conditions of saturation or near saturation. Studies by Luthin and Day (1955) and Kraijenhoff van deLeur (1962) have shown that both the hydraulic conductivity and flux may remain high for some vertical distance in the capillary fringe, depending on the nature of the materials. In general, the thickness of the capillary fringe is greater in fine materials than in coarse deposits.

An understanding of the height of the capillary fringe is important for several reasons. Clearly, any capillary fringe which extends up into the unprotected foundations of buildings will result in continuous mildew problems, resulting in esthetic and structural damage (Fig. 24.3). In addition, investigations related to petroleum hydrocarbon spills and leaks have shown that the water table will act as a barrier to the free product light non-aqueous phase liquids. Controversy exists, however, as to the extent of hydrocarbon movement through the capillary fringe. One school of thought implies that the capillary fringe acts as a barrier to the free hydrocarbon product. A second school of thought indicates that the concept of a capillary fringe is not consistent with the interpretation of hydrocarbon models. Current federal legislation related to monitoring underground storage tanks, however, dictates that the screened interval must extend 1.5 m above the seasonal high water table to account for capillary activity. It should be pointed out, however, that capillary rise which is determined by grain size distribution is indicative of a minimum capillary fringe, whereas, infiltration/percolation of water from the surface could result in a maximum capillary rise on top of which hydrocarbon contamination may exist. The estimation of minimum and maximum capillary rise measurements is poorly documented in the literature.

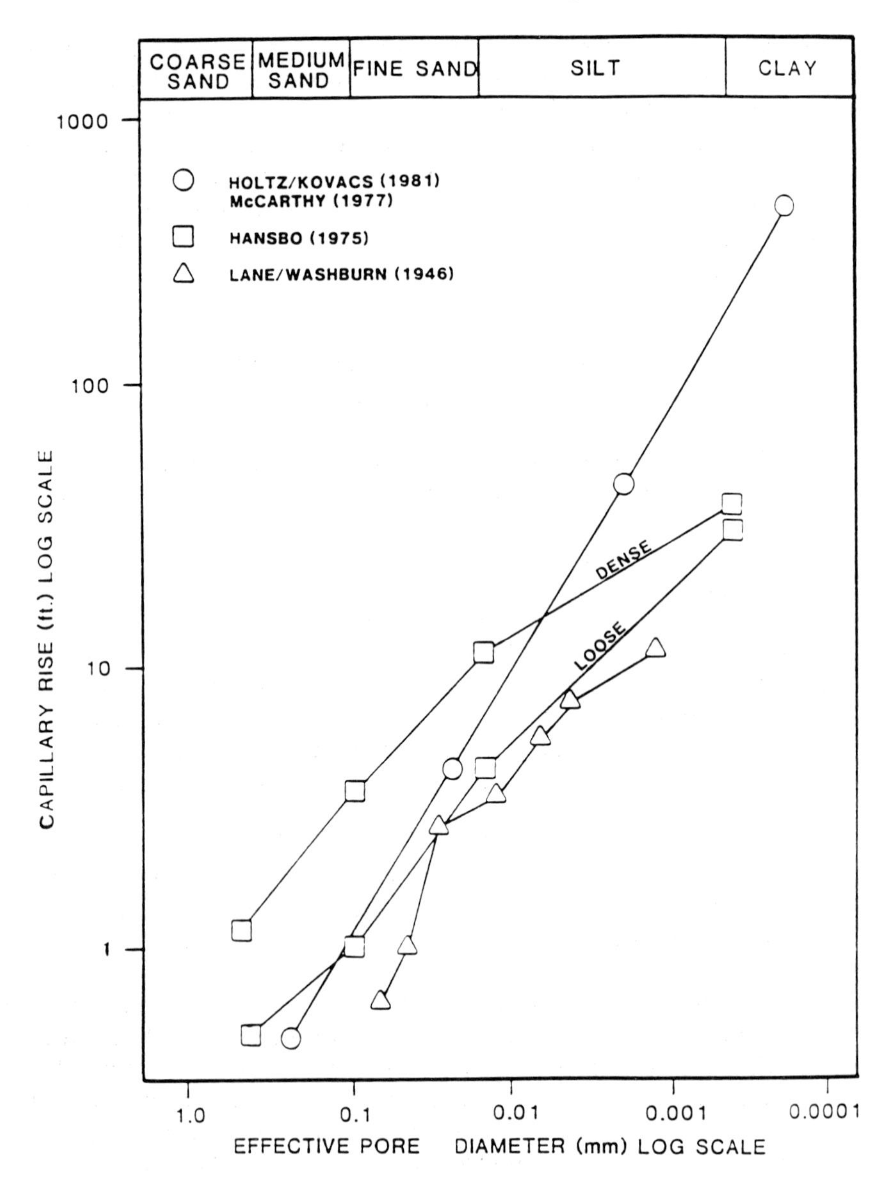

Fig. 24.3 Height of capillary rise.

Vadose zone flow regimes

The classical concept of infiltration depicts a distinct, somewhat uniform wetting front which slowly advances in a Darcian flow regime after a precipitation event. This wetting front infiltration concept was formally presented by Bodman and Coleman in 1944. The second flow regime is sometimes referred to as flow down macropore, non-capillary flow, subsurface storm flow, channel flow, and other descriptive names, but most often referred to as macropore flow. Unfortunately, this is a poor choice of terms since macropores typically are discontinuous, and therefore could not support the flows associated with them. A more accurate term should be secondary porosity or fracture flow. For a mathematical comparison of unsaturated Darcian flow and flow-through secondary porosity, see Hern and Mellencon (1986) and Mercer *et al.* (1983).

24.3 SOIL PORE-LIQUID MONITORING FOR HAZARDOUS WASTE SITES

Generally, we distinguish three ranges of water saturation between the limits of 0% and 100%. Figure 24.4 shows water in a water wet granular soil such as sand. At a very low saturation water forms rings, called pendular rings, around grain contact points. The air/water interface takes on the shape of a saddle under these conditions.

As seen in Fig. 24.4a, the water saturation is rather low, and the rings are isolated and do not form a continuous water phase. Although a very thin film of water, several molecules thick, which does not behave as ordinary liquid water, due to the very strong forces of attraction between the water and solid molecules, does remain on the solid surface, practically no pressure can be transmitted through it from one ring to the next. These conditions are typically too dry to use field tensiometers effectively. Figure 24.4b shows a pendular ring between two spheres. For this case, it is possible to relate the volume of the ring to the radius of the curvature of the air/water interface. The latter, in turn, is related, to the difference in pressures in the air and water across it.

As the amount of water saturation increases, the pendular rings expand until a continuous water phase is formed. The saturation at which this occurs is called equilibrium water saturation. Above this equilibrium saturation, the saturation is called funicular and flow of water is possible (Fig. 24.4c). Both the water and the air phases are continuous. As the water saturation increases, a situation develops in which the air (non-wetting phase) is no longer a continuous phase; it breaks into individual bubbles lodged in the larger pores (Fig. 24.4d). The air is then said to be in a state of insular saturation. These bubbles of air can only move if a pressure difference, sufficient to squeeze it through a capillary size restriction, is applied across it in the water. Obviously, if the air can escape from the void space (or is dissolved in the water) we have complete water saturation.

Classically, moisture in the vadose zone system is broken down into three components: gravity, pellicular, and hygroscopic waters. Gravity

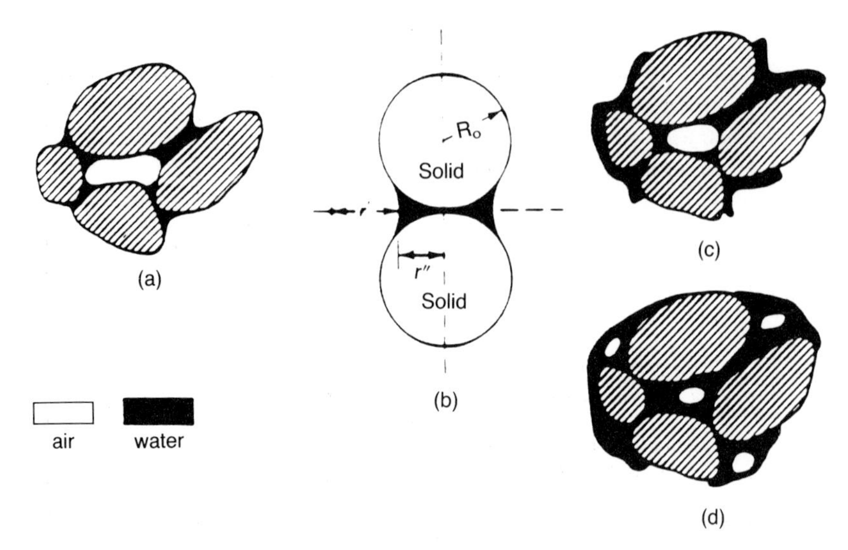

Fig. 24.4 Water and air saturation states.

water is the moisture in a soil which can be drained by gravitational forces; pellicular water is the moisture in a soil which cannot be drained by gravity forces, but can be lost to the atmosphere through natural evaporation; and hygroscopic water is the moisture which will never be lost through the above natural processes. This distribution of soil moisture can be mathematically approximated by a step-wise distribution composed of three components (Mercer *et al.*, 1983). Figure 24.5 shows a comparison of the step-wise wetness distribution and its corresponding original wetness distribution.

24.3.1 Soil moisture/tension relationships

Unlike water in a bucket, free, unlimited access to water does not exist in the soil. Soil water or, as it is frequently called, 'soil moisture', is stored in the small 'capillary' spaces between the soil particles and on the surfaces of the soil particles. The water is attracted to the soil particles, and tends to adhere to the soil. The smaller the capillary spaces between the particles, the greater the sticking force. For this reason, it is harder to get moisture out of fine clay soils than it is from the larger pores in sandy soils, even if the percentage of moisture in the soil, by weight, is the same.

Figure 24.6 shows the results of careful research work done with special extractors. As described by the Soilmoisture Equipment Corporation (1983), the graph shows the relationship of the percent of moisture in a soil to the pressure required to remove the moisture from the soil. These are called moisture retention curves. The pressure is

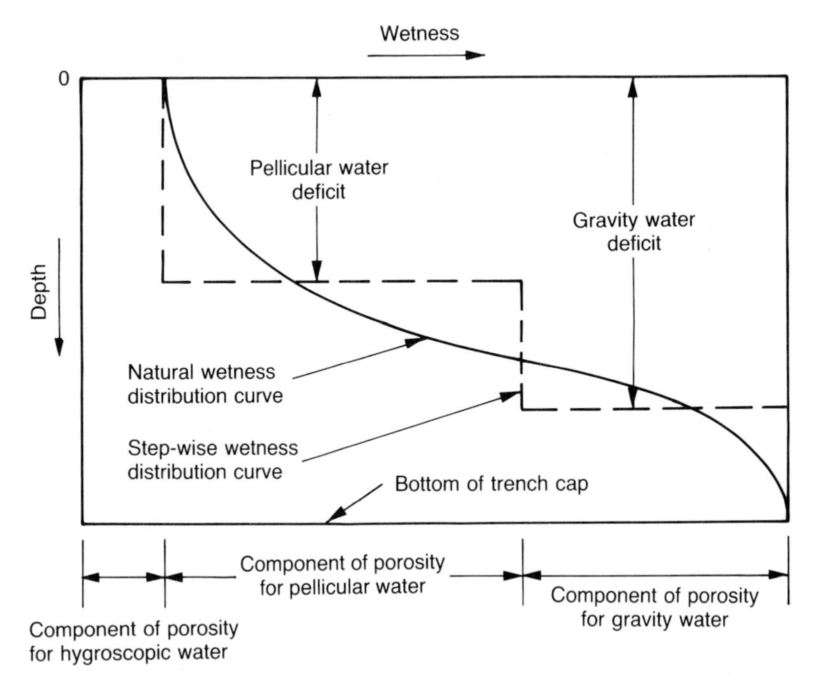

Fig. 24.5 Comparison of a schematic natural wetness distribution curve and its corresponding step-wise wetness distribution curve, after Mercer *et al.* (1983).

measured in bars* which is a unit of pressure in the metric system. Figure 24.6 clearly points out that two factors are involved in determining ease of water sampling: moisture content; and soil type.

Moisture in unsaturated soil is always held at suctions or pressures below atmospheric pressure. To remove the moisture, one must be able to develop a negative pressure or vacuum to pull the moisture away from around the soil particles. For this reason, we speak of 'soil suction'. In wet soils, the soil suction is low, and the soil moisture can be removed rather easily. In dry soils, the soil suction is high, and it is difficult to remove the soil moisture.

Given two soils (one clay and one sand) with identical moisture contents, it will be more difficult to extract water from the finer soil (clay) because water is held more strongly in very small capillary spaces in clays.

Another fact, brought out by the graphs on Fig. 24.6, is that silty clay soil with 30% moisture, if placed in contact with a sandy soil with only

* By definition, a bar is a unit of pressure equal to 10^6 dyne/cm^2. It is equivalent to 100 kPa (Kilopascals) or approximately 1 atmosphere, or 750 mm of mercury, or 1020 cm of water.

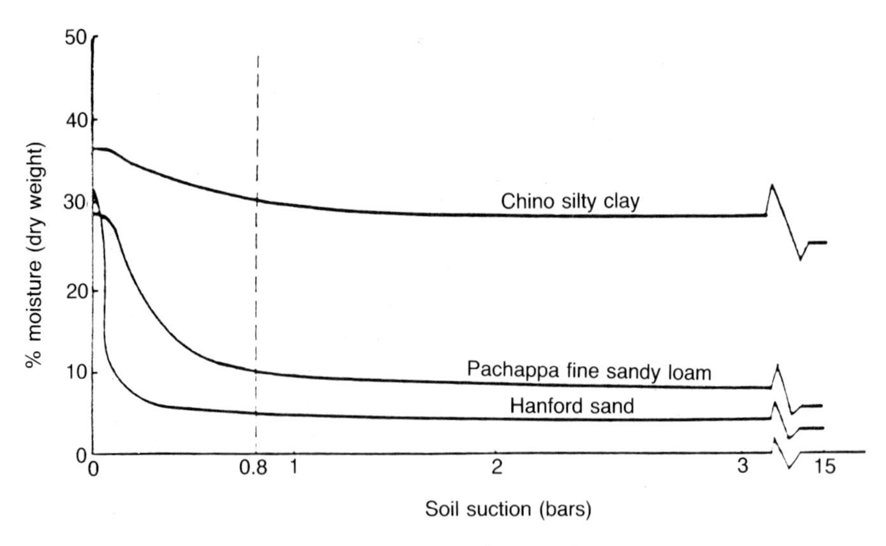

Fig. 24.6 Moisture retention curves – three soil types, after Soilmoisture Equipment Corporation (1983).

10% moisture will actually suck moisture out of the sandy soil until the moisture content in the sandy soil is only 5%. This is due to the greater soil tension in the fine clay texture.

24.3.2 Soil pore-liquid monitoring

The sampling of soil pore-liquid was reported in the literature in the early 1900s when Briggs and McCall (1904) described a porous ceramic cup which they termed an 'artificial root'. As shown in Fig. 24.7, different soils are capable of yielding different levels of water. At saturation, the volumetric water content is equivalent to the soil porosity, while in contrast, the unsaturated zone is usually found to have a soil moisture content less than saturation. For example, the specific retention curve on Fig. 24.7 depicts the percentage of water retained in previously saturated soils of varying texture after gravity drainage has occurred. Lysimeters are used to sample pore-liquids in unsaturated media because pore-liquid will not readily enter an open cavity at pressures less than atmospheric (the Richard's outflow principle).

Lysimeters are made up of a body tube and a porous segment. When placed in the soil, the pores become an extension of the pore space of the soil. Consequently, the water content of the soil and porous segment become equilibrated at the existing soil-water pressure. By applying a vacuum to the interior of the porous segment such that the pressure is slightly less inside the lysimeter than in the soil, flow occurs into the lysimeter. The sample is pumped to the surface,

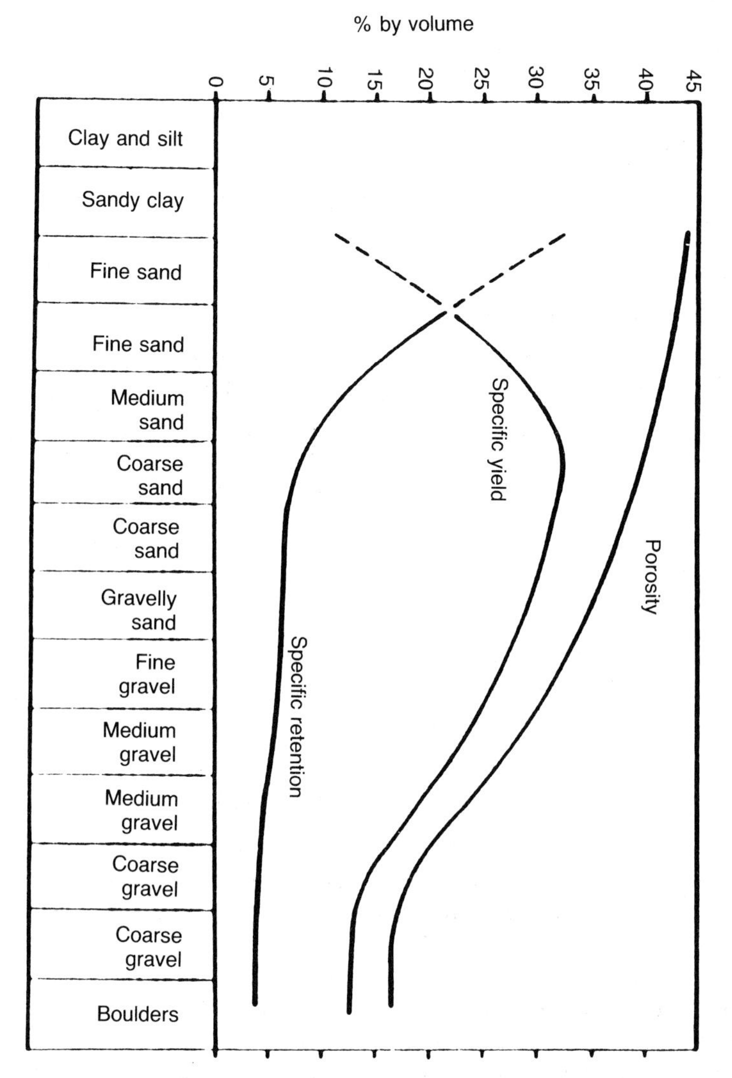

Fig. 24.7 Variation of porosity, specific yield, and specific retention with grain size, after Scott and Scalmanini (1978).

permitting laboratory determination of the quality of the soil pore-liquids.

It should be recognized, however, that situations may occur where the flow velocities in the unsaturated zone are higher than empirically demonstrated by Darcy's Law. As a result, the wetting front will not be uniform and most of the flow will occur through macropores. This type of gravity flow in highly structured soils will not be sampled effectively by suction lysimeters. The most promising technique for sampling soil pore-liquid in highly structured soils is pan lysimeters (e.g., free drainage samplers).

24.3.3 Pore-liquid sampling equipment

Well and open cavities cannot be used to collect solution flowing in the unsaturated zone under suction (negative pressures). The sampling devices for such unsaturated media are thus called suction samplers or lysimeters. Everett and Wilson (1983) provide an in-depth evaluation of the majority of unsaturated zone monitoring equipment. Law Engineering and Testing Company (1982) provides a description of some of the available suction lysimeters.

Suction lysimeters may be subdivided into three categories:

1. vacuum operated soil-water samplers;
2. vacuum-pressure samplers; and
3. high-pressure vacuum samplers with check valves.

Soil-water samplers generally consist of a ceramic cup mounted on the end of a small-diameter PVC tube, similar to a tensiometer. The upper end of the PVC tubing projects above the soils surface. A rubber stopper and outlet tubing are inserted into the upper end. Vacuum is applied to the system and soil water moves into the cup. To extract a sample, a small-diameter tube is inserted within the outlet tubing and extended to the base of the cup. The small-diameter tubing is connected to a sample-collection flask. A vacuum is applied via a hand vacuum-pressure pump and the sample is sucked into the collection flask. These units are generally used to sample to depths up to 2 m from the land surface. Consequently, they are used primarily to monitor the near-surface movement of pollutants from irrigation return flow.

To extract samples from depths greater than the suction lift of water (about 8 m), a second type of unit is available, the so-called vacuum-pressure lysimeter. These units were developed by Parizek and Lane (1970) for sampling the deep movement of pollutants from a land disposal project in Pennsylvania. The design of the Parizek and Lane sampler is shown in Fig. 24.8. The body tube of the unit is about 600 mm long, holding about 1 l of sample. Two polyethylene lines are forced through a two-hole rubber stopper sealed into a body tube. One line extends to the base of the ceramic cup as shown and the other terminates a short distance below the rubber stopper. The longer line connects to a sample bottle and the shorter line connects to a vacuum-pressure pump. All lines and connections are sealed.

In operation, a vacuum is applied to the system (the longer tube to the sample bottle is clamped shut at this time). When sufficient time has been allowed for the unit to fill with solution, the vacuum is

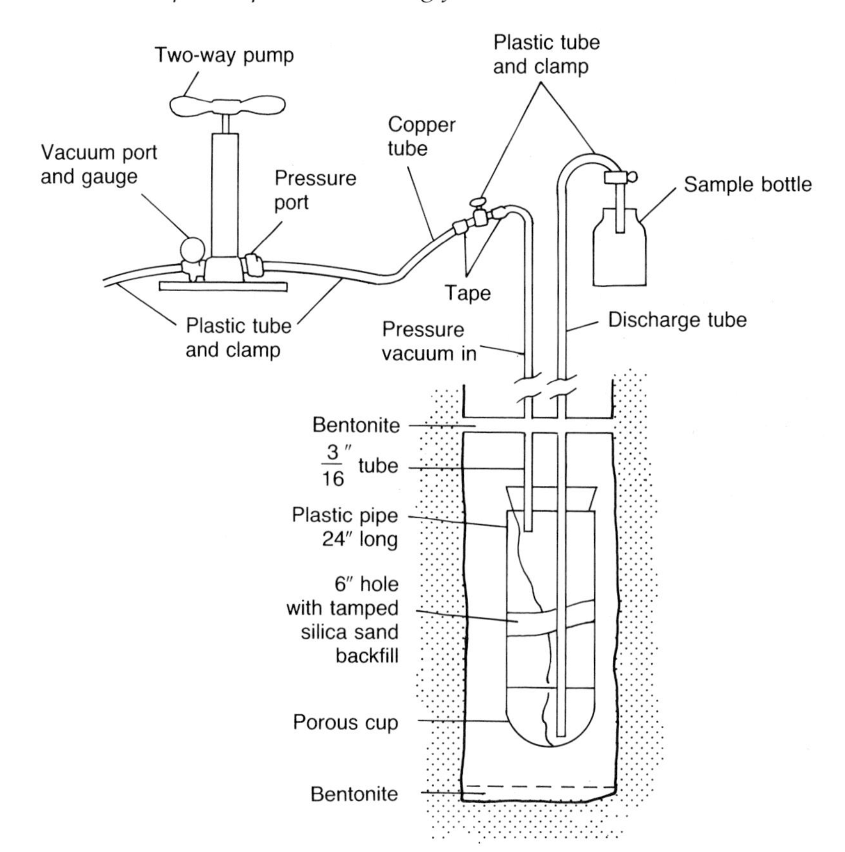

Fig. 24.8 Vacuum-pressure sampler (Parizek and Lane, 1970).

released and the clamp on the outlet line is opened. Air pressure is then applied to the system, forcing the sample into the collection flask. A basic problem with this unit is that when air pressure is applied, some of the solution in the cup may be forced back through the cup into the surrounding pore-water system. Consequently, this type of pressure-vacuum system is recommended for depths only up to about 15 m below land surface. In addition to the monitoring effort of Parizek and Lane, these units have been used extensively to sample leachate movement in the vadose zone underlying sanitary landfills.

Wood (1973) reported on a modified version of the design of Parizek and Lane. Wood's design, the third suction sampler to be discussed in this subsection overcomes the main problem of the simple pressure-vacuum system; namely, that solution is forced out of the cup during application of pressure. A sketch of the sampler is shown in Fig. 24.9. The cup ensemble is divided into lower and upper chambers. The two chambers are isolated except for a connecting tube with a check valve. A sample delivery tube extends from the base of the upper chamber to

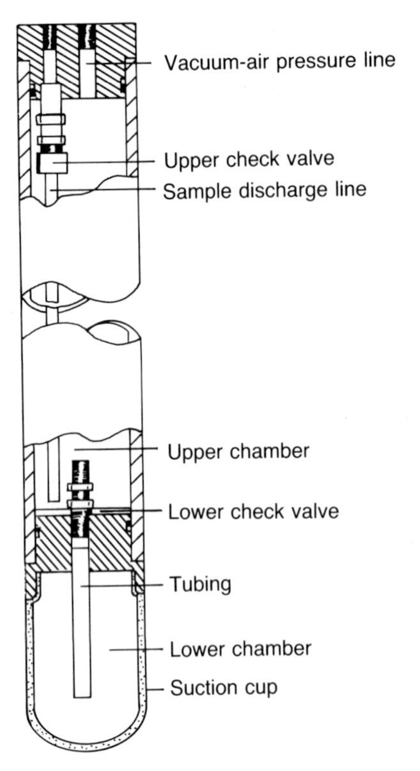

Fig. 24.9 'Hi/pressure-vacuum soil-water sampler'. Courtesy of Soilmoisture Equipment Corporation (1978).

the surface. This tube also contains a check valve. A second shorter tube terminating at the top of the sampler is used to deliver vacuum or pressure. In operation, when a vacuum is applied to the system, it extends to the cup through the open one-way check valve. The second check valve in the delivery tube is shut. The sample is delivered into the upper chamber, which is about 1 l in capacity. To deliver the sample to the surface, the vacuum is released and pressure (generally of nitrogen gas) is applied to the shorter tube. The one-way valve to the cup is shut and the one-way valve in the delivery tube is opened. Sample is then forced to the surface. High pressures can be applied with this unit without danger of damaging the cup. Consequently, this sampler can be used to depths of about 100 m below land surface (Soilmoisture Equipment Corporation, 1978). These samplers have been widely used in solid waste landfills.

24.3.4 Porous segments in lysimeters

The vadose (unsaturated) zone consists of a mixture of soil particles, water that is held on the surface of the particles and in small capillary spaces between the particles, and interconnecting air passages that are open to the atmosphere at the soil surface. Removing moisture for chemical analysis from the vadose zone requires the use of special porous materials. Simply exerting a suction on an open tube inserted into the vadose zone will not remove moisture since the interconnecting air passages in the soil will result only in the flow of air into the evacuated tube. However, by using a porous cup sealed to the end of the tube, samples can be removed by suction, providing the diameter of the individual pores in the porous cup do not exceed a critical value.

If the porous cup is fabricated from a hydrophilic material, such as ceramic, water will fill the pores of the cup completely. The water bonds to the porous ceramic and cannot be removed from the pores unless the air pressure differential across the wall of the cup reaches a critical value which is related to the pore size. If the porous cup is fabricated from a hydrophobic material such as PTFE (polytetrafluorethylene), water will fill the pores of the cup but the bonding of the water to the hydrophobic material will be less.

The air pressure required to force air through a porous cup which has been thoroughly wetted with water is called the 'bubbling pressure' or 'air entry value'. The smaller the pores in the cup the higher this pressure will be. The relation of the pore size to the bubbling pressure or air entry value is defined by the equation

$$D = \frac{30Y}{P} \tag{24.1}$$

where D is the pore diameter measured in microns, P is the bubbling pressure or air entry value measured in millimeters of mercury, and Y is the surface tension of water measured in dynes/cm.

In order to build a soil water sampling device which can be used successfully in the vadose zone to withdraw moisture from the soil, the device must incorporate a porous cup which has pores so small that the air in the soil, under atmospheric pressure, cannot enter even though a full vacuum is created within the sampler. Under these conditions, water from the capillary spaces in the soil will flow through the pores in the porous cup and into the sampler but air will not enter (Fig. 24.10).

With respect to Eq. (24.1), the maximum size of the pores that will permit this action is as follows. At 20 °C the surface tension of water is 72 dynes/cm. The maximum air pressure is 1 atmosphere or 760 mm of mercury. In accordance with the equation, the maximum pore size in the porous cup would be $D = (30)(72)/(76) = 2.8\,\mu m$. The pore size of ceramic cups is between 2 and 3 μm. If the pores of the wetted sampler cup do not exceed 2.8 μm in diameter, then a full vacuum can be maintained within the sampler and the water films in the pores of the porous cup will not break down. If the pore size of the cup is twice this amount, namely 5.7 μm, then the maximum vacuum that can

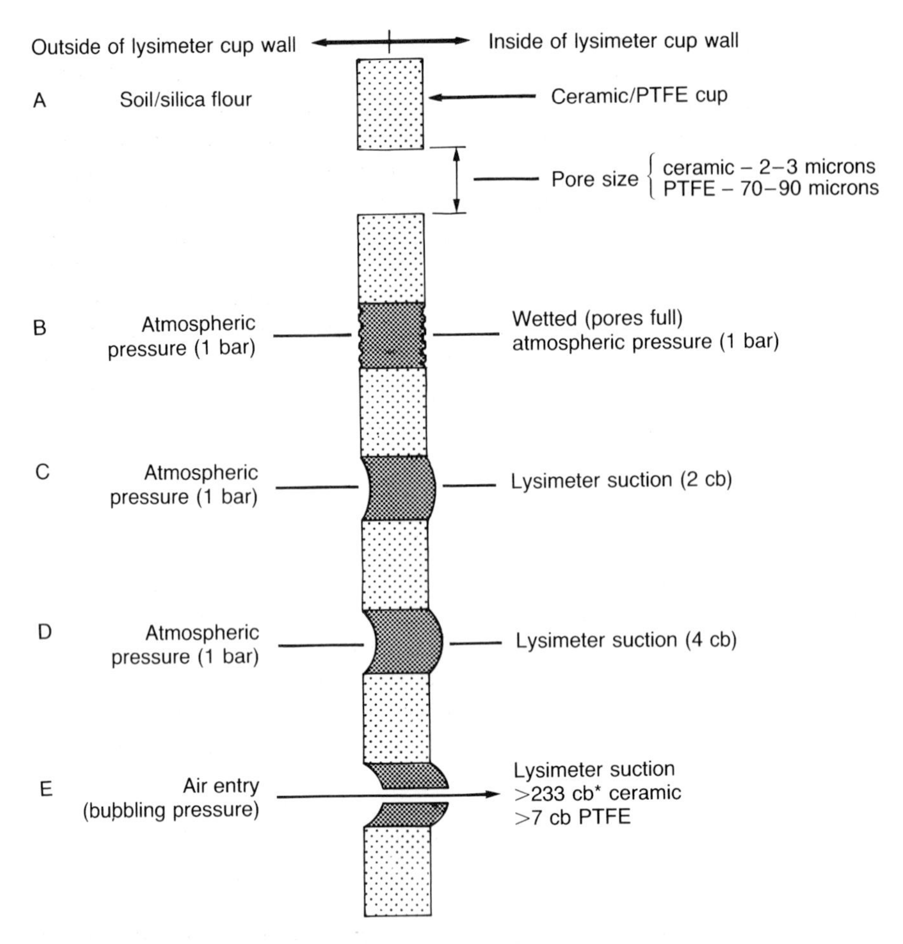

Fig. 24.10 Diagrammatic view of lysimeter cup wall.
* Vacuum > 100 cb not possible

be pulled within the sampler is 380 mm of mercury or 50% of an atmosphere. Likewise, if the pore size is twice again as large, namely 11.4 μm in diameter, then the maximum vacuum that can be pulled without the cup leaking air is 190 mm of mercury or 25% of an atmosphere. Since the majority of the pores used in PTFE suction lysimeters are about 40 μm, the bubbling pressure is only a few per cent of an atmosphere.

Where porous materials are being used in air-water systems, such as in suction lysimeters, the most direct method of evaluating the pore size of the material is through the use of air pressure. By thoroughly wetting the porous material and then exposing one side of it to increasingly higher air pressure values, with the other side under water, one can readily observe when the air pressure becomes high enough to enter the pores and cause bubbling on the opposite side. The specific air pressure at which this bubbling occurs is a direct measurement of the pore size as defined by the above formula and indicates directly the effectiveness of the porous materials to withstand air pressure differentials when in use. Evaluating pore size distribution by the mercury intrusion method or other means does not give direct information as to how the porous material will perform in the air-water system in which it is used. PTFE pores are generally round and symmetrical, while ceramic pores are of various ragged shapes. The strength of the water meniscuses in the individual pores are a function of pore shape as well as overall size, and for this reason an accurate measurement of the pressure at which the meniscus will break down and allow air to enter can only be made accurately by direct measurement of the bubbling pressure or air entry value of the wetted porous material.

24.3.5 Lysimeter failure confirmation

In the event that a sample cannot be retrieved from an installed suction lysimeter under conditions where the operator knows that the soil suction levels should be low enough to obtain a sample, such as after a major rainfall event, specific procedures should be followed. Adjacent to a suction lysimeter that appears to have failed, a soil suction determination must be made to determine if the available soil moisture is high enough to obtain a sample. Soil suctions are determined using tensiometers. Tensiometers are commercially available and are produced with various designs and lengths.

A tensiometer consists of a tube with a porous ceramic tip on the bottom, a vacuum gauge near the top, and a sealing cap. When it is filled with water and inserted into the soil, water can move into and out of the tensiometer through the connecting pores in the tip. As the soil dries and water moves out of the tensiometer, it creates a vacuum inside the tensiometer, which is indicated on the gauge. When the vacuum created equals the 'soil suction', water stops flowing out of the tensiometer. The dial gauge reading is then a direct measure of the force required to move the water from the soil. If the soil dries further, additional water moves out until a higher vacuum level is reached. When moisture is added to the soil, the reverse process takes place. Moisture from the soil moves back into the tensiometer through the porous tip until the vacuum level is reduced to equal the lower soil suction value, then water movement stops. If enough water is added to the soil so that it is completely saturated, the gauge reading on the tensiometer will drop to zero. Because water can move back and forth through the pores in the porous ceramic tip, the gauge reading is always in balance with the soil suction.

The effective operational range for suction lysimeters is between saturation and 60 centibars of suction as determined by the tensiometer.

Above 60 centibars of suction, a ceramic lysimeter will operate (Everett *et al.*, 1986). However, the flow rates will be so low that effectively one cannot get a sample. If the tensiometer readings are between 0 and 60 centibars of suction, the suction lysimeter should obtain a sample. If no sample is obtained under these soil suction ranges, the suction lysimeter will be deemed to have failed and should be excavated or abandoned.

24.4 INDIRECT MONITORING DEVICES

Monitoring units classically are grouped into direct sampling or measurement units and indirect measurement units. Soil suction samplers measure soil salinity directly, whereas measurements using the four-probe method must be converted to salinity by a calibration relationship. Similarly, tensiometers measure matric potential directly, whereas the output from heat dissipation sensors must be converted to matric potential using a calibration curve. Numerous types of vadose zone indirect measurement techniques are available which include the EC probe, salinity sensors, gravimetric measurements, neutron moderation, gamma ray attenuation, hygrometer/cycrometers, heat dissipation sensors, resistor/capacitor type sensors, remote sensing, heat pulse methods, etc. Each of the above indirect vadose zone monitoring techniques are extensively described by Everett *et al.* (1984).

The most promising of the indirect vadose zone monitoring techniques is called neutron moderation. Neutron moderation has been shown to be one of the most accurate and precise indirect predictors of soil moisture distribution. As a result, neutron moderation is rapidly gaining acceptance as a real-time, minimal cost indirect technique for measuring dissolved contaminant migration in the vadose zone.

Principle of the neutron moderation method

The principle of neutron moderation depends on two properties related to the introduction of neutrons with matter: scattering and capture (Gardner, 1965). High-energy neutrons emitted from a radioactive source may be slowed down or thermalized by collisions with atomic nuclei. The statistical probability for such thermalization depends on the scattering 'cross-section' of various nuclei. The scattering cross-section of hydrogen causes a greater thermalizing affect on fast neutrons than many elements commonly found in soils. This forms the basis for detecting the concentration of water in soil (van Bavel, 1963). The second property of interest in the neutron moderation method is capture of slow neutrons by elements present in the soil with the release of other nuclear properties or energy. Cadmium and boron have extremely large capture cross-sections compared with hydrogen. The property of energy released during capture serves as a means of detecting the concentration of slow neutrons.

In operation, when a source of fast neutrons is lowered into a soil through a suitable well or casing, a cloud of thermalized neutrons is established (Gardner, 1965). This cloud reflects the moderating effect of scattering cross-sections of nuclei in the soil mass on fast neutrons. If a suitable calibration is made to isolate the moderating effects of soil nuclei other than hydrogen, changes in the volume of the thermalized cloud will reflect changes in water content. In general, the wetter the soil, the smaller the volume of thermalized neutrons. A detector that relies on capture of thermalized neutrons is used in conjunction with suitable electronic circuitry to measure the water volume.

Instrumentation used to measure the water content of neutron moderation requires three principal components:

1. a source of fast neutrons, i.e., 50 millicuries of americium beryllium;
2. a detector of slow neutrons, i.e., boron detector; and
3. an instrument to determine the count rate from the detection equipment (Fig. 24.11).

Field implementation

Access tubes for neutron moisture logging may be constructed of seamless steel, aluminum, or PVC. In addition, access tubes can either use closed or perforated casing. Although the use of PVC casing slightly dampens the signal, more than sufficient thermalized neutrons can be detected to measure soil moisture over all ranges of concern (Keller *et al.*, 1989).

The application of neutron moderation at landfill sites is gaining acceptance. In addition, the cost advantages of neutron moderation are resulting in widespread application of the technology as a part of long-term post-closure monitoring strategies for closed landfills and contamination sites. Neutron probes cost less than $5000, provide real-time digital readout, have zero analysis cost, and have a lifetime which is substantially longer than most landfill or contamination sites.

With respect to landfills and contamination sites, the neutron moderation access tubes can either be installed vertically, on an angle, or horizontally beneath landfill cell expansion areas. Since neutron moderation operates over the range between zero matric potential (water saturation) and 90 bars of soil water tension, the neutron probe can be utilized under all conditions. Although the neutron probe was originally designed to measure indirectly soil moisture distribution, recent patents have expanded the application of neutron moderation to include soil gas analysis. In effect, patents held by Toxic Technology,

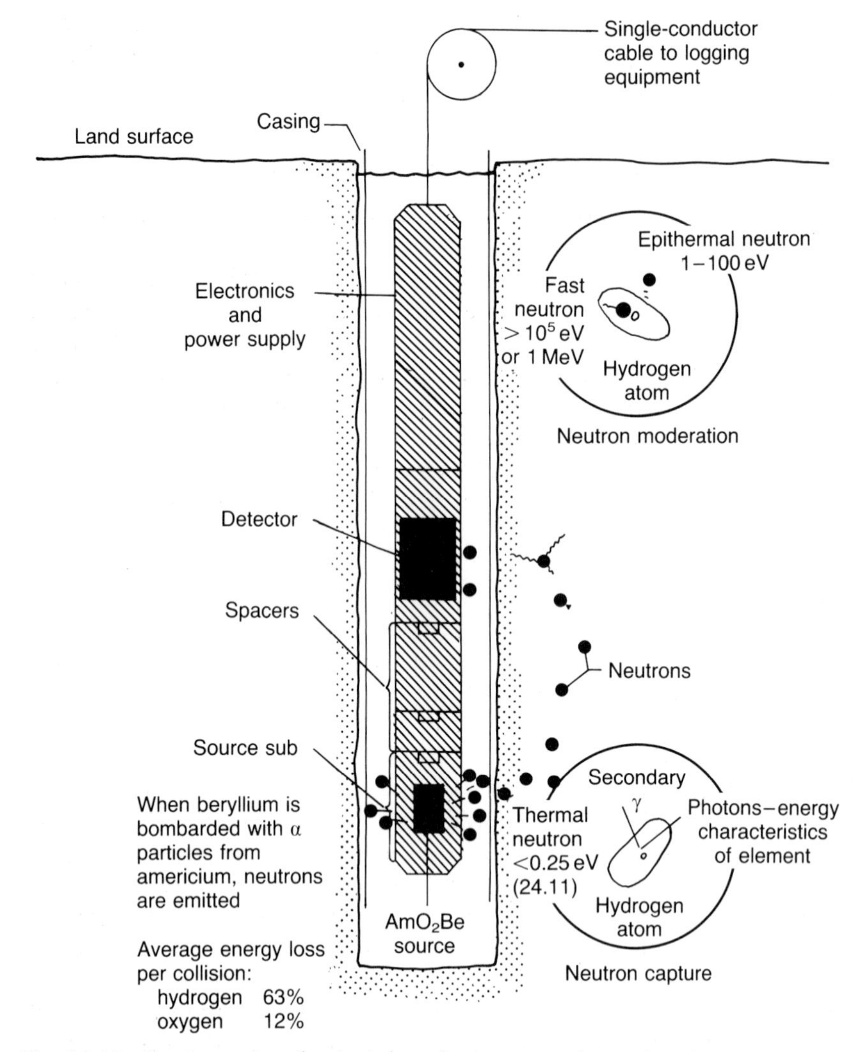

Fig. 24.11 Equipment and principles of neutron moisture logging.

Inc. of Costa Mesa, California utilize a perforated hydroprobe access casing wherein the neutron probe identifies the location of the contamination. Subsequently, a soil gas sample is taken adjacent to the depth identified by the neutron probe and analyzed using a field GC to confirm the presence of volatile organic contaminants. This approach is particularly attractive because the neutron probe is used as a least-cost screening tool, wherein confirmation of dissolved organic contamination can be obtained through soil gas analysis. Clearly, therefore, the installation of neutron probe access tubes below landfill liner sumps, beneath V-shaped landfill cells, or through any depth associated with contamination sites, provide a very accurate, and precise method of measuring not only soil moisture migration, but also dissolved contaminant transport.

Underground storage tank monitoring

One example of the application of neutron moderation to underground storage tanks is provided to show the multiple application of direct and indirect monitoring techniques. When a release takes place from a petroleum hydrocarbon underground storage tank, fluids will begin to pass through the vadose zone. This release results in an absorbed phase in the vadose zone, sometimes called residual oil saturation, a vapor phase, a dissolved phase in soil moisture, and a free product phase. The neutron moderation technique has been shown to definitively identify each of these phases. In addition, the application of neutron moderation satisfies tank integrity testing, tank and pipeline monitoring, risk analysis for adsorbed and dissolved phase, distinguishes between spills and leaks, and greatly enhanced passive and active remediation.

The neutron probe is a 50 mm by approximately 250 mm long probe which conveniently passes through a 50 mm subsurface casing which may be either closed or perforated. If perforated casing is used, the process is protected by a US Patent held by Toxic Technology, Inc. Based on Richard's principle of unsaturated flow, fluids will not move from a small pore to a large pore under unsaturated conditions, and, as such, water will not move into the perforated casing in the vadose zone. Hundreds of gasoline stations have been monitored using perforated casing and not one incident of fluids moving into the casing have been detected, thereby totally supporting Richard's principle. In effect, the neutron moderation technique uses the neutron probe to give an exact vertical distribution of soil moisture (dissolved phase in the vadose zone or absorbed phase of hydrocarbon in the vadose zone) (Fig. 24.12). Since the neutron probe reads hydrogen atoms from either water or hydrocarbons (or, for that matter, from any other source of hydrogen) a field gas chromatograph (GC) is used to confirm the presence of volatile hydrocarbons.

A typical installation for monitoring an underground storage tank involves four slotted PVC casings located in the corners of a tank excavation. When the access holes are drilled, a continuous core is obtained to determine the exact background condition. Regardless of background conditions, any deviation in terms of any of the flow conditions described above is precisely picked up by the neutron probe and confirmed by the GC. The time savings, the cost advantages, and the multiple application of the technique has enormous attraction to an underground storage tank program.

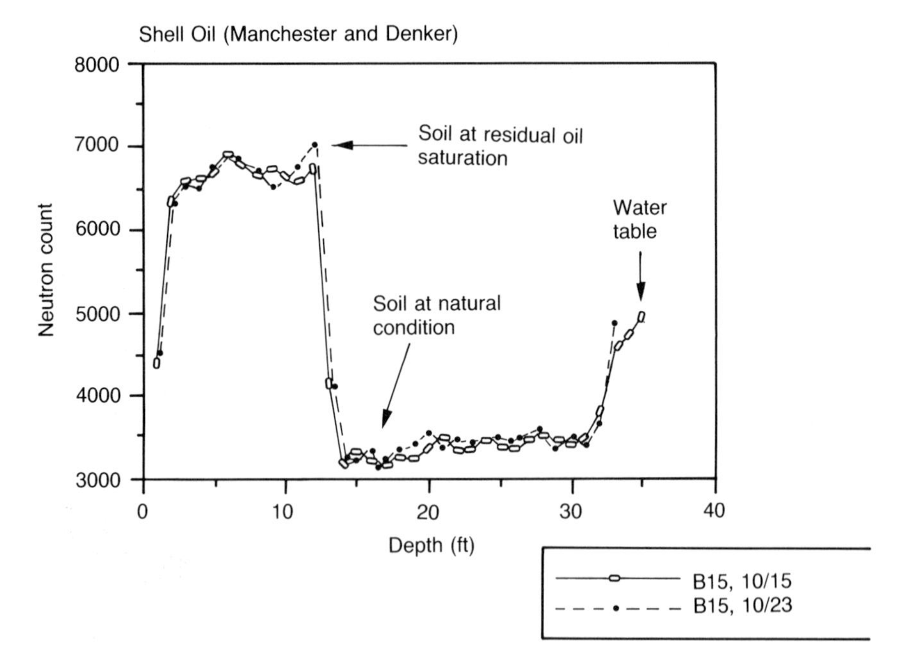

Fig. 24.12 Vertical profile of vadose zone contamination based upon neutron moderation.

Doctoral candidates at the University of California at Santa Barbara have demonstrated through artificial leakage conditions that a rate of 0.1 l/hr can be determined using the neutron probe system. Other researchers have routinely demonstrated that the technique can satisfy the 0.2 l/hr leak detection capability typically required in the US. If an anomaly is identified, successive runs with a tighter timeframe, i.e., measured in hours, can demonstrate whether dissolved product or free product is moving down through the excavation. This rate of flow for both the dissolved product, adsorbed product, and free product constitutes the basis for a definitive risk analysis. By tracking the movement of the contaminant front, one can easily determine the difference between a spill or a leak.

With respect to remediation, the perforated access holes offer immediate opportunities for minimal cost techniques. For example, small spills and leaks, which have been repaired, can be quickly remediated through the use of soil venting techniques using a blower system on the existing perforated access tubes. In addition, the opportunity to use an inlet air tube and to drive the volatiles towards the other three outlet perforated tubes can accelerate the process. For spills and leaks of larger proportion, further active approaches can include the use of the perforated access holes to introduce air, nutrients, soil moisture, bacteria, pH controls, etc., all of which can be easily introduced through the use of the perforated casing to accelerate either an active or a passive bioremediation program.

A major advantage of the technique is gaining widespread application in California relative to passive hydrocarbon remediation. In particular, the neutron probe provides a definitive risk analysis relative to whether the dissolved phase or free product is moving. The use of packers allows the taking of soil gas samples directly adjacent to the lens of contamination. In particular, the neutron probe provides a definitive risk analysis relative to whether the dissolved phase or free product is moving. The use of packers allows the taking of soil gas samples directly adjacent to the lens of contamination. The analysis of the gas distribution in the vapors obtained utilizes the ratio of carbon dioxide to oxygen to demonstrate passive bioremediation. The gas analysis also demonstrates the aerobic decomposition of the hydrocarbon, however, if the system should turn, anaerobic, air can easily be introduced through the perforated casing. The elevated levels of CO_2 relative to O_2 demonstrates that passive bioremediation is taking place and since neither the dissolved phase nor the adsorbed phase is moving, the contaminated soil can be left in place.

The application appears to be a combination of proven techniques which satisfy tank integrity testing, pipeline testing, leak detection monitoring, active remediation programs, and more importantly, allows for inplace passive remediation strategies. Since the radioactive source will certainly outlive the hydrocarbon contamination, the real time, zero analysis cost screening, is a major advantage to the technique.

REFERENCES

Bodman, G.B. and Coleman, E.A. (1944) Moisture and Energy Conditions During Downward Entry of Water into Soils, *Soil Sci. Soc. Amer. Proc.*, **8**, 116–22.

Bouwer, H. (1978) *Groundwater Hydrology*, McGraw-Hill, New York.

Brakensiek, D.L., Osborn, H.B. and Rawls, W.J. (1979) *Field Manual for Research in Agricultural Hydrology*, Agricultural Handbook No. 224, Sciences and Education Adm., US Dept. of Agric.

Briggs, L.J. and McCall, A.G. (1904) An artificial root for inducing capillary movement of soil moisture, *Science* **20**, 566–9.

Childs, E.C. (1969) *An Introduction to the Physical Basis of Soil Water Phenomena*, Wiley Interscience, New York.

Cooley, R.L., Harsh, J.F. and Lewis, D.C. (1972) *Principles of Ground-Water Hydrology*, Hydrologic Engineering Methods for Water Resources Development, Vol. 10, Hydrologic Engineering Center, US Army Corps of Engineers, Davis, California.

Davis, S.N. and De Wiest, R.J.M. (1966) *Hydrogeology*, John Wiley and Sons, New York.

Driscoll, F.G. (1986) *Groundwater and Wells*, (2nd edn), Johnson Division, St. Paul, Minnesota.

Everett, L.G. (1980) *Groundwater Monitoring*, General Electric Co. Technology Marketing Operations, Schenectady, New York.

Everett, L.G., McMillion, L.G. and Eccles, L.A. (1988) Suction lysimeter operation at hazardous waste sites, in *Groundwater Contamination – Field Methods*, (eds A.G. Collins and A.I. Johnson), ASTM, STP-963, Philadelphia, Pennsylvania.

Everett, L.G. and Wilson, L.G. (1983) *Unsaturated Zone Monitoring at Hazardous Waste Land Treatment Units*, Guidance Manual to EPA, Office of Solid Waste, Washington, DC.

Everett, L.G., Wilson, L.G., and Hoylman, E.W. (1984) *Vadose Zone Monitoring for Hazardous Waste Sites*, Noyes Data Corp., Park Ridge, New Jersey.

Everett, L.G., Wilson, L.G. and McMillion, L.G. (1986) Suction Lysimeter Operation at Hazardous Waste Sites, ASTM, Cocoa Beach, Florida.

Freeze, R.A. and Cherry, J.A. (1979) *Groundwater*, Prentice-Hall, Englewood Cliffs, New Jersey, 604pp.

Gardner, W.H. (1965) Water Content, in *Methods of Soil Analyses*, (ed. C.A. Black), Agronomy No. 9, Amer. Soc. Agron., Madison, Wisconsin, pp. 82–125.

Hern, S.C. and Mellencon, S.M. (1986) *Vadose Zone Modeling of Organic Pollutants*, Lewis Publishers, Chelsea, Michigan.

Hillel, D. (1971) *Soil and Water Physical Principles and Processes*, Academic Press, New York.

Keller, B.R., Everett, L.G. and Marks, R.J. (1989) *Efects of Access Tube Material and Grout on Neutron Probe Measurements in the Vadose Zone*. Ground Water Monitoring Review, NWWA, Fall Issue, Dublin, Ohio.

Klute, A. (1965) Laboratory measurement of hydraulic conductivity of unsaturated soil, in *Methods of Soil Analyses*, (ed. C.A. Black), Agronomy No. 9, Amer. Soc. Agron., Madison, Wisconsin, pp. 253–61.

Kraijenhoff van deLeur, D.A. (1962) Some Effects of the Unsaturated Zone on Nonsteady Free-Surface Groundwater Flow as Studied in a Scaled Granular Model, *J. Geophys. Res.*, **67**(11), 4347–62.

Law Engineering and Testing Co. (1982) Lysimeter evaluation, report to American Petroleum Institute.

Luthin, J.M. and Day, P.R. (1955) Lateral Flow Above a Sloping Water Table, *Soil Sci. Soc. Amer. Proc.*, **19**, 406–10.

Meinzer, O.E. (1942) Groundwater, in *Hydrology*, (ed. Oscar E. Meinzer), Dover Publications, New York, 385–477.

Mercer, J.W., Rao, P.S. and Wendell Marine, I. (1983) *Role of the Unsaturated Zone in Radioactive and Hazardous Waste Disposal*, Ann Arbor Science Publishers, Ann Arbor, Michigan.

Parizek, R.R. and Lane, B.E. (1970) Soil-water sampling using pan and deep pressure-vacuum lysimeters, *J. of Hydrology*, **11**, 1–21.

Scott, V.H. and Scalmanini, J.C. (1978) *Water Wells and Pumps: Their Design, Construction, Operation and Maintenance*, Bull. 1889, Div. Agric. Sci., U. of California.

Soilmoisture Equipment Corporation (1978) Operating instructions for the Model 1900 Soil Water Sampler, Santa Barbara, California.

Soilmoisture Equipment Corporation (1983) Internal memo on soil tension, Santa Barbara, California.

Todd, D.K., Tinlin, R.M., Schmidt, K.D. and Everett, L.G. (1976) *Monitoring Groundwater Quality: Monitoring Methodology*, EPA-600/4-76-026, USEPA,

Environmental Monitoring and Support Laboratory, Las Vegas, Nevada.
US Environmental Protection Agency (1986) *Permit Guidance Manual on Unsaturated Zone Monitoring for Hazardous Waste Land Treatment Units*, EPA/530-SW-86-040, OSW&ER, Washington, DC.
van Bavel, C.H.M. (1963) Neutron Scattering Measurement of Soil Moisture: Development and Current Status, *Proc. Int. Symp. Humidity and Moisture*, Washington, DC., 171–84.
Wood, W.W. (1973) A technique using porous cups for water sampling at any depth in the unsaturated zone, *Water Resources Res.*, **9**(2), 486–8.

Index